Lecture Notes in Mechanical Engineering

Lecture Notes in Mechanical Engineering (LNME) publishes the latest developments in Mechanical Engineering - quickly, informally and with high quality. Original research reported in proceedings and post-proceedings represents the core of LNME. Volumes published in LNME embrace all aspects, subfields and new challenges of mechanical engineering. Topics in the series include:

- Engineering Design
- Machinery and Machine Elements
- Mechanical Structures and Stress Analysis
- Automotive Engineering
- Engine Technology
- Aerospace Technology and Astronautics
- Nanotechnology and Microengineering
- Control, Robotics, Mechatronics
- MEMS
- Theoretical and Applied Mechanics
- Dynamical Systems, Control
- Fluid Mechanics
- Engineering Thermodynamics, Heat and Mass Transfer
- Manufacturing
- Precision Engineering, Instrumentation, Measurement
- Materials Engineering
- Tribology and Surface Technology

To submit a proposal or request further information, please contact the Springer Editor in your country:

China: Li Shen at li.shen@springer.com
India: Dr. Akash Chakraborty at akash.chakraborty@springernature.com
Rest of Asia, Australia, New Zealand: Swati Meherishi at
swati.meherishi@springer.com
All other countries: Dr. Leontina Di Cecco at Leontina.dicecco@springer.com

To submit a proposal for a monograph, please check our Springer Tracts in Mechanical Engineering at http://www.springer.com/series/11693 or contact Leontina.dicecco@springer.com

Indexed by SCOPUS. The books of the series are submitted for indexing to Web of Science.

More information about this series at http://www.springer.com/series/11236

Prabhakar V. Varde · Raghu V. Prakash ·
Gopika Vinod

Editors

Reliability, Safety and Hazard Assessment for Risk-Based Technologies

Proceedings of ICRESH 2019

 Springer

Editors
Prabhakar V. Varde
Reactor Group
Bhabha Atomic Research Centre
Mumbai, Maharashtra, India

Raghu V. Prakash
Department of Mechanical Engineering
Indian Institute of Technology Madras
Chennai, Tamil Nadu, India

Gopika Vinod
Probabilistic Safety Section
Bhabha Atomic Research Centre
Mumbai, Maharashtra, India

ISSN 2195-4356   ISSN 2195-4364   (electronic)
Lecture Notes in Mechanical Engineering
ISBN 978-981-13-9007-4   ISBN 978-981-13-9008-1   (eBook)
https://doi.org/10.1007/978-981-13-9008-1

This Springer imprint is published by the registered company Springer Nature Singapore Pte Ltd.
The registered company address is: 152 Beach Road, #21-01/04 Gateway East, Singapore 189721, Singapore

Preface

Advances in Risk Based Technology

Recent decades have seen an increased attention towards 'risk-based technology' by reliability and safety practitioners from industries like nuclear, defence, aerospace, oil and gas and chemical. Efficient data logging and trend analysis tools have helped in raising the confidence of risk-based approaches towards design, operation and maintenance of engineering systems in safety-critical industries. Economics also proceeds hand in hand with safety, which directs the research as well as academia to work on new and improved models and methods based on risk-based/risk-informed approach along with deterministic insights for decision-making. This provides a holistic framework for addressing the issues related to reliability, quality, safety and economics in a sustainable manner.

With the advancement of technology, reliability and risk practitioners are facing fresh challenges to adapt the existing modelling framework, especially in the fields of digital systems, human reliability, structural reliability, etc. The focus on research and development not only is limited to the realistic estimation of reliability, but also extends to understanding the failure mechanisms, dynamically changing failure modes of advanced intelligent systems, developing suitable models based on advanced computational techniques, etc. Considering the changing scenarios, the field of risk and reliability needs to cater to future requirements of the sustainable systems.

This book is a compilation of full-length paper presented during the proceedings of the International Conference on Reliability, Safety and Hazard 2019 (ICRESH-2019) held during 11–13 January 2019, at IIT Madras, Chennai, organized by Bhabha Atomic Research Centre (BARC), Indian Institute of Technology Madras (IITM), Society for Reliability and Safety (SRESA) and Centre for Advanced Life Cycle Engineering (CALCE). The primary focus of the conference was to discuss the research and development work being performed in research organizations, academic institutes and industries and chart out the road map for future directions in risk-based asset management of engineering systems. The topics covered include probabilistic safety assessment, digital system reliability, structural reliability, RAMS (reliability, availability, maintainability, safety), failure analysis, human reliability, nuclear safety and risk-based methods.

The conference attracted 170 registrants from five countries (India, USA, UK, Germany and France). The conference, which spanned for 3 days, had 16 keynote speakers, delivering expert talks on topics ranging from electronic reliability, material reliability, structural reliability, human reliability, etc. There were around 100 papers delivered in four parallel sessions. The conference was sponsored by several agencies such as Board of Research in Nuclear Sciences, Heavy Water Board, Electronics Corporation of India Ltd., Atomic Energy Regulatory Board, Board of Research in Isotope Technology and Defence Research and Development Organisation. Springer provided the support in the form of merit prizes for the best papers presented at the conference and is proud to be associated in bringing out this book.

The papers included in this book have been peer-reviewed by the domain experts to ensure the quality and correctness of the technical content. ICRESH-2019 Technical Committee and editors appreciate and thank the authors for their support during the review process. Looking at the spectrum of application domain covered in the conference, this book will serve as 'one-stop reference' for applying the reliability and risk-based technologies in power and process industries.

This conference with a theme 'risk-based technology' provided a platform for practising scientists and engineers to disseminate their knowledge and learn from other's experience. We hope this tradition will continue to play a key role in advancing knowledge on the development and application of risk-based engineering approach to complex systems.

Mumbai, India   Dr. Prabhakar V. Varde
Chennai, India   Dr. Raghu V. Prakash
Mumbai, India   Dr. Gopika Vinod

Contents

Reliability Methods

Nuclear Safety

About the Editors

Prof. Prabhakar V. Varde is an expert in the field of application of reliability and probabilistic risk assessment to nuclear plants and is currently working as Head of the Research Reactor Services Division and Senior Professor at Homi Bhabha National Institute, Bhabha Atomic Research Centre, Mumbai, India, where he also serves in advisory and administrative capacities in Atomic Energy Regulatory Board (AERB), India, and the Homi Bhabha National Institute, India. He is the Founder and President of the Society for Reliability and Safety (SRESA) and is one of the chief editors for its international journal—Life Cycle Reliability and Safety Engineering. He completed his B.E. (Mech) from Government Engineering College, Rewa, in 1983 and joined BARC, Mumbai, in 1984, where he worked as a Shift Engineer in the Reactor Operations Division until 1995. In 1996, he received his Ph.D. in Reliability Engineering from the Indian Institute of Technology, Bombay, Mumbai, following which he worked as a Postdoctoral Fellow at the Korea Atomic Energy Research Institute, South Korea, and a Visiting Professor at the Center for Advanced Life Cycle Engineering (CALCE) at the University of Maryland, USA. Professor Varde is also a consultant/specialist/Indian expert for many international organizations, including OECD/NEA (WGRISK), Paris; International Atomic Energy Agency, Vienna; University of Maryland, USA; Korea Atomic Energy Research Institute, South Korea. Based on his R&D work, he has published over 200 publications in journals and conferences, including 11 conference proceedings books.

Prof. Raghu V. Prakash is a Professor in the Department of Mechanical Engineering, Indian Institute of Technology Madras (IIT Madras); he specializes in the areas of fatigue, fracture of materials (metals, composites, hybrids), structural integrity assessment, remaining life prediction of critical components used in Transportation, Energy sectors, apart from new product design. He has more than twenty five years of professional experience in the field of fatigue and fracture and has more than 100 journals, chapter publications and 100 Conference publications and has edited 3 book volumes. He has developed test systems for use in academia, R&D and industry during his tenure as Technical Director at BiSS Research,

Bangalore; at IIT Madras, he teaches courses relating to Fracture Mechanics, Design with Advanced Materials, Product Design, DFMA. He is a voting rights member of ASTM International (Technical Committees, D-30, E-08 and E-28), vice-Chair of the Technical Committee on Materials Processing and Characterization of ASME. He serves in the Editorial boards of Journal of Structural Longevity, Frattura ed Integrità Strutturale (IGF Journal), Journal of Life Cycle Reliability and Safety Engineering.

Dr. Prakash received his Bachelor's degree in Mechanical Engineering from College of Engineering, Guindy, Madras (now Chennai); Master's degree (by Research) and Ph.D. from the Department of Mechanical Engineering, Indian Institute of Science, Bangalore. He is a member of several technical societies (Indian Structural Integrity Society, Society for Failure Analysis, Indian Institute of Metals). He has won several prestigious awards (Binani Gold Medal, Indian Institute of Metals), scholarships and Erasmus-Mundus Fellowships. He is the recipient of Distinguished Fellow of the International Conference on Computational and Experimental Engineering and Sciences (ICCES) 2015.

Dr. Gopika Vinod is a faculty member and the Head of the Probabilistic Safety Section in Bhabha Atomic Research Centre, Mumbai, India. She received her doctoral degree in Reliability Engineering from Indian Institute of Technology, Bombay and also been post doctoral fellow at Steinbies Advanced Risk Technologies, Germany. She is a recipient of DAE Young Engineer Award 2007. She has been to visiting scientist at Brookhaven National Laboratory. She has been actively involved in reliability, safety and risk analysis of Indian nuclear power plants, and nuclear and chemical facilities. She has been actively involved in reliability, safety and risk analysis of Indian Nuclear power plants, nuclear and chemical facilities. She has worked on the development of reliability based operator support systems such as risk monitor, symptom based diagnostic system, for Indian nuclear power plants. Her other areas of research activities include risk informed in-service inspection, reliability of computer based systems, dynamic reliability analysis, human reliability analysis, etc. She is in the editorial board of international Journal of System Assurance Engineering and Management, and the Journal of Life Cycle Reliability and Safety Engineering. She also serves as Journal Referee to IEEE Transactions on Reliability, Reliability Engineering and System Safety, Risk Analysis, Annals of Nuclear Energy, Nuclear Engineering and Design, etc.

Probabilistic Safety Assessment

Estimating Passive System Reliability and Integration into Probabilistic Safety Assessment

R. B. Solanki, Suneet Singh, P. V. Varde and A. K. Verma

Abstract Passive safety systems are being increasingly deployed into advanced designs of nuclear power plants (NPPs) with an objective of enhancing the safety. Passive systems are considered to be more reliable than the active systems as mechanical failures of active component and failure due to human errors are not contributing towards the system failure probability. The introduction of passive systems into NPPs on the one hand improves the reliability, and it poses challenges to the estimation of its reliability and integration into probabilistic safety assessment (PSA). The active system reliability can be estimated using the classic fault tree analysis technique. For estimating the passive system reliability, a different approach is required due to the presence of phenomenological failures apart from the mechanical component failures. In this paper, the approach is presented to estimate the passive system reliability of a typical isolation condenser system using artificial neural network (ANN)-based response surface method. The integration of passive system reliability into PSA is also demonstrated using the accident sequence analysis for a typical operational transient, which requires NPP to shut down from the full power and successfully remove the residual heat from the reactor core through isolation condenser system. The thermal-hydraulic behaviour of the isolation condenser system is analysed using thermal-hydraulic computer code RELAP 5/MOD 3.2 for different system configurations depending upon the initial conditions of the process parameters. The variability of the process parameters is represented by discrete probability distribution within

R. B. Solanki (✉)
Atomic Energy Regulatory Board, Mumbai, India
e-mail: rajsolanki@aerb.gov.in

S. Singh
Department of Energy Science, IIT, Mumbai, India
e-mail: suneet@iitb.ac.in

P. V. Varde
Bhabha Atomic Research Centre, Mumbai, India
e-mail: varde@barc.gov.in

A. K. Verma
Western Norway University of Applied Sciences, Haugesund, Norway
e-mail: ajitkumar.verma@hvl.no

© Springer Nature Singapore Pte Ltd. 2020
P. V. Varde et al. (eds.), *Reliability, Safety and Hazard Assessment for Risk-Based Technologies*, Lecture Notes in Mechanical Engineering,
https://doi.org/10.1007/978-981-13-9008-1_1

3

the given operating range. The reliability of the reactor protection system (RPS) is estimated using the fault tree analysis method using Risk Spectrum computer code. The core damage frequency induced through the operational transient is estimated using the classical event tree analysis approach. The open issues are also identified for integration of passive system reliability into PSA for future work.

Keywords Passive system reliability · Probabilistic safety assessment · Artificial neural network · Response surface method · Risk spectrum

1 Introduction

Passive systems are aimed at simplification in the design of the safety systems and enhancement of safety by relying on natural driving forces, which do not require external power sources. Two of the important attributes of failure of the 'active' safety systems, namely (i) 'Operator error' and (ii) 'Active component malfunction', do not contribute in the failure of 'Passive' safety systems. Hence, passive systems are theoretically considered to be 'more reliable' than an active system. However, adequate operating experience is not available to corroborate this. Further, considering the weak driving forces of passive systems based on natural circulation, careful design and analysis methods must be employed to assure that the systems perform their intended functions.

The increasing use of passive systems in the innovative nuclear reactors puts demand on the estimation of the reliability assessment of these passive systems. The passive systems operate on the driving forces such as natural circulation, gravity and internal stored energy, which are moderately weaker than that of active components. Hence, phenomenological failures (virtual components) are equally important as equipment mechanical failures (real components) in case of passive system reliability evaluation.

Substantial efforts are underway towards improving the reliability assessment methods for passive systems through coordinated research project by International Atomic Energy Agency (IAEA) involving different member countries; however, consensus is not reached [1]. The REPAS and APSRA methods are widely used reliability analysis methods for passive systems. REPAS methodology is based on estimating the reliability through parametric uncertainty analysis [2]. APSRA methodology is based on establishing failure surface and estimating the system reliability through classical fault tree approach [3]. The response surface-based methodology using a linear regression has been demonstrated for safety grade decay heat removal system [4].

In this paper, the approach is presented to estimate the passive system reliability of a typical isolation condenser system using artificial neural network (ANN)-based response surface method. The integration of passive system reliability into PSA is also demonstrated using the accident sequence analysis for a typical operational transient.

2 Description of Isolation Condenser System

There may be two or more identical, redundant trains of ICS depending on the design of NPP. Typically, each ICS train consists of seam drum, isolation condenser submerged into elevated water pool, 'normally open' steam supply valves, 'normally closed' condensate return valves and associated piping. Figure 1 provides a schematic of a typical ICS loop deployed in advanced NPP.

During the normal operation of reactor, steam produced in the steam drum is fed to the steam turbine through steam lines. The main condenser rejects the heat to atmosphere through condenser cooling system and feed the condensate back to steam drum through series of feed water heaters and feed pumps. Tap-off connections are made from the main steam lines to ICS loops. When reactor is shut down, the normal heat removal system is isolated and ICS is actuated to remove the decay heat produced in the reactor. Normally, steam side valve is kept open and condensate side valves get open on actuation signal. The steam rises into isolation condensers submerged into water pool generally located at a higher elevation for creating density difference in the loop, which is a driving force of the operation of ICS. The steam gets condensed into submerged isolation condenser, and the condensate is returned to steam drum due to gravity when the return valves open and when ICS is actuated.

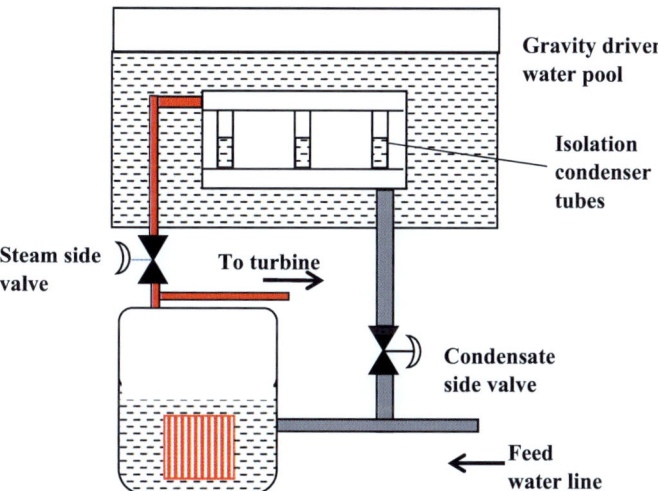

Fig. 1 Schematic of isolation condenser system [5]

3   Estimation of Passive System Reliability

The linear regression-based approach is adopted in most of the prevailing methods for simplifying the problem, which may not be necessarily providing the accurate prediction in all cases. The REPAS uses linear regression approach for generating mathematical relationship between input parameters and the output parameter. However, linear relationship between input parameters and output parameter may not be always true. The relationship could be nonlinear also. In such cases, more sophisticated machine learning tools are required. To make prediction about output parameter, there can be two approaches, namely (i) predict a complex nonlinear function (ii) and break the entire problem spectrum into multiple steps and solve for each step. The latter approach leads to the use of artificial neural network (ANN). Based on the Wilks' approach [6], sample size of 93 data set points was found to be adequate in order to affirm with 95% confidence coefficient that the output parameter would bind at least 95% of the population. The same data set is used for both REPAS method and ANN approach. The details on the input parameters and uncertainty characterization are provided in Table 1.

Initially, various ANN architectures were studied using different training data sets, hidden layers and hidden layer neurons for selecting the optimized ANN model. Finally, two-layer feed-forward ANN network with 10 numbers of 'sigmoid hidden layer neurons' and 'linear output layer neurons' is developed in MATLAB. The 93 sets of six numbers of input parameters were provided in ANN model as columns in a matrix; 93 sets of output vectors (i.e. integral power ratios obtained from thermal-hydraulic analysis of ICS) were also provided in another matrix. The 'Levenberg-Marquardt' back propagation algorithm is used as training function for the ANN model. Inside ANN model, the input vectors and target vectors are randomly divided into three sets as follows [7]:

- 70% (65 data) are used for training.
- 20% (18 data) are used to validate ANN.
- 10% (10 data) are used as a completely independent test of network generalization.

The trained ANN model was then used to estimate the output parameter values for the 'new' set of input parameters. The results obtained through ANN model were validated against the corresponding results obtained by thermal-hydraulic performance analysis carried out using the RELAP5 computer code. Table 2 provides the details on the validation of the ANN-based model as compared to linear regression-based REPAS method. It can be seen that ANN method outperforms the linear regression-based REPAS method. The results improve when the number of training data points increases.

In the present study, isolation condenser ratio (ICR) was used as a performance indicator parameter for the ICS. The ICR is defined as follows:

Table 1 Input parameters of ICS

Design parameters	Unit	Nominal value	Range	Discrete initial values and probabilities							
SD pressure (P1)	Bars	76.5	70–86	70.0		73.0	76.5	83.0	86.0		Value
				0.01		0.03	0.90	0.04	0.02		PDF
SD water level (L1)	m	2.165	0.6–3.0	0.8		1.5	2.0	2.165	3.0		Value
				0.01		0.02	0.07	0.85	0.05		PDF
GDWP water level (L3)	m	5.0	3.0–5.0	3.13		3.50		4.48	5.0		Value
				0.01		0.04	0.05	0.05	0.9		PDF
GDWP initial temperature (Tp)	°C	40	35–95	35		40	50	60			Value
				0.1		0.85	0.03	0.02			PDF
Non-condensable gas fraction (X1)		0.0	0–1	0.00	0.01	0.10	0.20	0.30	0.40	0.50	Value
				0.80	0.10	0.05	0.03	0.01	0.006	0.004	PDF
IC Tube water level (L2)	%	100	0–100	50.0		80.0		100.0			Value
				0.03		0.07		0.90			PDF

Table 2 Goodness of fit for ANN-based method and REPAS method

Data points	Coefficient of determination (R^2)	
	REPAS method	ANN method
50	0.670960	0.790234
70	0.706086	0.652558
100	0.722400	0.941200

$$\text{ICR} = \int_0^t W_i \, \mathrm{d}t \Bigg/ \int_0^t \dot{W}_{\text{Base}} \, \mathrm{d}t \qquad (1)$$

where W_i and W_{Base} are the cumulative heat rejected into GDWP during the specific system configuration and cumulative heat rejected into GDWP during the base case for the specified mission time.

If the estimated value of ICR falls below 0.6, the ICS is considered to have failed in its mission to successfully remove the decay heat from the NPP. This way, the ICR values were obtained using the trained ANN model for new set of input parameters for large number of configurations using Monte Carlo simulation approach. Total 20 sets of Monte Carlo simulations with the size of 10,000 were carried out (total nos. of simulations $= 2,00,000$) for estimating the passive system failure probability. Using the above-mentioned failure criteria (ICR less than equal to 0.6), the failure probability of ICS was estimated to be $7.08\text{E}{-}03 \pm 0.79\text{E}{-}03$.

4 Integration of Passive System Reliability into Probabilistic Safety Analysis

Probabilistic safety assessment (PSA) is a systematic and comprehensive methodology to evaluate risks associated with every life cycle aspect of a complex engineered technological entity such as a facility, a spacecraft or a nuclear power plant. The development of PSA over the years has led to three internationally accepted levels of analysis (i.e. Level 1, Level 2 and Level 3).

4.1 Description of Operational Transient

For the purpose of the demonstration of approach for integrating passive system reliability into probabilistic safety assessment (PSA), a hypothetical reactor design based on general engineering design concepts being deployed in NPPs is considered. In this paper, only those systems, which are required to function under the operational

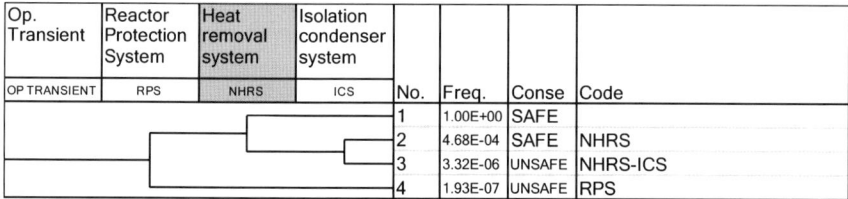

Op. Transient	Reactor Protection System	Heat removal system	Isolation condenser system	No.	Freq.	Conse	Code
OP TRANSIENT	RPS	NHRS	ICS	1	1.00E+00	SAFE	
				2	4.68E-04	SAFE	NHRS
				3	3.32E-06	UNSAFE	NHRS-ICS
				4	1.93E-07	UNSAFE	RPS

Fig. 2 Event tree model for operation transient

transient, are described. These include reactor protection system (RPS), normal heat removal system (NHRS) and passive isolation condenser system (ICS).

The reactor is operating at the rated power with all systems working normal. One of the process parameters is changed due to small deviation in the operating condition; the disturbance is beyond the capability of the control system, and reactor trips on process parameter reaching a pre-defined set point. The RPS actuates and brings reactor to shutdown state. If RPS fails to trip the reactor, core damage occurs.

Subsequent to reactor trip, decay heat is to be removed from the core to achieve the safe state. The NHRS continues to operate, takes the heat from the reactor core and transfers it to ultimate heat sink. If NHRS fails, the standby passive ICS actuates and provides the necessary cooling to the reactor coolant system. If both active and passive decay heat removal systems (i.e. NHRS and ICS) fail, then 'core damage' occurs. The event tree model is developed using RiskSpectrum® PSA software tool [8] which is shown in Fig. 2.

4.2 Assessment of System Unavailabilities of Active Systems

The frequencies for above-mentioned event sequences leading to core damage are required to be estimated in PSA. The probability of failure of passive ICS is estimated in Sect. 3.

The failure probabilities of active systems such as RPS and NHRS are estimated in this section. The system unavailabilities of these active systems are estimated using the classical fault tree analysis approach [9] using RiskSpectrum® PSA software tool. The generic failure data [10, 11] are used for the estimation of component failure rates and probabilities. Common cause failures are considered appropriately using alpha factor method [12]. For human reliability analysis, only latent human errors are considered, and human failure probabilities for latent actions are estimated using THERP method [13]. The dynamic human errors are not envisaged in the operational transient considered in this analysis.

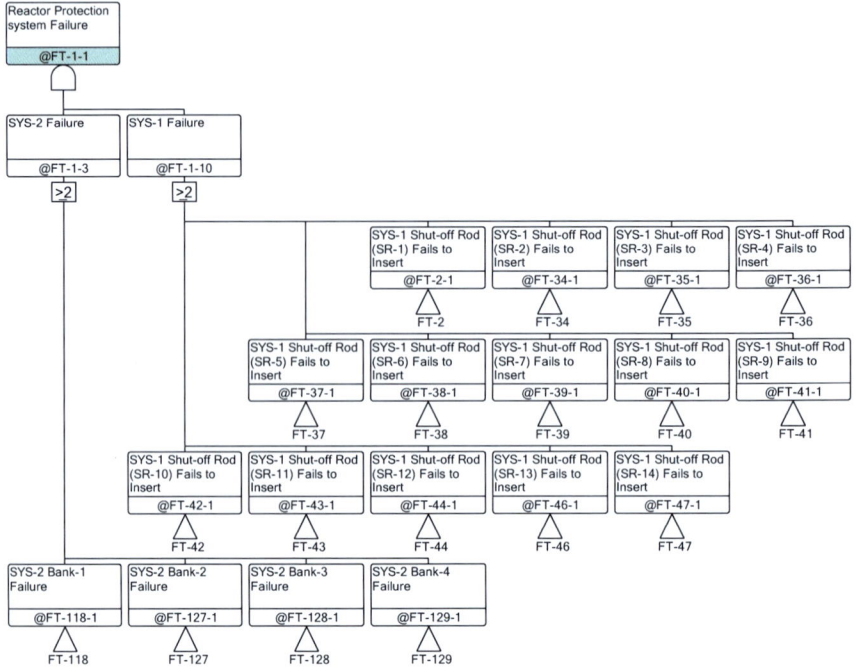

Fig. 3 Broad level fault tree model for RPS

4.2.1 Reliability Analysis of RPS

The RPS consists of two independent shutdown systems. The first system (SYS-1) consists of mechanical shutoff rods, which are parked outside the reactor and drops into reactor under gravity upon actuation. The second system (SYS-2) provides back-up to the first system, which consists of shut-off tubes in which liquid neutron absorber solution is added to provide negative reactivity. If both SYS-1 and SYS-2 fails to function, RPS fails to trip the reactor. The Broad level fault tree model for RPS is shown in Fig. 3 for illustration purpose.

4.2.2 Reliability Analysis of NHRS

The NHRS supplies feed water to steam drum during normal operation of the reactor. When reactor trips, the steam from the steam drum is dumped into main condenser through steam dump valves. The steam from the turbine exhaust is condensed in the main condenser. The condensate water is supplied to the deaerator storage tank through series of feed water heaters through condensate extraction pumps and the feed water pumps. Detailed fault tree is developed using RiskSpectrum® PSA software

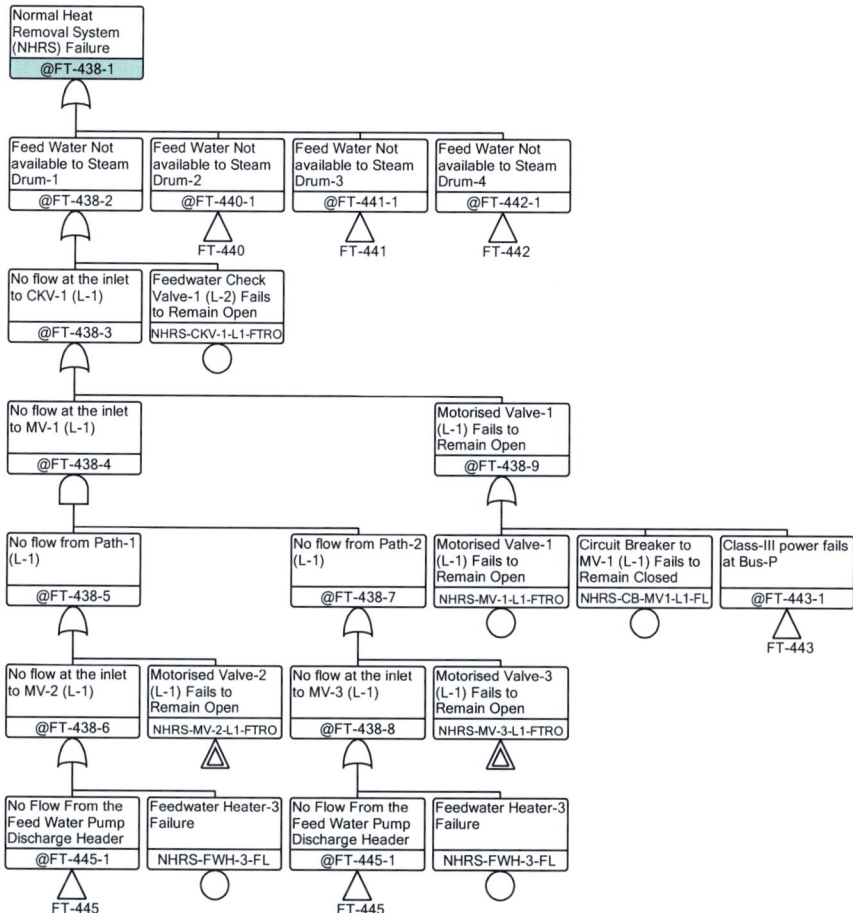

Fig. 4 Broad level fault tree model for NHRS

tool to estimate the system unavailability. Broad level fault tree model is shown in Fig. 4 for illustration purpose. The unavailability of NHRS is given in Table 3.

4.3 Estimation of 'Core Damage Frequency'

Each accident sequence is categorized with end-state category as 'Safe' or 'Unsafe'. The sequences leading to 'Unsafe' failures are considered to eventually result in 'Core damage'. For estimating core damage frequency induced due to operational transient (CDF_{OT}), the frequencies of all event sequences depicted in Fig. 2, which results into the end-state category of 'Unsafe' failure, are summed together. With this approach, the CDF_{OT} works out to be 3.51E−06/year.

Table 3 Active system unavailabilities

System	Unavailability
RPS	1.93E−07
NHRS	4.68E−04

5 Conclusion

The use of ANN approach for response surface generation provided improved results as compared to linear regression-based REPAS approach. This is due to the fact that the simplified assumptions made in linear regression-based approaches may not always be true, especially for complex systems. Considering the fact that passive systems are increasingly being deployed in the modern NPPs, estimation of their reliability with reasonable degree of confidence and its integration into PSA is essential for 'Risk-Informed' regulatory decision-making. Researchers have put efforts to address the open issues in reliability assessment; however, still there is a scope for improvement. As an extension of this work, efforts may be devoted to consider the dynamic effects of interactions of passive systems with other active systems in NPP, while integrating the reliability analysis results of passive systems into PSA.

References

1. P.E. Juhn, J. Kupitz, J. Cleveland, B. Cho, R.B. Lyon, IAEA activities on passive safety systems and overview of international development. Nucl. Eng. Des. **201**, 41–59 (2000)
2. F. Bianchi, L. Burgazzi, F. D'Auria, M.E. Ricotti, NEA/CSNI/R(2002)10, The REPAS approach to the evaluation of Passive Safety Systems Reliability, in *Proceedings of an International Workshop hosted by the Commissariat à l'Energie Atomique (CEA) held in Cadarache, France*, 4–6 Mar 2002
3. A.K. Nayak, M.R. Vikas Jain, Hari Prasad Gartia, A. Anthony, S.K. Bhatia, R.K. Sinha, Reliability assessment of passive isolation condenser system of AHWR using APSRA methodology. Reliab. Eng. Syst. Saf. **94**, 1064–1075 (2009)
4. T. Sajith Methews, M. Ramakrishnan, U. Parthasarthy, A. John Arul, C. Senthilkumar, Functional reliability analysis for safety grade decay heat removal system of Indian 500 MWe PFBR. Nucl. Eng. Des. **238**, 2369–2374 (2008)
5. R.B. Solanki, H.D. Kulakarni, S. Singh, A.K. Verma, P.V. Varde, Optimization of regression model using principal component regression method in passive system reliability assessment. Prog. Nucl. Energy **103**, 126–134 (2018)
6. S.S. Wilks, Determination of sample sizes for setting tolerance limits. Ann. Math. Stat. **12**, 91–96 (1941)
7. R.B. Solanki, H.D. Kulakarni, S. Singh, P.V. Varde, A.K. Verma, Artificial Neural Network (ANN) based response surface approach for passive system reliability assessment, in *Proceedings of 1st International and 4th National Conference on Reliability and Safety Engineering (INCRS-2018), PDPM IIITDM Jabalpur*, 26–28 Feb 2018
8. RiskSpectrum® PSA, Advanced fault tree and event tree software tool, A member of the Lloyd's Register Group
9. Atomic Energy Regulatory Board, *Probabilistic Safety Assessment Guidelines, AERB Safety Manual*, AERB/NF/SM-1, June (2008)

10. International Atomic Energy Agency, *Component Reliability Data for use in Probabilistic Safety Assessment*, IAEA-TECDOC-478 (1998)

11. Atomic Energy Regulatory Board, *Compendium of Standard Generic Reliability Database for Probabilistic Safety Assessment of Nuclear Power Plants*, AERB/NPP/TD/O (2006)

12. NUREG/CR-5801, *Procedures for Treating Common Cause Failures in Safety and Reliability Studies* (1993)

13. NUREG\CR. 1278, *Handbook of HRA with Emphasis on Nuclear Power Plant Application*, Final Report, Sandia National Laboratories, August (1989)

PSA Level-2 Study: Estimation of Source Term for Postulated Accidental Release from Indian PHWRs

Amit Kumar, Vageesh Shukla, Manoj Kansal and Mukesh Singhal

Abstract Source term is generally known as the amount of the radionuclides (fission products along with activation and actinides) that get released from a nuclear power plant in an accident. Since the release of radionuclides depends on accident scenario, various plant damage states which include design basis accident and severe accident with and without core meltdown have been considered in a systematic way as a part of PSA level-2 study with containment event tree during release categorization. Detailed study was done to estimate the source term for Indian 220 MWe PHWR with current understanding of radionuclide release from core and behavior of it in primary heat transport (PHT) system and containment. The major factors affecting potential radionuclide releases into environment as a result of nuclear power plant accidents are described. Engineering safety features play an important role to control and reduce the radionuclide release to environment. It is shown that during accident, release to environment is reduced if containment integrity is maintained. Well-established accident management guidelines exist for every Indian nuclear power plant to mitigate the accident progression. Using containment filtered venting system, containment integrity can be maintained if containment is over-pressurized, and effect of it to reduce the radionuclide release to environment is discussed. Based on the release of radionuclides to environment, final release category has been defined, which helps to know the severity of accident, and accordingly, emergency preparedness can be done. The estimated source term will be used for PSA level-3 study.

A. Kumar (✉) · V. Shukla · M. Kansal · M. Singhal
Reactor Safety & Analysis, Nuclear Power Corporation of India Limited,
Anushakti Nagar, Mumbai 400094, India
e-mail: kamit@npcil.co.in

V. Shukla
e-mail: vshukla@npcil.co.in

M. Kansal
e-mail: kansalm@npcil.co.in

M. Singhal
e-mail: singhalm@npcil.co.in

© Springer Nature Singapore Pte Ltd. 2020
P. V. Varde et al. (eds.), *Reliability, Safety and Hazard Assessment
for Risk-Based Technologies*, Lecture Notes in Mechanical Engineering,
https://doi.org/10.1007/978-981-13-9008-1_2

Keywords Accidental release · CET · CFVS · Containment · ESFs · LBLOCA · PHT · PSA · Radionuclides · SAMG · SBO · Source term

1 Introduction

Source term can be defined as the magnitude, timing, composition, physical and chemical form of release from core and containment, which in term depends upon the fissile content, operation power and accident consequences. The amount of radionuclides is a fundamental parameter to estimate the consequences of an accident on individuals and environment. A level-2 PSA quantifies the magnitude and frequency of radioactive release to the environment following core damage and containment failure. This level of analysis builds on various plant damage states which makes interface between PSA level-1 and level-2 studies. It evaluates accident phenomena with different containment failure modes that can lead to radioactive releases to environment from a nuclear power plant.

The aim is to find out what part of the radionuclide originally released from the core will be retained in different areas of the plant, and what will escape to the environment during accident scenario for 220 MWe Indian Pressurized Heavy Water Reactor (IPHWR). The estimation of source term has been done for different plant damage states with containment event tree for preliminary release categories. The in-house developed code 'STAR' [1] has been used to model the release from fuel for various plant damage states (PDS), and another developed code ACTREL [2] has been used for modeling the various phenomena/behavior of radionuclides in containment with engineering safety features (ESFs). Based on the release of radionuclides to environment, final release category has been defined. If quantum of release is similar to the particular postulated accident scenarios, it has been grouped into same final release category.

Further, the effect of containment filtered venting system (CFVS) on release of radionuclides to environment is described for extended station blackout (SBO)-initiated scenario and large break loss-of-coolant accident (LBLOCA)-initiated severe accident with core melt.

2 Factors Determining Importance of a Radionuclide

The importance of any radionuclide is determined by its potential to cause harm to the human being or to the environment after an accident. The main factors determining importance of a radionuclide are:

- Total inventory and its volatility;
- Half-life and nature of its radioactive emission;
- Chemical and physical properties that determine transport behavior;

- Biological characteristics (such as its uptake, biological half-life, specific effects on organs).

Some of the factors mentioned are inherent to the radionuclide involved (such as inventory, half-life, radioactive emissions), while others are dependent on the features of the reactor and the accident scenario. Further, the composition of the released radionuclide might change with the nature of accident. However, in general, the fission product noble gases and the volatile (iodine, cesium, and tellurium) are of importance in practically all accidents.

2.1 Radionuclide Groups

In a nuclear reactor, radioactive material is generated either due to fission of fissile and fissionable materials or due to activation of materials by neutron interaction. Due to the production of very large number of radionuclides (more than 800), the radionuclides are arranged in different groups based on their chemical nature and release rate from fuel. Due to similarities in chemical nature, similar release and transport behavior is employed. The grouping of it is given in Table 1 [3, 4].

2.2 Selection of Important Radionuclides for Accident Study

Due to difference in physical and chemical properties, all radionuclides present inside the core are not equally important from health effect viewpoint. WASH-1400 recommends that radionuclides having half-life less than 25.7 min should not be considered for PSA level-2 and level-3 studies [3]. Radionuclides having half-life less than 3 min are not considered as they will decay before coming out from containment. Apart from half-life, total inventory of radionuclide, fractional release, decay mode, and chemical interaction are the key features for selection of radionuclides.

Table 1 Grouping of radionuclides

Group	Element
1	Noble gases (Xe, Kr)
2	Halogens (I, Br)
3	Alkali metals (Cs, Rb)
4	Tellurium (Te, Se, Sb)
5	Alkaline earth metals (Ba, Sr)
6	Noble metals (Ru, Mo, Rh, Pd, Tc, Ag)
7	Lanthanides, actinides, and low-volatile elements (Y, Zr, Nb, La, Am, Cm, Pm, Sm, Ce, Np, Pu)
8	Tritium and carbon-14

As release of radionuclides from fuel mainly depends on the fuel temperature which in turns depends upon accident scenario, low-volatile radionuclides, e.g., zirconium, ruthenium, molybdenum, etc., and actinides are considered only in category B of design extension condition (DEC) scenario where core melting takes place. In design basis accident (DBA) and category A of DEC, where fuel does not reach to very high temperature, FPNG, volatiles, and semi-volatile radionuclides have been considered. Considering the above, 45 radionuclides have been chosen for DBA and DEC-A scenarios, and 84 for DEC-B scenario as listed below [5].

Accident scenario	Volatility	Radionuclides
DBA and DEC-A	FPNG	Kr-83m, Kr-85m, Kr-85, Kr-87, Kr-88, Kr-89, Xe-131m, Xe-133m, Xe-133, Xe-135m, Xe-135, Xe-137, Xe-138
	Volatile elements	I-131, I-132, I-133, I-134, I-135, Rb-86, Rb-88, Rb-89, Cs-134, Cs-136, Cs-137, Sb-127, Sb-129 Te-127, Te-127M, Te-129, Te-129m, Te-131m, Te-132, H-3, C-14
	Elements with intermediate volatility	Sr-89, Sr-90, Sr-91, Sr-92, Ba-139, Ba-140, Ru-103, Ru-105, Ru-106, Ag-110m, Ag-111
DEC-B	Low-volatile elements	Y-90, Y-91, Y-92, Y-93, Zr-95, Zr-97, Co-58, Co-60, Nb-95, Mo-99, Tc-99M, Rh-103m, Rh-105, Pd-100, Pd-107, Pd-109, La-140, La-141, La-142, Pr-143, Nd-147, Am-241, Am-243, Cm-242, Cm-244,Cm-245, Eu-154, Pm-148m, Pm-150, Sm-146, Sm-151, Ce-141, Ce-143, Ce-144, Np-239, Pu-238, Pu-239, Pu-240, Pu-241 and radionuclides considered in DBA/DEC-A scenario

3 Radionuclide Release During Accident

Design basis accident (DBA) and design extended condition (DEC) with and without core melt have been covered using various plant damage states (PDS) for the present study. A wide range of LOCA- and SBO-initiated accidents has been analyzed to determine the release into the environment.

3.1 Barriers to Fission Product Release

Fission products generated within the core are prevented by many barriers to reach the environment. The aim of source term estimation is to quantify the effectiveness of these barriers under accident condition. The first barrier to radionuclide release is the UO_2 fuel matrix, which normally has a huge capacity for the retention of fission products. The second barrier is the clad which hermetically seals the fuel, thus retaining the small fraction of volatile FPs which diffuse out from the fuel matrix and collect in the space within and outside the fuel matrix. It must be noted that the clad is subject to overheating during severe accident conditions and may fail at high temperature. The third barrier is the reactor coolant system, which depends strongly on the reactor type and the event sequences. The fourth barrier is the reactor containment, whose effectiveness again depends on the reactor type.

3.2 Release from Fuel

The volatile fission products, stable as well as radioactive, that are diffused out from the fuel matrix are accumulated in the fuel-clad gap during normal reactor operation. When the clad ruptures at high temperature, the entire gap inventory [calculated using WASH-1400 model] may get released to reactor coolant system (RCS). Subsequently, volatile, semi-volatile, and low-volatile fission products trapped inside the fuel matrix will get released due to overheating of fuel. Transient release at high temperature is calculated using CORSOR-M model. Due to the unavailability of moderator cooling in PDS-A2 and PDS-A3, moderator will start boiling and after some time uncovery of fuel channels will take place, which will cause heat up of fuel channels and finally falling of fuel channels on the bottom of calandria. Due to the availability of calandria vault water, further heating will transform the porous terminal debris bed into a homogeneous pool of molten corium sandwiched between top and bottom crusts [6]. Release from core for various plant damage states is given in Table 2.

Special Treatment of Tritium and carbon-14:

Most of the carbon-14 produced in moderator and coolant is efficiently removed by purification system, and only a small steady-state quantity is remaining in the systems. Carbon-14 generated inside the fuel (around 8.3% of total production) will be available inside the fuel and get released during anticipated accident. For the release of tritium, the activity present inside coolant and moderator is considered as total inventory. For tritium release, a part of the total activity present inside coolant and moderator will release to the environment. Actual release depends on the accident scenario [7].

Table 2 Release from core for various plant damage states

Plant damage state (PDS)	Description	Release percentage from fuel								
		I-131	Cs-137	Te-132	Xe-133	Kr-88	Sr-90	Ru-103	Cm-242	C-14
PDS-A1	Large break loss-of-coolant accident and shutdown system failure	100	100	100	100	100	100	–	–	100
PDS-A2	Station blackout and total failure of heat sink	100	100	100	100	100	60	0.2	5.0	100
PDS-A3	Large break loss-of-coolant accident and total failure of heat sink	100	100	100	100	100	60	0.2	5.0	100
PDS-B1	Large break loss-of-coolant accident and emergency core cooling system failure	2.73	11.8	2.06	0.82	0.18	0.047	0.031	–	22.5
PDS-B2	Small break loss-of-coolant accident and emergency core cooling system failure	2.0	8.53	1.54	0.61	0.16	0.022	0.010	–	16.4
PDS-B3	Critical break loss-of-coolant accident in reactor inlet header/outlet header	0.90	3.5	0.69	0.25	0.04	0.0005	0.0002	–	–
PDS-B4	Stagnation feeder break	0.10	0.14	0.10	0.097	0.095	–	–	–	–

3.3 Behavior in PHT System

It is clear that the thermal-hydraulic conditions, the timing and composition of the core material released, make a number of different released species possible. The species involved and history of their formation, including aerosol nucleation, will affect transport and deposition in the RCS because of the different physical and chemical properties involved. The determination of the chemistry associated with the release of fission products from fuel under accident is complex. Except noble gases and a small fraction of iodine, all other fission products would have condensed prior to entering the containment atmosphere and would be present as aerosol.

The behavior of iodine largely depends on the chemical form in which it is released. It is evident from experiment that CsI is a predominant form of iodine in reducing conditions. CsI is quite stable at temperature as high as 1600 °C and condense either forming aerosol by self-nucleation or depositing on RCS walls or on other aerosols. Usually, the process of conversion into aerosols is completed in the RCS. Credit of water-trapping factor for volatile release is considered in the PHT circuit.

3.4 Behavior in Containment

The behavior of radionuclides in containment largely depends on the chemical and physical forms of the radionuclides released to the containment. The initial chemical form is in most cases either elemental or oxide, and physical form is either gas or vapor and aerosol. As it has been discussed, except noble gases and a small fraction of iodine, all other radionuclides would enter into the containment in aerosol form.

In the specific case of iodine, if it enters containment in the volatile form, it would get readily deposited on metal and painted surfaces. The deposition rate would be higher for the surfaces where steam condensation occurs. Formation of organic iodide would occur due to the presence of various hydrocarbons present in the containment, which may dissociate under the high radiation conditions.

Partitioning of iodine in containment atmosphere would mainly depend on the concentration of iodine in the water, its pH and temperature. If ECCS fails to come initially, but established sooner or later through operator action, the iodine would finally accumulate in suppression pool water. The credit of partition factor is taken for that. The volatile plate-out halftime for deposition is credited in primary as well as secondary containments. Natural decay phenomenon for all radionuclides is considered.

4 Release Category

In the level-2 PSA, for given plant damage states (PDS), containment event trees (CETs) are used in modeling severe accident progression and containment response. Each branch point corresponds either to the availability of containment system function or to the likelihood of occurrence of some physical phenomenon. A CET serves to link PDS with radionuclide releases. The CETs produce a large number of end states, some of which are identical in terms of key release attributes. These end states are grouped together. These groupings are referred to as release categories/release bins. The release categories group CET end states that would be expected to have similar radiological characteristics and potential off-site consequences.

4.1 Preliminary Release Category

The source term estimation has been attempted in maximum detail possible with the available modeling capability. Preliminary following four source term release groups are considered to cover the entire spectrum of accident scenarios [8].

RC-A: This category encompasses event sequences with design extended condition with core melt scenarios (DEC-B) with early/late containment failure.

RC-B: This category involves event sequences involving design basis accidents/design extended condition (DEC-A) followed by failure of primary containment isolation.

RC-C: This category includes plant damage states related to design basis accidents/DEC-A along with various combinations of containment-related events like excess leakage/failure of instrument air isolation system followed by failure of one or more activity management systems (ESFs).

RC-D: It includes design basis accidents/DEC-A with successful containment isolation and isolation of instrument air related to non-LOCA valves, and all containment ESFs are activated as per design. Both containments leak as per specified leak rates at respective design pressures. The containment bypass events like SG tube rupture with successful actuation of all mitigation systems are also included in this group.

4.2 Final Release Category

The preliminary categorization has been done for identifying the accident sequences from each subgroup for release estimation in an organized way. Later on, after the release estimation, the accident sequences resulting in similar range of magnitude of release were merged and the characteristics of the final release categories were defined in terms of magnitude of release expected from this category. For all the final release categories, one event sequence is selected as representative of all sequences in that category in such a way that this sequence exhibits a release magnitude and consequent dose typical of events in that category and should be an expected significant contributor to that release category. The description of final release categories after merging the preliminary release categories is given in Table 3 in the subsequent section.

5 Assessment of Source Term (Release to Environment)

The effects of the engineered safety features (ESFs), in-leakage of instrument air and leak-tightness of the primary and secondary containments, are modeled in the estimation of radioactivity release (both at ground and at stack level) to the atmosphere depending upon the accident sequences. Following are the failure modes of containment and containment ESFs that can give rise to releases outside containment either independently or in combination.

(a) Containment isolation failure;
(b) Local hydrogen deflagration due to failure of post-accident hydrogen management system and passive catalytic recombiner devices (PCRDs);
(c) Pressurization due to compressed air in-leakages due to failure of isolation of instrument air to non-LOCA-related valves;
(d) Secondary containment recirculation and urge (SCRP) system failure;
(e) Primary containment filtration and pump back (PCFPB) system failure;
(f) Primary containment controlled discharge (PCCD) system failure.

Early containment failures are usually associated with high source terms. On the other hand, a delayed failure will ensure that a good part of the radionuclide reaching the containment is retained therein. During preliminary release category, effects of ESFs were considered. For final release categorization, containment isolation failure/late containment failure mode dominates and that representative case has been taken. The description of final release category and its release percentage with respect to core inventory has been described in Tables 3 and 4, respectively.

6 Observations

It can be seen that power excursion events with containment isolation failure (RC-1) give maximum release. During severe accident with DEC-B type scenario, if containment isolation fails, then significant release to environment occurs as shown in RC-2A and RC-2B. If late containment failure occurs, then release decreases significantly due to retention time and plate out as shown in RC-4A and RC-4B. For release category RC-4A, containment failure and excess leakage occur due to local hydrogen deflagration at 36 h of accident, and by this time, maximum radionuclides

Table 3 Description of final release category

Release category	Sub-release category	PDS	Release scenarios
RC-1		PDS-A1	Large break LOCA + failure of reactor shutdown system + containment isolation failure
RC-2	RC-2A	PDS-A3	Large break LOCA + total failure of ECCS + failure of moderator cooling + failure of primary containment isolation
	RC-2B	PDS-A2	Station blackout + failure of fire water system to steam generator + failure of primary containment isolation
RC-3		PDS-B1	Large break LOCA + total failure of ECCS + failure of primary containment isolation
RC-4	RC-4A	PDS-A3	Large break LOCA + total failure of ECCS + failure of moderator cooling + failure of PCRDs + late containment failure
	RC-4B	PDS-A2	Station blackout + failure of fire water system to steam generator + failure of PCRDs + late containment failure
RC-5		PDS-B2	Small break LOCA + total failure of ECCS + failure of primary containment isolation
RC-6	RC-6A	PDS-A3	Large break LOCA + total failure of ECCS + failure of moderator cooling + containment intact
	RC-6B	PDS-A2	Station blackout + failure of fire water system to steam generator + containment intact
	RC-6C	PDS-A3	Large break LOCA + ECCS + moderator cooling failure * Calandria vault cooling failure * Calandria vault hook-up successful [SAMG action] * PCRD successful * CFVS successful
	RC-6D	PDS-A2	Station blackout + failure of fire water system to steam generator * Calandria vault cooling failure * Calandria vault hook-up successful [SAMG action] * PCRD successful * CFVS successful
RC-7		PDS-B4	Stagnation feeder break + containment isolation failure

Table 4 Release percentage for final release categories

Release category	Sub-release category	Release from containment			
		I-131 (%)	Cs-137 (%)	Te-132(%)	Noble gases (%)
RC-1		10.09	10.04	10.04	53.09
RC-2	RC-2A	2.68	3.08	2.49	27.36
	RC-2B	2.62	2.84	2.35	12.78
RC-3		0.027	0.118	0.020	0.20
RC-4	RC-4A	0.007	0.004	0.004	1.34
	RC-4B	0.027	0.018	0.014	2.6
RC-5		0.015	0.079	0.011	0.15
RC-6	RC-6A	0.0053	0.0040	0.0038	0.58
	RC-6B	0.0061	0.0045	0.0037	0.84
	RC-6C	0.0051	0.0040	0.0038	4.42
	RC-6D	0.0097	0.0096	0.0077	3.91
RC-7		0.0000003	0.0000004	0.0000003	0.030

plate out except noble gases. For release category RC-4B, design pressure of containment reaches by 26 h, so excess leakage occurs after that. If calandria vault cooling fails, then release can be increased significantly and molten core–concrete interaction (MCCI) may occur. To stop the MCCI, accident is ceased using SAMG action to put water in calandria vault by injection through hook-up point. As extra water inventory is added, pressure within containment reaches very high from design pressure due to steam formation which can breach containment. Further, containment integrity is maintained using CFVS as per severe accident management guidelines, and release of radionuclides is limited as shown in the case of RC-6C and RC-6D. Due to excess pressure, release of noble gases increases in case of RC-6C and RC-6D. Decontamination factor of 100 is considered for non-noble gas radionuclides, while for noble gas radionuclides, retention in CFVS is not considered. If containment is intact, then release occurs very less even in severe accident (DEC-B), as can be seen in RC-6A and RC-6B. For DEC-A type of scenario, LBLOCA-initiated event (RC-3) gives more release than SBLOCA-initiated event (RC-5). During design basis accident (RC-7), no significant release occurs, even in case of containment isolation failure.

7 Conclusion

Using systematic approach for PSA level-2 study, different plant damage state was analyzed. For design basis accident and DEC-A type of scenario, ESFs play an important role to decrease and control the release which has been analyzed during preliminary release category. During severe accident, release to environment is very less if containment integrity is maintained. During late containment failure for severe accident, release decreases significantly due to retention and plate out of radionuclides in containment. Using accident management guidelines action, fuel melt can be avoided if water inventory is maintained in calandria using external hook-up point. Further, if accident progresses up to calandria vault cooling failure, then MCCI can be avoided using SAMG action of adding water to calandria vault through injection using hook-up. Containment integrity is maintained using CFVS, and release also decreases significantly. This quantitative estimation of release helps for assessing the effectiveness of safety design features and for the planning of post-accident emergency measures in the public domain.

References

1. User's manual of "Source term analysis and Release (STAR)" code
2. User's Manual of "Activity Release (ACTREL)" code
3. Appendix VI, WASH-1400 (NUREG-75/014), An assessment of Accident Risks in U.S. Commercial Nuclear Power Plants, United States Nuclear Regulatory commission (1975)
4. Accident Source Terms for Light-Water Nuclear Power Plants, NUREG 1465
5. AERB-Safety Guide on methodology for Radiological Impact Assessment of NPPs, Mar 2018
6. M.J. Brown*, D.G. Bailey, Fission-product releases from a PHWR Terminal Debris bed, Canadian Nuclear Laboratories Ltd, Chalk River, ON K0J 1J0, Canada (2015)
7. W. Sohn, D.-W. Kang, W.-S. Kim, An Estimate of Carbon-14 Inventory at Wolsong Nuclear Power Plant in the Republic of Korea (2003)
8. Level-2, Probabilistic Safety Assessment Report for KAPS-1&2, July 2012

Comprehensive Safety Analysis of Station Blackout Scenario in TAPS-1&2

Ritesh Raj, Venkata V. Reddy, Sameer Hajela and Mukesh Singhal

Abstract Tarapur Atomic Power Station-1&2 (TAPS-1&2) are the boiling water reactor units, commissioned in year 1969. TAPS-1&2 are well equipped to cater unavailability of offsite power, by having diverse route for two independent offsite power buses. Three emergency diesel generators (EDGs) of 100% capacity for a twin-unit station have been provided along with a dedicated station blackout diesel generator (SBO DG) with 100% capacity, which will provide a reliable and redundant electric power to continue operation of safety loads in case of loss of offsite power. Even in a very remote scenario of station blackout (SBO) (loss of all AC power including SBO DG) scenario, emergency condenser (EC) is provided to remove decay heat from the reactor system by thermosyphoning for around 8 h. During postulated design extension condition (DEC), time available for operator intervention to mitigate the accident progression of SBO is highly influenced by the EC operation/availability. This paper discusses the results obtained from study of postulated station blackout (SBO) scenarios with both the availability and unavailability of EC in TAPS-1&2. Following SBO, emergency condenser (EC) acts as heat sink by removing core decay heat through thermosyphoning. As a result, the reactor pressure and coolant temperature are reduced, effecting reactor core to cool down. Extended cooling of reactor system beyond 8 h will be done by adding the water in EC shell and reactor pressure vessel (RPV), whereas, in case of unavailability of EC, reactor pressure vessel (RPV) gets pressurized and pressure reaches relief valve at their set point and discharges the steam into wet well; consequently, the RPV level falls. Auto blow down system (ABDS) actuates automatically at reactor low level. Due to ABDS actuation,

R. Raj (✉) · V. V. Reddy · S. Hajela · M. Singhal
Reactor Safety & Analysis, Nuclear Power Corporation of India Limited,
Mumbai 400094, India
e-mail: riteshraj@npcil.co.in

V. V. Reddy
e-mail: vvreddy@npcil.co.in

S. Hajela
e-mail: shajela@npcil.co.in

M. Singhal
e-mail: singhalm@npcil.co.in

© Springer Nature Singapore Pte Ltd. 2020
P. V. Varde et al. (eds.), *Reliability, Safety and Hazard Assessment for Risk-Based Technologies*, Lecture Notes in Mechanical Engineering,
https://doi.org/10.1007/978-981-13-9008-1_3

27

the RPV depressurizes quickly. This action will make EC unavailable completely, and plant will lose its first layer of defence. In order to control the accident progression, the present analysis is done to estimate the time available for operator to resume emergency condenser service upon its failure to start on auto. In such a case, due to loss of heat sink and loss of coolant through relief valves, time available for operator to bring back the EC into service is very important. Using system thermal-hydraulic code RELAP-5, both the above scenarios have been modelled to assess the availability of time for operator intervention to mitigate the accident scenario. Findings of this study are used to develop accident management guidelines.

Keywords DEC · Emergency condenser · Operator intervention · RELAP · SBO

1 Introduction

Tarapur Atomic Power Station (TAPS) is a twin-unit BWR plant with an installed capacity of 2×210 MWe, presently operating at capacity of 2×160 MWe [1]. The reactor is a forced circulation boiling water reactor producing steam for direct use in the steam turbine. The fuel consists of uranium dioxide pellets contained in Zircaloy-2 tubes. Water serves as both the moderator and coolant. Coolant enters from the bottom of the reactor core and flows upwards through the fuel element assemblies where it absorbs heat and gets converted into steam. Steam separators located above the fuel assemblies separate the water particles from the steam–water mixture. This steam is then allowed to pass through a steam dryer to become dry saturated steam before it falls on turbine blades via steam line.

During gliding on the turbine blade, this steam loses its energy in moving the turbine. At the end of the series of turbine blade, a low-pressure steam is available, which is then passed through the condenser. Low-pressure and low-temperature water is then again allowed to pass through the reactor via two recirculation pumps to the tube side of two secondary steam generators and then to the reactor core. Figure 1 shows the general arrangement of reactor pressure vessel (RPV) and recirculation loops of the TAPS-1&2.

In the era of post-accidents at TMI, Chernobyl and Fukushima, IAEA Safety Guide [2] requires that NPPs should be equipped with independent hardware provisions in order to fulfil the fundamental safety functions, as far as is reasonable for design extension conditions including core melt scenarios (severe accidents), to enhance further defence-in-depth (DID). This Safety Guide also requires that equipment upgrades aimed at enhancing preventive features of the NPPs should be considered on priority. In the mitigatory domain, equipment upgradation is aimed at the preservation of containment function. In order to enhance preventive measures to avoid severe accidents, provisions in terms of hardware and facilities are incorporated in TAPS-1&2.

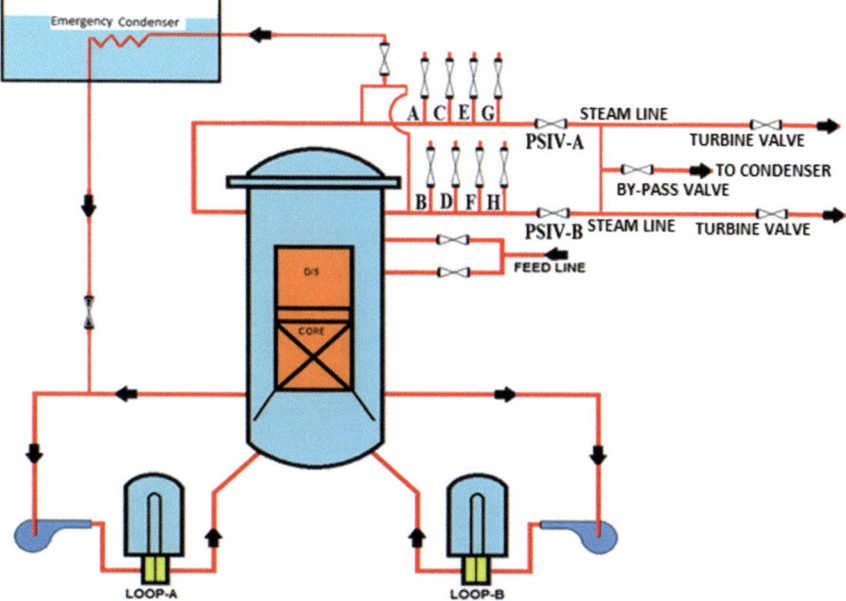

Fig. 1 General arrangement of reactor pressure vessel and recirculation loop

2 Description and Nodalization of SBO Scenario

The present analysis is carried out using commercially available international system thermal-hydraulic computer code RELAP-5 [3–5], which is applicable for LWR reactors. Plant components are subdivided into interconnected small control volumes using lumped parameter approach. The plant configuration is simulated as network of control volumes and junctions. In lumped parameter approach, the thermal-hydrodynamic properties such as pressure, temperatures, quality and voids are calculated at the centre of control volumes, and flows are calculated at the junctions connecting adjoining volumes. Nodalization scheme for the analysis is shown in Fig. 2.

Comprehensive safety analysis for station blackout scenario is done by modelling all the reactor components: RPV, fuel assemblies, emergency condenser, recirculation loops and detailed lines like steam line, feed line, etc. Pump, steam generators and valves are modelled as component-specific models by using realistic associated logics and trip conditions for realistic calculations. The reactor core of TAPS-1&2 contains 284 fuel assemblies (FAs), which are subdivided into three zones respectively. The central hot zone has one FA, the middle zone has 183 FAs and the outermost zone has 100 FAs. There is one bypass channel to simulate the leakage flow out of fuel assemblies. The essential trip parameters and associated control logics are also accounted for the simulation of SBO scenario.

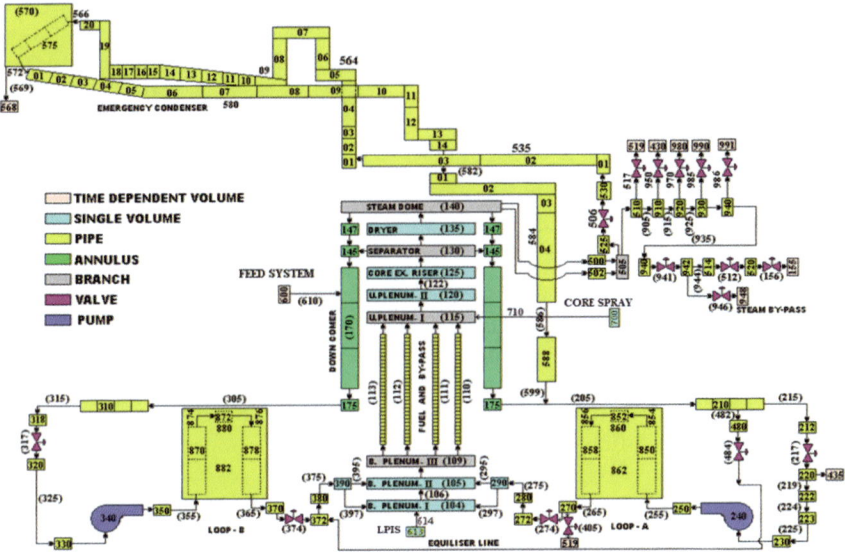

Fig. 2 TAPS-1&2 nodalization scheme for RELAP-5

3 Unavailability of Emergency Condenser

System-level modelling was carried out with plant-specific data for emergency condenser system during the development of full power internal event level-1 probabilistic safety assessment (PSA) [6] for TAPS 1&2. From PSA analysis, it is seen that unavailability of EC is in the order of 9.14 e-5 by considering EC failure modes due to common cause failures of motorized-operated valves (MOVs) fail to open, EC shell failure, EC tubes fail and relays failure. Out of all failures, the dominating sequence is found to be due to common cause failure of MOVs to open (more than 96%).

A number of symptoms such as reactor pressure high, EC shell water temperature not increasing/EC shell water level not depleting and no steam discharge from EC shell are available to draw operator attention about non-functioning of EC. These plant symptoms together with available procedures and training ensure that corrective actions can be initiated within the acceptable time frame. Time available to put EC into service to ensure core cooling is given in the subsequent section.

4 Handling of SBO

Following SBO, the emergency condenser (EC) system comes into service that removes core decay heat for initial six to eight hours; during the process, the reactor pressure and coolant temperature fall, effecting reactor core to cool down. During this period, EC shell side water reaches saturation and starts evaporating and RPV inventory also decreases due to system leakages. In order to cool down the reactor for extended period, replenishment of ECS shell and RPV is needed. For handling of prolonged SBO hook-up, systems are provided to add water inside RPV, EC shell side and fuel pool. Hook-up systems are provided to supply water to different systems from outside of the reactor building/plant buildings for providing the heat sink. These hook-up systems are to be used as part of accident management strategy. In addition to hook-up systems, a 200 kVA air-cooled mobile DG and containment filtered venting system (CFVS) were included as part accident management features during design extension conditions.

However, a series of systems are available to manage any unwanted situation during SBO. Still, it is most desirable to cool the reactor core through designed route of emergency condenser. Therefore, following SBO if EC is not coming into service, then it can be taken into line manually within a short period of time before ABDS actuation. This paper contains the results that estimate the maximum time available to operator to bring back the EC into service before ABDS actuation.

5 Results and Analysis

Two cases were analysed, in which first case is analysed to show the effectiveness of EC for decay heat removal through thermosyphoning and second case is analysed to show the effectiveness of EC if it comes by manual intervention after 30 min.

5.1 Case-1: SBO with Availability of EC on Auto

Steady-state run is from -300.0 to 0.0 s to obtain the normal plant parameters. At time 0.0 s, it is postulated that station blackout (SBO) occurs, resulting in total loss of AC power supply. Scram of the reactor takes place on Class IV power failure, and the reactor power goes down to decay power. Emergency condenser comes into action at its actuation logic (i.e. reactor pressure high for 10 s), thereby the reactor system starts depressurizing. RPV dome pressure and temperature are shown in Fig. 3.

Just after SBO, the RPV pressure shoots up from $71.0 \, \text{kg/cm}^2(\text{a})$ to $81.3 \, \text{kg/cm}^2(\text{a})$ due to the closure of primary steam isolation valves (PSIVs) on failure of Class IV power. After that, the EC quickly depressurizes the RPV and thereby cooling of the clad.

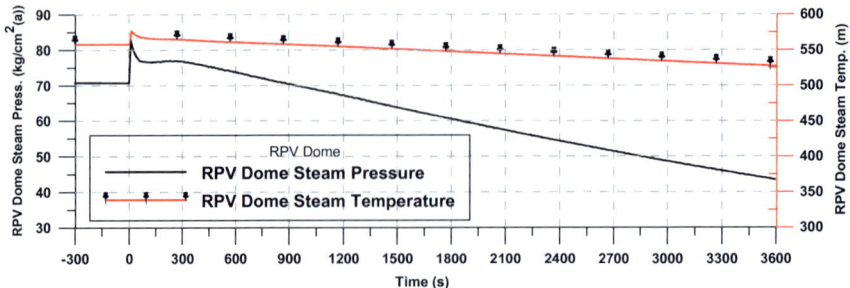

Fig. 3 Reactor pressure vessel dome steam pressure and temperature

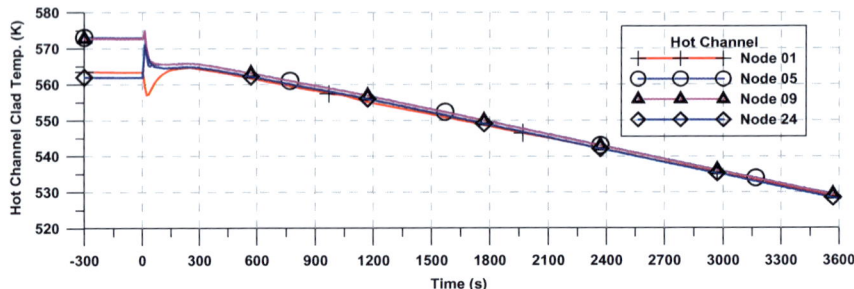

Fig. 4 Fuel clad temperature of hot channel

During an attempt to cool down the reactor, EC water depletes continuously by absorbing the heat of primary steam. Hence, in order to cool down the reactor for extended period of time, i.e. beyond 8 h, replenishment of EC shell and RPV is needed, as RPV inventory also depleted through system leakages. Therefore, water can be added in EC shell and RPV after 8.0 h of SBO with a flow rate of 2.0 kg/s each.

The clad temperatures of average channel-1, average channel-2 and hot channel become constant at around 558.0–573.0 K depending upon the axial power profile. As SBO occurs, clad temperature follows the reactor pressure in steam region. On actuation of emergency condenser, clad temperatures start decreasing and fall below 530.0 K in one hour.

Figure 4 is showing clad temperatures of hot channel. As in this whole study the temperature remains well below the threshold temperature for hydrogen generation 923 K, there is no hydrogen generation at all. Reactor water level is shown in Fig. 5. It is maintained at 13.0 m of level during steady-state run. On occurrence of SBO, it falls up to 11.8 m due to increase in pressure and discharge of water inventory through relief valve to maintain the pressure. As EC starts reducing the pressure, water comes from SSG shell into RPV through the leak paths (refer Fig. 6), which increases the reactor water level during first hour.

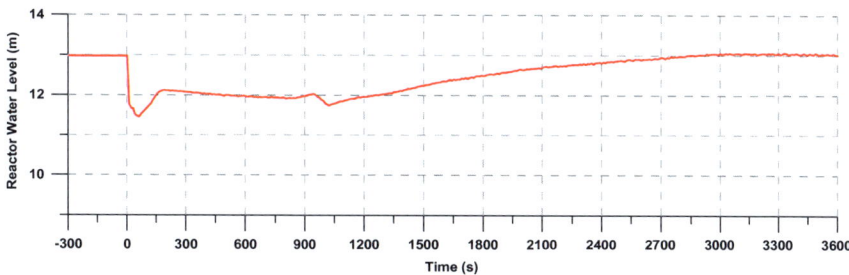

Fig. 5 Reactor water level

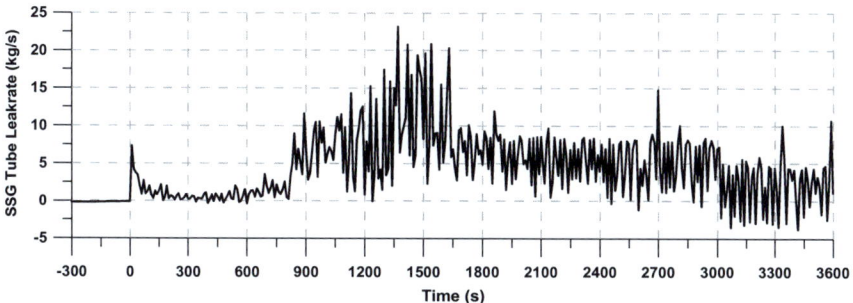

Fig. 6 Leakages in from SSG tubes

5.2 Case-2: SBO with Manual Initiation of EC After 30 min

Steady-state run is from −300.0 to 0.0 s to obtain the normal plant parameters. At time 0.0 s, it is postulated that station blackout (SBO) occurs, resulting in total loss of AC power supply. Scram of the reactor takes place on Class IV power failure, and the reactor power goes down to decay power The primary steam isolation valve (PSIV) also closes along with reactor scram. The closure of PSIV results in the pressurization of reactor pressure vessel (RPV). Because of unavailability of EC, in order to maintain the RPV pressure within design limit, the relief valves (RVs) come into service at their respective set point and discharge the steam into wet well; consequently, the RPV level falls. Auto blow down system (ABDS) actuates automatically at reactor low level after 35 min (2100 s) (refer Fig. 7). Due to ABDS actuation, the RPV depressurizes quickly. This action facilitates the replenishment of the RPV, with alternate water source to cool down the reactor core.

In order to save the first level of defence, operator can manually open the EC valve to cool down the reactor system during this duration of 35 min before actuation of ABDS. After a series of calculations, it is found that if operator is able to open the EC valve within 30 min of incident, then reactor system will be able to cool down by EC system.

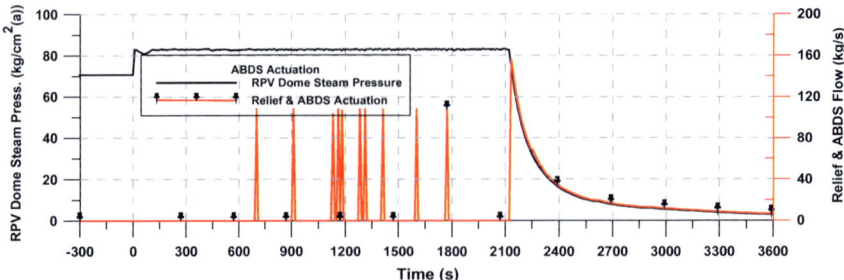

Fig. 7 Actuation of ABDS

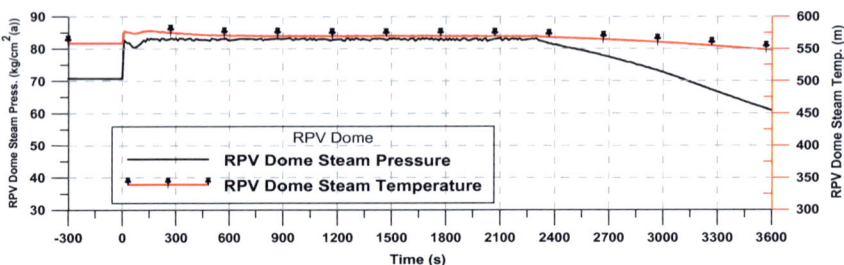

Fig. 8 Reactor pressure vessel dome steam pressure and temperature

RPV dome pressure and temperature are shown in Fig. 8. Reactor is operating at 71.0 kg/cm²(a) pressure during steady-state run. On SBO, the pressure increase rapidly up to 81.3 kg/cm²(a), i.e. set pressure of relief valves and maintained there as EC is assumed to fail on auto. After half an hour when EC is made available manually, the reactor pressure starts decreasing and becomes around 60.0 kg/cm²(a) after one hour. The RPV steam temperature is always maintained at the saturation temperature of RPV dome pressure.

The clad temperatures of average channel-1, average channel-2 and hot channel become constant at around 558.0–573.0 K depending upon the axial power profile. As SBO occurs, clad temperature follows the reactor pressure. On manual actuation of emergency condenser, after half an hour, clad temperatures start to decrease and fall below 550.0 K in one hour. Figure 9 shows the clad temperatures of hot channel. As in this whole study, the temperature of clad does not cross the 923 K; hence, there is no hydrogen generation at all.

Reactor water level is shown in Fig. 10. It is maintained at 13.0 m of level during steady-state run. On occurrence of SBO, it falls up to 11.8 m due to increase in pressure and discharge of water inventory through relief valve to maintain the pressure. As EC starts reducing, the pressure water comes from SSG shell into RPV through the leak paths (refer Fig. 11), which increases the reactor water level during first hour.

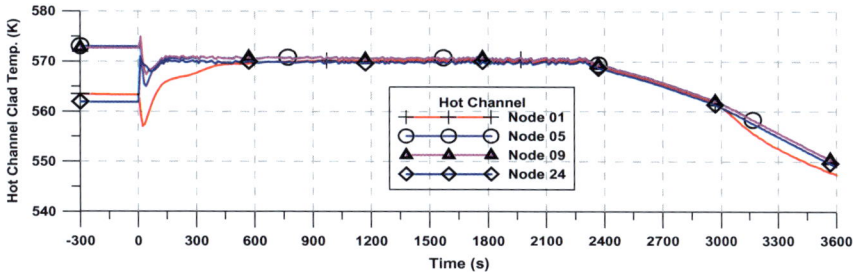

Fig. 9 Fuel clad temperature of average channel-1

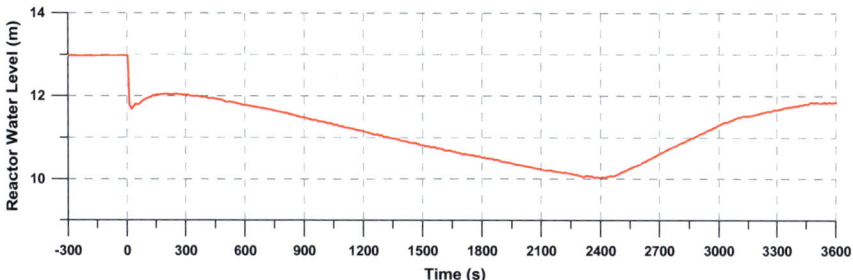

Fig. 10 Reactor water level

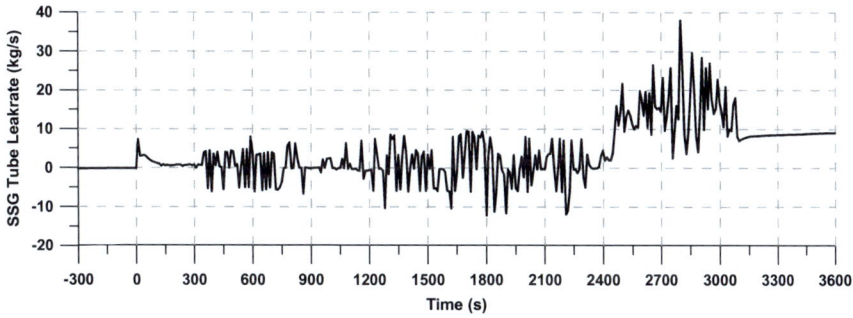

Fig. 11 Leakages in from SSG tubes

6 Conclusion

From this comprehensive safety analysis of station blackout scenario in TAPS-1&2, it can be seen that time available for operator to manually resume the emergency condenser is of the order of 30 min, if it is failed to actuate on auto. This time is considered adequate to manually actuate emergency condenser, and SBO scenario can be managed effectively.

References

1. R. Raj et al., Analysis of Anticipated Transient Without Scram For TAPS-1&2, Advances in Reactor Physics (ARP-2017), Mumbai, India, 6–9 Dec 2017
2. IAEA Safety Standards (Safety Guide NS-G-2.15) (2009)
3. SCDAP/RELAP-5/MOD-3.2 Code Manual Volume III: User's Guide and Input Manual, Idaho National Engineering Laboratory, NUREG/CR-6150, Revision 1, INEL-96/0422, Nov 1997
4. SCDAP/RELAP-5/MOD-3.2 CODE MANUAL VOLUME III-A: User's Guide and Input Manual, Idaho National Engineering Laboratory, NUREG/CR-6150, Revision 1, INEL- 96/0422, Nov 1997
5. RELAP5/MOD3 Code Manual, Volume I: Code Structure, System Models and Solution Methods, Idaho National Engineering Laboratory, NUREG/CR-5535, INEL-95/0174, June 1995
6. Probabilistic safety assessment level-1, internal event R-2 for TAPS-1&2

Probabilistic Safety Assessment of a Typical Chemical Process Plant

N. S. Ghuge, Pragati Shukla, S. Manivannan and D. Mandal

Abstract Probabilistic safety assessment (PSA) of an upcoming chemical process plant handling toxic materials was carried out. The PSA includes all possible accident scenarios and their quantification in terms of release frequency and consequences. Based on the comprehensive review of various storage facilities, process systems and safety systems, major PIEs are screened-in for further quantitative analysis. The facility may cause potential hazards mainly due to the three hazardous chemicals, viz. hydrogen, chlorine and a pyrophoric material. Hydrogen leakage causes flash fire or explosion hazards; chlorine gas leakage causes toxic hazards. Adequate measures have been considered to mitigate these hazards. PSA will help in probabilistic estimation. The analysis takes into account for development of accident scenarios, considering possible failure of safety functions and human interactions. Models available in ISOGRAPH software have been used for component modelling. The events considered in fault tree are failures that are associated with component hardware failures, human errors, maintenance or breakdown unavailability that may lead to undesired state. Standard formats were used for coding basic events, common cause failure (CCF) events and human error events in the fault trees with due care were considered to ensure that the basic event clearly relates to specific component/system identification, component failure mode and type. This work will be helpful in dealing with gaseous hazards and flash fire hazard and explosion hazard.

Keywords Safety · Hazard · Risk · PSA · HAZOP

N. S. Ghuge (✉) · P. Shukla · S. Manivannan · D. Mandal
Alkali Material & Metal Division, Bhabha Atomic Research Center, Mumbai, India
e-mail: gnagesh@barc.gov.in

P. Shukla
e-mail: pragati@barc.gov.in

S. Manivannan
e-mail: smani@barc.gov.in

D. Mandal
e-mail: dmandal@barc.gov.in

© Springer Nature Singapore Pte Ltd. 2020
P. V. Varde et al. (eds.), *Reliability, Safety and Hazard Assessment for Risk-Based Technologies*, Lecture Notes in Mechanical Engineering,
https://doi.org/10.1007/978-981-13-9008-1_4

1 Introduction

Hydrogen and chlorine generation by electrolysis is common process. Hydrogen is being generated in breaking of alkali, alkaline electrolyser system. Chlorine is generated in fused electrolyser salt electrolysis, viz. during production magnesium from used magnesium chloride. Such facility houses potential hazards mainly such as hydrogen leak and which may lead to flash fire hazard and explosion and chlorine leak may lead to toxic hazard. Adequate measures are in place to mitigate these hazards, and to ensure the sufficiency of these measures, HAZOP and qualitative risk analysis were carried out [1].

2 Overall Safety Approach

2.1 Hydrogen

Hydrogen gas is produced in electrolyser system, and this system should be leak free. Hydrogen gas generation rate is about 20 m^3/h is by product which gets sucked out through a common header with the help of water jet ejector to scrub out any carry-over of alkali traces before it is let out to any atmosphere. There are two separate ejector systems for hydrogen being produced at electrolyser as well as oxygen being produced at electrolysers to avoid flash fire. Main concern of handling of hydrogen is that it forms unsafe mixture with air in range of 4–74%. Hydrogen area monitor is installed to observe area concentration which should be less than 0.5% as per operational safety guideline of standing operating procedure (SOP).

2.2 Chlorine

Chlorine gas is generated in production of magnesium metal from electrolysis of fused salt electrolysis. The chlorine generation rate is about 0.5–10 m^3/h. Threshold limit value (TLV) of chlorine is 0.5 ppm.

Chlorine has several health hazards such as irritation to nose, throat, skin and eyes, also tearing, causes coughing and chest pain. High concentration may lead to burn the lungs and can cause a build-up of fluid in the lungs or death. Frequent exposure may permanently damage the lungs. Chlorine area monitor has been installed to observe area concentration which should be less than 0.1 ppm as per operational safety guideline of SOP.

3 Design Safety Features for Hydrogen Handling

Design safety features considered for hydrogen handling are as follows.

(a) Hydrogen gas is produced in the electrolyser system and is sucked out with the help of hydrogen ejector system. In order to avoid any hydrogen gas leakage in the plant area, it is ensured that:

- Hydrogen ejector system is switched ON before start of the electrolysis.
- Hydrogen ejector system always remains ON while the plant is in operation, by providing a suitable interlock with the rectifier system. In the event of failure of hydrogen ejector system, the interlock system gets activated and trips the rectifier.

(b) A minimum negative pressure of 25 mm water gauge is always maintained in the common header to ensure no leakage of hydrogen gas in the plant area.

(c) Surveillance of plant area is carried out at regular intervals with the help of portable hydrogen detectors.

(d) It has been ensured that even if complete hydrogen produced comes out in the plant area, hydrogen concentration will always be below 4% as 20 air changes per hour has been provided in the plant area.

(e) Hydrogen area monitor with 15 hydrogen sensors and with two annunciators will be installed in different areas of plant. An interlock will be provided to trip the rectifier at 2% hydrogen concentration in air.

(f) Other checks: Regular check to see that water seal is maintained in the condensate traps of the hydrogen suction line.

4 Design Safety Features for Chlorine Handling

Design safety features considered for chlorine handling are as follows.

(a) Chlorine gas is scrubbed in alkali scrubber.

(b) Chlorine gas is generated in electrolyser and sucked out through ventilation system and is passing out through the alkali scrubber. In order to avoid any gas leakage in the plant area, it is ensured that:

- Ventilation system is switched ON before start of the electrolysis.
- Scrubber system always remains ON while the plant is in operation, by providing a suitable interlock with the rectifier system. In the event of failure of scrubber system, the interlock system gets activated and trips the rectifier.

(c) Surveillance of plant area is carried out at regular intervals with the help of portable chlorine gas detectors.

5   Risk Assessment/Probable Accidental Condition

5.1   Failure of Hydrogen Jet Ejector System and Leakage of Hydrogen

5.1.1   Causes

Except for circulation pumps, there are no moving parts in the hydrogen jet ejector system. The system can fail, only if the circulation pump trips due to overload, electrical and mechanical failure.

5.1.2   Associated Risk

May lead to hydrogen leak, if no manual action is taken or safety system does not work.

5.1.3   Actions

The circulation pump associated with hydrogen jet ejector system is connected with rectifier tripping system. In the event of circulating pump getting tripped, relay in the inter-locking system gets activated and trips all the rectifier systems and which in turn stops hydrogen production. The suction pressure created by the hydrogen jet ejector system is monitored through a water-level manometer every hour. So, even in the event of hydrogen—rectifier interlock system gets failed, the 'Nil' suction pressure gets notified in an hour. In an hour, the total hydrogen produced is around 18 m^3, which corresponds to less than 0.5% (v/v) of air circulation in the plant area. Hence, even in the event of hydrogen that gets produced in an hour leaks to the atmosphere, the hydrogen concentration in the plant area is less than the explosive limits.

5.2   Leakage of Chlorine

5.2.1   Causes

Chlorine leakage in the plant area may happen due to failure of scrubbing pump, choke up in chlorine line and breakage in piping. Chlorine concentration may build up in plant area.

5.2.2 Associated Risk

May lead to chlorine leak possibility and health hazard.

5.2.3 Actions

Standby pump will be brought online in case of failure of scrubber pump. Required action will be taken to clear the choke up in line or to stop leakage. If not able to do so, action will be taken to stop electrolysis and increase the air change per hour in the area. Beyond area concentration of 0.5 ppm, area monitor will trip rectifier to stop electrolysis and further generation of chlorine.

6 Postulated Initiating Events and Its Consequences

6.1 PIE1: Hydrogen Leak

Hydrogen generated is about 20 m^3/h and hydrogen gas is generated and forms unsafe mixture with air in wide range 4–74% and can lead to flash fire in availability of ignition source.

6.2 PIE3: Leakage of Chlorine

Chlorine generated is about 0.5 m^3/h and chlorine leakage in the plant area may happen due to failure of scrubbing pump or choke up in chlorine line or breakage in piping. Chlorine concentration may build up in plant area more than TLV and possibility of health hazard.

7 Event Tree Analysis

The analysis takes into account for development of accident scenarios, considering possible failure of safety functions and human interactions. Models available in ISO-GRAPH software have been used for component modelling. The events considered in fault tree are failures that are associated with component hardware failures, human errors, maintenance or breakdown unavailability that may lead to undesired state. Standard formats were used for coding basic events, common cause failure (CCF) events and human error events in the fault trees with due care were considered to ensure that the basic event clearly relates to specific component/system identifica-

Table 1 Risk categories for various chemical products

State	Hydrogen safety limits	Chlorine safety limits
Safe	Below 4% concentration in air	$C < 0.5$ ppm 30-d time-weighted average (TWA)
Low risk		C 1.0 ppm 15 min—short-term exposure limit (STEL)
Medium risk	4–74% in air	More than 1.0 ppm
Hazards associated	Flash fire	Health hazard

tion, component failure mode and type. The risk category of each vent has been described in Table 1 [2–13].

7.1 Hydrogen Leak

7.1.1 Support Systems for Mitigation

Ventilation System (AHU)

It has been ensured that even if complete hydrogen produced comes out in the plant area, hydrogen concentration will always be below 4% as 20 air changes per hour has been provided in the plant area. System unavailability probability per demand is $5.81e^{-5}$, and this has been calculated by using fault tree in Fig. 3.

Hydrogen Area Monitors

Hydrogen area monitor with 15 hydrogen sensors (3 sensors in one zone) and with two annunciators will be installed in different areas of plant. An interlock will be provided to trip the rectifier at 2% hydrogen concentration in air. System unavailability probability per demand is $7.83e^{-3}$, and this has been calculated by using fault tree in Fig. 4. Annunciator will be used for human action to trip the rectifier from control room.

Auto Rectifier Trip

Rectifier trip will stop the electrolysis and pumping process which will stop further hydrogen production. System unavailability probability per demand is $1.343e^{-2}$, and this has been calculated by using fault tree in Fig. 5.

Human Action for Rectifier Trip

In this situation on hearing alarm of hydrogen area monitor, human has to trip the rectifier in case failure of auto-rectifier trip. He has to monitor hydrogen concentration in plant after shutdown of rectifier. Unavailability probability per demand is $5.0e^{-3}$, and this has been considered. In normal condition, operator has to stop the rectifier from control room only.

Frequency Quantification

The frequency for hydrogen leak has been taken from plant experience. The plant experience is about 20 years and till date this event has not happened. One event has been assumed in plant life. From this assumption, the initiating frequency is found to be 0.05/year. The event tree is shown in Fig. 1, and the consequence resulting from this event tress is hydrogen flash fire or explosion, if there is a failure in following SOP.

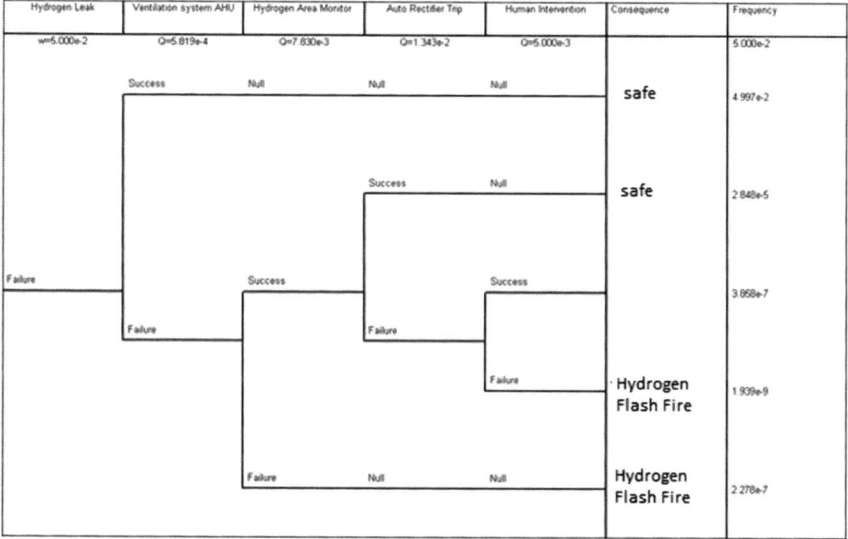

Fig. 1 Event tree for hydrogen leak

7.2 Chlorine Leak

7.2.1 Support Systems for Mitigation

Ventilation System (AHU)

It has been ensured that even if complete chlorine produced comes out in the plant area, chlorine concentration will always be below 0.5 ppm because 20 No. air changes per hour has been provided in the plant area. System unavailability probability per demand is $5.819e^{-5}$, and this has been calculated by using fault tree in Fig. 3.

Chlorine Area Monitors

Chlorine area monitor with 4 sensors in one zone and with two annunciators will be installed in different areas of plant. An interlock will be provided to trip the rectifier at 0.2 ppm chlorine concentration in air. System unavailability probability per demand is $7.83e^{-3}$, and this has been calculated by using fault tree in Fig. 6. Annunciator will be used for human action to trip the rectifier from control room.

Auto Rectifier Trip

Rectifier trip will stop the electrolysis which will stop further chlorine production. System unavailability probability per demand is $1.343e^{-2}$, and this has been calculated by using fault tree in Fig. 5.

Human Action for Rectifier Trip

In this situation on hearing alarm of chlorine area monitor, human has to trip the rectifier in case failure of auto-rectifier trip. He has to monitor chlorine concentration in plant after shutdown of rectifier. Unavailability probability per demand is $5.0e^{-3}$, and this has been considered. In normal condition, operator has to stop the rectifier from control room only.

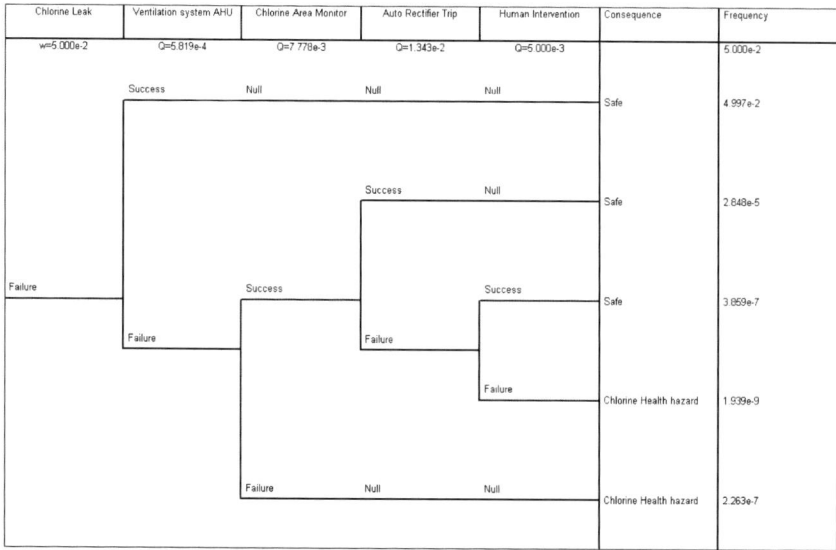

Fig. 2 Event tree for chlorine leak

Frequency Quantification

The frequency for chlorine leak has been taken from plant experience. The plant experience is about 20 years and till date this event has not happened. One event has been assumed in plant life. From this assumption, the initiating frequency is found to be 0.05/year. The event tree is shown in Fig. 2. The consequence resulting from this event tree is chlorine high concentration and will lead to health hazard, if there is a failure in following SOP.

7.2.2 Fault Tree of Ventilation System AHU

See Fig. 3.

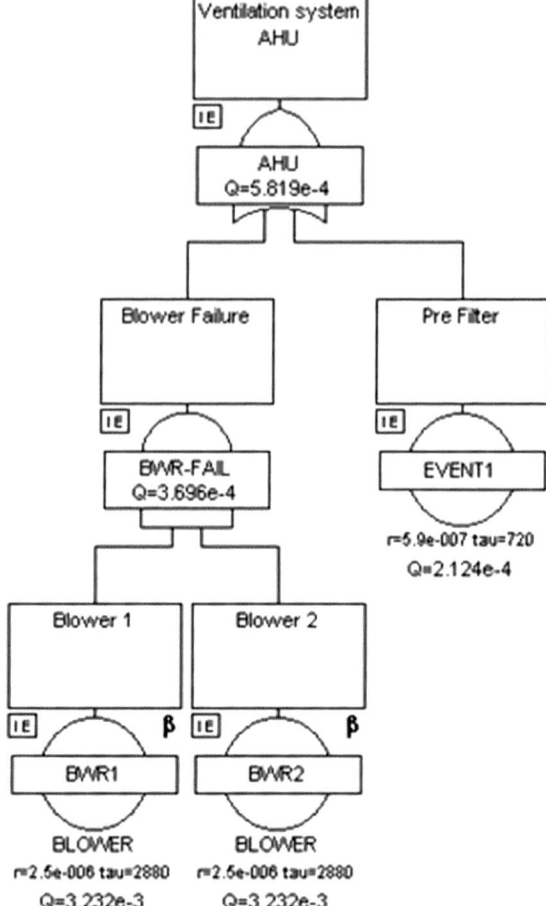

Fig. 3 Fault tree diagram for ventilation system AHU

7.2.3 Fault Tree of Hydrogen Area Monitor

See Fig. 4.

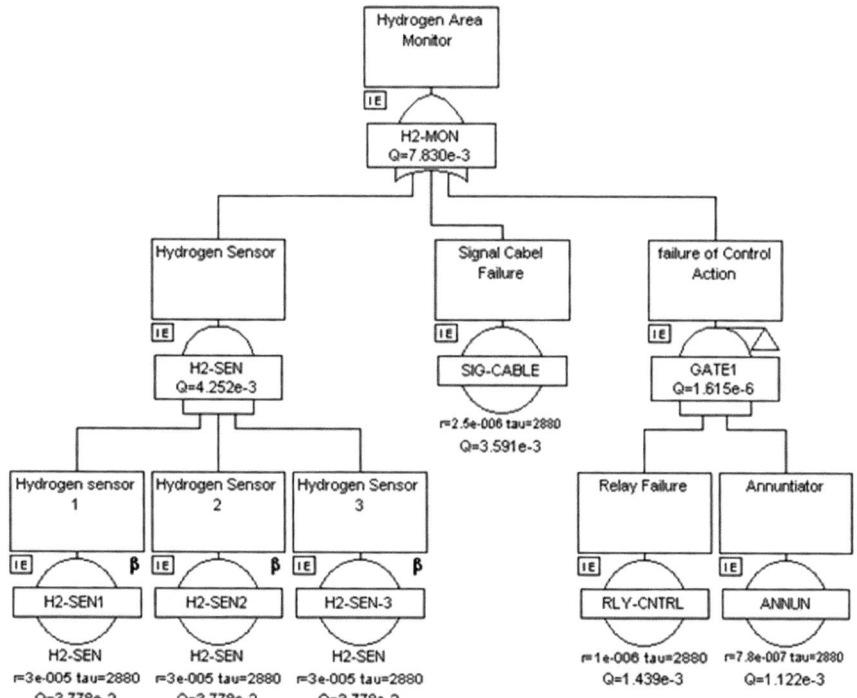

Fig. 4 Fault tree diagram for hydrogen area monitor

7.2.4   Fault Tree of Auto Rectifier Trip

See Fig. 5.

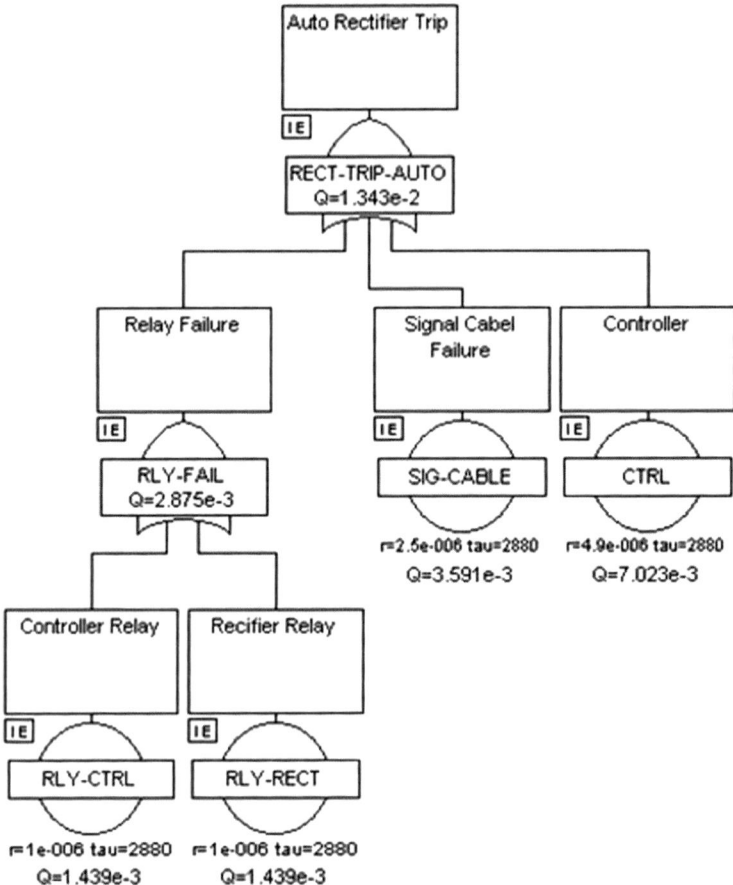

Fig. 5 Fault tree diagram for auto rectifier trip

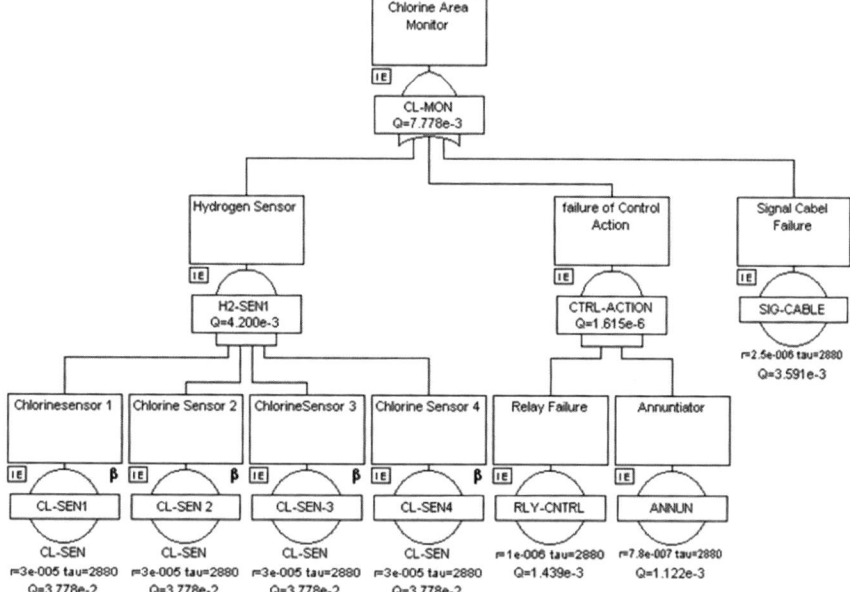

Fig. 6 Fault tree diagram for chlorine area monitor

7.2.5   Fault Tree of Chlorine Area Monitor

See Fig. 6.

8   Conclusions

The PSA carried out for chemical plant represents an integrated model of the safety of the plant. The study encompasses the design, operational safety practices, component reliabilities, dependencies and human reliabilities. This study identifies the dominant contributors to possible release states, in terms of initiating events, component failures and human errors.

With the current database on component reliabilities and the identified vulnerabilities to risk and hazard, the estimated frequency of various chemical compounds. Log-normal distribution has been considered for the components, and the 5000 simulations were carried out (Table 2).

Table 2 Risk frequency of various chemical compounds

Release state	Mean frequency (per year)
Hydrogen flash fire or explosion due to hydrogen leak	$2.27e^{-7}$
Health hazard due to chlorine high concentration	$2.26e^{-7}$

References

1. Procedures for Conducting Probabilistic Safety Assessments of Nuclear Power Plants, IAEA-Safety Series 50-P-4 (Level 1) (1992)
2. A.D. Swain, H.E. Guttmann, *Handbook of HRA with Emphasis on Nuclear Power Plant Application*. Final Report: NUREG\CR. 1278, Sandia National Laboratories, August. NUREG/CR-5695 (1989)
3. A Process for Risk-Focused Maintenance, U.S Nuclear Regulatory Commission, Washington, DC 20555, March (1991)
4. An Approach for Plant-specific, Risk-Informed Decision making: RG-1.177, Technical Specifications, U.S Nuclear Regulatory Commission, August (1998)
5. Applications of Probabilistic Safety Assessment (PSA) for Nuclear Power Plants, International Atomic Energy Agency, TECDOC-1200, February (2001)
6. Human Reliability Analysis in PSA for NPPs, International Atomic Energy Agency, Safety Series No: 50 P-10, IAEA, Vienna (1995)
7. Ralph R. Fullwood, Butterworth-Heinemann, Probabilistic Safety Assessment in Chemical and Nuclear Industries (2000)
8. FaultTree+V10.1, Copyright Isograph Reliability Software (1986)
9. Reliability Data for Probabilistic Safety Assessment of NPP, International Atomic Energy Agency, IAEA-TECDOC-478, Vienna (1986)
10. Atmospheric Dispersion and Modelling, AERB Safety Guide No. AERB/NF/SG/S-1, AERB (2008)
11. DBFL estimation report by NPCIL
12. J. Mishima, Airborne Release Fractions/Rates and Respirable Fractions for Nonreactor Nuclear Facilities, DOE-HDBK-3010-94, Change Notice 1, March (2000)
13. L.C. Cadwallader, Reliability Estimates for Selected Sensors in Fusion Applications, INEL-96/0295, ITER/US/96/TE/SA-16, September (1996)

PSA Level-3 Study—Estimation of Area and Persons Affected for Postulated Accidental Release from Indian PHWR After Implementing Urgent Protective Actions

P. Bhargava, Praveen Kumar, Brij Kumar, Manish Mishra, K. D. S. Singh, R. M. Tripathi, Amit Kumar, Pratima Singh, Vibha Hari, A. K. Vijaya, Manoj Kansal and Mukesh Singhal

Abstract A PSA level-3 study has been carried out for postulated accidental release from an Indian PHWR. Postulated release due to triple failures (LBLOCA with simultaneous failure of ECCS and containment isolation failure) in typical 220 MWe Indian PHWR is simulated in this study to know the impact of accident. The principal phenomenon for atmospheric transport of effluent is considered using a Gaussian plume model, MUSEMET which takes hourly variation of meteorological condition into account. Site-specific meteorological and demographic data has been considered for this study. Appropriate meteorological sampling has been done to represent 5-year meteorological data and severity of accident. COSYMA Code is used for calculation of projected doses for seven days to estimate the area and persons affected after implementation of urgent protective actions on the basis of IAEA GSR Part-7. As an element of emergency response planning, affected sector and area are determined under variety of meteorological conditions. The protection offered by iodine prophylaxis and sheltering has been emphasized in this study. Sheltering can be effective by keeping the public out of the plume exposure pathway during the time when the radioactive concentration of plume is high and iodine prophylaxis will avert thyroid dose significantly if taken at appropriate time. The effect of accident in public domain can be minimized effectively using proper countermeasures strategies. Further, risk to public and health effects has been found limited in this study.

Keywords Accidental release · MUSEMET · Projected dose · Health effects · Risk · Countermeasures · COSYMA · PSA

P. Bhargava (✉) · P. Kumar · B. Kumar · M. Mishra · K. D. S. Singh · R. M. Tripathi
Health Physics Division, HS&EG, BARC, Mumbai, India
e-mail: pradeepb@barc.gov.in

A. Kumar · P. Singh · V. Hari · A. K. Vijaya · M. Kansal · M. Singhal
Reactor Safety & Analysis, NPCIL, Mumbai, India

© Springer Nature Singapore Pte Ltd. 2020
P. V. Varde et al. (eds.), *Reliability, Safety and Hazard Assessment for Risk-Based Technologies*, Lecture Notes in Mechanical Engineering,
https://doi.org/10.1007/978-981-13-9008-1_5

1   Introduction

COSYMA (Code system from MARIA) [1] is applicable for probabilistic safety assessment (PSA) level-3 as well as deterministic calculations of the off-site consequences due to hypothetical accidental releases of radioactive material to atmosphere from nuclear facilities. The health effects, impact of countermeasures and economic costs of the releases can also be evaluated.

Impact of the implementation of the urgent protective action on the public domain has been carried out in case of hypothetical accident scenario as part of PSA level-3 [2] study. Aim of this study was to provide the support during emergency situation in preparation of the documents for planning and preparedness during emergency situation for target site having 220 MWe PHWR in India. Sub-system of COSYMA is Near Early (NE), Near Late (NL) and Far Late (FL). The NE sub-system is limited to calculate the early health effects and the influence of emergency actions to reduce those effects and is intended for use in the region near to the site. The NL sub-system is limited to calculate the late health effects and the impact of associated countermeasures and is intended mainly for use in the region near to the site. The FL sub-system is concerned with calculating late health effects and effect of the appropriate countermeasures at larger distances from the site.

2   Materials and Method

2.1   Accident Sequence

In this study, triple failure, i.e. LBLOCA + emergency core cooling system (ECCS) failure + containment isolation failure is considered as the accident scenario for 220 MWe Indian PHWR. The frequency of the accident is estimated to be in the range of 10^{-8}–10^{-9} per year.

LOCA with simultaneous failure of emergency core cooling system leads to the conditions where fuel cooling would only be possible by heat conduction and radiation to pressure tubes and calandria tubes and eventually through sub-cooled nucleate boiling heat transfer to moderator. Therefore, in such case the moderator in calandria serves as heat sink for decay heat removal. This mode of core heat dissipation requires moderator circulation.

The temperatures in this scenario are such that there would be no melting of uranium oxide. The pressure tubes may deform and touch calandria tubes but integrity of pressure tubes would be maintained. The Zircaloy fuel cladding would experience oxidation. In case of subsequent failure of this cladding, inventory in the gap between the pellet and clad of fuel elements would be released instantaneously with furthermore fission products releasing due to the diffusion process at high temperature.

This is one of the accident sequences considered for source term estimation in PSA level-2 study for Indian PHWR. Noble gases, iodine, cesium, strontium and tellurium are the five-release groups of radionuclides considered for estimation of projected does for seven days. Ground and stack level release has been considered.

2.2 Source Term

The core inventories of a typical Indian PHWR 220 MWe reactor are based on the results from ORIGEN-2.0 [3]. The inventories are shown in Table 1. These inventories correspond to an average incore burn up of 3700 MWD/t of uranium. Methodology adopted for the source term estimation resulting from the accident sequence considered above is summarized below.

For accident scenarios involving loss-of-coolant accident (LOCA) with failure of the ECCS, the activity would be released from core on sheath failure due to overheating of fuel which is estimated using wash-1400 and CORSOR-M model. Water-trapping factor of 2 for volatile release is considered in the PHT circuit. Part of the radionuclides gets released to the containment from the core and consequently releases to the environment due to high containment pressure through isolation dampers. Some additional fission products release would be there depending on instrument air in-leakage and increase in containment pressure. The volatile plate-out half-time for deposition in the containment is 1.5 h in primary envelope and 2.0 h in secondary envelope. Natural decay phenomenon for all radionuclides is considered. The activities released to the environment for postulated accident scenario are shown in Table 1.

Table 1 Core inventory and source term for Typical 220 MWe IPHWR

SN	Radio nuclide	Core inventory (Bq)	Release to environment (Bq)	SN	Radio nuclide	Core inventory (Bq)	Release to environment (Bq)
1	I-131	8.51E+17	2.32E+14	12	Xe-133m	5.55E+16	1.61E+14
2	I-132	1.22E+18	2.76E+13	13	Xe-133	1.67E+18	1.00E+16
3	I-133	1.74E+18	1.54E+14	14	Xe-135m	3.48E+17	1.48E+12
4	I-134	1.92E+18	1.83E+13	15	Xe-135	1.22E+17	3.69E+13
5	I-135	1.63E+18	7.55E+13	16	Xe-138	1.44E+18	5.37E+12
6	Cs-134	1.37E+16	1.96E+13	17	Kr-83m	1.11E+17	6.73E+12
7	Cs-137	2.33E+16	2.76E+13	18	Kr-85m	2.41E+17	4.48E+13
8	Sr-89	7.77E+17	2.20E+12	19	Kr-85	2.22E+15	4.41E+13
9	Sr-90	1.70E+16	3.52E+10	20	Kr-87	4.81E+17	1.88E+13
10	Te-132	1.22E+18	2.49E+14	21	Kr-88	6.66E+17	7.03E+13
11	Xe-131m	8.88E+15	9.41E+13				

2.3 Population Data

The basic village-wise population data [Census-2011] for the target site has been used to estimate the number of person affected by the urgent protective action [4]. The population data was extrapolated for 2019 using average growth rate of 0.68%/year. For areas beyond 32 km state-averaged population density of 325 persons/km^2 was used, up from a value of 308 persons/km^2 in 2011.

2.4 Atmospheric Dispersion

A segmented Gaussian plume model [1], which assumes that the atmospheric conditions affecting the plume are determined from those at a single meteorological station representing the site of release is considered. It assumes that the wind direction changes every hour throughout the release and the subsequent travel. In this case, dispersing material assumed to follow a trajectory, which consists of a series of straight segments.

Dry deposition is calculated using the deposition velocity, with the amount of material remaining in the plume calculated using the source depletion model, in which the vertical profile of the plume is assumed to remain Gaussian. Wet deposition is included using the washout coefficient, which increases with increasing rainfall rate.

The effects of down wash and mixing in the wake of a reactor building have been considered using a virtual source model based on the work of Hanna [5].

2.5 Meteorological Sampling

Probability safety assessment predicts the probability distribution of consequences, if an accident occurs in any of the wide range of atmospheric conditions which might occur at the site of interest during the period in which site operates. The sequences of conditions are obtained from the hourly spaced meteorological data over a period of few years, and assuming that the conditions during the future operation of the site will be similar to those observed in the past. It is impossible to undertake the calculations for every sequence of conditions over the operating period of the site in advance. Therefore, a representative sample of starting times or weather sequences must be used.

Stratified sampling [6], which is widely used for sampling weather sequences, was used to select the starting time or weather sequence from the target site having meteorological data for the year 2012–2016. Sequences of conditions from the meteorological data, for which the consequences are likely to be similar, are combined into groups. Sequences are chosen at random from those within each group and assigned a probability based on number of sequences allocated to each group and the number selected from within each group.

The grouping scheme distinguishes two main types of sequences; (1) Dry (without rain), (2) Wet (with rain) occurring within about 100 km of the site. Dry sequences were further divided into groups depending upon travel time of material to two distances (20 and 60 km) from release point, and the stability category in the first hour of release. Wet sequences can be defined by quantities related to the wet deposition density rather than just to rainfall rate. The distance at which high deposition occurs is also important. The sequences with rain in the scheme adopted here were grouped in terms of ratio of washout coefficient ($\wedge$) to wind speed (u). Meteorological sampling scheme adopted in this study is given in Table 2.

2.6 Exposure Pathways

Exposure pathways considered for estimating acute risk for adults are cloud shine, inhalation, skin doses from material deposited on skin, ground shine and resuspension of ground-deposited activity. The first three routes of irradiation only affect those people who are exposed to the plume of material as it passes overhead.

External irradiation from the cloud only occurs while the plume is present. Material is only inhaled while the plume is present, but some material can be retained in the body for long periods. Irradiation from third route stops once material is removed from the skin and clothing. The other two routes can lead to the source of exposure being present for long periods of time.

2.7 Countermeasure

Implementation of countermeasures for urgent protective action is based on the projected dose level for thyroid (50 mSv) for iodine prophylaxis and effective dose (100 mSv) for sheltering and evacuation for 7 days [7].

3 Results and Discussion

Results of estimated projected dose are presented in Figs. 1 and 2 with distance wise from the centre of reactor building with probability distribution of doses according to weather sequences (for the range atmospheric conditions that could occur at the time of accident). The number of person and area affected by the urgent protective action is given in Table 3.

Table 2 The sampling scheme used in this study

Group	Definition		
Groups are based on ratio of washout coefficient to wind speed			
1	$\wedge/u > 5 \times 10^{-5}$ within 10 km		
2	$\wedge/u > 5 \times 10^{-5}$ within 25 km		
3	$\wedge/u > 5 \times 10^{-5}$ within 50 km		
4	$\wedge/u > 5 \times 10^{-5}$ within 75 km		
5	$\wedge/u > 5 \times 10^{-5}$ within 100 km		
6	$\wedge/u > 5 \times 10^{-6}$ within 10 km		
7	$\wedge/u > 5 \times 10^{-6}$ within 25 km		
8	$\wedge/u > 5 \times 10^{-6}$ within 50 km		
9	$\wedge/u > 5 \times 10^{-6}$ within 75 km		
10	$\wedge/u > 5 \times 10^{-6}$ within 100 km		
Groups defined using initial stability category and travel time in hour for 20 km and 60 km			
	Category	Travel time (h)	
		20 km	60 km
11	A/B	<5	<8
12	A/B	<5	Any
13	A/B	Any	<13
14	A/B	Any	Any
15	E/F	<2	<4
16	E/F	<2	<6
17	E/F	<2	Any
18	E/F	Any	Any
19	C/D	<2	<4
20	C/D	<2	Any
21	C/D	<3	<4
22	C/D	<4	<6
23	C/D	<4	<7
24	C/D	<4	<8
25	C/D	<5	<9
26	C/D	Any	Any

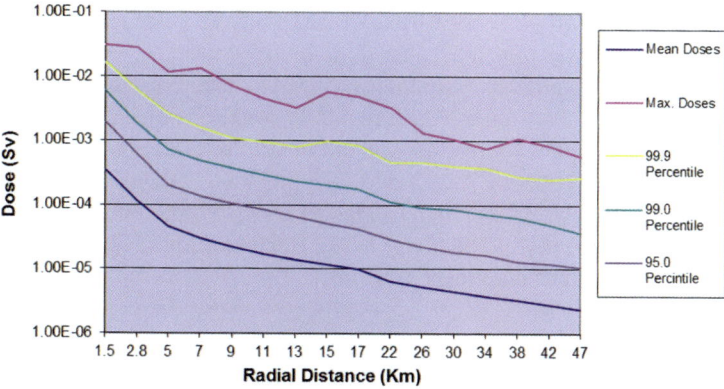

Fig. 1 Projected Effective dose for 7 days

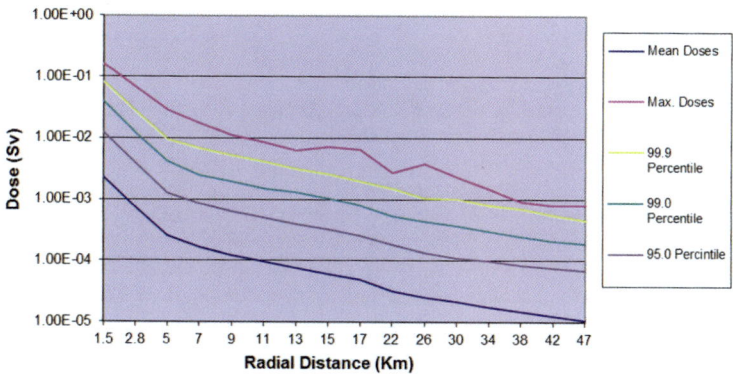

Fig. 2 Projected Thyroid dose for 7 days

Table 3 Number of person and area affected by the urgent protective action	Countermeasure implemented	
	Sheltering	Iodine prophylaxis
No. of affected persons	–	2410
Affected area (km^2)	–	8.91

4   Conclusion

It is observed from Fig. 1 that projected effective dose is below the 100 mSv considering all the weather sequences, hence sheltering and evacuation are not required in the public domain for the postulated accident scenario. It is observed from Fig. 2 that projected thyroid dose exceeds 50 msv of prescribed limit on the basis of maximum and 99.9 percentile of weather sequences. The iodine prophylaxis may be needed as urgent protective action on the basis of maximum and 99.9 percentile of projected dose received by the off-site person up to distance 4 and 2.1 km, respectively. Maximum number of person which may get thyroid dose more than prescribed limit is 2410, which can be averted using iodine prophylaxis. The projected thyroid dose is below the prescribed limit up to 99 percentile of weather sequences, hence iodine prophylaxis is not required during most of the weather sequences for this postulated accident scenario and none of the off-site personnel will be affected. Proper emergency preparedness plan exists around all the NPP sites to substantially reduce dose to public using iodine thyroid blocking (ITB). Hence no adverse health effects are expected for the postulated accident scenario.

References

1. I. Hasemann, J.A. Jones, COSYMA User Guide (Version 95/1); A joint report by NRPB and KfK, EUR 13045, KfK 4331 B
2. Procedure for conducting Probabilistic Safety Assessments of Nuclear Power Plants (Level 3); Safety Series No. 50 P-12, IAEA Vienna (1996)
3. A.G. Croff, A user's manual for the ORIGEN 2.0 computer code, ORNL/TM-7175, July 1980
4. Census of India 2011: Village data for Surat District
5. S.R. Hanna, G.A. Briggs, R.P. Hosker, *Handbook on Atmospheric Diffusion*. Technical Information Center, US Department of Energy (1982)
6. J.A. Jones, P.A. Mansfield, S.M. Haywood, I. Hasemann, C. Steinhaur, J. Ehrhardt, D. Faude, PC-COSYMA (Version 2); an accident consequences assessment package for use on PC. EUR report 16239 (1995)
7. GSR part 7: IAEA Safety Standard Series, Preparedness and response for nuclear and radiological emergencies. IAEA Vienna 2015

Shutdown Probabilistic Safety Assessment of Boiling Water Reactor

Manish Tripathi, Sonali Parmar, C. Bose, Vibha Hari, A. K. Vijaya, N. Mohan and Mukesh Singhal

Abstract The aim of the safety of the nuclear power plant is to assure that risk from nuclear power plant to public and plant personnel is insignificant during all plant operating states. Shutdown Probabilistic Safety Assessment (SPSA) is a systematic methodology for assessment of the safety of nuclear power plant at shutdown mode of operation. A comprehensive Shutdown PSA has been conducted for a typical boiling water reactor (BWR) to demonstrate the safety of the reactor during the shutdown mode of operation. The study is carried out in accordance with procedures outlined in various IAEA documents, NUREGs and AERB documents, etc. Refueling and associated maintenance is a dynamic process. It can have various configurations of the plant during the entire span of the refueling shutdown. In this study, a typical refueling shutdown (RSD) plan of BWR Plant at Tarapur Site has been used to derive the plant operating states. To reduce the effects of dynamic nature, refueling shutdown (RSD) duration is divided into 7 plant operating states (POS) and plant configuration is assumed constant during each POS. Six transients are considered as initiating events (IEs) in this study. For each IE, a separate event tree with various possible accident

M. Tripathi (✉) · S. Parmar · C. Bose · V. Hari · A. K. Vijaya · N. Mohan · M. Singhal
Nuclear Power Corporation of India Limited, B-3 Nabhikiya Urja Bhavan,
Anushaktinagar, Mumbai 400094, India
e-mail: mtripathi@npcil.co.in

S. Parmar
e-mail: sonaliparmar@npcil.co.in

C. Bose
e-mail: cbose@npcil.co.in

V. Hari
e-mail: vibhahari@npcil.co.in

A. K. Vijaya
e-mail: akvijaya@npcil.co.in

N. Mohan
e-mail: nmohan@npcil.co.in

M. Singhal
e-mail: singhalm@npcil.co.in

© Springer Nature Singapore Pte Ltd. 2020
P. V. Varde et al. (eds.), *Reliability, Safety and Hazard Assessment for Risk-Based Technologies*, Lecture Notes in Mechanical Engineering,
https://doi.org/10.1007/978-981-13-9008-1_6

sequences has been developed considering the availability of systems/components in the POS considered. Dynamic human actions modeled in different event trees are quantified using Accident Sequence Evaluation Program (ASEP) model based on NUREG/CR-4772. The Level-1 Shutdown PSA developed for TAPS-1&2, represents a complete model of the safety of the plant enclosing design, maintenance, and safety practices during refueling shutdown, component reliabilities, common cause failures (CCFs), and operator errors. This model finds out the main contributors to most likely severe accidents which can occur during refueling shutdown as a combination of postulated initiating events, component malfunction, and operator errors. The operation and maintenance procedure at TAPS-1&2 can be developed based on the outcome of the present study. With this integrated model available, the effect of proposed changes in maintenance practices can be examined by studying their impact on changes in accident frequencies. The results of the Shutdown PSA indicate that a fairly high level of defense-in-depth exists in TAPS-1&2 design during shutdown mode of operation. This is evident from the final CDF value, as well as from the predominantly higher order minimal cut-sets observed in the core damage sequences. The various tasks of SPSA are discussed in this paper.

Keywords Boiling water reactor (BWR) · Initiating event · Refueling shutdown · Tarapur atomic power station (TAPS-1&2)

1 Introduction

The aim of the safety of the nuclear power plant is to assure that risk from nuclear power plant to public and plant personnel is insignificant during all plant operating states. Shutdown Probabilistic Safety Assessment (SPSA) is a systematic methodology for assessment of the safety of nuclear power plant at shutdown mode of operation [1]. A comprehensive Shutdown PSA has been conducted for a typical boiling water reactor (BWR) to demonstrate the safety of the reactor during the shutdown mode of operation. The study is carried out in accordance with procedures outlined in various IAEA documents, NUREGs and AERB documents [2], etc. Refueling and associated maintenance is a dynamic process. It can have various configurations of the plant during the entire span of the refueling shutdown.

2 Objectives and Scope

The aim of SPSA of the BWR at Tarapur site is to give a comprehension of the conceivable vulnerabilities to reactor core emerging from equipment, human or procedural insufficiencies and conditions in the shutdown state of the reactor. The anticipated uses of the study include the following:

i. Presenting a complete model of the safety of the 160 MWe BWR that includes plant design, operational practices, component reliabilities, CCFs, and human behavior.

ii. Identification of major contributors to most likely severe core damage in terms of component malfunction, operator errors, plant design, and operation and maintenance practices in the shutdown state of the reactor.

iii. Identification of the weak links or imbalances in the plant design and operation and maintenance practices which have effect on the safety of the plant with respect to components malfunction and human behavior, which can be enhanced.

iv. Assessing the core damage frequency in terms of relative significance of major contributors to make comparative evaluation.

v. Reliability assessment of the available systems during shutdown with variable configurations as per refueling shutdown plan of TAPS-1 &2 which are important to plant safety.

vi. The source of radioactivity considered in the analysis is the reactor core and reactor is considered to be in refueling shutdown.

3 Plant Description

The site consists of two light water moderated and cooled boiling water reactors (BWRs) using slightly enriched uranium oxide fuel contained in Zircaloy-2 tubes, two turbine generator sets and associated auxiliary systems and services. The reactor was originally a dual cycle, forced circulation boiling water reactor producing steam for direct use in the steam turbine. However, since 1984 after isolation of secondary steam generators due to excessive tube leakage, the reactors are operated in single cycle at a load of 160 MWe [3].

Cooling water drawn from the Arabian Sea condenses the steam in the condenser. The condensate is pumped to the feed water heaters where its temperature is raised and it enters the reactor vessel through the feed water spargers. The feed water mixes with the downward flow from the steam separators in the annulus between the core shroud and the vessel wall. This flow is withdrawn from the vessel at the recirculation outlet, pumped through the secondary steam generators, and returns to the vessel at the recirculation inlet. The flow is then upward through the core where steam is produced. The water is separated from the steam in the separators and dryers. The separated water is returned to the annulus to repeat the cycle [3].

4 Methodology

The distinct plant configuration and operational conditions which occur during low power and shutdown operation are modeled by defining plant operational states (POSs) [1]. Pre-POSs are characterized as various procedural steps or actions that take

Table 1 List of pre-POS

Plant condition mode	Scope	Remarks
Full power	Covered in full power PSA	–
Hot shutdown	Covered in full power PSA	–
Cool down early (280–150 °C)	Covered in full power PSA	–
Cool down late (150–50 °C)	Covered in full power PSA	(150–50 °C) is covered under full power PSA as same plant and system configuration as in full power PSA
Refueling outage: Cold shutdown for maintenance (Rx water temp 55 °C, pressure atmospheric, Decay heat 2% in the beginning)	Present PSA study	–
Start-up/standby mode	Covered in full power PSA	During start-up, the plant and mitigating system configuration is same as during on power state of the reactor. And also the IEs covered are same as on power PSA hence this period 18 h for start-up is covered in full power PSA

place during the outage. The complete list of pre-POSs and their elucidations forms the basic input about the refueling outage which is utilized in the SPSA development. Following Table 1 gives an overview of pre-POS considered in the analysis.

In this study, a typical refueling shutdown (RSD) plan of BWR Plant at Tarapur Site has been used to derive the plant operating states. To reduce the effects of dynamic nature, refueling shutdown (RSD) duration is divided into 7 plant operating states (POS) and plant configuration is assumed constant during each POS.

4.1 Initiating Events and Their Frequency Estimation

A basic prerequisite for the PSA is the inclusive list of the initiating events for the plant under consideration which can end-up with core damage [4, 5]. For BWR SPSA, only the initiating events related to cold shutdown are considered those include faulty operation of plant components through random failures, human errors during maintenance, etc. To ensure that within the defined scope of this SPSA, the list of IEs is as comprehensive and complete as possible more than one approach is followed. Engineering evaluation of plant, INDIAN BWR Operational history, BWR Level-1 Internal Events PSA report, and IAEA-TEC-DOC–1144 were considered for the identification of the initiating events. The initiating events listed in the initial list were reviewed for their applicability with respect to procedures, administrative controls, etc. and screening was done. The following Sects. 4.1.1–4.1.7 describe the initiating events and their frequency estimation methods considered in the present work.

4.1.1 Small Break LOCA

There is possibility of Loss of Coolant when the system integrity is affected. During an outage, components of the primary system are taken out for the maintenance. The part of the system which is taken for the maintenance should be isolated. Upon the failure of the isolation, there is possibility of the LOCA, which in this study can be termed as maintenance-induced LOCA. During the cold shutdown, reactor recirculation pumps are not running and reactor recirculation loop valves are closed and in isolation condition. The temperature and pressure are low in the primary system. Therefore, possibility of large break LOCA and medium break LOCA is not envisaged. During cold shutdown, shutdown cooling system is running and is used for decay heat removal from the fuel. Some maintenance works are taken up on primary system. Hence there is probability of small break LOCA in the shutdown mode of the reactor operation. The initiating event frequency estimation for SB LOCA is based on IAEA-TEC-DOC–719 [6].

4.1.2 Loss of Station Power Supply

Failure of Class IV power supply will cause the failure of various pumps of various heat sinks. This will call for Class-III supply from EDGs and subsequent starting of essential equipment by EMTR. Therefore, it is considered and included in the list of initiating event. Loss of station power frequency is estimated based on plant-specific data.

4.1.3 Loss of Reactor Building Cooling Water System

Loss of RBCW will cause failure of shutdown cooling system and fuel pool cooling system. This will also result in cooling water failure to post-incident heat exchangers. Therefore, it is considered and included in the list of initiating events. Fault tree method was adopted for the initiating event frequency estimation for RBCW failure.

4.1.4 Loss of Salt Service Water System

Loss of SSW will cause ultimate heat sink failure and result in failure of shutdown cooling system and fuel pool cooling system. Therefore, it is considered and included in the list of initiating events. Fault tree method was adopted for the initiating event frequency estimation for SSW failure.

4.1.5   Loss of Shutdown Cooling System

Shutdown cooling system is the main source of decay heat removal during cold shutdown whenever fuel is in the reactor vessel. Therefore, it is considered and included in the list of initiating events. Fault tree method was adopted for the initiating event frequency estimation for SDC failure.

4.1.6   Loss of Fuel Pool Cooling System

Fuel pool cooling system is the main source of decay heat removal from spent fuel pools where unloaded core is stored during refueling shutdown. Therefore, it is considered and included in the list of initiating events. Fault tree method was adopted for the initiating event frequency estimation for fuel pool cooling system failure.

4.1.7   Loss of Fuel Cooling System

During refueling shutdown, fuel movement is made from reactor vessel to spent fuel pools and vice versa. At this time, reactor cavity basin is filled with water and gates in between reactor cavity basin and fuel pool are open. During this period, shutdown cooling system and fuel pool cooling system both remain in service for decay heat removal. Any of these two systems are sufficient for decay heat removal of fuel stored in fuel pool and reactor vessel. Failure of both the system simultaneously may occur. Therefore, it is considered and included in the list of initiating events. Fault tree method was adopted for the initiating event frequency estimation for the loss of fuel cooling system failure.

4.2   Event Sequence Modeling

Plant response modeling for each group of initiating events (IEs) is a major task of PSA. It will generate event sequences which can be safe state or unsafe state of the plant. Event sequences which lead to unsuccessful decay heat removal from the core are an unsafe state. These unsafe states may cause a core damage and called as core damage sates. Event sequences generated during plant response modeling are expressed in terms of IEs and successes or failures of mitigating functions and their system. Event tree method is widely used for event sequence modeling. An event tree consists of an initiating event and series of function events which are the mitigation functions or system by which plant will respond as per design and operational procedure. These mitigation functions or systems are modeled using fault tree technique.

4.3 System Modeling

The performance required from a frontline system depends in general on the IE. Required performance of system is that it will successfully execute the intended function during the accident condition arises due to initiating event. The success criteria of frontline systems are extremely important for the PSA model as they characterize the top events for the subsequent modeling of the system.

A frontline/support system failure depending upon its importance contributes to the core damage frequency (CDF). The reliability of the system is expected to be of high order to achieve a CDF as low as possible. The availability/reliability of a system in turn depends upon the reliability of its various components since the combination of certain component failures in a system causes the system failure.

System modeling has been carried out based on the success criteria identified during event tree development support systems, and frontline system models are developed separately and connected appropriately before quantification of event tree sequences. Thus, support system dependencies are included systematically in the quantification process. Any additional requirements for support system modeling, identified during the event tree development stage, are also reviewed, developed, and connected before quantification of event trees. Fault tree method has been used for system modeling and analysis [7]. The approach followed in this Shutdown PSA is the small event tree/large fault tree approach where the system models include the relevant support systems models, human errors, and common cause failures (CCF). CCF modeling was done using Alpha Factor Model (AFM). System modeling is recognized to acquire great importance in this kind of approach and thus it is essential to have an established quality system in this area to have consistency in the modeling of different systems.

4.4 Assumptions

Following is the summary of assumptions that are credited for Shutdown PSA for BWR. The assumptions made are general and applied for entire shutdown state of operation. Wherever necessary additional assumptions are made and mentioned with description of specific model.

- Reactor is in cold shutdown condition with primary system in cold and depressurized state.
- Fuel is in reactor vessel in POS-1 and decay heat is being removed by shutdown cooling system. Reactor basin cavity shield blocks will be removed. Drywell head unbolting and removal will follow. Subsequently reactor vessel head unbolting and removal will be done. Rubber bellow seal between drywell and reactor building is replaced prior to cavity flooding. Reactor water inventory is maintained by emergency feed system. Duration of this POS is for 5 days.

- Fuel is in reactor vessel in POS-2, and decay heat is being removed by shutdown cooling system. Gate in between fuel pool and cavity is in place. Reactor basin cavity is flooded and dryer separator assembly unbolting and removal is done. In-core sipping is done for 48 h to detect in-situ fuel failures. Emergency feed is available, and main feed is not available. Condensate service water and demineralized water system can be used to feed loss of water in reactor cavity basin. Duration of this POS is for 2 days.
- During POS-3, gate in between fuel pool and cavity is removed and fuel is transferred from reactor vessel to spent fuel pool. Unloading of the core activity takes 3–4 days. Decay heat is removed by shutdown cooling system and fuel pool cooling system. Duration of this POS is for 5 days.
- During POS-4, fuel is in fuel pool and decay heat is removed by shutdown cooling system and fuel pool cooling system. Control Rod Drive blade shuffling and replacement, LPRM R&R and Control Rod Drive R&R are done during this period. Gate in between cavity and fuel pool remains removed. Duration of this POS is for 4 days.
- During POS-5, fuel is in fuel pool and decay heat is removed by fuel pool cooling system. Fuel pool gate is closed and cavity basin water is drained to carry out drained vessel works. Duration of this POS is for 4 days.
- During POS-6, gate in between fuel pool and cavity is removed and fuel is transferred from fuel pool to reactor vessel. Loading of the core activity takes 3–4 days. RBCW and SSW system outage is availed for maintenance works during this POS. Duration of this POS is for 4 days.
- During POS-7, fuel is in reactor vessel, fuel pool gate in between fuel pool and cavity is in place. Dryer separator assembly installation is done. Reactor basin cavity is drained, and reactor vessel head is installed. Drywell head is installed, and cavity basin shield blocks are placed. Remaining works related to RBCW and SSW system outage are taken up during this POS. Duration of this POS is for 4 days.
- As per technical specification, at least two emergency diesel generators shall be operable including the associated automatic isolation and tie switchgear.
- As per technical specification, Core spray and post-incident system shall be operable when irradiated fuel is in either of the reactor vessels.
- Reactor water inventory makeup can be done through emergency feed water system, condensate service water system, demineralized water system, or core spray system.
- During shutdown cooling system, failures credit has been given to feed through emergency feed system or condensate service water or demineralized water or core spray. Bleed path will be provided by either bleed to main condenser or to drywell.
- Once reactor basin cavity is flooded, it contains $650\ M^3$ of water which acts as a large heat sink open to reactor building. This can accommodate large amount of heat with small increase in temperature.

Table 2 shows the different initiating events considered in shutdown PSA.

Table 2 Matrix of initiating events versus POS

Initiating Event	POS						
	1	2	3	4	5	6	7
Small LOCA	x	x	x			x	x
Loss of station power	x	x	x	x	x	x	x
Loss of shutdown cooling system	x	x					x
Loss of fuel pool cooling system				x	x		
Loss of fuel cooling system			x			x	
Loss Of RBCW	x	x	x	x	x		
Loss of SSW	x	x	x	x	x		

Choosing the type of data applicable to the plant components and operating conditions is one of the main challenges for PSA analysis. Most of the data used in shutdown case is obtained from the data of full power operation as the operating conditions during full power operations are more stressful than shutdown conditions. Therefore, use of failure data for full power operation in shutdown mode provides a conservative estimate.

Ideally, the best study would use plant-specific data, but this data would be sparse in nature due to the limited number of plant operating years. Hence, generic data which is been integrated with plant-specific data using Bayesian method has been used. Possibility of collecting and using the plant-specific data has been explored and used wherever feasible.

4.5 Human Error Probability

Adequate accounting for the potential contribution of human errors is essential in PSA analysis. This involves identification of operator errors as well as assessment of their probabilities of occurrence. By virtue of the capacity to amalgamate the treatment of both human reliability and component reliability, PSA provides a distinctive picture of contribution of specific human actions pertinent to the accident progression which contribute to the plant safety.

Malevolent behavior (deliberate acts of sabotage and the like) is not considered for analyzing human reliability. Only human errors that could be made in performing assigned tasks are considered. Human reliability analysis (HRA) in NPP Level-1 Shutdown PSA is Accident Sequence Evaluation Program (ASEP) methodology.

4.6   Uncertainty and Importance Analyses

Along with the Minimal Cutset Analysis, uncertainty analysis was also carried out for overall core damage frequency, the purpose of which was to estimate the range of uncertainty surrounding the mean frequency estimates for the core damage state. The analysis addresses the uncertainty which arises from the quantification of the frequencies and probabilities of the individual events that appear in the Minimal Cutsets. Uncertainty analysis was carried out for basic events, initiating events, and overall core damage frequency. As a final step, importance and sensitivity analysis were carried out for all basic events, CCF groups, various frontline and support systems and for all dominant event sequences for ranking the components to priorities corrective measures. This includes computation of risk reduction factor, risk increase factor, and Fussel-Vesely importance factors.

5   Results and Conclusions

The Level-1 Shutdown PSA developed for TAPS-1&2 represents a complete model of the safety of the plant enclosing design, maintenance and safety practices during refueling shutdown, component reliabilities, common cause failures (CCFs), and operator errors. This model finds out the main contributors to most likely severe accidents which can occur during refueling shutdown as a combination of postulated initiating events, component malfunction, and human errors. The operation and maintenance procedure at TAPS-1&2 can be developed based on the outcome of the present study. With this integrated model available, the effect of proposed changes in maintenance practices can be examined by studying their impact on changes in accident frequencies.

With the present database on component reliabilities and the identified vulnerabilities to core damage, the estimated mean core damage frequency (CDF) in shutdown is 10% of the at full power. Emergency Diesel Generator fails to run and fail to start common cause failure, Loss of station power are some examples of system and component failure modes which are high on the importance list. Following Fig. 1 shows the POS wise contribution to the CDF.

The results of the Shutdown PSA indicate that a fairly high level of defense-in-depth exists in design during shut down mode of operation. This is evident from the final CDF value, as well as from the predominantly higher order minimal cut-sets observed in the core damage sequences. Some of the strengths of plant systems which enhance the safety are identified as:

- The capability of the reactor shut down system to maintain the reactor in long-term sub-criticality by control rods.
- All the control rods are hydraulically isolated soon after the cold shutdown of the reactor to eliminate inadvertent withdrawal of control rods.

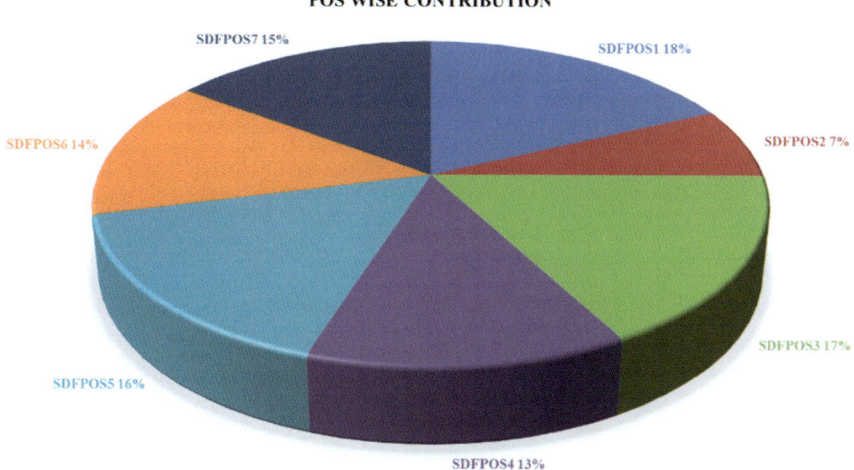

Fig. 1 POS wise contribution to core damage frequency

- All the maintenance works in primary system within first isolation valve are planned after complete unloading of the core into the fuel pool.
- Availability of large amount of water in reactor cavity and fuel pools during refueling operations. This provides sufficient time (more than 8 h) for making alternate arrangement for replenishment of inventory, before fuel gets uncover.
- Facility to inject water in reactor cavity and fuel pool during refueling from multiple sources like condensate service water and demineralized water apart from emergency feed water, core spray system for maintaining the coolant inventory.
- Provision of 3 × 100% frontline emergency diesel generators with independent cooling water systems in flood proof, seismically qualified separate rooms. Frontline DGs are backed up by a separate full capacity SBO DG which is located at a distance from EDGs.
- Stringent procedural control during fuel movement and emergency operating procedures (EOPs) for ensuring safety during shutdown.

As per of severe accident management guidelines (SAMG) provisions at the station for severe accident conditions, station has commissioned a 200 KVA truck-mounted mobile air cooled diesel generator (DG) and successfully demonstrated the running of equipment needed during station blackout (SBO) supplied by mobile DG. Station has incorporated several hookup points to inject water as a part of post-Fukushima modifications. Various hookup points for the water makeup are effective in shutdown state of the reactor, however credit have not been considered in the present study to show that: The existing plant design will meet with the general safety goals; to demonstrate that the current design is balanced in such a way that no initiating event or failure of mitigation system contributes disproportionately to the overall risk to the plant; identifying the systems, components, or human actions for which improvements or modifications to operational procedures could reduce or mitigate the likelihood of core damage. Moreover, considering the hookups in the analysis will mask the plant possible vulnerabilities, important human actions, and system weaknesses. Taking the credit of these hookups, the CDF value will reduce substantially.

Acknowledgements The authors express their gratitude to TMS Management for the support of the plant-specific data collection for the SPSA.

References

1. IAEA-TECDOC-1144, Probabilistic safety assessments of nuclear power plants for low power and shutdown modes
2. Atomic Energy Regulatory Board, "Design of light water reactor based nuclear power plants" No. AERB/NPP-LWR/SC/D-Atomic Energy Regulatory Board (2015)
3. Design Analysis Report, Volume-I and Volume-II, TAPS BWR, Dec 1968
4. J.W. Hickman, PRA procedures guide: a guide to the performance of probabilistic risk assessments for nuclear power plants. NUREG/CR-2300 (1983)
5. R.J. Breeding, J.C. Helton, E.D. Gorham, F.T. Harper, Summary description of the methods used in the probabilistic risk assessments for NUREG-1150. Nucl. Eng. Des. **135**(1), 1–27 (1992)
6. International Atomic Energy Agency. Defining Initiating Events for Purposes of Probabilistic Safety Assessment. IAEA-TECDOC-719, International Atomic Energy Agency (1993)
7. W.E. Vesely, F.F. Goldberg, N.H. Roberts, D.F. Haasl, *Fault tree Handbook*. No. NUREG-0492. Nuclear Regulatory Commission Washington DC (1981)

Probabilistic Safety Assessment of Process Plant Handling Hydrogen and Radioactive Fluid

Kandalam Venkatesh, Mistry Krunal, Bhanja Kalyan and Sadhana Mohan

Abstract The potential hazard in hydrogen handling plant containing radioactive fluid is mainly due to the activity from the radioactive fluid and flammability/explosion hazard from hydrogen. This study has focused on preliminary study of Probabilistic Safety Assessment of such a facility to identify the sufficiency of safety provisions/systems in the design and operation. As the availability of literature for the safety assessment of such facility is scarce, this study helps in providing an insight about the most critical system/equipment of the plant and the frequency of every consequence envisaged during the accident scenarios, thereby assisting in making design changes to the safety systems. For the assessment, potential accident sequences that dominate the risk were identified from design and safety analysis study and the features of the plant that contribute to the dominant accident sequences were analyzed in terms of their availability. Initiating events due to natural and energetic phenomena such as earthquakes, tornadoes, fires, floods, and explosions have not been considered in the analysis. Failure data was taken from both operational data and generic database. Appropriate Human Error Probability (HEP) was considered for the study but no modeling of HEP was carried out in the study. Fault tree analysis has been carried out for the major safety systems of the plant that are crucial for the mitigation of any consequence led by the initiating events viz. Gas Expansion System, Liquid Leak Detection System, Gas Leak Detection System, Vacuum Mopping System, Ventilation System, and Emergency Cleanup System. Similarly, event tree analysis has been carried out for the identification of postulated initiating events, leading to major release of activity in plant area. These events were large leakage in gaseous form, large leakage in

K. Venkatesh (✉) · M. Krunal · B. Kalyan · S. Mohan
Bhabha Atomic Research Center, Mumbai, India
e-mail: venky@barc.gov.in

M. Krunal
e-mail: krunalm@barc.gov.in

B. Kalyan
e-mail: kbhal@barc.gov.in

S. Mohan
e-mail: msadhana@barc.gov.in

© Springer Nature Singapore Pte Ltd. 2020
P. V. Varde et al. (eds.), *Reliability, Safety and Hazard Assessment for Risk-Based Technologies*, Lecture Notes in Mechanical Engineering,
https://doi.org/10.1007/978-981-13-9008-1_7

liquid form, loss of refrigeration system, and loss of Class IV Power. The PSA study was carried out in ISOGRAPH software version 10.1. Ground Release of activity was identified as the major consequence with the frequency of 1.09 × 10^{-10} per year which is very low. For further decreasing the frequency of the consequence, implementation of secondary containment for high activity handling process equipment is recommended.

Keywords Derived Air Concentration · Preliminary PSA · Radioactive liquid leak · Radioactive gas leak

1 Introduction

The main goal of radiological safety is to keep the toxicity and radioactive exposures from facilities to members of the public and workers as low as reasonably achievable (ALARA) during normal operational states (certainly below the limits set by the regulatory bodies) and in the event of an accident. In order to adhere to the safety goal, carrying out safety analysis has become almost mandatory for all industrial facilities handling toxic and radioactive material.

In the plants handling hydrogen and radioactive isotopes, the potential hazards are fire/explosion as well as radioactivity. The radioactivity exposure can be due to the radioisotopes in both gaseous and vapor form. Adequate measures are in place to mitigate these hazards and safety analysis has been carried out to ensure the sufficiency of these measures. However, to estimate probabilistically the risk from these hazards, Probabilistic Safety Assessment (PSA) has been carried out.

Probabilistic safety analysis was initially developed as a tool for quantitative risk assessment in nuclear power plant safety system design. However, it was soon realized that the same methodology can be adopted for evaluation of risks associated with chemical process plants as well. In the past, numerous PSA studies have been carried out for process plants handling hazardous chemicals. Sklet [1] identified accident sequences that may lead to hydrocarbon releases on offshore oil and gas production platforms. Rathnayaka et al. [2] developed fault tree and event trees for LNG processing facility. In all these studies, PSA was found to be an effective tool for comparing various process designs and their associated safety systems. The efficacy of these safety systems and their effect on consequence can serve as a useful guide to suggest focus areas. In the present analysis, accident scenarios are developed, considering possible failure of safety functions and human interactions. The PSA would give a quantitative insight of the consequences due to the accident scenarios. The work carried out is limited to identifying and assessing the risks at the facility during normal operation conditions and beyond design basis accident conditions. The identification of risks is focused on those emanating only from activity release and risks from fire/explosion hazard have not been considered in this study.

As the availability of literature for the safety assessment of such facility is scarce, this study helps in providing an insight about the most critical system/equipment of the plant and the frequency of every consequence envisaged during the accident scenarios, thereby assisting in making design changes to the safety systems. Component failure data has been taken from generic source such as IAEA TECDOC-478 [3], generic reliability database for components as well from operating experience. Component maintenance/repair times are considered based on operating experience. Test Intervals are as per Technical Specifications of the facility. Mission times considered are based on Safety Analysis.

The present study is a preliminary PSA and initiating events due to natural and energetic phenomena such as earthquakes, tornadoes, fires, floods, and explosions have not been considered in the analysis. Man-made events, specifically sabotage leading to large-scale fire/internal fire, are also not considered in the analysis.

2 Safety Concerns

The safety concern in the plant under study is the radioactive exposure to the plant personnel and the public. The main parameters to consider while handling radioactive isotope are Derived Air Concentration (DAC) and Annual Limit of Intake (ALI). The DAC for a radioactive isotope is the airborne concentration that, if inhaled over one-year work period (2000 h), will lead to Annual Limit of Intake (ALI) and produce approximately 20 mSv dose to an "average" worker [4].

DAC value is estimated considering that the person is taking the dose for the whole year, working for 2000 h in 8-h shifts. Thus, if a person is exposed to higher DAC for a shorter period of time, the total dose taken by the person should be less than ALI. If the dose is equal to ALI, the person should not be allowed to take any further dose in the remaining year. Thus, any event that can lead to more than ALI is a safety hazard.

3 Process Description

The contaminated water from the Nuclear Reactors is to be processed for removing the radioactivity before recycling back into the reactor to reduce the radioactivity exposure to the plant personnel. The contaminated water is generally transported to the facility in drums and stored in large storage tank for processing. The water is processed in a sequence of ion exchange beds and distillation columns before sending back to the reactor. The contaminated water is thus handled in both liquid, vapor, and gas forms and can lead to activity release in case of leakages in the facility.

4 Postulated Initiating Events

All the events that can lead to radioactivity exposure of 1 mSv to public or ALI
for plant personnel are identified based on the safety analysis. The basis to identify
such events was to consider any event leading to 250 DAC or more inside the plant
area even with the safety systems in operation. Below 250 DAC, the plant personnel
will not be receiving dosage above the permissible dosage for the year during such
accident even after considering that the personnel is operating the active area for the
whole shift (8 h). The following is the postulated initiating events arrived from the
safety analysis.

- Radioactivity leakage in liquid form due to tank rupture leading to activity more
  than 250 DAC inside plant
- Radioactivity leakage in Gas form due to tank rupture leading to activity more
  than 250 DAC inside plant
- Loss of refrigeration system
- Loss of Class IV Power.

5 Event Tree Development

5.1 Radioactivity Leakage in Liquid Form Leading to Activity More Than 250 DAC Inside Plant

The source for major radioactivity leakage in vapor form is storage tank for liquid. The
leakage of liquid inside the plant from tank can evaporate and lead to activity more
than 250 DAC. From the safety analysis, it is observed that rupture of drums/tanks
storing the liquid can lead to increase in DAC above 250 within shorter period of
time.

Liquid Leak Detection System (LLDS), Vacuum Mopping System (VMS), Emergency Cleanup System (ECS), and Ventilation System (VS) act as the safety systems
for the mitigation of above-initiating event. LLDS (beetle monitors) forms the first
step by detecting the liquid leakage in the plant. After the detection of leak, either
VMS or ECS needs to be manually initiated to avoid spreading the activity inside
plant area. In case of successful operation of VMS, ECS needs not to be started.
ECS needs to be initiated in case of VMS failure. In case of failure of both VMS
and ECS, the VS takes care of reducing the activity in the plant area by diluting and
releasing the exhaust through stack. If all the safety systems fail, the activity would
be isolated inside the plant area and not spread to the public domain, but the dosage
to the plant personnel will increase and can exceed ALI in the 8 h of shift operation.
Thus, **Ground Release** even with isolation inside the plant is a major consequence.

The frequency for liquid leakage by tank rupture was estimated from the operating
experience by considering the number of such events occurring in the total years of

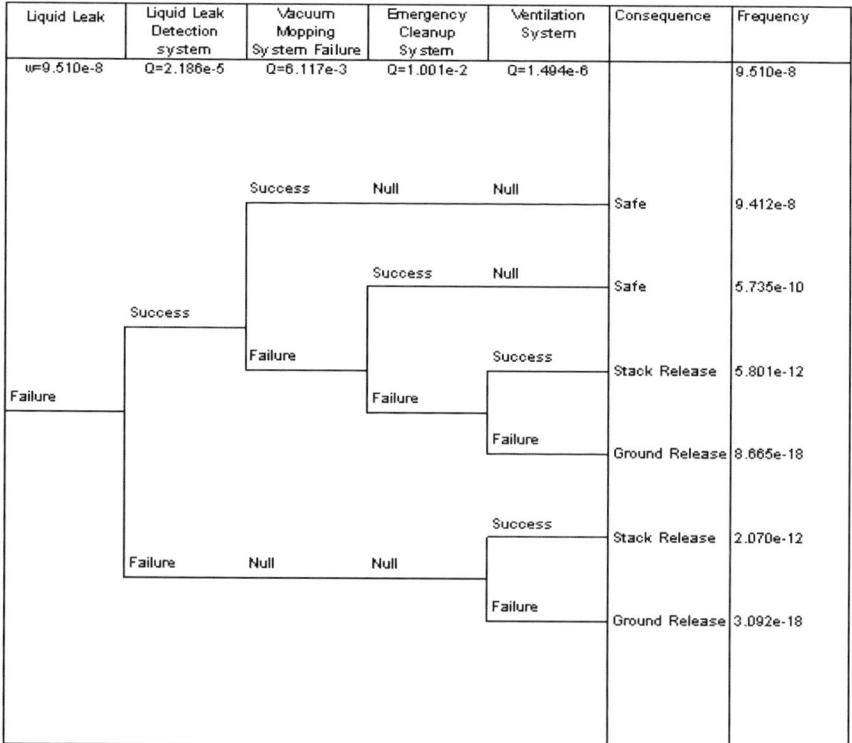

Liquid Leak	Liquid Leak Detection system	Vacuum Mopping System Failure	Emergency Cleanup System	Ventilation System	Consequence	Frequency
w=9.510e-8	Q=2.186e-5	Q=6.117e-3	Q=1.001e-2	Q=1.494e-6		9.510e-8
		Success	Null	Null	Safe	9.412e-8
			Success	Null	Safe	5.735e-10
	Success	Failure		Success	Stack Release	5.801e-12
Failure			Failure	Failure	Ground Release	8.665e-18
				Success	Stack Release	2.070e-12
	Failure	Null	Null	Failure	Ground Release	3.092e-18

Fig. 1 Event tree for radioactivity leakage in liquid form

operation. No such event has occurred in the past 30 years of operation of hydrogen handling plant which has a total of 40 number of liquid storage tanks. But for this study, one such event is considered. The initiating frequency is found to be 9.51×10^{-8}/h. The event tree is shown in Fig. 1. The consequences resulting from this event tree are **Ground Release** in case of failure of ECS, VMS, and VS together or failure of LLDS and VS together, **Stack Release** in case of ECS and VMS failure together with successful operation of VS or LLDS failure with successful operation of VS. The successful operation of LLDS and VMS together or LLDS and ECS together would result in **Safe Condition** (DAC in plant area below 250).

5.2 Radioactivity Leakage in Gas Form Leading to Activity More Than 250 DAC Inside Plant

As the leakage through any valve or tube would not lead to an activity increase to more than 250 DAC before identification through various safety and monitoring systems, only the failure of tanks is considered for the analysis as it can immediately release the whole inventory into the plant area and increase the DAC, giving no time in hand to avoid this scenario.

Gas Leak Detection System (GLDS), Emergency Cleanup System (ECS), and Ventilation System (VS) act as the support systems for the mitigation of above-initiating event. GLDS forms the first step by detecting the leakage in the plant through monitoring the increase in DAC. ECS is manually started in case of detection of DAC > 5. In case of failure of ECS, the VS takes care of reducing the activity in the plant area by diluting and releasing the exhaust through stack. If all the safety systems fail, the activity would be isolated inside the plant area and the dosage to the plant personnel will increase and can exceed ALI in the 8 h of shift operation.

The frequency for gas leakage through tank failure was estimated from the operating experience by considering the number of such events occurring in the total years of operation. No such event has occurred in the past 25 years of operation of hydrogen handling plant which has a total of six number of tanks. But for this study, one such event is considered. The initiating frequency is found to be 7.6×10^{-7}/h. The event tree is shown in Fig. 2. The consequence resulting from this event tree is **Ground Release** in case of ECS and VS failure together or GLDS and VS failure together, **Stack Release** in case of ECS failure with successful operation of VS or GLDS failure with successful operation of VS. The successful operation of GLDS and ECS together would result in **Safe Condition**.

5.3   *Loss of Refrigeration System*

The major utility for the plant is the Helium refrigeration system. For liquefaction of the process gas, functioning of Helium refrigeration system is crucial. During any loss of refrigeration coolant for the liquefaction column, the gas inside the system will expand due to the increase in the temperature and ultimately leading to increase in pressure. With any valve in the loop being closed, the pressure can increase to beyond the design pressure leading to the rupture of the equipment and release of activity above 250 DAC in the plant in gaseous form.

Gas Expansion System (GES), Gas Leak Detection System (GLDS), Emergency Cleanup System (ECS), and Ventilation System (VS) act as the support systems for the mitigation of above-initiating event. GES is provided to take care of expansion of hydrogen due to warm up caused by loss of refrigeration system. When GES successfully takes care of the expansion without any fail, no other safety system would be required as the activity release would not occur at all. GLDS forms the next step of safety system by detecting the leakage in the plant due to tank rupture. ECS is manually started in case of detection of DAC > 5. In case of failure of ECS, the VS takes care of reducing the activity in the plant area by diluting and releasing the exhaust through stack. If all the safety systems fail, the activity would be isolated inside the plant area and the dosage to the plant personnel will increase and can exceed ALI in the 8 h of shift operation.

The frequency for loss of refrigeration system was estimated by fault tree analysis. The failure frequency of refrigeration system was estimated considering the failure of either Class IV power or refrigeration compressor or expander. The initiating frequency was found to be 4.9×10^{-4}/h. The failure data of refrigeration compressor and expander was taken from the operational data. The event tree is shown in Fig. 3. The consequence resulting from this event tree is **Ground Release** in case of GES and VS failure together, **Stack Release** in case of GES and ECS failure with successful operation of VS or GES and GLDS failure together with successful operation of VS. The successful operation of GES alone or successful operation of GLDS and ECS during failure of GES would result in **Safe Condition**.

5.4 Loss of Class IV Power

The loss of power would lead to the loss of liquid nitrogen system. The liquid nitrogen acts as refrigeration system for gaseous adsorption and loss of the system during a

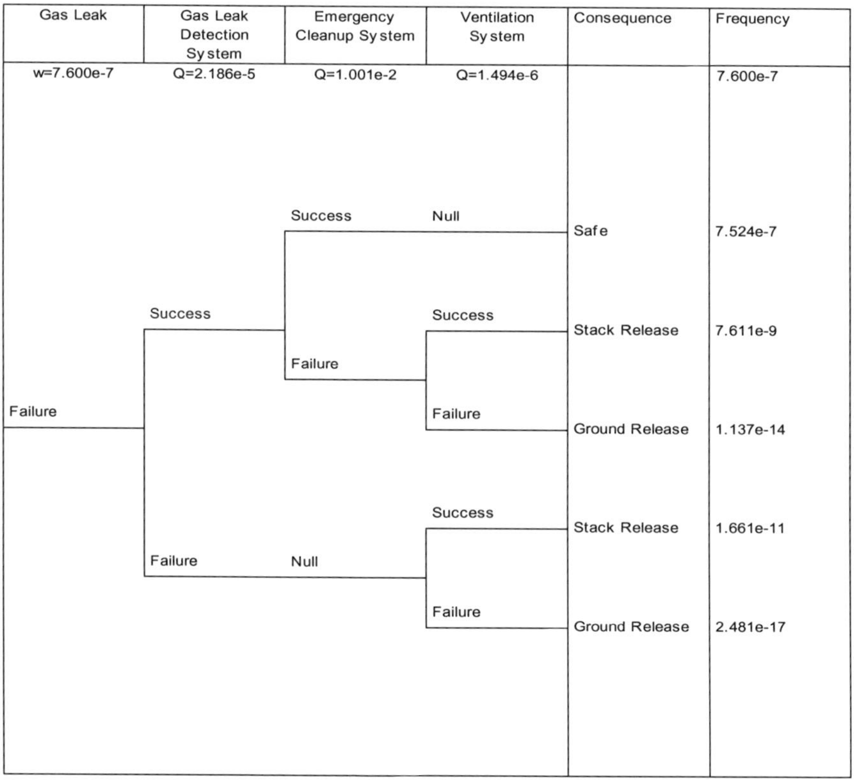

Gas Leak	Gas Leak Detection System	Emergency Cleanup System	Ventilation System	Consequence	Frequency
w=7.600e-7	Q=2.186e-5	Q=1.001e-2	Q=1.494e-6		7.600e-7
		Success	Null	Safe	7.524e-7
	Success		Success	Stack Release	7.611e-9
		Failure			
Failure			Failure	Ground Release	1.137e-14
			Success	Stack Release	1.661e-11
	Failure	Null			
			Failure	Ground Release	2.481e-17

Fig. 2 Event tree for radioactivity leakage in gas form

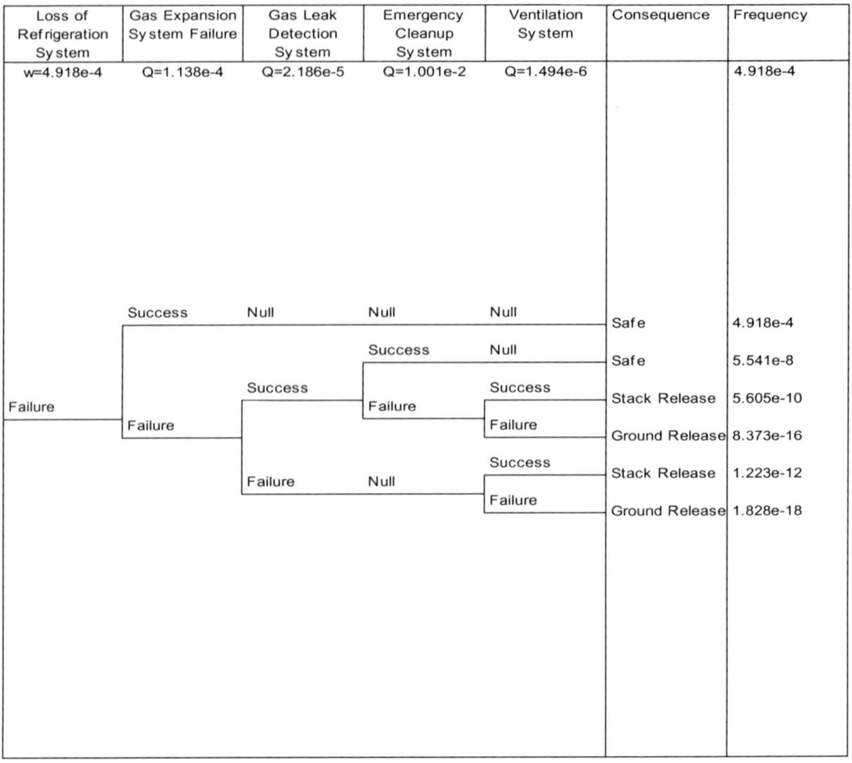

Loss of Refrigeration System	Gas Expansion System Failure	Gas Leak Detection System	Emergency Cleanup System	Ventilation System	Consequence	Frequency
w=4.918e-4	Q=1.138e-4	Q=2.186e-5	Q=1.001e-2	Q=1.494e-6		4.918e-4
	Success	Null	Null	Null	Safe	4.918e-4
			Success	Null	Safe	5.541e-8
		Success		Success	Stack Release	5.605e-10
Failure	Failure		Failure	Failure	Ground Release	8.373e-16
				Success	Stack Release	1.223e-12
		Failure	Null	Failure	Ground Release	1.828e-18

Fig. 3 Event tree for loss of refrigeration system

Class IV Power failure will lead to desorption of the gases and increase in the pressure inside the bed due to the gas expansion with temperature difference. With any valve in the loop being closed, the pressure can increase to beyond the design pressure leading to the rupture of the equipment and release of activity above 250 DAC in the plant in gaseous form.

Gas Expansion System (GES), Gas Leak Detection System (GLDS), Emergency Cleanup System (ECS), and Ventilation System (VS) act as the support systems for the mitigation of above-initiating event. The sequence of safety system is similar to that of the event "Loss of Refrigeration System."

The typical grid failure frequency of 1.4×10^{-4}/h has been considered in the analysis. The failure frequency has been taken from operational data of the plant site. The event tree is shown in Fig. 4. The consequence resulting from this event tree is **Ground Release** in case of GES and VS failure together, **Stack Release** in case of GES and ECS failure with successful operation of VS or GES and GLDS failure together with successful operation of VS. The successful operation of GES alone or successful operation of GLDS and ECS during failure of GES would result in **Safe Condition**.

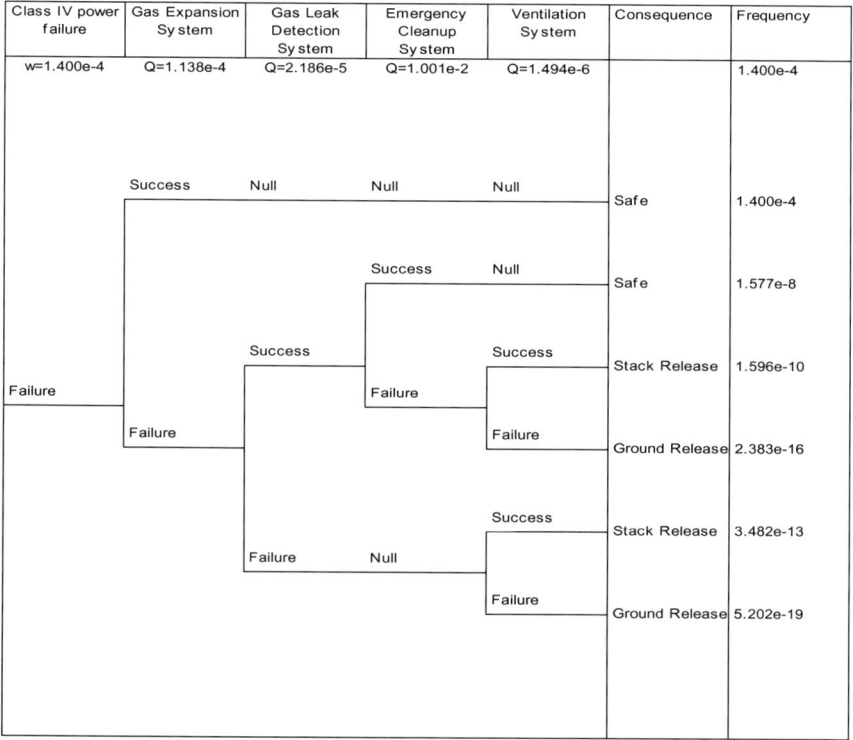

Class IV power failure	Gas Expansion System	Gas Leak Detection System	Emergency Cleanup System	Ventilation System	Consequence	Frequency
w=1.400e-4	Q=1.138e-4	Q=2.186e-5	Q=1.001e-2	Q=1.494e-6		1.400e-4
	Success	Null	Null	Null	Safe	1.400e-4
			Success	Null	Safe	1.577e-8
		Success		Success	Stack Release	1.596e-10
Failure			Failure	Failure	Ground Release	2.383e-16
	Failure			Success	Stack Release	3.482e-13
		Failure	Null	Failure	Ground Release	5.202e-19

Fig. 4 Event tree for loss of class IV power

6 Conclusion

With the current database on component reliabilities and the identified vulnerabilities to release, the estimated frequency of various release states along with the details of uncertainty analysis carried out using ISOGRAPH [5] is presented in Table 1. Lognormal distribution has been considered for the components, and the 5000 simulations were carried out.

The major consequence state from the plant is Ground Release, which is radioactivity inside the plant area greater than 250 DAC. The frequency of the consequence

Table 1 Release frequency of activity

Release state	Mean frequency (/h)	95% confidence value (/h)	Mean frequency (/year)
Safe	7.20e−4	1.18e−3	6.31
Stack release	8.36e−9	1.00e−8	7.32e−5
Ground release	1.25e−14	1.88e−14	1.09e−10

was found to be 1.09×10^{-10} per year which is very low. The consequence effects can be further reduced by isolating the high activity tanks inside secondary containment like glove box.

The second consequence state from the plant is Stack Release, which is safe gaseous release of radioactivity through stack. The consequence is safe as the total inventory release through stack during the initiating events under this study, does not exceed the annual dose limit for public within the exclusion zone due to the dilution. The frequency of the consequence was found to be 7.32×10^{-5} per year. The frequency of the consequence can be further reduced by implementing autoinitiation of ECS after leak detection and also including more redundancy in Detection Systems.

The highest mean release frequency among all the states for the plant is found to be 7.32×10^{-5}/h with 95% confidence for "Stack release." As the consequence of "Stack release" for the plant is not hazardous, the overall consequence is not of a major concern.

References

1. S. Sklet, Hydrocarbon releases on oil and gas production platforms: release scenarios and safety barriers. J. Loss Prev. Process Ind. **19**, 481–493 (2006)
2. S. Rathnayaka, F. Khan, P. Amyotte, Accident modeling approach for safety assessment in an LNG processing facility. J. Loss Prev. Process Ind. **25**, 414–423 (2012)
3. International Atomic Energy Agency, *Reliability Data for Probabilistic Safety Assessment of NPP*, Iaea-Tecdoc-478, Vienna (1986)
4. *Radiation Protection during Operation of Nuclear Power Plants,* AERB safety guide No. AERB/SG/O-5
5. *Fault Tree+V10.1*, Copyright ISOGRAPH Reliability Software (1986)

Source-Term Prediction During Loss of Coolant Accident in NPP Using Artificial Neural Networks

T. V. Santhosh, Akhil Mohan, Gopika Vinod, I. Thangamani and J. Chattopadhyay

Abstract Nuclear power plant (NPP) is a highly complex engineering system which experiences a number of transients such as equipment failure, malfunctioning of process and safety systems, etc. during its operations. Such transients may eventually result in an abnormal state of the plant, which may have severe consequences if not mitigated. In case of such an undesired plant condition, the chances of the release of source term (e.g. release of Iodine, Caesium, Krypton, Xenon, etc.) and subsequent dose to public and to the environment cannot be neglected. The early knowledge of the expected release of source term will help in planning the emergency preparedness program. In view of this, several computational intelligence techniques have been studied and employed to early prediction of source term based on containment thermal-hydraulic parameters and also taking into account the actuation/non-actuation state of associated engineered safety features (ESFs). This paper presents an integrated framework based on artificial neural networks (ANNs) for early prediction of the expected release of source term during the large break loss of coolant accident (LOCA) in 220 MWe pressurised heavy water reactors (PHWRs). A simulated data of fission product release up to 48 h from the beginning of the LOCA has been considered in the model development. Several neural networks with forward and reverse configuration of the hidden layer were tried to reach at an optimal network for this problem. As the range of the input data was significantly large, a data transformation to a logarithmic scale was also performed to improve the efficiency and accuracy in prediction. The developed ANN model has been validated with the blind case LOCA scenarios. The performance of the final model is found to be satisfactory with a percentage error between actual and predicted value being under 5% expect for a few cases for all the species.

Keywords Source term · Artificial neural networks · Nuclear power plants · Loss of coolant accident · Emergency preparedness

T. V. Santhosh (✉) · G. Vinod · I. Thangamani · J. Chattopadhyay
Reactor Safety Division, Bhabha Atomic Research Centre, Mumbai, India
e-mail: santutv@barc.gov.in

A. Mohan
Homi Bhabha National Institute, Mumbai, India

© Springer Nature Singapore Pte Ltd. 2020
P. V. Varde et al. (eds.), *Reliability, Safety and Hazard Assessment for Risk-Based Technologies*, Lecture Notes in Mechanical Engineering,
https://doi.org/10.1007/978-981-13-9008-1_8

81

1 Introduction

Nuclear power plant is a complex engineering system which experiences a number of transients during its operations. When faced with an unplanned event such as an equipment failure, the operator has to carry out necessary corrective actions within the specified time in order to mitigate the situation. However, in the case of a major accident scenario such as LOCA where there is a risk of radiation exposure to the public, the emergency response team has to take countermeasures to reduce the radioactive spread around the plant to protect public and environment. This demands an early detection of the accident scenario, expected source term and implementation of an effective countermeasures to mitigate radiation exposure to public. Hence, an intelligent system that will assist the emergency response team to identify the accident scenario at the early stages and predicts the expected radioactive release is very much essential.

The primary heat transport (PHT) system a typical of 220 MWe PHWRs is shown in Fig. 1. The main function of the PHT system is to circulate the heavy water coolant and transport the heat generated from the reactor core to the steam generator. The entire circuit consists of pumps, headers, feeder pipes, pressure tubes and steam generators. There are two steam generators and two heat transport pumps at each end of the reactor. The entire system is in the shape of a 'Figure of Eight' with each leg having a reactor inlet header (RIH) providing coolant to half of the channels of the reactor core and a reactor outlet header (ROH) receiving the hot coolant from the reactor channels. The reactor is also equipped with an emergency core cooling system (ECCS) designed to limit the consequences of events such as LOCA [1–3]. The break anywhere in the PHT loop triggers to the actuation of ECCS. If there is a loss of cooling, there can be the release of source term from the reactor core.

Source term is the amount and nature of the radioactive material released from a nuclear facility following a severe accident. Source term indicates information about the actual or potential releases of radioactive material from a given source which may include a specification of the amount, the composition and the rate and mode of release [4]. The probability of a major event at nuclear facilities leading to the release of large quantities of radioactivity into the environment is always ensured to be negligibly small. However, even in the event of a major release into the environment, the prompt and effective implementation of countermeasures can reduce the radiological consequences for the public. The response action post-accident scenario in a nuclear facility is the responsibility of the emergency preparedness team. This team will first calculate the hazard consequences due to such a scenario. Safety intervention is then done to reduce the radioactivity release to the public domain [5]. Brusque knowledge of source term is very relevant in safety applications of the reactor. It can be considered as a preliminary stage for accident consequence evaluation. It is also handy for the planning of accident mitigation and the assessment of safety features taken.

Development of analytical tools for a fast online accident diagnosis and subsequent source-term release prognosis at a nuclear power plant is highly desired by

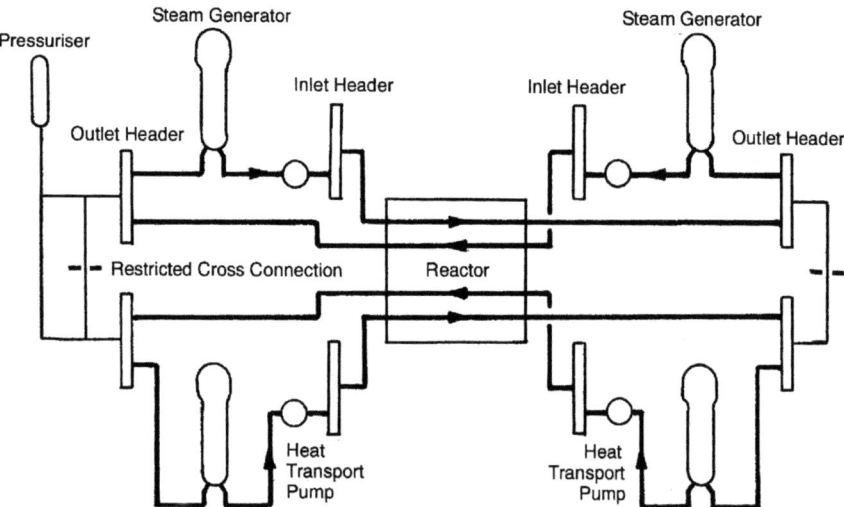

Fig. 1 PHT system of typical PHWR

all organisations dealing with this scenario across the globe. The uncertainties pertaining to accident scenarios are inherent to the plant model and its specifications. In this regard, several countries are undertaking individual research work on developing such models specifically for their plant accident scenarios. A computerised source-term prediction tool known as (RASTEP) Rapid source-term prediction by Nordic nuclear safety research organisation for Swedish nuclear power plants [6] and the German approach for source-term prediction—fast-pro [7] are similar researches in this area. Both these models make use of level-2 PSA and Bayesian belief networks (BBN) for accident predictions and then display the pre-calculated source term based on defined accident conditions. If the identified scenario is not available in the defined accident condition, these tools will not be able to display any source term as the pre-calculated source term is not readily available. The objective of this research work is to overcome this limitation by predicting the actual source term using advanced computational intelligence models.

A particular accident scenario known as loss of coolant accident (LOCA) and the consequent source-term release prediction has been considered in this paper. LOCA generally occurs due to a break initiation in the coolant channel. Commencement of such a break at inlet header leads to a sudden depressurisation of the primary heat transport (PHT) system. Several reactor trip/instrumented signals will be activated one after the other in a short period of time. Sequence of actuation of trip signals largely depends upon the break size and location of the event. Once a trained diagnostic system is actually accustomed to this behaviour of trip signals, it can then predict the real-time LOCA event based on these signals. Another prediction system known as the prognosis system can take inputs from this LOCA scenario and also available inputs from containment pressure and temperature sensors to predict the

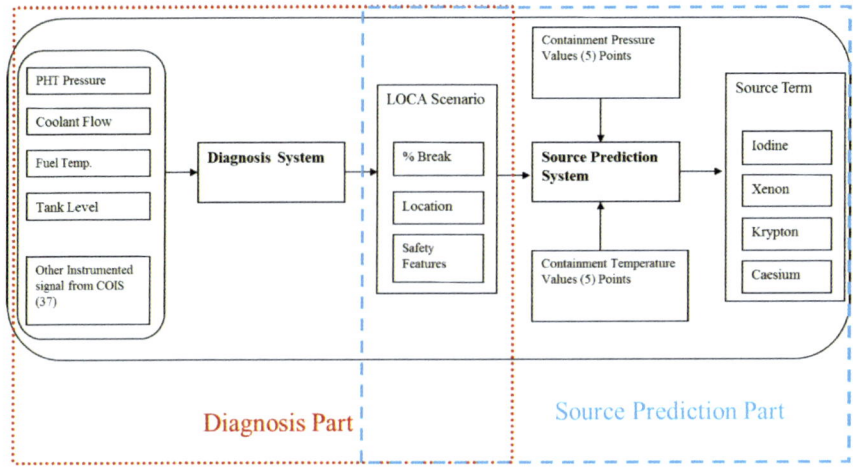

Fig. 2 Block diagram of source prediction system

source-term release into the environment. The prognosis part is considered in detail in this work and the source-term release has been predicted.

The overall system as a block diagram is shown in Fig. 2. The diagnosis part will take inputs from various instrumented signals from computer operated information system (COIS) and predict the LOCA scenario in the picture. The prognosis part will take inputs from the diagnostic part and also from pressure and temperature sensors at various points in containment to predict the source-term release into the environment. Since the major release will be in the form of four radioactive elements namely, Iodine, Caesium, Xenon and Krypton, they will collectively form the output part of the system.

In this study, the artificial neural network is employed to correctly predicts any given accident scenario that has occurred and ultimately relates it to the radioactivity spread to the environment. Here, the parameter sequence is characterised by various input parameters collected from sensors, and the considered accident scenario is LOCA which may lead to the release of radioactivity into the environment. The main performance criteria influencing this undertaking are the generalisation ability of the method and time taken before an accurate prediction of the event. Generally speaking, transient identification can be regarded as a pattern recognition problem of various input instrumented signal parameters. Any transient starts from the steady-state operation of the reactor. So, when an abnormal event occurs, the monitoring sensors register a varied pattern than the usual ones. These patterns are supposedly unique with respect to the type and severity of the accident, as well as to the initial conditions [1]. In our case, we can further define this pattern recognition of sensors to a classification problem for calculating the radioactivity release to the environment.

2 Methodology

2.1 Feature Selection

The foremost requirement of an ANN-based data-driven system is the modelling of the desired output. This system being used for the prediction of the source term, the output parameters were defined accordingly. The total dose emission to the environment of four major radioactive elements namely, Iodine, Caesium, Xenon and Krypton were defined as output parameters. The input feature selection process was based on their significant variation in their signal transforms when a radioactivity spread occurs due to LOCA. A total of 14 input feature parameters were selected for the system. Pressure and temperature signals from five different points of the containment were taken as the major input parameters and the rest being, elapsed time, percentage of break, location and the engineering safety features controlling the radioactivity release. Percentage break due to LOCA and its location was taken from the existing diagnostic system. Block diagram of the framework developed is shown in Fig. 3.

Simulated instantaneous data (per second wise) of the radioactivity release from the containment of the four output parameters was calculated and then matched with their corresponding 14 input parameters for obtaining the training and testing files for the neural network. A total of 172,800 data sets were taken for the learning phase of the network. The appropriate selection of input–output parameters are an important

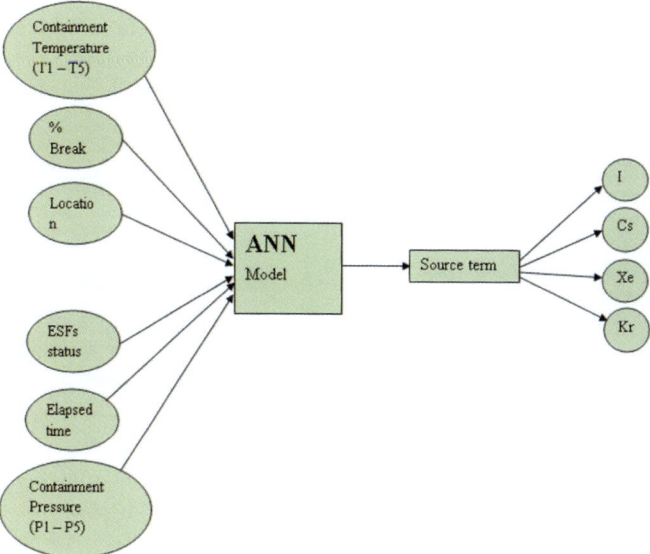

Fig. 3 Block diagram of source-term prediction

factor in neural networks algorithm. The spread of each input parameter distribution is more important than the number of parameters, because it decides the full range possibility of output.

2.2 Challenges Encountered

The major challenge during the study was the processing of featured data for the network algorithm. Vast records of instantaneous data features up to 48 h were generated from simulated studies. It was recognised that the release data was stagnant after 12 h and thus the initial data was reduced to 12 h. The idea of instantaneous prediction of radioactivity release was first envisaged. But its usage in a real-time scenario was found out to be minimal. A future prediction model was found to be a superior choice for such a scenario. Hence, a cumulative per hour wise prediction of the four radioactive elements up to 12 h was considered as final output.

Large variation in the output data without following any specific pattern as shown in Fig. 4 was a major concern. The variation was in terms of 10^8 for Iodine and 10^6 for Caesium.

The maximum and minimum values of both input and output are shown in Tables 1 and 2 to illustrate the spread of data values. The feasibility of any network to accurately predict this large variation from a comparatively lesser deviated input was found to be difficult [8]. To overcome this, the output values were then converted to their logarithmic values. The distribution spread of the logarithmic output reduced drastically and was in comparison with the input deviations. For a superior prediction system, it is important to have output deviation in comparison with the input deviation. Prediction system will not be able to produce better accuracy if the output is varying drastically with small deviation in input values. So, conversion of the output spread to logarithmic scale has drastically improved the efficiency of the neural network system.

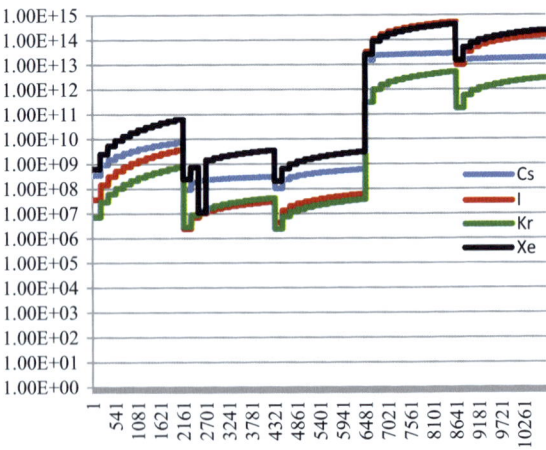

Fig. 4 Large variation of output data

Table 1 Minimum and maximum input values

Input data range	P1 (Pa)	P2 (Pa)	P3 (Pa)	P4 (Pa)	P5 (Pa)	T1 (°K)	T2 (°K)	T3 (°K)	T4 (°K)	T5 (°K)	Break (%)	Elapsed time (h)	Break location	Safety function actuation status
Min value	101.35	101.35	101.35	101.35	101.35	305.37	305.37	305.37	305.37	305.37	75	1	0	0
Max value	185.21	185.16	185.14	101.39	173.62	392.04	378.81	364.95	305.42	351.55	200	12	1	1

Table 2 Minimum and maximum values of output data

Output data range	Caesium (Bq)	Iodine (Bq)	Krypton (Bq)	Xenon (Bq)
Min value	9.09E+07	2.35E+06	2.83E+06	2.44E+08
Max value	2.82E+13	5.05E+14	5.01E+12	4.22E+14

2.3 Network Architecture

The artificial neural networks comprise of a large number of strongly connected neurons arranged in a particular fashion. The different neurons are interconnected with each other creating different layers [9]. The feed forward architecture is the most commonly adopted format in all related studies on other fields. So, in this work, we have also used that format to find out the radioactive release to the environment. The network can be trained to give a desired pattern at the output when a corresponding input data set is applied. The learning phase is accomplished with a large number of input and output target data. The training algorithm used in this study is the back-propagation algorithm. The initial output pattern is compared with the desired pattern and the weights are adjusted by the algorithm to minimise the error [10]. The iterative process finishes when the error becomes minimum. Full range of possible results was presented to the learning network for improving its prediction ability.

The number of layers to be used and the number of neurons in each layer are not known prior and are estimated by trial and error method. Each network was then compared for their average RMS error, to select the best network using BIKAS simulator [11].

In this work, this was calculated based on a systematic method of three phases. At each phase, a fixed number of iterations were considered with different neuron combinations in a progressive way.

(a) **First Phase**: In this phase, a fixed number of 5000 iterations were used for all the possible network structures developed with one and two hidden layer networks. This phase was then again divided into three stages based on hidden layers.

(i) *One hidden layer*: All one hidden layer network structures starting a single hidden neuron to 15 neurons were compared for a fixed iteration of 5000, in this stage. Figure 5 shows the results of this stage. It is understood from the figure that there is no specific pattern for error reduction in terms of the network architecture. It follows a random path and cannot be pre-decided. This proves that trial and error is the only method for finding the network architecture.

(ii) *Two hidden layers (Forward direction)*: A total of 23 network structures with two hidden layers in forward direction were tried in this phase. Here, forward direction means that the first hidden layer is having more number of neurons than the second one. Different permutations and combinations were tried in this selection and their outputs were then compared accordingly. Figure 6 shows this comparison in detail. Randomness of the error with respect to network architecture is very clear in this scenario also.

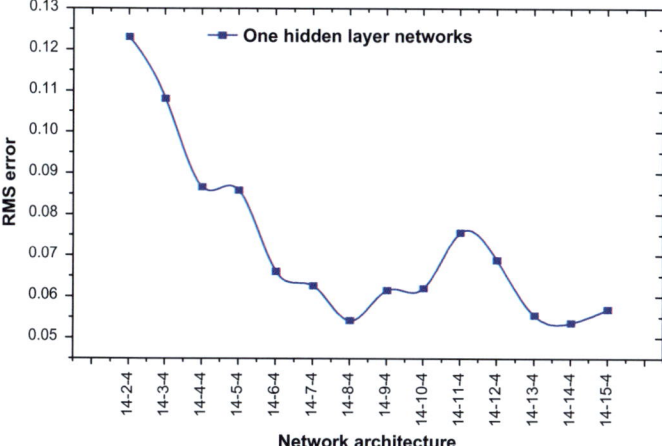

Fig. 5 Comparison of one hidden layer networks

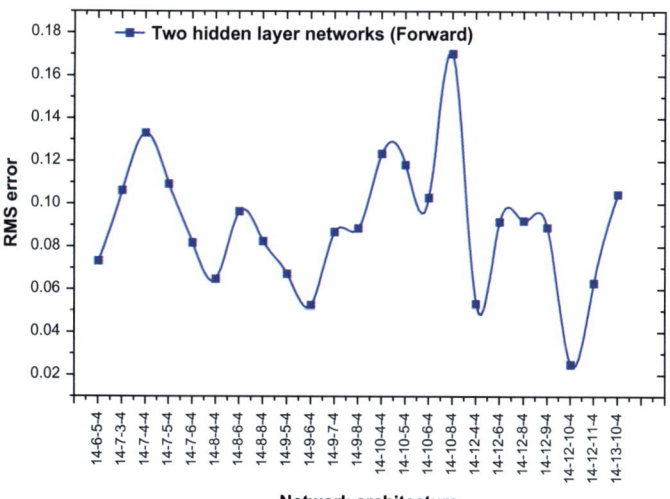

Fig. 6 Comparison of two hidden layer networks in forward direction

(iii) *Two hidden layers* (*Reverse direction*): Here, the first hidden layer is having less number of neurons than the second one. Hence the name, reverse direction. Eight such cases were compared for their RMS error for 5000 iterations. Comparison chart is shown in Fig. 7.

(b) **Second Phase**: All network structures in the first phase were compared, and the six selected ones with least error were taken up for the second phase. These selected architectures were run for 10,000 iterations. RMS error of each networks

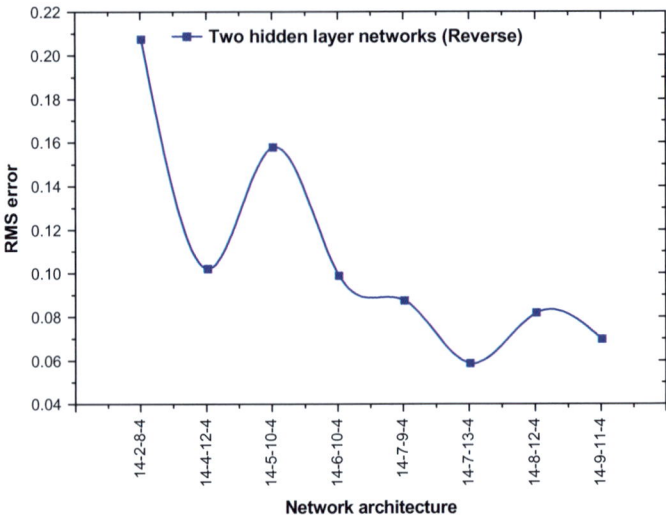

Fig. 7 Comparison of two hidden layer networks in reverse direction

for this much iterations was then found out. This was then compared for arriving at the final stage.

(c) **Third Phase**: At the end of the second phase, two architectures with minimum errors were finalised. Evaluation of these two for minimum error was carried out in the last phase of 50,000 iterations, and the final network architecture was selected. The neuron network configuration finalised for this study is 14-13-4 with a minimum RMS error of 0.045134. This architecture consists of 14 neurons in the input layer, one hidden layer with 13 neurons in it and an output layer consisting four neurons as outputs. Final network architecture and its error comparison with iterations are shown in Fig. 8.

The characteristic parameters of the networks, like the weight change factor, learning rate coefficient, etc., are the user-defined values for each networks. In this study, these parameters were kept constant throughout, for even comparison of all architectures. The values used are 0.7 for the learning rate parameter and max and min weight change factor being 1.2 and 0.5, respectively.

3 Results and Conclusions

The framework developed is to predict the source-term release to the environment using ANN algorithm, when an LOCA scenario occurs inside the containment of a pressure reactor. To reach at such a model, various network structures were tried and tested and the one with minimum error has been selected as the final architecture. Calculation of the actual source-term release in case of an accident scenario is an

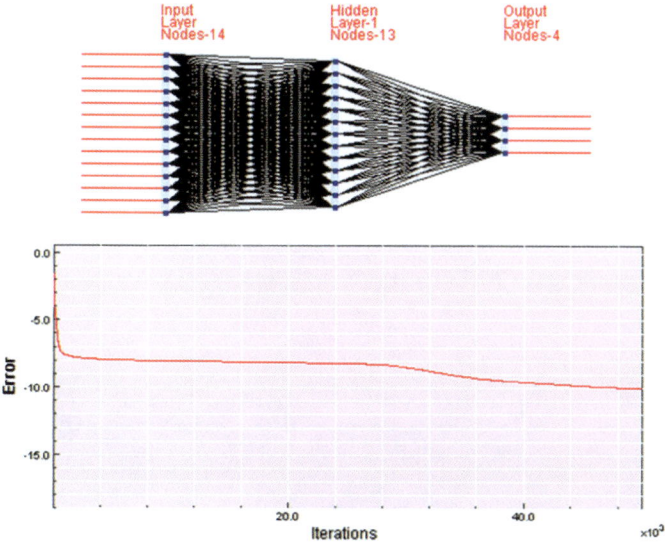

Fig. 8 Final network architecture

extreme computationally intensive task. The trained and tested model was then used for prediction of random scenarios.

3.1 Prediction of Fission Products

After the selection of suitable network architecture and its training procedure, the network was then tasked to produce predictions of random scenarios. Eleven random input scenarios with known output data were fed to the system for target prediction. The output obtained was then compared with the desired source-term data for all the four radioactive elements. The predicted output was exactly matching to the desired output for all the four cases. Figures 9, 10, 11 and 12 represent the comparison graph of the predicted and desired values of various fission products.

3.2 Effect of Noise

In a practical source-term prediction problem, the probability of the sensor data input containing noise is not negligible. So, it is important that the model is robust to give better accuracy even in the presence of noise. Both the pressure and temperature inputs were added with 1, 2, 5 and 10% noise, respectively, and their effect on predicted output was studied. It was found out that the RMS error increased with the

Fig. 9 Desired and predicted values of Iodine

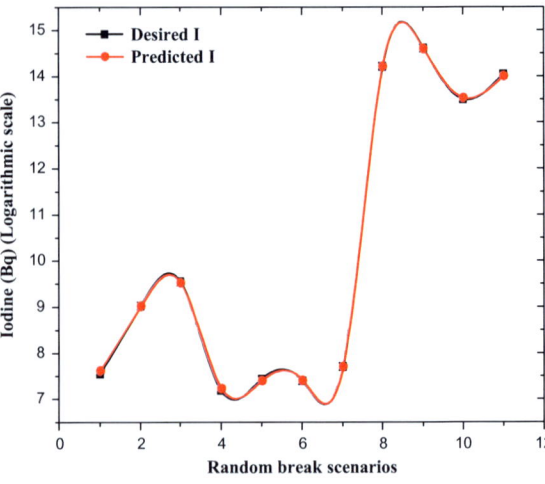

Fig. 10 Desired and predicted values of Caesium

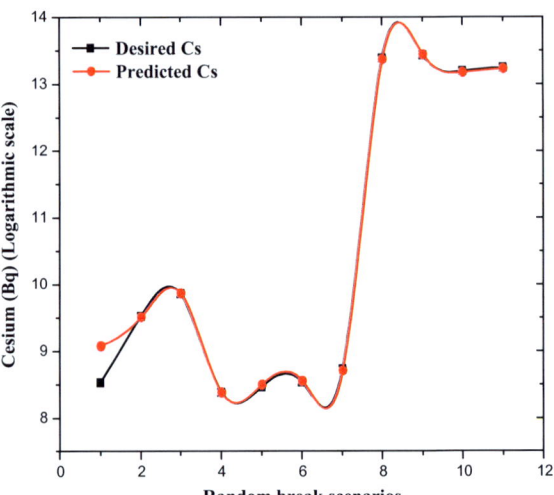

addition of noise, but its overall effect on the system was found to be minimal. Pressure and temperature signals from different containment positions were considered for this study.

It is important to consider strategic input values which have got maximum weightage in output decision making for this study. It was found out that the pressure at point 5 and temperature at point 1 have got large variations and affect the outputs phenomenally. Noise was added to these points and the system behaviour to this was found. Figures 13 and 14 show the effect of adding noise to pressure and temperature inputs, respectively. RMS error increases with the addition of noise in a limited manner in the system. Up to 1% noise addition has got a negligible effect on both

Fig. 11 Desired and predicted values of Krypton

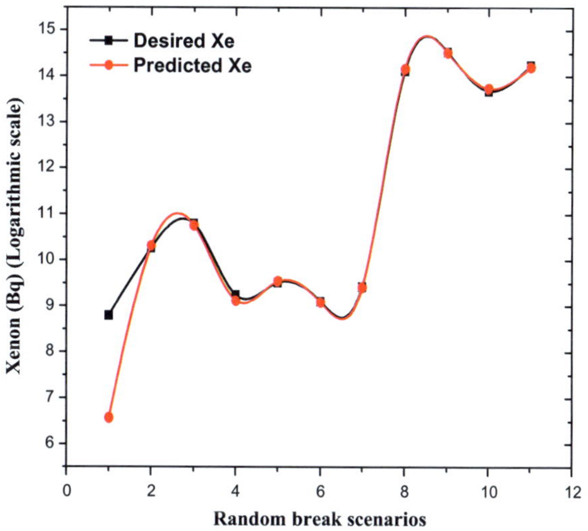

Fig. 12 Desired and predicted values of Xenon

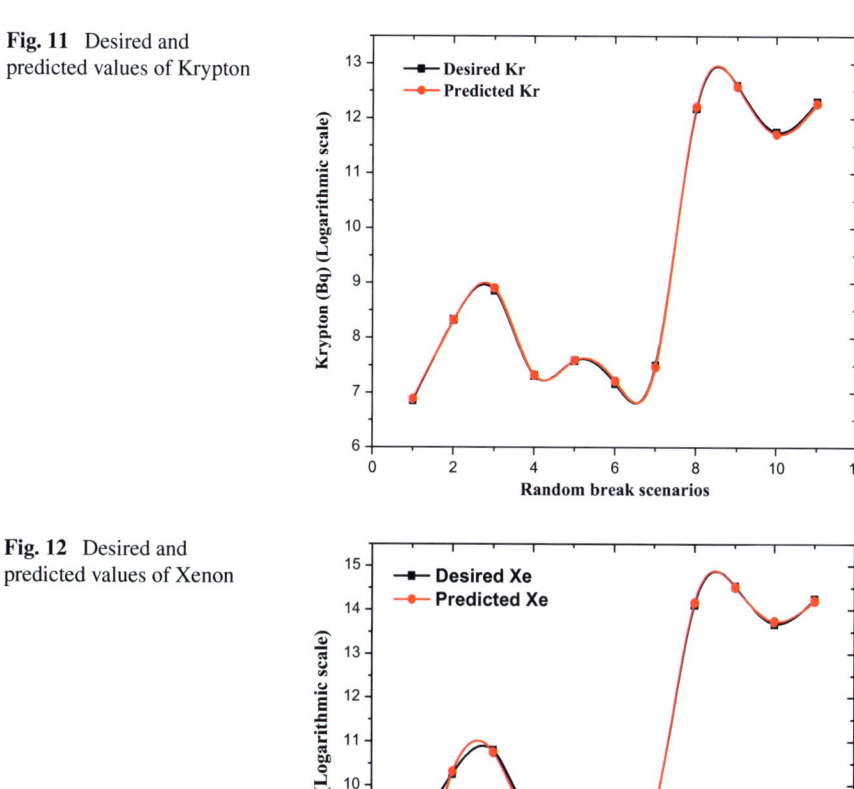

pressure and temperature inputs. The system is capable of predicting accurate source terms even in noisy input scenarios, which explains the robustness of the system.

3.3 Salient Features

The goal of future prediction of four radioactive elements release to the environment during LOCA scenario has been achieved by this model with an error level of almost null. It was also found that the system was having a dynamic response with better accuracy to all the random scenarios given for prediction. The effect of noise in

Fig. 13 Effect of noise in the pressure signal

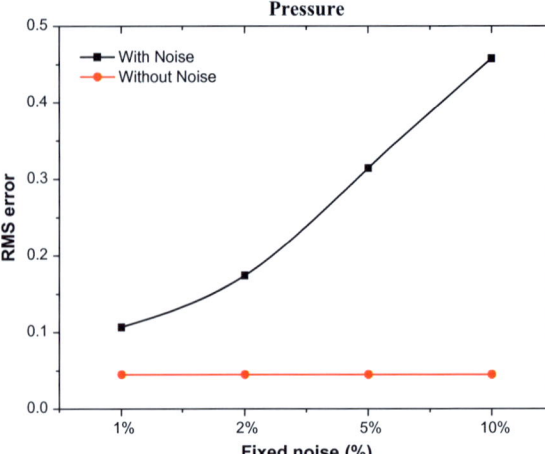

Fig. 14 Effect of noise in the temperature signal

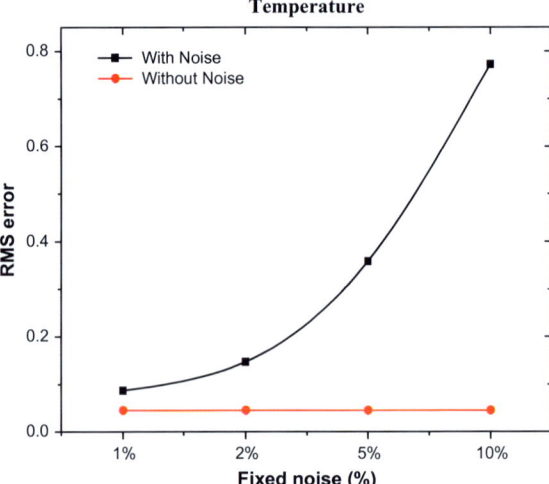

input data was also studied in detail and the error in prediction was found to be minimal. This explains the robustness of the system and its ability to work with better accuracy in any accident scenario. With proper training, it is understood that the system is capable of producing accurate results.

The present system is catered for only LOCA type scenarios. All other accident scenarios and their respective effect on the environment can be developed as a deep learning model as an extension of the same framework. Source-term release prediction cases of different accident scenarios can be developed individually and then added on to this existing system. It is pertinent to mention that the study so far was very satisfactory and results yielding. Further advancement of this model will be considered in the future.

References

1. T.V. Santosh, A. Shrivastava, V.V.S. SanyasiRao, A.K. Ghosh, H.S. Kushwaha, Diagnostic system for identification of accident scenarios in nuclear power plants using artificial neural networks. Reliab. Eng. Syst. Saf. **54**, 759–762 (2009)
2. Francesco Di Dedda, *Definition of a dynamic source term module within RASTEP* (Chalmers University of Technology, Sweden, 2013)
3. Samuel Glassstone, Alexander Sesonske, *Nuclear Reactor Engineering* (CBS Publications, New Delhi, 1998)
4. IAEA, Techniques and decision making in the assessment of off-site consequences of an accident in a nuclear facility, Safety series—88, Vienna (1987)
5. IAEA, Manual for First responder to a Radiological Emergency, Vienna (2006)
6. M. Knochenhauer, V.H. Swaling, F. Di Dedda, F. Hansson, S. Sjökvist, K. Sunnegård, Using Bayesian Belief Network (BBN) modelling for rapid source term prediction. Nordic Nuclear Safety Research, Sweden (2013)
7. M. Hage, H. Löffler, A probabilistic approach for source term prediction in case of severe accidents, in *PSAM 13*, Seoul, Korea (2016)
8. B. Yan, Y. Cui, L. Zhang, C. Zhang, Y. Yang, Z. Bao, G. Ning, Beam structure damage identification based on bp neural network and support vector machine. Hindawi Publishing Corporation-Mathematical Problems in Engineering (2014)
9. A.K. Jain, J. Mao, K.M. Mohiuddin, *Artificial neural networks: A Tutorial*. IEEE (1996)
10. S. Haykin, *Neural Networks: A Comprehensive Foundation* (MacMillan College Publishing Co., New York, 1994)
11. BIKAS, Neural Networks Simulator, Version 7.0.5, BARC (2005)

Digital Reliability

Time-Dependent Short Mode Failures in MLCCs Due to Cracks in the Device Bodies

S. K. Dash, Y. R. Bhanumathy and P. J. V. K. S. Prakasha Rao

Abstract Insulation resistance degradation is one of the most common failure modes in multilayer ceramic chip capacitors (MLCC). Degradation in insulation resistance facilitates short mode failures in long run due to certain failure mechanism. Therefore, it is important to understand, how these mechanisms resulted in short mode failures of MLCC. From many factors, those govern short mode failure in MLCC bias voltage is a major factor. There are two MLCC failure regimes defined by the bias voltage: low-voltage failure (<5 V) and normal voltage failure (>5 V). The normal voltage failures in multilayer ceramic chip capacitors initiated due to crack developed in them during assembly process are described in this paper. MLCCs are susceptible to cracking, if subjected to sudden change in temperature (temperature gradient of ~250 °C) during soldering process. These cracks facilitate moisture ingression which resulted in short mode failures in ceramic chip capacitors over a period of time due to metal migration. In this paper, controlled experiments were conducted to demonstrate short mode failure in MLCCX7RU capacitors initiated due to cracks developed in them during assembly fabrication. Effect of cracks on insulation resistance of the capacitors is also demonstrated by an innovative method wherein capacitors having cracks were soaked by solder flux mixed isopropyl alcohol. This method effectively simulates the cleaning process of the capacitors after mounting them on to the PCB. Subsequently, these capacitors were tested at normal voltage (50 and 70 V) under ambient conditions (25 °C, 50%RH). Both single and stacked type MLCCX7RU capacitors were considered for this study. Some capacitors were conformally coated to see its effect against further propagation of failure. From this test, it is inferred that mechanism of short mode failure in ceramic chip capacitors are due to (i) crack in the capacitor body resulted during soldering, (ii) moisture/contaminants penetration

S. K. Dash (✉) · Y. R. Bhanumathy · P. J. V. K. S. Prakasha Rao
Spacecraft Reliability and Quality Area, U R Rao Satellite Centre, Bangalore 560017, India
e-mail: sarat@isac.gov.in

Y. R. Bhanumathy
e-mail: bhanu@isac.gov.in

P. J. V. K. S. Prakasha Rao
e-mail: pjvks@isac.gov.in

© Springer Nature Singapore Pte Ltd. 2020
P. V. Varde et al. (eds.), *Reliability, Safety and Hazard Assessment
for Risk-Based Technologies*, Lecture Notes in Mechanical Engineering,
https://doi.org/10.1007/978-981-13-9008-1_9

99

during cleaning process, and (iii) potential difference across the capacitor during usage. With all these conditions, silver migration takes place between two adjacent electrodes resulting in short mode failure.

Keywords Conformal coating · Insulation resistance · MLCC · Metal migration · Moisture ingression

1 Introduction

Multilayer ceramic chip capacitors (MLCC) are one of the wide components in electronics circuits. Although these capacitors are reliable and quite robust they sometimes fail during hand soldering and assembly fabrication process (soldering on to PCB). MLCCs fail due to various, from which cracks on the body is most significant. Cracks on the MLCCs are not desirable as it limits the assembly yield and pose long-term reliability issue. Ceramic capacitors being fragile/brittle are prone to crack when mishandled during soldering and assembly process. The part experiencing differential temperature (gradient) across its body is considered detrimental factor for the crack initiation. Once assembled, the MLCCs are susceptible to develop a crack either because of printed circuit board flexing or capacitors experiencing higher than the rated temperature. These cracks manifest themselves as electrical defects such as intermittent contact, loss of capacitance, and reduction in insulation resistance/excessive leakage current. However, sometimes cracks on capacitor body do not affect any of the electrical parameters of MLCC such as capacitance (C), dissipation factor (DF), and insulation resistance (IR). Very often it is also observed that ceramic chip capacitors with cracks on the body as a result of soldering and assembly process could successfully withstand relevant high reliability component level screening like burn in as well system level environmental stress screening involving temperature cycling and vibration test. Such capacitors, while tested individually after removal from the package/hardware, revealed perfect electrical characteristics or slight degradation of insulation resistance (10^8 or 10^9 Ω against >10^{10} Ω). Usually ceramic chip capacitor failed in resistive short mode or dead short mode resulting in catastrophic effect such as burning/charring of capacitors. Hence, it is necessary to understand the effectiveness of cracks in terms of electrical behavior of MLCC and the mechanism of manifestation of crack into degradation of insulation resistance and further into short mode failure.

Reason of crack in MLCCs are well known and sufficient reference in literature is available regarding this [1, 2]; hence, it is not elaborated in this paper. However, efforts are made to understand the effect of such cracks on insulation resistance of MLCCs through a novel test method. Controlled experiments were conducted to demonstrate short mode failure in MLCCs initiated due to cracks developed in them during assembly fabrication. Ceramic chip capacitor MLCCX7RU (MIL part number) was chosen as representative sample for this study.

2 A Novel Method for Determination of Cracks in MLCC

Identification of crack on MLCCs body is always a matter of debate. Visual inspection using optical microscope is not a foolproof method for identification of crack because sometimes these cracks look superficial and sometimes they look like a demarcation line. Even x-ray radiography has a low success rate in revealing cracks [3–5]. Many times MLCCs with body crack meet electrical test specification in terms of capacitance, dissipation factor, and insulation resistance. Presence of crack usually not affect capacitance and dissipation factors of the MLCC substantially [6, 7] but can cause increase in leakage current (specified limit <2.5 µA) or reduction in insulation resistance (specified limit >10^{10} Ω). The majority of field failures of MLCCs are caused by low insulation resistance that is often due to cracks [3]. In low-impedance applications, a decrease in resistance might cause catastrophic failures. Although cracks in ceramic capacitors might not lead to immediate failures, they facilitate degradation in insulation resistance, which would degrade with time (hours to months) resulting eventually in field failures.

Various methods are described in the literature [3] which could reveal crack in ceramic chip capacitors such as visual inspection, x-ray radiography, scanning acoustic microscopy (SAM), environmental testing (temperature cycling and humidity testing), methanol testing, and saltwater testing [8]. From the above methods, visual inspection and x-ray radiography are more subjective and not much informative about nature or depth of crack. Scanning acoustic microscopy is a technique, which can show any crack (internal or external) quantitatively and provide its depth. Whereas environmental testing, saltwater testing and methanol testing are inductive methods, because in these methods induce extraneous material such as water, salt water, and methanol to determine if such crack as sensitive to such extraneous materials in terms of insulation resistance. Even though saltwater test and methanol test are quite effective in determining crack, they are no realistic, because these materials are not used in fabrication of MLCCs. However, several other factors may also lower the insulation resistance of MLCCs, thus evaluating the effect of cracks. These include contaminated cleaning solvents or large amounts of dissolved flux residue [9].

In this study, we have chosen more realistic method of determining cracks, which in turn resulted in reduction of insulation resistance. During fabrication process, the entire printed circuit board with assembled parts is subjected to IPA cleaning process to remove the flux and other contaminations. During such process, it is probable that the cleaning solution IPA mixed with solder flux could seep through the crack developed during soldering process. To evaluate effect of such process on insulation resistance of MLCC following experiment was carried out. Ceramic chip capacitors MLCC 2225X7RU, 1.2 µF, 5%, from a particular date code and a particular make were chosen for this study.

2.1 Test Results

Five MLCC 2225X7RU capacitors (sl. nos. C1 to C5), which had developed cracks on their body during manual soldering process, were chosen for study (refer Fig. 1 for optical photograph). Isopropyl alcohol was mixed with small amount of RMA flux (Rosin flux mildly activated), and the solution was thoroughly rinsed so that flux completely dissolved in IPA. The capacitors were soaked in the above solution for one hour at room temperature. Two good capacitors, which do not have any cracks on their bodies (sl. nos. C6 and C7), were also subjected to the above test (refer Fig. 2 for optical photograph). Then, the capacitors were air dried and subjected to insulation resistance (IR) measurement to evaluate for any change in the insulation resistance as compared to the initial IR readings. The capacitors were also baked at 125 °C for 1 h and post-baking IR measurement also carried out. The results are shown in Table 1.

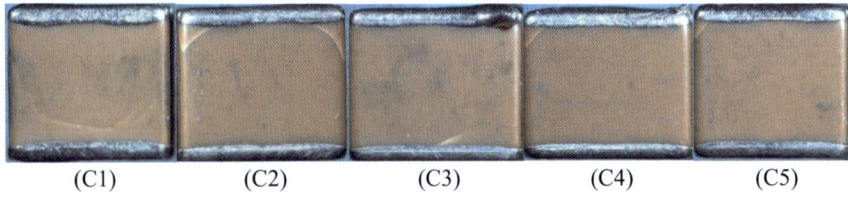

    (C1)   (C2)   (C3)   (C4)   (C5)

Fig. 1 Optical photographs of capacitors MLCC 2225X7RU, show cracks on their bodies, magnification: 5X

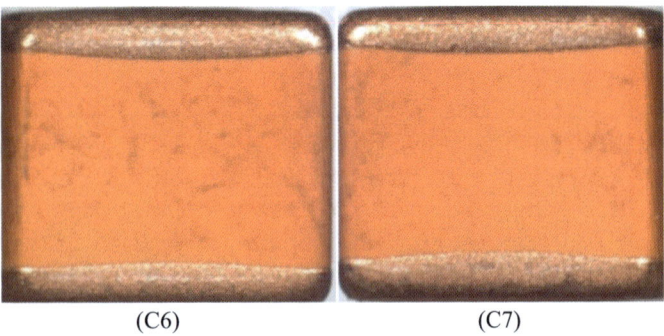

    (C6)   (C7)

Fig. 2 Optical photographs of capacitors MLCC 2225X7RU, show no cracks on their bodies, magnification: 5X

Table 1 Pre- and post-crack determination test electrical measurement of MLCCs

Capacitor sl. nos.	Pre-crack determination IR reading in Ω	Crack determination test	Post-crack determination IR reading in Ω	Post-baking IR reading in Ω	Specified limit of IR in Ω	Remarks
C1	3.93×10^{10}	Capacitors soaked in IPA mixed with solder flux for 1 h	8.31×10^{8}	3.12×10^{9}	$>1 \times 10^{10}$	IR decreased after soaking in flux mixed IPA solution and improved after baking
C2	4.75×10^{10}		2.44×10^{8}	5.21×10^{9}		
C3	3.93×10^{10}		5.67×10^{9}	2.32×10^{9}		IR decreased after soaking in flux mixed IPA solution and did not improve after baking
C4	2.62×10^{10}		1.19×10^{9}	2.59×10^{9}		
C5	4.85×10^{10}		7.62×10^{8}	7.45×10^{8}		
C6 (good capacitor)	3.98×10^{10}		3.94×10^{10}	3.92×10^{10}		No change in IR value was noticed
C7 (good capacitor)	2.84×10^{10}		2.88×10^{10}	2.82×10^{10}		No change in IR value was noticed

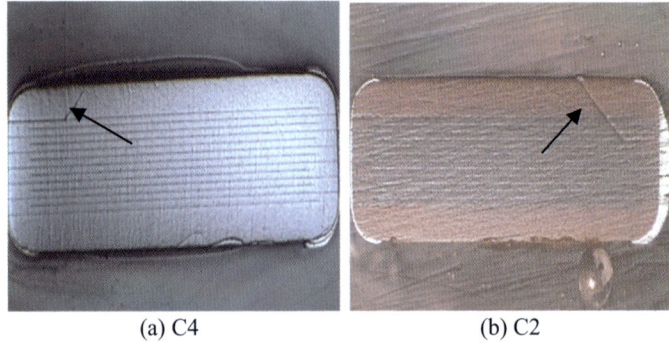

(a) C4   (b) C2

Fig. 3 Cross-sectioned view of cracked capacitors C2 and C4, magnification: 10X

2.2 Discussion

The above results showed that capacitors sl. nos. C1 and C2, which showed reduction in IR after soaking with solder flux mixed IPA, improved after baking at 125 °C for 1 h. Capacitor sl. nos. C3, C4, and C5, which showed reduction of IR after soaking with solder flux mixed IPA did not recover after baking. As the conductivity of solder flux mixed IPA is more than the ceramic materials, cracks filled with such solution increases conductivity of the overall dielectric material which in turn decreases insulation resistance of the capacitor. If the trapped solution is able to be evaporated and escaped after baking, then the insulation resistance value improves.

It is also observed that dimension of cracks in capacitors sl. nos. C3, C4, and C5 is lesser than those in capacitors sl. nos. C1 and C2. Hence, it may be presumed that lesser IPA solution might have seeped into capacitors sl. nos. C3, C4, and C5 than those of capacitors sl. nos. C1 and C2. The seeped liquid might have evaporated and escaped after baking at 125 °C in capacitors sl. nos. C1 and C2 resulted in improvement of insulation resistance. To validate the presumption, capacitor sl. no. C4 (lesser crack) and C2 (bigger crack) were molded and cross-sectioned. The optical microscopic photographs are shown in Fig. 3. As observed from the fig, depth of crack in capacitor sl. no. C4 is lesser and is up to the internal electrode layers, whereas depth of crack in capacitor sl. no. C2 is more and is penetrated into the internal electrode layers.

3 Mechanism of Short Mode Failure in MLCCs

As discussed in earlier sections, crack in MLCC itself may not create any electrical deviation. Also deviation in insulation resistance (10^8 or 10^9 against 10^{10} Ω) does not create any anomalies under the application environment of MLCC, unless otherwise there is a substantial degradation in insulation resistance which manifests into leakage

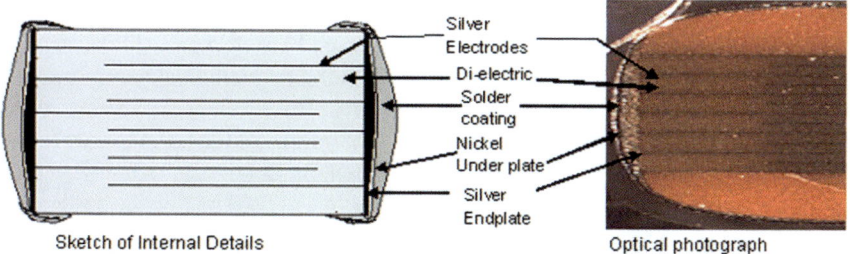

Sketch of Internal Details   Optical photograph

Fig. 4 Cross-sectioned ceramic chip capacitor and sketch of internal details

current in order of mill ampere (mA). It is also extensively discussed in literature [3] that crack developed on the capacitor in most cases would be dormant as the part does not experience stress which can propagate the developed crack. However, moisture or conductive contaminants (flux mixed IPA) penetrate inside then it can cause increase in leakage current, which on long run fail in short mode.

The basic structure of a multilayer ceramic capacitor MLCC consists of alternating thin layers of dielectric ceramic material and metal electrodes. Two metal end terminations function as electrical contacts and are used to solder the MLCC on a printed circuit board. For high reliability parts, the electrodes are basically silver mixed with palladium separated by dielectric materials Barium Titanate ($BaTiO_3$). Sketch of internal details along with photograph of cross-sectioned ceramic chip capacitor is shown in Fig. 4.

Generally, in MLCC the observation of increased leakage current manifested into short mode failure with time. Increased leakage currents in MLCCs having cracks might be due to various mechanisms. A virgin surface of the crack might have increased electron conductivity due to the presence of a high concentration of surface traps [3]; however, this conductivity is not great enough to cause failures. A substantial increase in leakage current occurs when a conductive media fills the crack that connects two adjacent electrodes layer (as shown in Fig. 3b) which is made of silver. This is possible either by moisture condensation, when the part is exposed to humid/contaminated environments or when silver migration of electrode metal occurs through the crack. In both cases, the presence of ionic contaminations substantially enhances the degradation process [3, 10, 11]. Finally, the capacitor fails in short mode. However, if the crack is on the external dielectric layer (as shown in Fig. 3a), then even if it filled with conductive media, no silver migration occurs. In such cases, capacitor shows only degradation in insulation resistance.

Metal migration is an electrochemical phenomenon that can occur under active condition in presence of moisture and temperature. Several factors influence metal migration, which includes temperature, relative humidity, voltage bias, conductor material, conductor spacing, contaminations type, and contamination amount [11]. The temperature ranges for metal migration to initiate is <100 °C, assuming the electrolytic solution is aqueous and current density is ~100 mA/cm^2. However, metal like silver can migrate at relatively high temperatures (more than 150 °C) by a field-induced diffusion mechanism [10] even in the absence of moisture. Voltage bias must be present for metal migration to occur. This electric field drives ionic migration by causing the positive ions to travel along the field lines from the anode to the cathode through an ion transport path provided by the aqueous/electrolytic medium. A higher bias increases the rate of propagation and the short mode failure gets precipitated faster [11]. The distance between the electrodes normally referred as conductor spacing in the MLCC, plays a major role in determining the time to failure. As this distance decreases, time-to-failure will decrease assuming that the rate of silver migration stays constant. Increased levels of contamination due to poor cleanliness can have a positive effect on metal migration behavior. Contamination usually increases the electrical conductivity of the electrolytic solution, thus decreasing the amount of time it takes for ions to migrate through the solution.

A popular mathematical model for mean time to failure (MTTF) due to metal migration growth based on Arrhenius equation was proposed by Hornung [12] as follows

$$T_f = \frac{\alpha G}{V} e^{\frac{Ea}{kT}}$$

where T_f is mean time to failure, α is proportionality constant, G is the gap between electrodes, V is the applied voltage, Ea is the activation energy, k is Boltzmann's constant, and T is the applied temperature. Hornung measured the dendrite growth of silver through borosilicate glass under an applied electric field and the growth rate was found to have an approximately linear dependence on the applied electric field. From the above equation, it is also evident that metal migration is directly proportional to voltage and indirectly proportional to gap between electrodes. For a normal testing condition, applied voltage (V) across the capacitor and temperature (T) environment is constant, gap (G) between the electrodes of capacitor is also constant. Hence, the only variable factor is activation energy (Ea), which varied from 0.6 to 1.2 eV for silver (metal) migration process [10].

From the above discussion, it is understood that time to failure in MLCCs under a normal testing environment (constant temperature and voltage) depends on activation energy, which manifested itself by moisture/contaminants entered through crack on the body of capacitors.

4 Failure Simulation Study on MLCCs

4.1 Background

Several MLCC 2225X7RU, 1.2 μF, 5%, 200 V (both single and stacked) failed during various phases of ground testing of ISRO satellites. Failure analysis on failed components revealed crack on all the failed capacitors, which indicates that all the failed capacitors have crack on them. In order to study crack induced time-dependent short mode failure in MLCCs, control simulation study was conducted on few MLCCs. Effect of conformal coating on capacitor having body crack was also studied.

4.2 Initial Visual and Electrical Examination of Chosen Components

MLCC 2225X7RU, 1.2 μF, 5%, 200 V (Single Capacitor 8 nos., Stack of Four Capacitors 8 nos.) were chosen for the study. All the components are mounted on PCB by hand soldering method by qualified operator. The components were visually inspected using optical microscope at 40X. No visual anomalies were observed. Refer Fig. 5 for single capacitors and Fig. 6 for stacked capacitors. The capacitors were tested electrical using a LCR meter and the parameters; C: capacitance, DF: Dissipation Factor, IR: Insulation Resistance are measured and shown in Table 2.

4.3 Thermal Stress Simulation on the Capacitors

Ceramic chip capacitors are susceptible to body crack, when they come across sudden temperature gradient of >100 °C. This fact was supported by numerous literatures [13–15] and proven by simulation study. In this study, crack was created on the capacitors by heating the body with touching the solder gun (at 280 °C) for ~1 min followed by sudden cooling by pouring little quantity of IPA on the heated capacitor

Fig. 5 Single capacitors C1–C8, pre-simulation optical photographs

Fig. 6 Stacked capacitor C1–C8, pre-simulation optical photographs

Table 2 Initial electrical test results of chip capacitors chosen for study

Capacitor type	Capacitor sl. nos.	Electrical parameters			Remarks
		C (Capacitance in μF): 1.14–1.26 for single 4.56–5.04 for stack of four	DF < 0.025	IR > 10^{10} Ω	
MLCC 2225X7RU, 1.2 μF, 5%, 200 V, **Single**	C1	1.1802	0.0123	8.76×10^{10}	Satisfactory
	C2	1.1745	0.0112	5.76×10^{10}	Satisfactory
	C3	1.1763	0.0156	7.14×10^{10}	Satisfactory
	C4	1.1937	0.0145	4.37×10^{10}	Satisfactory
	C5	1.2012	0.0138	6.46×10^{10}	Satisfactory
	C6	1.1523	0.0125	5.45×10^{10}	Satisfactory
	C7	1.1889	0.0118	9.16×10^{10}	Satisfactory
	C8	1.1922	0.0178	2.32×10^{10}	Satisfactory
MLCC 2225X7RU, 4.8 μF, 5%, 200 V, **Stack of Four**	C1	4.9602	0.0145	4.92×10^{10}	Satisfactory
	C2	4.7740	0.0118	3.45×10^{10}	Satisfactory
	C3	4.6359	0.0132	4.42×10^{10}	Satisfactory
	C4	4.8248	0.0122	5.98×10^{10}	Satisfactory
	C5	4.7452	0.0119	4.26×10^{10}	Satisfactory
	C6	4.8545	0.0117	3.72×10^{10}	Satisfactory
	C7	4.7860	0.0120	3.56×10^{10}	Satisfactory
	C8	4.7939	0.0138	4.45×10^{10}	Satisfactory

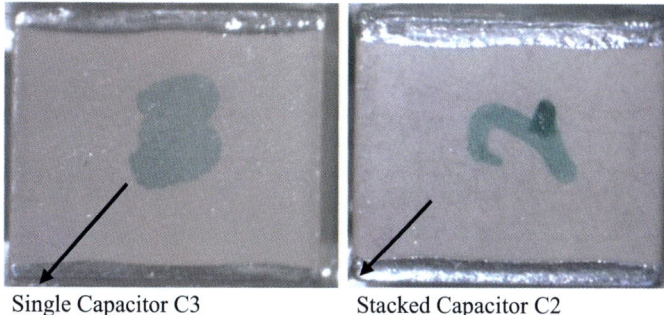

Single Capacitor C3   Stacked Capacitor C2

Fig. 7 Optical photograph of the capacitors those developed crack during thermal stress simulation

body. In this method, crack was created on single capacitor sl. no. 3 and stacked capacitor sl. no. 2 (refer Fig. 7).

4.4 Conformal Coating on Selected Capacitors

In order to study the effect of conformal coating on cracked capacitors, both single and stacked capacitors sl. nos. C3, C4, C7 and C8 were conformally coated prior to further testing.

4.5 Test Configurations

The capacitors were tested as per the following configurations:

Configuration-1: Single capacitors, C1 and C2 (normal) and C3 and C4 (with conformal coating) biased with 40 V
Configuration-2: Stacked capacitors, C1 and C2 (normal) and C3 and C4 (with conformal coating) biased with 40 V
Configuration-3: Single capacitors, C5 and C6 (normal) and C7 and C8 (with conformal coating) biased with 70 V
Configuration-4: Stacked capacitors, C5 and C6 (normal) and C7 and C8 (with conformal coating) biased with 70 V

4.6 Test Set-Up

Set-up was made as per circuit shown in Fig. 8.

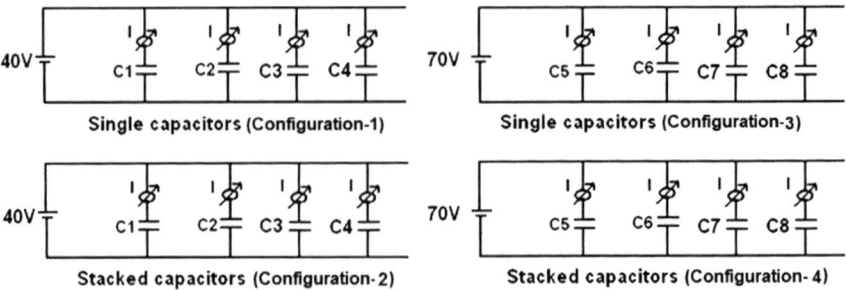

Fig. 8 Test set-up for all the four configurations

4.7 Test Conditions and Results

Capacitors were tested under biased condition (40/70 V) at room temperature as per the aforesaid configurations. The test was carried out for 1000 h or till any failure occurred. Supply current (I) across each capacitor was monitored in real time. Electrical parameters of each capacitor (C: capacitance, DF: Dissipation Factor, IR: Insulation Resistance) were measured offline in an interval of 250 h and recorded. Readings were tabulated in Tables 3, 5, 7 and 9 for test configurations 1–4, respectively. Once the supply current increases beyond 5 μA, measurement carried out at an interval of 1 h till complete short mode failure. Readings were tabulated in Tables 4, 6, 8 and 10 for test configurations 1–4, respectively.

4.8 Discussion

As per the aforesaid results on capacitors, observations can be summarized as follow

Case-1: Single capacitor, C3, which was having crack on the body but not conformally coated and biased with 40 V, developed leakage current from the beginning of testing (refer Tables 3 and 4). Leakage current increases from <1.0 to ~3.0 μA in 500 h and at the end of 750 h test leakage current increases to ~5 μA. Further testing of 16 h, leakage current increased from 5 μA to >1 A. Insulation resistance also gradually decreases from ~10^8 to ~10^5 Ω and finally to 3.2 Ω. As the capacitor was already having body crack, hence degradation of IR started as soon as the test started and the capacitor failed on resistive mode after 766 h of testing.

Case-2: Stacked capacitor, C2, which was having crack on the body but not conformally coated and biased with 40 V, developed leakage from beginning of testing (refer Tables 5 and 6). Leakage current increases from <1.0 to ~5.0 μA in 500 h and at the end of 525 h leakage current increases to ~27.0 μA. Further testing of 7 h, leakage current increased from ~27.0 μA to >1 A. Insulation resistance also gradually decreases from ~10^8 to ~10^6 Ω and finally to 1.5 Ω. As the capacitor was

Table 3 Result for test configuration-1 (I in µA, C in µF, IR in Ω, DF: no unit)

Capacitor sl. no. and type	Initial reading (0 h)				250 h				500 h				750 h				1000 h			
	I	C	DF	IR	I	C	DF	IR	I	C	DF	IR	I	C	DF	IR	I	C	DF	IR
C1, single	<1.0	1.16	0.012	8.76E10	<1.0	1.16	0.012	8.54E10	<1.0	1.16	0.012	8.82E10	<1.0	1.16	0.012	8.59E10	<1.0	1.16	0.012	8.59E10
C2, single	<1.0	1.17	0.011	5.76E10	<1.0	1.17	0.011	5.79E10	<1.0	1.17	0.011	5.69E10	<1.0	1.17	0.011	5.61E10	<1.0	1.17	0.011	5.61E10
C3, single	<1.0	1.17	0.015	**8.24E8**	<1.0	1.16	0.012	**9.12E7**	**~3.0**	1.16	0.012	**1.39E7**	**~5.0**	1.16	0.012	**7.55E6**	Refer Table 4			
C4, single	<1.0	1.19	0.014	4.37E10	<1.0	1.17	0.011	5.79E10	<1.0	1.17	0.011	5.69E10	<1.0	1.17	0.011	5.61E10	<1.0	1.17	0.011	5.61E10

Table 4 Test configuration-1 (capacitor sl. no. C3, single)

Test parameter	750 h	760 h	765 h	766 h	Remark
Leakage current (I)	**5.30 μA**	**37.9 μA**	**228.6 μA**	**>1 A**	Refer Fig. 9 for optical photograph of the failed capacitor
Capacitance (C)	1.16 μF	**0.8 μF**	**0.107 nF**	**13.41 nF**	
Dissipation factor (DF)	0.012	**0.170**	**0.210**	**−2.752**	
Insulation resistance (IR)	**7.55×10^6 Ω**	**9.48×10^5 Ω**	**1.75×10^5 Ω**	**3.2 Ω**	

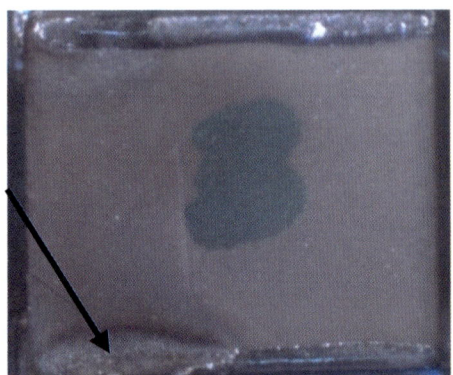

Fig. 9 post-simulation optical photograph of capacitor C3 (single) showing discolouration and crack on the body

having body crack, hence degradation of IR started as soon as the test started and the capacitor failed on resistive mode after ~532 h of testing.

Case-3: Single capacitor, C5, which was conformally coated and biased with 40 V, developed leakage current after 750 h of testing (refer Tables 7 and 8). Initially, there was no crack on the capacitor body. Leakage current increases from <1.0 to 10 μA in 750 h and at the end of 775 h leakage current increases to ~47 μA. Further testing of 30 h, leakage current increased from ~47.0 μA to 1.85 mA. Insulation resistance also gradually decreases from ~10^{10} to ~10^4 Ω. Total testing time taken for initiation of IR degradation is ~250 h and the capacitor failed on resistive mode after 805 h of testing.

Case-4: Stacked capacitor, C8, which was conformally coated and biased with 70 V, developed leakage current after 500 h of testing (refer Tables 9 and 10). Initially, there was no crack on the capacitor body. Leakage current increases from <1.0 to ~1.0μA in 500 h and at the end of 750 h leakage current increases to ~13.0 μA. Further testing of 30 h, leakage current increased from ~13.0 μA to 5.70 mA. Insulation resistance also gradually decreases from ~10^{10} to ~10^6 Ω and finally to ~10^4 kΩ. Total testing time taken for initiation of IR degradation is ~250 h and the capacitor failed on resistive mode after 780 h of testing.

Table 5 Result for test configuration-2 (*I* in μA, *C* in μF, IR in Ω, DF: no unit)

Capacitor sl. no. and type	Initial reading (0 h)				250 h				500 h				750 h				1000 h			
	I	*C*	DF	IR	*I*	*C*	DF	IR	*I*	*C*	DF	IR	*I*	*C*	DF	IR	*I*	*C*	DF	IR
C1, stack	<1.0	4.96	0.014	4.92E10	<1.0	4.96	0.012	4.81E10	<1.0	4.96	0.012	4.22E10	<1.0	4.96	0.012	4.77E10	<1.0	4.96	0.012	4.92E10
C2, stack	<1.0	4.77	0.011	**3.45E8**	<1.0	4.77	0.011	**7.22E7**	**~5.0**	4.77	0.011	**8.83E6**	Refer Table 6				–	–	–	–
C3, stack	<1.0	4.63	0.013	4.42E10	<1.0	4.63	0.013	4.72E10	<1.0	4.63	0.013	4.39E10	<1.0	4.63	0.013	4.56E10	<1.0	4.63	0.013	4.92E10
C4, stack	<1.0	4.82	0.012	5.98E10	<1.0	4.82	0.012	5.58E10	<1.0	4.82	0.012	5.67E10	<1.0	4.82	0.012	5.15E10	<1.0	4.82	0.012	5.22E10

Table 6 Test configuration-2 (capacitor sl. no. C2, stacked)

Test parameter	500 h	525 h	530 h	532 h	Remark
Leakage current (*I*)	**4.5 μA**	**27.4 μA**	**347.8 μA**	**>1 A**	Refer Fig. 10 for optical photograph of the failed capacitor
Capacitance (*C*)	4.77 μF	4.77 μF	**6.56 nF**	**55.45 pF**	
Dissipation factor (DF)	0.011	**0.18**	**0.712**	**−7.46**	
Insulation resistance (IR)	**8.83×10^6 Ω**	**1.45×10^6 Ω**	**1.15×10^5 Ω**	**1.5 Ω**	

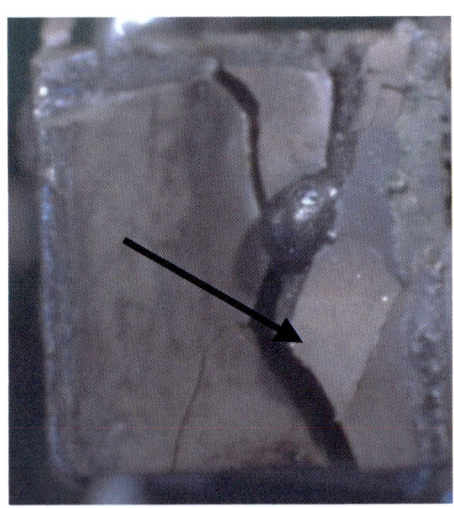

Fig. 10 post-simulation optical photograph of capacitor C2 (stack) showing discolouration and crack on the body

From the above discussions, following points can be noted:

(a) The capacitors which developed crack during fabrication process failed faster than those which do not have initial body crack. This indicates that with surface crack on the capacitor body moisture/contaminants entry is more than those do not have crack on the body.

(b) Capacitors, which do not have surface crack, still failed in resistive short mode after some duration. This indicates that manual soldering process, which induce sudden temperature gradient on the capacitor, facilitates propagation of internal crack and voids internal to the capacitors. This is long term absorb moisture and causes degradation of insulation resistance.

There is no effect on capacitance and dissipation factor of the capacitor till the insulation resistance in the order of 10^6 Ω. The capacitor degrades in a faster rate once the insulation resistance falls below 10^6 Ω or leakage current increases more than 100 μA.

Table 7 Result for test configuration-3 (I in µA, C in µF, IR in Ω, DF: no unit)

Capacitor sl. no. and type	Initial reading (0 h)				250 h				500 h				750 h				1000 h			
	I	C	DF	IR	I	C	DF	IR	I	C	DF	IR	I	C	DF	IR	I	C	DF	IR
C5, single	<1.0	1.20	0.013	6.46E10	<1.0	1.20	0.013	5.30E9	<1.0	1.20	0.012	8.42E8	~10.0	1.20	0.012	7.45E6	Refer Table 8			
C6, single	<1.0	1.15	0.012	5.45E10	<1.0	1.15	0.012	5.54E10	<1.0	1.15	0.012	5.39E10	<1.0	1.15	0.012	5.34E10	<1.0	1.15	0.012	5.38E10
C7, single	<1.0	1.18	0.011	9.16E10	<1.0	1.20	0.013	6.72E10	<1.0	1.20	0.013	6.52E10	<1.0	1.20	0.013	6.70E10	<1.0	1.20	0.013	6.49E10
C8, single	<1.0	1.19	0.017	2.32E10	<1.0	1.15	0.012	5.54E10	<1.0	1.15	0.012	5.39E10	<1.0	1.15	0.012	5.34E10	<1.0	1.15	0.012	5.38E10

Table 8 Test configuration-3 (capacitor sl. no. C5, single)

Test parameter	750 h	775 h	800 h	805 h	Remark
Leakage current (I)	9.5 μA	47.3 μA	92.7 μA	1.85 mA	Refer Fig. 11 for optical photograph of the failed capacitor
Capacitance (C)	1.20 μF	1.20 μF	3.20 nF	3.34 nF	
Dissipation factor (DF)	0.012	0.25	0.45	0.452	
Insulation resistance (IR)	7.45×10^6 Ω	1.48×10^6 Ω	7.55×10^5 Ω	3.77×10^4 Ω	

Fig. 11 Post-simulation optical photograph of capacitor C5 (single) showing rack on the body

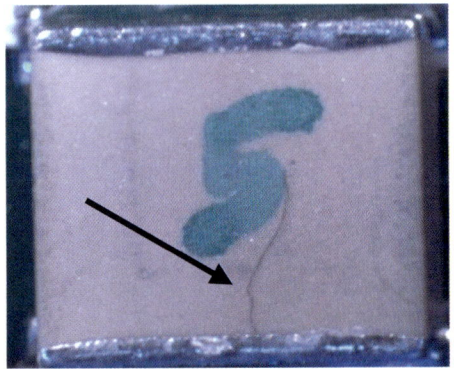

(d) Conformal coating did not prevent time-dependent failure of the capacitor if they are improperly soldered.

With the above feedbacks, alternate fabrication process (reflow soldering process) was adopted for chip capacitors and resistors at ISRO, which, in turn eliminated body crack-related issues and improved reliability of these components.

5 Conclusion

A novel method for conclusive termination of crack in MLCCs was discussed. This method also determines the depth of the crack in terms of insulation resistance. The simulation study on ceramic chip capacitor MLCC 2225X7RU, 1.2 μF, 5%, 200 V revealed that fabrication (hand soldering) induced crack resulted in time-dependent resistive short mode failure in the capacitors. The capacitors which developed crack during fabrication process failed faster than those which do not have body crack. MLCC capacitors, even though do not develop body crack during manual solder process, there is a possibility of aggravation of internal defects process, which in

Table 9 Result for test configuration-4 (I in µA, C in µF, IR in Ω, DF: no unit)

Capacitor sl. no. and type	Initial reading (0 h)				250 h				500 h				750 h				1000 h			
	I	C	DF	IR	I	C	DF	IR	I	C	DF	IR	I	C	DF	IR	I	C	DF	IR
C5, stack	<1.0	4.74	0.012	4.26E10	<1.0	4.74	0.013	4.69E10	0.001	4.74	0.013	4.51E10	0.001	4.74	0.013	4.33E10	0.001	4.74	0.013	4.26E10
C6, stack	<1.0	4.85	0.012	3.72E10	<1.0	4.85	0.012	3.82E10	0.001	4.85	0.012	4.12E10	0.001	4.85	0.012	4.02E10	0.001	4.85	0.012	3.72E10
C7, stack	<1.0	4.78	0.012	3.56E10	<1.0	4.78	0.012	4.19E10	0.001	4.78	0.012	4.02E10	0.001	4.78	0.012	4.33E19	0.001	4.78	0.012	4.26E10
C8, stack	<1.0	4.79	0.013	4.45E10	<1.0	4.79	0.013	7.45E8	~1.0	4.85	0.012	6.72E7	~13.0	4.85	0.012	5.09E6	Refer Table 10			

Table 10 Test configuration-4 (capacitor sl. no. C8, stacked)

Test parameter	750 h	775 h	780 h	Remark
Leakage current (I)	**13.75 µA**	**54.7 µA**	**5.7 mA**	Refer Fig. 12 for optical photograph of the failed capacitor
Capacitance (C)	4.852 µF	4.852 µF	**2.572 nF**	
Dissipation factor (DF)	0.012	**0.35**	**3.77**	
Insulation resistance (IR)	**5.09×10^6 Ω**	**1.28×10^6 Ω**	**1.23×10^4 Ω**	

Fig. 12 Post-simulation optical photograph of capacitor C8 (stack) showing discolouration on the body

turn resulted in resistive short mode failures. Post-body crack conformal coating on MLCCs did not prevent time-dependent failure in the capacitors.

Acknowledgements The authors are thankful to Ramesh Baburaj S, Head, Component Quality Control Division for all his supports for testing of capacitors during course of our experiment. Authors would also like to thank Shri P. Kunhikrishnan, Director URSC for his encouragement and guidance during the course of writing this paper.

References

1. C.R. Koripella, Mechanical behavior of ceramic capacitors. IEEE Trans. Compon. Hybrids Manuf. Technol. **14**(4) (1991) (December)
2. G. de With, Structural integrity of ceramic multilayer capacitor materials and ceramic multilayer capacitors. J. Eur. Ceram. Soc. **12**, 323–336 (1993)
3. A. Teverovsky, Effect of manual soldering induced stress in ceramic capacitors, NASA Electronic Parts and Packaging (NEPP) Program, December, 2008
4. M. Tarr, Failure mechanisms in ceramic capacitors, in *Proceedings of Online Postgraduate Courses for the Electronics Industry* (2007), http://www.ami.ac.uk/courses/topics/0179_fmcc/index.html

5. D.S. Erdahl, I.C. Ume, Online-offline laser ultrasonic quality inspection tool for multilayer ceramic capacitors. IEEE Trans. Adv. Packag. **27**, 647–653 (2004)
6. B. Sloka, D. Skamser, A. Hill, M. Laps, R. Grace, J. Prymak, M. Randall, A. Tajuddin, Flexure robust capacitors, in *Proceedings of the 27th Symposium for Passive Components, CARTS'07*, Albuquerque, NM (2007)
7. J. Prymak, M. Prevallet, P. Blais, B. Long, New improvements in flex capabilities for MLC chip capacitors, in *Proceedings of the 26th Symposium for Passive Components, CARTS'06*, Orlando, FL (2006), pp. 63–76
8. B.S. Rawal, et al., Reliability of multilayer ceramic capacitors after thermal shock, AVX Technical Information
9. M. Tarr, Failure mechanisms in ceramic capacitors, in *Proceedings of Online Postgraduate Courses for the Electronics Industry* (2007), http://www.ami.ac.uk/courses/topics/0179_fmcc/index.html
10. H.C. Ling, A.M. Jackson, Correlation of silver migration with temperature-humidity-bias (THB) failures in multilayer ceramic capacitors. IEEE Trans. Compon. Hybrid Manuf. Technol. **12**(1), 130–137 (1989)
11. E. Bumiller, C. Hillman, White Paper, A review of models for time-to-failure due to metallic migration mechanisms, www.DFRSolutions.com
12. A. Hornung, Diffusion of silver in borosilicate glass, in *Proceedings of the 1968 Electronic Components Conference* (IEEE, New York, NY, USA, 1968), pp. 250–255
13. D. Wilson, S. Walker, C. Ricotta, B. Scott, Development of Low Resistance Conductive Paths in Ceramic Chip Capacitors, FACTS 2000
14. M.J. Cozzolino, B. Wong, L.S. Rosenheck, Investigation of Insulation Resistance Degradation In BG Dielectric Characteristic, MIL PRF 55681 Capacitors, CARTS 2002
15. A. Teverovsky, Thermal-shock testing and fracturing of MLCCs under manual-soldering conditions. IEEE Trans. Dev. Mater. Reliab. **12**(2), 413–419 (2012)

Local Orbit Bump Control in Indus-2 Using Intelligent Agents-Based Control for Enhanced System Operability

Rahul Rana, R. P. Yadav and P. Fatnani

Abstract Indus-2 is a 2.5 GeV 200 mA Synchrotron Radiation Source (SRS) available to researchers all over the country and being operated in round the clock mode as a national facility. The position and stability of electron beam in the circular orbit determine the critical position and size of photon beam spot in a typical beamline of tens of meter long. Precise orbit control is therefore extremely critical. Application of local electron orbit bump for various purposes is often needed in an SRS. It is applied in such a manner that the desired local electron orbit profile in the ring is achieved with least affect elsewhere. This in turn requires accurate system model. However, in accelerators, parameters change with the operating point and also an accurate model of the system is not possible for such a dynamic system. Even slight deviation of model from system dynamics may result in drastic side effects on the beam. So, to minimize the efforts put into orbit correction and to minimize disturbance outside local bump every time the manual tuning activity is required, an intelligent agent-based solution for finding bump parameters, with self skill update capability, is planned to be implemented into Indus-2 orbit control though. The criticality of the application demands a stable, reliable, and robust operation of the proposed software. Checks are adopted into the application and verification and validation tests would be carried out before final deployment.

Keywords Intelligent agents · Local bump · Orbit control · SRS

R. Rana (✉) · R. P. Yadav · P. Fatnani
Raja Ramanna Centre for Advanced Technology, Indore, India
e-mail: rahulr@rrcat.gov.in

R. P. Yadav
e-mail: rpyadav@rrcat.gov.in

P. Fatnani
e-mail: fatnani@rrcat.gov.in

© Springer Nature Singapore Pte Ltd. 2020
P. V. Varde et al. (eds.), *Reliability, Safety and Hazard Assessment for Risk-Based Technologies*, Lecture Notes in Mechanical Engineering,
https://doi.org/10.1007/978-981-13-9008-1_10

121

1 Introduction

A wide variety of experiments are performed using SR from Indus-2 at experiment stations located at the commissioned beamlines (BL). These experiments are sensitive to the electron beam position and angle at source point [1]. The magnitude of dependence is such that for few 100 μm changes in electron beam position there is an mm change at the target sample. In advanced SRS, the orbit stability requirement is more stringent, i.e., 10% of beam size which comes out to be micron to sub-micron level. Various schemes like Slow Orbit Feedback (SOFB) and Fast Orbit Feedback (FOFB) systems are employed for achieving these stability requirements [2]. This stability is with reference to a predefined orbit, but sometimes a need to change this orbit altogether or locally at some place arises. For this, until now SOFB was used interactively while continuously consulting BL users after each modification in the orbit. After some iterative exercises, the beam orbit is attained that satisfies user requirement. Machine time is generally taken specifically for this activity as disturbance propagates throughout the ring while finding feasible orbit. Hence machine downtime increases. This iterative exercise is proposed to be replaced by self-learning application which, according to operating conditions or when commanded, can adjust the orbit locally. This application is particularly useful when one or two of the BL users are not getting the beam, and local correction is required without affecting other BL users.

Intelligent agents are software entities placed in a dynamic environment with a well-defined goal which the entity tries to accomplish by dynamically interacting with its environment. While interacting autonomously, these entities update their beliefs based on the result of their actions. When they give deteriorated performance and old skill is not good enough for achieving the assigned goal, they learn new skills. Agents are generally modeled to perceive their environment through properties like, their autonomy, intentions, beliefs, reactivity, and proactivity [3, 4]. Different types of agents are proposed and used by researchers and engineers for their applications. Multi-agent-based schemes have also been explored for control of large distributed plants. The work on multi-agent-based control has also been done toward minimization of closed orbit distortion (COD) of Indus-2 electron orbit, in simulated environment [5]. The simulation studies were performed earlier in the way that different agents were deployed at different layers of control to leverage the computational capability of each layer while locally handling most control tasks. The work presented here is an extension to it in a way that for the ease of development and deployment, the agents are first deployed only at application layer for initial test trials and scheme qualification for local orbit bump and COD of Indus-2 machine. This paper discusses the deployment of an intelligent algorithm core into already existing SOFB infrastructure for COD and control of local bump for any particular BL. It also peeks into the area of reliability and constraint incorporation into algorithm of this application for overall system stability.

2   Intelligent Agent-Based Orbit Control for Local Bump

Local bumps in electron orbit in SRS are generally provided for alignment of SR at BL experiment stations. Corrector magnets or steering magnets are used on top of other magnet optics. The same set of correctors are used by SOFB. Figure 1 shows three corrector scheme. Corrector one (CV1) opens bump, CV2 applies kick toward CV3 and CV3 closes it and the relation between them for closed orbit distortion (COD) [6] is given by Eq. (1). BPI1 and BPI2 are responsible for measuring beam position in this bump. Any mismatch in the corrector ratios would give rise to bump leakage and orbit distortion all over the ring. The skill set is defined as the ratio of correctors: λ:C1:C2:C3, where λ is a scaling factor for 1 mm bump at BPI. Application of local bump involves some amount to bump leakage; however, efforts are put to minimize this leakage by fine trimming the corrector value iteratively.

Equation (1) is used for theoretically calculating corrector strengths at three correctors ($\theta_1, \theta_2, \theta_3$) for closed orbit distortion. β is Twiss Parameter and depends on the location of the corrector. $\Psi_{i,j}$ denotes the phase difference between corrector i and j. Practically this ratio is calculated through measuring system response to step excitation of all correctors to generate response matrix (RM).

$$\frac{\theta_1 \sqrt{\beta_1}}{\sin \psi_{2,3}} = \frac{-\theta_2 \sqrt{\beta_2}}{\sin \psi_{1,3}} = \frac{\theta_3 \sqrt{\beta_3}}{\sin \psi_{3,1}} \tag{1}$$

For control of electron orbit locally three and four kicker schemes are used. While three kicker scheme controls only position at BPIs placed in between these three correctors, four kicker scheme controls angle at them as well. Until now, local bump was provided using SOFB system. Predictive facilities were provided to the operator in the SOFB application for easy visualization of the effects of changing reference orbit [7]. However, the process becomes iterative and a need of local, automatic, and intelligent algorithm was felt. The agents are developed with the aim of minimizing orbit bump leakage outside the bump zone while achieving the desired bump height. It involves updating their existing skill sets in case bump leakage is found to be outside the defined band of tolerance. Next sections describe the bump control system overview, proposed application overview, COD agents and their training and system constraints.

Fig. 1 Three kicker bump scheme

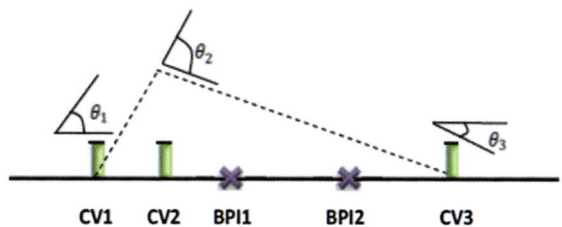

2.1   Bump Control System Overview

This section gives an overview of the Indus-2 orbit system and its control. The electron beam is made to circulate in the Indus-2 vacuum chamber using various magnets while continuously monitoring its position at various locations through beam position indicators (BPIs). The position is acquired from BPIs through data acquisition system and passed to orbit control system. The control system calculates the correction, in the form of steering magnets' current settings, required to mitigate the disturbances in the beam. The corrections are passed to the steering magnets that correct the beam position variations. Current settings for SOFB are calculated using inverse response matrix (RM). Inverse RM is evaluated from RM using singular value decomposition (SVD). The SOFB control system is implemented through state machine-based algorithm in LabVIEW real-time environment. For local bump correction, the same controller is used with an additional state. Here, the corrections for local correctors are received from application layer and passed to power supply (PS) interface for application to corrector magnets. Figure 2 shows the block diagram of the local bump control scheme highlighting different components of the control system.

2.2   Application Overview

The new orbit control application is embedded inside the existing SOFB application by inserting a new state. The system model for local bump controller is RM of SOFB system itself as same hardware is used. The new state has the facility to communicate with the client GUI through ethernet network for sending and receiving the corrector values of steering coils and BPI data. The local bump control application

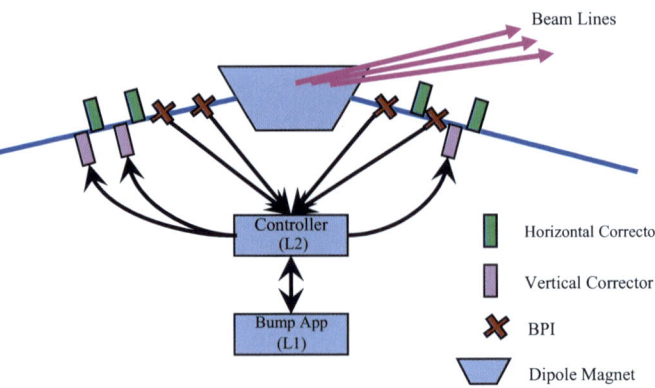

Fig. 2   Block diagram of the local bump scheme

incorporating agent-based control algorithm is deployed and integrated w client GUI in the control room. All the actions are calculated at this applic the corrector offsets are passed to the new state of SOFB controller. Han mechanism is adopted for reliable communication between RT application a application.

2.3 COD Agents, Skill Learning, and Algorithm

For achieving a common goal of closed orbit distortion (COD) minimizatio orbit correction agent work to find best corrector settings using hill climbing rithm. The initial skill set is calculated by genetic algorithm. It finds new sett by evaluating cost of applying incremental changes to all three correctors separa and then applying the most effective step. Here, the cost is the bump leakage. bump leakage is measured as rms orbit deviation outside bump. Most effective is with least cost. It also updates the skill set after application of most effective st Three corrector schemes are used for providing local bump. Training of the ag involves iteratively finding a corrector set, i.e., by applying a small delta step in p itive and negative direction to each of the three correctors one by one, which cou minimize the deviation outside local bump while achieving desired bump in the bun region. The training continued till a corrector set is achieved that keeps the leakag deviations in a band of tolerance, named as stop learning band. The training agai starts when the leakage increases beyond a band of tolerance, named as start learnin band. Figure 3 highlights various states of a typical learning cycle at different leakag bump costs. The numerical values of these bands are such that the absolute value o stop learn is less than start learn band. This ensures better learned skills compared to earlier and avoids going to learn mode frequently. Several iterative exercises were performed to finalize the values of stop learn band and start learn band to be 6 and 8 μm, respectively.

Applying a higher value of correctors while achieving a bump (not in learn mode) may lead to high leakage bump despite applying corrections with correct skill set,

Fig. 3 Learning state at different bump levels

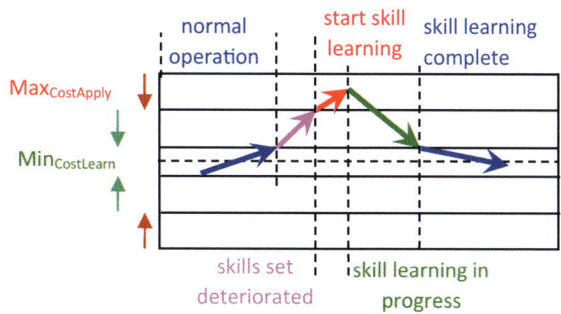

so this has to be avoided. Too low value of application will reduce the bump achieve speed to impractically low, so a compromise has to be made. It is seen that noise floor in the beam position is 2–3 μm rms. A peak-to-peak deviation of 30 μm is tolerable for BL users. Also system measurements highlight that a maximum deviation of up 1.2–1.5 mm is possible with 1A change in corrector current. So for interactive learning of the system with actuators excitation, a current of 15 mA is taken. While being in learn mode, corrector excitation should be such that it exceeds the noise floor inside the electron beam but do not much affect the orbit, i.e., COD should be intolerable limits even in the worse case. The cost of stop learn should be such that the COD becomes minimum but greater than the noise floor so that the system does not oscillate due to noise dominance. The stop learn should also be selected such that the BL users do not get affected by the perturbation caused in the beam due to skill learning. The start learn value should be selected higher than stop learn band so that learning starts only when it is beyond the learned skill limit. The delta step for applying correction using skill set, delta step for learn cycle, cost selection for start learn, and stop learn band are very crucial and are to be chosen wisely. Table 1 gives the details of the final values of system parameters used for implementation of the agents.

Flow graph of the COD agent for control of bump in "*Apply Request Interactively*" mode of operation is shown in Fig. 4. *Apply Request Interactively* mode of operation is the core of the application. Initially, simulation request is raised with offset requirement at BL sample end. This request uses genetic algorithm to find current settings for the designated corrector sets. User may initialize skills with this set of corrector current or may keep the previously learned skill set. At first step, the corrections are applied based on this skill set. The cost is evaluated for this applied step. If the cost is greater than a predefined value, i.e., $Max_{CostApply}$, learning mode gets activated otherwise bump achievement condition is checked. If bump is achieved, it stops, otherwise it again applies correction with same skill set. In learn mode, hill climbing algorithm is used to find the corrector which gives minimum cost on excitation. Each corrector is excited in both directions with small current (Δ_{Learn}) and cost is recorded. The step with minimum cost is applied and the skill set is updated. The cost is compared with the cost of stop learn ($Min_{CostLearn}$). If it is less than this value, it comes out of learn mode and checks for bump achieve condition. If not, then it again repeats the process.

Table 1 System parameters

Δ_{Apply} (max current at one step for any corrector)	1 mA
Δ_{Learn} (current applied for skill learning)	15 mA
rms Noise (nost without application bump)	2–3 μm
$Max_{CostApply}$ (cost start Learn)	8 μm
$Min_{CostLearn}$ (cost stop Learn)	6 μm

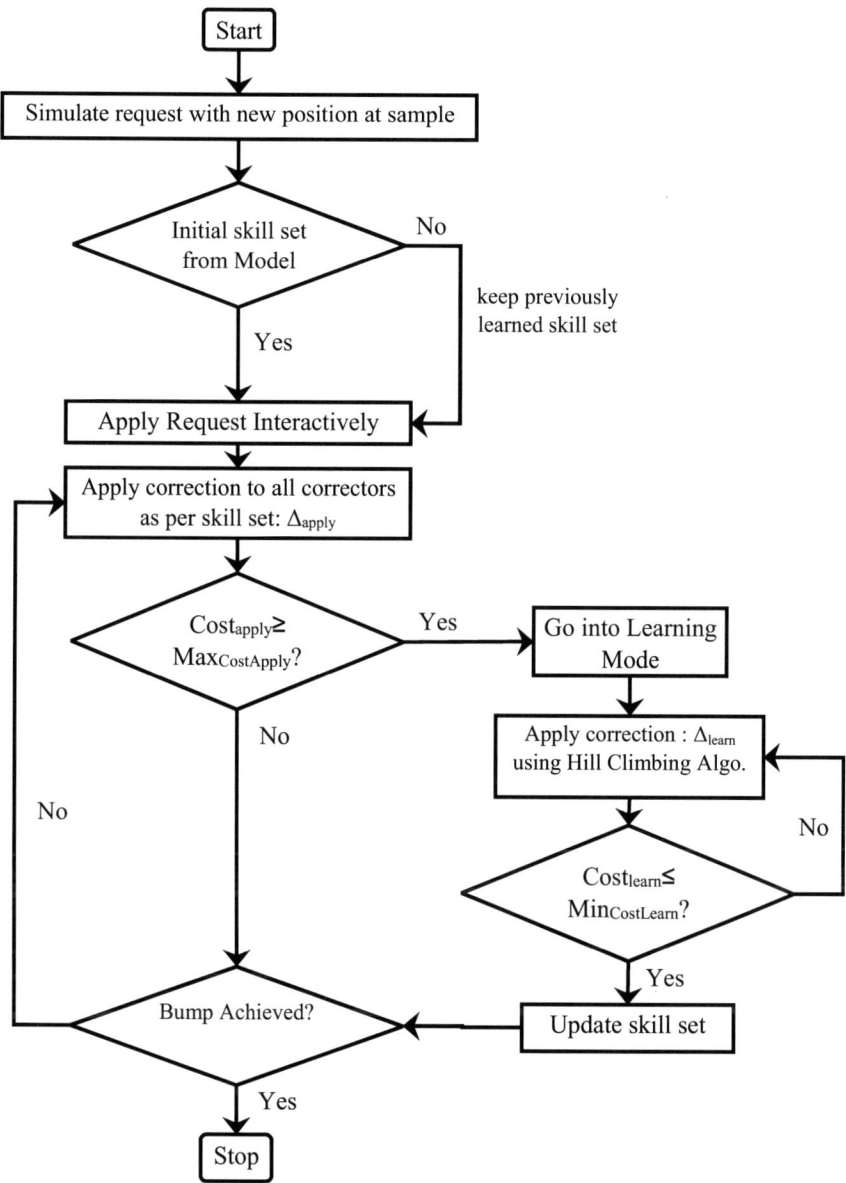

Fig. 4 Algorithm for *Apply Request Interactively* mode of operation

2.4   System Constraints and Validation Criteria

The application is equipped with system constraints so that no instability arise. The constraints are embedded into the controller. The application of corrector setting through the application is permitted only when the final value of the corrector does not exceed its safe limit which is kept at 6 A for initial trials and would finally be kept at 7 A. The correctors are capable for sustaining 10 A current through. While using manual and quick application modes, the maximum rate at which the simulated corrector current may be applied is kept at 0.1 A/s that is below corrector PS rate limit of 0.13 A/s. This avoids the system to get into nonlinear zone. Sufficient time is given for settling of the response before evaluation of the applied step for cost evaluation. While evaluating incremental apply and learn steps to correctors, it was considered that the perturbation introduced by the learning and application of correction would be within tolerance band of 30 μm.

3   Results of Initial Trials of the Scheme

The proposed scheme has been implemented and tested for skill learning and overall functionality of the application. Some of the implementation results are given next, in the form of application interface GUI and skill learning capability.

Figure 5 is a snap of the client application GUI highlighting its various parameters. For the selected BL, i.e., BL 12, skills set graph, local corrector current graph, bump and leakage graph and interface button for activating various modes of operation are highlighted. By simulating offset requirement at BL sample, corrector current requirements, and effects on BPIs can be seen. Three modes of operation for applying the simulated request are *apply request interactively* (with learning, if required), *apply request quickly* and a*pply request manually* (step by step). The *apply request quickly* and *manually* modes are performed only when the skill set is known to be good as these modes does not involve learning new skills and the correctors are incremented gradually with previously known skill sets. The *apply request interactively* mode works on the algorithm as explained earlier, which applies bump while also ascertaining that the leakage is minimum, if it is not, then the algorithm goes for learning new skills. Figure 5 shows, the local bump is achieved with a bump leakage of 6.3 μm rms for achieving a bump height of around 250 μm, and corrector values are within constraint boundary of $\pm6A$. As per system model, a position change of 640 μm was required at BL 12 sample point which was achieved by the application without any learn cycle. The RM for system model was measured recently, so it was possible to achieve the bump without any need of learning new skills. In Fig. 5, the center graph shows skill set, blue is initial and red is new learned skill set. Since no learning took place the red dot was not updated and is same as it was in previous bump condition. For the sake of verifying the functionality, the initial skill set was modified to make it poor and then bump request was placed. Figure 6 gives the details

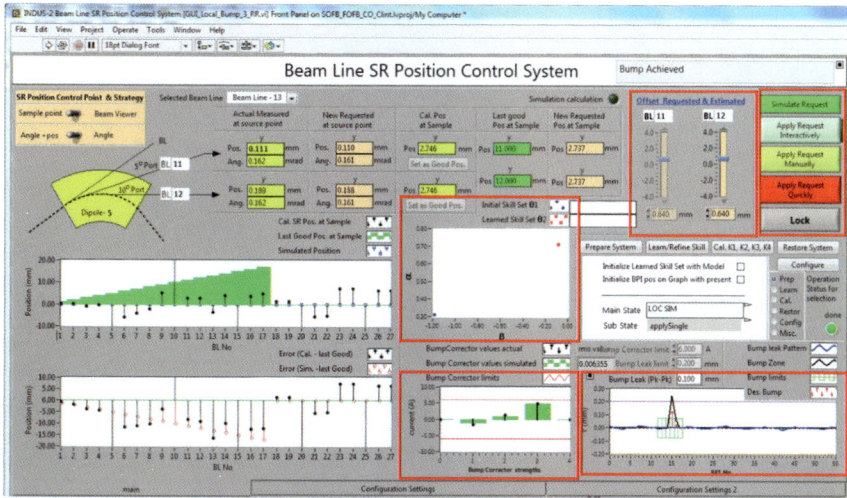

Fig. 5 GUI snap of the client application

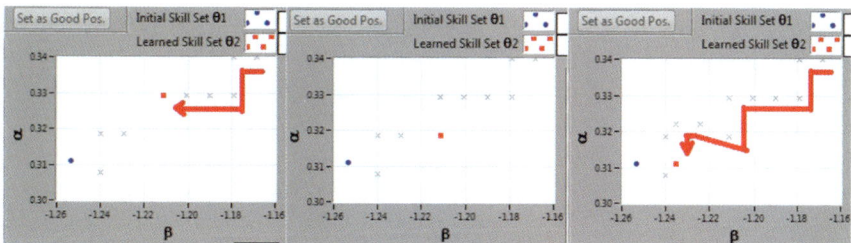

Fig. 6 COD Agent in process of learning new skill set while providing bump for BL 1

of the track made by the skill set while achieving the bump at BL1. The α and β on the skill graph are the ratio of correctors C2:C1 and C3:C1.

Since the initial skills, skills from system model, were good enough, and we altered the skills to see the movement of the learned skill set, the learned skill set tries to come back to the same point while achieving the desired goal of bump application. Similar procedure was repeated for BL 6 to get response shown by Fig. 7, while applying bump. The skill learning pattern also confirms with the simulation studies pattern.

Fig. 7 COD Agent in
process of learning new skill
set while providing bump for
BL 6

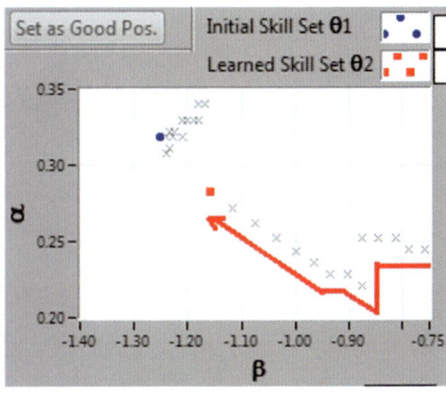

4   Conclusions

The proposed scheme has been validated, and initial trials were performed to evaluate
various parameters skill learning and overall functionality of the application in Indus-
2. With the chosen parameters and the intentionally degraded skill set, the application
tries to achieve desired bump while learning new skill sets that are approaching the
earlier known good skill set. Along with system constraint implementation into the
application, its parameters are fine trimmed to have optimum system performance.
The application is able to apply bump locally at any BLs with tolerable disturbances
at other locations. The initial results are in line with the simulation studies carried
out for the same system model. A GUI has been developed for remote operation in
control.

References

1. R.P. Yadav, P. Fatnani, Electron beam orbit control systems for Synchrotron Radiation Sources
   and Indus-2: a perspective. RRCAT Newsl. **27 T.2**(2) (2014)
2. A.D. Ghodke, A.A. Fakhri, G. Singh, Beam position stability in INDUS-2 storage ring, in
   International Symposium on Nuclear Physics, Mumbai, India, 18–22 Dec 2000, vol. 43 B,
   pp. 542–543
3. A. Rao, M. Georgeff, BDI Agents: from theory to practice, in *Proceedings of the First International Conference on Multi-Agent Systems*, Cambridge, MA, USA (1995), pp. 312–319
4. W. Shen, D.H. Norrie, J.P. Barthes, *Multi-Agent Systems for Concurrent Intelligent Design and
   Manufacturing* (Taylor and Francis, London, UK, 2001)
5. R.P. Yadav, P.V. Varde, P.S.V. Natraj, P. Fatnani, C.P. Navathe, Intelligent agent based operator
   support and beam orbit control scheme for synchrotron radiation sources. Int. J. Adv. Sci.
   Technol. Gatlinburg Tenn. **52**, 11–33 (2013)

6. R. Husain, A.D. Ghodke, S. Yadav, A.C. Holikatti, R.P. Yadav, P. Fatnani, T.A. Puntambekar, P.R. Hannurkar, Measurement, analysis and correction of the closed orbit distortion in Indus-2 synchrotron radiation source. PRAMANA J. Phys. **80**(2), 263–275 (2013)
7. R. Rana, R.P. Yadav, P. Fatnani, Slow orbit feed back system of Indus-2 with predictive control concepts, in *Indian Particle Accelerator Conference (InPAC-2018)*, Madhya Pradesh, India, 9–12 Jan 2018

Reliability Analysis of Smart Pressure Transmitter

Vipul Garg, Mahendra Prasad, Gopika Vinod and J. Chattopadhyay

Abstract In recent years, the advent of technology has necessitated the use of 'smart' transmitters. A transmitter is described as 'smart' if it incorporates signal conditioning and processing functions that are carried out by embedded microprocessors. Smart transmitters generally have the features like self-diagnosis, fault detection, digital communication, etc. However, with the advancement in the features and capabilities, the complexity of the smart transmitter has increased and also the estimation of its reliability. In this paper, we have carried out reliability analysis of a smart pressure transmitter. Incorporation of the smart features requires use of many electronic items/components. These components of the system may or may not fail independently, and their failure may lead to unavailability or degraded performance of the system. This has been modelled using the goal tree (GT) and success tree (ST) methodology. GT defines system objective which is a set of functions that shall be fulfilled to achieve the goal. Success tree ST defines the structure of the system and comprises system components used to achieve the GT functions. The system performance is considered as a function of its components. The interdependency between the system performance and its components is modelled using the Master Logic Diagram (MLD) wherein the potential faults/failures are introduced into the system. Nine potential faults/failure modes were identified, and their impact on the system capability to perform was studied. The cause and effect relationship captured using the MLD is then translated into the mathematical model. The evaluation of this model is carried out using fault tree, which provides an estimate of the unavailability of the system. The unavailability of the system for a mission time of one year has been found to be 4.617E−2.

Keywords Reliability analysis · Smart transmitter · Goal tree · Success tree · Fault tree

V. Garg (✉) · M. Prasad · G. Vinod · J. Chattopadhyay
Reactor Safety Division, Bhabha Atomic Research Centre, Mumbai, India
e-mail: vipulgarg@barc.gov.in

© Springer Nature Singapore Pte Ltd. 2020
P. V. Varde et al. (eds.), *Reliability, Safety and Hazard Assessment for Risk-Based Technologies*, Lecture Notes in Mechanical Engineering,
https://doi.org/10.1007/978-981-13-9008-1_11

133

1 Introduction

Smart transmitters are the advanced versions of the conventional analogue transmitters. A typical analogue transmitter senses some process parameter, which could be temperature, pressure, flow, etc., and produce an output signal proportional to the input (measured signal). Smart transmitter comprises microprocessor that helps it perform calculations and is more accurate than its analogue counterpart. Also, it has a digital communication protocol which is useful in reading the transmitter's measured output, in configuring various setting in transmitters, to know the device status, to have its diagnostic information, etc. Because of these advantages, smart transmitters are now finding their application even in the sophisticated industries like nuclear power plants (NPPs). However, in order to make the transmitter 'smart' requires incorporation of electronic devices which increases system complexity. Hence, it becomes necessary to estimate their reliability in order to ensure that their performance meets the industrial criteria, especially in the safety-critical systems. In this paper, goal tree–success tree (GT–ST) methodology has been adopted to perform the reliability analysis of smart transmitters. GT–ST model can completely and rigorously describe a system and its operations. It incorporates a structured approach that shows how a specific objective in a plant is achieved [1]. Firstly, the objective of the system is described. This is referred to as 'goal' in GT–ST methodology. In the GT, the objective or goal is divided/broken down into sub-goals or main functions that should be fulfilled in order to achieve the objective. Similarly, the sub-goals are further broken down into their supporting functions and the process continues until no more breaking down of the sub-functions is possible. However, in order to achieve the system goal by performing various functions as covered in the GT, some system hardware is required. This forms the basis for success tree (ST). Similar to the GT, at the first level the main hardware systems are described that are necessary to achieve the system goal. Then, the main hardware systems are further broken down into the supporting hardware systems that must be in healthy state for the main hardware system to be active. This breaking down process continues until further breaking down is not possible. Because of the large number of equipment, people and software in very complex systems such as nuclear power plants, development of a complete success tree becomes a major and often limiting task. In addition, because of the huge number of interacting parts, the GT–ST representation becomes very complex. In order to present the success logic of a very complex interacting system in a compact and transparent fashion, the MLD can be used [2]. The MLD explicitly shows the interrelationships among the independent parts of the systems, including all of the support items [2].

2 System Architecture of the Smart Transmitter

ELPRT-100SPT is a compact and lightweight 24 V DC powered 2-wire pressure transmitter with HART communication protocol. It contains a piezoresistive type sensing element for pressure measurement. Output of piezoresistive sensor is converted electronically to 2-wire 4–20 mA signal. The system architecture is depicted as shown in Fig. 1.

Earlier, reliability prediction of smart pressure transmitter for use in NPP [3] was performed, wherein the total failure rate of the transmitter was calculated as the cumulative of the failure rates of different components in the transmitters. Relex Architect software tool [4] was used to predict the failure rate of different electronic components. The results obtained from the analysis [3] are shown in Table 1.

3 MLD Model for Smart Pressure Transmitter

The GT–ST methodology along with the MLD has been utilized to perform the reliability analysis of the smart pressure transmitter [5].

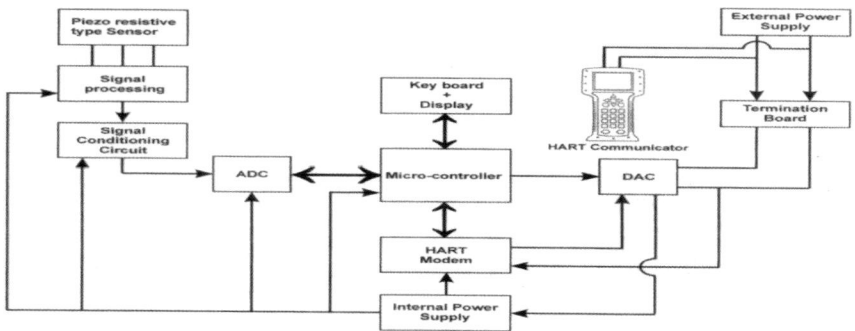

Fig. 1 System architecture of the smart transmitter

Table 1 Results obtained from the reliability prediction model of smart transmitter

Module	Failure rate ($/10^6$ h)	MTBF (h)
PCB-CPU board	0.211823	4.72E+06
PCB-signal conditioning board	2.510037	398,400.5
PCB-display board	1.386153	721,421.0147
PCB-input board	0.529118	1.89E+06
PCB-termination board	0.003044	3.29E+08

3.1 Goal Tree Model

The GT describes the objective/goal of the system. The objective/goal function is decomposed into sub-functions at increasing levels of detail [6]. The main purpose of the decomposition is to define physically meaningful functions, realization of which assures that the designated objective can be attained. The goal function of the smart pressure transmitter for this case study has been considered as 'to obtain the measurements'. This goal function can be decomposed into two main functions:

1. Obtain the process data (henceforth called 'measured data')
2. Process measured data.

These main functions are linked with the 'AND' logic, i.e. both these functions shall be attained in order to achieve the system goal.

3.2 Success Tree Model

The ST focuses on the physical aspects of the system and is developed from top to bottom, looking at all levels at which the system can be analysed. In essence, the physical elements collect all the components of the system necessary to achieve any of the functions present in the GT [6]. The necessary hardware components required for the successful operation of the main functions thereby ensuring attainment of the goal function are as follows: Input board, Signal conditioning board, Display board and CPU board.

3.3 Faults and Failures

For performing the reliability analysis, faults and failures are introduced which provide the dysfunctional aspects [5]. A fault is an abnormal condition that may cause a reduction in, or loss of, the capacity of an entity to perform a required function; a failure is the termination of the ability of an entity to perform a required function or in any way other than as required [7]. Nine significant/potential faults/failures were identified for the smart pressure transmitter as shown in Fig. 2. The data for some of these failures is taken from Table 1. Some other failures, for e.g. wrong power supply connections, incorrect calibration are human errors and the corresponding Human Error Probability (HEP) has been taken from THERP handbook [8].

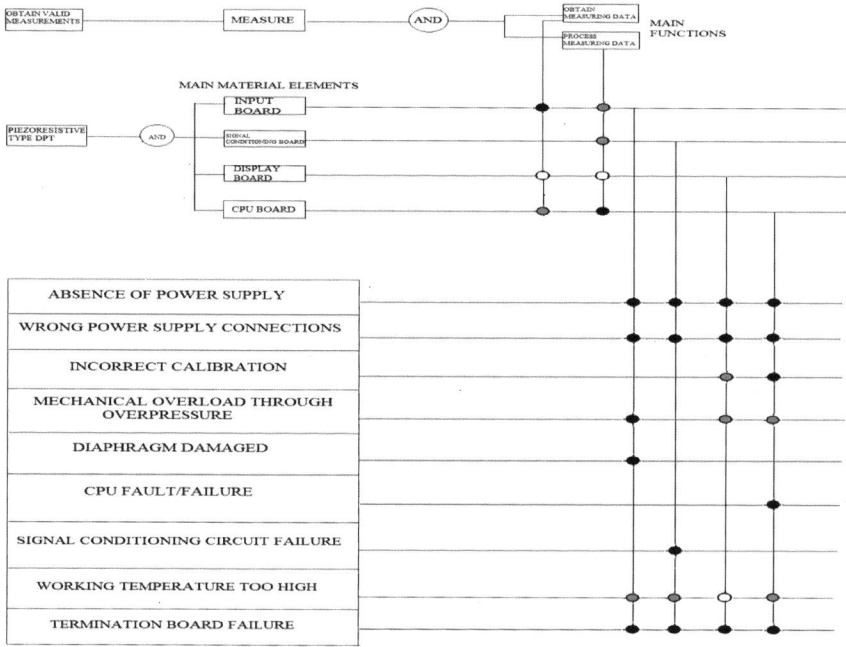

Fig. 2 MLD for smart pressure transmitter

3.4 *Master Logic Diagram*

Due to the huge number of interacting parts, the GT–ST representation becomes very complex. In order to present the success logic of a very complex interacting system in a compact and transparent fashion, the MLD is a good choice [2]. The hierarchy of the MLD is displayed by using the interdependency matrix in the form of a lattice. The bottom left side of the MLD constitutes the success tree. The top right side of the MLD constitutes the goal tree. The relationship/dependency between the entities is displayed by using the interdependency matrix in the form of a lattice. This relationship is represented in the MLD by filled dots. The degree of relationship is further specified by using a suitable colour of the dot. Strong relationship is represented by a black dot. Medium relationship is represented by grey dot. No dot is placed if there is no relationship between the elements. The MLD for this study is given in Fig. 2.

4 Reliability Analysis

To perform the reliability analysis, the methodology presented in [5] has been followed.

To do so, the relationship between the elements of the systems needs to be understood. To do so, some events need to be defined first.

D_d: {fault or failure d occurs} with $d = 1, \ldots, n_d$
P_p: {supporting material element p is in a failed state} with $p = 1, \ldots, n_p$
M_m: {unit (main material element) m is in a failed state} with $m = 1, \ldots, n_m$
S_s: {supporting function s malfunctions} with $s = 1, \ldots, n_s$
F_f: {basic function (main function) f malfunctions} with $f = 1, \ldots, n_f$
G_g: {global function (main function) g malfunctions} with $g = 1, \ldots, n_g$.

The direct relationship events are defined as follows:

$AB_{a,b}$: {event A_a directly implies (i.e. without request for any other event) event B_b}.

That is, the direct relationship events are given as follows:

$DP_{d,p}$: {an occurrence of fault or failure d directly implies a failed state of supporting material element p}.

Similarly, other direct relationships may be derived.

All these direct relationship events are assumed to be independent, and their probabilities of occurrence are denoted $P[AB_{a,b}]$. These values of relationships (probabilities) depend on the dot colours given in the relationship matrices. For example, Table 2 can be used to translate a degree of relationship into a probability and vice versa [5].

Total relationships take all the direct and indirect relationships into account.

The total relationship events between the faults and failures and units are defined by
$DMtot_{d,m} = $ {an occurrence of fault or failure d (directly or indirectly) implies a failed state of unit m}, with

$$DMtot_{d,m} = \left\{ DM_{d,m} \cup_p \left(DP_{d,p} \cap PM_{p,m} \right) \right\} \tag{1}$$

Table 2 Relationship analysis: input values and graphical representation [5]

Relationship	Input value (probability)	Result translation for graphical representation
Full/high	1.00	0.833–1.00
Medium	0.667	0.50–0.833
Low	0.333	0.167–0.50
Nil/very low	0	0–0.167

$$\text{Similarly, } \mathbf{MFtot_{m,f}} = \left\{ MF_{m,f} \cup_s \left(MS_{m,s} \cap SF_{s,f} \right) \right\} \tag{2}$$

$$\text{Therefore, } DF_{d,f} = \left\{ \cup_m \left(DMtot_{d,m} \cap MFtot_{m,f} \right) \right\} \tag{3}$$

$DG_{d,g}$ = {an occurrence of fault or failure d (directly or indirectly) implies a malfunction of global function g}, with

$$\mathbf{DG_{d,g}} = \left\{ \cup_{f(g)} DF_{d,f(g)} \right\} \tag{4}$$

and $f(g)$, the set of basic functions f which have to be fulfilled to achieve the global function g.

It is then possible to express the probability of global function g malfunctioning, as follows:

$$P[G_g] = P\left[\cup_d \left(D_d \cap DG_{d,g} \right) \right] \tag{5}$$

4.1 Events for Case Study

d, number of potential faults or failures = 9 or $d = \{1, 2, 3, ..., 9\}$
m, materials/modules in the system = 4 or $m = \{1, 2, ..., 4\}$ and f, number of main functions = 2 or $f = \{1, 2\}$
g, number of global functions = 1, or $g = \{1\}$ and G_g = global function 'g' malfunctions

$$\mathbf{DG_{d,g}}\{\text{Fault 'd' causing failure of global function 'g'}\} = \left\{ \cup_{f(g)} DF_{d,f(g)} \right\} \tag{6}$$

$$\mathbf{DF_{d,f}}\{\text{Fault 'd' causing failure of main function 'f'}\}\left\{ \cup_m \left(DM_{d,m} \cap MF_{m,f} \right) \right\} \tag{7}$$

$DM_{d,m}$ = Fault 'd' leads to the failure of material 'm' and
$MF_{m,f}$ = Failure of material 'm' leads to the failure of basic function 'f'.

A typical case of a fault and its impact on the system through the relationships modelled using MLD are presented.

4.2 Absence of Power Supply, d = 1

As, $DG_{d,g} = \left\{ \cup_{f(g)} DF_{d,f(g)} \right\}$
  Therefore, $DG_{1,1} = DF_{1,1} \bigcup DF_{1,2}$

$$DF_{d,f} = \left\{ \cup_m \left(DM_{d,m} \cap MF_{m,f} \right) \right\}$$

$$DG_{1,1} = DF_{1,1} \bigcup DF_{1,2} = \left\{ \left(DM_{1,1} \bigcap MF_{1,1}\right) \bigcup \left(DM_{1,3} \bigcap MF_{3,1}\right) \bigcup \left(DM_{1,4} \bigcap MF_{4,1}\right) \right\}$$
$$\cup \left\{ \left(DM_{1,2} \bigcap MF_{2,2}\right) \bigcup \left(DM_{1,4} \bigcap MF_{4,2}\right) \bigcup \left(DM_{1,1} \bigcap MF_{1,2}\right) \bigcup \left(DM_{1,3} \bigcap MF_{3,2}\right) \right\}$$

Similarly, the system has been analysed for the remaining eight potential faults, and their equations have been developed and modelled in the fault tree.

4.3   Results

Equation 5 has been solved by developing its corresponding fault trees in the Isograph software [9]. The unavailability of the smart pressure transmitter for a system lifetime of one year has been found to be 4.617E−2.

Figure 3 shows the fault tree developed to solve the system equations.

The performance/unavailability of the system also depends upon the time duration for which it has been operating. The unavailability values for the global function were calculated at the end of different period of operation (different mission times) as shown in Fig. 4.

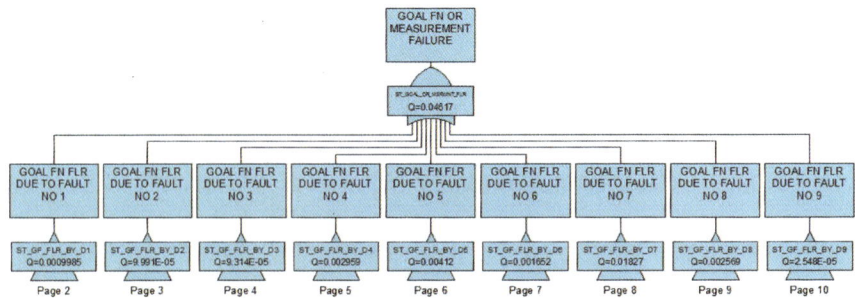

Fig. 3 Fault tree of the smart pressure transmitter

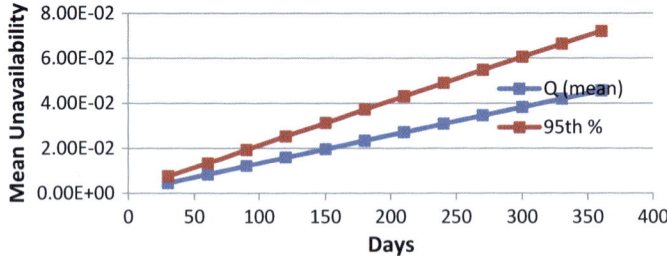

Fig. 4 Probability of the global function malfunctioning

References

1. I.S. Kim, M. Modarres, Application of goal tree-success tree model as the knowledge-base of operator advisory systems. Nucl. Eng. Des. **104**, 67–81 (1987)
2. M. Modarres, S.W. Cheon, Function-centered modeling of engineering systems using the goal tree–success tree technique and functional primitives. Reliab. Eng. Syst. Saf. **64**, 181–200 (1999)
3. Y.M. Reddy, R. Narne, T.V. Santhosh Reliability prediction of smart pressure transmitter for use in NPPs (2016) (Springer)
4. Windchill Quality Solutions (formerly Relex) (Online). http://www.crimsonquality.com/products/. Windchill Quality Solutions (formerly Relex)
5. F. Brissaud, A. Barros, C. Bérenguer, D. Charpentier, Reliability analysis for new technology based transmitters. Reliab. Eng. Syst. Saf. **96**(2), 299–313 (2011)
6. Y.F. Li, S. Valla, E. Zio, Reliability assessment of generic geared wind turbines by GTST-MLD model and Monte Carlo simulation. Renewable Energy **83**, 222–233 (2015). ISSN 0960-1481
7. IEC, Functional safety of electrical/electronic/programmable electronic safety-related systems. [s.l.]: International Electrotechnical Commission, vol. 61508 (2010)
8. A.D. Swain, H.E. Guttmann, *Handbook of Human Reliability Analysis with Emphasis on Nuclear Power Plant Applications*, NUREG/CR-1278, Sandia National University, August 1983
9. *Isograph Reliability Workbench*. [Online] Isograph. https://www.isograph.com/

Accelerated Failure Time Models with Corrective and Preventive Maintenance for Repairable Systems Subject to Imperfect Repair

A. Syamsundar, V. N. A. Naikan and V. Couallier

Abstract Repairable systems used in an industry work under differing environmental conditions such as usages, loads, stresses, temperatures, pressures and under dissimilar operating and maintenance conditions at diverse sites. These factors have an effect on their working and play a role in hastening or slowing down system deterioration leading to failure. Identifying the significant factors which effect the failure process and understanding their impact will help in improving the reliability of these systems. In this paper, accelerated failure time models, incorporating both corrective and preventive maintenance for repairable systems subject to imperfect repair, are proposed to quantify the effects of the significant factors which influence their failure process and assess how this information can be used to improve the reliability of the system.

Keywords Accelerated failure time models · Repairable system · Imperfect repair · Preventive maintenance

1 Introduction

Repairable systems function under varying operational, maintenance and environmental conditions especially in industrial use. The reliability characteristics of the same repairable system vary widely when subjected to different levels of these conditions. When subjected to extremes of temperatures, to higher loading/stress levels, when undergoing more usage, with improper maintenance or upkeep, the condition

A. Syamsundar (✉)
Research and Development Centre, Visakhapatnam Steel Plant, Visakhapatnam, India
e-mail: syamsundar.annamraju@gmail.com

V. N. A. Naikan
Subir Chowdhury School of Quality and Reliability, Indian Institute of Technology, Kharagpur, India

V. Couallier
Institut de Mathematiques de Bordeaux, Universite de Bordeaux, Talence, France

© Springer Nature Singapore Pte Ltd. 2020
P. V. Varde et al. (eds.), *Reliability, Safety and Hazard Assessment for Risk-Based Technologies*, Lecture Notes in Mechanical Engineering, https://doi.org/10.1007/978-981-13-9008-1_12

143

of the system deteriorates more rapidly than it would otherwise leading to a higher intensity of failures than normal.

To understand the effect of varying conditions on the performance of the repairable systems this additional information, i.e. the usage, loading, stress levels and maintenance information can be included as covariates to form regression models for repairable systems. Regression models where the covariates act through a link function to modify the time scale of the intensity of the failures are termed accelerated life testing or accelerated failure time models depending on their application, i.e. life testing or industrial usage, respectively.

These models have initially been applied to repairable systems motivated by life testing applications. Guida and Giorgio [1] and Guerin et al. [2] have used homogeneous and nonhomogeneous Poisson accelerated life testing models with constant covariates. Srivastava and Jain [3] have used linearly increasing stresses as time-varying covariates with nonhomogeneous accelerated life testing models. Schabe [4] constructed various accelerated life testing models using nonhomogeneous processes with constant and time-varying covariates. Yun et al. [5, 6] have used accelerated life testing models with imperfect repair and constant covariates.

These models have just begun to be extended to repairable systems in the field. Martorell et al. [7] use accelerated failure time models with imperfect repair and constant covariates for motor operated valves of a nuclear power plant. Novak [8] uses an accelerated failure time model with a renewal process and time-varying covariates for oil pumps. Syamsundar et al. [9] use accelerated failure time models with imperfect repair and global usage rate as a constant and time-varying covariate for cars. Studies of multiple factors affecting the failure process of a repairable system subject to scheduled preventive maintenance along with corrective maintenance have not been carried out so far.

In this paper, the failure data of a cement plant roller mill with several factors affecting its failure process and subject to corrective and scheduled preventive maintenance is investigated with the help of accelerated failure time models. This is carried out to determine which of these factors are significant in causing failures and to assess and quantify the effects of the significant factors. This will help to ascertain what is to be done to improve roller mill's reliability and probably decide its maintenance strategy.

The paper is divided as follows. Section 2 describes the models, their inference and simulation using these models. In Sect. 3, analysis of the cement plant roller mill failure data with several factors is carried out to identify significant factors and their influence and help improve its reliability and maintenance strategy. Section 4 provides the conclusion.

2 Accelerated Failure Time Models, Inference and Simulation for Repairable Systems

A repairable system is subjected to various conditions in use like different usages and loads, number of corrective and preventive maintenance actions, stresses, temperatures, pressures and working at different sites. Considering these factors as covariates, the chronological time is transformed using these covariates through a link function to generate accelerated failure time models for modelling the failure processes of repairable systems used under different conditions.

2.1 Accelerated Failure Time Models

2.1.1 Basic Model

The conditional intensity of the accelerated failure time (AFT) model with a time-varying covariate is given by:

$$\lambda(t|H_{t^-}) = f(\mathbf{Z}(t))\lambda\left(\int_0^t f(\mathbf{Z}(u))du\right) \tag{1}$$

where $\lambda(t|H_{t^-})$ is the intensity function given the past history of the model and $\mathbf{Z}(t)$ the time-varying covariates $\mathbf{Z}(t) = (z_1(t), z_2(t), \ldots, z_m(t))$ where the failure time t is transformed by the covariate function $f(\mathbf{Z}(t))$ where $f(.)$ is known as the link function. The effect of the covariates is to transform (accelerate or decelerate) the time scale of the intensity function depending on the nature and strength of the covariates.

It is assumed that the values of the time-varying covariates are those prevalent at the failure times or corrective/preventive maintenance (CM/PM) action times only. It is also assumed that the values remain the same during the sojourn since the previous failure or CM/PM times and change only after repair at the next failure or CM/PM time.

This causes the covariate function to be taken out of the integral leading to mathematical tractability without affecting the model properties. The conditional intensity of the accelerated failure time (AFT) model with time-varying covariates becomes:

$$\lambda(t|H_{t^-}) = f(\mathbf{Z}(t))\lambda(f(\mathbf{Z}(t)t) \tag{2}$$

2.1.2 AFT Model with Corrective Maintenance and Scheduled Preventive Maintenance

Preventive maintenance (PM) can be of two types scheduled or condition based. Here scheduled preventive maintenance is considered. Scheduled PM can also be of two types, time based carried out at specified times and age based carried out at a fixed age after the previous failure. Scheduled PM censors the failure mechanism. However, it is independent of the failure process. U_i is used as an indicator variable to indicate PM and CM at time t_i, with $U_i = 0$ indicating a CM and $U_i = 1$ indicating a PM.

Imperfect repair being the more general case of maintenance of repairable systems, imperfect repair models are considered for AFT models. For an imperfect repair PM and CM model, two factors indicating the repair factors or degrees of repair are considered as: ρ_C for CM and ρ_P for PM.

For a PM and CM AFT model having Arithmetic Reduction of Age with memory one (ARA$_1$) imperfect repair process with a power-law baseline intensity, the conditional intensity of the PM-CM process as given in Doyen and Gaudoin [10] with covariates forming an AFT model is given by:

$$\lambda(t|H_{t-}) = f(\mathbf{Z}(t))\alpha\beta\left[\left(t - \left(\sum_{i=1}^{K_{t-}} \rho_P^{U_i}\rho_C^{1-U_i} w_i\right)\right) f(\mathbf{Z}(t))\right]^{\beta-1} \tag{3}$$

where w is the inter-event time being a CM or PM event and K the total number of events including CM and PM events.

For a PM and CM AFT model having Arithmetic Reduction of Intensity with memory one (ARI$_1$) imperfect repair process with power-law baseline intensity, the conditional intensity of the PM-CM process with covariates forming an AFT model is given by:

$$\lambda(t|H_{t-}) = f(\mathbf{Z}(t))\alpha\beta\left[\left(t^{\beta-1} - \left(\sum_{i=1}^{K_{t-}} \rho_P^{U_i}\rho_C^{1-U_i} w_i\right)^{\beta}\right)(f(\mathbf{Z}(t)))^{\beta-1}\right] \tag{4}$$

2.2 Estimation of AFT Models

Estimation in these models is carried out using the method of maximum likelihood. The likelihood function for the failure process of a repairable system observed on $[0, t]$ with $i = 1, 2, \ldots, n$ failures as given in Lindqvist [11] is:

$$L(\theta|\text{data}) = \prod_{i=1}^{n} \lambda(t_i|H_{t-}) \exp\left(-\sum_{i=1}^{n} \int_{t_{i-1}}^{t_i} \lambda(y_i|H_{y-})dy\right) \tag{5}$$

where θ represents the parameter vector.

Taking the logarithm of both sides we get:

$$\ln L(\theta|\text{data}) = \sum_{i=1}^{n} \ln(\lambda(t_i|H_{t^-})) - \sum_{i=1}^{n} \int_{t_{i-1}}^{t_i} \lambda(y_i|H_{y^-}) dy \tag{6}$$

The log likelihood equation is solved either analytically or using a numerical procedure to obtain the estimates of the parameters of the models.

2.3 Hypothesis Testing of AFT Models

Since the method of maximum likelihood is used for estimation of model parameters, the model with the maximum likelihood value is considered as the model with better fit, as it will provide a better fit to the data set of failures.

A penalised likelihood criterion such as the Akaike Likelihood Criterion (AIC) as given in Akaike [12] can be used to counter the better fit observed for models of higher complexity with more data. The criterion is given by:

$$\text{AIC}(k) = -2 \ln L + 2k \tag{7}$$

where $k =$ the number of parameters of the model.

The model with the minimum AIC estimate is considered as the model with the better fit.

Graphical method is used to check the accuracy of fit by plotting the observed number of failures and the estimated cumulative failure intensities and/or estimated expected number of failures versus failure times.

2.4 Simulation with AFT Models

The times to failures given covariate information can be obtained through simulation using the inverse transform method. Consider t^z as the internal age of the system under consideration. Then:

$$t^z = \int_0^t f(\mathbf{Z}(u)) du = f(\mathbf{Z}(t)) t \tag{8}$$

The probability $F\left(t_{i+1}^z|t_i^z\right)$ is generated as a uniform random variable in $(0-1)$ and the t_i^z values are obtained by inverting the distribution which can be used to obtain t values given the covariate information.

These values can be used to estimate the expected times to next failure as well as the expected number of failures approximately. Getting average t_i^z values from say five hundred simulations and obtaining t values from these will give the estimated expected times to next failure. Counting the number of failures up to a specific failure time and taking the average of these from say five hundred simulations will provide the estimated expected number of failures till that failure time. Five hundred simulations will provide a sufficient number of trajectories to smooth out the variation in the average values bringing them closer to the actual values.

3 Application to Cement Plant Roller Mill Failure Data

In this study the failure data of the roller mill of a cement plant with CM times, PM times, and the data pertaining to five factors affecting the failure process, given in Love and Guo [13] is investigated. The five factors are instantaneous change in pressure on the main bearing, recirculating load on the bucket elevator, power demanded by the main motor, corrective maintenance (CM) actions, preventive maintenance (PM) actions. It has been stated in Love and Guo [13] that the covariate value instantaneous change in pressure on the main bearing is not significant and the power demanded by the main motor and recirculating load on the bucket elevator values are highly correlated. Hence only three covariate values, recirculating load on the bucket elevator, number of CM and number of PM actions are considered in this study. The idea is to understand the effect of these factors on the failure process of the roller mill.

The PM-CM AFT models given in Eqs. (3) and (4) are considered for modelling the failure intensity of the roller mill failure data with the three covariates. The most commonly used link function $f(.)$ is the exponential $\exp(.)$ which leads to easy tractability and provides only positive values to $f(.)$ The covariate function $f(.)$ with the three covariates is $f(\mathbf{Z}(t)) = \exp(\gamma_1 z_1(t) + \gamma_2 z_2(t) + \gamma_3 z_3(t))$: $z_1(t)$ being the number of corrective maintenance actions, $z_2(t)$ the number of preventive maintenance actions and $z_3(t)$ the recirculating load on the bucket elevator.

To obtain the failure intensity, the PM-CM times without covariates and with all three covariates are fitted with ARA_1 and ARI_1 imperfect repair AFT models given in Eqs. (3) and (4), respectively with a PLP baseline. The parameters of the model are estimated using the method of maximum likelihood as given in Sect. 2.2. The log likelihood and AIC values with these models are given in Table 1. Hypothesis testing and model fit are carried out as given in Sect. 2.3. It can be seen that the AIC Ln L values of the failure intensity with the covariates are much lower than that of the intensity without covariates indicating that the AFT model with three covariates is a better fit to the failure intensity. Thus, the three covariates have an effect on the failure intensity of the cement plant roller mill failure data.

Table 1 Values of log likelihood and AIC with ARA_1 and ARI_1 imperfect repair models with PLP baselines fitted to the roller mill CM and PM times data with single covariate and three covariates together

Likelihood values	Models					
	No covariate		With covariates			
			Three covariates recirculating load, No. of CMs, No. of PMs		Single covariate recirculating load	
	ARA_1	ARI_1	ARA_1	ARI_1	ARA_1	ARI_1
ln L	106.15	106.14	94.88	95.40	94.89	95.57
AIC ln L	220.30	220.28	203.76	204.80	199.78	201.14

To understand whether any of these covariates individually have a major effect on the failure intensity, the PM-CM times with a single covariate, the recirculating load on the bucket elevator which is more likely to effect the failure intensity is first fitted with ARA_1 and ARI_1 imperfect repair AFT models with a PLP baseline. The log likelihood and AIC values with these models are given in Table 1. It can be seen that this model has marginally lower AIC Ln L values than that of model fitted with the three covariates indicating a better fit to the failure intensity. Thus, only the single covariate recirculating load on the bucket elevator has an effect on the failure intensity. The other covariates, number of PM times and number of CM times have no effect on the failure intensity.

Graphical model fit is carried out as given in Sect. 2.3. Plot of estimated cumulative intensity with ARA_1-AFT model and observed number of failures with recirculating load on the bucket elevator as a time-varying covariate versus the times to failure is given in Fig. 1. Estimated expected number of failures is obtained using simulation as per Sect. 2.4. Plot of estimated expected times to failure with the ARA_1-AFT model and observed number of failures with recirculating load on the bucket elevator as a time-varying covariate versus observed times to failure is given in Fig. 2. It can be seen that there is a reasonably good fit in Fig. 1 and there is a good fit in Fig. 2.

The parameter estimates with this model are given in Table 2.

The conditional intensity of the AFT ARA_1 PM-CM model with power-law baseline intensity for the failure data of the roller mill of cement plant is:

$$\lambda(t|H_{t^-}) = \exp(9.78\, z_3(t))(5.57e{-}10)(2.72)$$

$$\left[\left(t - \left(\sum_{i=1}^{K_{t^-}} 1^{U_i} 0.99^{1-U_i} w_i\right)\right) \exp(9.78\, z_3(t))\right]^{1.72} \tag{9}$$

The covariate recirculating load on the bucket elevator has a transforming effect on the failure times with a value equal to exp (9.78 times covariate measure) for the ARA_1-AFT model. To reduce the number of failures and improve the reliability of the cement plant roller mill, the recirculating load on the bucket elevator is to be lowered/optimised. The values of repair effect factor for CM and PM are one and indicate perfect repair for both CM and PM. As both CM and PM actions are optimal, only CM can be considered as a maintenance strategy for the roller mill. This however can be reviewed/optimised based on cost considerations which are also a very important factor in maintenance.

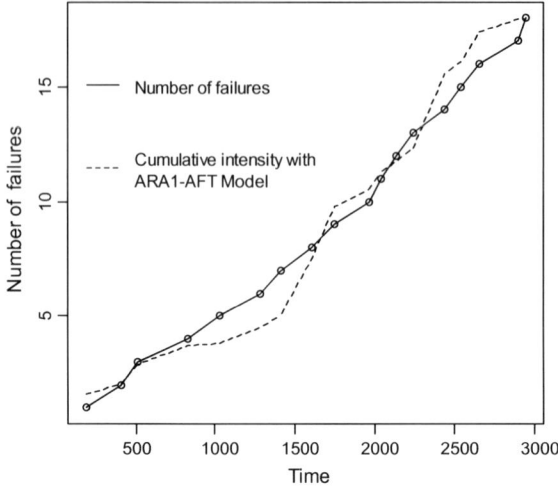

Fig. 1 Plot of the observed number of failures and the estimated cumulative intensity of ARA_1-AFT model vs times to failures for roller mill failure data

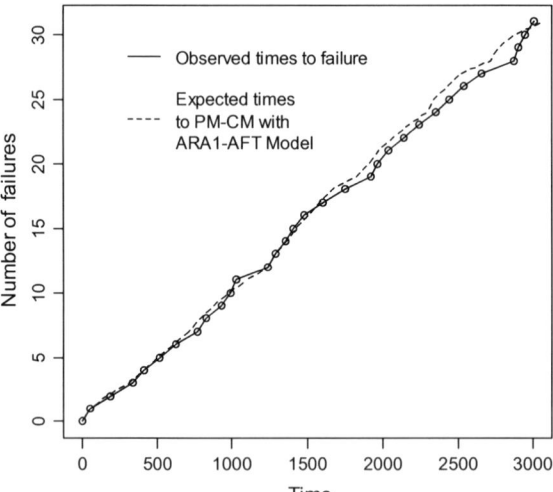

Fig. 2 Plot of expected times to failure with ARA_1-AFT model versus observed times to failure for roller mill failure data

Table 2 Estimated values of the parameters of AFT models with better fit for roller mill failure and PM times data

Model	ARA$_1$ with single covariate Recirculating load						
Parameter	$\hat{\alpha}$	$\hat{\beta}$	$\hat{\rho}_C$	$\hat{\rho}_P$	$\hat{\gamma}$	ln L	AIC ln L
Estimated value	5.57e−10	2.72	0.99	1	9.78	94.89	199.78

4 Conclusion

Accelerated failure time models have been used in this paper to model the failure processes of repairable systems in the field with data on PM and CM times and covariate information consisting of data on the factors affecting the failure process. These models are very useful in assessing whether these factors affect the failure process and to estimate/quantify their effects. This information can be used to improve the reliability of the system by addressing the influential factors and help to properly plan its maintenance strategy.

References

1. M. Guida, M. Giorgio, Reliability analysis of accelerated life-test data from a repairable system. IEEE Trans. Reliab. **44**, 337–346 (1995)
2. F. Guerin, B. Dumon, P. Lantieri, Accelerated life testing on repairable systems, in *Proceedings of Reliability and Maintainability Symposium* (2004), pp. 340–345
3. P.W. Srivastava, N. Jain, Optimum ramp-stress accelerated life test for m identical repairable systems. Appl. Math. Model. **35**, 5786–5793 (2011)
4. H. Schabe, Accelerated life testing models for non homogeneous Poisson models. Stat. Pap. **39**, 291–312 (1998)
5. W.Y. Yun, E.S. Kim, Estimating parameters in repairable systems under accelerated stress, in *Proceedings of the International Conference on Computational Science and its Application, ICCSA, Part IV* (Springer, Berlin, 2005), pp. 558–565
6. W.Y. Yun, E.S. Kim, J.H. Cha, Accelerated life test data analysis for repairable systems. Commun. Stat. Theory Methods **35**, 1803–1814 (2006)
7. S. Martorell, A. Sanchez, V. Serradell, Age-dependent reliability model considering effects of maintenance and working conditions. Reliab. Eng. Syst. Saf. **64**, 19–31 (1999)
8. P. Novak, Regression models for repairable systems. Methodol. Comput. Appl. Probab. **17**(4), 963–972 (2015)
9. A. Syamsundar, D.E. Vijay Kumar, V.N.A. Naikan, Imperfect repair accelerated failure time models with time varying covariates for maintained systems, in *Proceedings of the 7th International Conference on Modelling in Industrial Maintenance and Reliability* (Sidney Sussex College, University of Cambridge, UK, 2011)
10. L. Doyen, O. Gaudoin, Modelling and assessment of aging and efficiency of corrective and planned preventive maintenance. IEEE Trans. Reliab. **60**(4), 759–769 (2011)
11. B.H. Lindqvist, On the statistical modelling and analysis of repairable systems. Stat. Sci. **21**, 532–551 (2006)

12. H. Akaike, Information theory and an extension of the maximum likelihood principle, in *International Symposium on Information Theory* (Akademia Kiado, Budapest, 1973)
13. C.E. Love, R. Guo, Using proportional hazard modelling in plant maintenance. Qual. Reliab. Eng. Int. **7**, 7–17 (1991)

Reliability Improvement of Single-Shot Device

Manmeet Singh and S. Nandula

Abstract Reliability demonstration and reliability improvement of a single-shot device (SSD) mainly depend on quality processes involved during selection and manufacturing of components, assembly/integration, environmental testing (vibration, bump, temperature conditioning, water immersion, etc.) and functional tests under different operating conditions (ambient, high altitude and desert). After a thorough root-cause analysis based on the Fault Tree Analysis and critical examination of the delayed functioning mechanism, the corrective actions were implemented. Further, the reliability requirements and the delay functioning requirements were met completely during developmental trials. Failures during developmental trials reveal that root cause for defects is due to human error, insufficient quality process, inadequate training and failure in addressing reliability aspects during design stage, etc. To conclude, the lesson learnt from the study was that the stringent reliability analysis of the product should be done, during initial design stage of the product, by identifying potential failure modes of the product and by addressing them prudently.

Keywords Failure · Single-shot device · Reliability improvement · Root-cause analysis · Fault tree analysis

1 Introduction

The term "single shot" refers to an apparatus which is destroyed when it is effectively functioned. Here the outcome from a single-shot device (SSD) is dichotomous (go-no-go or success-failure) rather than data acquired for analysis, by measuring a continuous variable parameters like stress level, pressure, temperature, response/activation time, etc. in a standard life testing situation [1].

M. Singh (✉) · S. Nandula
DRDO, New Delhi, India
e-mail: manmeet@hqr.drdo.in

S. Nandula
e-mail: s_nandula@hqr.drdo.in

© Springer Nature Singapore Pte Ltd. 2020
P. V. Varde et al. (eds.), *Reliability, Safety and Hazard Assessment for Risk-Based Technologies*, Lecture Notes in Mechanical Engineering, https://doi.org/10.1007/978-981-13-9008-1_13

153

Functional reliability of an SSD could be defined as the probability that a system (SSD) or product will perform in a satisfactory manner for a given period of time when used under specific operating conditions. By definition, single-shot devices are utilized and consequently fully operational, once within their lifetime. Due to this fact, typical reliability approaches must be tailored to answer for this distinctive challenge. The reliable performance of single-shot device is due to inherent robust design and the functionality of the device being successfully and safely demonstrated, such as a rocket, bomb, etc., or having indispensable components which will no longer be able to function appropriately after each individual event, such as an automobile airbag, pyro igniters, etc. In the case of a rocket, it is very expensive to test its full functionality over a statistically significant sample size because of the self-destructive nature of the system. Due to this constraint of testing, accurate reliability modelling and analysis are essential to predict its optimal operational reliability and understand any potential reliability issues [2].

The reliability testing of SSD can be regarded as attribute testing. The results of an attribute test are classified as one of two possible states which include success/failure, acceptable/unacceptable and conforming/nonconforming [3]. Attribute testing is used to evaluate the reliability of SSDs that must function when called on to function but have no active mission time. For example, if several operating SSDs are placed inside an environmental test chamber and at the end of the test, each SSD is checked to see if it is still operating then the test result recorded for each SSD is success or failure. The SSDs are referred to as single-shot items which could be battery (thermal or primary), pyros (squibs, cartridges, pin pullers), separation or ejection systems, sensors, or igniters.

Binomial distribution is normally used to estimate the reliability of SSDs subjected to attribute testing. An essential condition for using the binomial distribution as a model is that the probability of success remains same from trial to trial irrespective of environment. This entails that each SSD on test has the equal probability of success. Therefore, reliability can be estimated by attribute testing in which the test could be described as testing "n" SSDs and recording "r" failures [3]. The reliability of the SSD for the test conditions can be estimated by the Eq. 1.

$$R = (n - r)/n \quad \text{for} \quad r \geq 1 \tag{1}$$

where "R" is the reliability, "n" denotes the number of SSDs tested and "r" denotes the number of SSDs failed.

The challenges encountered are plenty in predicting/estimating and demonstrating the reliability of a single-shot device. The most prominent are contemplation of the effects of extended periods of dormancy (storage) and exposure to an array of potential environments and operating phases. In reality, many single-shot devices may be dormant for decades and experience many environmental transitions. However, they will still be relied upon to function effectively when called upon during lifetime or they may never get utilized throughout their life span.

2 Reliability Demonstration

Reliability estimation/verification tests are performed to demonstrate conformity to stated reliability parameters such as MTBF or failure rate at a given confidence level. Reliability testing is formal, statistical in nature and requires standard procedures. Reliability tests are classified based on the way the tests are performed and on the type of results that are recorded. The type of data to be taken and the way in which they are reported must be a part of the overall reliability test plan. Reliability tests can be continuous (recorded as variables data) or dichotomous (recorded as success/failure attribute data). Attribute data are recorded as success or failure for a given test and each SSD is numbered separately for traceability. The strategy for reliability test is to produce failures typical of the product use or to eliminate design weaknesses from the product.

Reliability demonstration and reliability improvement of a single-shot device (SSD) mainly depend on design methodology, quality processes involved during selection of raw material and manufacturing of components, assembly/integration and environmental testing. In the case of trials, the strategy to create failure conditions typical of product usage was adopted, and the test was conducted in an environment that reflected the environment of exploitation or deployment. The SSDs were subjected to various tests viz shock, acceleration, bump, temperature conditioning, EMI/EMC, water immersion and functional tests under different operating conditions like ambient temperature, high altitude and desert.

3 Reliability Improvement

The SSD has four main components and out of which two components are redundant or hot standby to each other. The reliability block diagram for SSD is shown in Fig. 1. The root cause for two specific events such as non-functioning of SSD and premature functioning of SSD are identified using Fault Tree Analysis (FTA).

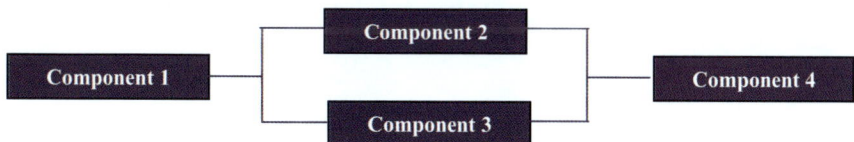

Fig. 1 Reliability block diagram of single-shot device failure

3.1   Fault Tree Analysis for Non-functioning of Single-Shot Device

The Fault Tree Analysis (FTA) was first introduced by Bell Laboratories and is one of the most widely used methods in reliability, availability, maintainability and safety analysis. FTA is a top-down, deductive failure analysis in which a system failure condition is simulated and analysed using a Boolean logic to combine a series of lower level events that are a combination of hardware and software failures and human errors that could cause the undesired event. Here the FTA is used as a diagnostic tool to identify and correct causes of the top event, i.e. non-functioning of SSD. Fault Tree for non-functioning of SSD is shown in Fig. 2. Failures during developmental trials reveal that the root cause for defects is due to human error, insufficient quality process, inadequate training and failure in addressing reliability aspects during design stage, etc.

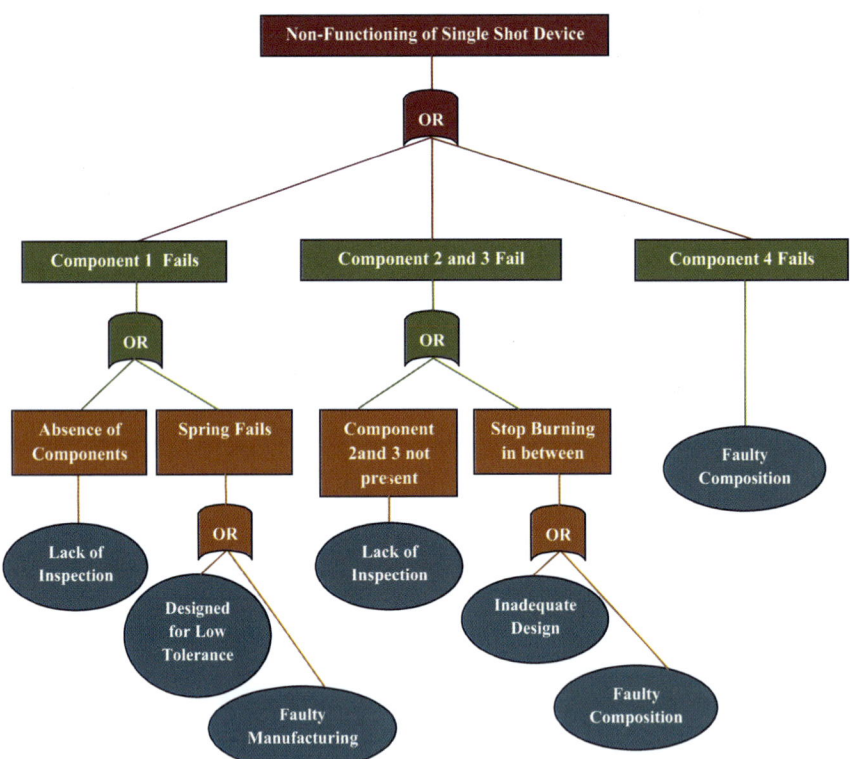

Fig. 2 Fault tree analysis diagram for non-functioning of single-shot device

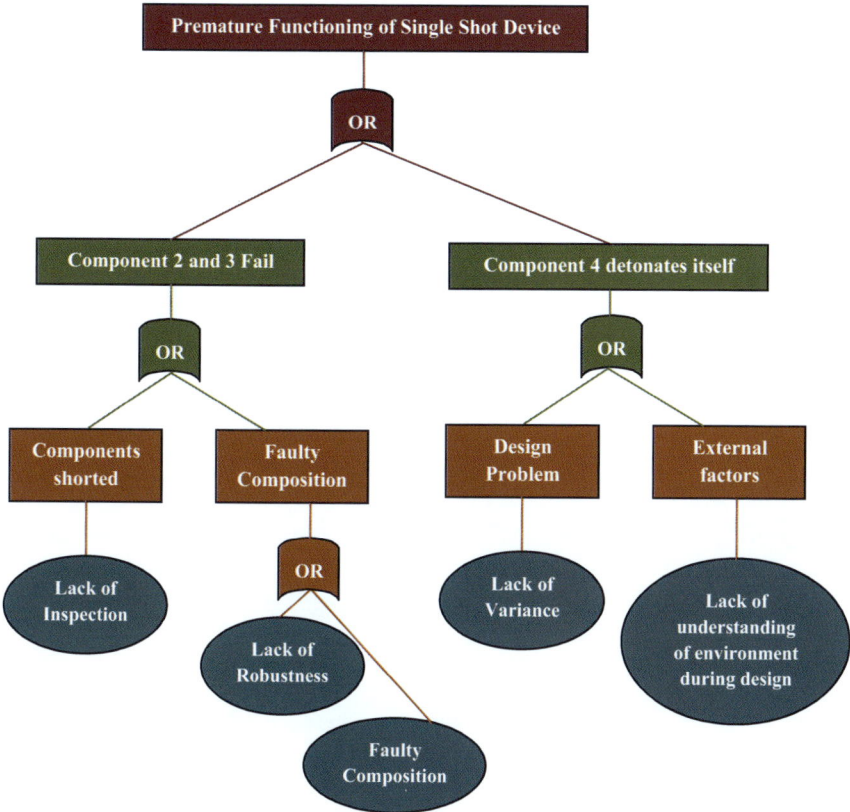

Fig. 3 Fault tree analysis diagram for premature functioning of single-shot device

3.2 Fault Tree Analysis for Premature Functioning of Single-Shot Device

The Fault Tree for premature functioning of SSD is shown in Fig. 3. The FTA analysis reveals that root cause for premature functioning of SSD is due to human error, insufficient quality process and failure in addressing reliability aspects during design stage, etc.

4 Results and Discussion

Considering the sensitivity of the device, SSDs usually demand the reliability of the order of 0.99 and above (Ideally 1.0). Further, the premature delay function-ing requirement also imposes challenges on the designers for ensuring the reliable functioning of every component of SSD including delayed functioning. During first developmental trials of the SSD, the desired reliability and delay functioning requirements were not achieved. Extensive technical analysis was undertaken to address the failure. The root cause for the failure was identified using the Fault Tree Analysis, and the corrective action was implemented and subsequently the environmental tests and functional tests were performed. During the repeat developmental trials although the reliability of the SSD had improved to 0.99, the delay functioning requirement could not be achieved. After a detailed root-cause analysis based on the post-trial data and critical examination of the delayed functioning mechanism, the design of the delay functioning mechanism was modified and further improvements in the quality processes during assembly were introduced. The third developmental trial has achieved the SSD reliability of 1.0, and the delay functioning requirements were also completely met [4].

5 Conclusion

FTA is the tool used for identifying the root causes of the top event, and then by addressing them judiciously, the reliability of the product was improved from less than 0.99 to 1.0. Further, the SSD had met the qualitative requirements necessary for its introduction or deployment in the desired environment. To conclude, the message to comprehend from this activity was that Design for Reliability (DfR) has to be resorted to rigorously and all potential failure modes have to be addressed during the initial design stage of the product development by using reliability tools (e.g. FMEA, FMECA, FTA, etc.), following design standards, selection of standard material and components, etc. This will result in the design, development and manufacturing of robust SSD product which would withstand adverse environmental conditions during actual operation/deployment for the given time period.

Acknowledgements The authors would like to thank Dr. Chitra Rajagopal, Distinguished Scientist and Director General (SAM) for providing consistent guidance and support to continue this study.

References

1. L.J. Bain, M. Engelhardt, Reliability test plans for one-shot devices based on repeated samples. J. Qual. Technol. (American Society for Quality Control, Rolla, Missouri, USA) (1991), pp. 304–311
2. D.J. Foley, D.W. Kellner, Single-shot device reliability challenges, in *2017 Annual Reliability and Maintainability Symposium(RAMS)* (IEEE, Orlando, FL, USA, 2017)
3. D.W. Benbow, H.W. Broome, *The Certified Reliability Engineer Handbook* (ASQ Quality Press, Milwaukee, WI, 2014)
4. M. Singh, S. Nandula, Reliability norms for research and development (R&D) activity, in *Proceedings of 11th National Conference on Aerospace and Defence Related Mechanisms (ARMS-2018)*, BITS Pilani Hyderabad Campus, November, 2018

Reliability Evaluation of Silicon Carbide (SiC) Boost Converter Using Stochastic Model

D. Umarani, R. Seyezhai and R. Sujatha

Abstract In recent years, power electronics has turned as an inevitable part in aerospace, renewable energy conversion, and automotive industry applications. Although it expands in all dimensions with full potential, the critical applications demand reliability from the safety point of view. So, efforts are being devoted towards the improvement of reliability of power electronic (PE) systems. Field experiences show that power electronic systems are usually one of the most vital parts in terms of lifetime, failure rate, and maintenance cost. There are three important queries to be considered for evaluating its reliability: reason for failure, how to build a reliable power electronic system, and the methods to test and monitor the reliable operation of the system. Taking into account these constraints, in this article, we have brought the detailed reliability evaluation of silicon carbide (SiC)-based boost converter. Compared to silicon devices, SiC power devices have many more advantages such as higher blocking voltages, higher switching frequencies, and higher junction temperatures. The boost converter acts as an interface in photovoltaic systems and regulates the input voltage to the subsequent parts of the power conversion system. A SiC-based boost converter incorporates the advantages of SiC material in the boost converter. Still, the reliability of the system in terms of lifetime, system availability, failure rate, and mean time to failure needs to be analysed before using it in any critical application. In this paper, the components prone to degradation/failure due to operating frequency, operating temperature, and variation in the input voltage have been investigated. A stochastic model called the Markov model has been implemented to find out the system availability. A reliability workbench called isograph has been used to model the entire system. The first-order equations have been

D. Umarani (✉) · R. Seyezhai
Renewable Energy Conversion Laboratory, Department of EEE, SSN College
of Engineering, Chennai, India
e-mail: umaranid@ssn.edu.in

R. Seyezhai
e-mail: seyezhair@ssn.edu.in

R. Sujatha
Department of Mathematics, SSN College of Engineering, Chennai, India
e-mail: sujathar@ssn.edu.in

© Springer Nature Singapore Pte Ltd. 2020
P. V. Varde et al. (eds.), *Reliability, Safety and Hazard Assessment
for Risk-Based Technologies*, Lecture Notes in Mechanical Engineering,
https://doi.org/10.1007/978-981-13-9008-1_14

solved in MATLAB using Runge–kutta method. The prototype of the SiC boost converter has been built and operated in both closed and open loops at various operating frequencies. The thermal study has been carried out to find out hot spots, and the system performance has been monitored under various operating conditions. The results obtained for various test cases are analysed, validated, and presented.

Keywords Boost converter · Markov model · Reliability · Silicon carbide and stochastic model

1 Introduction

The eminence of distributed energy sources like solar photovoltaic (PV) systems that produce DC power has increased the demand for DC to DC power converters. They not only provide high efficiency but also provide fast dynamic response with good control [1, 2]. Battery-operated systems need stack of cells to obtain higher DC voltage. However, due to space constraints, heaping of cells is not acceptable in high-voltage applications. Therefore, DC-DC converters can be used for solar photovoltaic and fuel cell applications, wherein the applied input voltage can be boosted from a low-level to the required high-level voltage [3, 4]. These converters generally use silicon (Si) MOSFET devices operated under high frequency [5]. The pulse width is adjusted by varying the duty ratio so as to maintain the boost factor. Recently, silicon carbide devices have gained so much attention due to its attracting features like high-temperature, high-voltage handling capability, and wider bandgap compared to silicon devices [6, 7]. In this power electronic era, SiC emerges to be an alternative to Si devices that downsize the converters and reduce the weight as well. SiC can be operated at very high temperature, and thus, the thermal profile needs to be investigated to ensure the reliable operation of the equipment [8, 9]. However, except SiC, all other components of the converter need to be cooled down with appropriate cooling systems to maintain the reliability. This paper deals with a SiC-based DC-DC boost converter. The reliability of the converter plays an important role as it predicts the lifetime of the converter in critical applications such as medical devices or military equipment. So, a reliability model should be developed to analyse and figure out the system availability [10–12]. There are various modelling available for reliability prediction. This paper has used the stochastic modelling of the SiC boost converter using the Markov state transitions. It is one of the important techniques of evaluating reliability. The model comprises possible states of the system, state transition paths, and the rate of transition such as failure or repair. The paper is organised as follows: Sect. 2 discusses the design and implementation of SiC boost converter. Section 3 discusses the stochastic modelling of SiC boost converter. Section 4 describes the thermal analysis of the proposed converter. Section 5 describes the hardware validation, and Sect. 6 provides the comparison with Si boost converter. The paper ends with conclusion and acknowledgement.

2 Design and Implementation of SiC Boost Converter

A DC-DC converter is analogous to an AC transformer where the input voltage can be stepped up or stepped down to meet the requirement. It works on the principle based on the inductor's property to resist any change in current. The duty cycle decides the boost factor of the converter. The SiC-based boost converter has been considered for study as it has high-temperature withstanding capability, high-voltage handling capability, and wider bandgap. It is also a fast switching device with ultrafast free-wheeling diode. All these features add advantage to the SiC boost converter compared to the Si-based converter and ensure reliable operation. The SiC-based DC-DC boost converter operating principle, design parameters, and simulation results are discussed in this section.

2.1 SiC Boost Converter and Its Operating Principle

The basic configuration of SiC boost converter is shown in Fig. 1. This topology comprises an inductor, diode, MOSFET switch, and output capacitor.

The boost converter operates in two modes. When the switch is closed, the input current rises and flows through inductor and switch. When the switch is open, the input current flows through L_i, C_o, diode, and load. The inductor-stored energy flows through the load until the next cycle. The equivalent circuits in mode 1 operation and mode 2 operation are shown in Figs. 2 and 3, respectively.

2.2 Design Parameters

The inductor and capacitor for DC-DC converter have been designed using the parameters listed in Table 1 and in Eqs. (1) and (2).

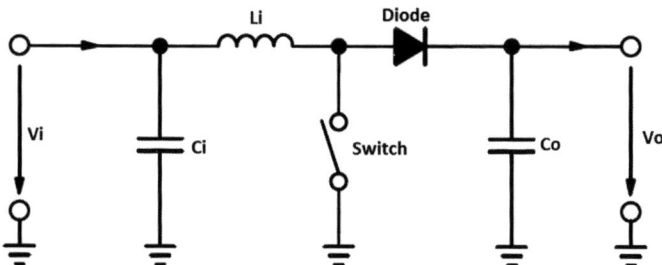

Fig. 1 SiC boost converter

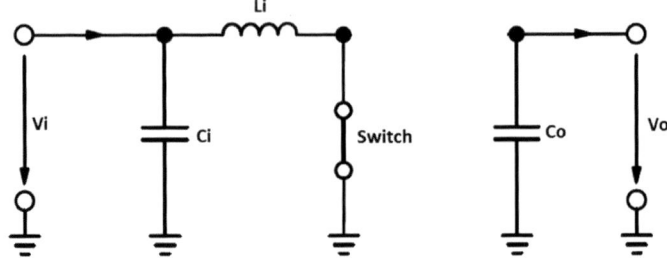

Fig. 2 Mode 1—switch ON

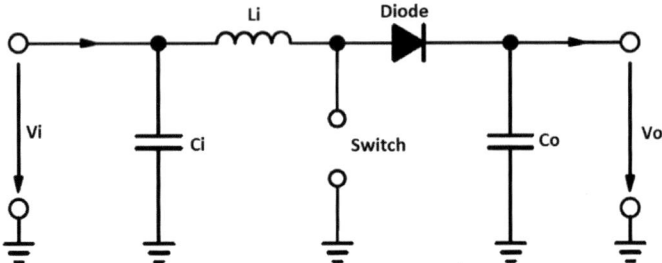

Fig. 3 Mode 2—switch OFF

Table 1 Design parameters

Specifications for boost converter	
Input voltage (V_i)	10 V
Output voltage (V_o)	20 V
Duty ratio (D)	50%
Voltage gain (G)	2
Switching frequency (f_{sw})	100 kHz
Current ripple (Δi)	3%
Voltage ripple (Δv)	10%

$$L_i = \frac{V_i D}{f_s \Delta I_L} \tag{1}$$

where V_i is the minimum input voltage, D is the duty cycle, f_s is the minimum switching frequency, and ΔI_L is the inductor current ripple.

$$C_o = \frac{I_o D}{f_s \Delta V_o} \tag{2}$$

where ΔV_o is the capacitor voltage ripple, and I_o is the output current.

The value of input inductor, L_i has been calculated as 2 mH, and the output capacitor, C_o is calculated as 330 μF.

2.3 Simulation of the Proposed Topology

The SiC boost converter has been modelled using MATLAB/Simulink and simulated with the parameters listed in Table 1. Switching frequency of 100 kHz and duty ratio of 50% have been considered for triggering the MOSFET switch. The output voltage is shown in Fig. 4.

For an applied input voltage of 10 V, an output voltage of 20 V has been obtained.

The input inductor current ripple and output capacitor voltage ripple are shown in Figs. 5 and 6, respectively.

The input inductor current ripple is found to be 1.20 and 5.73% for SiC boost converter. The SiC model of the MOSFET switch has been used in the simulation of the boost converter.

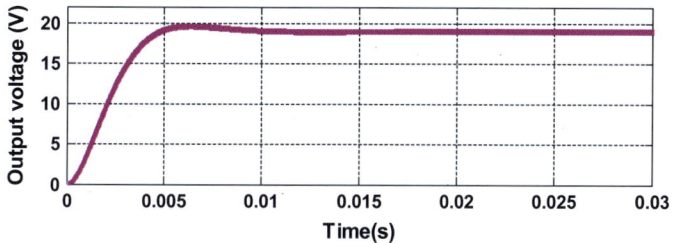

Fig. 4 Output voltage waveform of boost converter

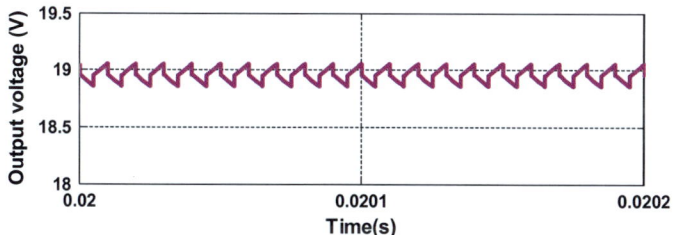

Fig. 5 Input inductor current ripple

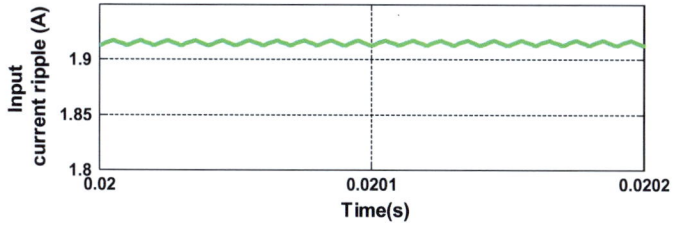

Fig. 6 Output capacitor voltage ripple

3   Stochastic Modelling of Boost Converter

3.1   Reliability Analysis

The reliability analysis of a power electronic converter is required to identify whether the converter meets the desired specifications. It helps in smooth operation and maintenance of the converter. Thus, we need a model either at component level or at system level to study the reliability of the DC-DC converter. Reliability analysis based on component level focuses on failure rate of the key components in power circuits such as semiconductors, capacitors, and inductors. There are various reliability models for these electronic components under component level [13, 14]. The empirical-based models depend on collected failure data to quantify model variables. They are mostly employed to study the component reliability. This paper focuses on stochastic modelling of boost converter using the Markov model. The state transitions are identified, and the operating states are fixed. Through the state transition diagram, failure and repair rates also taken into account, and the system availability has been calculated. The thermal model of the SiC boost converter has been developed for reliability prediction. The failure and repair rates have been calculated with MIL-HBK 217 [15].

3.2   Markov Modelling of SiC Boost Converter

Markov's model is a combinatorial model based on graphical representation of operating states that corresponds to a system topology. The system will reach a failed state from the failure-free state through various transitions. There are two types of states in the Markov models: (1) absorbing states with failed architecture and (2) non-absorbing states in which the system delivers full or partial functionalities. Figure 7 represents the state Markov's transition states of an SiC boost converter. In this case, three subsystems are considered. They are DC source, inductor, diode, and switches. These three subsystem failures may result in partial output of the system or total failure of the system. The subsystem failure rates are represented by λ_1, λ_2, and $\lambda3$, and the repair rate of the three subsystems is represented by μ_1, μ_2, and μ_3, respectively. The fully operating states are X, Y, and Z. The partial operating states are used as X', Y', and Z'. The failed states are used as x, y, and z.

From the state diagram, the first-order probability differential equation for each Markov state is as follows.

$$P_{1'}(t) = \mu_1 P_2(t) + \mu_3 P_3(t) + \mu_2 P_4(t) - (\lambda_1 + \lambda_2 + \lambda_3) P_1(t) \qquad (5)$$

$$P_{2'}(t) = \lambda_1 P_1(t) - \mu_1 P_2(t) \qquad (6)$$

Fig. 7 State transition diagram of SiC boost converter

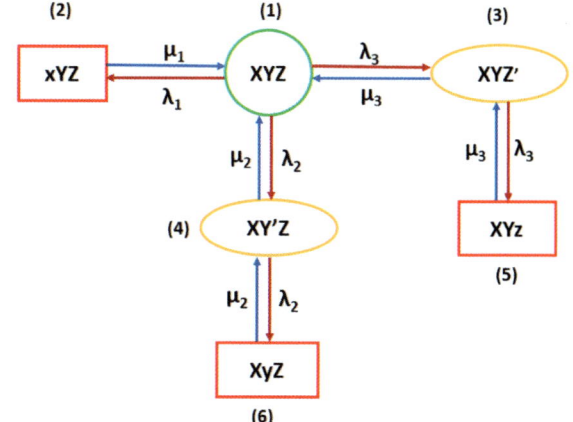

$$P_{3'}(t) = \lambda_3 P_1(t) + \mu_3 P_5(t) - (\mu_3 + \lambda_3) P_3(t) \qquad (7)$$

$$P_{4'}(t) = \lambda_1 P_1(t) + \mu_2 P_6(t) - (\lambda_2 + \mu_2) P_4(t) \qquad (8)$$

$$P_{5'}(t) = \lambda_3 P_3(t) - \mu_3 P_5(t) \qquad (9)$$

$$P_{6'}(t) = \lambda_2 P_4(t) - \mu_2 P_6(t) \qquad (10)$$

The rate transition matrix is given as follows.

$$
\lambda
$$
$$
-(\,|\, + \lambda_2 + \lambda_3)
$$
$$
\mu_1 \ \ \mu_3 \ \ \mu_2 \ 0 \ 0
$$
$$
\lambda_1 \ -\mu_1 \ \ 0 \ \ 0 \ 0
$$
$$
\lambda_3
$$
$$
0
$$
$$
-(\lambda_3 + \mu_3)
$$
$$
0
$$
$$
\mu_3
$$
$$
0
$$
$$
\lambda_1
$$
$$
0
$$
$$
0
$$
$$
-(\lambda_2 + \mu_2)
$$
$$
0
$$
$$
\mu_2
$$

$$\begin{bmatrix} P_1'(t) \\ P_2'(t) \\ P_3'(t) \\ P_4'(t) \\ P_5'(t) \\ P_6'(t) \end{bmatrix} = [\ |0| \ |0| \ |\lambda_3| \ |0| \ |-\mu_3| \ |0| \ |0| \ |0| \ |0| \ |\lambda_2| \ |0| \ |-\mu_2| \ ] \begin{bmatrix} P_1(t) \\ P_2(t) \\ P_3(t) \\ P_4(t) \\ P_5(t) \\ P_6(t) \end{bmatrix}$$

The system availability can be calculated as $P_1(t) + P_3(t) + P_4(t)$. The constant failure rate and repair rates have been calculated as $\lambda_1 = 0.01$; $\lambda_2 = 0.001$; $\lambda_3 = 0.001$; $\mu_1 = 0.01$; $\mu_2 = 0.02$; and $\mu_3 = 0.02$. The initial probability vector is taken as $[1 \ 0 \ 0 \ 0 \ 0 \ 0]$. By solving the rate transition matrix, we get the component availability as 0.999976999652514. The system availability plotted against time of operation is shown in Fig. 8.

The system availability of the SiC boost converter with change in time of operation shows that, after a period of 10^6 h of operation, the system availability reaches a probability of 0.73.

3.3 Thermal Analysis of SiC Boost Converter

SiC MOSFET is the main switching device in PV boost converter. The on and off period of the device contributes to increased switching loss at high switching frequency. Therefore, the thermal profile of the boost converter has to be observed under operating conditions like varying switching frequency, varying applied voltage both in closed-loop and in open-loop operations. The switching device is subjected to

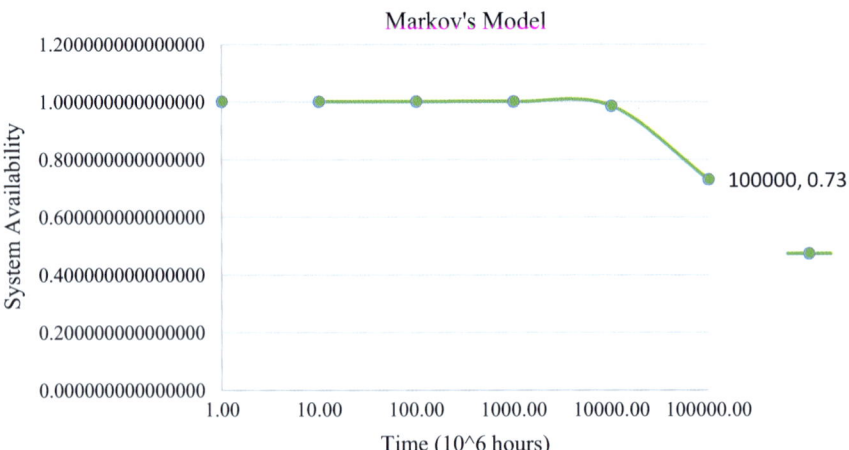

Fig. 8 Time of operation (10^6 h) versus system availability

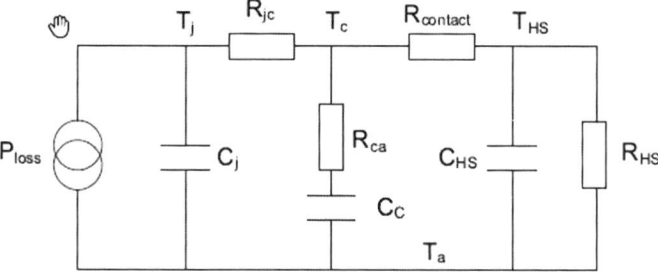

Fig. 9 Thermal model of SiC MOSFET and heatsink

more stress in closed-loop operation until it reaches the desired output. This results in change in $R_{ds}(ON)$ of the device and further increases the junction temperature [16, 17]. As the SiC device can withstand high temperature and high voltages, it can handle abnormalities during closed-loop operation. The thermal behaviour of the SiC device has been discussed in this section. The simple thermal equivalent of SiC MOSFET and heatsink is shown in Fig. 9.

From the datasheet of the device SCT2080KE, thermal resistances and capacitances have been taken. The heatsink parameters have been determined from the geometry of the heatsink. Thermal study has been performed for SiC boost converter with parameters such as switching loss, $R_{ds}(ON)$, and variation in applied voltage. The failure of the MOSFET depends on base failure rate, junction temperature, and case temperature [18–20]. It is given by the relation

$$\lambda_{MOSFET} = \lambda_b \pi_T \pi_Q \pi_A \pi_E \tag{11}$$

where

$$\pi_T = \exp\left[-1925\left(\frac{1}{T_j + 273} - \frac{1}{298}\right)\right] \tag{12}$$

$$T_j = T_C + \theta_{JC} P_D \tag{13}$$

$$T_C = T_a + \theta_{CA} P_D \tag{14}$$

where T_j is the junction temperature of the semiconductor device, T_c is the case temperature and T_a is the ambient temperature, θ_{JC} is the junction to case thermal resistance, and θ_{CA} is the case to ambient thermal resistance.

Fig. 10 Switching
frequency versus switching
loss

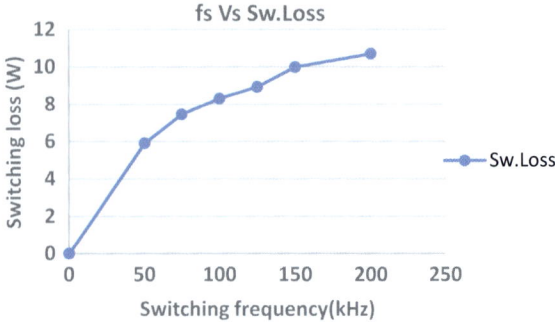

3.3.1 Switching Frequency Versus Switching Loss

Switching loss plays a significant role in increasing the junction temperature at higher operating frequencies. So, the switching frequency versus switching loss has been studied for SiC MOSFET-based boost converter.

The switching loss of the boost converter is given by

$$P_{SW} = (t_1 + t_2)\left(\frac{V_i I_O}{2}\right) f_s \tag{15}$$

where t_1 and t_2 are turn on and turn off period, V_i is the input voltage, I_o is the output current, and f_s is the switching frequency. When the switching frequency increases, the switching loss also increases proportionally. The switching frequency and the switching loss are shown in Fig. 10.

From Fig. 10, it can be inferred that the switching loss of the SiC boost converter increases with increasing switching frequency, but the temperature rise with respect to this switching loss will be handled by the SiC device as it has high thermal withstanding capability.

3.3.2 Temperature Versus R_{ds}(ON)

Increase in on-state drain source resistance results in increased conduction loss which in turn increases the junction temperature of the device. By using the thermal model of the SiC device, the effect of R_{ds}(ON) has been studied and presented in Table 2. Required data have been taken from the datasheet of the device. The following equations provide the relationship between calibration factor and conduction losses of the device.

$$P_{cond} = K \times P_{cond_{cal}} \tag{16}$$

$$P_{cond_{cal}} = V_{DS} \times I_D \tag{17}$$

Table 2 R_{ds}(ON) versus junction temperature

R_{ds}(ON) Ω	T_j (°C)
0	62.3
0.1	84
0.2	105.5
0.3	125

Table 3 Applied voltage versus temperature

Input voltage (V)	100 kHz	125 kHz	150 kHz
Closed-loop operation			
10	33.4	33.6	33.7
12	33.6	33.75	33.9
15	33.9	34	34.1
20	34	34.1	34.5
Open-loop operation			
10	33	33.2	34
12	33.3	33.3	34.8
15	33.4	33.6	35
20	33.5	33.9	35.2

where V_{DS}—transistor drain source voltage, I_D—drain current, K—calibration factor, P_{cond_cal}—calculated conduction loss.

The results from Table 2 show that the increment of the R_{ds}(ON) shows visible increment in conduction loss and the junction temperature has also increased. Still, the SiC MOSFET can withstand the temperature rise and ensures continuous operation.

3.3.3 Applied Voltage Versus Temperature for Various Switching Frequencies

The variation of temperature for different drain source voltage under varying switching frequencies with different input voltages has been listed in Table 3 for closed-loop and open-loop operation.

Table 3 shows that the junction temperature increases with increase in applied input voltage. But that is more prominent at higher switching frequencies. It is observed that the temperature change is more significant in open-loop operation. The devices that are stressed for producing boosted output voltage without maintaining nodal voltages are the reason for increasing junction temperatures.

From the above analysis, it is observed that SiC MOSFET has better thermal profile and good performance compared to Si-based boost converter.

4   Hardware Validation of SiC Boost Converter

A hardware circuit has been built using SiC MOSFET SCT2080KE for DC-DC boost converter. The pulse pattern has been generated with PIC microcontroller and embedded with the system. The open-loop and closed-loop operations were performed at varying input voltage with different switching frequencies. The hardware parameters have been listed in Table 1, and hardware circuit is shown in Fig. 11. The thermal images have been taken using FLIR thermal imaging camera.

4.1   Open Loop Operation of SiC Boost Converter with Varying Input Voltage and Switching Frequency

The boost converter has been operated in open loop with varying input voltage and switching frequency. The input voltage of 10, 12, 15 and 20 V has been applied with switching frequencies of 100, 125 and 150 kHz. The temperature variations have been observed using FLIR thermal imaging camera. The variation of temperature with input voltage change is shown in Fig. 12.

From the plot, it can be observed that the temperature increases with increase in input voltage and operating frequency under open-loop operation of SiC boost

Fig. 11 SiC boost converter circuit

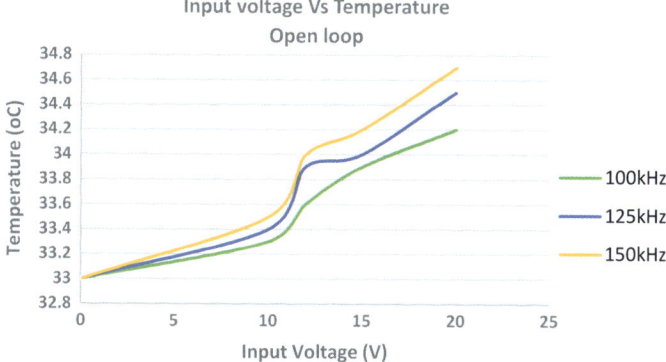

Fig. 12 Input voltage versus temperature for different frequencies in open-loop operation

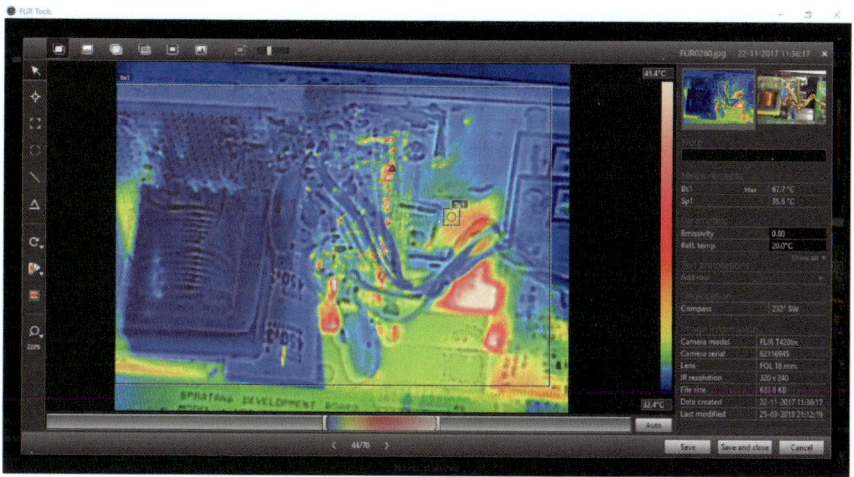

Fig. 13 Thermograph of open-loop SiC boost converter @ 20 V, 150 kHz input

converter. The thermal image during open-loop operation at 20 V input and 150 kHz frequency is shown in Fig. 13.

4.2 Closed-Loop Operation of SiC Boost Converter with Varying Input Voltage and Switching Frequency

The boost converter has been operated in closed loop with varying input voltage and switching frequency. The input voltage of 10, 12, 15 and 20 V has been applied with switching frequencies of 100, 125 and 150 kHz.

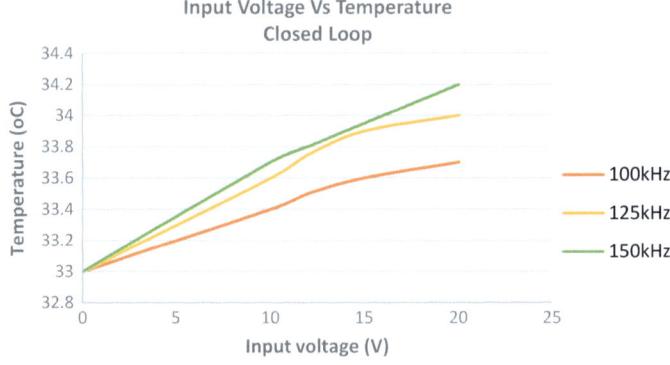

Fig. 14 Input voltage versus temperature for different frequencies in closed-loop operation

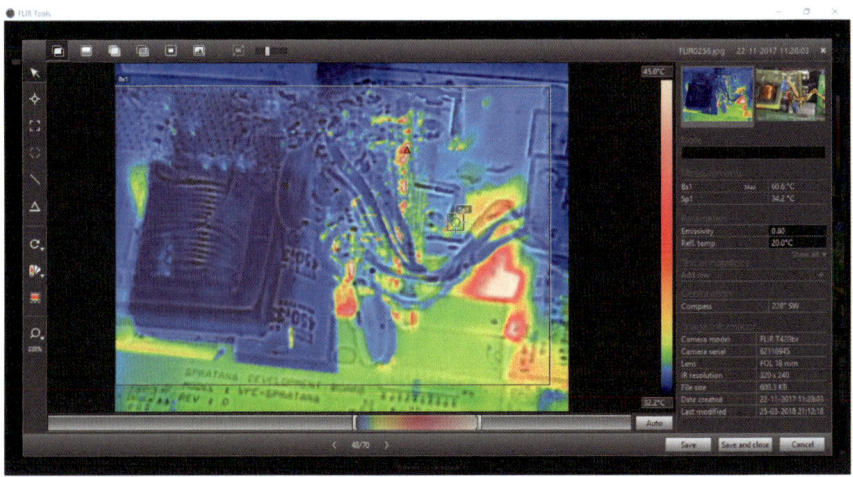

Fig. 15 Thermograph of open-loop SiC boost converter @ 20 V, 150 kHz input

The temperature variations have been observed using FLIR thermal imaging camera. The variation of temperature with input voltage change is shown in Fig. 14.

From the plot, it can be observed that the temperature increases with increase in input voltage and operating frequency under closed-loop operation of SiC boost converter. The temperature change is very less in closed-loop operation compared with open-loop operation. The thermal image during closed-loop operation at 20 V input and 150 kHz frequency is shown in Fig. 15.

Fig. 16 Temperature versus on-state resistance $R_{ds}(ON)$

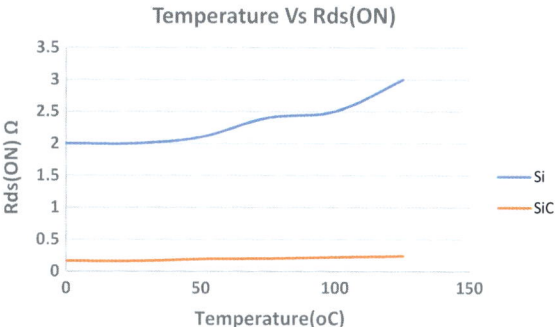

5   Comparison of SiC and Si Boost Converter

5.1   *Temperature Versus On-State Drain Source Resistance $R_{ds}(ON)$*

The on-state drain source resistance of the Si and SiC-based boost converter has been analysed with varying temperature. The effects are shown as graphical result in Fig. 16.

From the graph, it can be seen that the temperature of Si boost converter is showing rapid change in $R_{ds}(ON)$ compared to the $R_{ds}(ON)$ of the SiC-based boost converter. The Si device shows 3 Ω $R_{ds}(ON)$ at 125 °C, whereas the SiC device shows only 0.2 Ω at 125 °C.

5.2   *Switching Frequency Versus Switching Loss*

The performance of Si and SiC boost converters has been studied with varying switching frequency. The switching losses were calculated and then compared. Figure 17 shows the variation of switching loss with respect to change in switching frequency. The switching losses were calculated using Eq. (18).

$$P_{SW} = (t_1 + t_2) \frac{V_i I_o}{2} f_s \qquad (18)$$

where t_1 is the turn ON period, t_2 is the turn OFF period, f_s is the switching frequency, V_i is the applied input voltage, and I_o is the output current.

The plot clearly shows that the switching loss is more for Si-based boost converter up to a maximum of 16 W at 150 kHz. But the SiC-based boost converter shows the switching loss of 10 W at the same frequency. Hence, SiC-based boost converter has better efficiency.

Fig. 17 Variation of
switching loss with
switching frequency

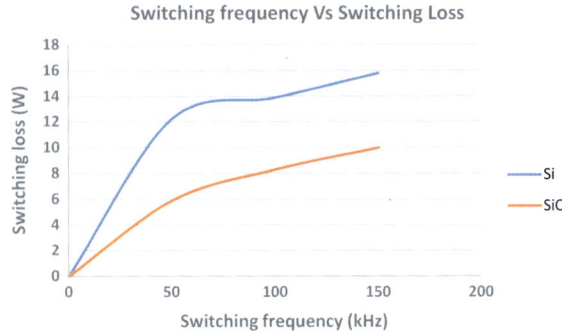

Switching frequency Vs Switching Loss

Table 4 Ripple comparison
for Si and SiC boost converter

	Si boost converter	SiC boost converter
Voltage ripple (%)	2.90	1.20
Current ripple (%)	9.38	5.73

Table 5 Theoretical and
hardware results for SiC
boost converter

	R_{ds}(ON) Ω	T_j (°C)
Theoretical	0.2	105.5
Practical	0.2	125
	fs (kHz)	T_j (°C)
Theoretical	100 kHz	33.7
Practical	100 kHz	33.9

5.3 Current and Voltage Ripple

The capacitor voltage ripple and inductor current ripple have been listed in Table 4
for Si boost converter and SiC boost converter.

The SiC boost converter has reduced voltage ripple and current ripple. The values
lie within the accounted values in the design of boost converter. Table 5 shows the
comparison of theoretical and practical results of the thermal study with 100 kHz.

From Table 5, it can be inferred that the SiC boost converter has better performance characteristics compared to SiC boost converter under 100 kHz operation.
The practical results are validated with the theoretical results.

6　Conclusion

A stochastic modelling for SiC boost converter for reliability assessment has been
done in this paper. The state transition-based Markov model is framed for the proposed converter, and the system availability has been calculated using Runge–Kutta

method. The proposed converter has been operated under varied input voltages and frequencies under open loop and closed loop to analyse the thermal behaviour. The thermal model of the device has been done, and the thermal profile of the SiC boost converter is better than the Si boost converter. The switching loss is reduced by 7% for the SiC boost converter. The $R_{ds}(ON)$ with 0.2 Ω of the SiC MOSFET has shown a temperature change of 125° C which could not be withstand by the Si MOSFET at a switching frequency of 100 kHz. The graphical results are presented for various cases such as temperature versus on-state drain resistance, temperature versus applied voltage, and switching frequency versus switching loss. A comparison is given between SiC- and Si-based boost converter. A hardware circuit was built for boost converter using SiC device SCT2080KE. The test results show that SiC boost converter has better performance in real time in closed-loop and open-loop operation.

Acknowledgements The authors would like to thank SSN management for funding this research work. We would also extend our gratitude to Dr. R. Damodaram, Associate Professor, SSNCE, and his team for helping in taking thermal images.

References

1. K.I. Hwu, C.F. Chuang, W.C. Tu, High voltage-boosting converters based on bootstrap capacitors and boost inductors. IEEE Trans. Ind. Elec. **60**(6) (2013)
2. D. Nicolae, C. Richards, J. van Rensburg, Boost converter with improved transfer ratio, in *Proceedings of IEEE IPEC*, 2010, pp. 76–81
3. N. Mohan et al., *Power Electronics* (Wiley, New York, 1989)
4. P.S. Bhimra, *Power Electronics* (Khanna publishers, New Delhi, India)
5. M.H. Rashid, *Power Electronics Circuits, Devices and Application*, 3rd edn. (Pearson Education, Inc, New Delhi, 2004)
6. X. Huang, G. Wang, Y. Li, A.Q. Huang, J. Baliga, Short-circuit capability of 1200 V SiC MOSFET and JFET for fault protection, in *2013 28th Annual IEEE Applied Power Electronics conference and Exposition (APEC)*, pp. 197–200, March 2013
7. A. Agarwal, H. Fatima, S. Haney, S.-H. Ryu, A new degradation mechanism in high-voltage SiC power MOSFETs. IEEE Electron Device Lett. **28**, 587–589 (2007)
8. K. Sheng, Maximum junction temperatures of SiC power devices. IEEE Trans. Electron Devices **56**, 337–342 (2009)
9. A.S. Alateeq, Y.A. Almalaq, M.A. Matin, Modeling a multilevel boost converter using SiC components for PV application, in *Wide Bandgap Power Devices and Applications*, vol. 9957 (International Society for Optics and Photonics, 2016), p. 99570J
10. A. Khosroshahi, M. Abapour, M. Sabahi, Reliability evaluation of conventional and interleaved DC–DC boost converters. IEEE Trans. Power Electron. **30**(10), 5821–5828 (2015)
11. H. Wang, K. Ma, F. Blaabjerg, Design for reliability of power electronic systems, in *IECON 2012-38th Annual Conference on IEEE Industrial Electronics Society* (IEEE, New York, 2012), pp. 33–44
12. S.V. Dhople, A. Davoudi, A.D. Domínguez-García, P.L. Chapman, Unified approach to reliability assessment of multiphase DC–DC converters in photovoltaic energy conversion systems. IEEE Trans. Power Electron. **27**(2), 739–751 (2012)
13. Y. Song, B. Wang, Survey on reliability of power electronic systems. IEEE Trans. Power Electron. **28**(1), 591–604 (2013)

14. S. Yang, D. Xiang, A. Bryant, P. Mawby, L. Ran, P. Tavner, Condition monitoring for device reliability in power electronic converters: a review. IEEE Trans. Power Electron. **25**(11), 2734–2752 (2010)
15. MIL-HDBK-217F "Military Handbook: Reliability Prediction of Electronic Equipment", Department of Defense, Washington, DC, 2 Dec 1991
16. A.H. Ranjbar, B. Fahimi, Helpful hints to enhance reliability of DCDC converters in hybrid electric vehicle applications, in *Record of IEEE Vehicle Power and Propulsion Conference*, 2010, pp. 1–6
17. C. Buttay, D. Planson, B. Allard, D. Bergogne, P. Bevilacqua, C. Joubert, M. Lazar et al., State of the art of high temperature power electronics. Mater. Sci. Eng.: B **176**(4), 283–288 (2011)
18. D.-P. Sadik, J. Colmenares, D. Peftitsis, J.-K. Lim, J. Rabkowski, H.-P. Nee, Experimental investigations of static and transient current sharing of parallel-connected silicon carbide MOSFETs, in *Proceedings of the 2013–15th European Conference on Power Electronics and Applications (EPE)*, pp. 1–10, Sept 2013
19. J. Lutz, H. Schlangenotto, U. Scheuermann, R. De Doncker, *Semiconductor Power Devices: Physics, Characteristics, Reliability* (Springer, Berlin, Heidelberg GmbH & Co. K, 2011)
20. F. Xu, B. Guo, L.M. Tolbert, F. Wang, B.J. Blalock, Evaluation of SiC MOSFETs for a high efficiency three-phase buck rectifier, in *2012 Twenty- Seventh Annual IEEE Applied Power electronics Conference and Exposition (APEC)*, pp. 1762–1769, Feb 2012

Application of HALT for Standardization in Aerospace Systems: A Case Study of Relay Contactors

Shivanand Tripathi, Mukesh Kumar, Yashwanth Deshmukh
and S. Giridhar Rao

Abstract There is practically no activity nowadays that is not framed, whether partly or totally, by standards viz. Government standards, Industry standards, Company standards, etc. Standardization not only helps in maximizing the compatibility, interoperability, repeatability, safety, and quality but also in minimizing the efforts and requirements of procurement, inventory, logistics, and maintenance activities. This paper deals with a case study in which it was required to use the same relay contactor part number in several different systems for use in the aerospace applications. Due to non-availability of aerospace-grade contactors, two COTS (Commercial-off-the-Shelf)-grade contactors having different part numbers were short-listed on the basis of technical specifications, make, availability, accessibility, and past usage history. However, both contactor types were found falling short to the end-use environmental specifications of some of the systems. Highly Accelerated Life Testing (HALT) was successfully utilized by applying various types of environmental stresses on several samples of the contactors to discover their design margins viz. operational and destruct margins. HALT results revealed that the discovered design limits of one contactor model were meeting the requirements of the system having the most harsh field use conditions and were much beyond its (contactor's) datasheet specifications and the same was selected to be used across all the systems. Over a period of time, the contactor has been integrated into many of these systems and has been successfully used for aerospace applications validating the HALT results.

Keywords Contactor · HALT · COTS · Design margins · Destruct margins · Load strength model · Operational margins · Organization standards · Relay

S. Tripathi (✉) · M. Kumar · Y. Deshmukh · S. Giridhar Rao
Advanced Systems Laboratory (ASL), DRDO, Hyderabad 500058, Telangana, India
e-mail: shivanand@asl.drdo.in

M. Kumar
e-mail: mukeshkumar@asl.drdo.in

Y. Deshmukh
e-mail: yashwanth@asl.drdo.in

S. Giridhar Rao
e-mail: sgiridhar@asl.drdo.in

© Springer Nature Singapore Pte Ltd. 2020
P. V. Varde et al. (eds.), *Reliability, Safety and Hazard Assessment
for Risk-Based Technologies*, Lecture Notes in Mechanical Engineering,
https://doi.org/10.1007/978-981-13-9008-1_15

1 Introduction

Standardization is the implementation of guidelines, rules, and specifications for common and repeated use, aimed at achieving optimum degree of order or *uniformity* in a given context, discipline, or field. Practically, every activity is framed, whether partly or totally, by standards. These can be government standards or industrial standards or company or organization standards [1]. The first two are externally imposed while the later one is normally internal to an organization.

Standardization has been important to industrial progress since the dawn of mass production, and it will continue to be so. Standardization affects every aspect of manufacturing. It is part of the glue holding things together [1]. Bills of material and lead times are standards essential to both engineering systems and materials systems [1] and hence plays a crucial role in optimizing the parts management program and finally brings consistency, visibility, and predictability in the production of the end product and as it reduces the variation and risks associated with quality, reliability, and production schedule.

In any application, the final products or systems can be of different types and capacity. For example, in commercial motor vehicle manufacturing the products can be a passenger car, an SUV, a Bus, or a truck. Similarly in X-ray machines manufacturing for industrial applications, different class, or capacity of the machines are produced targeting specimens of different materials (such as metal, ceramic, plastic), different thicknesses, and different specimens-dimensions. Similarly, in an aerospace or defense applications the products manufactured are of different capacity, range, payload, speed, dimensions, etc. However, it is always desired the common parts and subsystems of those products or systems be exactly identical (i.e. same part or type number) as this standardization helps in high level of compatibility and interchangeability across the systems and thus helps in minimizing the time and efforts in procurement activities and inventory management leading to high order of maintainability and system availability. In a complex mission aerospace system, the requirements of such standardization are much more than in commercial or industrial applications as the whole process of *identification, selection, testing, verification, validation, and acceptance of a new part* is a long and complex process consuming a good amount of time, money, and efforts.

To achieve the standardization of parts in an aerospace systems, the chosen parts must be able to meet the technical and functional specifications and also withstand the environmental stresses during its entire life cycle consisting of handling, storage, transportation, and field deployment. For standardization of a part across various systems, it is necessary that not only the functional requirements across the systems are same but it is also necessary that the selected part must be evaluated and validated for the system going to see the harshest environmental stresses. So that if a part is qualified for the system which is going to experience the highest levels of environmental stresses expected to be encountered by it during its life cycle then it need not be qualified separately for the other systems and can be used across all the systems.

HALT is normally known as a rapid design improvement tool and a good number of literature is available on the idea. But, this paper puts forward a different idea of how to utilize the tool even if the objective is not improvement of design. The HALT tool can be made use of in evaluating and selecting a bought out Commercial off-the-shelf (COTS) components by utilizing the design margins discovery during the process.

This paper explains with a case study, how HALT technique was successfully used to select, evaluate and ultimately meet the requirement of using a single type and model of non-Mil grade relay contactors across several different systems for an aerospace applications.

2 HALT—Highly Accelerated Life Testing

2.1 Introduction

"HALT" is an acronym and was coined by *Dr. Gregg Hobbs*. This *Highly Accelerated Life Test* is basically a "Design Ruggedization" test [2] and its significance lies in its ability to "rapidly" find defects in electronic products.

The goal of HALT is to quickly break the product and learn from the failure modes the product exhibits. The key value of the testing lies in the failure modes that are uncovered and the speed with which they are uncovered [3]. The basic philosophy behind HALT can be understood with the help of the load-strength model (refer Fig. 1).

With passage of time, the strength of a product gradually reduces due to fatigue caused over a period of time by the ambient load or stress. And, ultimately the products fail due to the fatigue of its weakest parts, i.e., *the weakest links* (Fig. 2).

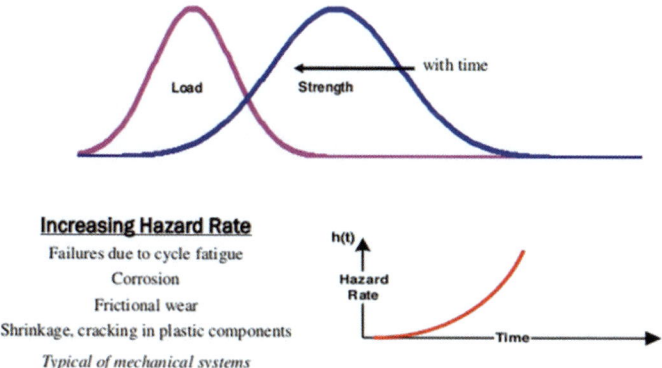

Fig. 1 Load-strength model

Fig. 2 Failure of the
weakest links due to fatigue

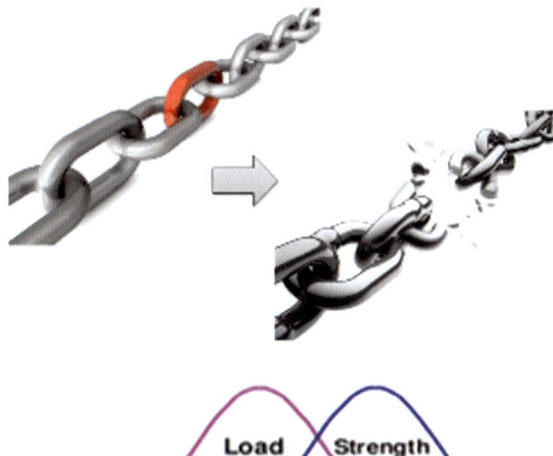

Fig. 3 HALT approach

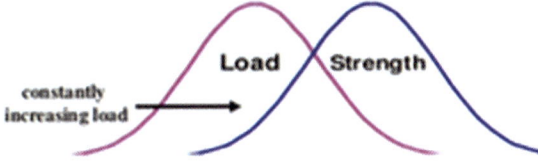

2.2 HALT's Approach

HALT tries to find out this weakest links by increasing the load *rapidly* (refer Fig. 3) starting from smaller stresses and thus increasing the probability of seeing an "existing" but hidden, latent, or potential failure mode.

To understand the design capability and the hidden weakness in the product, stress is applied in gradual manner till a failure is precipitated and detected. The failures may be investigated to understand the root cause(s). HALT in present case was used for finding the design capability of the contactors.

2.3 HALT—Stress Application

The stresses are applied starting with the least destructive and going to the most destructive. If the selected stresses are thermal and vibration, then the sequence normally followed is:

cold step stress → hot step stress → rapid thermal cycling stress → vibration step stress → combined thermal cycling/vibration stress.

2.4   HALT—Stress Level of Testing

The level of testing can be decided on case-to-case basis considering factors such as type or complexity of product, capability of functional test equipments, HALT equipments limitations, fixtures availability, cost, time, etc. and hence, the levels of stresses must be decided on case-to-case basis only.

2.5   HALT—Where to Stop

At what point to stop testing is an engineering decision. The stopping point may be either when the fundamental technological limits (FTL) or HALT test equipment's limit is reached. This FTL is the point where multiple failures begin to occur with small increases in stress and failure analysis reveals fundamental and catastrophic failures across several devices, with corrective action being prohibitive or impossible.

2.6   HALT—Margin Discovery, at the Core of Halt

HALT uncovers the operational and destructs limits of the product as HALT stresses the product quite beyond the product specifications.

It should be noted that some engineers incorrectly think that HALT only consists of determination of operational and destruct limits. However, operating margins are important indicators of product robustness (and therefore reliability) [4].

During the testing when the stresses are increased gradually, the "soft failure" limits called operational limits and "hard failure" limits called destruct limits are found. Soft failure implies that a failed product returns to its normal operational state when the stress is removed, whereas if the product fails permanently and does not return to its operational state ever after removal of the applied stress it is called "hard failure" [5]. The difference between these limits (operational/destruct) and the product specifications is called *operational margin* and *destruct margin,* respectively. Thus, HALT is a *Design Margin discovery tool* wherein the hitherto unknown margins available in the product are discovered and this knowledge gives the confidence about the strength or capability of the product and hence informed decisions can be taken whether to release the product in the field.

As the failure modes responsible for these limits are found and eliminated, the limits are pushed further and further out, maximizing the margins and thus increasing the product's useful life and reliability. Figure 4 illustrates the various margins discovered during the HALT process when a number of samples are HALT tested [5].

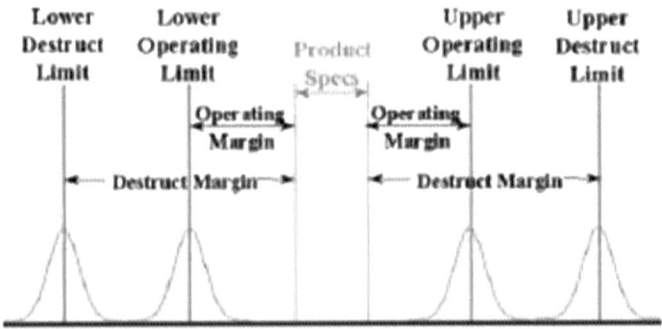

Fig. 4 HALT: margin discovery curves

3   A Case Study of Halt Application Leading to Meeting Standardization Requirement

3.1   Introduction

HALT is known generally as a design ruggedization technique to rapidly improve the design in the developmental phase of a product. However, there may be cases where the goal is not to improve the design by increasing the design margins *but mainly to know the margins of the existing design to understand the suitability and robustness* of the product to a particular application. The best example is frequent cases of parts uprating [6], wherein components are used in a product the specification of which is wider than the datasheet specification limits of the components. In the case study also, HALT tool was used to discover the design capability.

This case study explains a case study wherein it was required to harness the benefits of standardization by using the same relay-contactor model/part number across several systems.

The DC relay contactor was required to be used across several systems for aerospace applications. However, an appropriate aerospace-grade part wasn't available meeting both the electrical as well as environmental specifications. The solution was then to find a suitable COTS component with the matching or equivalent specifications and to use the same for our requirements. In this background, a number of relay contactors were studied and out of the available options, two different contactors bearing different models/part numbers were short-listed on the basis of technical specifications, make, availability, accessibility, past usage history, lead time, etc. The two models (part numbers) of contactors henceforth would be referred to as **Contactor-1** and **Contactor-2**. As per the datasheets, the electrical specifications of both the models were appropriate for all the systems, however, the environmental (dynamic) specifications were found falling short to the end-use environmental specifications of some of them. To meet the standardization requirement of the project, it was decided to use that contactor model across all the systems which can meet the end-use environmental specifications of the system having the most severe and stringent end-use environmental requirements.

3.2 Reason for Using HALT

To make an objective and informed decision whether to use any of the two contactors for all the systems the task at hand was to generate scientific and reliable data. In this backdrop, the HALT technique was used to generate the required data. However, it may be noted that HALT was utilized here not for design development as it is generally known for but for the discovery of the margins of the (existing) design; in other words, to find and understand the inherent capability of the contactors' design.

3.3 Selection of Environmental Stresses for HALT Application

In the HALT plan, only dynamic tests were included [such as sine vibration (SV), random vibration (RV), mechanical shock, and acceleration tests] as the climatic specifications [temperature, humidity, and pressure] of both the contactors were in line with the requirements/specifications of all the systems.

3.4 Implementation of HALT

3.4.1 Sampling Plan

Based on the availability at the time of execution of HALT plan, six samples of each of Contactor-1 and Contactor-2 were selected.

3.4.2 Functional Test Setup

The functional test setup diagrams for Contactors 1 and 2 are given in Fig. 5.

A DC power supply PS-I set at 28 V was used to energize the coil circuit and another one PS-II set at 32 V, 5 A was used for the main contacts circuit. The supply to the contactor coil was controlled using a switch provided on the power supply. As the main intention here was to ascertain whether the contactors are functioning or not at elevated environmental stresses, the load connected to the contactor contacts was kept low; a 10 Ω wire-wound resistive load across PS-II resulting in 3.2 A load current. A 16 channel recorder was used for data acquisition. The fastener used for

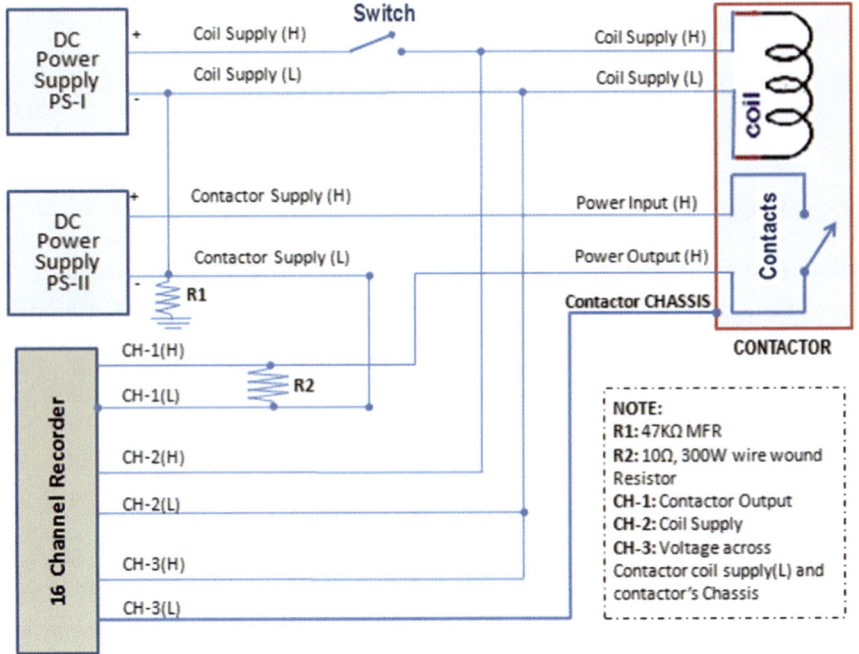

Fig. 5 Test setup for contactors functional checks during HALT

the purpose of mounting the contactor on the fixture was used to represent the chassis (body). A 48 kΩ metal film resistor was used as bleeder resistance between low and ground to suppress ground noise.

3.4.3 Measurements for the Contactors

The following measurements were carried out.

(a) *Output of main contact of the contactor*: in Channel-1 of the recorder,
(b) *Coil supply reaching to the contactor coil*: in Channel-2 of the recorder,
(c) *Low to contactor chassis (body) voltage*: in Channel-3 of the recorder.

3.4.4 HALT Stress Method

The HALT tests were carried out in *Step-stress* manner and in the sequence of sine vibration → random vibration → linear acceleration → mechanical shock test.

- The sine vibration test was carried out in the 20–2000 Hz frequency band. The test was carried out in step-stress mode by increasing the amplitude in steps of ±5g

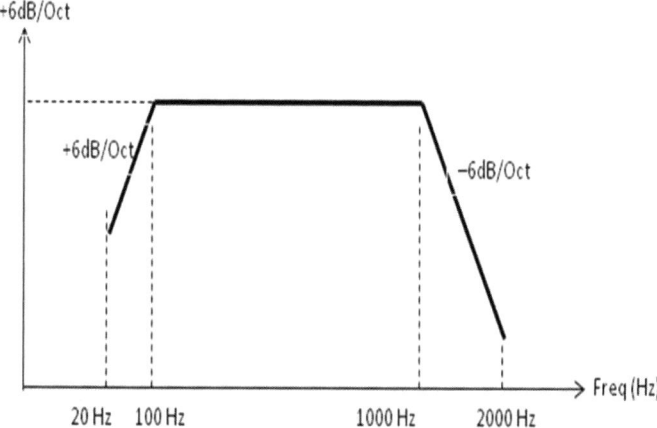

Fig. 6 Random vibration test profile

(peak-to-peak). The test was carried out at sweep rate of 1 octave/min along all three axes for a duration of 500 s per axis.

- The random vibration test was carried out as per the profile given in Fig. 6 in the 20–2000 Hz frequency band as per guidelines in Mil-std-810 spectrum for aerospace application. The test was carried out in step-stress mode by increasing the amplitude in steps of $5g_{rms}$. The test was carried out along all three axes for a duration of 500 s per axis.
- The linear acceleration test was carried out by increasing the amplitude in steps of 5g. The test was carried out along all six directions and for a duration of 1 min per direction.
- The classical shock test (half sine wave, 11 ms period) was carried out by increasing 5g in every step. The test was carried out along all five directions with four shocks per direction.

3.4.5 Determination of Operating and Destruct Limits

While applying the environmental stresses during the HALT exercise the contactors were functionally operated as explained earlier. The coil supply was manually made ON and OFF several times to see whether the contactors were displaying their normal functionality.

Any misbehavior or fault observed would mean abnormal operation of contactors. The following would indicate faulty operation of the contactors:

(a) If contactor supply (H) and contactor coil supply (H) shorts then both Ch-1 and Ch-2 will go to 32 V. This fault means Ch-2 (monitoring the coil supply) would be showing ON even after switching OFF the supply to the contactor coil.

(b) If contactor supply (H) touches the contactor chassis (body) then Ch-3 would go to −32 V.
(c) If contactor coil supply (H) touches the chassis (body) then Ch-3 would go to −28 V.
(d) If low lines of any power supply shorts to chassis, then it is also a fault condition.
(e) Fluctuation/chattering in contactor output (Ch-2).
(f) Both coil supply (Ch-2) and contactor output (Ch-1) fluctuating even when all supplies are kept ON and constant.
(g) Output (Ch-1) dropping to 0 V even though coil supply (PS-I) being ON.

Detection of any of the above-mentioned faults would mean abnormal operation of the contactor which in turn would mean the applied stress level being beyond the operating limit of the contactor. The minimum stress level at which at least one sample of the contactors starts misbehaving could be the operational limit or the destruct limit. If the contactor resumes normal operation after removing the HALT stress applied, the stress limit is taken as the operational limit. The destruct limit is the minimum stress at which at least even one contactor sample displays an operational fault or some physical damage of a permanent nature, i.e., the contactor fails to resume its normal operation even after removing the applied stress.

3.5   HALT—Result

- *Sine vibration test*: (1) Only one sample of Contactor-2 shown permanent damage. (2) There was no permanent fault or damage observed in Contactor-1 in the entire range of applied stress.
- *Random vibration*: (1) Only one sample of Contactor-1 failed permanently. (2) The discovered operational limits of both Contactors 1 and 2 were found to be the same.
- *Classical shock test*: Operating limit of Contactor-1 was found better however, the destruct limit of Contactor-2 was higher.
- *Linear acceleration test*: (1) Both operating as well as destruct limits were found to be same for both the contactors. (2) There was only one failure mode observed for Contactor-1 which is the output remains high even when the coil supply is withdrawn. *The behavior of Contactor-1 was found to be more deterministic.* In those aerospace applications where power needs to be kept permanently ON for loads, this contactor is more safe and reliable to use.
- Operating margins were comfortably higher for both the contactors in all the dynamic tests.
- From the HALT exercise, operating and destruct margins of both the contactors were found. Though the margins exceeded much beyond the requirements, Contactor-1 was selected as it displayed better operational as well as destruct limits except in case of classical shock test where the destruct limit of Contactor-1 was higher (Table 1).

Table 1 Design margins discovered for the two contactors

Test	Operating margins		Destruct margins	
	Contractor-1	Contractor-2	*Contractor-1*	Contractor-2
Sine vibration (pk-pk)	±10g	±5g	±15g	±10g
Random vibration (g_{rms})	5 g_{rms}	5 g_{rms}	15 g_{rms}	10 g_{rms}
Classical shock (pk)	20g	10g	25g	30g
Linear acceleration (g)	10g	10g	15g	15g

3.6 Outcome

Based on the results obtained in the HALT exercise, Contactor-1 was selected and over time it was integrated into several systems and have been successfully used for aerospace applications.

4 Conclusions

Standardization simplifies the efforts in meeting organizational and project requirements. Standardization in bill of materials is crucial for optimizing the parts management program and finally for project management in a complex aerospace application because the associated variation in lead time and quality and reliability of parts will have a bearing on the quality and reliability of the final products and the project schedule. The part manufacturer provides electrical specifications for a given environmental range of operation. If mismatches exist between the manufacturer specified environmental ranges and the local environment, the parts selection and management team must take additional steps to assure proper operation of the part in the product [6]. And for standardization of the part across all the systems, it is required to be evaluated for the highest level of stress expected to be encountered by the systems. Traditionally, it has been considered that "the goal of HALT is to improve the design margin" [5]. However, the tool can be used even for just discovering the operating and destruct margins before using the component as it is in a system. This is the case when need is to select, evaluate, and use a bought out commercial off-the-shelf (COTS) components, and this situation is not very infrequent these days.

In the case study of two contactors, we have shown that when the suitable aerospace (Mil) grade relay contactors for power switching were not available, COTS-grade contactors were identified matching the electrical specifications. However, their environmental specifications in the datasheet were not meeting the environmental stress specifications of some of the systems. Here, the popular HALT tool was successfully exploited in evaluation of the two contactor types for discovering the operating and destruct margins of the contactors' design beyond the datasheet specifications. It helped in generating the necessary scientific data, and in turn, it enabled the management to meet the standardization requirements of the project, i.e., to use the same relay contactor part number across multiple systems characterized by different levels of end-use environmental specifications.

References

1. R.W. Hall, *Continuous Improvement Through Standardization*
2. G.K. Hobbs, *HALT & HASS, The New Quality and Reliability Paradigm* (2002)
3. D. Neill, QualMark Corporation, *Highly Accelerated Life Testing—Testing With a Different Purpose*
4. A. Barnard, Lambda Consulting, Pretoria, South Africa (2012) Ten Things You Should Know About HALT & HASS, *RAMS Symposium*
5. J. Cooper, *Introduction to HALT—Making Your Product Robust*, Microelectronics symposium, Pan pacific (2017)
6. CALCE (Center for Advanced Life Cycle Engineering), Performance Assessment and Part Uprating

Failure Mode Effect Analysis of Analog Alarm Trip Unit Using Simulation Technique

Subhasis Dutta, Sushil B. Wankhede, Kaustubh V. Gadgil and S. Panyam

Abstract This study is focused on identifying and recognizing the failures associated with analog alarm trip units (ATUs). Electronic components of ATU were identified and simulation was carried out to study the effects of component failures. Different modes of failure of each electronic component were considered. Component failures leading to safe and unsafe states of ATU were found out and documented. This study would help to identify unsafe failure modes of ATU and their effect on plant safety.

Keywords Analog alarm trip units · Modes of failure · Electronic components · Unsafe state

1 Introduction

ATUs are an important part of control and instrumentation system of nuclear research reactors. They are used to compare signals from a measuring device located in the field and generate outputs for various alarms, interlocks, and trips which are required for proper operation of safety-critical systems of nuclear research reactors. Typically around hundred ATUs are used in a nuclear research reactor facility. ATU consists of electronic components, and failure of these components drives the output to unknown states. As the ATUs were used in safety systems like reactor trip logic system, detail analysis of the effects of component failures was required. Analysis (FMEA) reports

S. Dutta (✉) · S. B. Wankhede · K. V. Gadgil · S. Panyam
Bhabha Atomic Research Center, Mumbai, India
e-mail: sdutta@barc.gov.in

S. B. Wankhede
e-mail: sushilw@barc.gov.in

K. V. Gadgil
e-mail: kaustubh@barc.gov.in

S. Panyam
e-mail: spanyam@barc.gov.in

© Springer Nature Singapore Pte Ltd. 2020
P. V. Varde et al. (eds.), *Reliability, Safety and Hazard Assessment for Risk-Based Technologies*, Lecture Notes in Mechanical Engineering,
https://doi.org/10.1007/978-981-13-9008-1_16

189

such as failures were not provided by vendors. Thus, a study was carried out to identify the states which the output of ATU may take, upon failure of components and to identify the unsafe states.

2 Analog Alarm Trip Unit

ATUs compare the current and voltage input with set-point values and generate two independent contact outputs [1]. Current signal is converted into voltage, and two independent set-point comparisons are performed to actuate two independent relay outputs. Set points and hysteresis are adjustable through individual potentiometers. Each output relay can be configured as high or low alarm through mode selection DIP switches. Output relay can also be configured to be energized or de-energized during alarm state and in power ON state (normal operating state) by appropriate selection of DIP switches. ATUs have additional features like status indicating LEDs for the status of output relays and terminals for monitoring voltages corresponding to input current and individual set points.

Electronic components of an ATU were segregated into different blocks depending on their functionality for the ease of analysis as mentioned below.

2.1 Blocks of Analog Alarm Trip Unit

2.1.1 Power Supply Block

Power supply block receives 24 VDC power supply from an external source and performs supply isolation, conditioning, and voltage regulation before supplying power to other electronic components of the ATU. It consists of fuse, reverse supply protection diode, isolating transformer, astable multi-vibrator, and voltage regulator ICs.

2.1.2 Input Amplifier Block

Input amplifier receives current and voltage signals from field instruments and generates amplified voltage output for further computation. It consists of an input filter, range selection DIP switch [1], and amplifier circuit consisting of the operational amplifier, resistors, and capacitors.

2.1.3   Set-Point Generating Block

Voltage from the input amplifier block is compared with two set-point voltages generated by the set-point generating block. It consists of a reference voltage unit, amplifier unit, and two set-point potentiometers. The set points can be adjusted independently through these two potentiometers.

2.1.4   High/Low Selection Blocks

An ATU can be set for high or low alarm operations [1] through these blocks. It consists of switch IC and a DIP switch used to configure the unit for high or low alarm operations. Switch IC consists of four switches which are controlled by individual externally applied control voltages. Alarm trip unit has two high/low selection blocks for two independent relay outputs.

2.1.5   Set-Point Comparator Blocks

Set-point comparator block compares set-point voltages with the input signal and drives the output relay. It comprises an amplifier and a network of resistances and capacitors. There are two comparator blocks for two set-point comparisons.

2.1.6   Hysteresis Block

This block generates a voltage which adds to the input of the comparator blocks, causing a hysteresis during the change of output state of the comparator. The value of the hysteresis is adjustable by potentiometers. This block consists of a network of resistances, operational amplifiers, and potentiometers. Alarm trip unit has two hysteresis blocks, one for each set-point comparison.

2.1.7   Output State Configuration Blocks

The output relay can be configured as energized or de-energized mode during an alarm condition through this block. This block consists of an XOR gate, DIP switch, and resistors. When the DIP switch is in the closed condition, the output relay is energized during an alarm or trip condition and vice-versa. There are two output state configuration blocks and one for each output relays.

2.1.8 Output Relay Driving Block

This is the final block comprising of the output driving transistor, relay status indi-cating LED, output relay, and a network of resistors and capacitors. Output relay is driven by the comparator through this block. Each output relay has independent driving blocks.

2.2 Working of ATU

ATU receives input current in the range 4–20 mA from a transmitter located in the field and generates relay outputs. In the case shown in Fig. 1, the ATU is configured in high alarm mode for both output relays. Relay 1 is configured to be in the energized condition in power ON state and in the de-energized condition in alarm state and Relay 2 is configured to be in the de-energized condition in power ON state and in the energized condition in alarm state. Alarm set point is set as 10 mA as shown in the figure. When the input current is below the set point (normal operating state or power ON state), Relay 1 is in energized state and Relay 2 is de-energized state. When input current exceeds the set point, Relay 1 de-energizes and Relay 2 energizes till the input

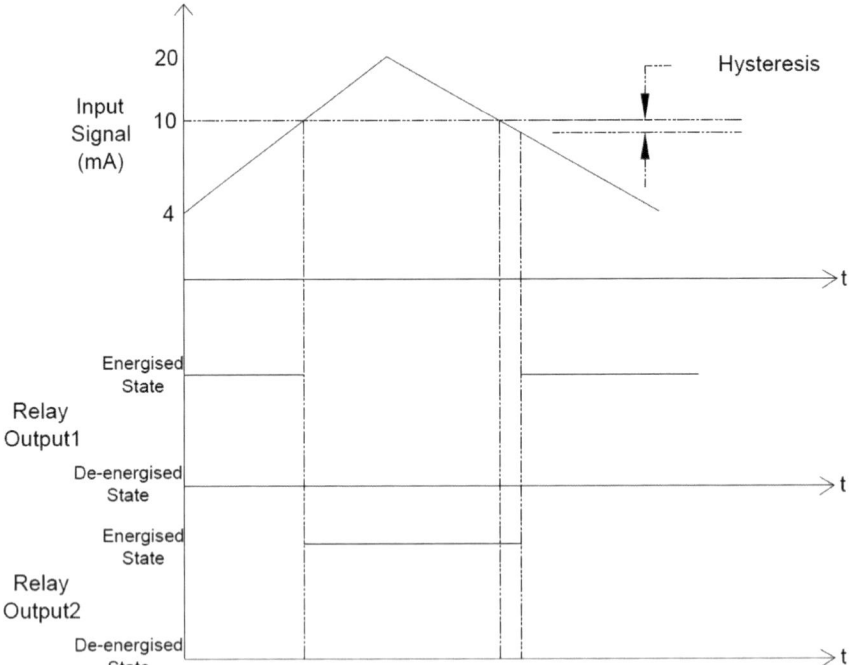

Fig. 1 Working of ATU

is greater than the set point. As the input current goes below the set point, the state of the relays do not change until the input goes below a certain value, the difference is called hysteresis and is shown in the figure. Hysteresis can be set independently for both the relays through potentiometers as discussed in Sect. 2.1.6. The changes in states of the relay contacts can be used for generating alarms, interlocks, and trips.

3 Failure Mode Effect Analysis

Failure modes and effects analysis (FMEA) is a systematic method of evaluating an item or process to identify the ways in which it might potentially fail, and the effects of the mode of failure upon the performance of the item or process and on the surrounding environment and personnel [2]. FMEA and its techniques are well established. The most commonly used techniques for a failure effects analysis are the tabular Failure Modes and Effects Analysis (FMEA), fault tree analysis, matrix FMEA, and sneak circuit analysis. This paper is an application of tabular Failure Modes and Effects Analysis technique with additional treatment of simulation.

Failure of components drives the system to either a safe state or unsafe state. A safe failure is defined as a failure that causes the module or system to go to a defined fail-safe state without a demand from the process [3]. An unsafe failure is defined as a failure that does not respond to a demand from the process and does not go to the defined fail-safe state [3]. Some unsafe failures are detectable and the system can be brought to the normal operating state through repair. Different states a system might take due to failure of components are shown in Fig. 2.

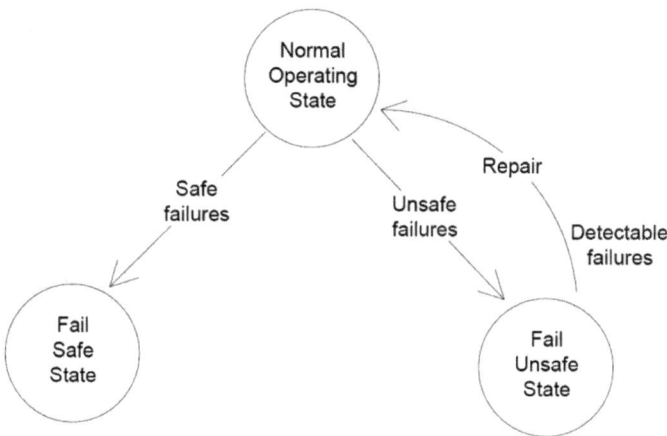

Fig. 2 States of a system

3.1   Assumptions Consider During FMEA

The following assumptions were made during the study [4]:

1. The effect of component failure on the loss of function is considered at the subsystem level.
2. The analysis assumes that only a single component fails at a time.
3. Temperature is considered to be within the specified levels of operation.
4. The system under consideration is assumed to be in a state of normal operation, with all inputs and outputs at their nominal values.

3.2   Methodology

Failure Mode Effect Analysis was carried out to identify different unsafe states of ATU with the help of Simulation software. Electronic components of ATU were traced, and their functionality was simulated as shown in Fig. 3. Effects of component failures on final relay outputs were observed. The observations for component failures in different blocks were tabulated in a worksheet.

Configuration of ATU considered during simulations was such that the output relay was energized during a normal state of input current and in the de-energized state during an alarm or trip condition. This configuration ensures the ATU to be

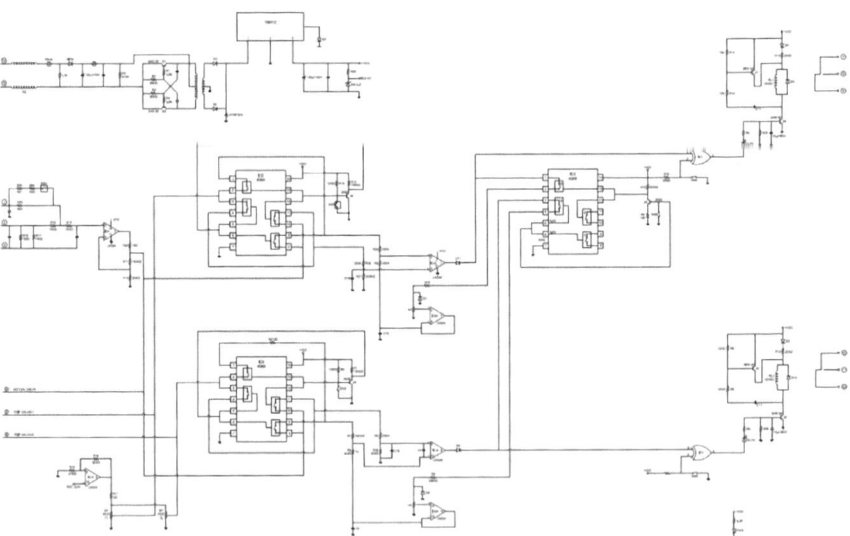

Fig. 3 Circuit of ATU

in alarm or trip state during power supply failures. Thus, de-energized state of the output relays was considered as the fail-safe state throughout the simulation.

The following configurations of the ATU were considered during the simulation.

1. Output relay was configured to be de-energized (fail-safe) during an alarm or trip condition.
2. Input of ATU was configured to be current in the range of 4–20 mA.
3. Relays were configured to be independent of each other.
4. Simulations were carried out with alarm tip unit in both high and low alarm mode for each component failure cases.

ATU was configured as discussed above. Values for set point, hysteresis, and input current were set, and faults in electronic components of each block were introduced in different modes as per Table 1. Simulations were carried out for both normal and alarm states of input, and changes in the status of output relays were observed.

3.3 Failure Modes

Failure of electronic components of ATU was considered to be in various modes [5] as shown in the table below.

3.4 Results of Simulations

Simulations were carried out to identify the safe and unsafe states of ATUs due to failure of components in different modes. In case of unsafe failures, detectable and undetectable failures were identified. Unsafe and undetected failures are considered to be a dangerous state as they cannot be rectified through repair or maintenance [3].

3.4.1 Case Study: Worksheet of Output Relay State Selection Block and Output Relay Driver Block

Worksheet of output relay state selection block and output relay driver block is shown in Fig. 4. In the worksheet, components, their functionality, different modes of component failures, and effects of failures are tabulated with the help of software simulation. The effects of the failures were studied, and safe/unsafe failures were recognized. In some cases, the unsafe failures were detectable, like in this worksheet two cases were found in which the functionality of the ATU was found to be opposite to that in a normal operating condition, i.e., during normal input state, the output goes into fail-safe de-energized state and in unsafe energized state during an alarm condition. Although the case is not fail-safe during an alarm or trip condition, it can be detected during the normal operating condition.

Table 1 Failure modes

Sl No.	Components	Failure modes
1	Resistors	Resistance becomes open
		Resistance becomes short
2	Inductor	Inductor becomes open
		Inductor becomes short
3	Capacitor	Capacitor becomes open
		Capacitor becomes short
4	Protection fuse	Protection fuse becomes open
		Protection fuse becomes short
5	Diode	Diode becomes open
		Diode becomes short
6	Voltage regulator IC	IC damage
		Input and output terminals become short
7	Reference voltage	Reference voltage is equal to supply voltage
		Reference Voltage is zero
8	DIP switch	Switch contact becomes open
		Switch contact becomes short
9	Switch IC	Switch contact outputs become open
		Switch contact outputs become short
10	Transistors	Base and emitter terminals become short
		Base and collector terminals become short
		Emitter and collector terminals become short
		Base terminal becomes open
		Emitter terminal becomes open
		Collector terminal becomes open
11	Operational amplifier	Inverting and non-inverting terminals become short
		Output and non-inverting terminals become short
		Output and inverting terminals become short
		Inverting terminal becomes open
		Non-inverting terminal becomes open
		Output terminal becomes open
		IC damaged and output is driven to supply voltage
		IC damaged and output is driven to zero voltage
12	Potentiometer	Terminals of potentiometer become open
		Terminals of potentiometer become short
13	XOR gate	Input terminals become short
		Input terminals become open
		Output terminal becomes open

Results of FMEA of ATU

Component	Function	Failure	Effect	Fail Safe	Detectable in case of not fail Safe
Output relay RL2 state selection block					
Switch (SW5)	De energised or energised relay RL2 in case of alarm	May become short	Opposite of Normal Functionality	No	Yes
		May Become Open	Normal Functionality	Yes	-
Resistor (R)	Limiting current to the input of XOR(IC1)	May become open	Opposite of Normal Functionality	No	Yes
		May become short	Normal Functionality	Yes	-
XOR(IC1)	De energised or energised relay RL2 in case of alarm	May become open	Relay always de-energised	Yes	-
		May become short	Relay always de-energised	Yes	-
Output relay RL2 driver block					
LED (OUT2)	Indictaion of relay RL2 status	May become open	Relay always de-energised	Yes	-
		May become short	Normal Functionality	Yes	-
Resistor (R5)	Limiting Current to base of transistor Q6	May become open	Relay always de-energised	Yes	-
		May become short	Normal Functionality	Yes	-
Resistor (R29)	Driving Output relay RL2	May become open	Normal Functionality	Yes	-
		May become short	Relay always de-energised	Yes	-
Capacitor (10uF)	Filter	May become open	Normal Functionality	Yes	-
		May become short	Relay always de-energised	Yes	-
Transistor(Q5)	Driving Output relay RL2	B/E short	Relay always de-energised	Yes	-
		B/C short	Relay always energised	No	No
		C/E short	Relay always energised	No	No
		C open	Relay always de-energised	Yes	-
		E open	Relay always de-energised	Yes	-
		B open	Relay always de-energised	Yes	-
		Transistor Short	Relay always energised	No	No
		Transistor Open	Relay always de-energised	Yes	-
Capacitor (C11)	Driving Output relay RL2	May become open	Normal Functionality	Yes	-
		May become short	Relay always de-energised	Yes	-
Resistor (R6)	Driving Output relay RL2	May become open	Normal Functionality	Yes	-
		May become short	Normal Functionality	Yes	-
Resistor (R8)	Driving Output relay RL2	May become open	Relay always de-energised	Yes	-
		May become short	Normal Functionality	Yes	-
Transistor(Q3)	Driving Output relay RL2	B/E short	Relay always de-energised	Yes	-
		B/C short	Normal Functionality	Yes	-
		C/E short	Normal Functionality	Yes	-
		C open	Relay always de-energised	Yes	-
		E open	Relay always de-energised	Yes	-
		B open	Relay always de-energised	Yes	-
		Transistor Short	Normal Functionality	Yes	-
		Transistor Open	Relay always de-energised	Yes	-
Resistor (R10)	Driving Output relay RL2	May become open	Normal Functionality	Yes	-
		May become short	Normal Functionality	Yes	-
Relay (RL2-12VDC)	Output relay RL2	May become open	Relay always de-energised	Yes	-
		May become short	Relay always de-energised	Yes	-
D10 diode	Free wheeling diode	May become open	Normal Functionality	Yes	-
		May become short	Relay always de-energised	Yes	-
D5 diode	for protection	May become open	Normal Functionality	Yes	-
		May become short	Normal Functionality	Yes	-

Fig. 4 Worksheet for output relay state selection block and output relay driver block

Table 2 Effects of component failures

Total numbers of failures of components	350
Total number of cases of unsafe failures of alarm trip unit	50[a]
Total number of detectable unsafe failures of alarm trip unit	06

[a]In some cases of undetected unsafe failures (dangerous failures), it was observed that component failures drive the unit to dangerous state only in high or low configuration mode, and thus, one of the relays may be configured in the opposite mode to drive the output of ATU to a safe failure state during the component failure

Effects of failures of components obtained through simulation studies can be summarized as shown in Table 2.

From the simulations, it was concluded that in order to prevent common cause failures, both relays should not be used for a single safety function as they have common components, namely input amplifier block and set-point generation block.

4 Conclusion

FMEA is a flexible tool that was tailored to meet a specific requirement. Specialized worksheets requiring specific entries were adapted. This work done (FMEA) has indicated the strengths and weaknesses in the system design of ATUs and provided information for allied analyses in maintainability, testability, logistics, reliability prediction, and safety. This study helps in configuring ATUs and applies C&I design techniques to prevent unsafe failures.

References

1. User manual of Current/Voltage Trip Amplifier, Reputed make
2. Standard IEC 60812: Failure modes and effects analysis (FMEA and FMEC)
3. Failure Modes, Effects and Diagnostic Analysis, exida GmbH, Stephen Aschenbrenner, Germany
4. MIL-STD-338B:Electronic reliability design Standard
5. Safety Analysis Report (Hardware) of Analog Comparator/Trip Amplifier, Reputed make

Ageing Model for Electrolytic Capacitors Under Thermal Overstress

Arihant Jain, Archana Sharma, Y. S. Rana, Tej Singh, N. S. Joshi
and P. V. Varde

Abstract Electrolytic capacitors are widely used components in various applications such as power supplies in avionics, DC–DC converters, and regulation and protection system of a nuclear reactor. These capacitors are frequently responsible for system failures. Ageing of the capacitor, influenced by operating thermal and electrical conditions, leads to failure, indicated by a change in capacitance and equivalent series resistance (ESR) beyond acceptable limits. As part of an experimental programme towards the development of degradation models for electronic components, we have performed accelerated ageing experiments on electrolytic capacitors under thermal overstress. Capacitance and ESR values are recorded with ageing time for various predefined levels of thermal overstress conditions. Based on the experimental results, a degradation model for electrolytic capacitors has been developed. This paper presents a detailed account of the experiments and the degradation model developed.

Keywords PHM · PoF · Reliability · Capacitor · Thermal overstress

A. Jain · A. Sharma · Y. S. Rana (✉) · T. Singh · N. S. Joshi · P. V. Varde
Bhabha Atomic Research Center, Mumbai, India
e-mail: yps@barc.gov.in

A. Jain
e-mail: arihantj@barc.gov.in

A. Sharma
e-mail: archphy@barc.gov.in

T. Singh
e-mail: t_singh@barc.gov.in

N. S. Joshi
e-mail: nsjoshi@barc.gov.in

P. V. Varde
e-mail: varde@barc.gov.in

© Springer Nature Singapore Pte Ltd. 2020
P. V. Varde et al. (eds.), *Reliability, Safety and Hazard Assessment*
for Risk-Based Technologies, Lecture Notes in Mechanical Engineering,
https://doi.org/10.1007/978-981-13-9008-1_17

1   Introduction

Physics-of-failure (PoF) models are derived using first principles of physical science and account for the dominant failure modes and underlying failure mechanisms for a given operating condition or stress [1]. We have started a PoF-based programme on prognostics and health management (PHM) for reliability prediction of electronic components. A database on PoF models for different electronic components available in the literature has been prepared. It is planned to carry out accelerated ageing experiments and determine the model constants. For these experiments, to start with, we have chosen electrolytic capacitor first.

Electrolytic capacitors are widely used components in various applications such as power supplies in avionics, DC–DC converters and regulation and protection system of a nuclear reactor. These capacitors are frequently responsible for system failures. Ageing of the capacitor, influenced by operating thermal and electrical conditions, leads to failure, indicated by a change in capacitance and equivalent series resistance (ESR) beyond acceptable limits. We have performed experiments on electrolytic capacitors under thermal overstress conditions. Capacitance and ESR values are recorded with ageing time for various predefined levels of thermal conditions. The experimental results are used for the development of a degradation model for electrolytic capacitors. In this report, we present the details of the ageing experiments and the degradation model developed.

2   Degradation in Electrolytic Capacitors

2.1   Mechanisms

The two main precursors used for monitoring the health of electrolytic capacitors are capacitance and ESR. With degradation, ESR of a capacitor increases and capacitance decreases. There are many factors responsible for the degradation [2], such as an increase in temperature due to heating and degradation of the oxide layer. The heating can occur either due to external operating conditions or internal processes such as charging/discharging and power dissipation due to ripple current and leakage current. As a result of an increase in temperature, evaporation of electrolyte takes place which leads to degradation of the capacitors. Another important factor which may lead to degradation or failure of a capacitor is the increase in internal pressure caused by higher rate of chemical reactions at increased temperatures.

2.2   Degradation Models

Based on the evaporation of electrolyte, the first principle models for degradation of electrolytic capacitors have been developed by Kulkarni et al. [2–4] and Celaya et al. [5]. The models are supported by accelerated ageing experiments with thermal overstress (TOS) and electrical overstress (EOS). Other studies on the subject have been reported by Darnand et al. [6], Wang et al. [7] and Shukla et al. [8].

   The model developed by Kulkarni et al. [2] relates the reduction in volume (due to evaporation) of the electrolyte to degradation in capacitance by considering structural details of capacitor design. Thermal overstress experiments have been performed by the authors in support of the model. Using the experimental data on decrease in capacitance with ageing time, decrease in volume of the electrolyte is estimated. Fitting the electrolyte volume versus ageing time data to a polynomial, the model constants are derived. The model constants are employed to estimate the reduction in volume as a function of ageing time for a test capacitor. The information is then used to predict a decrease in capacitance.

2.3   Variation of Capacitance Under Thermal Overstress

In line with the model of Kulkarni et al. [2], the degradation of capacitance can be related to the distance or separation, d (i.e. the dielectric thickness) between the two conductive plates. Capacitance is related to 'd' in the following manner:

$$C = \frac{2\varepsilon_o \varepsilon_r A}{d} \tag{1}$$

   As the electrolyte evaporates due to heating, its concentration reduces. The reduction in concentration can be visualized as the equivalent increase in dielectric thickness, which thus leads to the reduction in capacitance. Using the above relation, the value of capacitance (in per cent) as a function of ageing time 't' at any given temperature 'T' can be derived as:

$$C(t, T) = \frac{100}{\frac{D(t,T)}{100} + 1} \tag{2}$$

where $D(t, T)$ is effective increase in dielectric thickness (in per cent with respect to the initial value) due to evaporation of the electrolyte.

3 Experimental Validation

Based on our model, we planned experiments to first establish the nonlinear behaviour of capacitance with a change in concentration of electrolyte, i.e. change in mass of capacitor. There is no way to measure the reduction in volume of electrolyte; thus, we took mass of capacitor as a measuring parameter. As the electrolyte evaporates, the mass of capacitor will decrease. As proposed by Shukla et al. [8], in order to accelerate the evaporation rate of electrolyte, the capacitors were punctured. A set of eight punctured capacitors were subjected to thermal overstress in a thermal shock chamber at temperatures of 100 °C. The capacitance and mass of each capacitor were recorded at regular intervals. Figure 1 shows the variation of capacitance with mass of eight capacitors. The data fits well to a quadratic polynomial. Figure 2 shows the comparison of the fitted data against the observation on a test capacitor 5.

In another experiment, a set of capacitors was subjected to thermal overstress (temperature 95 °C) and variation of capacitor mass was recorded with time at regular intervals. Figure 3 shows the variation of capacitor mass with time. The mass of a capacitor reduces continuously with ageing under thermal overstress.

From the above results, it is observed that (i) capacitance varies nonlinearly with mass of the capacitor and (ii) mass of a capacitor reduces with ageing time under thermal overstress. Thus, it is evident that the capacitance will reduce with ageing time. As mentioned in Sect. 2.3, the reduction in mass of a capacitor can be related to an increase in hypothetical thickness 'd' of dielectric. With this background, thermal overstress experiments were carried out to obtain a degradation model. The experiments are described in the following sections.

Fig. 1 Variation of capacitance with mass

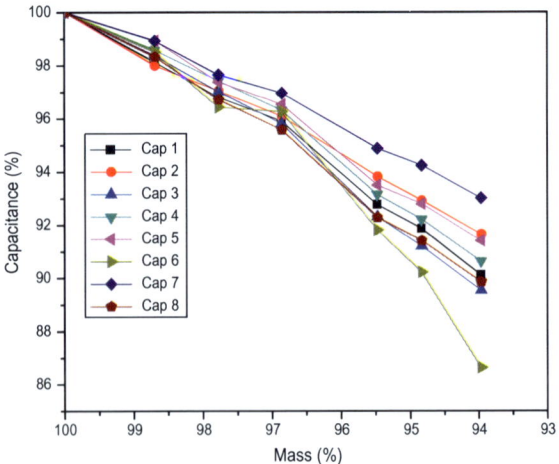

Fig. 2 Predicted and
observed values of
capacitance with mass

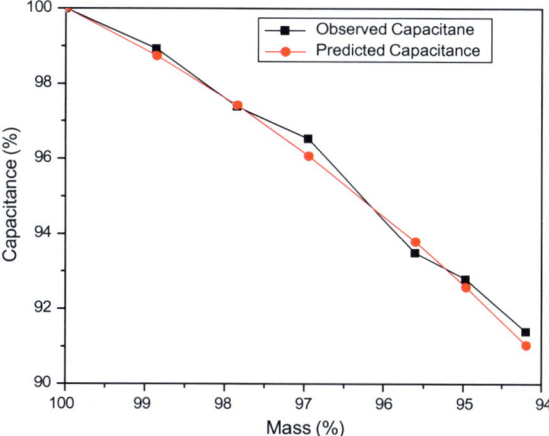

Fig. 3 Variation of capacitor
mass with time

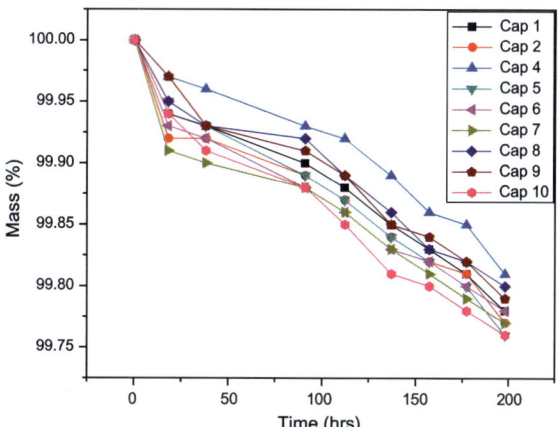

4 Thermal Overstress Experiments

We carried out accelerated ageing experiments on AECs by subjecting them to thermal overstress. The high thermal stress was applied by putting them in a thermal shock chamber whose temperature was raised slowly up to the desired level. Figure 4 shows the experimental set-up.

The AECs considered for the experiment were chosen randomly from a particular lot with rated capacitance of 2200 μF, rated voltage of 10 V and maximum operating temperature of 85 °C. The electrolytic capacitors under test were characterized (measurement of capacitance and ESR) in detail before the start of the experiment at an ambient temperature of 30 °C. Wayne Kerr 6500B Series Precision Impedance Analyser was used for the purpose. The 6500B series of Precision Impedance Analysers provide precise and fast testing of components at frequencies up to 120 MHz with

(a)   (b)

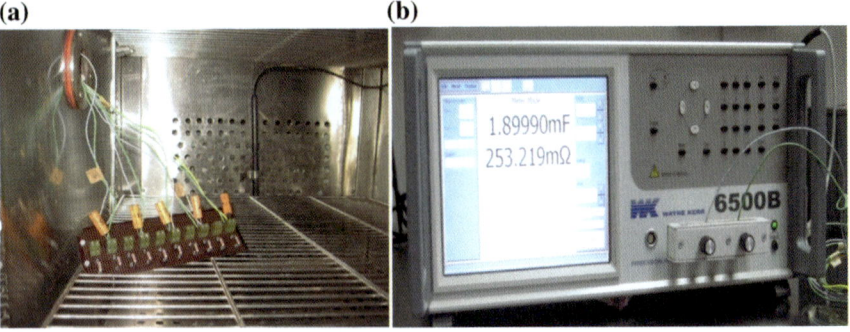

Fig. 4 Experimental set-up of thermal overstress (TOS)—**a** capacitors in thermal chamber, **b** measurements by precision impedance analyser

basic measurement accuracy of ±0.05%. Four sets of thermal overstress experiments were carried out at temperature levels of 85, 95, 105 and 115 °C. The temperature levels were chosen by optimizing temperature range (from near the operating conditions to 20–30% higher than rated value) to observe degradation within reasonably short time. The experiments were performed by subjecting the capacitors to the desired temperature level in a thermal chamber. In line with the work performed by Kulkarni et al. [2], the parameters (capacitance and ESR) were measured at regular intervals of few hours at ambient temperature. The temperature was raised and lowered slowly at a rate of 2 °C/min without causing any thermal shock to the capacitors.

4.1 Thermal Overstress Experiment—1

The first thermal overstress experiment was performed at 95 °C. The variation of C (% of initial value) of ten capacitors with ageing time is shown in Fig. 5. To obtain an ageing model, we have plotted the effective increase in thickness of dielectric (in per cent) with respect to ageing time, in Fig. 6, using the relation $C \alpha (1/d)$. The average behaviour of all the capacitors except capacitor 5 is used for obtaining an ageing model. The fifth capacitor is considered as a sample case to validate the model. Based on the studies presented in Sect. 1, the experimental data points are fitted to a nonlinear polynomial. The following model fits well to the data:

$$D(t) = a(1 - \exp(-t/b))$$ (3)

where $a = 5.37748$, $b = 102.79409 \pm 9.53663$.

Using the above model, variation of D with ageing time is predicted for the fifth capacitor.

Figure 7 shows a comparison of the predicted value of capacitance using Eq. 3 and the measured values obtained from the experiment for capacitor 5. It is seen from

Fig. 5 Variation of C with
ageing time

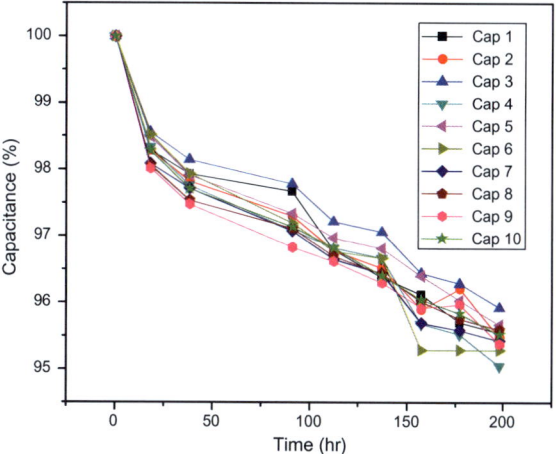

Fig. 6 Variation of D with
ageing time

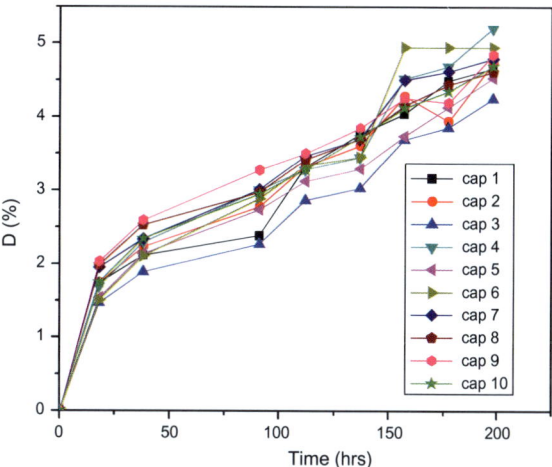

the figure that the model predicted and experimental values are in good agreement
barring a few points.

4.2 *Thermal Overstress Experiment—2*

The second thermal overstress experiment was performed at 105 °C. The variation
of C (% of initial value) with ageing time is shown in Fig. 8. The plot of the effective
increase in thickness of dielectric (in per cent) with respect to ageing time is shown
in Fig. 9. The average behaviour of the four capacitors (no. 1, 2, 4, 5) is used for
obtaining an ageing model. The third capacitor is considered as a sample case to

Fig. 7 Comparison of the model predicted and measured values

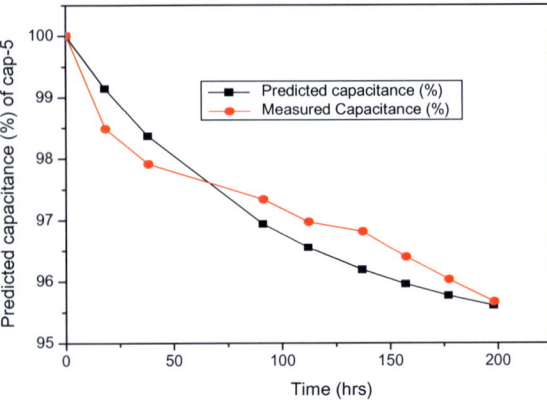

Fig. 8 Variation of C with ageing time

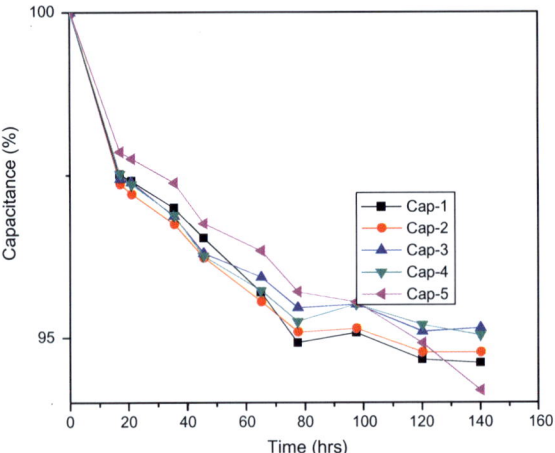

validate the ageing models. The experimental data points are fitted to a nonlinear polynomial. The same model of the experiment—1 (Eq. 3) fits well to the data. The model constants are:

$$a = 5.55519, b = 37.86181 \pm 2.05299$$

Using the model of Eqs. 2 and 3 with the above model constants, the values of C with ageing time are predicted for the third capacitor.

Figure 10 shows a comparison of the predicted value of capacitance using Eq. 3 and the measured values obtained from the experiment for capacitor 3. It is seen from the figure that the model predicted and experimental values are in good agreement barring a few points.

Fig. 9 Variation of D with ageing time

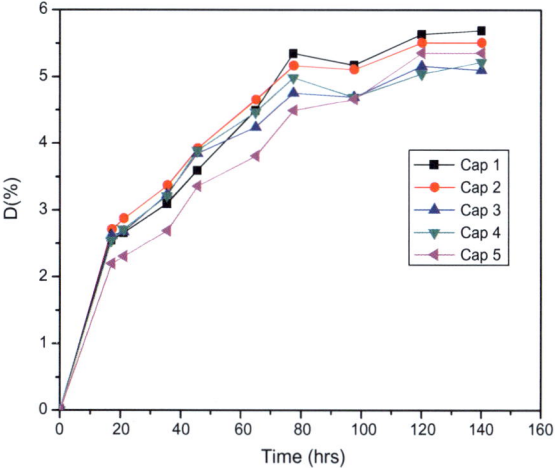

Fig. 10 Comparison of the model predicted and measured values

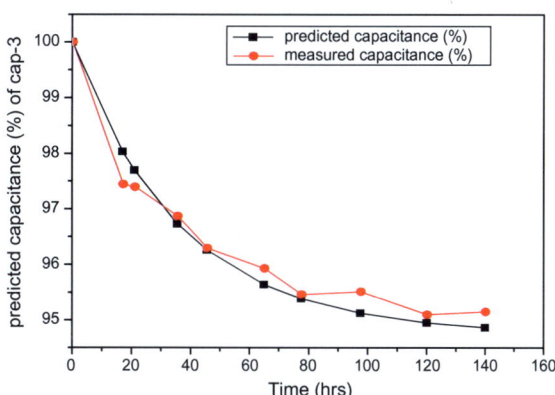

4.3 Thermal Overstress Experiment—3

The third thermal overstress experiment was performed at 115 °C. The variation of C (% of initial value) with ageing time is shown in Fig. 11. The plot of the effective increase in thickness of dielectric (in per cent) with respect to ageing time is shown in Fig. 12. The average behaviour of the first three capacitors is used for obtaining an ageing model. The fifth capacitor is considered as a sample case to validate the ageing models. Again, the experimental data points fit well to the model given in Eq. 3. The model constants are:

$$a = 7.05896, b = 26.83738 \pm 2.92788$$

Using the model of Eqs. 2 and 3 with the above model constants, the values of C with ageing time are predicted for the third capacitor.

Fig. 11 Variation of C with ageing time

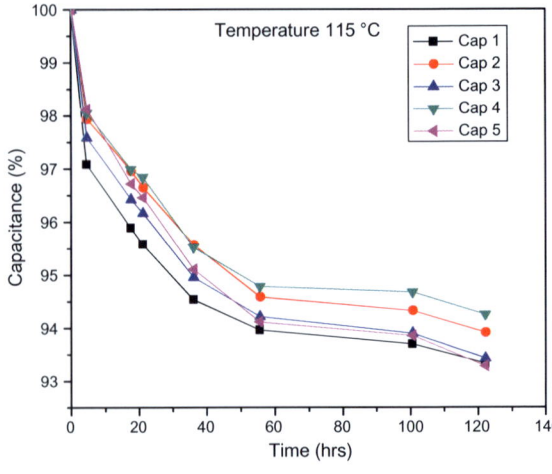

Fig. 12 Variation of D with ageing time

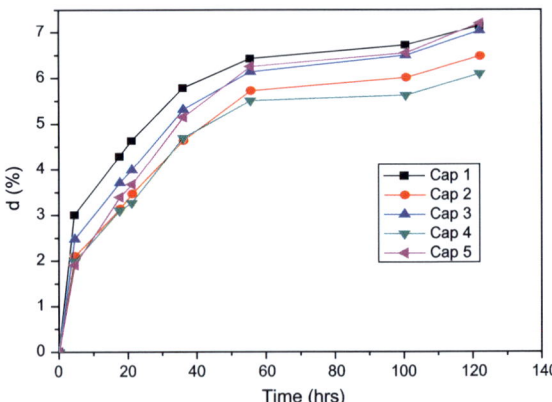

Figure 13 shows a comparison of the predicted value of capacitance using Eq. 3 and the measured values obtained from the experiment for capacitor 3. It is seen from the figure that the model predicted and experimental values are in good agreement barring a few points.

4.4 Thermal Overstress Experiment—4

The fourth thermal overstress experiment was performed at 85 °C. The variation of C (% of initial value) with ageing time is shown in Fig. 14. The plot of the relative thickness of dielectric with respect to ageing time is shown in Fig. 15.

The average behaviour of the three capacitors (no. 1, 3, 4) is used for obtaining an ageing model. The third capacitor is considered as a sample case to validate the

Fig. 13 Comparison of the model predicted and measured values

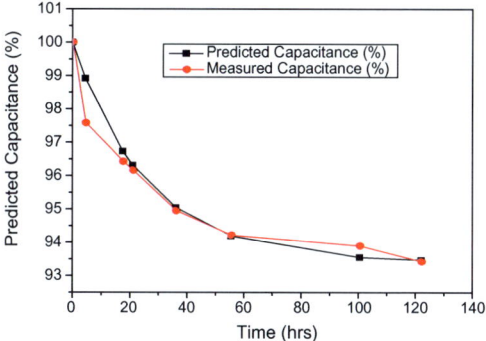

Fig. 14 Variation of *C* with ageing time

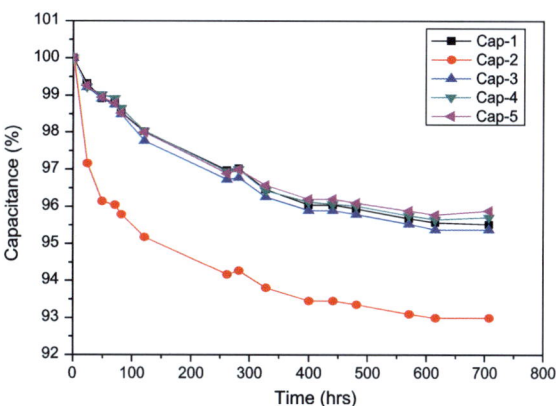

Fig. 15 Variation of *D* with ageing time

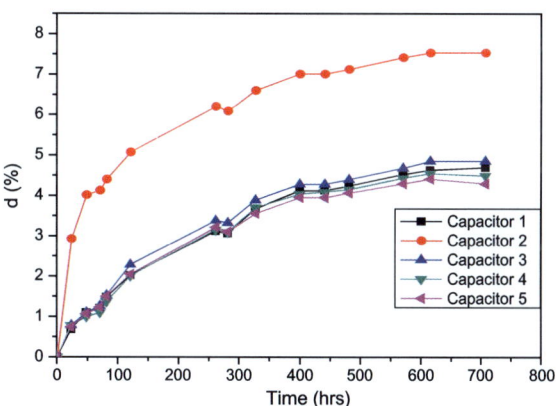

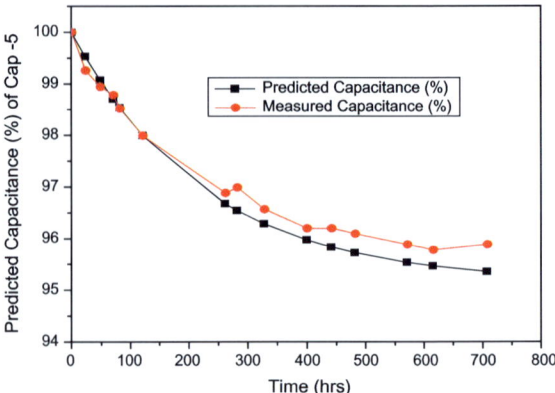

Fig. 16 Comparison of the model predicted and measured values

ageing models. The experimental data points are fitted to the model given in Eq. 3, with the model constants being:

$$a = 4.85527, b = 226.07829 \pm 6.4807$$

Using the model of Eqs. 2 and 3 with the above model constants, the values of C with ageing time are predicted for the fifth capacitor.

Figure 16 shows a comparison of the predicted value of capacitance using Eq. 3 and the measured values obtained from the experiment for capacitor 5. It is seen from the figure that the model predicted and experimental values are in good agreement barring a few points.

5   The Ageing Model

In order to obtain an ageing model for capacitors under any given level of thermal overstress condition, it is required to have functional from of the model constants with temperature, i.e.

$$d(t, T) = a(T)\big[1 - \exp\{-t/b(T)\}\big] \tag{4}$$

The model constants estimated for the three experimental results are given in Table 1.

The experimental data points result in the following functional forms of the model constants.

$$a(T) = 23.15521 - 0.42289 \times T + 0.00245 \times T^2 \tag{5}$$

$$b(T) = 2 * 102312.11897(\exp(-T/12.42556)) + 7.271 \tag{6}$$

Table 1 Model constants at different temperatures

Model constant	Value			
	$T = 85\,°C$	$T = 95\,°C$	$T = 105\,°C$	$T = 115\,°C$
a	4.85527	5.37748	5.55519	7.05896
b	226.07829	102.79409	37.86181	26.83738

6 Summary and Conclusions

Accelerated ageing experiments have been performed on aluminium electrolytic capacitors under thermal overstress conditions. Capacitance is considered as the failure precursor. Based on the observations, an ageing model has been developed which predicts capacitance with ageing time under any given thermal stress condition. The model is nonlinear in time.

Acknowledgements The authors are thankful to Instrument Maintenance Section (NI), Research Reactor Maintenance Division, Reactor Group, for their help in preparing the test circuit and providing electronic instruments as and when required. We thank Shri Parag Punekar, Research Reactor Maintenance Division, Reactor Group, for his help in arranging the required instruments as and when required. We also thank to Shri P. Sumanth, Head, Research Reactor Maintenance Division, Reactor Group, for his support. Thanks are also due to Shri Hari Balakrishna and Shri Ankit Agarwal of Reactor Control Division for their help through technical discussions on the subject from time to time.

References

1. M. Pecht, J. Gu, Physics-of-failure-based prognostics for electronic products. Trans. Inst. Measur. Control **31**(3/4), 309–322 (2009)
2. C.S. Kulkarni et al., Physics based electrolytic capacitor degradation models for prognostic studies under thermal overstress, in *European Conference of the Prognostics and Health Management Society* (2012)
3. C.S. Kulkarni et al., Physics of failure models for capacitor degradation in DC-DC converters. http://www.isis.vanderbilt.edu/sites/default/files/KulkarniBiswasKoutsoukos_GoebelCelaya_MARCON2010.pdf
4. C.S. Kulkarni et al., Physics based modeling and prognostics of electrolytic capacitors. http://www.isis.vanderbilt.edu/sites/default/files/AIAA20AIAA2012_Final_Kulkarni_Celaya_Biswas_Goebel1.pdf
5. J.R. Celaya et al., A model-based prognostics methodology for electrolytic capacitors based on electrical overstress accelerated ageing, in *Annual Conference of the Prognostics and Health Management Society* (2011)
6. H. Darnand, P. Venet, G. Grellet, Detection of faults of filter capacitors in a converter application to predictive maintenance, in *INTELEC Proceeding,* vol. 2, pp. 229–234 (1993)
7. H. Wang et al., Transitioning to physics-of-failure as a reliability driver in power electronics. IEEE J. Emerg. Sel. Top. Power electron. **2**(1) (March 2014)
8. R. Shukla et al., Accelerated ageing of aluminium electrolytic capacitor. https://www.ee.iitb.ac.in/npec/Papers/Program/NPEC_2015_paper_55.pdf

Analytical Modelling of Distributed File Systems (GlusterFS and CephFS)

Akshita Parekh, Urvashi Karnani Gaur and Vipul Garg

Abstract With the exponential increase in digital data worldwide, the need to provide efficient, reliable and scalable storage solution has become a significant concern. Also, there has been exceptional growth in network-based computing recently and client-/server-based applications have brought revolutions in this area. An efficient solution is required for enhancing the storage capacity of existing systems without migration of server software and with minimum or zero downtime. The industries like software services, health care, education, research, etc., require a reliable storage infrastructure that relies on distributed environment to store and process large amounts of data. With the advent of various distributed file systems (DFSs), the focus is on a solution over off-the-shelf commodity servers which provide ease of scalability and reliability. In this paper, we will discuss analytical modelling of two major DFSs—GlusterFS and CephFS—which are being used in the industry.

Keywords Availability · CephFS · GlusterFS · Petri nets

1 Introduction

A distributed file system (DFS) is a file system that spans over distributed clustered nodes and is transparent to the user or client application. The data is accessed and processed as if it is stored on the local client machine [1]. DFS can span over off-the-shelf commodity servers and can be given as a storage solution for various services like cloud computing, email, database, etc. Various file system benchmarks

A. Parekh (✉)
Homi Bhabha National Institute, NFC, Hyderabad, India
e-mail: akshita@nfc.gov.in

U. K. Gaur · V. Garg
Bhabha Atomic Research Centre, Mumbai, India
e-mail: urvashi@barc.gov.in

V. Garg
e-mail: vipulgarg@barc.gov.in

© Springer Nature Singapore Pte Ltd. 2020
P. V. Varde et al. (eds.), *Reliability, Safety and Hazard Assessment for Risk-Based Technologies*, Lecture Notes in Mechanical Engineering,
https://doi.org/10.1007/978-981-13-9008-1_18

are available to test the performance (throughput and input–output per second) of DFS, but to run them, file systems have to be deployed on commercial hardware or laboratory set-up requiring actual physical hardware. Hence, studying behaviour of DFS and analysing its scalability and availability using generalized stochastic Petri nets (GSPNs) will provide a better insight to the service provider without actual deployment of the hardware and DFS [2, 3].

In our research work, we have selected two open-source DFSs—GlusterFS and CephFS—and compared them based on their behaviour and availability. GlusterFS is a scalable distributed file system, which can be scaled up to petabytes of storage. GlusterFS aggregates disc storage resources from multiple servers into a single global namespace [4]. Ceph uniquely delivers object, block and file storage in one unified system. CephFS is an object-based file system that treats object storage as its foundation and provides block and file system capabilities as layers built upon that [5]. GlusterFS and CephFS were studied in depth, and their behaviours were modelled using Unified Modelling Language (UML) sequence diagrams. Time equations derived from sequence diagrams help in predicting the performance of the file systems without running actual file system benchmarks. Availability of GlusterFS and CephFS was modelled using GSPN, and simulations were carried using Pipe tool.

2 Modelling

To model the behaviour of GlusterFS and CephFS, both the DFS were studied. After in-depth survey and study of the working paradigms, sequence diagrams were used to model their behaviour and the control structure of the systems [6]. Figures 1 and 2 illustrate the architecture of GlusterFS and CephFS, respectively [4].

2.1 GlusterFS Behavioural Modelling for Read and Write Operations

GlusterFS volumes use directories as bricks to store data which are mentioned at the time of creating volumes [7]. The sequence diagrams for read and write file operations over GlusterFS are shown in Figs. 3 and 4, respectively. For each file operation, the GlusterFS Virtual File System (VFS) request is forwarded to Filesystem in Userspace (FUSE) kernel module. The GlusterFS translators perform their respective function like mapping the request to the correct brick containing data, handling replication, etc. [4]. In case of write operation, the data is written to all the replicated bricks of the volume, and then, an acknowledgement is sent from the server to the application. GlusterFS replication factor of two is used for the following sequence diagrams.

$$T_{\text{read}} = T_{\text{appn\_read}} + T_{\text{fuse}} + T_{\text{trans\_client}} + T_{\text{trans\_server}} + T_{\text{readblock}} \qquad (1)$$

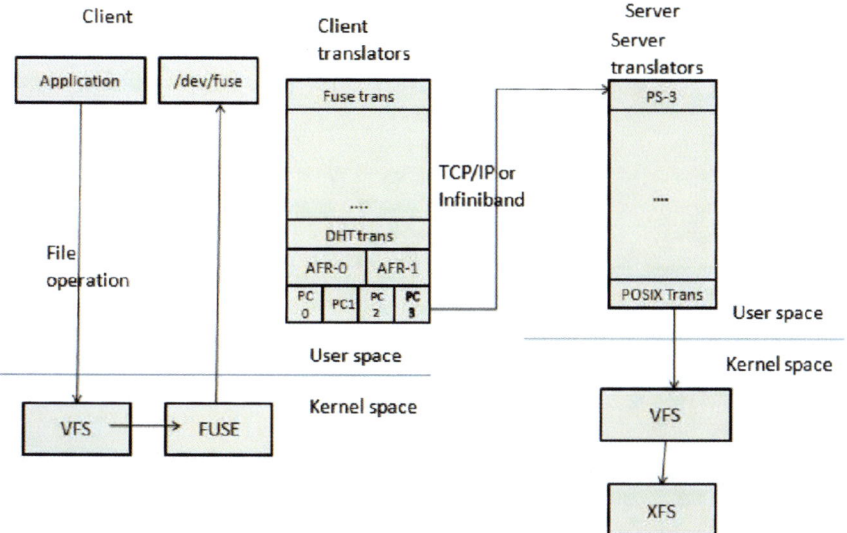

Fig. 1 Architecture of GlusterFS replicated volume

Fig. 2 Ceph architecture

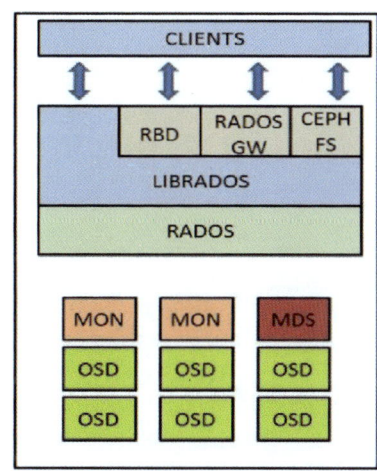

$$T_{\text{write}} = T_{\text{appn\_write}} + T_{\text{fuse}} + T_{\text{trans\_client}} + T_{\text{trans\_server}} + M * T_{\text{write}} \qquad (2)$$

where

M	Number of replicated servers,
$T_{\text{appn\_read}}$	Processing time at application level,
T_{fuse}	Time taken by kernel or system calls (VFS-FUSE-GlusterFS module),

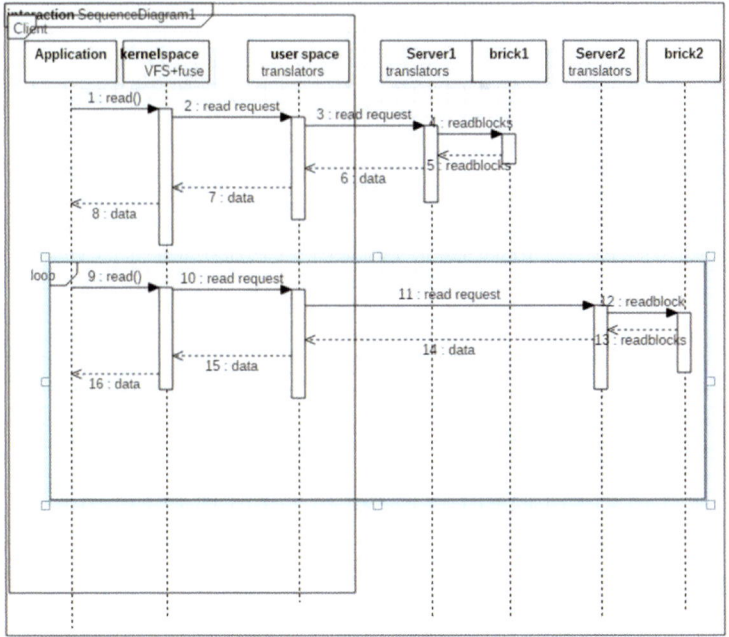

Fig. 3 GlusterFS read operation

$T_{\text{trans\_client}}$ Time taken by the translators in user space at client side to perform their functions,

$T_{\text{trans\_server}}$ Total time taken by the translators in user space at server side to perform their functions,

$T_{\text{readblock}}$ Time to read a single block from the file system,

$T_{\text{appn\_write}}$ Processing time at application level,

T_{write} Time to write a block of data by the file system.

2.2 CephFS Behavioural Modelling for Read and Write Operation

Figure 5 shows the behaviour of CephFS consisting of two object storage daemons (OSDs)—one monitor node, one Metadata Server (MDS) node and one admin node. The cluster name and a monitor address of Ceph storage cluster were provided at the client side. All file operations in CephFS include establishing of connection of client with the monitor that syncs recent copy of the cluster map containing information about OSDs and storage pools. To read/write data, an I/O context to a specific pool is

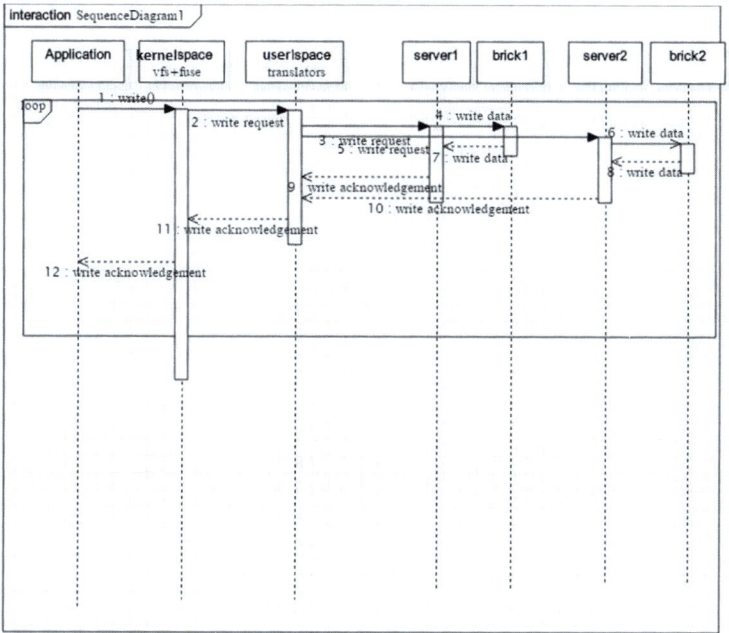

Fig. 4 GlusterFS write operation

created in Ceph storage cluster. The data is read or written to pool if the permissions exist. Ceph OSDs use Controlled Replication Under Scalable Hashing (CRUSH) algorithm to compute the location to store replicas of objects [5]. The Ids of the secondary OSDs (which are equal to the number of replicas defined) are computed by primary OSD for the acting set. The data is written to secondary OSDs by primary OSD. After the acknowledgement from secondary OSDs and its own completion of write operation, an acknowledgement is sent to the Ceph client by the primary OSD.

$$T_{\text{write}} = T_{\text{appn\_read}} + T_{\text{crush\_map}} + M * T_{\text{write\_osd}} \tag{3}$$

$$T_{\text{read}} = T_{\text{appn\_read}} + T_{\text{crush\_map}} + T_{\text{read\_block}} \tag{4}$$

where

M	Number of replicated OSDs,
$T_{\text{appn\_read}}$	Processing time at application level,
$T_{\text{crush\_map}}$	Time taken by the CRUSH map algorithm to compute object location,
$T_{\text{read\_block}}$	Time taken to read a block of data from an OSD,
$T_{\text{write\_osd}}$	Time taken to write a block of data on an OSD.

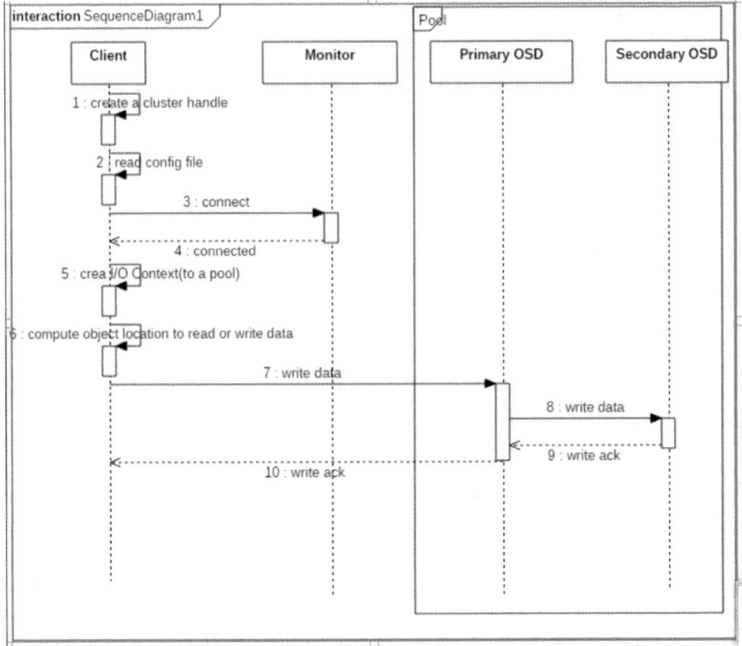

Fig. 5 CephFS read/write operation

2.3 Validation of Behaviour Model of GlusterFS and CephFS

To validate the time equations derived from UML sequence diagrams, we deployed both the file systems on a cluster of five commodity servers having x86 architecture, 3.3 GHz CPU frequency and 8 GB RAM connected over 1 Gbps network. T_{write} for GlusterFS and CephFS was found to be 86.31 μs and 1.58 ms, respectively.

3 Modelling Availability of File Systems using Petri Nets

Availability is the duration of uptime for operations and is a measure of how often the system is up and healthy [8]. It is basically a probability that a given system is operating appropriately when it is requested for use. It is a function of both reliability (probability that a system will not fail) and maintainability (probability that a system is successfully restored after failure). It is often measured as:

$$Availability = (uptime)/(uptime + downtime) \tag{5}$$

$$= MTBF/(MTBF + MTTR) \tag{6}$$

where MTBF is mean time between failure and MTTR is mean time to repair.

In the following Petri nets, we have assumed that the connections in the network topology are perfect. Failure and repair rates of server are assumed to follow exponential distribution.

3.1 Petri Net for GlusterFS with Replication Factor 2

$$\text{Downtime} = P[\#P4 > 0] \tag{7}$$

$$P = \{P1,\ P2,\ P3,\ P4\}$$

$$T = \{T1,\ T2,\ T3,\ T4,\ T5,\ T6\}$$

Initial State: M_o ($P1 = 2$) Failure states $= P2, P3$
$T1$: Input $P1$, Output $P2$ (Hardware Failure of Server)
$T2$: Input $P2$, Output $P1$ (Repair)
$T3$: Input $P1$, Output $P3$ (Software Failure of Server)
$T4$: Input $P3$, Output $P1$ (Repair).

Figure 6 represents the modelling of availability of GlusterFS using GSPN using Pipe tool [9]. From the above figure, it is evident that the main components of the file system are m clients, n identical servers and network module (collection of switch, router, communication media, etc.). Here, $m = 1$ and $n = 2$. Disc/network/server failure can be categorized as hardware failure. Software failure can be like OS failure or file system corruption. The rates of transitions $T1$ and $T3$ are defined using minimum of mean time to fail (MTTF). The mean time to repair (MTTR) of the hardware component and that of software component are shown by the transition $T2$ and $T4$. GlusterFS is available even if one of the storage servers is up. No token in $P1$ indicates unavailability of all the servers. This duration is equal to the duration for which there are tokens in $P4$. So, duration of tokens in $P4$ is the time or which the system is unavailable. The model is scalable as increased number of servers can be easily modelled with the help of tokens in place $P1$.

3.2 Petri Net for CephFS with three OSDs, Replication Factor 2

$$\text{Downtime} = P[\#P4 > 0] + P[\#P5 > 0] + P[\#P8 > 0]$$
$$+ P[\#P9 > 0] + P[\#P11 > 0] \tag{8}$$

$$P = \{P1,\ P2,\ P3,\ P4,\ P5,\ P6,\ P7,\ P8,\ P9,\ P11\}$$

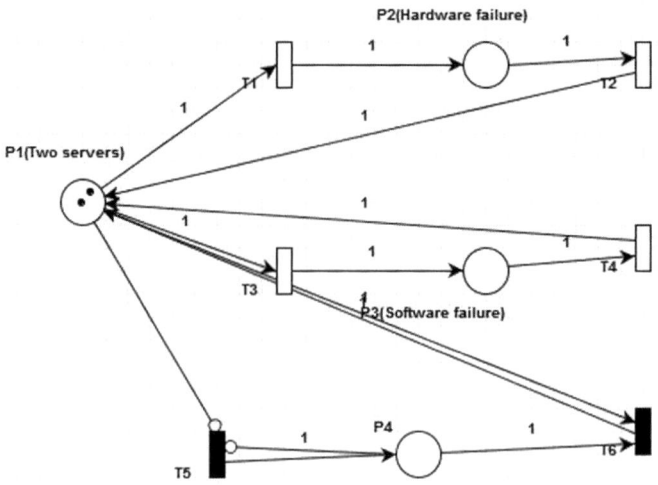

Fig. 6 Modelling of GlusterFS using Petri net

$$T = \{T1, T2, T3, T4, T5, T6, T7, T8, T9, T10, T11, T12, T13, T14\}$$

Initial State: M_o $(P1 = 1, P2 = 3, P3 = 1)$
Failure states $= 4, P5, P6, P7, P8, P9$
$T1$: Input $P1$, Output $P4$ (Hardware Failure of Monitor)
$T2$: Input $P4$, Output $P1$ (Repair)
$T3$: Input $P1$, Output $P5$ (Software Failure of Monitor)
$T4$: Input $P5$, Output $P1$ (Repair).

Similarly, hardware and software failures for other components are depicted using remaining transitions.

Figure 7 represents the modelling of availability of CephFS using GSPN using Pipe tool. From the above figure, it is evident that the main components of the file system are l identical monitors, m identical OSDs and n identical MDS (Metadata Servers) and network module (collection of switch, router, communication media, etc.). Here, $l = 1$, $m = 3$ and $n = 1$. The read/write requests by the client are made on the OSDs. Transitions $T1, T3, T5, T7, T9$ and $T11$, are timed according to mean time to failure of the respective components. Transitions $T2, T4, T6, T8, T10$ and $T12$ are timed according to mean time to repair of the respective components. Here, k out of n logic is used to implement logic of availability of OSDs (with k being 2 and n, 3). Transition $T13$ will fire if there are less than k tokens in $P2$. The Petri nets drawn above are scalable in nature, and as the number of servers is represented by tokens, it can be increased with more number of tokens.

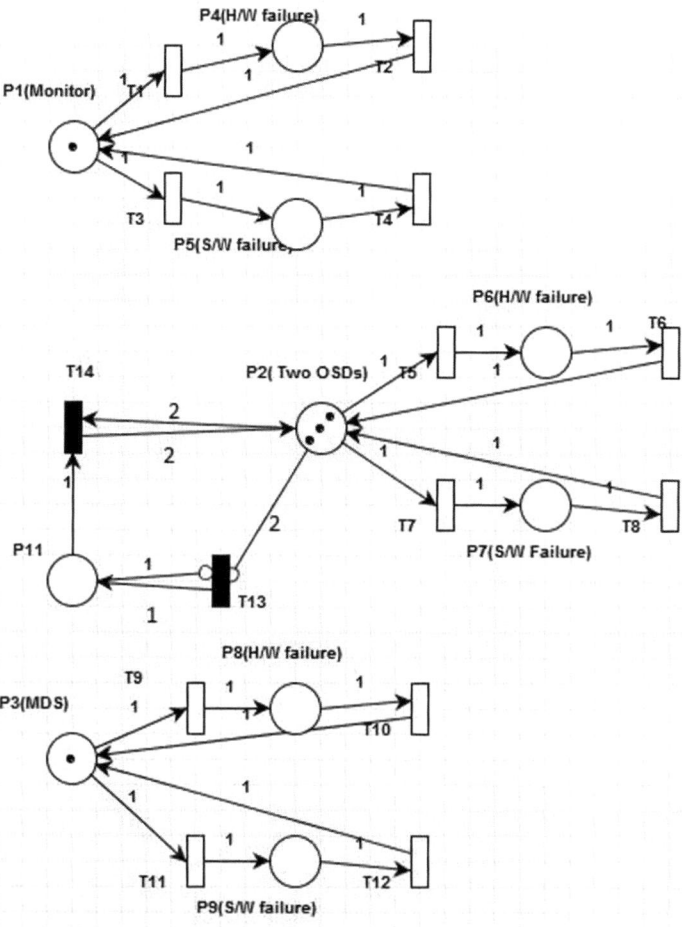

Fig. 7 Modelling of CephFS using Petri net

3.3 Analysis of Availability

The simulation for Petri net for both the file systems was done. After consulting an expert, the approximate failure rates for file system and OS (software) is taken as once in 7 months, while disc and network (hardware) failure rate as once in a year. Repair rate for hardware is taken as 48 h and software as 24 h, for both the file systems. This can be changed according to production infrastructure, and simulations can be carried out on the modelled Petri nets (Table 1).

Table 1 Comparison of availability of GlusterFS and CephFS in %

Number of servers	GlusterFS	CephFS Considering (1 MDS, n OSDs, 1 MON)	Considering availability of only OSDs
2 servers	99.35	81.17	99.29
3 servers	99.95	81.18	99.22
5 servers	99.99	81.92	99.95

4   Conclusion

A lot of research work is done by comparing performance of various DFSs by running benchmarks like Iozone, specs, etc. In our work, we have tried to construct behavioural models using sequence diagrams for GlusterFS and CephFS. Model-driven performance analysis is an important technique to analyse any system's performance. So, the internals of the distributed file systems were studied in depth and file operations were modelled. The performance modelling is dependent on various internal parameters like $T_{\mathrm{appn\_read}}$, T_{crushmap} and $T_{\mathrm{translators}}$ that in turn depend on the server's hardware specification like CPU frequency, IPC, etc. GSPN models were designed, and simulations were run using Pipe tool to analyse availability of GlusterFS and CephFS with respect to increase in number of servers. It was observed that, with same number of servers and replication factor, the availability of GlusterFS was better than that of CephFS.

References

1. A. Khazanchi, A. Kanwar, L. Saluja, *An Overview of Distributed File System*
2. T. Murata, Petri nets: properties, analysis and applications. Proc. IEEE **77**(4), 541–580 (1989)
3. M. Ajmone Marsan, Stochastic petri nets: an elementary introduction, *in European Workshop on Applications and Theory in Petri Nets* (Springer, Berlin, 1988), pp. 1–29
4. http://docs.gluster.org/en/latest/
5. http://docs.ceph.com/docs/master/architecture/
6. Y. Wu, F. Ye, K. Chen, W. Zheng, *Modelling of Distributed Filesystems for Practical Performance Analysis*
7. M. Selvaganesan, M.A. Liazudeen, An insight about GlusterFS and its enforcement techniques, *in 2016 International Conference on Cloud Computing Research and Innovations*
8. H. Paul, *Availability, Reliability, Maintainability, and Capability* (Barringer, P.E)
9. PIPE2: Platform Independent Petri Net Editor by Nadeem Akharware, MIEE

Vulnerability Assessment of Authorization System for USB-Based Storage Devices

Ajay Kumar, C. S. Sajeesh, B. Vinod Kumar, Vineet Sharma, Gigi Joseph and Gopika Vinod

Abstract USB-based storage devices such as flash drives and external hard disks are very popular for data transfer because of their high speed and portability. But they also give rise to information security threats like data loss and spread of malware. In this paper, we present an authorization system to allow only authorized devices to be accessed on a computer for reducing the chances of data loss via USB-based removable storage media. The unique identification of devices is based on device parameters which are part of the USB standard. These parameters have been used for authorizing USB-based storage devices connected to a computer. A vulnerability assessment of the authorization system as well as device authorization parameters has been carried out.

Keywords Device authorization system · Device identification · Unauthorized data loss · USB-based storage devices · Vulnerability assessment

A. Kumar (✉) · C. S. Sajeesh · B. Vinod Kumar · V. Sharma · G. Joseph · G. Vinod
Bhabha Atomic Research Centre, Mumbai, India
e-mail: gargajay@barc.gov.in

C. S. Sajeesh
e-mail: sajeesh@barc.gov.in

B. Vinod Kumar
e-mail: bvinod@barc.gov.in

V. Sharma
e-mail: vineets@barc.gov.in

G. Joseph
e-mail: gigi@barc.gov.in

G. Vinod
e-mail: vgopika@barc.gov.in

© Springer Nature Singapore Pte Ltd. 2020
P. V. Varde et al. (eds.), *Reliability, Safety and Hazard Assessment for Risk-Based Technologies*, Lecture Notes in Mechanical Engineering,
https://doi.org/10.1007/978-981-13-9008-1_19

223

1 Introduction

Universal Serial Bus (USB)-based storage devices like pendrives and external hard disks are the popular form of storage available for the transfer of data. There are various reasons for this popularity like small physical size, high-speed data transfer, large data storage, low prices, and plug and play. Unfortunately, this portability, convenience, and popularity also pose threats to an organization's information like data theft, data loss, unauthorized access to data, and spread of malware and viruses. These further aggravate the information security risks whose mitigation has become one of the most pressing challenges for all kind of organizations today. Therefore, it is imperative to find a way of using USB-based storage devices so that its impact on security is minimal.

As part of this work, we have developed an authorization system which is used to authorize the access of USB storage devices as per the authorization policy defined by an organization based on its security requirements. These requirements may vary from very stringent (no USB storage device allowed) policies to flexible policy (allow only selected list of USB storage devices) to a relaxed policy (all USB storage devices allowed). The proposed system caters to all these requirements.

The following sections explain about various popular USB device authorization systems, identification of USB storage device, framing of authorization policy, and vulnerability assessment of authorization system.

2 Authorization System

Authorization system performs the identification of USB devices plugged on an end-system and enforces the application of authorization policy on it. Authorization system uniquely identifies all USB storage devices by collecting device authorization parameters. When a device is plugged into the system, device's parameters are matched with the white list of authorized USB devices. If they match the device can be accessed, otherwise it gets blocked and considered as unauthorized. Unauthorized devices are blocked by unloading the corresponding device driver which renders the device unusable from the point of view of file-system access. Devices remain in blocked state until they are added to the white list. After addition in white list, the blocked devices are unblocked. The updated white list is retrieved from the server on a regular basis. The system ensures that only USB storage devices which are included in the policy white list are authorized to work and other USB devices are blocked.

2.1 USB Device Authorization Techniques

Two of the popular approaches for USB device authorization systems are described here:

Password-based—Password-based authorization systems are available in the market and provided by vendors at the commercial level, e.g., USB Block and USB lock [1]. In such systems; password needs to be set which is further used for accessing the USB storage device. If the password entered is correct, it will unblock the USB storage device else it remains blocked. In this way, a USB storage device is authorized. The major disadvantage of password-based authorization system is that the strength of the model depends on the password strength. Passwords may be guessed, shared or stolen easily.

Trusted device list—Such systems keep a trusted device list which is used to authorize the plugged USB storage device. This system allows a user to add a particular USB device to the trusted device list so that whenever this device is plugged on the system it will work. Disadvantage of such systems is that local device lists can be manipulated easily.

We have developed a new authorization system which installs an agent application on end-system and fetches authorization policy from a centralized server. It is a user-mode application authorizing plugged USB storage device as per the policy. The advantage is that devices need to be registered only once but they can be used across multiple systems seamlessly. The USB device owner has full control to decide where his device may be used. At the same time, a PC owner also has full control to decide which devices are allowed to work to his/her PC.

The agent application is responsible for USB authorization policies enforcement and collection of data. It works on Windows as well as Linux platform. The agent identifies device parameters from USB-based storage devices (e.g., pendrives and external hard disks) connected to an end-system. Information from other USB-based devices like printers, scanners, webcams, and smartphones was ignored for the current work. The dataset of USB-based storage device's parameters collected by the agent is further used for the vulnerability assessment of USB authorization parameters and the updation of organization policy.

2.2 USB Storage Device Identification

USB devices are identified using device type, vendor ID, product ID, and serial number. For USB-based storage devices, the device type is DISK. Vendor ID is specific to a manufacturer and must be purchased from USB Implementers Forum (USB-IF) by the manufacturer. Product ID is assigned by the vendor to distinguish various product lines (e.g., 1 or 2 GB variants of pendrives). The serial number is used by vendors to distinguish between multiple devices of the same variety. As per the USB standard, it is recommended that these parameters taken together should uniquely distinguish a device. It is also recommended that serial number should have at least 12 Hexadecimal characters. It is observed that these recommendations may or may not be honored by the device manufacturers. Identification parameters and their size are shown in Table 1.

2.3 Authorization Policy

Authorization policy is framed to facilitate the authorization of USB storage devices on end-system. Policy contains white list of USB storage devices which are registered. An organization may also choose to allow devices of a particular make (e.g., all Sony devices) or a particular model to work on the end-systems irrespective of matching of other parameters and without the need of registration. A sample policy is shown in Table 2. On the agent, USB device is authorized after matching its authorization parameters with the policy. Once the USB storage device is registered, it does not imply that it will be accessible in every user's system. Users can authorize the use of USB devices on their systems, i.e., users can define their own policies which are categorized into three levels as shown in Fig. 1.

(i) Global Policies—It define the accessibility of registered USB device in owner's system as well as other user's system. If a user registers USB device in global pool, he/she is able to use that USB device on all end-systems. This facilitates a user to roam around and access data from any system using globally registered USB device. It may not be acceptable to other users that devices registered by other users will work on their system. To avoid this, end-system owner can either disable or selectively enable globally registered devices for use on his/her system.

Table 1 USB storage device identifiers

Identification parameters	Size
Device type	2 hex digits
Vendor ID	4 hex digits
Product ID	4 hex digits
Serial number	12 hex digits

Table 2 Authorization policy sample

Parameter	Operator	Value	Allow	Block
Device type	Matches exact	DISK	Yes	No
Serial number	Matches exact	Agent authorization policy	Yes	No
Vendor ID	Matches exact	Agent authorization policy	Yes	No
Product ID	Matches exact	Agent authorization policy	Yes	No

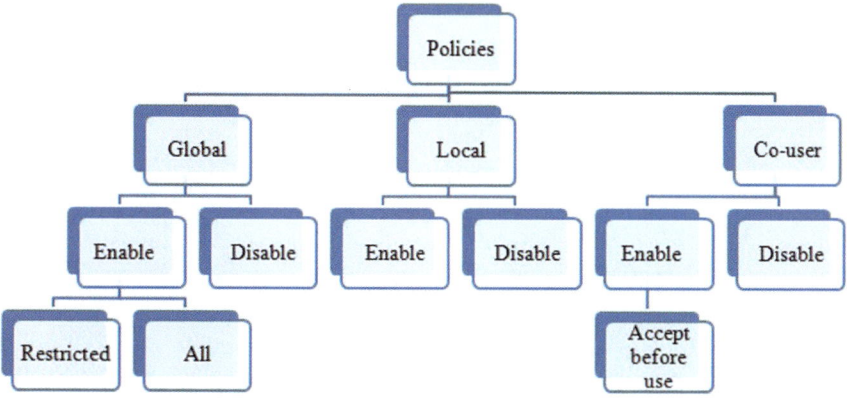

Fig. 1 User control over use of USB storage devices authorization

(ii) Local Policies—In local policies, device is registered to be used only on the owner's system. User can delete such devices also in order to stop working of devices in their own systems.

(iii) Co-user Policies—In co-user policies, the owner of USB device can send a co-user to add request to other users (e.g., his colleagues), to allow his local device to work on their system. But devices will work on other's system only if the co-user request is accepted.

3 Vulnerability Assessment

This section discusses the vulnerability present in the agent-based system for USB device authorization. Authorization system identifies a plugged USB device based on its authorization parameters. After identification, the authorization policy is applied on the device and it is enabled or disabled based on the policy. If a device publishes incorrect data for any reason, then it leaves the whole system in a vulnerable state. E.g., if devices are authorized based on the matching of serial number and two devices are having identical serial number and one of them is registered, then both will be enabled irrespective of the registration of the second one. We need to identify such cases present in the system. Furthermore, if any of the parameters does not follow the standard, it may lead to vulnerabilities in the system. Vulnerable USB storage devices may be considered as unauthorized devices.

The system was deployed in a network of around 3000 end-systems. Agent system identified a total of 7030 USB storage devices out of which 3618 were registered and 3412 were unregistered. Data obtained from the agent system has been analyzed using various metrics. The vulnerability assessment is done at two levels—first at authorization parameter level and another one at authorization system level. For the parameter level, two metrics are proposed: measure of security of device registration and ease of user for device registration. Similarly, system level data was analyzed using classification and information retrieval measures. These measures could provide indication of the vulnerability in the proposed system, which are discussed below.

It is to be noted here that the vulnerability assessment as part of this study limits itself to the study of device parameters only. There are some use cases which allow an unauthorized device to be accessed on the end-system, e.g., non-installation of agent application on the end-system, installation of custom device drivers on end-system to hide the device class, and forced manipulation of authorization policy. As part of the current work, assessment of vulnerabilities pertaining to such cases has not been done.

3.1 Authorization Parameters Level

As per USB-IF (Implementers Forum) guidelines, authorization parameters should follow standards but it does not attempt to enforce and police the implementation of proprietary serial numbers of all product lines of all manufacturers. In the absence of enforcement of standard, it has been observed that some of the manufacturers do not follow the standards and leave authorization parameters in vulnerable states. Based on the dataset, the USB authorization parameters may be present in the following states:

1. **Serial Number**—Matching characters in serial number are different from total number of characters and counted after left and right trimming. E.g., 0000000000W0000000, left trim of 0000000000W0000000 gives W0000000 and right trim of W0000000 gives W, so the final serial number is W. In this case, total number of characters is 18 while matching character is only 1. Like this, there are various states of serial number which are given as follows:

   - **Case 1** Total number of character ≥ 12 and actual matching character ≥ 12 (follow standard) (Total 2189 devices), e.g., 000AEBFFB507SK88040C036B
   - **Case 2** Total number of character ≥ 12 and actual matching character < 12 (Total 225 devices), e.g., 00000000000000000000061
   - **Case 3** Total number of character < 12 (Total 1133 devices), e.g., 0AC69B23
   - **Case 4** All the characters are zeroes (Total 7 devices), e.g., 0000000
   - **Case 5** Temporary serial number generated by Operating System, i.e., having second char '&' (Total 59 devices), e.g., 6&11979AA7
   - **Case 6** Same serial number for 2 or more USB storage devices (Total 5 devices) Case 1 is considered as a valid state since it follows USB standards while Case 2 through Case 6 is considered as invalid states. Data on instances of various states of serial number are collected and given along with each case. In further discussion, the use of term "number of characters" in serial number implies "number of matching characters".

Table 3 Instances for VendorID and ProductID

VendorID	ProductID	No. of instances
Not-present	Not-present	25
Not-present	Present	24
Present	Not-present	9
Present	Present	3560

2. **VendorID**—It has two states which are as follows:

   - **Present**—Vendor Id is present in USB storage devices.
   - **Not-Present**—Vendor Id is not present in USB storage devices.

3. **ProductID**—It has two states which are as follows:

   - **Present**—Product Id is present in USB storage devices.
   - **Not-Present**—Product Id is not present in USB storage devices.

Data on instances of various states of VendorID and ProductID were collected and given in Table 3.

Vulnerability Measures

Based on vulnerability found in authorization parameters, we have defined two measures for this study—security of device registration (S) and ease of user for device registration (E) which are given in Eqs. (1) and (2). These measures depend upon the authorization policy set by an organization. The authorization policy may define criteria for the presence of VendorID and ProductID and minimum length of serial number. From Table 3, it can be observed that VendorID and ProductID are found in most of the USB devices and do not significantly affect the system vulnerability. Therefore in further analysis, the comparison using VendorID and ProductID is ignored.

$$\text{Security of device registration } (S) = \frac{\text{Number of USB devices that follow organization policy}}{\text{Number of USB devices allowed to be registered}} \tag{1}$$

$$\text{Ease of user for device registration } (E) = \frac{\text{Number of USB devices allowed to be registered}}{\text{Total number of USB devices}} \tag{2}$$

$S < 1$ means all the devices allowed to be registered do not follow organization policy for device registration.

$S = 1$ means all the devices allowed to be registered follow organization policy for device registration.

$S > 1$ means there are some devices which follow organization policy but are not allowed to be registered.

E is between 0 and 1.

$E = 0$ means no device is allowed to be registered.

$E = 1$ means all the devices are allowed to be registered.

A stringent USB authorization policy, based on USB standard, may state that USB device's serial number must have at least 12 characters. From the data set, it is found that

Number of USB devices following policy = 2197
Number of USB devices allowed to be registered = 2197
Total number USB devices = 3618

For this policy, S and E, defined above, are calculated as:
Security of device registration $(S) = 2197/2197 = 100\%$
Ease of user for device registration $(E) = 2197/3618 = 60.7\%$

For the above policy, S is 100% but E remains only 60.7% which implies that only 60.7% of devices are allowed to be registered. In this case, an administrator may choose to relax his organization policy so that more devices can be registered by the users without significant compromise on the security of device registration. Suppose it is decided to relax the restriction on the serial number to 8 characters or more (from 12 characters or more), it would result in the following changes in the parameters:

$S = 2197/3434 = 64\%$ which means only 64% of the devices allowed to be registered follow organization policy.

$E = 3434/3618 = 95\%$ which means 95% of the devices out of total devices are allowed to be registered.

This implies that if organizational policy is changed to allow devices with 8 or more characters in serial number, then there is a 36% decrease in security of device registration (because of registration of vulnerable devices) but nearly all (95%) of the user devices are allowed to be registered. This demonstrates the trade-off between the two metrics, and it serves as a guideline to the system administrator to choose a policy appropriate for his organization. In Fig. 2, we have modeled various policies by varying the criteria for USB device registration and the values of S and E for a particular policy have been observed.

3.2 Authorization System Level

Vulnerability assessment of the system investigates the possibility of existence of defects which has given access to unauthorized devices. Typically, system can exist in four different states, as per classification definitions. There are two types of devices: authorized devices (+) which are registered and unauthorized devices (−) which are not registered. For the system, we have defined:

True Positives (TP): Correctly allow access to an authorized USB storage device.

True Negatives (TN): Correctly deny access to an unauthorized USB storage device.

False Positives (FP): Incorrectly allow access to an unauthorized USB storage device.

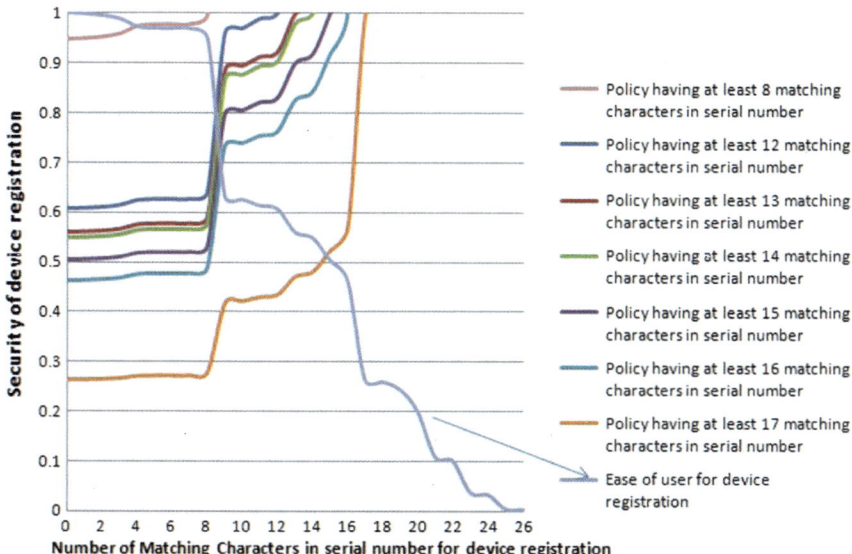

Fig. 2 Security of device registration and ease of user for device registration (based on number of characters in serial number)

		Classification (Detection)	
		+	-
Registered Device	+	TP	FN
Rogue Device	-	FP	TN

Fig. 3 Confusion matrix for classification

False Negatives (FN): Incorrectly deny access to an authorized USB storage device.

Using the above classification states, confusion matrix is drawn as shown in Fig. 3.

The whole database of USB storage devices is divided based upon the number of matching characters in the Serial Number (denoted as n). From the dataset, it is observed that for $n = 4, 5$, and 6, number of instances of USB devices are very less so they are combined into one dataset. In the same manner number of instances for $n = 9$ and 10 also are combined. Finally, there are total 10 datasets which are given in Table 4 which shows authorization system state for each dataset.

Vulnerability Measures

There are two types of measures existing for vulnerability assessment: (i) Classification measures and (ii) Information Retrieval measures. These measures were applied to the dataset for assessing the system vulnerability. Confusion matrix can be drawn

Table 4 Dataset of USB storage devices—matching characters, authorization system state for each dataset, classification, and information retrieval measures

Sr. No.	(Dataset) Number of matching character n	Number of instances	Authorization system state				Classification measures					Information Retrieval (IR) measures		
			TP	TN	FP	FN	Accuracy	Weighted Accuracy	TPR	FPR	Specificity	Precision	Recall	F1 Score
1	$n = 0$	7	3	0	4	0	0.43	0.24	1	1	0	0.43	1	0.6
2	$n = 1$	52	12	40	0	0	1	1	1	0	1	1	1	1
3	$n = 2$	74	25	49	0	0	1	1	1	0	1	1	1	1
4	$n = 3$	127	56	71	0	0	1	1	1	0	1	1	1	1
5	$n = 4, 5, 6$	41	13	28	0	0	1	1	1	0	1	1	1	1
6	$n = 7$	143	71	72	0	0	1	1	1	0	1	1	1	1
7	$n = 8$	2057	1145	908	1	3	0.998	0.998	0.997	0.001	0.999	0.999	0.997	0.998
8	$n = 9, 10$	87	13	24	2	48	0.43	0.57	0.21	0.077	0.92	0.87	0.21	0.34
9	$n = 11$	90	25	65	0	0	1	1	1	0	1	1	1	1
10	≥ 12	4352	2189	2155	0	8	0.998	0.999	0.996	0	1	1	0.996	0.998

Classification (Detection)

		+	-
Registered Device	+	0	C_{12}
Rogue Device	-	C_{21}	0

Fig. 4 Cost matrix for classification

for each dataset shown in Table 4. Traditional measures of binary classification are discussed below:

Accuracy is defined as the total number of correct classifications to the total number of classifications. Mathematically, it is given in Eq. (3). While accuracy gives a measure of test's effectiveness, it does not take into account the cost of misclassification imbalances. If the cost matrix, shown in Fig. 4, is considered, then Cost Ratio (CR) is defined as C_{21}/C_{12} and **Weighted Accuracy** (WA) is given by Eq. (4). In this work, cost ratio is chosen as 0.7/0.3. FP is given more weightage to FN because unauthorized USB-based storage devices pose a greater risk.

Specificity measures the ability of the classifier to identify irrelevant USB storage devices. Mathematically, it is given in Eq. (5).

$$\text{Accuracy} = \frac{\text{TP} + \text{TN}}{\text{TP} + \text{FP} + \text{TN} + \text{FN}} \quad (3)$$

$$\text{WA (Weighted Accuracy)} = \frac{\text{CR} \cdot \text{TN} + \text{TP}}{\text{CR}(\text{TN} + \text{FP}) + \text{TP} + \text{FN}} \quad (4)$$

$$\text{Specificity} = \frac{\text{TN}}{\text{TN} + \text{FP}} \quad (5)$$

Accuracy and weighted accuracy curves are plotted in the same plot shown in Fig. 5. Both accuracy and weighted accuracy curves have two dips (i) at $n = 0$ reason is lots of FP and (ii) at $n = 9$ reason is lots of FN; these two points are outlier points. Apart from these two dips, the accuracy and weighted accuracy are high which means the system is mostly working in an accurate manner.

Receiver Operator Characteristics (ROC) space originally used in signals theory, ROC space is a useful measure of a binary classifier's ability. It is a plot of TPR, given in Eq. (6) versus FPR, given in Eq. (7) as the discriminating threshold is varied. A ROC space is defined by FPR and TPR as *x*- and *y-axes,* respectively, which depicts relative trade-offs between true positive (benefits) and false positive (costs). It is shown in Fig. 6.

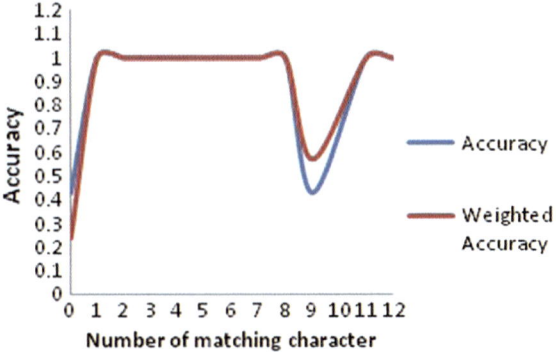

Fig. 5 Accuracy and weighted accuracy curve for authorization system

Fig. 6 ROC space for
authorization system

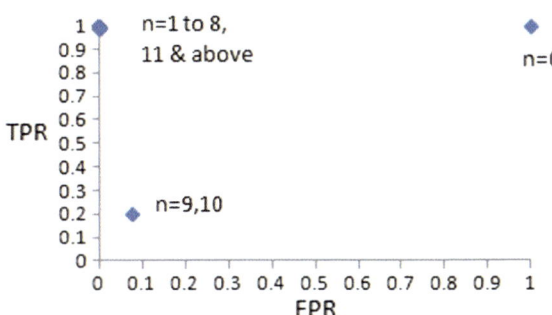

$$\text{TPR (True Positive Rate)} = \frac{TP}{TP + FN} \tag{6}$$

$$\text{FPR (False Positive Rate)} = 1 - \text{specificity} = \frac{FP}{TN + FP} \tag{7}$$

Mathews Correlation Coefficient (MCC) is widely used in machine learning to evaluate the quality of two-class classifiers. It is a balanced measure that works well even in cases where the classes are highly skewed. The MCC is in essence a correlation coefficient between the observed and predicted binary classifications; it returns a value between -1 and $+1$. A coefficient of $+1$ represents a perfect prediction, 0 an average random prediction and -1 an inverse prediction. Mathematically, it is given in Eq. (8).

$$\text{MCC} = \frac{(TP \times TN) - (FP \times FN)}{\sqrt{(TP + FP)(TP + FN)(TN + FP)(TN + FN)}} \tag{8}$$

MCC is calculated for the whole system.
   TP $= 3552$, FP $= 7$, FN $= 59$, TN $= 3412$, MCC $= 0.98$.

As MCC tends to $+1$, it represents a perfect prediction. Hence, the proposed authorization system is working ideally.

Information Retrieval (IR) Measures

Precision is defined as the fraction of retrieved results that are relevant. For classification, it is analogous to positive predictive value, i.e., the proportion of relevant devices, which are registered and are correctly detected. Precision gives a measure of the **exactness** or fidelity of the classification. Mathematically, it is given in Eq. (9).

   Recall is the fraction of relevant results that are successfully retrieved. For classification, it is called the true positive rate and denotes the fraction of correctly classified relevant alerts among all the relevant alerts. Recall gives a measure of completeness of the classification. Mathematically, it is given in Eq. (10).

   F1-Score is the harmonic mean of precision and recall; this is considered to be a good measurement of the accuracy of a method. The best value of $F1$ score is 1 and the worst is 0. $F1$ score combines the precision and recall elements and gives a composite view of the performance of the system. Mathematically, it is given in Eq. (11).

   Information Retrieval measures for authorization system are given in Table 4.

$$\text{Precision} = \frac{\text{TP}}{\text{TP} + \text{FP}} \tag{9}$$

$$\text{Recall} = \frac{\text{TP}}{\text{TP} + \text{FN}} \tag{10}$$

$$F1 = 2 \times \frac{(\text{precision} \times \text{recall})}{\text{precision} + \text{recall}} \tag{11}$$

PR Space viewing precision and recall values in isolation does not have much value. From the visualization viewpoint, it is better to view the precision-recall space (or the PR curve) of the system. The graph plots precision as a function of various recall values and shows us the relationship between the two. The aim of a good system is to be in the region of high precision and high recall.

   Figure 7 for PR space shows most of the points 8 out of 10 fall in the upper right corner of PR space which shows the proposed system is behaving ideally. There are two outlier points in PR space corresponding to $n = 9$ and $n = 0$ datasets. For these two points, system does not behave ideally. The reason is for $n = 0$ there are many FP cases and for $n = 9$ there are many FN cases.

Fig. 7 PR space for authorization system

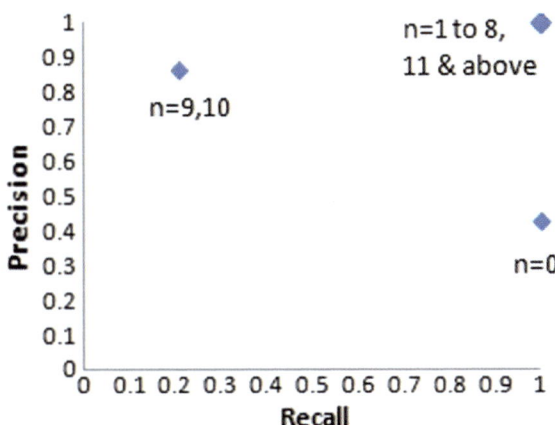

Table 5 Recommendation for purchasing a new USB storage device

Manufacturer	No. of USB devices following standard	No. of total USB devices found	% of USB devices following standard
Manufacturer 1	120	120	100
Manufacturer 2	180	181	99.4
Manufacturer 3	781	789	98.9
Manufacture 4	674	707	95.3
Manufacturer 5	195	207	94.2
Manufacturer 6	113	122	92.6
Manufacturer 7	423	566	74.7
Manufacturer 8	1163	2711	42.9
Manufacturer 9	1	472	0.21

4 Conclusions

In this work, agent-based authorization system is presented which authorizes the device by making a unique profile considering three authorization parameters which are serial number, vendor ID, and product ID. Vulnerability assessment at authorization parameter shows serial number as the most vulnerable parameter and analysis shows that there is a compromise between security of device registration and ease of user for device registration. Vulnerability assessment at authorization system level shows the system behaves like a perfect classification working ideally. Based on the vulnerability assessment of authorization parameters and system authorization policy can be redefined as per the organization needs.

Further, the authorization system gives recommendation for the procurement of new USB-based storage device after analyzing more than 7000 devices. Table 5 shows some of the manufacturers who follow standard as laid by USB-IF.

References

1. Existing Authorization System for USB device available at http://www.newsoftwares.net/usb-block/
2. J. Larimer, Exploiting vulnerabilities with removable storage, available at https://media.blackhat.com/bh-dc-11/Larimer/BlackHat_DC_2011_Larimer_Vulnerabiliters_w-removeable_storage-wp.pdf
3. P. Krzyzanowski, "Client Server Communication" available at https://www.cs.rutgers.edu/~pxk/rutgers/notes/content/02-networking.pdf
4. White paper on Data Leakage Worldwide White Paper: The High Cost of Insider Threats is available at http://www.cisco.com/c/en/us/solutions/collateral/enterprise-networks/data-loss-prevention/white_paper_c11-506224.html
5. C. Goutte, E. Gaussier, A probabilistic interpretation of precision, recall and *F*-score, with implication for evaluation, in *Proceedings of 27th European conference on IR research,* 2005, pp. 345–359
6. A.P. Bradley, The use of the area under the ROC curve in the evaluation of machine learning algorithms. J. Pattern Recognit. **30**(7), 1145–1159 (1997)

Crypto-Ransomware Detection Using Behavioural Analysis

Parth S. Goyal, Akshat Kakkar, Gopika Vinod and Gigi Joseph

Abstract In recent years, there has been an increasing number of crypto-ransomware attacks that encrypt files and ask for money in exchange of decryption key causing huge loss of information and money to the affected organizations and users. A major reason for this increasing number of attacks is large financial gains with benefits of anonymity provided by crypto-currency like bitcoin. Traditional security solutions are incapable to detect such crypto-ransomware. This is mainly because they use signature-based detection methods, whereas crypto-ransomware keeps on evolving and use techniques like code obfuscation that can bypass these security solutions. This paper discusses the working mechanism of crypto-ransomware, limitations of signature-based detection methods and how behaviour-based detection method is an effective mechanism for the detection of crypto-ransomware. The paper investigates the distinctive behaviour of crypto-ransomware as compared to genuine applications. This behaviour analysis is further utilized to propose and validate a classification model to detect and classify ransomware from genuine applications.

Keywords Behavioural analysis · Bitcoin · Crypto-ransomware detection · Cryptographic ransomware

P. S. Goyal (✉) · A. Kakkar · G. Vinod · G. Joseph
Bhabha Atomic Research Center, Mumbai, India
e-mail: parth@barc.gov.in

A. Kakkar
e-mail: akshat@barc.gov.in

G. Vinod
e-mail: vgopika@barc.gov.in

G. Joseph
e-mail: gigi@barc.gov.in

© Springer Nature Singapore Pte Ltd. 2020
P. V. Varde et al. (eds.), *Reliability, Safety and Hazard Assessment for Risk-Based Technologies*, Lecture Notes in Mechanical Engineering,
https://doi.org/10.1007/978-981-13-9008-1_20

1 Introduction

There are many types of malware that ask for ransom in one or other way around, for example, asking for fake antivirus subscription, asking for penalty fees by acting as fake government authority [1], etc. But these types of malware just threaten user of consequences, whereas in actual, they rarely do anything to the underlying data of system or system itself. However, **crypto-ransomware** is a type of malware that encrypts the user files in such a way that it is very difficult to get the original files back without decrypting the files. For decryption, a key/string is needed which is revealed by attacker to victim only against huge amount of money, mostly in the form of crypto-currency such as bitcoin, which cannot be tracked easily. Many organizations and users have become the victim of crypto-ransomware attacks, causing loss of millions of dollars and affected their business operations. According to Kaspersky [2], the impact of WannaCry ransomware was so severe that a leading car maker had to close its factory in France and hospitals in the UK had to turn away patients. German transport giant, West Bengal power distribution company and the Russian Interior Ministry were all hit too.

This paper presents behaviour-based detection of crypto-ransomware based on a set of behaviours that are exhibited by crypto-ransomware. For dynamic behaviour analysis, more than 200 crypto-ransomwares [3–6] have been executed in different test environments. In order to capture behaviours that distinguish the crypto-ransomwares, many genuine applications including those that have some functionality related to crypto-ransomware have been executed and analysed. Finally, set of ten behaviours have been selected to identify the crypto-ransomware such as high generation rate of encrypted files, high file write operations, high CPU utilization, deleting shadow copies, changing registries, renaming files, increasing size of files, changing wallpaper and network activity.

This paper is structured as follows. Section 2 describes the working mechanism of crypto-ransomware. Section 3 discusses about the detection mechanisms and limitations of signature-based detection mechanism. Section 4 presents a detailed dynamic behaviour analysis of crypto-ransomware. Section 5 presents the model for the detection of the crypto-ransomware and finally, conclusions are drawn in Sect. 6.

2 Crypto-Ransomware: How It Works

Crypto-ransomware infection generally involves five stages:

Payload Delivery is the first stage. There are multiple ways for the delivery of the crypto-ransomware in the system: sending malicious emails, hosting malicious advertisement, using exploit kits, etc.

Execution of crypto-ransomware is the second stage, attackers while sending the crypto-ransomware payload looks for the various vulnerabilities in order to execute their crypto-ransomware payload, for example, embedding the payload into a pdf file and using vulnerability of pdf readers where opening the pdf file will lead to the execution of the payload.

Command and Control (C&C) server communication is the third stage. Crypto-ransomware tries to communicate with the command and control server to send the information about the infected host and exchange cryptographic keys, which will be used in the process of encryption. Addresses of these servers are either hard-coded or generated using the domain-generating algorithm. There are two scenarios.

- Crypto-ransomware contact to C&C servers for exchanging host information and encryption keys. It will proceed with encryption process only after communication with C&C server.
- Crypto-ransomware will proceed with the encryption process even if it is not able to contact C&C server for information exchange.

Encryption is the fourth stage, according to Symantec [1], crypto-ransomware use a combination of symmetric and asymmetric encryption. The advantage of symmetric encryption is that it encrypts files very fast. This is an important aspect as attackers want to complete the encryption before being detected. The disadvantage of symmetric encryption is that if the key is discovered during encryption or communication to C&C, then the victim can use it to decrypt files. Asymmetric encryption uses two encryption keys: public and private. It is more secure, but the process of encryption becomes slow. So, attackers combine the strength of both methods in such a way that victim's files can be encrypted using symmetric encryption and then symmetric encryption key can be encrypted using asymmetric encryption.

Extortion is the fifth stage. Once the encryption process is completed, crypto-ransomware displays a set of instructions on the screen regarding ransom payment procedure to get decryption key. Attackers asked ransom payment in form of crypto-currencies such as bitcoins. As bitcoin wallets are free and disposable, this encourages the attackers to create a new, unique wallet for each infection.

3 Detection Mechanisms

3.1 Signature-Based Detection

Signature is a string of bytes that is unique to known crypto-ransomware. Signature detection is the process of analysing the program based on known crypto-ransomware signatures and identifies it as crypto-ransomware if it is matched with the previously observed signature. Signature-based detection relies on the enormous repository of malicious code signatures and repository has to be frequently updated. Advantages of this method are simple to use, provides good protection and fast detection from the known crypto-ransomware. Limitations [7] are as follows:

- The main issue with signature-based detection is its inability to detect new crypto-ransomwares, which are yet to be turned into a signature.

- Attackers use various tools to generate large crypto-ransomware variants of the original crypto-ransomware using code obfuscation techniques where each variant will have different signatures.
- Crypto-ransomwares might not also be readily available to derive a signature of it, as it can be either a targeted attack is some specific location or just a trial run of a potential crypto-ransomware.

3.2 Behaviour-Based Detection

As per [7], there are two approaches to behaviour-based detection. One is a static analysis and another is dynamic analysis. **In static analysis**, behaviours are extracted from the binary of an unknown executable. This extracted behaviour then checked using matching algorithm to see if it matched with malicious behaviour. **In dynamic analysis**, behaviours of executable are captured while executable is running on the system. This captured behaviour then checked using matching algorithm to see if it matched with malicious behaviour. In this paper, dynamic behaviour analysis is used for detecting the crypto-ransomware. Advantages of dynamic behaviour-based detection method are (1) detection of crypto-ransomware whose signatures are not known (2) detection of zero-day crypto-ransomwares (3) requires less maintenance as no need to store a large number of signatures in the repositories (4) immune to code obfuscation.

4 Dynamic Behaviour Analysis

For behaviour analysis, crypto-ransomware and genuine applications have been executed and analysed. It was observed that ransomware extensively use the registry database to get complete information about the system and made changes to some of the system information through the registry. It was also observed that crypto-ransomware made a lot of file system and network calls in the system. As encryption is CPU-extensive job, so some of the crypto-ransomware put the limit on the CPU utilization to avoid detection, whereas other use CPU at their full capacity in order to complete encryption as fast as possible. Some of the distinctive and frequently exhibited behaviours that were observed are listed below:

Generation of encrypted files: Basic functionality of crypto-ransomware is the encryption of file so that file cannot be recovered without the decryption key. Encrypted files tend to have high entropy, where entropy is a measure of uncertainty of the data [8]. Entropy of a file can be calculated using Eq. 1 as follows:

$$e = -\sum_{i=0}^{255} P_{Xi} \log_2\left(\frac{1}{P_{Xi}}\right) \tag{1}$$

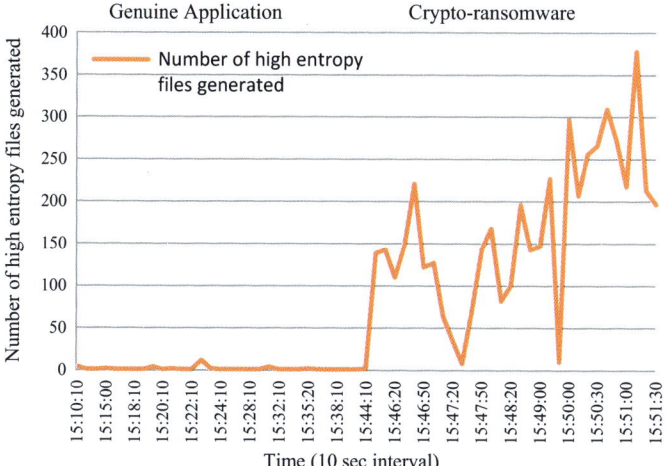

Fig. 1 Graph representing number of high entropy files generated by genuine applications and crypto-ransomwares with respect to time

For $P_{Xi} = F_i$/total bytes and F_i, the number of instances of byte value "i" in the array. This produces a value from 0 to 8 bits/bytes, where 8 represent a perfectly even distribution of byte values in the array. Encrypted files will tend to approach the upper bound 8, since each byte in a ciphertext should have a uniform probability of occurring [8]. Figure 1 shows a graph that represents a number of high entropy files generated by genuine applications and crypto-ransomware with respect to time (in interval of 10 s). Figure 2 shows a graph that represents a number of high entropy files generated per minute on an average by some of the crypto-ransomwares and genuine applications.

Write to large number of files: Crypto-ransomware creates encrypted files using one of two methods. First one is that crypto-ransomware reads the original file and overwrites its content with the encrypted one. The second one is that crypto-ransomware creates a new file, reads the content from the original file, writes encrypted content to the new file and after that, it deleted the original file. The first method was observed in most of the crypto-ransomwares. However, in both methods, ransomware writes the encrypted contents to files. In order to evade detection and maximize damage, crypto-ransomware tries to encrypt maximum files possible within minimum time. Apart from writing encrypted content to files, crypto-ransomware also writes instruction files (for payment) in every directory of the system where it encrypted the files. Figure 3 shows a graph that represents a number of files written by genuine applications and crypto-ransomware with respect to time (in interval of 10 s). Figure 4 shows number of file written per minute on an average by some of crypto-ransomwares and genuine applications.

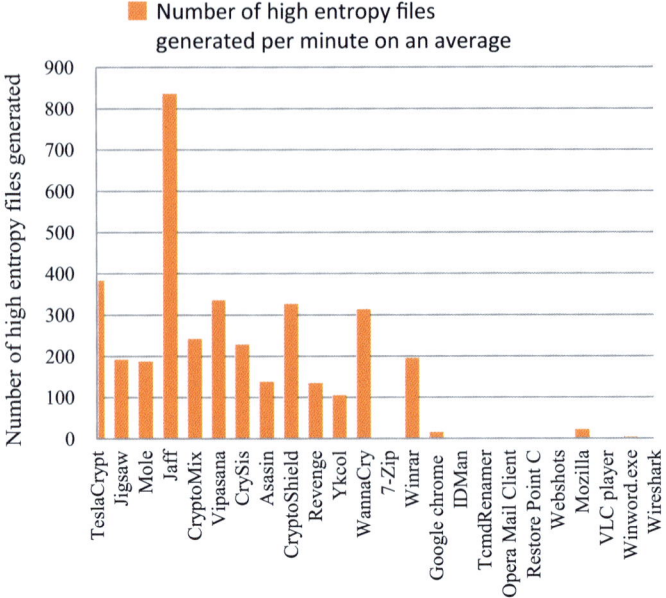

Fig. 2 Number of high entropy files generated per minute on an average by crypto-ransomwares and genuine applications

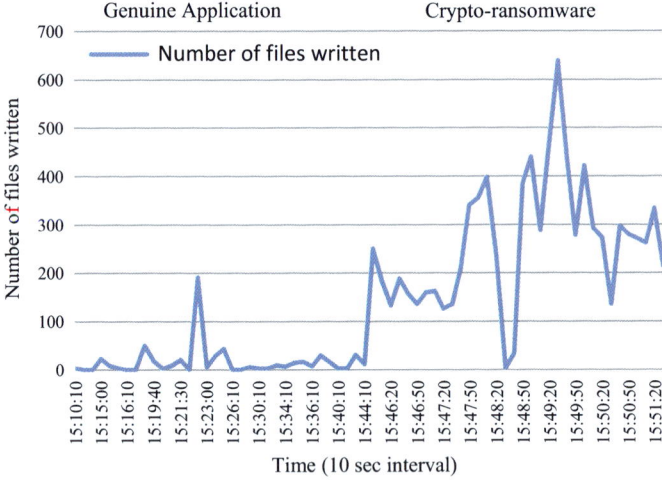

Fig. 3 Graph representing number of files written by genuine applications and crypto-ransomwares with respect to time

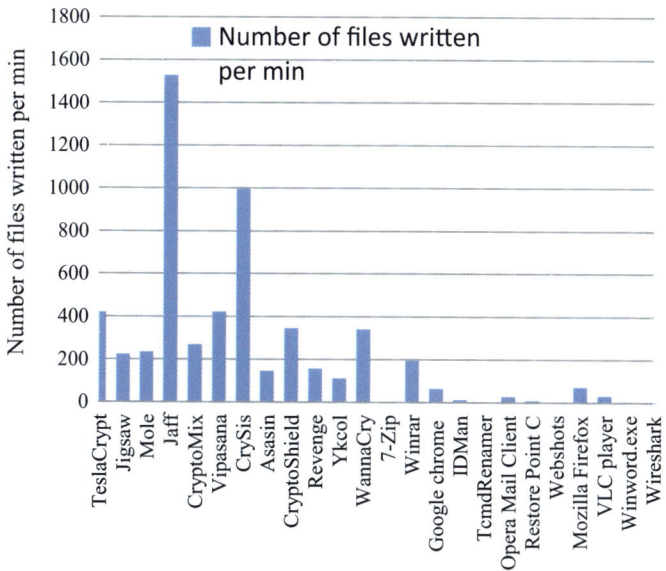

Fig. 4 Number of files written per minute on an average by crypto-ransomwares and genuine applications

Increase of file size: Crypto-ransomware writes some additional information such as original file name, file size and encrypted symmetric key in the encrypted file. This increases the size of the file.

Change of file name: Crypto-ransomware renames the file either by adding particular extension or to some random name or both. This may help in protecting the encrypted files from being encrypted again either by same crypto-ransomware or some other crypto-ransomware. It was observed that crypto-ransomwares changed the file names with high rate as shown in Fig. 5, which shows a graph that represents a number of files renamed by genuine applications and crypto-ransomware with respect to time (in interval of 10 s).

Deletion of volume shadow copies: Volume shadow copies are the snapshots of the Microsoft Window's volumes, which are used for backups. In order to ensure that encrypted files are not recovered, crypto-ransomware deletes the volume shadow copies. This is done using vssadmin.exe tool, which is provided by Microsoft to manage (create/delete) the volume shadow copies.

Persistence: In order to ensure that encryption process would continue across reboots, crypto-ransomware creates autorun registry key that means a new registry key is created under HKCU\Software\Microsoft\Windows\ CurrentVersion\Run and the value is set to the path of the executable of the crypto-ransomware.

Set wallpaper: Crypto-ransomware puts scary images or instructions to pay ransom or both as desktop wallpaper. They did so by setting the registry key "HKCU\Control\Panel\Desktop\Wallpaper" value to the path of image.

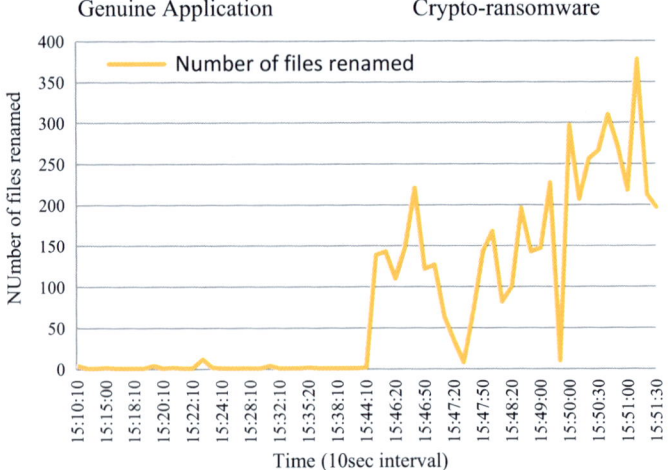

Fig. 5 Graph representing number of files renamed by genuine applications and crypto-ransomwares with respect to time

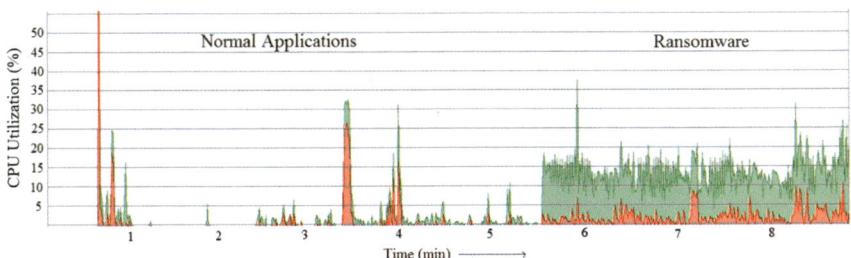

Fig. 6 Graph representing CPU utilization by genuine applications and crypto-ransomwares with respect to time

Execution of process from the C:\User\<CurrentUser>\AppData\: Most of crypto-ransomwares executed itself from the C:\User\<CurrentUser>\AppData\Local\temp, while few executed from C:\User\<CurrentUser>\AppData\Roaming, as it might provide them user's permissions.

High CPU Utilization: File encryption is CPU-intensive task and encryption of thousands of files makes CPU utilization more visible. Same happens in the case of crypto-ransomware as it encrypts a large number of files. It has been observed that crypto-ransomware continuously utilized the CPU on an average of 10% per minute, whereas generally any normal genuine application utilized CPU on an average of 2–3% per minute as shown in Fig. 6.

Network Activity: Crypto-ransomwares try to contact their command and control servers by using various domain names. In cases where domain names do not

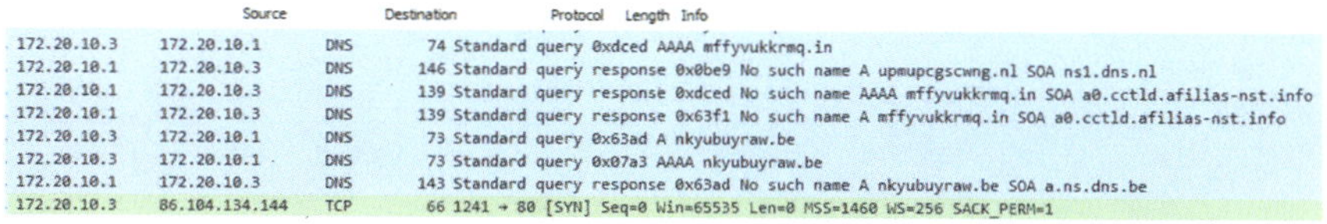

	Source	Destination	Protocol	Length	Info
172.20.10.3	172.20.10.1	DNS		74	Standard query 0xdced AAAA mffyvukkrmq.in
172.20.10.1	172.20.10.3	DNS		146	Standard query response 0x0be9 No such name A upmupcgscwng.nl SOA ns1.dns.nl
172.20.10.1	172.20.10.3	DNS		139	Standard query response 0xdced No such name AAAA mffyvukkrmq.in SOA a0.cctld.afilias-nst.info
172.20.10.1	172.20.10.3	DNS		139	Standard query response 0x63f1 No such name A mffyvukkrmq.in SOA a0.cctld.afilias-nst.info
172.20.10.3	172.20.10.1	DNS		73	Standard query 0x63ad A nkyubuyraw.be
172.20.10.3	172.20.10.1	DNS		73	Standard query 0x07a3 AAAA nkyubuyraw.be
172.20.10.1	172.20.10.3	DNS		143	Standard query response 0x63ad No such name A nkyubuyraw.be SOA a.ns.dns.be
172.20.10.3	86.104.134.144	TCP		66	1241 → 80 [SYN] Seq=0 Win=65535 Len=0 MSS=1460 WS=256 SACK_PERM=1

Fig. 7 Crypto-ransomware try to contact C&C servers using various domain names and hard-coded ip address

Table 1 Final set of behaviours

Behaviour notation	Behaviour description
B1	Generation of files with high entropy at high rate
B2	Increase of file size at high rate
B3	Change of file name at high rate
B4	Creation of autorun registry key
B5	Write large number of files at high rate
B6	Deletion of volume shadow copies
B7	Set the wallpaper registry key
B8	Execution of process from the C:\User\<CurrentUser>\AppData\
B9	Network activity
B10	High CPU utilization

get resolved, crypto-ransomware tries to connect to command and control server through hard-coded IP addresses. Figure 7 shows this scenario in which crypto-ransomware tried to contact to servers using domain names such as mffyvukkrmq.in, upmupcgscwng.nl and nkyubuyraw.be. When these domain names did not get resolved, then it sent the TCP request to hard-coded IP address "86.104.134.144".

In order to distinguish behaviours of crypto-ransomware from genuine applications, various genuine applications were executed and their behaviours were analysed. It has been observed that crypto-ransomwares generate the encrypted files, write to large number of files, increase the file size and rename the files at very high rate in comparison with genuine application. Table 1 contains the final set behaviours for the detection of crypto-ransomware.

5 Modelling

Identification of a set of behaviours for the crypto-ransomware is the first step towards its detection. Next step is utilizing these behaviours to design a model to distinguish crypto-ransomwares from the genuine application. Bayesian belief network (BBN) classification model has been selected for the same. BBN classification model has

Fig. 8 Threshold
calculation for BBN
classification model

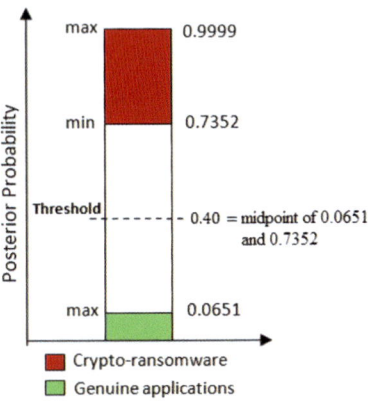

been trained using data set that consists of behaviours shown by crypto-ransomwares
and genuine applications. For generating data set, more than 200 crypto-ransomware
samples have been executed. Out of these, 67 crypto-ransomwares samples have successfully encrypted the files. This is due to the fact that these 67 crypto-ransomwares
have been able to encrypt the files without communicating with C&C servers. Threshold T for flagging the process as crypto-ransomware was calculated from the posterior
probabilities of crypto-ransomwares and genuine applications of training data set.
Threshold T for the BBN classifier has been calculated as middle value of lowest
probability of process being a ransomware given that process is ransomware and highest probability of process being ransomware given that process is not ransomware as
shown in Fig. 8. After calculation, threshold is 0.4, which means if posterior probability of any process calculated by BBN classification model is greater than 40%,
then the process is flagged as crypto-ransomware.

BBN classification model has been validated using crypto-ransomwares and genuine applications, which were not part of the training data set. Validation results for
some of the crypto-ransomwares and genuine applications are shown in Table 2. Performance evaluation of BBN classification has been done using performance metrics
such as accuracy, sensitivity, precision and $F1$ score. Table 3 shows the values of
these parameters for BBN classification model, which were calculated using Eqs. 2,
3, 4 and 5, respectively.

$$Accuracy = \frac{TP + TN}{TP + TN + FP + FN} \tag{2}$$

$$Sensitivity = \frac{TP}{TP + FN} \tag{3}$$

$$Precision = \frac{TP}{TP + FP} \tag{4}$$

$$F1\ score = 2 * \frac{precision * sensitivity}{precision + sensitivity} \tag{5}$$

Table 2 BBN classification validation results for some of the crypto-ransomwares and genuine applications

| S. No | Test samples | Features | | | | | | | | | | | P (R|B1, B2, …, B10) (%) | BBN Classifier (Threshold = 40%) |
		B1	B2	B3	B4	B5	B6	B7	B8	B9	B10		
1	Shade Ransomware	✓	✓	✓		✓	✓	✓		✓	✓	99.99	Ransomware
2	Chircrypt Ransomware	✓	✓	✓		✓				✓	✓	99.350	Ransomware
3	Satana Ransomware	✓	✓	✓	✓	✓	✓		✓		✓	99.99	Ransomware
4	Hermes Ransomware	✓	✓			✓	✓		✓	✓	✓	99.9	Ransomware
5	Bart Ransomware	✓				✓		✓		✓	✓	30.811	Not ransomware
6	Kruptos 2	✓				✓					✓	1.293	Not ransomware
7	Xps Viewer					✓						0	Not ransomware
8	Logitech webcam								✓			0	Not ransomware
9	Winzip										✓	0	Not ransomware
10	Dummy file creator	✓				✓					✓	1.293	Not ransomware

Table 3 Performance parameters for BBN classification model

	BBN classifier
Accuracy (%)	97.5
Sensitivity (%)	95
Precision (%)	100
$F1$ score (%)	97.4

TP True Positive, i.e., Crypto-ransomware correctly classified as crypto- ransomware

FP False Positive, i.e., Genuine application incorrectly classified as crypto-ransomware

TN True Negative, i.e., Genuine application correctly classified as not crypto-ransomware

FN False Negative, i.e., Crypto-ransomware incorrectly classified as not crypto-ransomware.

6 Conclusion

This paper presents the limitation of signature-based detection methods and how behaviour-based detection mechanism can be effective for crypto-ransomware detection. For dynamic behaviour analysis, more than 200 crypto-ransomwares have been executed in different test environments, out of which 67 crypto-ransomwares successfully encrypted the files despite not being able to communicate with any C&C. In order to capture behaviours that distinguish the crypto-ransomwares, many genuine applications have been executed and analysed. After thorough analysis, ten behaviours have been selected to identify the crypto-ransomware such as high file write operations, deletion of shadow copies, high cpu utilization, renaming and increasing the size of files, changes in registries, network activity, etc. Even though a total number of crypto-ransomware and genuine applications tested were comparatively less due to the various experimental challenges in downloading, executing and storing the crypto-ransomwares and their logs, still we get a good number of behaviours for the detection of crypto-ransomware. For the detection of crypto-ransomware, BBN classification model has been selected and trained. Model has been able to detect crypto-ransomwares and genuine application correctly with good accuracy, sensitivity, precision and $F1$ measure.

References

1. Symantec, *An ISTR Special Report: Ransomware and Businesses 2016* (2016). Retrieved from https://www.symantec.com/content/en/us/enterprise/media/security_response/whitepapers/ ISTR2016_Ransomware_and_Businesses.pdf
2. Kaspersky Lab, *Kaspersky Security Bulletin: Story of the year 2017* (2018). Retrieved from https://cdn.securelist.com/files/2017/11/KSB_Story_of_the_Year_Ransomware_FINAL_ eng.pdf
3. Virustotal. Malware Samples. Retrieved from https://www.virustotal.com/
4. Github. Malware Repository. Retrieved from https://github.com/ytisf/theZoo/tree/master/ malwares/Binaries
5. Virusshare. Malware Samples. Retrieved from https://virusshare.com/
6. Malware-Traffic-Analysis. Retrieved from https://www.malware-traffic-analysis.net
7. D. Nieuwenhuizen, A behavioural-based approach to ransomware detection. *Whitepaper. MWR Labs Whitepaper* (2017)
8. N. Scaife, H. Carter, P. Traynor, K.R. Butler, Cryptolock (and drop it): stopping ransomware attacks on user data, in *2016 IEEE 36th International Conference on Distributed Computing Systems (ICDCS)* (IEEE, New York, 2016, June), pp. 303–312

Software Reliability Growth as an Offshoot of Verification and Validation Process

Avijit Das, Manish Kr. Tiwari and D. R. Nayak

Abstract Software developed for mission-critical systems employ rigorous verification and validation techniques to achieve good quality. Traditional software reliability models use execution time during software testing for reliability estimation. Although testing duration is an important factor in reliability, it is possible to improve the accuracy of estimation by factoring other techniques that influence the final software quality. In this paper, we propose a novel method to utilize the effectiveness of the verification and validation process to predict reliability. The main idea is that failure detection is not only related to the time that the software experiences under testing, but also to different levels of review rigors the software has undergone. Our experimental results with multi-version software used in mission-critical systems show our model achieves substantial reliability estimate.

Keywords Verification and validation · Software reliability · Static analysis · Dynamic testing · Code walkthrough

1 Introduction

Software deployed in mission-critical systems has well-defined safety and reliability concerns. The main difference between mission-critical software and conventional software is that the development process is bounded by quality norms and should meet the safety requirements of internationally accepted standards [1]. Realization of highly reliable software for mission-critical systems is a challenging and complex

A. Das (✉)
Electronics & Radar Development Establishment, DRDO, Bengaluru, India
e-mail: avijitdas@lrde.drdo.in

M. Kr. Tiwari · D. R. Nayak
Advanced Systems Laboratory, DRDO, Hyderabad, India
e-mail: manishkumar@asl.drdo.in

D. R. Nayak
e-mail: ramzanayakd@asl.drdo.in

© Springer Nature Singapore Pte Ltd. 2020
P. V. Varde et al. (eds.), *Reliability, Safety and Hazard Assessment for Risk-Based Technologies*, Lecture Notes in Mechanical Engineering, https://doi.org/10.1007/978-981-13-9008-1_21

task. Hence, software verification and validation (V&V) becomes an integral part of the development life cycle for such systems. V&V encompasses analysis and testing activities across the full life cycle and complements the efforts of other quality-engineering functions [2].

Software reliability is one of the key factors for the determination of software quality. It is defined as the probability of failure-free operation of a computer program for a specified time and in a specified environment [3]. Many analytical models [4, 5] have been proposed for software reliability estimation based on the execution time of software. These models use the failures collected in testing phases to predict the failure occurrences in the operational environment. However, it is possible to improve the accuracy of estimation by integrating the exhaustive V&V process that organizations follow to improve the software quality.

We propose a software reliability modeling technique that attempts to measure the effectiveness of the different methodologies incorporated to review the correctness of software. It shows an incremental growth in reliability with quantitative proof of each methodology contributing toward improving the final software quality. To assess the effectiveness of our approach, we have experimented with multi-version software and estimated approximate reliability.

The rest of the paper is organized as follows. Section 2 surveys the related work. Section 3 and 4 describe our approach in detail. Section 5 presents our experimental results. Section 6 concludes this paper.

2   Related Work

Software reliability estimation using test coverage is getting considerable research attention. In literature, several models [6, 7] have been proposed to formulate the relationship between number of faults or failures and test coverage achieved. Chen et al. [7] proposed an approach to estimate the reliability of software from code coverage metric. The basic premise of this measurement process is that code coverage will ensure that no part of code is untested before deployment. Senthil [8] further extends the concept by integrating functional profile with code coverage. It gives weightage to the functions in the code which are called multiple times during execution.

Recent researches also show that static analysis has a significant impact on software reliability [9]. Static analysis as a part of reliability assessment provides a qualitative and quantitative description of software attributes. Rallis et al. [10] have estimated reliability from sequential reviews of the code using binomial distribution and Bayesian conditional probability. The model considers a review to be successful if it results in a latent fault to be detected. The probability that all the faults are detected at the end of the last review is the measure of software reliability. Automated static analysis tools reason about runtime properties of program code without actually executing it. Approaches [11, 12] use these concepts of static analysis into quantitative reliability estimates.

3   V&V Process of a Mission Critical Software

In this section, we will present the V&V process being followed in our organization for assessing the quality of software. This process is generic in nature and has been used successfully for different mission-critical software.

A schematic representation of the aforesaid V&V process is shown in Fig. 1. As shown in the figure, the review activities follow an iterative method. The errors identified in one iteration are reported for error correction and the corrected software is again verified with the same rigorous process. The process continues till no more errors are detected and sufficient confidence is gained that the software will function properly in its intended environment. Each phase of the V&V process is explained in detail in the following subsections. A quantitative study on the effectiveness of each phase and the entire process is presented in Sect. 4.

3.1   Code Walkthrough

Code walkthrough is a classic review technique that relies on the visual examination of source code to detect errors. Code walkthrough is an informal code analysis technique where the main objective is to discover the algorithmic and logical errors in the code. A member of V&V team who undertakes code walkthrough traces the execution of code through different statements and functions unearthing majority of the logical errors before the commencement of testing. This is a very tedious and rigorous task as the size of software, especially in case of mission-critical system scales from kilo lines of code (KLOC) to million lines of code (MLOC).

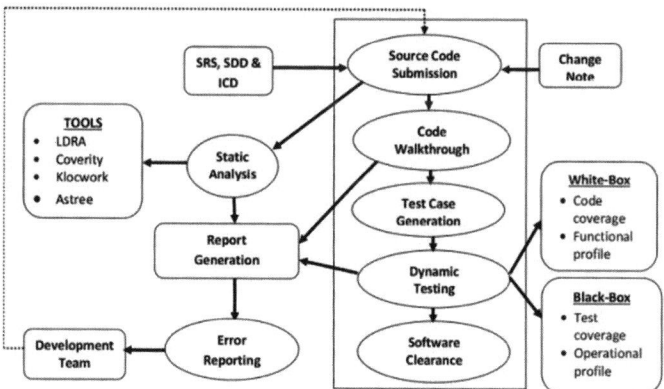

Fig. 1 Schematic representation of V&V process of mission-critical software

3.2 Automated Static Analysis

No single fault-detection technique is capable of addressing all fault-detection concerns. Automated static analysis (ASA) is one such complementary technique to source code walkthrough which does not require program execution. The principal aim of ASA is to check the presence of some common types of semantic and runtime errors that usually creep into code and to check the adherence to coding standards. ASA tools utilize the control, data, information flow, path, and inter-procedural analyses of software code [11]. Also, empirical studies have emphasized the fact that different tools seem to find different sometimes non-overlapping errors [13]. The reason behind this inference is that not every tool employs all the analysis techniques. Also, the undecidability of runtime properties implies that it is impossible to have an analysis which always finds all defects and produces no false positives [14]. Hence, it is recommended to use a multitude of tools for ASA for better fault identification results.

3.3 Dynamic Testing

The aim of program testing is to identify all defects in a program. It is the symbolic or physical execution of a set of test cases with the intent of exposing errors in the program. A given testing strategy may be good for exposing certain kinds of faults but not for all possible kinds of faults in a program. Testing strategy may be based on functional specification (black box), or it may be based on internal program structure (white box). Testing strategy from the viewpoint of software reliability also known as statistical testing involves developing functional and operational profiles. The objective of statistical testing is to determine the reliability of the product rather than discovering errors.

3.3.1 White Box Testing and Functional Profile

White box test case design requires a thorough knowledge of the internal structure of a program. White box testing strategy can be either coverage-based or fault-seeding-based. The complete set of functions with corresponding occurrence probabilities make up the functional profile of software. The design of functional profile from the structure of the code involves assigning probabilities to each function. The probabilities are assigned based on the frequency of function calls.

3.3.2   Black Box Testing and Operational Profile

In the black box approach, test cases are designed using the functional specification of the software. Testing is carried out to test each functionality for both pass and fail criteria to ensure that software does not behave arbitrarily in case of an exception condition.

A complete set of operations of a system or software with occurrence probabilities makes up the operational profile. Operations, as opposed to functions, are actually implemented tasks of a system, while functions are tasks of a system at the design level. Normally, the number of operations is higher than the number of functions, as a single function may be implemented by multiple operations [15].

4   Process Effectiveness

In this section, we will present the quantitative modeling of the V&V techniques discussed in the earlier section. The models will provide a comprehensive analysis of each technique and their degree of impact on the overall effectiveness.

4.1   White Box Adequacy

White box testing employs code coverage and fault injection techniques. We use both the strategies to measure the adequacy of our testing process. Code coverage adequacy is calculated as a weighted average giving importance to large, complex and frequently called functions. We extend the model proposed by Senthil [8] for embedded software by factoring the interrupt service routines (ISR) which was not supported in [8].

$$\text{Code coverage adequacy} = \frac{\sum w_i t_i}{\sum w_i} \qquad (1)$$

where t_i is the code coverage measure achieved for each function during testing and is given by

$$t_i = \text{minimum(Statement, Branch, MCDC)} \qquad (2)$$

and

$$w_i = \text{no. of statements} * \text{cyclomatic complexity} * \{(\text{function call frequency})$$
$$\text{or (interrupt frequency in 100 ms cycle)}\} \qquad (3)$$

Mutation testing is a fault-based testing where realistic faults are induced intentionally into the source code. The fault-induced program is known as a mutant. In
our approach, we induce mutants into the code and then test the mutated code with
the test cases generated for code coverage.

$$\text{White box adequacy} = \text{Code coverage adequacy} * \text{Mutation score}$$
$$= \frac{\sum w_i t_i}{\sum w_i} * \frac{\text{no. of mutants killed}}{\text{total no. of mutants}} \tag{4}$$

4.2 Black Box Adequacy

Black box testing is a method that examines the functionality of an application
without peering into its internal structures or workings. The adequacy of the black
box test suite is a measure of the completeness of the test suite. A test suite is
considered complete if for each requirement of the software there exits at least one
test case in the test suite that tests the correctness of the software with respect to
that requirement. For mission-critical systems, testing for the correctness of each
requirement may not be sufficient as the software may not be designed to handle
exception conditions and system may behave inadvertently. Hence, it is mandated to
test each functionality for correctness as well as for failure conditions. In addition,
during black box testing, it is necessary to design the operational profile of the
software if the objective to achieve a measurable reliability estimate is desired.

$$\text{Black box adequacy} = \frac{\sum pf_i cf_i}{\sum pf_i} \tag{5}$$

where f_i is the ith functionality of the software as specified in the SRS and pf_i is
the probability of occurrence of the functionality f_i. The set $\{f_i\}$ along with the set
$\{pf_i\}$ represents the operational profile of the software. Also, cf_i is the functionality
coverage score. The value of cf_i is 1, if the functionality is tested for both positive
and negative criteria, is 0.6, if the functionality is tested only for correctness, is 0.4,
if the functionality is tested only for failure conditions and is 0, if the functionality
is completely untested. The expression $\sum pf_i$ represents the operational profile of
the software.

4.3 Code Walkthrough Score

Code walkthrough as discussed earlier is an essential technique to detect algorithmic
and logical errors. Since code walkthrough is a very tedious activity, it is usually not
possible to review the complete code in a single attempt. Hence, it is usually carried

out in an incremental manner and errors are reported as they are detected. We use the term "review" for each iteration of code walkthrough. Rallis et al. [10] have proposed that the probability of all faults getting detected by the end of the last review is a measure of software reliability. The model considers a review to be successful if it results in a latent fault to be detected. We have adopted the model to calculate the effectiveness of our code walkthrough cycles.

Let Q_i = The probability of detecting a fault during the ith review, N_i = no. of faults detected in the system during the ith review, $E_i = N_1 + N_2 + \cdots N_i$ and $K =$ probable no. of faults in the system before observing any data for the system. The detection probability Q_i is assumed to be the same for all faults. If K numbers of algorithmic and logical faults are present at the beginning of the first review, then applying Binomial distribution $B(N_i; K - E_{i-1}, Q_i)$ is the probability that exactly N_i faults are detected during the ith review is expressed as follows:

$$B(N_i; K - E_{i-1}, Q_i) = \frac{(K - E_{i-1})!}{(K - E_{i-1} - N_i)!N_i} Q_i^{N_i}(1 - Q_i)^{K - E_{i-1} - N_i} \quad (6)$$

Hence, after n reviews, Code walkthrough score is:

$$\sum_{i=1}^{n} B(N_i; K - E_{i-1}, Q_i) = \sum_{i=1}^{n} \frac{(K - E_{i-1})!}{(K - E_{i-1} - N_i)!N_i} Q_i^{N_i}(1 - Q_i)^{K - E_{i-1} - N_i} \quad (7)$$

The unknown variable in the above expression K is taken from the software reliability prediction model which we previously designed in collaboration with an academic institute [16]. The model can predict the numbers of faults in software before V&V activities are started using fuzzy logic methods. The detection probability Q_i is taken to be 0.7 based on our empirical experience of code walkthrough of several software over a period of ten years.

4.4 Automated Static Analysis Score

Automated static analysis tools can identify a wide range of probable software anomalies such as non-compliance with coding standards, unused and dead code, uninitialized variables, buffer overflows, null pointer dereferences, infinite loops, division by zero, and floating-point arithmetic problems. As discussed in earlier sections, different ASA tools employ multiple analysis techniques ranging from simple pattern matching logics to complex and sound techniques like abstract interpretation. A framework for static analysis is said to be sound if all defects checked for are reported, i.e., there are no false negatives but there may be false positives. Hence, performance evaluation of the tools becomes a necessity to gain confidence in analysis results produced by these tools. We have categorized ASA tools into three categories based on the analysis techniques incorporated, viz, *Cat 1(Pattern matching), Cat 2(Unsound dataflow analysis),* and *Cat 3(Sound dataflow analysis).*

The four ASA tools that have been used in our V&V process belong to these three categories. We have classified *LDRA* as a *Cat 1* tool, *Coverity & Klocwork* as *Cat 2* tools and *Astree* as *Cat 3* tool. These classifications are based on our empirical experience with these tools from historical data as well as from literature survey available on these tools [17–20]. To measure the effectiveness of ASA tools, two measures are commonly used in literature, viz, *Precision & Recall. Precision* is the ratio of a number of true positives over the number of reported errors. *Recall* is the ratio of a number of true positives over the actual number of errors in the source. It is quite obvious that for any ASA tool, *Precision & Recall* should be high. The code walkthrough score model that we have discussed in the earlier subsection can now be extended to measure the effectiveness of ASA with a few changes in the attributes.

Let Q_i = The probability of detecting a fault by the ith tool, N_i = no. of faults detected in the system by the ith tool, and K = probable no. of runtime faults and coding rule violations in the system. $N_i = \text{TP}_i + \text{FP}_i$, $E_i = \text{TP}_1 + \text{TP}_2 + \cdots \text{TP}_i$, where TP_i is the no. of actual faults detected by the ith tool and FP_i is the no. of false positives reported by the ith tool. $\text{FN}_i = K - \text{TP}_i$, where FN_i is the no. of false negatives, i.e., errors latent in the system and not reported by the ith tool. In ideal scenario, FN_i should be zero but as we have discussed earlier it is infeasible to design such a sound tool. Hence, we use *Recall* to represent the relationship between K and TP_i. $\text{TP}_i = \mathcal{R}_i * K$, where $\mathcal{R}_i$ is the *Recall* measure for the ith tool. Empirical studies have shown the value of $\mathcal{R}$ (*Recall*) is 0.5 ± 0.05 for *Cat 1*, 0.75 ± 0.05 for *Cat 2* and 0.9 ± 0.05 for *Cat 3* tools. The problem with tools with high $\mathcal{R}$ value is that their $\wp$ value (*Precision*) is less. This is due to over-approximation techniques employed by the Cat 3 tools. The over-approximation though reduces the precision but ensures that the results are safe. The safeness of analysis comes from the fact that over-approximation ensures that no false negatives are reported which is the problem with Cat 1 and Cat 2 tools. These two categories of ASA tools employ under-approximation to improve the efficiency and precision of analysis but increase the probability of false negatives. The application of code walkthrough score model in this case $B(\text{TP}_i; K - E_{i-1}, Q_i)$ is the probability that exactly TP_i faults are detected by the ith tool and is represented as:

$$B(\text{TP}_i; K - E_{i-1}, Q_i) = \frac{(K - E_{i-1})!}{(K - E_{i-1} - \text{TP}_i)!\text{TP}_i} Q_i^{\text{TP}_i}(1 - Q_i)^{K - E_{i-1} - \text{TP}_i} \quad (8)$$

The approximate value of K can be calculated as $K = \frac{\text{TP}_i}{\mathcal{R}_i}$. The fault-detection probability Q_i is assumed to be 0.99 as any automated static analysis tool invariably reports errors. The errors may be true positives or false positives and requires analysis to filter the results. Hence, for n tools used in the V&V process,

$$\sum_i^n B\left(\text{TP}_i; \frac{\text{TP}_i}{\mathcal{R}_i} - E_{i-1}, Q_i\right) = \sum_i^n \frac{\left(\frac{\text{TP}_i}{\mathcal{R}_i} - E_{i-1}\right)!}{\left(\frac{\text{TP}_i}{\mathcal{R}_i} - E_{i-1} - \text{TP}_i\right)!\text{TP}_i} Q_i^{\text{TP}_i}(1 - Q_i)^{\frac{\text{TP}_i}{\mathcal{R}_i} - E_{i-1} - \text{TP}_i}$$

$$(9)$$

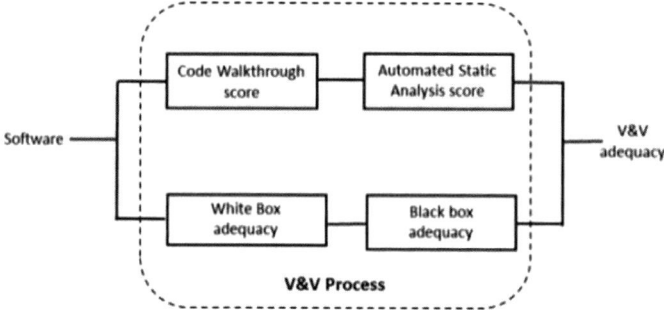

Fig. 2 V&V process adequacy in detail

The above expression is the ASA score for our V&V process using the three categories of tools.

4.5 V&V Adequacy Score

In the earlier subsections, we have explained the design of process adequacy models for each individual process. Now, we are going to present a cumulative adequacy model for the entire V&V process. The basic premise of this formulation is that each individual review technique contributes independently and equally for the overall process improvement. The schematic representation of the adequacy model is shown in Fig. 2.

As shown in Fig. 2, the overall V&V process can be visualized as two static testing technique blocks in series in one line which is parallel to other line containing the dynamic testing technique blocks in series. This is in fundamental principle of software verification wherein static and dynamic testing complements each other and ensures the development of more reliable software compared to only code-execution-based testing. The representation of blocks in series and parallel connections resemblances reliability block diagram (RBD) modeling. RBD is used to estimate the system reliability when subsystems are connected in series or in parallel or in a mixed fashion [21]. The application of RBD formula for mixed systems in our process is expressed as follows:

$$\text{V\&V adequacy score} = 1 - \big[\{1 - (\text{White box adequacy}) * (\text{Black box adequacy})\} \\ * \{1 - (\text{Code walkthrough score}) * (\text{ASA score})\}\big] \quad (10)$$

5 Experimental Study

We have conducted experiments to quantitatively validate our model on multi-version software used in mission-critical systems. The primary objective of our experimental evaluation is to arrive at a quantifiable figure in reliability growth based on the effectiveness of the V&V process.

5.1 Experiment Design

In our experiments, we have used embedded and GUI software. The embedded software is developed in C for 8051 microcontroller and the GUI software is developed in C++ in RHEL 5.3. Both the software are used in mission-critical systems and has been put through the rigorous V&V process. At the completion of each stage, the effectiveness of the particular process stage is computed. Finally, an overall process adequacy score is calculated for both the software.

5.2 Results

We present our experimental results in the subsequent paragraphs. Table 1 represents the characteristics of the software on which experiments were carried out. In Table 1, *Cum_Cyclo_complexity* column indicates the sum of individual cyclomatic complexities of all functions in the software. Tables 2 and 3 represent the White box and Black box adequacy results, respectively.

Table 4 represents the code walkthrough results. Column K shows the predicted numbers of algorithmic or logical faults present in the software as computed from fuzzy-logic-based software reliability prediction model [15]. Table 5 represents the automated static analysis results.

Table 1 Statistics of software under test

Module	KLOC	Files	Functions	*Cum_Cyclo_complexity*
Embedded	6	11	85	599
GUI	35	19	302	3740

Table 2 White box adequacy results

Module	Code coverage adequacy	Mutants	Mutants killed	White box adequacy
Embedded	0.77	24	21	0.68
GUI	0.82	40	35	0.72

Table 3 Black box adequacy results

Module	f_i	$\sum p f_i$	$\sum p f_i c f_i$	Black box adequacy
Embedded	15	12.5	10.3	0.83
GUI	28	25.2	19.91	0.79

Table 4 Code walkthrough results

Module	KLOC	K	Reviews	Faults detected	Code walkthrough score
Embedded	6	15	7	11	0.78
GUI	35	26	10	21	0.84

Table 5 Automated static analysis results

Module	LDRA	Coverity	Klocwork	Astree	ASA score
Embedded	35	56	53	68	0.92
GUI	63	78	82	110	0.73

Table 6 Software reliability estimate

Module	*Rel1*	*Rel2*	*Rel3*	*Rel4*	*Rel5*
Embedded	0.8981	0.9124	0.9361	0.9216	0.9422
GUI	0.864	0.881	0.875	0.907	0.928

We apply the RBD formula on the four process adequacies to arrive at a quantifiable estimate of reliability. The number of iterations through the V&V process uncovers more numbers of errors as the software is corrected and repaired. This results in software reliability to improve as the number of latent defects in the software reduces. The results have been tabulated in Table 6. Columns *Rel1*, *Rel2*, *Rel3*, *Rel4,* and *Rel5* depict the software reliabilities attained after the completion of first, second, third, fourth, and fifth iteration of V&V process cycles, respectively.

Figure 3 shows the growth in software reliability for the modules under test. It can be observed from the plot that there is a decrement in the curve in one cycle for the modules under test. The reason for this decrement is attributed to the fact that as errors get fixed some new errors creep into the software during the fixation process. Also, due to changes in the requirements, modifications and upgrades are carried out in the software which induces newer errors. Once, the additional errors are detected and corrected, the reliability continues to improve.

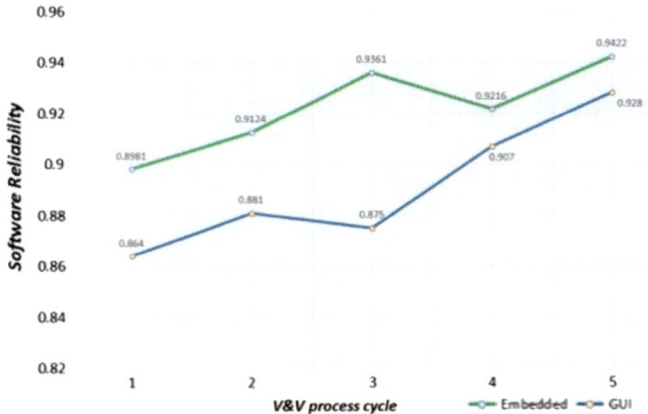

Fig. 3 Software reliability growth

6 Conclusion and Future Work

In this paper, we have presented an approach wherein the rigorous V&V process of mission-critical systems is assessed for its efficacy. The model depicts quantitative proof that each V&V process stage contributes significantly to improving the final software quality. Our experimentation on mission-critical software modules substantiates our model with satisfactory estimates. The software modules used in our experiments are operational software modules deployed and used in mission-critical systems. However, we will like to improve the accuracy of our model on open-source benchmark software.

Our model does not utilize the concept of formal verification and model-based techniques. Formal verification is highly recommended V&V philosophy to improve the reliability of any type of software. In our future work, we plan to augment our V&V process by integrating formal verification on critical logics of software. Hence, our objective for the future study will be to identify the critical logics in each software module and formally specify them at the outset. Once, the critical logics are identified and specified mathematically we can verify them formally using model-based techniques. This in turn will result in growth of the software reliability as the basis of formal verification is to prove the absence of bugs.

References

1. RTCA, Inc., December 1992, RTCA/DO-178B, *Software Considerations in Airborne Systems and Equipment Certification*, Washington, DC
2. D.R. Wallace, R.U. Fujii, Software Verification and Validation: An Overview. *IEEE Software* (May, 1989)

3. I. Musa, K. Okumoto, *Software Reliability Engineering: Measurement, Prediction, Application* (McGraw Hill, 1987)
4. A.L. Goel, Software reliability models: assumptions, limitations, and applicability. IEEE Trans. Software Eng. **11**(12), 1411–1423 (1985)
5. M.R. Lyu (ed.), *Handbook of Software Reliability Engineering* (McGraw Hill, 1996)
6. Y.K. Maliya, N. Li, J. Bieman, R. Karcich, Software reliability growth with test coverage. IEEE Trans. Reliab. **51**(4), 420–426 (2002)
7. M.H. Chen, M.R. Lyu, Effect of code coverage on software reliability measurement. IEEE Trans. Reliab. **50**(1), 165 (2001)
8. C. Senthil Kumar, Hybrid approach for estimation of software reliability in nuclear safety systems, in *AERB Newsletter*, vol. 26, No. 1 (June 2013)
9. M. Kersken, The role of static analysis in software reliability assessment, in *11th Advances in Reliability Technology Symposium*, pp 169–182 (1990)
10. N.E. Rallis, Z.F. Landsdowne, Reliability estimation for a software system with sequential independent reviews. IEEE Trans. Software Eng. **27**(12), 1057 (2001)
11. J. Jheng, N. Nagappan, J.P. Hudepohl, M.A. Vouk, On the value of static analysis on fault detection of software. IEEE Trans. Software Eng. **32**(4), 240 (2006)
12. W.W. Schilling, M. Alam, Modeling the reliability of existing software using static analysis, in *Proceedings of the. IEEE International Conference on Electro/Information Technology*, pp. 366–371 (May 2006)
13. P. Emanuelsson, U. Nilsson, A comparative study of industrial static analysis tools, in *Technical reports in Computer and Information Science* (Linkoping University, Sweden, January 2008)
14. N. Rutar, C.B. Almazan, J.S. Foster, A comparison of bug finding tools for java, in *Proceedings if the ISSRE,* pp. 245–256 (2004)
15. H. Koziolek, Operational profiles for software reliability, in *Seminar of Dependability Engineering* (Carl von Ossietzky University of Oldenburg, July 2005)
16. A. Das, M. Venkat, Missile software early reliability prediction using fuzzy logic, in *Proceedings of the DRDO Technical Seminar* (2008)
17. LDRA. http://www.ldra.com
18. Coverity: Software Testing and Static Analysis Tools. http://www.coverity.com
19. Klocwork: Source Code Analysis Tools for Security & Reliability. http://www.klocwork.com
20. Astree Runtime Error Analyzer-AbsInt. https://www.absint.com/astree
21. M. Rausand, A. Hoyland, *System Reliability Theory Models, Statistical Methods, and Applications*, 2nd edn. (Wiley Series)

A Review of Recent Dynamic Reliability Analysis Methods and a Proposal for a Smart Component Methodology

Darpan Krishnakumar Shukla and A. John Arul

Abstract The next-generation nuclear power plants are designed to have inherent safety features and passive safety systems and use advanced digital instrumentation and control (IC) systems to achieve required operational performance and to meet the safety goals. Digital IC systems often perform complex tasks while interacting with process dynamics. Static reliability methodologies have been shown to be inadequate for modeling and accurately estimating the reliability of such systems due to the lack of close correspondence between the model and the system. Though a number of new dynamic methods have been developed for the reliability evaluation of such systems, they have not reached a stage of maturity, like that of the traditional event tree/fault tree methods. Therefore, an effort is made to review the diverse and recent development in the field, with a view to identify suitable attributes and methods which are necessary for it to be widely adopted and for which a general-purpose software tool could be developed. The qualitative analysis is performed for the attributes among the reviewed methods for comparison and identifying the best method. A qualitative comparison is given for the major attributes. The paper recommends and outlines the development of smart component methodology (SCM) for reliability modeling of dynamic safety systems.

Keywords Dynamic reliability · Probabilistic dynamics · Reliability · Smart component methodology

D. K. Shukla (✉)
Indira Gandhi Centre for Atomic Research,
Homi Bhabha National Institute, Kalpakkam 603102, Tamil Nadu, India
e-mail: darpanks@igcar.gov.in

A. John Arul
Indira Gandhi Centre for Atomic Research, Kalpakkam 603102, Tamil Nadu, India
e-mail: arul@igcar.gov.in

© Springer Nature Singapore Pte Ltd. 2020
P. V. Varde et al. (eds.), *Reliability, Safety and Hazard Assessment for Risk-Based Technologies*, Lecture Notes in Mechanical Engineering,
https://doi.org/10.1007/978-981-13-9008-1_22

267

1 Introduction

Real systems are complex interacting entities, where the performance requirements, system configuration and failure parameters may change with time. With the advent in digital instrumentation and control (IC) technologies, the next-generation nuclear power plants are expected to be built with inherent safety features, passive performance and self-diagnostics. Static reliability methods, though widely used, do not adequately model such systems exactly using its pure combinatorics. This gap between the reliability model and the system is claimed in the literature, will be reduced if we carry out *dynamic reliability* evaluation for the system, and will get the results of our interest. While doing dynamic reliability evaluation of the system, physical process, hardware, software, human performance, etc., can be considered precisely in the system model for failure analysis; a methodology is followed for system structuring and for quantifying it, which is called dynamic reliability method. In this section, we summarize some of the reported definitions of dynamic reliability and then mention some of the challenges to be checked for reviewing the methods. Some views on the definition of dynamic reliability and dynamic reliability methodologies found in the literature are presented below.

Devooght, J. [1]: Dynamic reliability accounts the dynamic nature of system, i.e., the evolution of the system, change in dynamics due to failures, repairs, maintenance, etc., and hence, change in failure rates due to new dynamics.

Labeau, P. E. et al. [2]: Dynamic reliability methods provide a framework for explicitly capturing the influence of time and process dynamics on scenarios.

Aldemir, T. [3]: Dynamic methodologies for PSA defined as a time-dependent phenomenological model of system evolution along with its stochastic behavior to account for possible dependencies between failure events.

Babykina, G. et al. [4]: In the context of safety analysis, dynamic reliability can be seen as an extension of system reliability assessment methods to the case in which the structure function changes in time with a discrete evolution (e.g., in the case of a phased mission system) and with a continuous evolution (e.g., in the case of dependency between the reliability of components and continuous physical variables). The structure function is meant to be a function describing the link between the states of system components, usually represented by their failures and repairs, and the state of the system itself. The dynamic reliability accounts for the following phenomena: (1) dynamic behavior of system components (aging and fatigue of different types), (2) importance of the order of occurring events (ordered sequences of events are considered in place of cut sets) and (3) multiple natures (stochastic and deterministic) of events driving the transitions between states.

To summarize the definitions, *dynamic reliability methods* must be able to model

1. Time dependence of system structure function.
2. Time ordering of basic events.
3. Time dependence of the reliability parameters (failure rate, mission time, test interval).

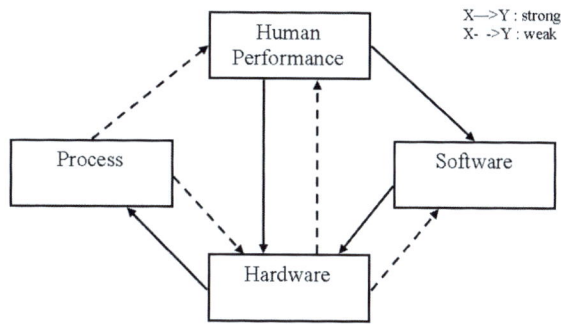

Fig. 1 Possible interaction between entities

4. Interaction effects: inter-dependence between process variables, hardware states, software function and human performance.

Hence, dynamic reliability methods give detailed modeling capability for the dynamic systems having interactions as described in Fig. 1, where the dotted and strong arrows indicate weak and strong interactions between the entities, respectively. In contrary to the benefits offered by the dynamic reliability methods of taking care of the interactions and other dynamic aspects in the method rather than the dependency on the analyst, which is for his careful approximation in static reliability analysis, the dynamic reliability analysis is challenged with the handling of complexity of the system, computational intensive nature and easiness of usage of the method. Various dynamic reliability methods have been developed in the last three decades and applied to many dynamic scenarios. In the literature, there is no consensus on the acceptance of a method for dynamic reliability assessment; the newly developed methods have not reached the maturity of application, like that of the traditional event tree/fault tree methods.

Therefore, the methods available in the literature are reviewed and compared with a view to identify suitability with respect to the attributes of accuracy, simplicity, scalability, fidelity and generality of modeling. However, we declare that the review does not aim to be exhaustive and complete but to find out a learned recommendation which would be a promising step toward a solution for all the above-mentioned problems. Based on the review, the paper recommends and outlines the peculiar qualities of smart component-based methodology along with Monte Carlo simulation for addressing the problems of dynamic reliability.

The paper is organized as follows. Section 2 presents the review of some of well-known dynamic reliability methods and approaches. Section 3 proposes a smart component methodology. Section 4 presents a comparative evaluation of the various methods presented in Sect. 2. The paper is concluded in Sect. 5.

2 Existing Methods, Approaches and Tools for Dynamic Reliability

A number of dynamic methodologies have been developed for application ranging from small- to large-scale systems. A large number of review papers have been published in the recent past on the subject of dynamic reliability and its methodologies [1–3, 5]. In the last decade, significant reports have been published by USNRC on the reliability analysis of digital I&C systems. The report Ref. [6] presents a study of a number of methods for reliability modeling of digital I&C systems and its incorporation into existing probabilistic risk assessment (PRA) models. The study identifies two types of dynamic interactions that need to be considered while modeling digital I&C system reliability: Type I—interactions between the reactor protection and control systems and the controlled plant process and Type II—interaction between the components of the reactor protection and control systems itself. According to the report, the required minimum criteria for acceptance of dynamic methodologies (only seven major requirements are reproduced out of eleven) are [6]:

1. The methodology should account for both Type I and Type II interactions.
2. The model must be able to predict encountered and future failures well and cannot be purely based on previous experience.
3. The model must make valid and plausible assumptions, and the consequences of violating these assumptions need to be identified.
4. The data used in the quantification process must be credible to a significant portion of the technical community.
5. The model must be able to differentiate between a state that fails one safety check and those that fail multiple ones.
6. The model must be able to differentiate between faults that cause function failures and intermittent failures.
7. The model must have the ability to provide uncertainties associated with the results.

From the qualitative analysis provided for some of the methods against the eleven criteria in Ref. [6], it is concluded that none of the methods meet all the criteria; however, each of the different methods may be better suited for specific problems. USNRC-6901 has ranked Markov/cell-to-cell mapping technique (CCMT) method and dynamic flowgraph methodology (DFM), as the top two methods, with most positive features and least negative or uncertain features (using subjective criteria based on reported experience) for dynamic probabilistic safety assessment. The report also mentions that there is no regulatory requirement for a single methodology to be applicable to all digital I&C systems. However, availability of such a methodology will be a convenience [6].

Subsequent report USNRC-6942 (2007) [7] and 6985 (2009) [8] proof checked the two identified methods. The study also applied the methods to benchmark systems and demonstrated the incorporation of the methods into PRA procedures. Report USNRC-6962 (2008) [9] presented the limitations of traditional reliability methods for digital system reliability evaluations. However, we observe a state explosion problem for Markov/CCMT method and doubt the solvability of dynamic flowgraph methodology when applied to industrial-scale systems.

The following section briefly reviews some of the significant methods with a view to assess its suitability for general applicability to dynamic reliability analysis, flexibility for deriving static reliability analysis and suitability for industrial-scale problems.

2.1 Methods Based on Chapman–Kolmogorov Equation

2.1.1 Chapman–Kolmogorov Equation

A number of methods are based on the general mathematical description of the time evolution of the joint probability of hardware state and process variables. The *Chapman–Kolmogorov equation (C-K equation)* for probability mass transport is described in several Refs. [1, 10–13], but first introduced in Refs. [14] as a dynamic approach for modeling failure in process control systems using Markovian assumption. Then, the duo of probabilistic and deterministic behavior of the reactor dynamics is mathematically supported by Ref. [11, 12] as a *theory of probabilistic dynamics (PD)/continuous event tree*. Briefly, the C-K equation is described; herein, integro-differential form and its more useful integral forms for reliability and availability problems for application to dynamic PSA are found in Refs. [11, 15].

Let, the reactor process variables be denoted by $X = [x_1, x_2, x_3, \ldots x_p]$, where p is the number of process variables under consideration which are functions of time. The system hardware (H/W) logic state is determined from the component states of the system. Let us denote the hardware states by i, $i = 1, 2, \ldots q$, then the evolution of the process variables of the system is given by the following differential equation:

$$\frac{\mathrm{d}X(t)}{\mathrm{d}t} = f_i(X), \quad \text{with} \quad X(0) = X_0, \quad \text{and} \quad X \in R^p \tag{1}$$

This is a complex nonlinear functional equation, the solution of which is given by

$$X = g_i(t, X_0), X_0 = g_i(0, X_0) \tag{2}$$

In the equation, function $f_i(X)$ describes the dynamic evolution of the process variables for a given H/W state i, which may have differential dependence on X. Hence, the overall state of the reactor dynamics is represented by (X, i). The system evolves either due to the change in state of the components (stochastic) or due to automated control action (deterministic). Generally, the probability of transition from one state to another depends mainly on X. Hence, $P(i \rightarrow j|X)$ is the conditional probability of transition from H/W state i to the H/W state j per unit time. Let $\Pi(X, i, t)$ be the probability density that the system is in state (X, i) at time t. Then, the C-K equation can be written as follows:

$$\frac{\partial \Pi(X, i, t)}{\partial t} = \sum_{j \neq i} P(j \rightarrow i | X) \Pi(X, j, t)$$

$$- \sum_{j \neq i} P(i \rightarrow j | X) \Pi(X, i, t) - \nabla_X f_i(X) \Pi(X, i, t) \qquad (3)$$

Or

$$\frac{\partial \Pi(X, i, t)}{\partial t} = \sum_{j \neq i} P(j \rightarrow i | X) \Pi(X, j, t) - \lambda_i \Pi(X, i, t) - \nabla_X f_i(X) \Pi(X, i, t)$$

$$(4)$$

where $\lambda_i(X) = \sum_{j \neq i} P(i \rightarrow j | X)$ is defined as the total probability of outgoing transition from state i. It is assumed here that the probability Π and function f_i are well behaved, so that the divergence operation exists. There are situations when considering nonlinear systems, where the divergence may not exist and such special conditions are addressed in [16]. The special issues are not addressed in this paper as the focus here is on the development of an overall framework.

The integral version of the above equation is as follows [15]:

$$\Pi(X, i, t) = \int \Pi(u, i, 0) \delta(X - g_i(t, u)) e^{- \int_0^t \lambda_i(g_i(s, u)) ds} du$$

$$+ \sum_{j \neq i} \int P(j \rightarrow i | u) du \int_0^t \delta(X - g_i(t - \tau, u)) e^{- \int_0^{(t-\tau)} \lambda_i(g_i(s, u)) ds} \Pi(u, j, \tau) d\tau$$

$$(5)$$

The integral equation describes that the probability density that the system is in state (X, i) at time t is the sum of two probabilities [15]: The first term represents the evolution of system in the same state i for all the time t without any transition, while the process variables are changing deterministically according to the dynamics. The second term describes all the transition into state i taking place before time t.

The C-K equation takes as input, transition probabilities (failure rates, repair rates), a set of process variables and related dynamics, domain of safe working of the system and initial distribution of probability density. The dynamic methodologies can further be classified according to input representation and solution methods for C-K equation. Numerical treatment of the C-K equation invokes inevitable discretization schemes. To reduce the complexity in solving the complete C-K equation, the process variables are divided into discrete cells and the most famous method applying this technique is the cell-to-cell mapping technique method. Some of the well-known solution schemes to the C-K equation are Monte Carlo algorithms [12], cell-to-cell mapping technique [17] and dynamic event tree techniques, which are reviewed in forthcoming sections. Dynamic event trees and classical event trees are also derived from continuous event tree which are essentially discrete time versions of continuous event tree [10] and are compared with each other in Refs. [15, 18].

2.1.2 Cell-to-cell Mapping Technique (CCMT) Methods

The *cell-to-cell mapping technique (CCMT)* for reliability and safety studies was first developed by Aldemir et al. [14, 17, 19, 20]. In the CCMT, the continuous process variables, present in the C-K equation, are divided into cells of smaller intervals. The representative C-K equation for CCMT is obtained from the general C-K equation by integrating over each of the cells. The probability density is assumed to be uniform in these cells. The discrete hardware states and the discretized process variables form a discrete state space. The interval boundaries of cells are preferentially placed at the limit values of the control action which will subsequently be useful for considering non-temporal distribution. The resulting Markov chain coupled with dynamic evolution equations is solved with time step Δt assuming that system will not change its state during the time step. The method can be used both in inductive mode, to identify system evolution with given initial conditions, and in deductive mode, to identify sequence of paths that lead to the top event/undesired system states [19, 21–23].

The Markov/CCMT method is attributed to have some inherent limitations from memory and computational time requirements. For storing and handling of the transition matrix, large memory is required. The required memory space is proportional to the number of process variables times the number of cells (into which each variable is discretized) times the number of states. Reducing the number of cells partitioning the safety domain will reduce the matrix size, but consequently accuracy will be reduced. Therefore, attempts have been made to reduce memory requirements for this method in [17, 24] by using the vectorization method and sparse matrix technique. Hassan [19] has provided improvements to the method for storage requirements by using database for storing process evolution. The limitation of the number of acceptable process variables for tractable treatment of the dynamics remains unsolved till now [2]. Next is the treatment of the transition rates which are dependent on the values of the process variables and the choice of the time step. The time step in Markov/CCMT should be so small that not more than one cell crossing should occur in that step. But, keeping a small time step will, first, underestimate the transition probability due to the small probability of leaving the cell boundary and, second, slow down the computations [25]. Reference [25] describes the process-dependent time step limitation as a modeling problem of CCMT; i.e., the dynamics underlying the whole process is incorrectly represented if the size of the cells is not reduced in correspondence to the value of the time step. Reference [26] has addressed the time step problem of CCMT by treating time as a continuous variable in continuous time cell-to-cell mapping technique (CCMT). Reference [27] treats the stiffness of the transition matrix which arises due to the difference in scales of the time constants of the process dynamics. Recently, Ref. [23] addresses the limitations of computational time and storage using backtracking algorithm with pruning mechanism. The reference also mentions that accuracy can be maintained by more sampling in the cells, and computational time can be managed by going for parallel computation. The Markov/CCMT is applied for modeling of BWR/6 SBLOCA [19, 28] and a benchmark system of digital feedwater control system of a PWR [29, 30], and hence it is a strong candidate for large-scale application [23].

2.2 Monte Carlo Simulation

2.2.1 Direct Monte Carlo Simulation

The first paper on the *Monte Carlo method (MC method)* was published in 1949 by N. Metropolis and S. Ulam. The MC method was initially applied to neutron transport calculations, and then the usage of the term Monte Carlo became synonymous with the term simulation, i.e., Monte Carlo simulation (MCS), while its applications in particle physics, communication network traffic, models of conflicts, computation of multiple integral, etc., have been explored in a wide range of the literature. For use in probabilistic dynamics, three slightly different MCS approaches, viz. discrete event simulation [5], dynamic Monte Carlo availability model (DYMCAM) [31] and analog MC [12], are used. The analog MC is the first application of Markov MC [32]. The analog MCS algorithm described in Ref. [12] is as follows:

Step-1   The initial configuration i and a value of the process variables X are sampled from $\Pi(X, i, 0)$.
Step-2   The next transition time t out of i is sampled.
Step-3   The evolution of X in state i is computed up to t; if a border of 'working' domain is crossed, or if the end of the accident duration is reached, the current history is stopped.
Step-4   A new state j is sampled; if j is an unacceptable configuration, the history is stopped.
Step-5   This procedure is repeated from Step-2, j being the new value of i.

The well-known advantages of MC method over other methods for application to probabilistic dynamics include: The method is insensitive to the dimensionality of the problem, MCS can treat all types of distributions for modeling the system, and MCS directly estimates the safety and reliability characteristics.

In contrast to the advantages, (i) the accuracy of results from the MCS method depends on the number of simulations performed, which may lead to large computational cost when it applies to PD, because after each sampling of the sequence of events the simulation calls the evolution function of the system according to the new configuration; (ii) in case of very low probability of failure events, most of the histories will not result in failures; hence, the number of histories required for good results could be huge. Consequently, increasing the number of simulations increases the computational time due to repeated calculations of deterministic evolutions, and decreasing the number of simulations decreases accuracy; hence, strong solution schemes are required.

2.2.2   Methods for Improving MCS

To improve the computational efficiency of MCS, two aspects need to be addressed: (i) reducing the number of samples required for a given accuracy and (ii) reducing the simulation time for each sample. Several methods have been developed for increasing the efficiency of statistical sampling for rare event simulation and improving deterministic simulation efficiency for repeated calculations. In the reliability estimation domain for reduction of variance of the result, importance sampling, subset simulation, biased simulation, forced transition, memorization are some of the acceleration techniques used Refs. [33–37]. The first time use of extreme value theory for rare event probability estimate is presented recently in Ref. [38]. The generalized extreme value distribution can be used to estimate the probability of crossing the safety boundary. Since there are only three asymptotic forms for extreme value distributions, they can be determined from limited simulations and can be used to extrapolate the estimate of reliability and failure frequency.

For reducing the process simulation time per sample, *pre-simulation memorization*-based schemes introduced in Refs. [39–42] are used. *Cell-to-boundary (CTB)* and *most probable evolution (MPE)*-based methods provide significant acceleration to the simulation. In CTB-based memorization method, the safety domain, D, is divided into cells and the evolution of system from all nodes is performed for all system configurations up to the closest boundary of the safety domain. The characteristics of these system evolutions are memorized. The characteristics include time and the value of X at the intersection with the closest boundary, the type of event encountered (control action/failure) at the boundary and a measure of the probability of the system to stay in the current state i up to the boundary. During the simulation, the time to the next transition is sampled from the current point in the process variable space. From the memorized time to the closest boundary from the neighboring nodes, since probability of reaching to the boundary is high, the dynamic calculations utilizing the memorization are brought to the boundary through an interpolation procedure [43]. The drawback of the memorization-based method is that if the size of the problem is large then memory requirements increase and the memorization becomes difficult.

In MPE, one defines a most probable set of initiating events; then for each of them, the dynamics of the system is integrated and main characteristics are stored till the end of the event sequence. The memory required to store is less compared to CTB method as least probable evolution paths are not calculated in MPE method [40]. Moreover, the use of a combination of CTB and MPE seems to be a more advantageous scheme. In a way, MPE is used to know the most probable evolution path, and cell in the neighborhood of the MPE paths is evaluated for CTB characteristics. The advantage of both the methods is brought together in Ref. [39]. In addition to these techniques, there is also the possibility of using response surfaces and metamodels for reducing the process evolution computation time [44].

The development of software tools based on MC method for dynamic reliability is based on hybridization of the MCS technique with other techniques which solve the problems highlighted in the previous paragraph. For example, the combination of CTB and MPE is described in Ref. [39].

Piecewise deterministic Markov process (PDMP) was first introduced by Davis in 1984 for Markov processes consisting of deterministic processes and random jumps. PDMP can be used for treating multiple failures [45] and preventive maintenance. The advantage of PDMP is that it is applicable to most types of dynamic reliability problems. However, PDMP is not easy to manipulate. It is difficult both to specify and to solve by methods other than MCS. The general PDMP algorithm for MCS is presented in Ref. [46] for dependability analysis of the famous heated tank system benchmark [14].

2.3   Dynamic Event Tree (DET)

In comparison with conventional event tree, dynamic event tree synthesizes branching at different time points and physical process evolution. It is often called as *discrete dynamic event tree (DDET)* to distinguish DET and CET. The first ever dynamic method introduced in 1981 for probabilistic transient analysis for PRA is called: event sequence and consequence spectrum using logical analytical methodology in Ref. [47], the first version of dynamic logical analytical methodology (DYLAM) [48]. Subsequently, many variants of DET-type methodologies and software tools that use DETs have been developed in USA and Europe, such as DYLAM-3 [48], dynamic event tree analysis method (DETAM) [49], accident dynamic simulator (ADS) [50], integrated safety assessment (ISA) [10], Monte Carlo dynamic event tree (MCDE-T) [51], analysis of dynamic accident progression tree (ADAPT) [21] and recently reactor analysis and virtual control environment (RAVEN) by INL [52–54]. All the methods differ in treatment of branching rules, stopping rules, branching generations (automatic), interfacing with existing deterministic dynamic code for physical parameters for an accident sequence, model of operator behavior. In other words, treatment of aleatory and epistemic uncertainties is different in all the methods.

DET is an event tree model in which branchings are allowed to occur at different points in time and each system can have more than two states. Initially, a fixed time interval was used for branching and then Ref. [55] introduced probabilistic branching. Reference [49] defines DETAM with five characteristic sets that define the dynamic event tree approach. These are: (i) Set of variables included in the 'branching set.' (ii) Set of variables defining the 'plant state.' (iii) Set of 'branching rules.' (iv) Set of sequence expansion rules. (v) Quantification tools. The 'branching set' is the set of variables that determine the space of possible branches. The plant state at any node in the tree is defined by the value of the variables in the branching set. The 'branching rules' are the rules used to determine when a branching should take place. Most straightforward branching rule could be a constant time step or based on hardware failure time, etc. The 'sequence expansion rules' are the rules used to limit the number of sequences, hence tree expansion. Sequence termination could be based on simulation time or reaching user-defined absorbing states or on sequence frequency falling below a user-specified lower limit. The quantitative tools are those used to compute the process variables as well as the branching frequencies.

There were various challenges in the use of DET [2, 3], but later it has been tackled in recently developed DET-based tools: first, the explosion of number of branches while allowing all possible branches in simulation, second the computational resource-intensive nature of the tools based on DET and third processing of the amount of data produced for a single initiating event. The first and second problems can be solved by increasing the time step, decreasing the probability threshold, limiting branches according to the probability cut-off or terminating the branches which exceed the user-defined number of failures or grouping of similar scenarios. The second problem can also be solved by using parallel processing [54, 56]. The main advantage of DET, which makes it a well-developed method, includes its capability to provide complete investigation of the possible accident scenarios. There is no assumption on the type of the distribution for branching, and the method provides readable results to the risk analyst. Because analysis of all the scenarios at once is difficult, many developments have been carried out in the literature for classification of similar scenarios, called scenario clustering, for more understanding of the DET results, and hence addressed the third problem of processing of the data generated for a single initiating event [57].

2.4 . Dynamic Flowgraph Methodology (DFM)

The method was introduced by S. Guarro and D. Okrent, for process failure modeling in 1984. DFM accounts for the interaction between hardware and software within embedded digital system. DFM is a digraph-based technique, and it uses graphical symbols like (i) process variable node (circle), (ii) condition edge (dotted arrow), (iii) causality edge (continuous arrow), (iv) condition node (square box), (v) transition box and (vi) transfer box [2, 3, 6, 58]. A brief description of the elements of DFM is as follows. A process variable node represents a process variable which is discretized into a finite number of states. There is no criterion for discretization of the process variable, but boundary/threshold crossing and logic of the system is preferentially used for discretization. The system dynamics is represented by a cause-and-effect relationship between these states, and states are connected by causality edges. Condition edges are used to connect condition node and transfer box. The transfer box selects the transfer function according to the value present in the condition node. The transition box allows system to evolve in time by providing time lag between input and output parameters.

DFM can be used for both deductive (top to bottom) and inductive (bottom to top) tracing of fault propagation. DFM yields the prime implicants for the system.

The advantages of DFM are that it has ability to change system configuration in discrete time. This can be used for acquiring information for the possible accident sequence leading to top event. In addition to that, DFM models the system and top event separately, which is useful in case where various top event probabilities ought to be determined. Reference [6] mentions DFM method as the most preferable methodology for the reliability modeling of digital l&C systems. Application of DFM includes: control software in advanced reactors [59, 60], modeling human performance and team effects [59] and PSA modeling of a digital feedwater control system similar to an operating PWR [29].

In DFM due to the discretization of process variable, there exists inevitable discretization error. Proper discretization of each variable can be achieved by progressively refining the discretization. Hence, there is a trade-off between the accuracy of the model and the size and complexity of the model [2]. Reference [61] suggested DFM and Markov/CCMT can be used in a complementary manner, finding prime implicants in DFM, i.e., the combination of basic events necessary and sufficient to cause top event, and quantifying the prime implicants using Markov/CCMT, to achieve comprehensive modeling and evaluation of digital I&C systems. The methods to find prime implicants from the DFM include the transformation of DFM into FT [62] or into binary decision diagram (BDD) [63]. In Yadrat code, the later approach is used to evaluate DFM [58]. The other computer program based on DFM is Dymonda [61].

2.5   Event Sequence Diagram (ESD)

This is a self-explanatory diagram mainly used for understanding of accident sequences. The ESD method with dynamic capabilities is the extension of continuous event tree methodology with visual graphics. The ESD uses 6-tuple of events, conditions, gates, process parameter set, constraint and dependency rules to represent dynamic scenarios like conditions, competitions, concurrent processes (output AND gate), mutually exclusive outcomes (output OR gate), synchronization processes (input AND gate) and physical process (physical variable condition). The objective of the method is to find failure frequency and availability of the system to be in a particular state at a particular time. The ESD models the time implicitly, in a way branching in ESD occurs based on the crossing of a boundary by a process variable. The method can be utilized to predict future failure, cut sets, etc. For availability of a system in some states, all the possible states of the system need to be declared and, of course, the transition rates between them. The drawbacks of ESD include: The methodology is not a component-based approach but a scenario modeling framework. The sequence identification is analyst dependent as the sequence delineation has to be performed by the analyst and not performed automatically. In a highly dynamic system, the ESD could become very large [2]. The application of ESD includes Refs. [64, 65] for Cassini mission and transient analysis of europa reactor.

A semi-analytical solution is possible using the ESD approach [2]. The procedure is summarized as follows [2]: first, constructing the ESD using the symbolic ESD language, second deriving mathematical representation for all components in the ESD using the governing equations, third obtaining analytical approximations to the relevant process variables and finally solving the integrals analytically or numerically to obtain end-state probabilities. The job of the analytical approximations could be laborious. The semi-analytical approach reduces the problem of dimensionality of the system evolution by solving a sequence of multidimensional integral equations.

2.6 Dynamic Fault Tree (DFT)

This is the extension of the traditional fault tree with the time element. DFT was first developed by Dugan et al. [66]. One type of dynamic fault tree uses *generalized dynamic gates* for representing interaction of the basic events in time [66, 67]. For instance [67] uses the following gates: functional dependency gate (FDEP), spare gates (CSP: cold spare gate, HSP: hot spare gate, WSP: warm spare gate), priority AND gate (PAND) and sequence enforcing gate (SEQ). If $c = a$ PAND b, then c becomes true when a and b are true but only if a becomes true before b. The SEQ gate is an expression of a constraint that components can only fail in the specified order and the top event occurs when all basic events have occurred. The FDEP gate represents the generation of dependent events once the trigger event (a failure) occurs. The dependent events are multiple inputs to this gate and can be connected to inputs of other gates. This gate does not have an output. The SPARE gate is used to model different types of redundancy, i.e., cold, warm and hot. CSP means the spare component does not fail, HSP means the spare component may fail at its full rate, and warm SPARE means the spare component may fail at a rate equal to the full rate reduced by dormancy factor. However, it is not clear as to how many of these gate types are essential to model the real systems and there is no systematic analysis to our knowledge that these gates are complete to model dynamic situations.

The other type of dynamic fault tree uses *house events* with a house event matrix handling the multiple operation mode changes in time and the configuration changes with time [68]. In a house event matrix, columns describe system configuration at the time periods and rows describe the number of the house events present in the model, which timely switch on and off in accordance with the status of the modeled system. The dynamic fault tree and static FT have common disadvantages: complexity in modeling and simultaneously handling the sequential dependent failure and multiple operation mode evolution of the system. Adversely, there is no model for process evolution in these DFTs. Intuitive and convenient *reliability graph with generalized gate* (RGGG) has been improved to account for these challenges using dynamic generalized gates [69] and reliability matrix [70], respectively, where dynamic generalized gates solve problem of sequence-dependent failure and reliability matrix able to account for the limitation of multiple operation modes of dynamic fault trees. The quantification of the dynamic fault tree can be performed either by time-dependent Boolean logic, Markov models, Bayesian network [71, 72], MCS [67] or numerical integration [73].

2.7   GO-FLOW Methodology

The *GO-FLOW* [74] methodology was developed as a success-oriented methodology; however, failure states were later included in the methodology by introducing NOT gates in the method. In this method, an analyst needs to make a chart of the physical layout of the system. The chart has all possible operational states of the system. Signal lines and operators are used to make the system chart. The operators are connected by signal lines. The operator presents the functioning or failure of the component or a logical gate or a signal generator. The operator also represents transition or any Boolean logic (AND, OR, NOT). The input/output signal lines represent the process variable or time or any other information.

Using GO-FLOW method, reliability and availability can be evaluated for the complex systems. The assumptions in the modelling include constant failure rate models, two state components and the time of failure of various components are independent. The GO-FLOW methodology is useful for phased mission analysis as it provides compactness of the representation of the phased mission situation in comparison with event trees. The GO-FLOW methodology can be used as quantitative reliability evaluation of process systems after utilizing static methods (such as failure mode and effect analysis (FMEA)) for qualitative analysis for identifying failure modes [75].

There are some drawbacks in the method. The method cannot model redundant systems easily [2]. The method possesses the state space explosion problem in the case of a large number of system components. Also, the concept of the method may be hard to learn and implement for the analyst [6]. The available computer program for GO-FLOW methodology includes in Ref. [74]. The GO-FLOW methodology has been applied to nuclear systems and control system of bullet trains in Ref. [2], and passive safety system of AP1000 reactor for dynamic reliability evaluation [75].

2.8   Stochastic Hybrid Automaton (SHA)

Though a component failure can occur gradually, it is often treated as a discrete event. Following this approximation, a system under consideration can be treated as a *discrete event system (DES)*, i.e., a system whose behavior is not controlled by time but by the occurrences of events. The *SHA theory* is an extension of the theory of FSA, considering stochastic events such as failure of components and deterministic evolution of the process variables [76, 77].

Here, the procedure for dynamic risk assessment is a two-step process involving the development of an SHA model of the studied system and then MCS of the system modeled by SHA for either a predefined mission time or a stopping criterion. The SHA method is able to work with different continuous modes of the operation of the system which are generated due to control action upon threshold crossing and the deterministic or stochastic transition events due to component failure. The later events are incorporated through probability distributions, and the former is defined by ordinary differential equations in the method. The SHA model does not require preliminary simplifying hypotheses, and thus it is able to consider all types of events such as failure, repairs, maintenance, aging, control actions, demand-dependent aging and future behavior dependent on the past [4]. The SHA method shows to be a suitable tool to model large complex systems operating in a dynamic environment. The method is implemented in Scilab/Scicos environment in Ref. [77] and is applied to test cases of the heated tank system and an oven with a temperature control system. In the validation study of the SHA method for the two systems, the features of the SHA method were tested for the following cases: (1) dependencies/interaction between stochastic events and continuous variables, (2) multiple aging modes according to repairable/reconfigurable components and (3) treatment of non-temporal behavior of components such as failure on demand. References [4, 78, 79] address the application of the SHA for the dynamic scenarios of a controlled steam generator, a feedwater control system of steam generator and a decanter unit of a chemical plant, respectively. The only disadvantage of the SHA method is the size of the model and consequently the simulation time.

The available computer programs which are based on SHA are *Dynamic Reliability and Assessment (DyRelA)* [80], PythoniC-Object-Oriented Hybrid Stochastic Automata (PyCATSHOO) [81] and *Stochastic Hybrid and Non-repairable Dynamic Fault Tree Automaton (SHyFTA)* [79]. The SHA is used as input representation for PDMP in the PyCATSHOO [81] tool. Recently, PyCATSHOO has been applied to a DHR system of sodium fast reactor in Ref. [82] for dynamic PRA-informed design. The SHyFTA, is a stochastic model of dynamic reliability uses *Hybrid Basic Event (HBE)*, and it is a combination of SHA and non-repairable dynamic FT. HBE is defined as the evolution of the basic events; i.e., basic event is not simply characterized by a static cumulative distribution function of probability, but also depends on the deterministic evolution of the system. The approach of SHyFTA is based on the separation of two mutually dependent processes, the deterministic and the stochastic. Compared to DyRelA and PyCATSHOO, SHyFTA includes *reliability, availability, maintainability and safety (RAMS)* formalism and also additional feature of MATLAB framework since it has been developed in Simulink. The Simulink architecture provided in Ref. [79] is generalized and can be used for any other dynamic reliability problems.

2.9 Petri Net (PN)

This is a graphical modeling method introduced by C. A. Petri in 1966 for system modeling. PN is similar to finite state machine. PN explicitly models time using a set of elements like nodes/places, tokens, arcs and transition. The nodes, represented by circles, are used to describe system state or process variables; transition is represented by a rectangular box, describe events; and arcs are used to connect nodes/places to transition or transition to nodes/places. The token, represented by a dot in the circle, describes the current state of the system. When PNs are executed, the tokens are consumed by transition in each of the places where it is connected from and tokens are produced by transition in each of the places where it is connected to provided that while triggering all the places that are connected from are having at least one token each. Initially, firing of timed transition was having only exponential distribution or deterministic times. The times derived from constant failure rate model can be used as mapping between PN and Markov model. Then, Ref. [83] has extended the *stochastic PN (SPN)* to *generalized stochastic Petri net (GSPN)* which represents a semi-Markov process (allowing any probability density function for time sampling). The overview of PN theory and SPN is presented in Ref. [84]. In *interpreted stochastic PN*, messages (Boolean variables (0/1, true/false)) are passed along with firing of the PN. Interpreted stochastic PNs cannot be proof checked, and hence the verification process is non-intuitive and time consuming [2]. Other extensions of PNs include fuzzy PN, colored PN, timed PN, stochastic PN [85, 86].

For quantitative evaluation of dynamic system reliability using PN, process variable models can be coupled with PN and the resulting MCS can be used to perform quantitative analysis [61]. Also, PNs can be used to simulate FTs directly and hence can be readily included in PRA [87]. A standard PN with inhibitor arcs and condition places can be used to determine missing safety requirements, uncertainties in safety requirements and inconsistencies in safety requirements.

The *stochastic reward Petri net (SRPN)* has been developed for reliability analysis of communication networks by [88]. Reference [89] presented a generalization of PN to mode automata. A mode automaton is an input/output automaton with a finite number of states called modes. The mode automata can be compiled into fault trees. Mode automata are a superset of PNs with inhibitor arcs and can model everything a computer can compute [6].

PN allows explicit representation of time and is able to simulate concurrent activities, dynamic activities and time delay. It also allows studying systems that are characterized as being asynchronous, distributed, parallel, non-deterministic and/or, stochastic. The major limitation of PN is combinatorial explosion when modeling large systems.

2.10 Dynamic Bayesian Network (DBN)

As mentioned earlier, in a dynamic system, system configuration can change, which, in turn, can change the failure rate of the corresponding components. That describes the conditional dependence of failure probability on dynamics or system configurations. In these situations, Bayesian updating of conditional probability can be directly utilized. The Bayesian network (BN) is a probabilistic approach, and it explicitly represents the relationships/interactions between the system components. For graphical representation of the interactions between components and process variables of a system, BN uses nodes for describing variables/events/system components and arcs/edges connecting to nodes for describing relationships (child and parent) between the nodes. BN graphs are directed acyclic graphs. The nodes having only incoming arcs are called leaf nodes. The nodes with only outgoing arcs are independent nodes (root nodes). The root nodes are defined with marginal prior probabilities. The prior conditional probabilities are stored in *conditional probability table (CPT)* for each node with success and failure instantiations except root nodes. The probabilities of each node are updated using Bayes formula. In a complex system having n number of components, a component may not depend on all other components but only on m of them, the number of parent nodes. Hence, there are 2^m entries instead of 2^n in the conditional probability table. This reduces computational workload substantially [90]. The drawbacks of BN model include the following: BN models require prior knowledge of probability of the interactions. Further, the BN model developed is strongly expert dependent. These disadvantages can be addressed by the development of suitable algorithms for automatic construction of BN [90].

Explicit representation of time in BN is referred to as *timed Bayesian network (TBN)*. Two approaches are presented in modeling time in BN, i.e., *time-sliced approach* and *event-based approach*. Early work on the application of BNs to dynamic domains has led to formalisms known as *temporal Bayesian networks* [91], *dynamic Bayesian networks (DBNs)* [92], network of 'dates,' and modifiable temporal belief networks (MTBNs) are based on the time-sliced approach. The temporal node Bayesian network (TNBN) and net of irreversible events in discrete time (NIEDT) are based on the event-based approach.

The dynamic extension of BN model, i.e., DBN, is a powerful and flexible tool to model dynamic system behavior and update reliability and uncertainty analysis with life cycle data [93]. In DBN, the BN is defined at n number of time steps called n-time-sliced BN and the nodes representing component/event/variables of each time step are connected to each other using edges. The relationships between time steps are described by interconnecting edges. The state of the variables at time step n depends on time step $n-1$, time step $n-2$ and so on. Generally, two time slices are considered to model the system time evolution, hence the model called 2-TBN or 2-time slice temporal BN. As the state variables depend only on previous time step variables, the 2-TBN model is Markovian model. For quantification, DFTs are translated into DBNs in Refs. [71, 72]. References [90, 94–97] show recent developments in the DBN based reliability analysis.

3   Proposal of a Combination of Smart Component Methodology and Monte Carlo Simulation

In this section, we propose and briefly describe a smart component methodology (SCM) in combination with MCS as a dynamic reliability analysis method. However, the quantitative validation and any theoretical development are out of the scope of the paper. One of the requirements for a dynamic reliability modeling is a system representation. In SCM, the system representation is done through object-oriented architecture. Here, the components of a system and their interrelationship (between entities that are presented in Fig. 1) are described using abstract objects (drawn from the paradigm of object-oriented analysis and design of software systems). The objects will have attributes and rules. The attributes capture all information of the individual component, for example, the functional and reliability variables. The rules capture the behavior of the object in the system, and it will be responsible for object evolution in time. The interrelationship between components is defined in the connector, and they capture the dependencies among them. This way of representing each component or subsystem for reliability analysis by an abstract software object (with attributes, interrelation and rules) is termed smart components (SCs). Next, after the system structuring using the SC model, the SC model of the system is operated with a suitable reliability quantification algorithm. Component-based MCS is selected for reliability analysis, where MC sampling of the component states and transition time is carried out to explore the system state space and then, along with deterministic system simulation, calculate reliability measures based on the number of visits to the system failed state. This representation is perhaps the most foolproof [2]. However, the challenge lies in the development of suitable MCS algorithms or methods that translate the object-oriented representation to, for example, a DFM.

To implement the SCM, a relational database architecture is selected for representing a given system which contains all the component-as-an-object. The components would include all entities of Fig. 1, i.e., process, hardware, software and human performance. The objects contain the information of the components with the following attributes:

1. Input parameters.
2. Output parameters.
3. Parameters describing the state of the component (hardware, software, etc.).
4. Reliability parameters (failure rate, repair rate, an indicator of the reliability model, the probability of failure).

In addition to the components, the database file must contain the 'connector' object describing the connections between the components' attributes which defines the hardware wiring of the system. The function of each component, manifested by the rules, may represent its functional dependence on other components or input variables. The process variables of an object are treated as a functional dependence through rules of the component. Using all this information about the system, MCS can be performed using the procedure described elsewhere.

In SCM, the analyst makes the SC model of the system and uses SC simulators for quantification. The building of the SC model is intuitive and user-friendly. Hence, the proof-of-correctness of the method is transferred to the simulation algorithm rather than the analyst. MCS handles the continuous time, multiple failure modes and aging separately, and the unification of all in one MCS algorithm for SCM is a challenge. The scalability of SCM depends on MCS algorithm, and it is not addressed here, but the presence of parallel computers and advanced computing techniques would alleviate the problem of computational intensive nature of the MC method. An advantage of SCM is that the same SC system model can be used for static reliability evaluation because the system represented using the connector and component objects can be translated to a reliability block diagram (RBD). For instance, static reliability evaluation can be done by carrying out a reachability check on the SC model. Hence, the method would be considered as a method with backward compatibility, which is expected from any dynamic reliability method.

In the next section, the dynamic reliability methods are qualitatively compared.

4 Qualitative Comparison of Various Methods

The qualitative attributes considered for this study are listed below. A 'Yes' implies the method is suitable for use in dynamic reliability assessment concerning that attribute, and 'No' implies otherwise.

A. The method provides improved accuracy compared to static reliability methods. The method takes care of the timing of occurrence of events, process variable evolution, control action, multiple failure modes, more importantly dependence of failure rate on both process parameters and time.

    I. Time declaration (continuous = Y/discrete = N).
    II. Treatment of process variable (Y/N).
    III. Multiple failure modes and aging (Y/N).

B. Burden of Proof: Equivalence of reliability model to the system model (fidelity), requirement of skillful analyst and level of approximations in the method (Y=burden of proof requirement is less).

S. Whether the method is scalable to large problems, including complexity of representation (memory) solution time (time) as a function of problem size (less than exponential growth or there is a way to handle exponential growth within the method = Y/exponential growth = N).

F. Whether the method is user-friendly or intuitive for modeling of system (Y/N).

G. Whether general-purpose software is available based on the method or suitable for being developed into one (Y/N).

I. Possibility of interfacing of the results of the analysis with FT/ET (Y/N).

H. Whether the method is able to incorporate software and human reliability model (Y/N).

Table 1 Qualitative comparison

Methods	A.I	A.II	A.III	B	S	F	G	I	H
Dynamic fault tree	Y	N	N	N	N	Y	Y	Y	N
Cell-to-cell mapping technique	Y	Y	Y	Y	N	N	N	N	N
Dynamic event tree	Y	Y	Y	Y	Y	Y	Y	Y	Y
Dynamic flowgraph methodology	Y	Y	Y	N	N	N	Y	Y	Y
GO-FLOW methodology	Y	N	Y	N	N	N	N	N	Y
Petri nets	Y	N	Y	Y	Y	N	Y	N	Y
Stochastic hybrid automaton	Y	Y	Y	Y	Y	N	Y	N	Y
Event sequence diagram	Y	Y	N	N	N	Y	N	Y	Y
Dynamic Bayesian network	Y	N	Y	N	N	Y	Y	N	Y
Piecewise deterministic Markov processes	Y	Y	Y	Y	N	N	N	N	N
Smart component methodology	Y	Y	Y	Y	Y	Y	Y	Y	Y
Direct Monte Carlo simulation	Y	Y	Y	Y	N	N	N	Y	Y

From Table 1, the most favorable method is the method with the most Y's. That includes DET, SHA, SCM and direct MCS. In DET, the deductive analysis is difficult (top event to basic event). SHA theory is then the second best method in the list with two disadvantages of not being able to interface the result with fault tree methodology and being less user-friendly and intuitive. Though the direct MC method is scalable and in principle can model anything, it lacks a system-to-simulation model translation mechanism. Hence, it is not readily usable by non-experts. This problem is precisely addressed in the SCM and hence is promising regarding most of the attributes.

We believe that one of the principal reasons for non-availability of a general-purpose tool for dynamic reliability assessment is the disconnect between representation schemes and simulation power. Many of the easy and intuitive methods do not scale up. Similarly, MCS which in principle can solve any problem requires a general-purpose representation scheme and interface. With this perspective, when the methods are evaluated for the attributes B, S, F, G and I, smart component method emerges as a very likely candidate for a general-purpose dynamic reliability analysis method.

5   Conclusion

We have presented a brief review of some available dynamic reliability methods with a view of selecting and developing the best among them for use in reliability analysis of dynamic systems. The attributes required for an ideal dynamic method are elucidated, and the reviewed methods are inter-compared. In the inter-comparison, we emphasize the general applicability and suitability of the methods for generic tool development, as these attributes mainly have contributed to the lack of widespread use of dynamic reliability methodology. Based on the current review, SCM is identified to have excellent potential for the widespread application, while capable of faithfully modeling dynamic scenarios including software and human reliability aspects. The framework provides the structure for the dynamic system representation, while MCS is the key computational method.

6 Future Developments

A framework for the implementation of the SCM, utilizing object-oriented design and relational database concepts, is in progress. The MCS algorithms need to be explored for unifying various MC reliability models and at the same time acceleration of results. Completeness of MC state space exploration is to be studied for verifying the completeness of results since the size of state space in dynamic systems is huge. Demonstration of application of the SCM for small- to an industrial-scale system is to be studied.

Acknowledgements The authors thank Martin B. Sattison and Diego Mandelli for their valuable feedback on the draft manuscript. The authors thank Director, Reactor Design Group, IGCAR, for the encouragement and support for completing the work. First, the author thanks the Board of Research in Nuclear Studies, Mumbai, India, and Department of Atomic Energy, India, for supporting through DGFS-Ph.D. fellowship.

References

1. J. Devooght, Dynamic reliability. Adv. Nucl. Sci. Technol. **25** (1997)
2. P.E. Labeau, C. Smidts, S. Swaminathan, Dynamic reliability: towards an integrated platform for probabilistic risk assessment. Reliab. Eng. Syst. Saf. **68**, 219–254 (2000)
3. T. Aldemir, A survey of dynamic methodologies for probabilistic safety assessment of nuclear power plants. Ann. Nucl. Energy **52**, 113–124 (2013)
4. G. Babykina, N. Brînzei, J.-F. Aubry, G. Deleuze, Modeling and simulation of a controlled steam generator in the context of dynamic reliability using a stochastic hybrid automaton. Reliab. Eng. Syst. Saf. **152**, 115–136 (2016)
5. N.O. Siu, Risk assessment for dynamic systems: an overview. Reliab. Eng. Syst. Saf. **43**(1), 43–73 (1994)
6. T. Aldemir, D. Miller, M.P. Stovsky, J. Kirschenbaum, P. Bucci, A.W. Fentiman, L.T. Mangan, Current State of Reliability Modeling Methodologies for Digital Systems and Their Acceptance Criteria for Nuclear Power Plant Assessments (NUREG/CR-6901), Technical report (2006)
7. T. Aldemir, M. Stovsky, J. Kirschenbaum, D. Mandelli, P. Bucci, L.A. Mangan, D. Miller, X. Sun, E. Ekici, S. Guarro, M. Yau, B. Johnson, C. Elks, S. Arndt, Dynamic Reliability Modeling of Digital Instrumentation and Control Systems for Nuclear Reactor Probabilistic Risk Assessments (NUREG/CR-6942), Technical report (2007)
8. T. Aldemir, S. Guarro, J. Kirschenbaum, D. Mandelli, L.A. Mangan, P. Bucci, M. Yau, B. Johnson, C. Elks, E. Ekici, M. Stovsky, D. Miller, X. Sun, S. Arndt, Q. Nguyen, D.J, A

Benchmark Implementation of Two Dynamic Methodologies for the Reliability Modeling of Digital (NUREG/CR-6985), Technical report (2009)

9. T.L. Chu, G. Martinez-Guridi, M. Yue, J. Lehner, P. Samanta, Traditional Probabilistic Risk Assessment methods for digital system (NUREG/CR-6962), Technical report (2008)

10. J.M. Izquierdo, E. Melendez, J. Devooght, Relationship between probabilistic dynamics and event trees. Reliab. Eng. Syst. Saf. **52**, 197–209 (1996)

11. J. Devooght, C. Smidts, Probabilistic reactor dynamics-I: the theory of continuous event trees. Nucl. Sci. Eng. **240**, 229–240 (1992)

12. C. Smidts, J. Devooght, Probabilistic reactor dynamics-II: a Monte Carlo study of a fast reactor transient. Nucl. Sci. Eng. **111**, 241–256 (1992)

13. J. Devooght, C. Smidts, Probabilistic reactor dynamics-III. A framework for time-dependent interaction between operator and reactor during a transient involving human error. Nucl. Sci. Eng. **3**(112), 101–113 (1992)

14. T. Aldemir, Computer-assisted markov failure modeling of process control systems. IEEE Trans. Reliab. 36(1), 133–144 (1987)

15. J. Devooght, C. Smidts, Probabilistic dynamics as a tool for dynamic PSA. Reliab. Eng. Syst. Saf. **8320**(95), 185–196 (1996)

16. T. Aldemir, Some measure theoretic issues in probabilistic dynamics. Nucl. Sci. Eng. **155**, 497–507 (2007)

17. M. Belhadj, T. Aldemir, The cell to cell mapping technique and Chapman-Kolmogorov representation of system dynamics. J. Sound Vib. **181**(4), 687–707 (1995)

18. C. Smidts, Probabilistic dynamics: a comparison between continuous event trees and a discrete event tree model. Reliab. Eng. Syst. Saf. **44**, 189–206 (1994)

19. M. Hassan, T. Aldemir, A data base oriented dynamic methodology for the failure analysis of closed loop control systems in process plants. Reliab. Eng. Syst. Saf. **27**(3), 275–322 (1990)

20. T. Aldemir, M. Belhadj, L. Dinca, Process reliability and safety under uncertainties. Reliab. Eng. Syst. Saf. **52**(3 SPEC. ISS.), 211–225 (1996)

21. A. Hakobyan, R. Denning, T. Aldemir, S. Dunagan, D. Kunsman, A Methodology for Generating Dynamic Accident Progression Event Trees for Level-2 PRA, in *PHYSOR-2006, ANS Topical Meeting on Reactor Physics*, Vancouver, BC, Canada (2006), B034 1–9

22. P. Bucci, J. Kirschenbaum, L.A. Mangan, T. Aldemir, C. Smith, T. Wood, Construction of event-tree/fault-tree models from a Markov approach to dynamic system reliability **93**(2008), 1616–1627 (2015)

23. J. Yang, T. Aldemir, An algorithm for the computationally efficient deductive implementation of the Markov/cell-to-cell-mapping technique for risk significant scenario identification. Reliab. Eng. Syst. Saf. **145**, 1–8 (2016)

24. M. Belhadj, T. Aldemir, Some computational improvements in process system reliability and safety analysis using dynamic methodologies. Reliab. Eng. Syst. Saf. **52**, 339–347 (1996)

25. P.E. Labeau, Modeling PSA problems-II: a cell-to-cell transport theory approach. Nucl. Sci. Eng. **150**, 140–154 (2005)

26. B. Tombuyses, T. Aldemir, Continuous cell-to-cell mapping. J. Sound Vib. **202**(3), 395–415 (1997)

27. B. Tombuyses, J. Devooght, Solving Markovian systems of O.D.E. for availability and reliability calculations. Reliab. Eng. Syst. Saf. **48**(1), 47–55 (1995)

28. M. Belhadj, M. Hassan, T. Aldemir, On the need for dynamic methodologies in risk and reliability studies. Reliab. Eng. Syst. Saf. **38**(3), 219–236 (1992)

29. T. Aldemir, S. Guarro, D. Mandelli, J. Kirschenbaum, L.A. Mangan, P. Bucci, M. Yau, E. Ekici, D. Miller, X. Sun, S. Arndt, Probabilistic risk assessment modeling of digital instrumentation and control systems using two dynamic methodologies. Reliab. Eng. Syst. Saf. **95**(10), 1011–1039 (2010)

30. J. Kirschenbaum, P. Bucci, M. Stovsky, D. Mandelli, T. Aldemir, M. Yau, S. Guarro, E. Ekici, S.A. Arndt, A benchmark systems for comparing reliability modeling approaches for digital instrumentation and control systems. Nucl. Technol. **165**, 53–95 (2009)

31. D.L. Deoss, N.O. Siu, A simulation model for dynamic system availability analysis, Technical report (1989)
32. E.E. Lewis, F. Boehm, Monte Carlo simulation of Markov unreliability models. Nucl. Eng. Des. **77**, 49–62 (1984)
33. M. Marseguerra, E. Zio, Optimizing maintenance and repair policies via a combination of genetic algorithms and Monte Carlo simulation. Reliab. Eng. Syst. Saf. **68**, 69–83 (2000)
34. M. Marseguerra, E. Zio, Nonlinear Monte Carlo reliability analysis with biasing towards top event. Reliab. Eng. Syst. Saf. **40**(1), 31–42 (1993)
35. P.E. Labeau, E. Zio, Procedures of Monte Carlo transport simulation for applications in system engineering. Reliab. Eng. Syst. Saf. **77**, 217–228 (2002)
36. M. Marseguerra, E. Zio, Monte Carlo biasing in reliability calculation with deterministic repair times. Ann. Nucl. Energy **27**(2000), 639–648 (1999)
37. H. Cancela, P.L. Ecuyer, M. Lee, G. Rubino, B. Tuffin, Analysis and improvements of path-based methods for Monte Carlo reliability evaluation of static models. Improv. Monte Carlo Reliab. Eval. static Model. (Chap. 1) (2009), 1–20
38. M. Grigoriu, G. Samorodnitsky, Reliability of dynamic systems in random environment by extreme value theory. Probabilistic Eng. Mech. **38**, 54–69 (2014)
39. P.E. Labeau, E. Zio, The cell-to-boundary method in the frame of memorization-based Monte Carlo algorithms. Math. Comput. Simul. **47**, 347–360 (1998)
40. P.E. Labeau, A Monte Carlo estimation of the marginal distributions in a problem of probabilistic dynamics. Reliab. Eng. Syst. Saf. **52**(1), 65–75 (1996)
41. M. Marseguerra, E. Zio, J. Devooght, P.E. Labeau, A concept paper on dynamic reliability via Monte Carlo simulation. Math. Comput. Simul. **47**, 371–382 (1998)
42. P.E. Labeau, Monte Carlo estimation of generalized unreliability in probabilistic dynamics-I: application to a pressurized water reactor pressurizer. Nucl. Sci. Eng. **126**(2), 131–145 (1997)
43. M. Marseguerra, E. Zio, The cell-to-boundary method in Monte Carlo-based dynamic PSA. Reliab. Eng. Syst. Saf. **48** (1995)
44. J.M. Lanore, C. Villeroux, F. Bouscatie, N. Maigret, Progress in methodology for probabilistic assessment of accidents: timing of accident sequences, in *ANS - ENS - International ANC/ENS Topical Meeting Probabilistic Risk Assessment*, Port-Chester, NY, USA (1981)
45. Y.H. Lin, Y.-F. Li, Dynamic reliability models for multiple dependent competing degradation processes, in *Safety and Reliability: Methodology and Applications—Proceedings of the European Safety and Reliability Conference ESREL 2014*, pp. 775–782 (1984)
46. H. Zhang, F. Dufour, Y. Dutuit, K. Gonzalez, Piecewise deterministic Markov processes and dynamic reliability. Proc. Inst. Mech. Eng. Part O J. Risk Reliab. **222**(4), 545–551 (2008)
47. A. Amendola, G. Reina, Event sequences and consequence spectrum: a methodology for probabilistic transient analysis. Nucl. Sci. Eng. **77**, 297–315 (1981)
48. G. Cojazzi, The DYLAM approach for the dynamic reliability analysis of systems. Reliab. Eng. Syst. Saf. **52**, 279–296 (1996)
49. C.G. Acosta, N.O. Siu, *Dynamic Event Tree Analysis Method* (DETAM) for accident Sequence Analysis, Technical report, 1991)
50. K.-S. Hsueh, A. Mosleh, The development and application of the accident dynamic simulator for dynamic probabilistic risk assessment of nuclear power plants. Reliab. Eng. Syst. Saf. **52**, 297–314 (1996)
51. K. Martina, P. Jörg, MCDET: a probabilistic dynamics method combining monte carlo simulation with the discrete dynamic event tree approach. Nucl. Sci. Eng. **153**(2), 137–156 (2006)
52. A. Alfonsi, C. Rabiti, D. Mandelli, J.J. Cogliati, R.A. Kinoshita, A. Naviglio, S.C. Columbia, Erin, F. Garrick, A. Hughes, I. Maracor Technical Services, et al., Dynamic Event Tree analysis through RAVEN, in *International Topical Meeting Probabilistic Safety Assessment Analysis 2013, PSA 2013*, 3 (October 2015) (2013) pp. 1697–1709
53. A. Alfonsi, C. Rabiti, D. Mandelli, J. Cogliati, C. Wang, P.W. Talbot, D.P. Maljovec, C. Smith, *RAVEN Theory Manual* (Technical Report March, Idaho National Laboratory, 2016)
54. A. Alfonsi, C. Rabiti, J. Cogliati, D. Mandelli, S. Sen, R. Kinoshita, C. Wang, P.W. Talbot, D. Maljovec, A. Slaughter, C. Smith, *Dynamic Event Tree Advancements and Control Logic Improvements* (Technical Report September, Idaho National Laboratory, 2015)

55. A. Amendola, Accident sequence dynamic simulation versus event trees. Reliab. Eng. Syst. Saf. **22**(1–4), 3–25 (1988)
56. A. Hakobyan, T. Aldemir, R. Denning, S. Dunagan, D. Kunsman, B. Rutt, U. Catalyurek, Dynamic generation of accident progression event trees **238**, 3457–3467 (2008)
57. D. Mandelli, A. Yilmaz, T. Aldemir, K. Metzroth, R. Denning, Scenario clustering and dynamic probabilistic risk assessment. Reliab. Eng. Syst. Saf. **115**, 146–160 (2013)
58. K. Björkman, Solving dynamic flowgraph methodology models using binary decision diagrams. Reliab. Eng. Syst. Saf. **111**, 206–216 (2013)
59. A. Milixi, R.J. Mulvihill, S. Guarro, J.J. Persensky, *Extending the Dynamic Flowgraph Methodology (DFM) to Model Human Performance and Team Effects*, US NRC (2001)
60. M. Yau, S. Guarro, G. Apostolakis, Demonstration of the dynamic flowgraph methodology using the Titan II space launch vehicle digital flight control system. Reliab. Eng. Syst. Saf. **49**(3), 335–353 (1995)
61. K. Björkman *Digital Automation System Reliability Analysis - Literature survey* (Technical report, 2010)
62. S. Guarro, D. Okrent, The logic flowgraph: a new approach to process failure modeling and diagnosis for disturbance analysis applications. Nucl. Technol. **67**, 348–359 (1984)
63. T. Tyrväinen, Prime implicants in dynamic reliability analysis. Reliab. Eng. Syst. Saf. **146**, 39–46 (2016)
64. S. Swaminathan, C. Smidts, The event sequence diagram framework for dynamic probabilistic risk assessment. Reliab. Eng. Syst. Saf. **63**, 224 (1999)
65. S. Swaminathan, J.-Y. Van-Halle, C. Smidts, A. Mosleh, S. Bell, K. Rudolph, R.J. Mulvihill, B. Bream, The Cassini Mission probabilistic risk analysis: comparison of two probabilistic dynamic methodologies. Reliab. Eng. Syst. Saf. **58**(1), 1–14 (1997)
66. J.B. Dugan, S.J. Bavuso, M.A. Boyd, Dynamic fault-tree models for fault-tolerant computer systems. IEEE Trans. Reliab. **41**(3), 363–377 (1992)
67. K.D. Rao, V. Gopika, V. Sanyasi Rao, H. Kushwaha, A. Verma, A. Srividya, Dynamic fault tree analysis using Monte Carlo simulation in probabilistic safety assessment. Reliab. Eng. Syst. Saf. **94**(4), 872–883 (2009)
68. M. Čepin, B. Mavko, A dynamic fault tree. Reliab. Eng. Syst. Saf. **75**(1), 83–91 (2002)
69. S.K. Shin, P.H. Seong, A quantitative assessment method for safety-critical signal generation failures in nuclear power plants considering dynamic dependencies. Ann. Nucl. Energy **38**(2–3), 269–278 (2011)
70. S.K. Shin, Y.G. No, P.H. Seong, Improvement of the reliability graph with general gates to analyze the reliability of dynamic systems that have various operation modes. Nucl. Eng. Technol. **48**(2), 386–403 (2016)
71. S. Montani, L. Portinale, A. Bobbio, D. Codetta-raiteri, Radyban: a tool for reliability analysis of dynamic fault trees through conversion into dynamic Bayesian networks. Reliab. Eng. Syst. Saf. **93**(7), 922–932 (2008)
72. A. Bobbio, L. Portinale, M. Minichino, E. Ciancamerla, Improving the analysis of dependable systems by mapping fault trees into Bayesian networks. Reliab. Eng. Syst. Saf. **71**(3), 249–260 (2001)
73. S.V. Amari, G. Dill, E. Howald, A new approach to solve dynamic fault trees. Reliab. Maintainab. Symp. 374–379 (2003)
74. T. Matsuoka, M. Kobayashi, GO-FLOW: a new reliability analysis methodology. Nucl. Sci. Eng. **98**, 64–78 (1988)
75. M. Hashim, H. Yoshikawa, T. Matsuoka, M. Yang, Quantitative dynamic reliability evaluation of AP1000 passive safety systems by using FMEA and GO-FLOW methodology. J. Nucl. Sci. Technol. **51**(4), 526–542 (2014)
76. J.-F. Aubry, N. Brînzei, *Systems Dependability Assessment: Modeling with Graphs and Finite State Automata*, risk management edition (Wiley, New York, 2015)
77. G.A.P. Castañeda, J.-F. Aubry, N. Brinzel, Stochastic hybrid automata model for dynamic reliability assessment. IMechE 28–41 (2015)

78. G. Babykina, N. Brînzei, J.-F. Aubry, G. Deleuze, Modelling a feed-water control system of a steam generator in the framework of the dynamic reliability, *ESREL-2013* (2014), pp. 3099–3107

79. F. Chiacchio, D.D. Urso, L. Compagno, M. Pennisi, F. Pappalardo, G. Manno, SHyFTA, a stochastic hybrid fault tree automaton for the modelling and simulation of dynamic reliability problems. Expert Syst. Appl. **47**, 42–57 (2016)

80. G.A.P. Castañeda, J.-F. Aubry, N. Brînzei, DyRelA (Dynamic Reliability and Assessment), pp. 39–40 (2010)

81. H. Chraibi, Dynamic reliability modeling and assessment with PyCATSHOO: application to a test case, in *PSAM* (2013)

82. V. Rychkov, H. Chraibi, Dynamic probabilistic risk assessment at a design stage for a sodium fast reactor, in *IAEA-CN245-042*, Yekaterinburg (2017), pp. 1–8

83. A. Marson, *Modeling with Generalized Stochastic Petri Nets* (Wiley, New York, 1995)

84. C. Cordier, M. Fayot, A. Leroy, A. Petit, Integration of process simulations in availability studies. Reliab. Eng. Syst. Saf. 8320 (96)

85. L. Gomes, J.P. Barros, Addition of fault detection capabilities in automation applications using Petri nets, in (IEEE, New York, 2004), pp. 645–650

86. L. Zhenjuan, P. Bo, L. Hongguang, Batch process fault diagnosis based on fuzzy petri nets, in *Proceedings of the First International Conference Innovations and Computer Information Control* (2006)

87. T.S. Liu, S.B. Chiou, The application of Petri nets to failure analysis. Reliab. Eng. Syst. Saf. **57**, 129–142 (1997)

88. M. Balakrishman, K. Trivedi, Stochastic Petri nets for reliability analysis of communication network applications with alternate routing. Reliab. Eng. Syst. Saf. (53), 243–259 (1996)

89. A. Rauzy, Mode automata and their compilation into fault trees. Reliab. Eng. Syst. Saf. (78), 1–12 (2002)

90. O. Doguc, J.E. Ramirez-Marquez, A generic method for estimating system reliability using Bayesian networks. Reliab. Eng. Syst. Saf. **94**(2), 542–550 (2009)

91. K. Kanazawa, Reasoning about Time and Probability Keiji Kanazawa, Ph.D. thesis, Brown University (1992)

92. A.E. Nicholson, J.M. Brady, Dynamic belief networks for discrete monitoring. IEEE Trans. Syst. Man. Cybern. **24**(11), 1593–1610 (1994)

93. J. Zhu, M. Collette, A dynamic discretization method for reliability inference in dynamic Bayesian networks. Reliab. Eng. Syst. Saf. **138**, 242–252 (2015)

94. M. Kalantarnia, F. Khan, K. Hawboldt, Dynamic risk assessment using failure assessment and Bayesian theory. J. Loss Prev. Process Ind. **22**(5), 600–606 (2009)

95. H. Boudali, J.B. Dugan, A discrete-time Bayesian network reliability modeling and analysis framework. Reliab. Eng. Syst. Saf. **87**(3), 337–349 (2005)

96. P. Weber, L. Jouffe, Complex system reliability modelling with Dynamic Object Oriented Bayesian Networks (DOOBN). Reliab. Eng. Syst. Saf. **91**(2), 149–162 (2006)

97. A. Ben Salem, A. Muller, P. Weber, Dynamic Bayesian networks in system reliability analysis, in *6th IFAC Symposium Fault Detection, Supervision and Safety Technical Processes*, pp. 481–486

Cell-to-Cell Mapping Technique Application for Dynamic Reliability Estimation

Mahendra Prasad, Vipul Garg, Gopika Vinod and J. Chattopadhyay

Abstract Digital systems have replaced analog because they can help increase plant safety and reliability. Conventional event-tree (ET)/fault-tree (FT) method have been used for reliability analysis of digital systems in nuclear power plants (NPPs). The static ET/FT approach does account the dynamic interaction between the digital system and the NPP process systems. Among the PSA analyst and regulatory body, there is limited consensus on the use of dynamic models for estimating risk from NPP. In this work, cell-to-cell mapping technique (CCMT) along with Markov model is used to estimate time-dependent reliability of water tank-level control problem. The Markov model is developed for inlet and outlet control valves, while the CCMT equations are solved with predefined time steps. These two are coupled to get the time-dependent probability of top event.

Keywords CCMT · Dynamic reliability · Digital system · Markov model

1 Introduction

In probabilistic safety assessment (PSA), dynamic methods consider the components failure behavior (stochastic model) in conjunction with the system model. When these two are coupled, then a time-dependent system evolution model is obtained. There are concerns on using the static methodology to adequately account for the impact of process/hardware/software/firmware/human interactions on the stochastic system behavior. references [1, 2] give a detailed list and description of dynamic PSA methods. Challenges in dynamic reliability are as follows:

- Assume specific relation between process variables (proportional change due to control action);

M. Prasad (✉) · V. Garg · G. Vinod · J. Chattopadhyay
Reactor Safety Division, Bhabha Atomic Research Centre, Mumbai, India
e-mail: mprasad@barc.gov.in

V. Garg
e-mail: vipulgarg@barc.gov.in

© Springer Nature Singapore Pte Ltd. 2020
P. V. Varde et al. (eds.), *Reliability, Safety and Hazard Assessment for Risk-Based Technologies*, Lecture Notes in Mechanical Engineering,
https://doi.org/10.1007/978-981-13-9008-1_23

- Handling of non-coherence;
- Modeling of human behavior;
- Uncertainties resulting from arbitrary discretization of the state space in modeling the process.

2   Need for Dynamic PSA

Dynamic PSA tends to increase the realistic modeling of systems risk quantification when the time-dependent interactions are considered. These methods should not be taken as an alternative to the PSA methodology that is currently used in NPP risk analysis. They can be a complementary tool for the improved modeling and useful insights into risk.

The dynamic methods can produce a large amount of system states and hence require computation power. The classical combination of fault-tree (FT) and event-tree (ET) analysis is used to develop risk models in PSA. FT and ET analysis are static and Boolean logic-based approaches. There are challenges in including the dynamic time-dependent models into PSA. These challenges can be in the fields of operator interactions, passive equipment, etc.

3   Types of Interactions in Dynamic PSA

3.1   Type 1 Interaction

Type 1 interaction refers to the coupling between the digital system and process system. The failure mode of the digital system may be dependent on the NPP processes, wherein the digital system is assigned for control/monitoring of the parameters such as pressure, temperature, and level.

3.1.1   Methodologies for Modeling Type I Interactions

There are broadly three methods, namely discrete time, continuous time, and the one which has visual interaction that are used to model Type 1 interactions. NUREG/CR-6901 [3] provides the details for the same. An example of continuous-time method is continuous cell-to-cell mapping (CCCM). Few discrete-time methods include dynamic event-tree analysis method (DETAM). The visual interface methods include Petri nets and 'GO-FLOW.'

3.2 Type 2 Interactions

Type 2 interaction concerns the interaction among components of the digital system.

3.2.1 Methodologies for Modeling Type 2 Interactions

Few of the methods to model Type 2 interaction are Petri net, dynamic flow graph, Markov methods, etc.

4 Markov/CCMT Methodology

Markov/CCMT methodology integrates the component Markov model and process dynamics to get time-dependent probabilities.

4.1 CCMT Methodology

In CCMT, the first requirement is the knowledge of the top cell (where the system can be considered as fail/success or as the case may be as per the analysis) so that state space is discretized (or partitioned) into U_j ($j = 1, …, J$) cells. Discrete time steps are considered. The system evolution is modeled through the probability that the controlled variables are in cell U_j in the state space at time $t = n\Delta t$ ($n = 0, 1, …$), while the components in the system are in combination condition $l = 1, …, L$ (modeled by the Markovian model). The transitions between cells depend on the dynamic laws (in terms of differential equations) governing the system and state of system components. The cells designated as 'absorbing cells' are those from which the system cannot move out, and hence, the transition probabilities from these cells to other cells in the state space are equal to 0. The top event cell is designated as 'absorbing cell.' The partitioning has to done such that all cells are disjoint and cover the whole state space.

4.2 Markov Methodology

In Markov method, component transition to different states is represented graphically by directional links. This method is used to get the probabilities of system in various configurations at each time instant. The method has the following constraints:

- States are mutually exclusive; state 'lm' ($m = 1,\ldots, M; l = 1,\ldots, L$) is defined for component 'm,' i.e., probability $P_{lm}(t)$ that a component 'm' is in state 'l' at time 't.'
- The probability of transition between states is known through solution of the state equations.

4.3 Markov/CCMT Methodology

When the conditions, such as process variable in cell U_j at time $t = n\Delta t$ and the components of system are in a combination l' at time $t = n\Delta t$, are met, then the $w(j|j', l', n)$ represents the cell-to-cell transfer probability that the controlled variables are in the cell $U_{j'}$ at time $t = (n + 1)\Delta t$.

The transfer probability $w(j|j', l', n)$ is derived as follows:

- An origin cell j' is partitioned into Z sub-cells.
- The midpoint of a sub-cell is taken as starting point for the governing equations for system dynamics. These are integrated over the time interval $n\Delta t \leq t \leq (n + 1)\Delta t$. The component state combination remains the same in this time interval.
- Let the number of times the process parameter of interest arrives in the in destination cell j be Y in the time interval $n\Delta t \leq t \leq (n + 1)\Delta t$.
- The probability $w(j|j', l', n)$ is derived as $w(j|j', l', n) = Y/Z$.

Thus, we get the probability of the system being in U_j at $t = (n + 1)\Delta t$, when it was in $U_{j'}$ at time $t = n\Delta t$. The sum of the probabilities over the whole space of control variable is 1 since the cells cover the whole space of variable (cells are non-overlapping, mutually exclusive, and exhaustive). Since the top cell defines the event where failure occurs, the system failure probability can be obtained at any time instant.

5 Application of the Markov/CCMT Model to the Adapted System

This section describes how the Markov/CCMT methodology can be applied on physical system. Here, this methodology is applied on the test case of water-level control system. This is shown in Fig. 1.

All physical relations of control variables and different components have been considered in the analysis. The initial level of water is 15 ft, and the initial position of control valve 1 and 2 is 5%. In this case, for simplification, failure of only two control valves has been assumed and that all other components are working properly. So in further discussion and simulation, this assumption should be taken care of. Figure 2 shows the Markov transition diagram for the control valves.

Notations in the figure have the following meanings:

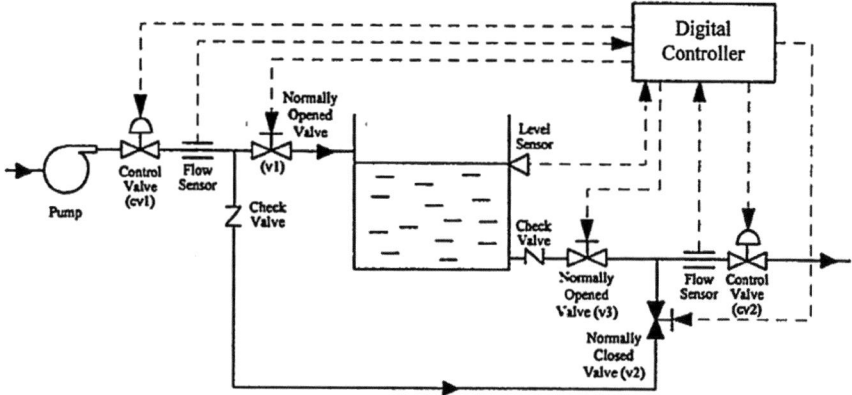

Fig. 1 Tank water-level controller [3]

Fig. 2 Markov transition
diagram for the control
valves

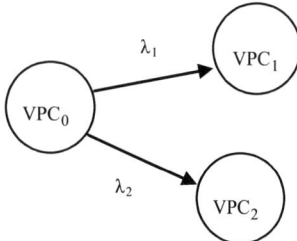

- VPC$_0$: normal working condition;
- VPC$_1$: failed open;
- VPC$_2$: failed close.

The state probabilities were obtained analytically by solving the coupled differential equations from the Markov diagram (Fig. 2). The time step chosen for the dynamic analysis must adhere to the following condition:

$$\Delta t < 1/\lambda \text{(so that the components keep same state '}n\text{' to '}n + \Delta t\text{')}$$

The time step chosen in the analysis is 500 ms. The controlled variable state space (CVSS) partitioning is the first step that must be accomplished during this stage. Table 1 shows the cells of the control variable state space that were chosen for this study. The shaded cell represents the top event. These cells are subsequently divided into $P = 8$ sub-cells.

Software program was written to solve the system dynamic equations along with the Markov modeling. Whole system dynamics are modeled together with the logic to find out the transition probability and incorporating Markov/CCMT methodology. Figure 3 shows the probability of top event for three time steps each of 500 ms.

Table 1 Cells for level discretization

$j = 5$	$x > 20$ ft	Failed high
$j = 4$	12 ft $< x \leq 14$ ft	Low
$j = 3$	14 ft $< x \leq 16$ ft	Normal–low
$j = 2$	16 ft $< x \leq 18$ ft	Normal–high
$j = 1$	18 ft $< x \leq 20$ ft	High

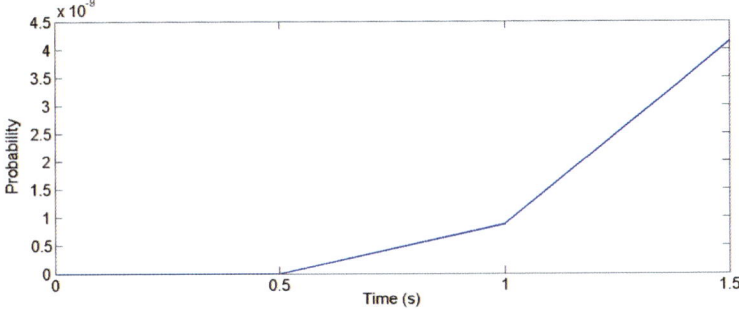

Fig. 3 Time-dependent probability of top event ($x > 20$ ft)

6 Conclusion

Markov/CCMT has been used to estimate dynamic reliability. The objective of this study was to obtain a better understanding of the Markov/CCMT methodology, in order to make future uses of it easier. In this study, the probability of the system to exceed 20-ft level in the tank was estimated at three discrete-time points. The code complexity would increase if more time steps were included in the study. Hence, for simplification, three time steps were used to demonstrate the dynamic reliability of the water tank problem.

References

1. T. Aldemir, M.P. Stovsky, J. Kirschenbaum, D. Mandellil, P. Bucci, L.A. Mangan, D.W. Miller, X. Sun, E. Ekici, S. Guarro, M. Yau, B. Johnson, C. Elks, S.A. Arndt, Dynamic reliability modeling of digital instrumentation and control systems for nuclear reactor probabilistic risk assessments. NUREG/CR/6942, US Nuclear Regulatory Commission (US NRC)
2. T. Aldemir, A survey of dynamic methodologies for probabilistic safety assessment of nuclear power plants. Ann. Nucl. Energy **52**, 113–124 (2013)
3. T. Aldemir, D.W. Miller, M.P. Stovsky, J. Kirschenbaurr, P. Bucci, A.W. Fentiman, L.T. Mangan, Current State of Reliability Modeling Methodologies for Digital Systems and Their Acceptance Criteria for Nuclear Power Plant Assessments, U.S. Nuclear Regulatory Commission (2006)

ARM SoC-Based VME CPU Board Development Using Open-Source Framework for Intelligent Accelerator Control System

R. P. Yadav, R. Rana and P. Fatnani

Abstract Accelerator control systems comprise of different off the shelf bought out and custom-built electronic hardware and software components. Mostly the accelerator control systems are of layered architecture and many times the components cost forces the control engineers to opt for simpler and low computation power CPU for lower layer control hardware. This restricts the use of computation-intensive algorithm at higher layers in overall control system. This situation is non-optimal toward implementation of intelligent agents-based control schemes in accelerator control as well as in similar industrial control systems. To overcome this, the low-cost ARM SoC-based VME controller card suited to accelerator control system as well as general industrial control system is being developed. This paper presents the overall VME CPU card hardware scheme, application software framework, RTOS porting, BSP development, methods for testing along with the initial test results.

Keywords SoC · VME · ARM · Intelligent agents

1 Introduction

The accelerator machines are comprised of different subsystems like injector, transport lines, storage ring, and beamlines. Also according to the different technologies and devices, this grouping further divides into RF, vacuum, power supply, timing, interlock and machine safety protection. To cater to the control requirements of all these, the control system of accelerator machines is of distributed layered architecture with interconnected software and hardware components of different layers spread

R. P. Yadav (✉) · R. Rana · P. Fatnani
Raja Ramanna Centre for Advanced Technology, Indore, India
e-mail: rpyadav@rrcat.gov.in

R. Rana
e-mail: rahulr@rrcat.gov.in

P. Fatnani
e-mail: fatnani@rrcat.gov.in

© Springer Nature Singapore Pte Ltd. 2020
P. V. Varde et al. (eds.), *Reliability, Safety and Hazard Assessment for Risk-Based Technologies*, Lecture Notes in Mechanical Engineering,
https://doi.org/10.1007/978-981-13-9008-1_24

over a large area. The operation challenge here is that all the subsystems are needed to be operated in a synchronized and sequential manner along with interlocking and pre-qualification of actions at different machine operation states.

Intelligent objects are software programs called as agents that are situated in a given environment and are capable of acting in this dynamic environment toward achieving their goal. An agent program is modeled in terms of mentalistic notions such as beliefs, desires, and intentions so that the developed software entity can pose some of the basic properties such as autonomy, social ability, reactivity, and proactivity generally exhibited by humans [1, 2]. With these attributes, agents provide a high abstraction level for developing software and thereby potentially simplify the control system design for complex systems. But this poses the requirements of large computation power at the controller level. Also, the distributed nature of computation involving the manipulation of control output signals and inference drawing from control input signals puts strict requirement on the timing synchronization among different CPUs that implement the agent framework.

The studies on the use of artificial intelligence (AI) methods for accelerator control [3–5] have shown that such systems if implemented on machines can bring out the significant improvement in overall system reliability and will enhance operation ability. The research study in this field as well as for hardware-in-loop (HIL) simulation studies and qualification of developed code requires a large number of CPU with high processing power with hard real-time operating system (RTOS) like VxWorks and OS-9-based platforms. This is a costly affair. For catering to these requirements at low costs, the scientists are trying to use the low-cost open-source hardware-based single-board computer (SBC) like RPi, Beaglebone, Cubieboard-based cluster [6, 7]. The comparative study done by Cloutier et al. [7] on 17 different SBC boards available from different vendors showed that the choice of Raspberry Pi SBC boards is the optimum for this purpose. Also using these bards, the High Performance Computing Division at Los Alamos National Laboratory has built 750 Rpi-based CPU clusters that are in operation now [6].

To fulfill such control system requirements, the accelerator control systems mostly use standard VME bus-based hardware. VME is a computer bus and standardized by the IEC as ANSI/IEEE 1014-1987. A typical VME bus system consists of one or more CPU card and different types of I/O cards. For building the low-cost HIL simulator, it is decided to develop VME CPU cards with high computation power using the open-source Raspberry Pi3 B-based SBC. Along with the hardware, the software platform is another important parameter that decides the overall price of such systems. The off-the-shelf industrial VME CPU boards rely on proprietary RTOS like VxWorks and OS-9 for providing the real-time operating system requirements.

For this card to keep the overall ownership cost low, it is decided to use the open-source RTOS, FreeRTOS. FreeRTOS is the outcome of collaborative efforts among the world's leading chip companies over a 15-year period. It is professionally developed, strictly quality controlled, robust, well supported, free to use in commercial products without a requirement to expose proprietary source code, and its IP is carefully managed [8].

The RTOS platform is needed to fulfill the timing and synchronization requirements of control loops, but such platforms often do not support the direct execution of high-level computation software (like MATLAB, LabVIEW, and Python) codes and logic programming languages (like Prolog, Answer set programming (ASP), and Datalog) codes. Support for high-level programming languages and logic programming languages in VME CPU is a must for implementing the agent framework. Also, the open-source SCADA supporting feature is required for easy and fast code development in an incremental manner. For this, it is decided to install the CPU with Linux OS and port Scilab, SWI-Prolog, and EPICS on it.

This paper presents the overall VME CPU card hardware scheme, application software framework, RTOS porting, BSP development, methods for testing along with the initial test results.

2 VME CPU Board Features

To fulfill the requirements of envisioned VME CPU board that can support the agent framework in the distributed control system environment, the overall hardware is designed with two CPUs where the first CPU is the ARM Cortex-A53 SoC (System on Chip) Quad core running at 1.2 GHz, on which Linux kernel (version 4.4.15 -v7+) is ported. This CPU system provides the basic framework for executing the SCADA and general programming/logical programming languages-based codes. This is the main software layer for implementing the agent framework and provides the networking and supervisory control features. The second CPU is ARM Cortex-M3 on which FreeRTOS (version 8.2.3) is ported. This CPU system provides the real-time response capability to the CPU board. For both OS, the device driver and board support package (BSP) with important features have been developed, tested, and integrated. The shared memory-based inter-CPU communication logic and VME32 interface logic has been implemented in FPGA. About 40% of the FPGA resources are available for users. Table 1 lists some of the important features provided in the CPU hardware. Table 2 lists some of the features for which the board firmware is being developed. Till now many of the firmware features have been developed, tested, and integrated.

3 Hardware Design

To have two types of OS (one for general computing needs and other for real-time computing needs) on one board simultaneously, the board is designed with two CPUs. The ARM Cortex-A53 series (BCM2837)-based Raspberry Pi 3B board has been used for CPU1. The ARM Cortex-M3-based CPU (SAM3X8E) has been used for CPU2. Figure 1 shows the block diagram for the CPU hardware configuration. There are three main blocks (a) SAM3X8E module, (b) BMC2837 module, and

Table 1 VME CPU board hardware features

Board parameter/features	Description/comments
VME standard and features supported	Up to VME 32 standard supported DTB Master: A32/A24/A16/D32/D16/D8/BLT32/BLT16 VME IRQ supported: Total 5 out of 7 (VME IRQ 3, 4, 5, 6, 7) AM supported: 28 (predefined) +16 (user-defined) codes
FPGA resources	Spartan-6 LX75 (40% resources available for user)
Processors	Two processors are provided CPU1: ARM 64Bit, Cortex -A53, Quad core, 1 GHz CPU2: ARM 32bit, Cortex-M3, 84 MHz
Memory	1 GB RAM, 16 GB ROM
Time synchronization methods	GPS, PPS UTC, 40 ns resolution. 1us accuracy, time stamping at source, cyclic event generation, future time event generation in distributed system environment
Additional onboard I/O	2 Analog Input, 12 bit, 1 Msps, 2 Analog outputs, 12 bit, 1 Msps, 4 Status I/O 3.3 V
Interface	2 USB 2.0, 1 LAN 10/100base-T, 1 RS232 port
Diagnostics	Onboard readout for +3.3, +5, +12, −12 V, CPU1 Temperature, CPU2 Temperature, FPGA Temp, 4 Temperature measuring devices and 12 status LED.
Special features	True Random Number Generator (TRNG), WatchDog, Power On Reset (POR)

Table 2 VME CPU board firmware features

Board parameter/features	Value/comments
RTOS support	FreeRTOS
General OS supported	Linux
SCADA supported	EPICS
BSP developed for	Linux and FreeRTOS
Inter-CPU communication methods	Hardware Que, Virtual com port, and Streaming Socket, total 27 bidirectional channels are possible
General programming language/logical programming language support	C, C++, ARDUINO, Scilab/SWI-Prolog

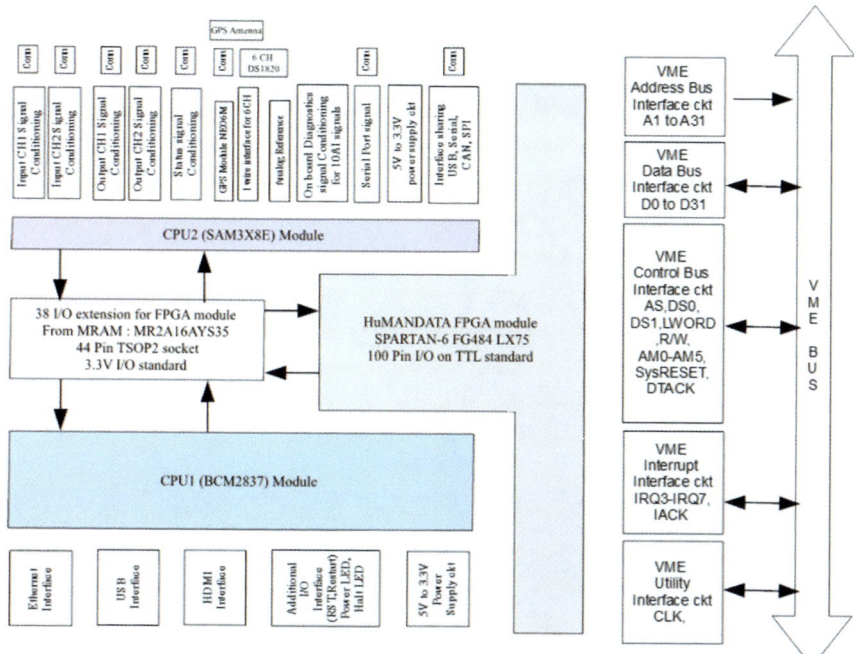

Fig. 1 Block diagram of VME CPU Hardware design

(c) FPGA module. These blocks are surrounded with the interfacing and signal conditioning blocks that handle the interfacing circuits and the I/O to these three main blocks. All the I/O listed in Table 1 are interfaced with the CPU2. All the VME bus signal conditioning blocks are interfaced with the FPGA module. The communication between CPU1–FPGA and CPU2–FPGA has been done using the CMOS 3.3 logic-based custom-defined parallel data bus that supports up to 6 MB/s data transfer rate in asynchronous bus mode and up to 40 MB/s in synchronous bus mode (under testing not fully qualified). Two interrupts have been provided from FPGA to CPU1 and CPU2. Presently, the token-based arbitration has been implemented and tested for communication between CPU1, CPU2 and FPGA. The complete VME controller logic for features listed in Table 1 has been implemented in FPGA that is capable of handling 40 MB/s data transfer rate on VME bus in BLT32 bit mode.

The RTC with 40 ns resolution has been implemented in FPGA that syncs with pulse per second (PPS) signal of GPS (u-Blox NEO-6) module. The glue logic for future time tag event generation, cyclic event generation, and event time tagging has been associated with the RTC. Figure 2 shows the assembled CPU unit along with the interface details and different mechanical parts, 6U 8T fascia, custom heat sinks for CPU1 and FPGA and GPS antenna. Figure 3 shows the thermal image of CPU card installed in VME crate with forced cooling. It can be seen that the CPU region

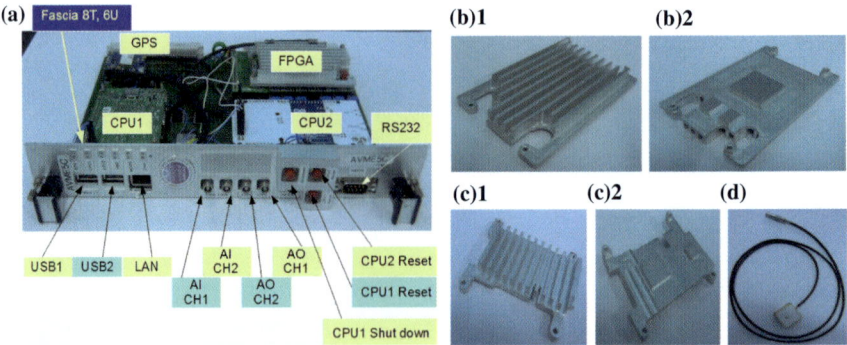

Fig. 2 Picture of VME CPU card with different parts **a** interface details, **b** FPGA Heat sink, **c** CPU1 Heat sink, **d** GPS antenna

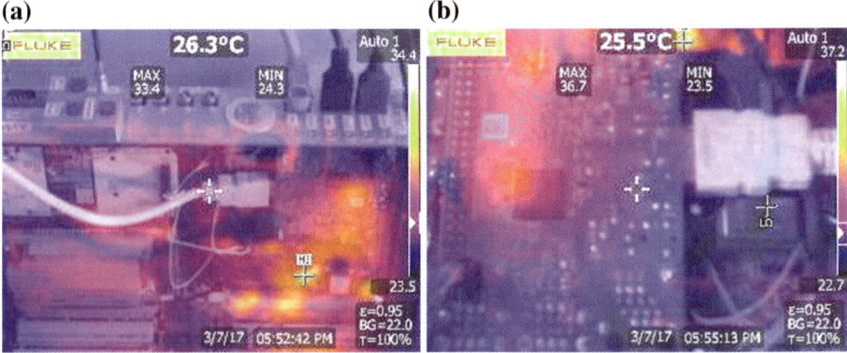

Fig. 3 Thermal Image **a** Showing the overall temperature distribution of CPU board, **b** CPU1 area temperature distribution

and the FPGA area are free from any high-temperature effects. The high-temperature marks observed are of VME driver IC and termination resistances only and that is also within limits. For handling thermal damage due to failure of air circulation, onboard diagnostics feature to monitor all the power rail (+3.3, +5, +12, −12 V) voltages, CPU1 temperature, CPU2 temperature, CPU1 heat sink temperature, FPGA heat sink temperature, and temperature measurement at four points on PCB has been provided. The shared memory-based inter-CPU communication has been provided. Both the CPU's can access VME bus. Out of five VME IRQs, three IRQs are routed to CPU1 and two VME IRQs are routed to CPU2. CPU1 can respond to VME IRQ with interrupt latency's of milliseconds, whereas the CPU2 responds to VME IRQ with in tens of microseconds. The time tagging of events and future time tag event generation are done at FPGA level.

4 Software Design

The software development activity for this card is divided into five main sub-activities (a) Linux kernel porting and kernel driver development for CPU1–FPGA interface (b) FreeRTOS porting and kernel driver development for CPU2–FPGA interface (c) FPGA code development for VME controller and board glue logic (d) Driver development for different board features and making of board BSP for both CPU1 and CPU2 (e) Development of multilayer agent framework for the board.

Figure 4 shows the block diagram of board firmware design. The inner dotted line box shows the logic blocks implemented inside the FPGA. Inside FPGA, all the logic is implemented in the form of state machines grouped into three main engines, (a) BMC2837 Engine (b) SAM3X8E Engine, and (c) VME Engine. This engine-based development methodology has been adopted for ease of development. First starting from the empty engine framework, the state machine for one feature has been developed tested and integrated. Then one by one in incremental manner state machines are developed and integrated till all the desired features have been achieved. There is 24 KB of dual port RAM arranged in 12 blocks of 2 KB for inter-engine communication. Command response-based feature serving has been implemented through 16-bit wide 128 multi-port general-purpose registers. The BMC2837 Engine and SAM3X8E Engine implement the interface logic for custom protocol to communicate with CPUs and handle the interrupt routed to CPU1 and CPU2, respectively. VME engine implements the VME and timing-related functionality listed in Table 1.

The Linux base 4.4.15 v7+ and kernel compilation toolchain for the board has been prepared. The OS is trimmed for remote login on telnet and command-line interface (CLI) for this board. CPU1 bridge protocol has been formulated, synthesized, tested, deployed, and integrated on both the sides CPU1 side and FPGA side. Linux kernel driver for the bridge has been developed and integrated with the system. Using this kernel driver, the device drivers have been developed for VME access (57 different modes including BLT and User AM), CPU2 communication, VME IRQ handling, and board diagnostics. Using the device drivers, the board BSP is made. EPICS version R7.0.1 (latest version) has been ported and compiled along with support modules. Asyn drivers are written for accessing VME and board diagnostics information. IOC has been developed for the demo application. CSS OPI (Fig. 5) has been developed for the demo application IOC testing.

CPU2 bridge protocol has been formulated, synthesized, tested, deployed, and integrated on both the sides CPU2 side and FPGA side. Driver codes have been developed for onboard devices (GPS, timing, CPU2 Bridge, temperature, IRQ, queues, and CLI) for Arduino mode and tested. FreeRTOS version 8.2.3 has been ported on the board. The board-specific device variables have been exported across the project. The RTOS is trimmed for serial port CLI interface for debugging and testing of this CPU board. FreeRTOS kernel driver for the bridge has been developed and integrated with the system. Using this kernel driver, the device drivers have been developed for VME access (basic modes), CPU1 communication (127 channels in each direction for inter-application intra-CPU [CPU1 and CPU2]), VME IRQ handling, and board

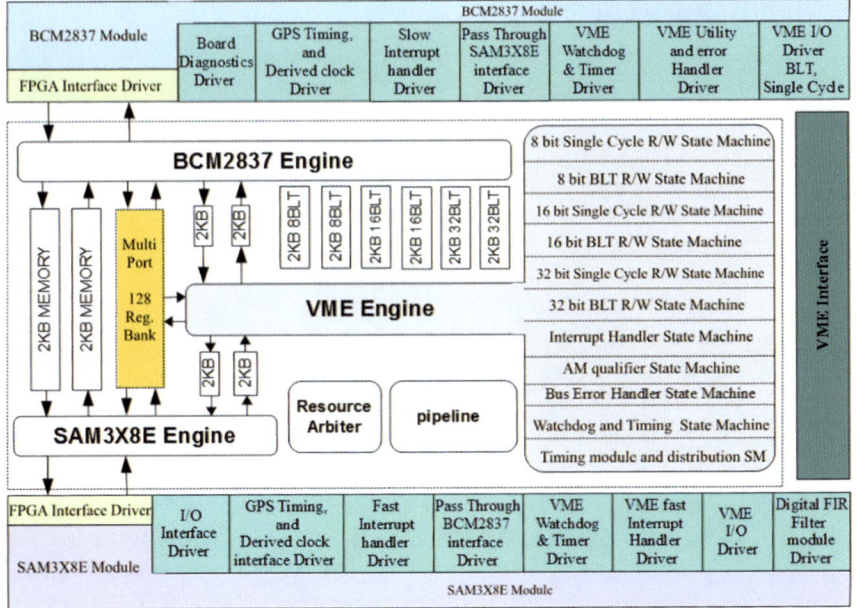

Fig. 4 Block diagram of board firmware design

diagnostics. Using the device drivers, the board BSP is made for board diagnostics devices (temperature and voltages) and GPS and timing modules.

The work is presently going on for implementing the agent framework at different layers. The resource demanding parts (memory, network, system model, and supervisory control) are going to be put on the Linux. The time-critical part is going to be put on the RTOS part. The fast execution demanding part can be put on the direct FPGA logic.

5   Test Results

Figure 6 shows the interrupt latency histogram for the developed VME CPU board for CPU1 running Linux over it and CPU2 running FreeRTOS on it. Figure 7 shows the interrupt latency histogram for the developed CPU with the best commercial VME CPU board (PPC7D) available at Accelerator Control Systems Division, RRCAT, for comparison. From Fig. 6, it can be seen that under Linux the interrupt latency can be as large as of ~600 μs. Further, it is observed that under medium computing load conditions, this value can rise up to 1.6 ms. Under FreeRTOS, the maximum interrupt latency value measured is ~32 μs. From Fig. 7, it can be observed that the interrupt latency of the developed CPU under FreeRTOS is comparable with that of the commercial VME CPU boards running industry leading RTOS VxWorks.

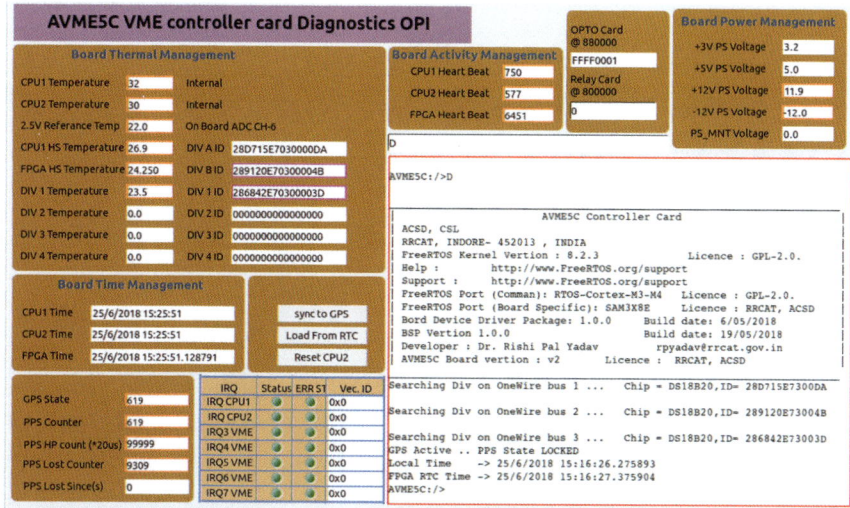

Fig. 5 Snapshot of EPICS OPI for the VME CPU board for remote diagnostics

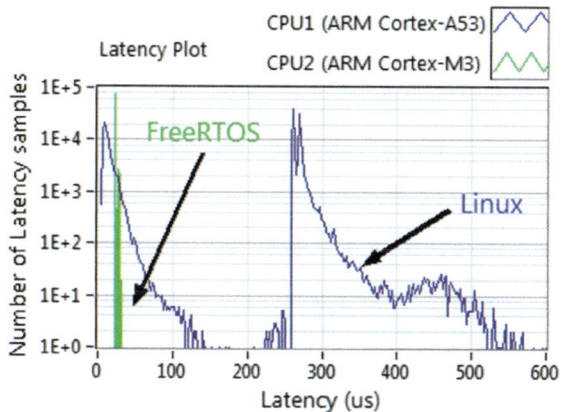

Fig. 6 Interrupt latency plot for CPU1 and CPU2 of VME CPU board

6 Conclusion

A VME32 CPU controller board has been developed based on the open-source software and hardware concept. It has two ARM processors. The first one is ARM Cortex-A53 SoC on which, Linux Kernel (version 4.4.15 -v7+) has been ported. The second one is ARM Cortex-M3 on which, FreeRTOS (version 8.2.3) has been ported. For both the OS, the device driver and BSP have been developed, tested, and integrated. VME32 interface logic has been implemented in FPGA. The EPICS version R 7.0.1 has been ported on the board, and the Asyn device drivers have been

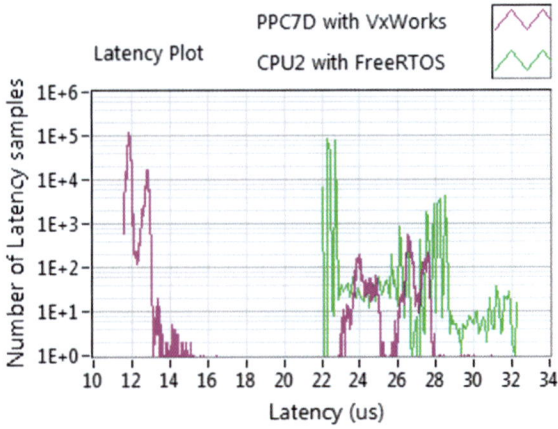

Fig. 7 The interrupt latency plot of CPU2 and Commercial VME CPU board PPC7D

developed. The command-line interface has been developed and configured for both Linux and FreeRTOS for easy development and debugging. This board is specifically designed for implementing the agent framework for accelerator control environment (VME I/O + GPS-based 40 ns Resolution time stamping of events + hard real-time performance with RTOS + soft real-time performance for SCADA + Fast processing on FPGA for feedback control). As this board is also of general nature, it can be used in almost all the industrial VME SCADA systems. The future work involves the testing and validation of high-throughput interface between CPU1 and FPGA (40 MB/s data transfer rate) and implementation of agent framework distributed on different board layers wise CPU1, CPU2 and FPGA logic.

References

1. A. Rao, M. Georgeff, BDI Agents: from theory to practice, in *Proceedings of the First International Conference on Multi-Agent Systems,* Cambridge, MA, USA, (1995), pp. 312–319
2. W. Shen, D.H. Norrie, J.P. Barthes, *Multi-Agent Systems for Concurrent Intelligent Design and Manufacturing* (Taylor and Francis, London, 2001)
3. R.P. Yadav, P. Fatnani, P.V. Varde, P.S.V. Nataraj, A multi-agent based control scheme for accelerator pre-injector and transport line for enhancement of accelerator operations. Online J. Elixir. Comp. Sci. Eng. **44**, 7405–7410 (2012)
4. R.P. Yadav, P.V. Varde, P.S.V. Nataraj, P. Fatnani, C.P. Navathe, Model-based tracking for agent-based control systems in the case of sensor failures. Int. J. Autom. Comput. **9**(6), 561–569 (2012)
5. R.P. Yadav, P.V. Varde, P.S.V. Nataraj, P. Fatnani, Intelligent agent based operator support and beam orbit control scheme for synchrotron radiation sources. Int. J. Adv. Sci. Technol. **52**, 11–34 (2013)
6. Article System with thousands of nodes brings affordable testbed to supercomputing system-software developers (2018, 28 August). Retrieved from https://www.lanl.gov/discover/news-release-archive/2017/November/1113-raspberry-pi.php

7. M.F. Cloutier, C. Paradis, V.M. Weaver, A Raspberry Pi cluster instrumented for fine-grained power measurement. Electronics **5**(61), 1–19 (2016)
8. FreeRTOS website *home page* (2018, 28 August). Retrieved from https://www.freertos.org/

Architecture-Centric Dependability Analysis for I&C Safety Systems in NPP: A Case Study

Amol Wakankar, Ashutosh Kabra, A. K. Bhattacharjee
and Gopinath Karmakar

Abstract System architecture plays a major role in achieving the safety, availability and reliability requirements of the system. Any violation of these requirements detected at a later stage of the system development life cycle may call for architectural modification, which can be expensive in terms of both cost and time. This paper presents a case study of architecture-centric dependability analysis based on rigorous model checking techniques. A large computer-based safety system of a PWR is considered for this study. This work demonstrates that architectural dependability analysis is feasible for large computer based system (CBS) and it increases the level of confidence in meeting the system dependability requirements at the early stage of development.

Keywords Architecture modeling and analysis · Dependability analysis · Compositional analysis · Discrete-Time Markov Chain · Engineered safety feature actuation system · Probabilistic model checking

1 Introduction

Dependability of computer-based instrumentation and control (I&C) safety systems for nuclear power plant (NPP) is the most sought after goal of both designers and the regulatory bodies. It is observed that system architecture has a major impact on

A. Wakankar (✉) · A. K. Bhattacharjee
Homi Bhabha National Institute, Mumbai, India
e-mail: amolk@barc.gov.in

A. K. Bhattacharjee
e-mail: anup@barc.gov.in

A. Wakankar · A. Kabra · A. K. Bhattacharjee · G. Karmakar
Bhabha Atomic Research Center, Mumbai, India
e-mail: kabra@barc.gov.in

G. Karmakar
e-mail: gkarma@barc.gov.in

© Springer Nature Singapore Pte Ltd. 2020
P. V. Varde et al. (eds.), *Reliability, Safety and Hazard Assessment for Risk-Based Technologies*, Lecture Notes in Mechanical Engineering,
https://doi.org/10.1007/978-981-13-9008-1_25

meeting the dependability requirements such as safety, availability, and reliability. Despite widespread applicability of standard design practices established by industrial standards and regulatory guides [1–4], the conformance of the systems w.r.t. to these requirements cannot be established until system integration stage.

Any design deficiency propagated to the later stages of the development life cycle will result in increased development cost and time. Hence, in order to evaluate the design w.r.t. the target dependability goals during architectural design phase, we used the advances in the area of architectural modeling and the model checking techniques to carry out dependability analysis of large computer-based system (CBS). The analysis is performed to establish the component-level specifications (failure rate, diagnostic coverage, repair rate, and surveillance test interval) for all the components of the safety system that achieve the target dependability goals.

We use Architectural Analysis & Design Language (AADL) [5], which has a formally defined semantics, for modeling the system architecture and associated error model. Quantitative analysis of system dependability is performed using a probabilistic model checking technique, where probabilistic model checker tool PRISM is used as a back-end analysis engine in the analysis. Although the probabilistic model checking-based analysis has been used by many researchers [6], it suffers from state-space explosion problem associated with the model checking techniques [7], when applied to analyze large systems.

In this paper, we present a case study to demonstrate the architecture-centric analysis of Engineered Safety Feature Actuation System (ESFAS) of a pressurized water reactor (PWR), which is a CBS. ESFAS is large in size and its associated state space is beyond the capabilities of symbolic model checking. In order to overcome this difficulty, we used the compositional analysis technique for the dependability analysis. Compositional analysis methodology decomposes the analysis into multiple steps based on the system architecture, which allows us to deal with large state spaces associated with industrial systems.

2 Preliminaries

2.1 Error Model

The error model of a system/component characterizes its fault behavior envisaged during the preliminary safety analysis. The error model of a component is represented by a state machine consisting of the *failed* and *operational* states along with the transitions with associated probabilities. A transition is triggered on the occurrence of an event, such as failure or repair of the component.

Let λ denote the failure rate of a component and DC denote the diagnostic coverage. λ is subdivided into safe failures rate (λ_S) and unsafe failures rate (λ_{US}). Further, unsafe failures can be categorized as unsafe detectable (USD) or unsafe undetectable (USUD) failures based on the self-diagnostics features of system/component and the

Fig. 1 Generic error model

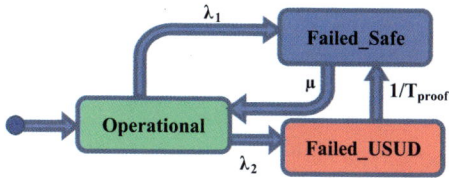

corresponding failure rates are denoted by λ_{USD} and λ_{USUD} respectively, as defined below.

$$\lambda_{USD} = (DC)\lambda_{US}$$

$$\lambda_{USUD} = (1 - DC)\lambda_{US}$$

We consider the USD failure also as a safe failure because operator can drive the system/component into a safe state on the basis of diagnostic information, while USUD failures represent the actual unsafe failures, which cause system's failure on demand and make plant unsafe. Hence, failure rate λ_1 for detectable or safe failure (which leads the system into the safe state either automatically or manually) and failure rate λ_2 for unsafe undetectable failures are defined as follows.

$$\lambda_1 = \lambda_S + \lambda_{USD}$$

$$\lambda_2 = \lambda_{USUD}$$

Figure 1 shows the generic error model used in our analysis. The component remains in *Operational* state till it is working as expected. The component undergoes a transition from *Operational* to *Failed_Safe* state with a failure rate of λ_1 and to *Failed_USUD* state with a failure rate of λ_2. The *USUD* failures can be detected during surveillance test, which is conducted periodically with an interval of T_{proof}. Once *USUD* failures are detected during surveillance test, the operator will drive system/component into a *Failed_Safe* state, where repair activities can be carried out. System/component undergoes transition from *Failed_Safe* to *Operational* state on repair with a rate denoted by μ. This error model is used for all the systems/components for dependability analysis throughout the paper. The attributes defined here, viz. λ, DC, μ, and T_{proof}, are collectively referred as *Component-level specification* (CLS) is an unordered set of attributes defined as CLS = {λ, DC, μ, T_{proof}}.

2.2 System Dependability

Dependability of a system is mainly characterized by its safety, availability and reliability attributes, and this paper focuses on these attributes only. Maintainability and testability are the other attributes, which are also considered during dependability analysis [8].

Computer-based I&C safety systems have two modes of operations: *standby* mode and *active* mode [10]. In the standby mode, system dependability is mainly defined by its availability, which is the probability of system being in the operational state on demand. Once the system is actuated on demand, it operates in active mode. In this mode, system reliability is one of the most important attributes, which quantifies the system's ability to provide the required functionality for the specified duration.

Our analysis consists of the following two steps.

- In *standby* mode of operation, we analyze the system architecture to determine its safety and availability as described in Sect. 4.3. Safety and availability are quantitatively represented by two attributes, viz. Probability of Failure on Demand (PFD) and Spurious Operation Probability (SOP). PFD and SOP should be minimized in order to increase the system safety and availability.
- System availability data is then used to calculate the system reliability during active mode of operation as described in Sect. 4.4.

We define an attribute called *target dependability attribute* TDA = {PFD, SOP}, which will be used in the rest of the paper. It is important here to note that TDA of a component is derived from its CLS. For our experiments, we have calculated the TDAs for various combinations of CLSs and stored in a lookup table, which is referred to as *cls-to-tda-lookup-table*.

3 Analysis Methodology

3.1 Architectural Modeling and Analysis Methodology

The Architecture Analysis & Design Language (AADL) is defined by Society of Automotive Engineers International (SAE) standard, which provides a unifying framework for model-based system engineering. The architectural model of a system consists of a hierarchy of components. A component can either be an *atomic* or a *composite* one. *Atomic components* are the basic building blocks of a system architecture, which are indivisible. *Composite components* are constituted using the existing atomic or composite components.

For dependability analysis, we build the system architectural model in the OSATE framework, which allows us to specify AADL model of architectural components and its associated error model [11]. DTMC model is generated from the architectural specifications and analyzed using a symbolic model checking tool PRISM with respect to the specification of a formal property given in Probabilistic Computation Tree Logic (PCTL) [9]. Dependability attributes, viz., PFD and SOP of the system are calculated by determining the steady-state probabilities of the system being in *Failed_USUD* and *Failed_Safe* states, respectively.

3.2 Probabilistic Model Checker PRISM

Probabilistic model checking is a formal verification method for analyzing quantitative properties of systems that exhibit stochastic behavior. PRISM facilitates probabilistic analysis of Markov chain models—discrete-time Markov chain (DTMC) and continuous-time Markov chain (CTMC). The properties relevant to the dependability analysis are specified in a formal notation called PCTL accepted by the PRISM model checker. An example of PCTL properties, used in this work, is given below.

S=? [F *Failed_USUD*]: For this specification, the model checker shall compute the steady-state probability (S) of a system eventually (F) going into the failed state.

4 Case Study

In this section, we present a case study in order to demonstrate the system dependability analysis of the proposed architectures of ESFAS of a PWR. Based on the design basis requirements of ESFAS [15], the required TDA is as follows:

- Probability of Failure on Demand (PFD) $\leq 10^{-5}$
- Spurious Operation Probability (SOP) $\leq 10^{-5}$

The goal of the dependability analysis is to determine the CLS of all atomic components of ESFAS that achieves the required TDA.

4.1 Proposed Architecture of ESFAS

ESFAS is a computer-based I&C safety system, which identifies the accident conditions and actuates Emergency Safety Systems (ESS) to mitigate the effect of postulated initiating event (PIE), such as loss-of-coolant accident (LOCA), a steam line break etc. ESFAS consists of four redundant, physically separate, and functionally independent trains as shown in Fig. 2 and performs its safety function based on *2-out-of-4* (2/4) voting logic. Each train of ESFAS consists of two diverse controllers (e.g., safety train *ESFAS-A* has two diverse controllers *A*1 and *A*2) and interfaces with all four trains of RPS (*RPS-A, RPS-B, RPS-C,* and *RPS-D*) over redundant P2P data communication links.

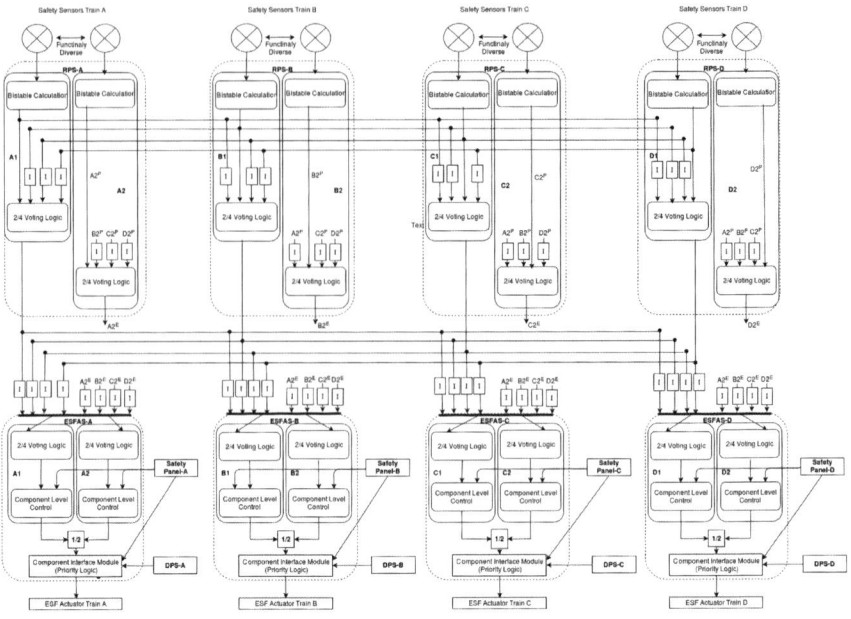

Fig. 2 Architecture of engineered safety feature actuation system

In the proposed I&C systems architecture, the same safety sensors and RPS controllers are used for both reactor trip system (RTS) and ESFAS. This architecture has an advantage of lesser penetrations and reduced hardware requirements. RPS controllers monitor all the parameters, which are required for ESFAS processing. After required signal conditioning and analog-to-digital conversion, measurements are compared with set points (corresponding to ESFAS actuation) to generate bi-stable signal. Bi-stable signals are communicated among all RPS controllers and *parameter actuation signal* is generated based on *2-out-of-4 (2/4)* voting logic on bi-stable signals for each parameter. Both the controllers of each train of RPS send the generated *parameter actuation signals* to all the four trains of ESFAS. Both the controllers of each train of the ESFAS receive the parameter actuation signals from all four RPS trains. Each controller of ESFAS combines parameter actuation signals (for same parameter) using 2/4 voting logic to determine the accident condition and generates *train-level ESF actuation signals*. Component-level control (CLC) module of both the ESFAS controller processes the *train-level ESF actuation signals* and generates *component-level actuation signals*, which are used to control of plant components of the corresponding ESS based on the *1-out-of-2 (1/2)* voting logic. It may be noted that ESF controllers are to be designed *fail as is* to reduce the spurious operation probability of ESF system.

For the simplicity of exposition, the following assumptions were made for dependability analysis of the ESFAS:

- All redundant trains of EFSAS are independent of each other.
- CLSs of both the diverse controllers of a ESFAS train are the same.
- Failure rates and repair rate of all the components/ systems are constant with respect to time.
- Surveillance test detects all unsafe undetectable (USUD) failures.
- Voting logics are failure-free, and hence, their failure rates are considered as *zero*.
- Standby mode lasts for a long time; hence steady-state probabilities are considered for computation of SOP and PFD.
- For simplicity of analysis, we have not considered the effect of common cause failure (CCF). However, dependability analysis can be extended to include CCF effect by using multiple beta factor (MBF) model [12] for error model of 2/4 voting logic. Generic Markov model for a safety system, which includes CCF of more than one train using MBF, is given in [13].

4.2 Architectural Model of ESFAS in AADL

Architectural model of ESFAS is specified in AADL language using OSATE framework, where the atomic components of the system are defined first and then these atomic components are gradually integrated to build the large hierarchical system/sub-system. ESFAS consists of four ESFAS safety trains, where each ESFAS safety train has two ESFAS controllers and one ESF actuator. Each ESFAS controller performs its function based on 2/4 voting logic on the input received from four RPS controllers over P2P links. Individual ESFAS controller also includes CLC module to processes *ESF-level actuation signal* and generate required *component-level auction signals*. The hierarchical model of ESFAS is shown in Fig. 3.

The total number of states in this model is 3^{404} ($\approx 10^{192}$), and it is infeasible to analyze it with state-of-the-art model checking technique even using high-end resources (multi-core CPU with 250 GB of RAM). To overcome this difficulty, we

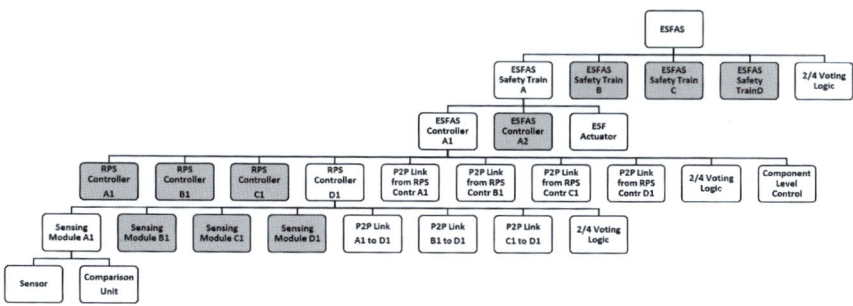

Fig. 3 Hierarchical model of ESFAS

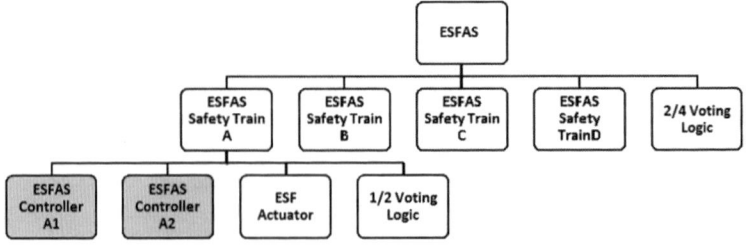

Fig. 4 M1—Hierarchical model of ESFAS with ESFAS controller abstracted

decompose the system architecture into multiple abstract models as described in the following subsections and carry out the dependability analysis compositionally.

4.2.1 Abstract Model of ESFAS M1

The model M1 is an abstract model of ESFAS, where an ESFAS controller is abstracted as atomic component. M1 consists of four trains of ESFAS and 2/4 voting logic. Formal description of 2/4 voting logic and the associated Markov model are given in [14]. Each ESFAS safety train consists of two ESFAS controllers with 1/2 voting logic for controller's output. This model is shown in Fig. 4.

Error Model of ESFAS

- ESFAS fails to provide safety action on demand (Probability of occurrence of this event is termed as *PFD*) if

  a. At least three ESFAS safety trains are in *Failed_USUD* state **OR**
  b. Two ESFAS safety trains are in *Failed_USUD* state **WITH** at least 1 train is in *Failed_Safe* state.

- ESFAS spuriously actuates ESS (Probability of occurrence of this event is termed as *SOP*) if

  a. At least three ESFAS safety trains are failed in *Failed_Safe* state.

Error Model of ESFAS Safety Trains

- ESFAS safety train is in *Failed_USUD* state if

  a. ESF actuator is in *Failed_USUD* state **OR**
  b. ESF actuator is in *Operational* state AND both ESFAS controllers are in *Failed_USUD* state.

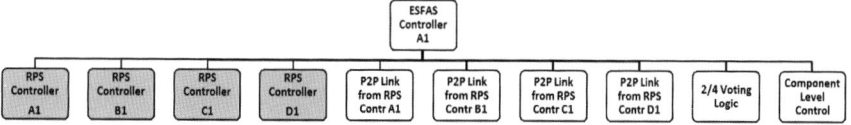

Fig. 5 M2—Hierarchical model of ESFAS controller with RPS controller abstracted

- ESFAS safety train is in *Failed_Safe* state if

  a. ESF actuator is in *Failed_Safe* state **OR**
  b. ESF actuator is in *Operational* state **AND** any of the ESFAS controllers is in *Failed_Safe* state.

4.2.2 Refined Model of ESFAS Controller M2

The model M2 is the refinement of ESFAS controller, where RPS controller is abstracted as an atomic component. M2 consists of four RPS controller with associated four P2P links, a 2/4 voting logic, and a CLC. This model is shown in Fig. 5.

Error Model of the ESFAS Controller

- ESFAS controller fails in Failed_USUD state if

  a. CLC is in *Failed_USUD* state **OR**
  b. CLC is in *Operational* state **WITH**
     i. At least three RPS controllers (including corresponding P2P link failure) are in *Failed_USUD* state **OR**
     ii. Two of the four RPS controllers (including corresponding P2P link failure) are in *Failed_USUD* state **AND** at least one is in *Failed_Safe* state.

- ESFAS controller is in *Failed_Safe* state if

  a. CLC is in *Failed_Safe* state **OR**
  b. CLC is in *Operational* state **AND** at least 3 RPS controllers (including corresponding P2P link failure) are in *Failed_Safe* state.

Error Model of the RPS Controller M3

The model M3 is the refined model of RPS controller. It consists of four sensing modules, three P2P link and a 2/4 voting logic. Sensing module consists of sensor and comparison unit (CU). This model (shown in Fig. 6) does not have any abstract component.

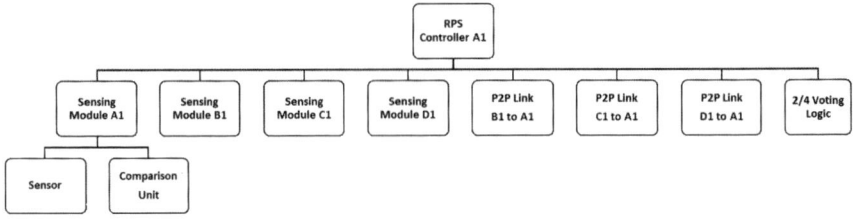

Fig. 6 M3—Hierarchical model of RPS controller

Error Model of the RPS Controller

- RPS controller fails in *Failed_USUD* state if

  a. At least three sensing modules (including corresponding P2P link failure) are in *Failed_USUD* state **OR**
  b. Two sensing modules (including corresponding P2P link failure) are in *Failed_USUD* state **AND** at least one is in *Failed_Safe* state.

- RPS controller fails in *Failed_Safe* state if

  a. At least three sensing modules (including corresponding P2P link failure) are failed in *Failed_Safe* state.

Error Model of the Sensing Module

- Sensing Module is in *Failed_USUD* state if

  a. CU is in *Failed_USUD* state **OR**
  b. CU is in *Operational* state **AND** Sensor is in *Failed_USUD* state.

- Sensing Module is in *Failed_Safe* state if

  a. CU is in *Failed_Safe* state **OR**
  b. CU is in *Operational* state **AND** Sensor is in *Failed_Safe* state.

4.3 Availability Analysis and Results

4.3.1 Analysis of Model M1

In this step, DTMC model in PRISM syntax [9] is generated corresponding to the abstract model of ESFAS (shown in Fig. 4) in OSATE. The generated DTMC model is analyzed using PRISM to determine the CLSs of ESFAS controller and ESF actuator

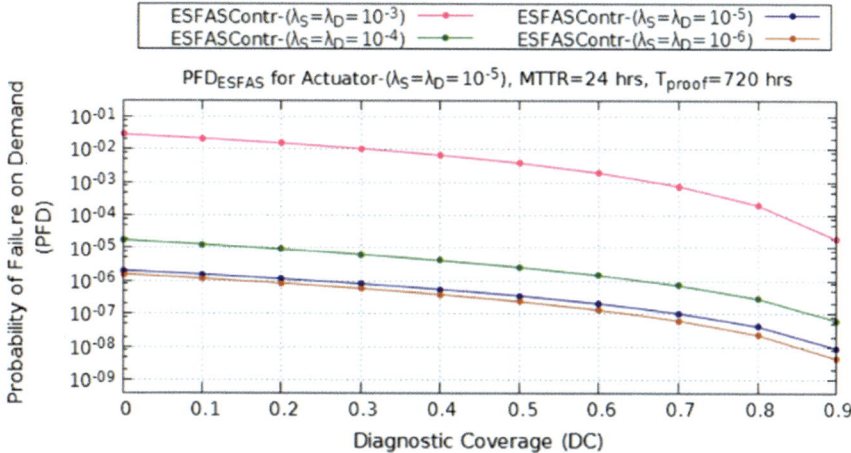

Fig. 7 PFD of ESFAS model M1

that achieve the required TDA of ESFAS. The PCTL specification to compute PFD and SOP of M1 is given by Eqs. 1 and 2, respectively.

$$S = ? [ESFAS\_Is\_Failed\_USUD] \tag{1}$$

$$S = ? [ESFAS\_Is\_Failed\_Safe] \tag{2}$$

We carried out the experiments to evaluate the PCTL specifications given in Eqs. 1 and 2 with respect to various values of DC, where $0 \le DC \le 0.9$. While doing so, parameters, viz. failure rates of ESAFS controller and ESF actuator, were varied with a step size of 10^{-1}. Based on the plant operating experience, the most used configuration for MTTR $(1/\mu)$ and T_{proof} are 24 and 720 h, respectively, and hence, the fixed values of these parameters were used for all ESFAS components in our experiments. The results of corresponding to PFD and SOP were plotted, which were analyzed and only the relevant plots that achieve desired TDA (PFD < 10^{-5} and SOP < 10^{-5}) are shown in Figs. 7 and 8. The results are summarized in Table 1 and the result relevant to achieve the required TDA is highlighted.

This analysis shows that actuator design with 10^{-5}/hour failure rate is the optimum for achieving PFD < 10^{-5} for ESFAS. No significant gain is observed if failure rate of the actuator is further decreased. Since ESF actuator is an atomic component, the obtained CLS can be used for its design. However, ESFAS controller being a composite component has to be analyzed further using its refined model M2 to obtain the CLS of its subcomponents. For this purpose, we used the obtained CLS (i.e., $\lambda_S = \lambda_{US} = 10^{-4}$ with DC = 0.2), to compute the TDA (PFD = 2.79×10^{-2} and SOP = 2.32×10^{-3}) using the *cls-to-tda-lookup-table* for ESFAS Controller discussed in Sect. 2.2.

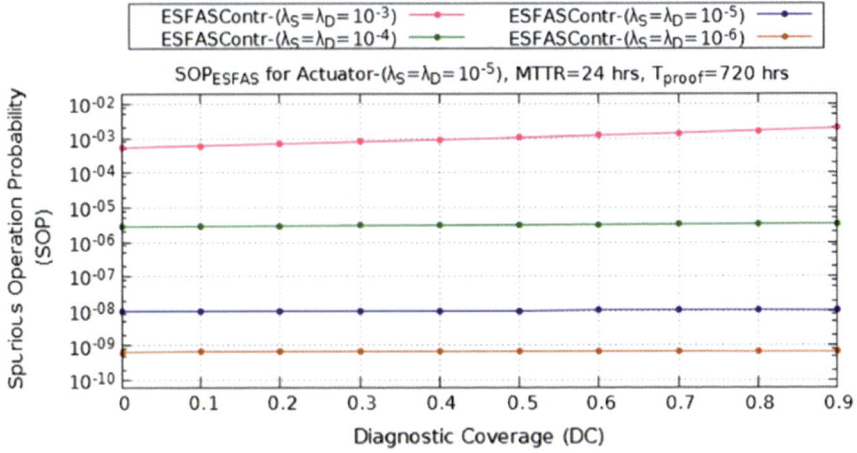

Fig. 8 SOP of ESFAS model M1

Table 1 PFD and Sop of ESFAS

S. No.	ESFAS component failure rate (/h)		DC	MDT (h)	T_{proof} (h)	PFD of ESFAS	SOP of ESFAS
	ESFAS controller	Actuator					
1	10^{-3}	10^{-7}	0.9	24	720	1.35×10^{-5}	1.97×10^{-3}
2	10^{-4}	10^{-4}	≈ 0.88	24	720	1.00×10^{-5}	9.72×10^{-6}
3	10^{-4}	10^{-5}	≈ 0.2	24	720	1.00×10^{-5}	2.94×10^{-6}
4	10^{-5}	10^{-5}	0	24	720	2.06×10^{-6}	1.01×10^{-8}
5	10^{-4}	10^{-6}	0	24	720	2.75×10^{-6}	2.5×10^{-6}
6	10^{-5}	10^{-6}	0	24	720	7.77×10^{-9}	3.5×10^{-9}
7	10^{-4}	10^{-7}	0	24	720	2.04×10^{-6}	2.47×10^{-6}
8	10^{-4}	10^{-8}	0	24	720	1.98×10^{-6}	2.47×10^{-6}

4.3.2 Analysis of Model M2

We carried out the experiments and obtained the plots for PFD and SOP of Model M2 with respect to various values of DC, where $0 \leq DC \leq 0.9$. While doing so, parameters, viz. failure rates for RPS controller and P2P link, were varied with a step size of 10^{-1}.

From the plots (shown in Figs. 9 and 10), it can be concluded that the failure rates of RPS Controller, P2P link and CLC are to be kept $\leq 10^{-4}$, $\leq 10^{-4}$ and $\leq 10^{-5}$, respectively, with $DC \geq 0.2$. Since P2P link and CLC are atomic components (refer Fig. 5), no further analysis is required for these components. However, for RPS controller, which is a composite component, the dependability analysis is further carried

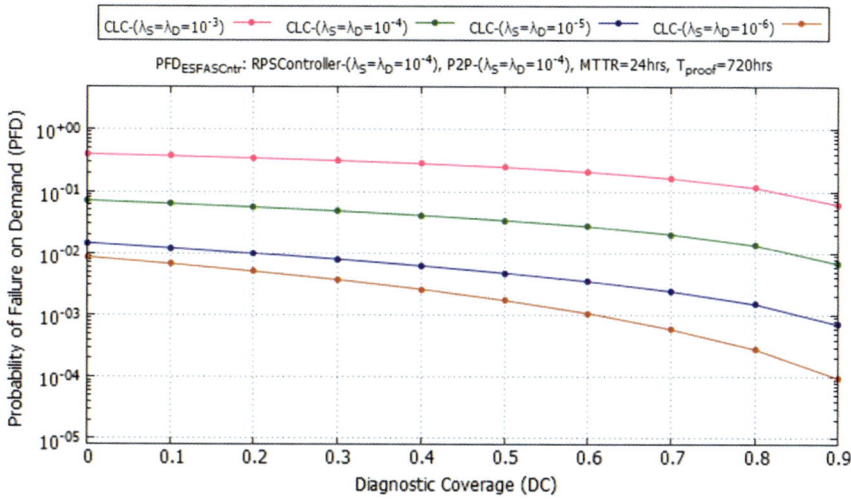

Fig. 9 PFD of ESFAS controller model M2

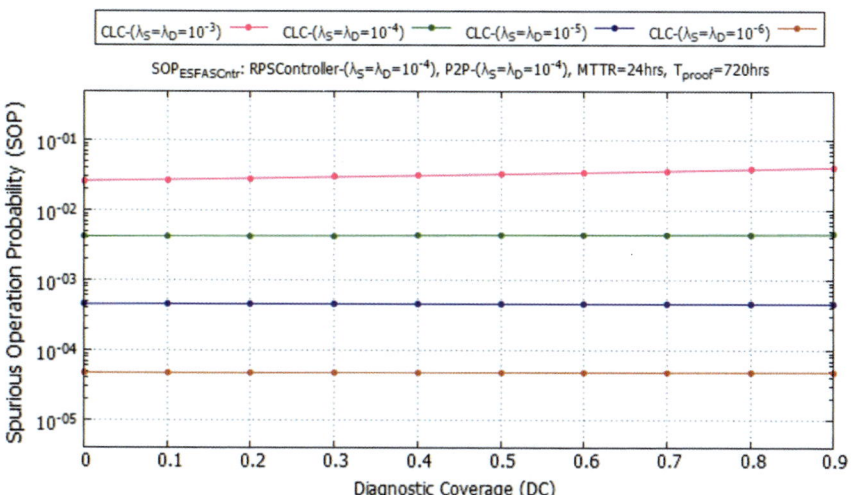

Fig. 10 SOP of ESFAS controller model M2

out to obtain the CLS of its subcomponents. From the calculated CLS requirement (i.e. $\lambda_S = \lambda_{US} = 10^{-4}$ with DC = 0.2), we obtain the TDA (PFD = 2.79×10^{-2} and SOP = 2.34×10^{-3}) for RPS Controller using *cls-to-tda-lookup-table*. The calculated TDA is used in the next step for analyzing RPS controller model M3.

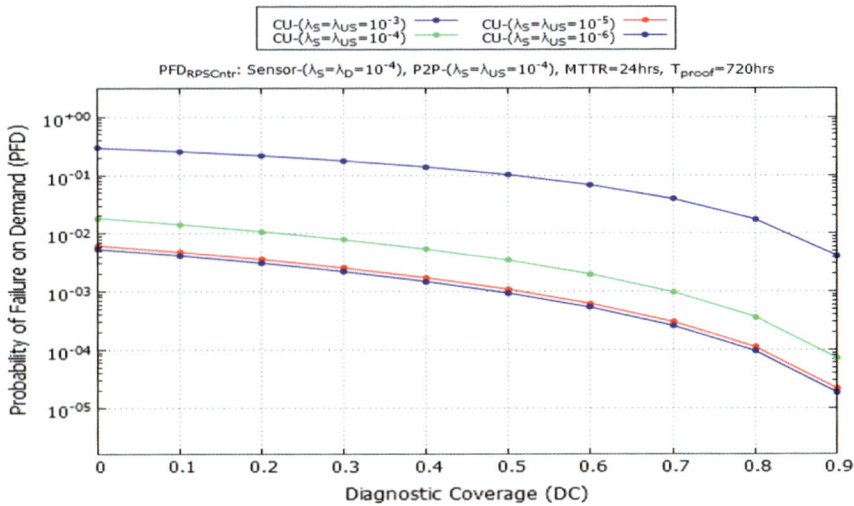

Fig. 11 PFD of RPS controller model M3

4.3.3 Analysis of Model M3

We carried out the experiments and obtained the plots for PFD and SOP for model M3 with respect to various values of DC, where $0 \leq DC \leq 0.9$. While doing so, parameters, viz. failure rates of Sensor, P2P link, and CU, were varied with a step size of 10^{-1}. It can be concluded from the plots shown in Figs. 11 and 12 that the failure rates of sensors, P2P link, and CU are to be kept $\leq 10^{-4}$, $\leq 10^{-4}$ and $\leq 10^{-4}$, respectively, with $DC \geq 0.2$.

4.4 Reliability Estimation of EFSAS

In the proposed architecture of ESFAS, sensors and RPS Controllers are required till initiation of the safety action. As ESS is to be operated for at least 72 h [16] after the occurrence of any accident condition, CLC of both the controller is required to be operational for at least 72 h once ESF actuation signal is generated. The reliability of CLCs of any ESFAS controllers with $\lambda_S = \lambda_{US} = 10^{-5}$/h and $t = 72$ h is expressed as

$$R_{CLC} = e^{-(\lambda_S + \lambda_{US}) \times t} = 0.99856$$

The overall ESFAS reliability is represented by the availability of ESFAS on demand and correct operation of ESFAS for at least 72 h. As two independent ESFAS controller performs safety function based on 1/2 voting logic, the reliability of ESFAS can be expressed as

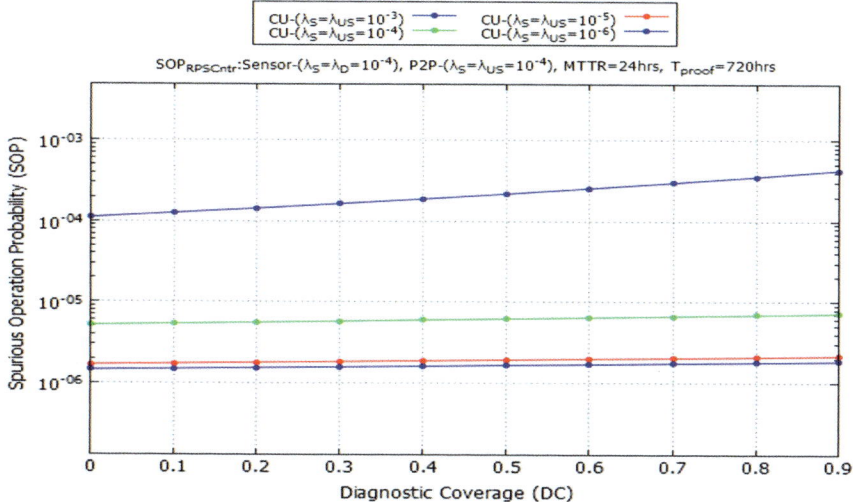

Fig. 12 SOP of RPS controller model M3

Table 2 Experimental results for ESFAS

Component of reactor trip system	$\lambda_S = \lambda_{US}$	DC	MTTR ($1/\mu$)	T_{proof}
Sensor (connected to RPS)	10^{-4}/h	0.2	24 h	720 h
CU of RPS controller	10^{-4}/h	0.2		
P2P Link between redundant RPS controller	10^{-4}/h	0.2		
P2P Link between RPS controllers and ESFAS controller	10^{-4}/h	0.2		
CLC module	10^{-5}/h	0.2		
ESFAS Actuator	10^{-5}/h	0.2		

$$R_{ESFAS} = A_{ESFAS}\left(1 - (1 - R_{CLC})^2\right) = (1 - (PFD + SOP))\left(1 - (1 - R_{CLC})^2\right)$$
$$= 0.999977$$

4.5 Results

Based on the experiments, the components of ESFAS are to be designed with CLSs given in Table 2 to achieve the required TDA (PFD < 10^{-5} and SOP < 10^{-5}). With these design specification, overall availability and reliability of ESFAS will be 0.999980 and 0.999977, respectively.

5 Conclusion

In this paper, we have demonstrated the architecture-centric dependability analysis of a large CBS with the help of a case study. This kind of analysis is useful in

1. Establishing the required component-level specification, viz. failure rate, diagnostic coverage, MTTR, and surveillance test interval at the early stage of system development life cycle.
2. Selecting an optimal design solution by exploring all the available options of component-level specification (for different system components) that can achieve the target dependability goals. This in turn reduces the overall system development time and cost.
3. Validating suitability of any proposed architecture design for achieving the desired reliability and availability goals for large CBS. Such an analysis strengthens the preliminary safety analysis report.

Acknowledgements Authors wish to thank Mr. Y. S. Mayya, formar Head, Reactor Control Division & Group Director, E&IG for his kind support and encouragement when this work was carried out. Authors also wish to thank Mr. Debashis Das, Director, E&IG, BARC for his continual encouragement.

References

1. IEEE Standard criteria for safety systems for nuclear power generating stations. *IEEE Std-603* (2009)
2. IEEE Standard criteria for programmable digital devices in safety systems of nuclear power generating stations. *IEEE Std 7-4.3.2-2016* (2016)
3. AERB Safety guide: Safety systems for pressurized heavy water reactors, *AERB/NPP-PHWR/SG/D 10*, (2005)
4. AERB Safety guide: Computer based systems of pressurized heavy water reactors, *AERB/NPP-PHWR/SG/D-25*, (2010)
5. P. Feiler, Open source AADL tool environment (OSATE), in *AADL Workshop*, Paris (2004)
6. M. Huth, M. Ryan, *Logic in Computer Science: Modelling and Reasoning about Systems* (Cambridge University Press, Cambridge, 2004)
7. E. Clarke, O. Grumberg, S. Jha, Y. Lu, H. Veith, *Progress on the State Explosion Problem in Model Checking* (Springer, Berlin Heidelberg, Berlin, Heidelberg, 2001), pp. 176–194
8. IAEA Nuclear Energy Series: Dependability Assessment of Software for Safety Instrumentation and Control Systems at Nuclear Power Plants. *IAEA Nuclear Energy Series NP-T-3.27*, International Atomic Energy Agency, Vienna (2018)
9. M. Kwiatkowska, G. Norman, D. Parker, PRISM 4.0: Verification of probabilistic real-time systems, in *Proceeding of 23rd International Conference on Computer Aided Verification (CAV'11)*, vol. 6806 of LNCS (Springer, 2011), pp. 585–591
10. L. Meshkat, J.B. Dugan, J.D. Andrews, Dependability analysis of systems with on-demand and active failure modes, using dynamic fault trees. IEEE Trans. Reliab. **51**(2), 240–251 (2002)
11. SAE Architecture Analysis and Design Language (AADL) Annex Volume 1: Annex a: Graphical AADL notation. Annex C: AADL MetaModel and Interchange Formats, Annex D: Language Compliance and Application Program Interface Annex E: Error Model Annex, AS5506/1 (2011)

12. P. Hokstad, K. Corneliussen, Loss of safety assessment and the IEC 61508 standard. Reliab. Eng. Syst. Saf. **83**, 111–120 (2004)
13. M. Kumar, A.K. Verma, A. Srividya, Modeling demand rate and imperfect proof-test and analysis of their effect on system safety. Reliab. Eng. Syst. Saf. **93**, 1720–1729 (2008)
14. A. Kabra, G. Karmakar, M. Kumar, P. Marathe, Sensitivity analysis of safety system architectures, in *Proceeding of International Conference on Industrial Instrumentation and Control (ICIC)*, India (2015), pp. 846–851
15. BARC Design basis report: Engineered Safety Feature Actuation System of Indian Pressurized Water Reactor, *BARC/E&IG/IPWR/DBR/ESFAS/Rev0* (2016)
16. AERB Safety Code: Design of light water based Nuclear Power Plants, AERB/NPP-LWR/SC/D (2015)

Modeling Inhomogeneous Markov Process in Smart Component Methodology

Darpan Krishnakumar Shukla and A. John Arul

Abstract Dynamic reliability analysis methods can account for the interactions present between physical process, control systems' hardware and software, and human actions in reliability analysis of dynamic systems. They provide scope for high fidelity in reliability modeling. A smart component-based methodology is developed recently to serve as a generic method for dynamic reliability analysis while solving existing challenges of dynamic reliability analysis, such as state space explosion, easy system structuring. The method is based on object-oriented representation of the dynamic systems' structure and interactions, and Monte Carlo simulation for reliability simulation. The method can account for the dynamics generated from the above-mentioned interactions. In addition to that, modeling and demonstration of the aging and wear in processes through time-dependent reliability parameters is needed. In this paper, we demonstrate time-dependent reliability parameters in the framework of smart component methodology (SCM) using inhomogeneous Markov process. The generality of SCM for inclusion of Weibull distributed failure rates and various repair schemes is validated with example systems and the reliability results from the literature, and numerical and fault tree methods. An acceleration scheme is also implemented within the SCM framework and results are found to be consistent.

Keywords Aging · Dynamic reliability analysis · Smart component methodology · Time-dependent failure rate

D. K. Shukla (✉) · A. John Arul
Indira Gandhi Centre for Atomic Research,
Homi Bhabha National Institute, Kalpakkam 603102, Tamil Nadu, India
e-mail: darpanks@igcar.gov.in

A. John Arul
e-mail: arul@igcar.gov.in

© Springer Nature Singapore Pte Ltd. 2020
P. V. Varde et al. (eds.), *Reliability, Safety and Hazard Assessment for Risk-Based Technologies*, Lecture Notes in Mechanical Engineering,
https://doi.org/10.1007/978-981-13-9008-1_26

1   Introduction

Dynamic reliability analysis provides reliability evaluation of realistic systems accurately [1]. The features of dynamic systems that raised a need for the development of dynamic reliability analysis method include interactions of the physical process with control systems' hardware and software, and human actions. The other challenges for developing dynamic reliability methods are scalability, intuitiveness, widely used and user-friendly interface. Recently, a smart component (SC)-based methodology is developed at Indira Gandhi Centre for Atomic Research (IGCAR) for dynamic reliability analysis which is based on object-oriented representation for system structuring and Monte Carlo simulation (MCS) for quantifying the structured system for the results of the user interest. The MC method having the capability of modeling any scenario in general is chosen as an evaluation method. The SCM has been demonstrated for the reliability analysis of the tested system, m-out-of-n system, and several dynamic systems elsewhere [5]. In these applications, MCS algorithms of SCM have considered constant failure rates. In the dynamical reliability evaluation of the realistic system, the kinematic behavior of probability fluid is needed to be studied due to its importance in fine aspects of reliability studies. They are generated due to the inhomogeneity in the reliability parameters such as time-varying failure rates of the hardware components or the physical process-dependent rates, dirac delta functions introduced because of the periodic testings/repairs, preventive maintenance (PM) and component dependencies. One part of the challenges lies in the generic development of an MCS procedure for including time-dependent features of a dynamic system in reliability evaluation [2]. These aspects are to be treated in the MCS procedures to be used in SCM-based dynamic reliability analysis (is the objective of the paper).

In literature, the MC treatment of time-dependent reliability parameters in Ref. [3] is based on mode sampling and self-transition MC sampling techniques for the inhomogeneous Markov processes. The reference is treating PM, revealed and unrevealed failures. In Ref. [4], MC sampling methods for time-dependent failure rates are improved for efficient calculations by introducing biased MC sampling schemes based on an exponentially and uniformly distributed biasing parameter, and they treat revealed failure. The MC techniques are not demonstrated clearly for unrevealed failures and PM, and the benefits of PM over corrective maintenance is not illustrated. Therefore, in this paper, MCS procedures to incorporate PM, revealed and unrevealed failures, and tested and repair systems for transient as well as mission time availability analysis within SCM are presented.

In this paper, in Sect. 2, various component models for inhomogeneous Markov processes of SCM are described briefly along with MCS for quantification for time-dependent reliability analysis. A single-component system is analyzed with the various component models using SCM and numerical method, and the comparison and the differences in the results are provided in Sect. 3. Section 4 presents the validation of the accelerated MC procedures of SCM for two-component series and parallel system with that from literature. Section 5 compares SCM with fault tree results for the three-component series–parallel system. The paper is concluded in Sect. 6.

2 Component Models and Monte Carlo Sampling in Smart Component Methodology

In this section, first, SCM is described in brief with its key methods and elements. The component reliability models and their modeling within SCM are presented next. Subsequently, the procedure for availability quantification of MC method is presented.

2.1 Smart Component Methodology

SCM is based on object-oriented system representation and MCS for reliability evaluation. The two-step procedure for reliability evaluation using SCM is as follows: one, building SC model of the system, and two, evaluation of the SC model of the system using the MCS algorithm. The SC model of the system consists of three parts.

1. Components-as-objects: While structuring SC model of the system, the components of the system are represented as an abstract object. The objects are defined with their attributes and local rules, each of them helps to describe the components' functional behavior in the SC model of the system. The attributes include process parameter, state of hardware, reliability parameters, etc.
2. Connector: The dependencies among the components are depicted here. These dependencies can be in the various forms such as logic dependencies, process control thresholds, message passing, testing.
3. Global rule: The global rules are defined for the system, to manifest a system's success criteria/failure criteria, simulation parameters, and system type.

These three together form a structure for system simulation. After building the SC model of the system, reliability evaluation is performed with a suitable but general MCS algorithm. Readers are referred to [5] for more information on the specific elements of the method.

2.2 Component Models

For reliability evaluation of a system, the reliability of each component and their combinations have to be evaluated to know completely the probabilistic behavior of the system [6]. In this section, existing component reliability models are mentioned for its application to find the system reliability/availability of interest. The existing two types of the components are repairable and nonrepairable. The nonrepairable component is the one that remains in the failed state after the failure due to the impossibility of repair or unaffordable cost of the repair. There exist two types of repair strategies for repairable component [7]: (1) corrective maintenance (CM) and

(2) preventive maintenance (PM). In CM, the component is repaired/replaced after its failure. The existing two models of the corrective maintenance are revealed failures (when the components are subjected to continuous monitoring) and unrevealed failures (where the failed components are not continuously monitored but periodically tested for finding its working condition). In PM, the components are replaced or repaired periodically to avoid failure. Further, there are two options for PM: (i) instantaneous and (ii) delayed. This assumption has implicated for the unavailability. It is assumed that PM is instantaneous in the paper, wherever mentioned next.

Let 1 and 0 states of the hardware of the component describe the working and failed states of the component, respectively, and state 2 describes the component being in repair state. The transitions from state 1 to 0 represent the failure of the component. Similarly transition from state 0 to 1 correspondence to the repair/replacement of the component. In the repair model of the revealed failures, the repair process is started immediately after the failure. In the unrevealed failures, the failed component remains in the failed condition until a test conducted, because the failure is unrevealed. The tests being conducted periodically, and based on the result of the test, whether the component is in a failed or working state, the repair process is initialized. Therefore in the revealed failure case, the repair process is the transition between the states 0 to 1, and in the unrevealed failure, the transition between the state 0 to 2 and subsequently 2 to 1 is considered as repair process. These two repair models are represented as state diagrams in Fig. 1 as (a) and (c), respectively. The repair rate, μ, is assumed to be exponentially distributed.

The failure rate, $\lambda(t)$, is represented as the sum of Weibull distributed failure rates as given in Refs. [3, 4] for the bathtub curve. Since Weibull distribution is widely used to represent time-dependent behavior and according to authors' understanding, it is easy to sample the time from the Weibull distributed time, Weibull distribution is considered for modeling. In case of the components with aging, where the failure rate and, hence, the unavailability are increasing with time, to maintain acceptable availability, PM of the component is carried out periodically. For example, the component is replaced with the new one or minimally repaired if found in the failed state. The two types of PM are as-good-as-new (replacement), and, as-good-as-old (mini-

Fig. 1 Component models **a** revealed repairable without PM, **b** revealed repairable with periodic PM, **c** unrevealed without PM, and **d** unrevealed with the PM. Dashed arrow suggests as-good-as-new replacement of PM

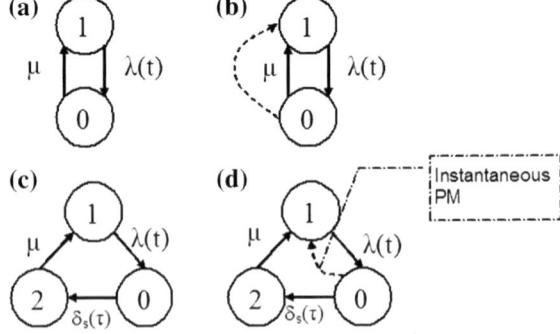

mal repair). It is assumed that PM is as-good-as-new in the calculations that follow. The options are illustrated as (b) and (d) in Fig. 1, respectively, for the revealed and unrevealed failures with PM. The dotted line represents the instantaneous PM.

In revealed failure case with PM, the as-good-as-new PM is carried out periodically, with a period τ, while the repair process starts instantaneously after failure. In the case of conflict of the PM occurring before finishing of the repair, the PM carried out at the periodic interval which replaces the component with the new one. In the unrevealed failure case, where the test is also periodic (represented in Fig. 1 as $\delta_s(\tau)$), the as-good-as-new PM is carried out along with testing (i.e., with period $\tau_{PM} = \tau$). Note that the PM for (B) and (D) is assumed to be instantaneous, which brings the component back to the working state instantaneously. The repair models discussed here, we believe, have closer correspondence to the practical scenario compared to the one given in Ref. [3], since in the reference, the repair time assumed to be zero for the unrevealed failures in inhomogeneous Markov processes.

2.3 Component Modeling in SCM

To model a system and its components in SCM for reliability analysis, as mentioned in Sect. 2.1, each component is modeled as an object and the system is modeled as a collection of the objects. The component model of the object is defined as an attribute which is used as an identifier. The summary of the available reliability models and the corresponding required reliability parameters are given in Table 1. The quantification algorithms of SCM, in background, take care of all the types of the reliability models while sampling time and state for simulating a history. The SCM provides the simple representation of the component for reliability analysis. For example, here, SC model of a single-component system is given as a tabular form Component 1 in Table 3. The model is defined with parameter (attribute), type of parameter and value of the parameter (data) in the table. In the model, the two-parameter Weibull distributed failure rate for three modes of failure is defined as $[[\lambda_1, \alpha_1], [\lambda_2, \alpha_2], [\lambda_3, \alpha_3]]$, where α_i are the Weibull parameters. The corresponding failure rate is given by $\lambda(t) = \lambda_1\alpha_1 t^{\alpha_1-1} + \lambda_2\alpha_2 t^{\alpha_2-1} + \lambda_3\alpha_3 t^{\alpha_3-1}$.

Table 1 List of reliability models available in SCM

Reliability model	Identifier	Required reliability parameters
Nonrepairable	1	$\lambda(t)$
PFD	2	P_f = probability of failure on demand (PFD)
Repairable		
Revealed w/o PM	3	$\lambda(t), \mu$
Revealed with PM	4	$\lambda(t), \mu, \tau_{PM}$
Unrevealed w/o PM	5	$\lambda(t), \mu, \tau_{test}$
Unrevealed with PM	6	$\lambda(t), \mu, \tau_{test}, \tau_{PM}$

2.4 Quantification in SCM

For quantifying the unavailability for the repairable component, which is defined as the ratio of time during which the component is not in working state to the total operating time, the downtime is calculated as the sum of the time during which the component is in state 0 and/or 2. The exact modeling of the periodic nature of tests, PM is elucidated in the MCS as a delta function, $\delta_s(t)$. The MCS procedure for simulating a history of a system for reliability evaluation consists of two steps. [8] The two-step procedure is modified here to incorporate the three states of the components of the SC model of the system. The first step of MCS procedure is sampling of time to transition, t_s, using (1) inversion of cumulative distribution function (CDF) based on the current state of the components of the system, $t_{sp} = F^{-1}(s, \lambda(t))$, where s is state of the system components and F is the CDF of transition rates. And, (2) deterministic periodic test, τ. The sampling time to transition, when incorporates unrevealed failures, has these two times. The first time sampling is due to the stochastic nature of the failure and repair rates. Moreover, the second deterministic times are due to the deterministic process of periodic testing. The minimum of the two times is selected as the transition time in the first step of MCS, i.e., $t_s = \min([t_{sp}, \tau])$. The elapsed time t_{elapsed} in the simulation history is updated to the transition time, i.e., $t_{\text{elapsed}} = t_{\text{elapsed}} + t_s$.

The second step of the MCS procedure is the sampling of state to which the system is making the transition. The sampling state is based on probabilistic or deterministic depending on the selected minimum time in the previous step. In the case of probabilistic transition, the new state is determined from the transition rates, $\lambda(t)$, at the new time t_{elapsed}. For the deterministic transition due to the test/PM in unrevealed failure, the transitions are from state 0 (failed) to state 2 (repair). The simulation history is continued until the stopping criteria reached. The stopping criteria for an MC history are based on the problem system, i.e., either system failure for mission time (MT) model or system regeneration for steady-state (SS) unavailability. A large number of histories are simulated to get statistically good average estimate. The reliability/availability tallies are carried out after each sampling of transition time for estimating the parameter of interest.

In the next section, the unavailability calculations of a single-component system with the various reliability models are presented.

3 Single-Component System

In this section, the SC model of a single-component system is constructed and transient unavailability is evaluated using SCM MCS algorithm. The analytical formulation for the transient unavailability calculations of the single-component system is presented. Results from both the methods are compared and discussed subsequently. The single-component system studied here is adapted from the literature [3]. Hence, results are also compared with that from the literature.

3.1 SC Model of the Single-Component System

The failure rate and repair rate of the single-component system are defined below. Failure rate, $\lambda(t) = \lambda_0 + \frac{\alpha}{\theta}(\frac{t}{\theta})^{(\alpha-1)}$ (per year), repair rate, $\mu = 10$ (per year), mission time, $MT = 5$ (year), $\tau = 1$ (year). $\lambda_0 = 0.013$ (per year), $\theta = 7.5$ (year), $\lambda_1 = \frac{1}{\theta}$, $\alpha = 2.5$ [3]. However, τ is test interval, defined only for simulating the unrevealed failure cases. In the single-component system, the system failure criteria are failing of the single component. The SC model of the system is consisted of only one component having defined the above-mentioned information of the component. The SC model of component 1 is given in Table 3. The SCM simulation algorithms, which are based on MCS, are used to estimate the transient and steady-state unavailability of the system.

3.2 Analytical Validation

To validate the results from the MC method, we use state equations for the corresponding system. In this section, the equations are presented for the single-component system with unrevealed failures. Readers are referred to state diagram (C) of Fig. 1. State equations are as follows:

$$\frac{dP_1(t)}{dt} = -\lambda(t)P_1(t) + \mu P_2(t) \tag{1}$$

$$\frac{dP_0(t)}{dt} = \lambda(t)P_1(t) - \delta_s(\tau)P_0(t) \tag{2}$$

$$\frac{dP_2(t)}{dt} = \delta_s(\tau)P_0(t) - \mu P_2(t) \tag{3}$$

where $\delta_s(\tau) = \sum_0^n \delta(n\tau - t)$, $n = \frac{t}{\tau} \in N$ and $P_i(t)$ represents probability of finding the system in state i at time t. $\delta_s(\tau)$ represents dirac delta function and it models accurately the instantaneous nature of tests. Unavailability, $\bar{U}(t)$, of the system is defined as the system being in state 0 and 2. Hence, $\bar{U}(t) = P_0(t) + P_2(t)$.

3.3 Results and Discussions

The unavailability of the single-component system is calculated from the two methods, SCM and numerical methods. The results are generated for the unrevealed failure cases one with the PM and another without the PM. The similar combinations of results are generated for model 1 (the constant failure rate) and model 2 (increasing failure rates). For the system with revealed failures, results are compared with that from the literature.

3.3.1 Case 1: Increasing Failure Rates and Unrevealed Failures

The instantaneous and interval unavailability for the increasing failure rates are calculated using SCM and numerical method for without PM. However, only SCM is used to estimate the unavailabilities for the case with the PM. The comparison is shown in Fig. 2a, b. From the results, it is concluded that the transient unavailability of the SCM follows closely with that from the numerical calculations. When PM is applied explicitly to the unrevealed failure cases, where at every test interval, as-good-as-new PM is carried out, i.e., replacement of the component with the new one, instantaneous unavailability is fallen to zero as expected, and interval unavailability is reduced. The interval unavailability is eventually reached steady-state value less than one for the increasing failure rates. As this is expected, this is pronounced in Fig. 2b.

3.3.2 Case 2: Constant Failure Rate and Unrevealed Failures

The transient unavailability for the cases where the system is being tested periodically is shown in Fig. 3a, b for the constant failure rate case. It is noticed that when PM is applied to the unrevealed failure case, irrespective of the failure rate time-varying form, i.e., either increasing or constant, the interval unavailability is improved as it is pronounced in Fig. 3b.

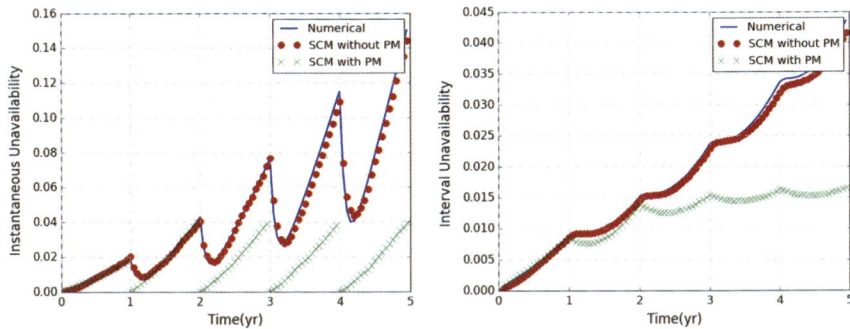

Fig. 2 Increasing failure rate case, **a** Comparison of the instantaneous unavailability results from the SCM and numerical for two cases: (1) with PM and (2) without PM. **b** Comparison of the interval unavailability results from the SCM and numerical for two cases: (1) with PM and (2) without PM

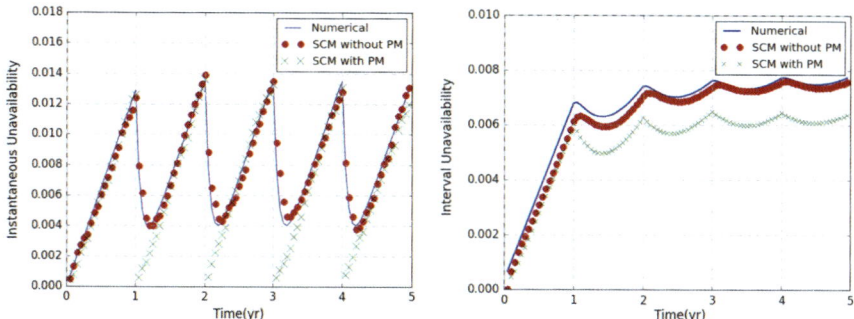

Fig. 3 Constant failure rate case, **a** Comparison of the instantaneous unavailability results from the SCM and numerical for two cases: (1) with PM and (2) without PM. **b** Comparison of the interval unavailability results from the SCM and numerical for two cases: (1) with PM and (2) without PM

3.3.3 Case 3: Increasing Failure Rates and Revealed Failures

For the system with revealed failures and with the increasing failure rates, the mission time interval unavailability from the Ref. [3] is compared with that from the SCM. The results in Table 2 show that the interval unavailability and mission unreliability results are matching well with that from the reference results. In the table, number of histories, unreliability, unavailability, variance, fractional error, coefficient of variation, average time per history (seconds), figure of merit are defined, respectively, as $N, \bar{R}, \bar{A}, \sigma^2, \delta = \frac{\sigma}{\bar{X}}, \Delta = \sqrt{N} * \delta, \bar{t}, \text{FOM} = \frac{1}{\sigma^2 \bar{t}}$.

In summary, the single-component system having various models of inhomogeneous nature is modeled in SCM with the distinction between the unrevealed failure and PM. The results also profoundly elaborate the differences in the models. The results match well with the numerical deterministic results.

4 Two-Component Parallel and Series Systems

In this section, two-component system connected in parallel and series is analyzed for reliability evaluation having the time-dependent failure rates. First, parallel system of Ref. [3] is analyzed, and subsequently, the series system of Ref. [4] is analyzed.

4.1 SC Model of the Two-Component System

The construction of the two-component parallel system in SC framework is carried out as shown in Table 3. The two-component object as shown in the table are created

Table 2 Comparison of interval unavailability and mission time unreliability results of Ref. [3] and SCM for the single-component and two-component parallel systems

System	Ref. results (N = 1E+4)	SCM results	$\bar{R}$	$\bar{A}$
Single component	$\bar{R}$ = 3.49E−1 (+/− 0.23E−2)			
		N = 1E+4	3.41E−1	8.04E−3
		σ^2	5.83E−5	8.19E−8
		δ	2.24E−2	3.56E−2
	$\bar{A}$ = 8.09 E−3 (+/− 1.1E−4)	Δ	2.24E+0	3.56E+0
		FOM	1.77E+7	1.26E+10
		$\bar{t}$	9.67E−4	9.67E−4
Two parallel components	$\bar{R}$ = 9.29E−3 (+/− 1.8E−4)	N = 1E+5	9.43E−3	9.16E−5
		σ^2	1.06E−7	1.29E−12
		δ	3.45E−2	1.24E−2
	$\bar{A}$ = 9.17E−5 (+/− 2.9E−6)	Δ	1.09E+1	3.92E+0
		FOM	5.40E+9	4.43E+14
		$\bar{t}$	1.75E−3	1.75E−3

with its attributes - hardware state, reliability parameter, and input–output. Connector for the parallel system shows that connections from the output of both the components are connected to a dummy-component result. The use of connector table is to identify system functioning using reachability checking. However, global rules are also given in the table for both the series and parallel connections for the completeness. For the parallel system, failure criteria are failing of both the systems, and for the series system, failure criteria are failing of either components. These criteria are presented in the global rules explicitly.

4.2 Results and Discussions

4.2.1 Two-Component Parallel System

The interval unavailability and the unreliability for mission time of five years are calculated for the two-component parallel system is calculated from the SC model of Table 3. The two components have the same failure and repair rates as defined in Sect. 3.1. The results are reported in row 2 of Table 2. The unreliability and unavailability results of SCM match well with that from the reference.

Table 3 SC model of two-component parallel and series system

Component 1			Component 2		
Parameters	Type	Data	Parameters	Type	Data
Hardware	State	1	Hardware	State	1
Rel_Model	Rel	4	Rel_Model	Rel	4
Failure Rate	Rel	[[0.013,1], [0.00649,2.5]]	Failure Rate	Rel	[[0.013,1], [0.00649,2.5]]
Repair Rate	Rel	10	Repair rate	Rel	10
Test interval	Rel	1	Test interval	Rel	1
CompInput	Input	1	CompInput	Input	1
CompOutput	Output	1	CompOutput	Output	1
Connector for parallel system					
Component 1	CompOutput		Result	CompInput1	
Component 2	CompOutput		Result	CompInput2	
Connector for series system					
Component 1	CompOutput		Component 2	CompInput	
Component 2	CompOutput		Result	CompInput1	
Global rule for parallel system:			*Global rule for series system:*		
If			If		
Component 1 == 1 or Component 2 == 1:			Component 1 == 1 and Component 2 == 1:		
Then			Then		
system success			system success		
Else			Else		
system fail			system fail		

4.2.2 Two-Component Series System

The two-component series system of Ref. [4] is analyzed for transient unavailability for the mission time of 30 yrs. The reliability parameter for the two-component series system unavailability analysis is: $\lambda(t) = 10^{-6} + 10^{-6}t$ (per year), $\mu = 10^{-2}$ (per year), $MT = 30$ (year) . The results are shown in Fig. 4 with analogue MC and with different values of a biased parameter of exponentially distributed biased MC techniques of Ref. [4]. The results show consistent behavior with that from the literature. When the failure rates are biased with the exponentially distributed failure rate, $\lambda*$, when the difference between the original and the biased rate is low, the nonlinear MC is not benefiting. When the difference is kept very high, then due to decrease in weight of the history in the later stage, results are not converging. But when the difference-biased and difference-unbiased rates are kept moderate, the acceleration is achieved successfully. It is also confirmed that the similar type of the behavior is observed in Fig. 1 of the Ref. [4].

Fig. 4 Instantaneous unavailability results from the SCM for the two-component series system with direct and biased MCS algorithms

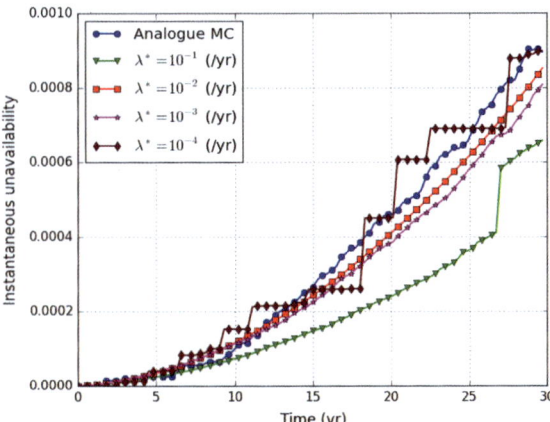

In conclusion, the SC model of the two-component system is constructed for analysis in SCM. The unavailability results for the two-component series and parallel systems agree well with the reference results. The optimized acceleration parameter is also achieved similar to that of the reference results.

5 Three-Component Series–parallel System

A three-component series–parallel system consists of same components as defined in Sect. 3.1 is analyzed in this section using SCM and fault tree method. The mission time is same as that used in Sect. 3. The fault tree of the three-component system is shown in Fig. 5. The results are shown in Fig. 6 for the transient unreliability of the system generated from SCM and Reliability Workbench V10.2 and they are matched well.

Hence, a three-component series–parallel system is analyzed using SCM. The results show that SCM with the inhomogeneous Markov reliability model can be used in general for any configuration of the system. In future work, a dynamic system consists of process, hardware, and software along with the inhomogeneous reliability parameters needs to be explored, and acceleration methods for the system with the unrevealed failures to be investigated.

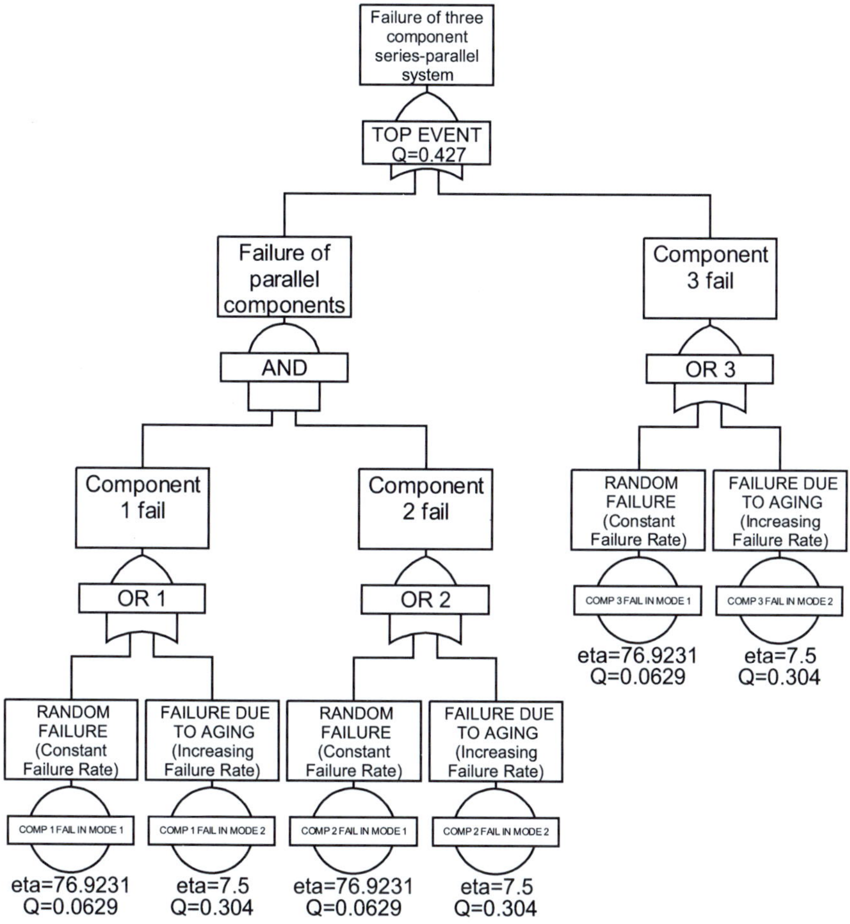

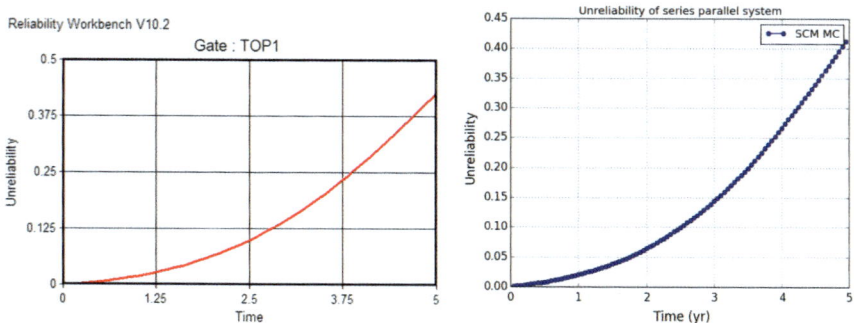

Fig. 5 Fault tree of series–parallel system having generated in Reliability Workbench V10.2

Fig. 6 Comparison of unreliability results from FT and SCM for the series–parallel system

6 Conclusion

SCM is being developed for dynamic reliability evaluation. One of the needs of a dynamic reliability method is to treat the kinetic effects on probability generated due to the time-dependent transition rates. In the paper, SCM is reported with its capability to treat the inhomogeneous Markov processes. PM, unrevealed and revealed failures are modeled clearly in the SCM. Results are agreeing well with that from the literature and the numerical methods. Results of unrevealed failures with and without PM show the clear disjoint between PM and unrevealed failures, and the benefits of PM in unavailability results are elucidated.

Acknowledgements The authors thank Director, Reactor Design Group, IGCAR, for their encouragement and support for completing the work. The first author thanks the Board of Research in Nuclear Studies, Mumbai, India, and Department of Atomic Energy, India, for supporting through DGFS-PhD fellowship.

References

1. D.K. Shukla, A.J. Arul, A review of recent dynamic reliability analysis methods and a proposal for a smart component methodology, in *Proceedings of 4th International Conference on Reliability, Safety and Hazard (ICRESH-2019)*, Indian Institute of Technology Madras, 10–13 Jan 2019
2. P.E. Labeau, C. Smidts, S. Swaminathan, Dynamic reliability : towards an integrated platform for probabilistic risk assessment. Reliab. Eng. Syst. Saf. **68**, 219–254 (2000)
3. E.E. Lewis, T. Zhuguot, Monte Carlo reliability modeling by inhomogeneous Markov processes. Reliab. Eng. **16**, 277–296 (1986)
4. M. Marseguerra, E. Zio, F. Cadini, Biased Monte Carlo unavailability analysis for systems with time-dependent failure rates. Reliab. Eng. Syst. Saf. **76**(0951), 11–17 (2002)
5. D.K. Shukla, A.J. Arul, Development of smart component based framework for dynamic reliability analysis of nuclear safety systems, in *International Conference on Fast Reactors and Related Fuel Cycles: Next Generation Nuclear Systems for Sustainable Development*, Yekaterinburg, Russia (2017), p. 603
6. W.E. Vesely, A time-dependent methodology for fault tree evaluation. Nucl. Eng. Des. **13**, 337–360 (1970)
7. E.E. Lewis, *Introduction to Reliability Engineering*, 2nd edn. (Wiley and Sons Ltd., New Delhi, 1994)
8. E.E. Lewis, F. Boehm, Monte Carlo simulation of Markov unreliability models. Nucl. Eng. Des. **77**, 49–62 (1984)

Estimation of Residual Lifespan of Instrumentation & Control (I&C) Cables of Nuclear Power Plants by Elongation Measurement

A. K. Ahirwar, P. K. Ramteke, N. B. Shrestha, V. Gopika and J. Chattopadhyay

Abstract Nuclear power plants (NPPs) contain myriads of electrical cables of all sizes, voltage ratings and lengths delivering electrical power to much vital, as well as non-vital equipment. These cables are insulated with some form of polymeric insulation. These insulating materials will gradually age from exposure to heat, radiation, moisture and chemicals. Usually, the evidence for satisfactory performance over a period of designed life is provided by the results of the qualification tests, simulating operational ageing and the postulated design basis accident. Experience has shown that the failure of nuclear cables is primarily due to the hardening and embrittlement of the insulation resulting in the formation of micro cracks, a loss of dielectric strength and high leakage currents. Percentage elongations-at-break (E-at-B) are derived from measurements by tensile tests on cable insulation materials to establish qualified life. Monitoring cable degradation by measuring E-at-B is in many cases not feasible as it requires the periodic removal of cable samples from the field for destructive testing. Condition monitoring (CM) is an effective technique to assess the functional capability/operational readiness of cables during the plant operation.

Keywords Ageing research · Elongation at break · Accelerated testing

1 Introduction

Cable is a vital component of the nuclear power plant. The magnitude of the current flowing through a control cable may vary from mA to few amperes. Any break in the path of current flow or leakage of current will constitute a failure. Loss in the

A. K. Ahirwar (✉) · P. K. Ramteke · N. B. Shrestha · V. Gopika · J. Chattopadhyay
Reactor Safety Division, Reactor Design & Development Group, Bhabha Atomic Research Centre, Trombay, Mumbai 400 085, India
e-mail: babulk@barc.gov.in

P. K. Ramteke
e-mail: pkram@barc.gov.in

© Springer Nature Singapore Pte Ltd. 2020
P. V. Varde et al. (eds.), *Reliability, Safety and Hazard Assessment for Risk-Based Technologies*, Lecture Notes in Mechanical Engineering, https://doi.org/10.1007/978-981-13-9008-1_27

flexibility/increase in hardness or stiffness, decrease in elongation and electric insulation resistance are good indices of ageing. When cables approach their end of life, they start creaking on bending or even may become brittle. It has been observed that cable terminations are most susceptible to ageing effects by virtue of their chemical reaction with oxygen. Invariably aged cable termination is small length, sometimes even only 15 cm; ageing of cables can be managed either by keeping some extra length at the time of installation or by providing junction boxes so that entire cable need not be replaced.

In general, a cable is not discarded as long as it shows continuity (electrical resistance of conductor almost zero) and insulation resistance (IR) greater than a set failure criteria limit which is normally 50 MΩ [1, 2]. It has been observed that it is the mechanical properties which degreed earlier than the electrical properties. However, the variation in electrical property such as IR is sudden and drastic. For example, when the cable is approaching brittleness stage, IR may decrease from 10^{10} to 10^{6} Ω in a short span of time.

2 Ageing Test Methodology

One of the most important tasks in ageing research is the selection of appropriate test temperature and gamma radiation dose rate such that item ages/degrades in the same manner as in the case of natural ageing and test is completed in 2–3 months. This is essential because the validity of extrapolated life under normal working environment is meaningful only if no new failure mechanisms are introduced due to accelerated stress. In order to find out test temperature for an engineering hardware, cable test specimen is kept in a thermal oven, and temperature of the oven is increased in steps after every day until failure. The temperature at which test item fails is called limit temperature. Accelerated test temperature has to be lower than limit temperature. Then, three to four batches of the test specimens are kept at different temperature less than limit temperature for 1000 h each to determine rated temperature. Rated temperature is the highest temperature at which there is no noticeable degradation at the end of 1000 h of test. Once the rated temperature is established, accelerated ageing tests are conducted at least at two temperatures above rated value. Performance parameters are measured initially and periodically after every ten days. Testing is to be continued until failure. Arrhenius equation is used for extrapolation of lifespan at nominal environmental temperature. In case of multiple environments, their individual and interaction effects have to be taken into consideration. In such a situation, design of experiment becomes complex. Lifespan under an operating environment can be extrapolated by using the Eyring model. Both the models are empirical mathematical equations. As far as radiation ageing is concerned, it has been observed that lower dose rates are more damaging than higher dose rates for the same integrated dose. This is probably due to more time being available for chemical reaction with oxygen. Normally, hardware is expected to receive a total lifetime and accident dose of about 50–100 Mrad. Dose rates of 50–100 Krad/hr are quite reasonable and practical [3, 4].

3 Brief Description of the Ageing Research Test Facilities Available in RSD, BARC

Nuclear power plants (NPPs) contain myriads of electrical cables of all sizes, voltage ratings and lengths delivering electrical power to much vital, as well as non-vital equipment. These cables are insulated with some form of polymeric insulation. Failure of cables is primarily due to the hardening and embrittlement of the insulation resulting in the formation of micro cracks, a loss of dielectric strength and high leakage currents. Percentage elongations-at-break (E-at-B) are derived from measurements by tensile tests on cable insulation materials to establish qualified life. Monitoring cable degradation by measuring E-at-B is in many cases not feasible as it requires the periodic removal of cable samples from the field for destructive testing. Correlation study of the E-at-B with the physical/chemical deterioration of the insulation and jacket materials is an important element and it may help in predicting the cables life. Condition monitoring (CM) is the assessment of continued ability to maintain functional capability under specified service conditions including design basis event environments. CM techniques are supporting tools for equipment qualification programme. Elongation-at-break CM method is used to estimate the life/residual life of the cables (Fig. 1) [5].

Fig. 1 Universal testing machine

4 Elongation-at-Break/Tensile Test Facility

As explained above, tensile property (E-at-B and Tensile Strength) measurements are basic tool for correlation. Tensile test is performed in accordance with ASTM-D2633-82, using a universal testing machine (UTM) equipped with pneumatic grips and having an extensometer clamped to the sample. Special tensile specimens (dumbbells for larger cables or cylinders for smaller cables) of the insulation or jacket materials without the copper conductor are used for these tests. These tests are destructive, and therefore, many samples are required if tests are conducted regularly.

The sample of cable sheath material was made according to ASME D638-10 standard and the specimen sample of sheath as shown in the Fig. 2.

Accelerated thermal ageing test was carried out in oven with forced air circulation with an accuracy of ±1 °C. During the accelerated thermal ageing test, test parameters monitored were the visual inspection and cracks on bending. The samples were taken out according to the test matrix for EAB measurement (Table 1).

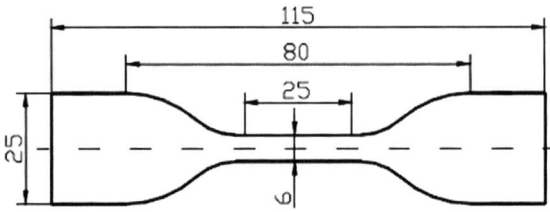

Fig. 2 Specimen of sheath (dumbbell shaped)

Table 1 Test matrix

EPR, HTXLPE and LTXLPE cable insulation			
120 °C		130 °C	
80 samples (All three types) (Dumb-bell samples)		50 samples (All three types) (Dumb-bell samples)	
Test duration days	Removing no. of samples	Test duration days	Removing no. of samples
Initial	5	Initial	5
10	5	10	5
…	…	…	…
…	…	…	…
150	…	80	…

Fig. 3 Thermal ageing test chambers

5 Thermal Ageing Test Chamber

Thermal chambers of various temperature ranges (ambient to 300 °C) and different dimensions are designed and fabricated for carrying out the accelerated thermal ageing studies of C&I components and equipments. Provision for on-line performance monitoring and process air connection of the items being tested has also been added in these chambers (Fig. 3).

6 Thermal Ageing Model

The Arrhenius model is used for addressing the time–temperature ageing effect. Other models have also been used. For example, a simple model, the 10 °C rule, states that chemical reaction rates double and the material life decreases by one-half for every 10 °C increase in the temperature. However, Arrhenius model is preferred over the 10 °C rule, since the latter is a rough approximation of the former. This equation is used in chemical kinetic to predict the rate process. Its application has been extended to reliability studies. This equation takes into account not only the chemical reaction but also physical and physicochemical reactions. This model is represented by the following equation [6–8].

$$t = Ae^{\phi/kT} \tag{1}$$

where

t Life of the insulating material
ϕ Activation energy (ev),

1 eV 23.06 kcal/mole = 1.6 × 10⁻¹² ergs

T Testing temperature in degree Kelvin

k Boltzmann's constant = 8.617 × 10⁻⁵ eV/K = 1.38 × 10⁻¹⁶ ergs/K.

To find the value of A and ϕ, accelerated thermal ageing tests have to be conducted at least two temperature levels. Accelerate tests reduce the test durations. The accelerated temperature for these tests should be fixed. The ϕ can be calculated using the following equation derived from Eq. (1), using two test temperatures.

$$\frac{t_1}{t_2} = e^{\frac{\phi}{k}\left(\frac{1}{T_1} - \frac{1}{T_2}\right)} \tag{2}$$

t_1 and t_2 are thermal ageing time at temperatures T_1 and T_2 (in Kelvin), respectively.

7 Case Study and Test Results

Tensile testing has been carried out on different types of cable insulation material namely, EPR, HTXLPE and LTXLPE. To establish the ageing characteristics of the cable insulation materials, all the three types of cable specimens were thermally aged at 120 and 130 °C separately. Tensile test on insulating specimens was carried out periodically till the elongation-at-break is observed to be 65% or below, which is the end life criterion. The test results are shown in the table.

An ageing curve was drawn for % elongation versus ageing test duration for EPR cable in Fig. 4; Similar ageing curves can be drawn for other cable materials. The criteria to establish the end of life was chosen as 65% elongation. From the ageing curves, test duration in days at 120 and 130 °C was obtained, and these values were used to draw the Arrhenius plot to extrapolate the life at normal operating temperature, i.e. 60 °C. Arrhenius plot for EPR cable is shown in Fig. 5. The constants A and $\emptyset$ were evaluated from this Arrhenius plot (Tables 2, 3, 4 and 5; Figs. 4, 5, 6 and 7).

Equations (1) and (2) can also be used to evaluate the constant A and $\emptyset$. Using these, the life was estimated at normal operating temperature, i.e. 60 °C.

For EPR cable insulation material:

$$\frac{t_1}{t_2} = \exp\left[\frac{\emptyset}{k}\left(\frac{1}{T_1} - \frac{1}{T_2}\right)\right] \tag{3}$$

where t_1 and t_2 are the test duration in days to get 65% elongation at 120 and 130 °C temperature, respectively, and were obtained from ageing curves-(Figure-).

$T_1 = 120 + 273 = 393$ K

$T_2 = 130 + 273 = 403$ K

$t_1 = 131$ days $t_2 = 66$ days

$\emptyset$ = Activation energy (kCal/mole)

Fig. 4 Ageing characteristics cable insulation material EPR

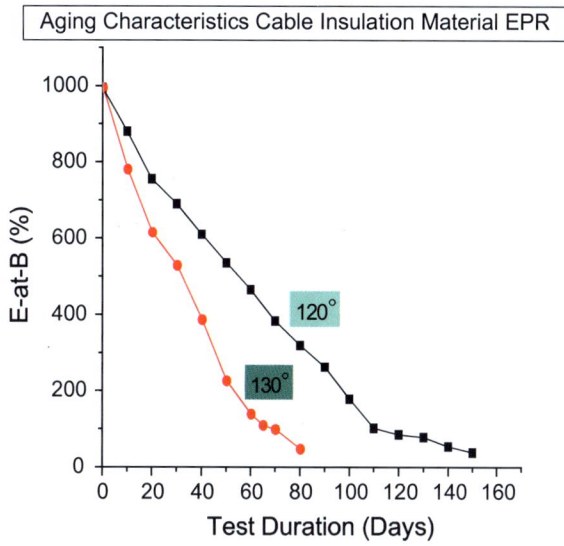

Fig. 5 Arrhenius plot for EPR cable

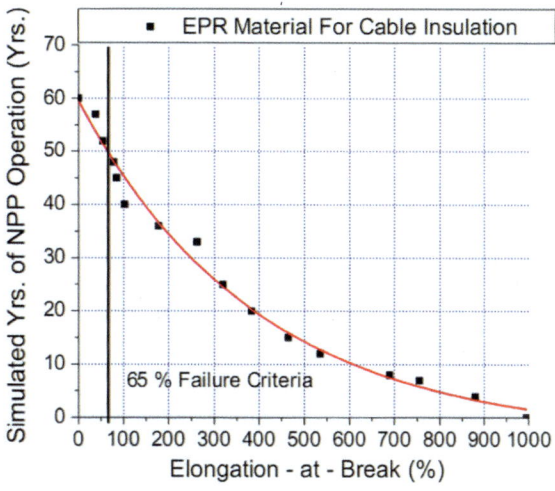

Table 2 Thermal ageing test results of EPR cable insulation materials

EPR cable insulation			
120 °C		130 °C	
Test duration days	E-at-B %	Test duration days	E-at-B %
Initial	995	Initial	995
10	880	10	780
20	755	20	615
30	690	30	528
40	610	40	386
50	535	50	225
60	465	60	138
70	383	65	109
80	319	70	98
90	262	80	47
100	178		
110	102		
120	85		
130	78		
140	54		
150	38		

Table 3 Thermal ageing test results of LTXLPE cable insulation materials

LTXLPE cable insulation			
120 °C		130 °C	
Test duration days	E-at-B %	Test duration days	E-at-B %
Initial	965	Initial	965
10	856	10	737
20	796	20	678
30	708	30	466
40	624	40	324
50	572	50	255
60	438	55	192
70	382	60	111
80	291	65	85
90	205	70	54
100	177	75	41
110	145		
120	112		
130	89		
140	29		

Table 4 Thermal ageing test results of HTXLPE cable insulation materials

HTXLPE cable insulation

120 °C		130 °C	
Test duration days	E-at-B %	Test duration days	E-at-B %
Initial	942	Initial	942
10	894	10	722
20	819	20	649
30	772	30	605
40	718	35	514
50	669	40	412
60	602	45	234
70	518	50	182
80	458	55	144
90	382	60	118
100	321	65	72
110	258	70	53
120	211	75	35
130	115	80	26
140	65		
150	55		

Table 5 Estimation of lifespan using the Arrhenius equation

Insulation material	Test duration for 65% Elongation-at-break		Estimated lifespan at 60 °C (Years)
	At 120 °C	At 130 °C	
EPR	131	66	52
LTXLPE	128	65	48
HTXLPE	140	69	55

$k = $ Boltzmann's constant $= 2 \times 10^{-3}$ kCal/mole/K

$$\frac{131}{66} = \exp\left[\frac{\emptyset}{k}\left(\frac{1}{393} - \frac{1}{403}\right)\right] \tag{4}$$

Ø 21.71(kCal/mole).

Constant A of the Arrhenius Eq. (1) can be determined using any one of the accelerated temperature data:

$$t = Ae^{\phi/KT}$$

where

Fig. 6 Ageing
characteristics cable
insulation material LTXLPE

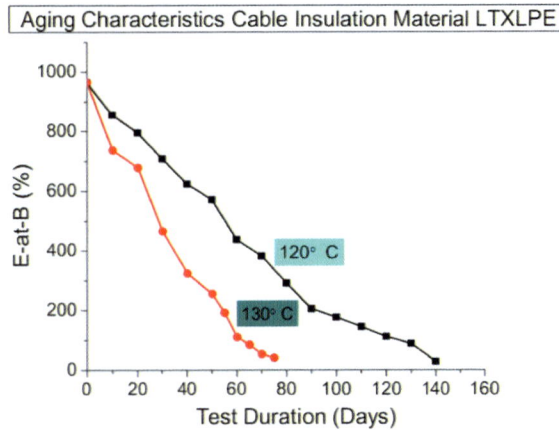

Fig. 7 Ageing
characteristics cable
insulation material HTXLPE

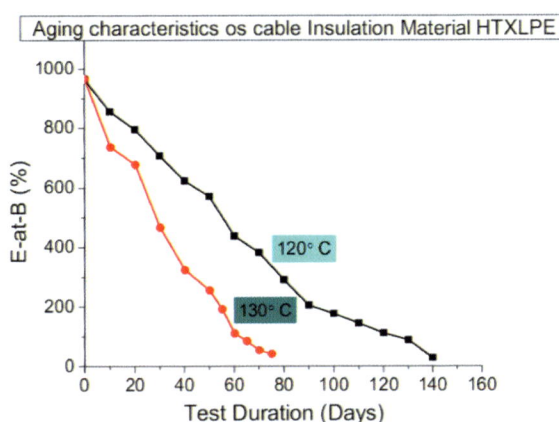

t 131 days at 393°K
A 1.31×10^{-10} days.

After determining the values of A and Ø, life at a normal operating temperature of 60 °C (333°K) can be determined as follows using Eq. (1):

$$t = Ae^{\phi/kT}$$

t 52 Years.

Similar evaluations for other cables were also carried out and estimated the lifespan. Test results are shown in Table 2.

8 Conclusion

The measurement of elongation-at-break for the cable insulation materials is the requirement for monitoring the ageing characteristics of the cables. By measuring the elongation-at-break, lifespan of the cable insulation materials can be calculated. The estimated life can be used as a guideline for the selection and standardization of the materials. Some uncertainties are also associated due to approximation made during designing of the accelerated thermal ageing tests. Condition monitoring (CM) techniques can be used to get an assurance of the residual life of the cables as discussed earlier part of this paper.

References

1. IEEE Standard for Qualifying Class 1E Equipment for Nuclear Power Generation Stations, IEEE Std. 323 (1974)
2. IEEE Standard 11-1974, Standard Test Procedure for Evaluation of Systems of Insulating Materials
3. S.G. Burnay, J. Cook, N. Evans, "Round–robin" testing of cable materials (IAEA Co-ordinate Research Programme on Management of Ageing of In-containment I&C cables. Report on the JAEA Research Co-ordination Meeting, 8–9 June 1998, EDF, Botdeaux, France
4. K. Anandakumaran, W. Seidl, P.V. Castaldo, Condition assessment of cable insulation system in operating nuclear power plants. IEEE Trans. Dielectr. Electr. Insul. 6, 376–384 (1999)
5. K.T. Gillen, R.A. Assink, R. Bernstein, Condition monitoring approaches applied to a polyhloroprene cable jacketing material. Polym. Degrad. Stab. 84, 419–431 (2004)
6. IEEE Std. 383-1974, IEEE standard for qualifying Class 1E Equipment for Nuclear Power Generating Stations, Institution of electrical and electronic engineers, New York (1974)
7. ASTM D638, Standard Test Mathod for Tensile Properties of Plastic. ASTM International
8. IS 10810 Part7, Method of test for Cables, Indian Standard (1997); S.G. Burnay, An overview of polymer ageing studies in the nuclear power industry. Nucl. Instrum. Methods Phys. Res. B 185, 4–7 (2001)

Structural Reliability

Finding Reciprocities Between the Intellectual Frames of Resilience Engineering and High Reliability Organizations

Vivek Kant

Abstract With the growth in new technologies, disparate sociotechnical systems, ranging from aviation to infrastructure, are rapidly getting interconnected. The scope, level and complexity of these novel sociotechnical systems present a renewed challenge for safety as a discipline. In this regard, in recent times, two major research traditions have emerged for addressing safety—high reliability organizations (HRO) and resilience engineering (RE). Both have emerged from different research traditions, have different origins and are sustained by different practices. While both HRO and RE have similarities and differences, they simply cannot be juxtaposed together due to their epistemic cultures. To this extent, researchers have emphasized the diversity and at the same time recognized the need for building bridges across these two research traditions. A particular insight provided by researchers has been to demonstrate that the intellectual frames shape epistemic discourses and concepts within these two traditions. Intellectual frames are a shared epistemological basis on which practitioners and researchers, dispense, negotiate and shape research traditions, as well as their own identities. Based on this insight, a particular manner in which HRO and RE can be partially comprehended together is in terms of the finding reciprocities between these frames and reconsidering them according to the requirements of the field of safety science. While the intellectual frames are a product of historical development, their joint comprehension would require a reflexive engagement with the key aspects that constitute these traditions. Toward this end, the article presents three epistemic linkages that provide possible reciprocities between the two intellectual frames of CE and HRO in sociotechnical systems. These three linkages include comprehending the social dimension of high-risk systems; perspectival nature of the lived experience of people in high-risk systems; and the designed dimension and engineering knowledge involved in sociotechnical systems. Building and assessing these pathways will allow for an all-rounded discussion of safety science for sociotechnical systems.

V. Kant (✉)
Indian Institute of Technology Bombay, Mumbai, India
e-mail: vivek.kant@iitb.ac.in

© Springer Nature Singapore Pte Ltd. 2020
P. V. Varde et al. (eds.), *Reliability, Safety and Hazard Assessment for Risk-Based Technologies*, Lecture Notes in Mechanical Engineering,
https://doi.org/10.1007/978-981-13-9008-1_28

357

Keywords Resilience engineering · Safety · High reliability organizations ·
Intellectual frames

1 Introduction

One of the most fundamental challenges for human factors and safety is the development of conceptual tools and frameworks to support practitioners in their everyday work. In order to achieve this end, over an epoch of safety discourse, two prominent threads of thought have developed: high reliability organizations (HRO) and resilience engineering (RE). These two schools of thought have different traditions, origins and ways of thinking about systems safety [1–8].

In the article on the epistemic cultures related to HRO and RE, Le Coze [9] highlighted various aspects of intellectual frames such as the identity of practitioners, purpose and origins of disciplinary discourses, as well as ways in which these shape individual researchers. Le Coze notes that there are certain aspects in the intellectual frames that allow for shared epistemological orientations. HRO in general has emphasized the organizational basis of high-risk systems, with an ethnographic orientation. This tradition has extensively debated the principles required for studying organizational reliability as well as comprehending these principles into concepts such as "collective mindfulness." In contrast, RE has had a marked emphasis on the deconstruction of conceptual notions such as "human error," both from a theoretical and empirical standpoint. Further RE has emphasized the systems view of accidents and also the development of actionable frameworks and models. RE prominently draws upon a natural science orientation toward addressing humans in sociotechnical systems, while HRO uses a more socially oriented approach. The difference between the two viewpoints of HRO and RE is quite nuanced and cannot be simply relegated to a dichotomous relation. Nevertheless, there is still the need for finding reciprocities between the intellectual frames of HRO and RE to develop safety as a holistic discipline. While these two threads are simultaneously sustained, a unified approach will require that issues related to safety are viewed in light of challenges related to the safety of systems; thus, requiring a reflective viewpoint that integrates rather than separates disciplinary strands of HRO and RE [9].

This article presents possible reciprocities of these two research foci in terms of comprehension and deconstruction of their epistemological basis starting from a bottom-up view of people and technology interaction in sociotechnical systems. Using a strand of RE/CE (Cognitive Engineering focus) emanating from Jens Rasmussen and engineered systems along with ecological psychology, this article links the RE-based approaches to HRO in terms of meta-theoretical similarities [9]. This article presents three epistemic linkages between HRO and RE. Specifically, it delves into the conceptual basis of RE emanating from cognitive engineering (CE) from the approach of Jens Rasmussen (henceforth RE/CE). While these three linkages are not a final answer for a complete merger, they nevertheless present the steps toward addressing the two traditions metatheoretically. The aim of these meta-theoretical reciprocities is to support researchers in developing conceptual frameworks and tools for the everyday work of practitioners.

The current article begins with a discussion of the challenges involved in addressing sociotechnical systems that need to be highlighted as precursors for building reciprocities at a meta-theoretical level between RE/CE and HRO (Sect. 2). The article then highlights three epistemic linkages in terms of a common metric of "social" (theme: sociological, Sect. 3), emphasis on a perspectival view of human knowing and acting (theme: cognitive, Sect. 4), and finally, an emphasis on the designed dimension and engineering knowledge for sociotechnical systems (theme: engineering-based philosophy, Sect. 5).

2 Fundamental Challenges for Epistemically Addressing Sociotechnical Systems and Safety

The first fundamental challenge of safety is that it requires a combined understanding of people, processes and technologies. More broadly, these are together often characterized as sociotechnical systems [10]. Sociotechnical systems are chimerical in nature, partly, because they are hybrid. These systems are neither completely social nor completely technical. Designed, yet emergent. Since they involve people, technologies and processes, they are not tractable by conventional divisions between subjective and objective; science and engineering; theory-driven and practice-driven [11]. Therefore, the creation of conceptual schemes to support practitioners is one of the foremost challenges in making sociotechnical systems safe and resilient. The standard engineering solutions to architecting technical systems do not completely address these hybrid actor-based systems (sociotechnical systems). In sociotechnical systems, the entities under study not only stare back but also act back in ways that thwart all well-intentioned designed approaches. The uneasiness increases for engineers when they try to comprehend the ways in which they want to *design the social* as well as *design for the social*. The malaise deepens even further when the engineers and managers faced with the sociotechnical challenge become unsure about the practical ways to handle and optimize these systems. Addressing sociotechnical systems requires appropriate analysis and design methods to comprehend and design for such systems. The challenges are to strike an optimal balance between design and emergence; constrain and flexibility; system growth and sustainability.

Following from the above, the second basic challenge of safety is that design, operation, management, regulation, mitigation and prevention have to be taken together for these systems. In addition, conceptual schemes for practitioners must be such that they cover these basic aspects of the heterogeneous entities of sociotechnical systems, in terms of a common basis. Currently, social scientists are involved in two prominent ways in engineering research and design. These include providing policy recommendations for systems as well as social analysis of such systems including wider ways of comprehending technology. To these broad ways, the third manner of engagement with engineering research would be to actively engage in providing design and analysis tools that will help the engineers to analyze and comprehend sociotechnical systems, and in turn provide the sustainable design of these systems. Since engineering is not simply an application of science, the existing concepts and methods of social science cannot be simply "applied" to sociotechnical safety research and modeling. The concepts and methods have to be framed in light of social science thought and engineering-based conceptions and aims.

Along with the challenges posed in the construal of sociotechnical systems, another basic challenge is one posed by epistemic dichotomies of realism and constructivism, subjectivity and objectivity, theory and praxis [7, 11, 12]. The dichotomy between constructivism presents a slippery slope for a safety practitioner due to the inherent drawing of the boundaries which is not occupationally palatable. Constructivists emphasize the socially construed nature of accidents and vulnerabilities that exist due to historically contingent development trajectories. In contrast, realists eschew the subjective viewpoint and argue for a "true" description. Constructivism and realism, while they present important viewpoints as debates on the nature of "what exists," they are not able to support the safety practitioner completely. This is because the safety professionals oftentimes find themselves in the midst of large technological setups, having historical trajectories and dealing with disgruntled workers, apathetic managers with pecuniary interests, and in many cases governmental regulations that do not completely recognize the damage they cause due to the lack of awareness about safety. In these circumstances, the safety practitioner deals with systems that are very much objective in behavior while having a subjective dimension in the form of non-apparent vested interests, embodied in its forms, functions, operations and services. The practitioner also deals with people, organizations and institutions that have primary subjective reality but are instantiated in an objective material basis of people, goods, laws and regulations. Therefore, the entities that the safety practitioners deal with are both subjective and objective; rather, they are inter-subjective and inter-objective; lying at the nexus of people, processes, technologies and systems.

One manner of approach is to define the subject matter under consideration as science and proceed to see if the existing practice fits with it or not, i.e., "nothing goes, even if it works in practice [11]." A different approach is to begin with practice as well as scrutinize the basic assumptions that lies behind the science or scope of knowledge. This current approach is adopted in this article. Therefore, out of the two camps of "anything goes as long as it works in practice" and "nothing goes, even if it works in practice" there is a mid-level position that this article adopts: *"only some things go, given that they work in practice."* The aim of this article is to scrutinize the basic viewpoints and find reciprocities related to HRO and RE/CE in terms of the key epistemological basis for addressing sociotechnical systems.

3 Epistemic Linkage 1: "Social" as a Common Metric for Comprehending People and Technologies in Sociotechnical Systems (Theme: Sociological)

One of the first challenges in bridging the divide between RE/CE and HRO is the conceptualization of the human in sociotechnical systems. Le Coze [9] highlights that the epistemic frames of reference of HRO derive from the social sciences; whereas, for RE, it is primarily situated in cognitive science and engineering knowledge. As a result, HRO delves on topics that are amenable to social scientists; whereas, RE/CE begins its discussion of issues and is not limited to those of perception, attention, among other topics. While this distinction is correct, digging a bit deeper into the theoretical and conceptual basis of RE will provide a solution toward linking this gap. Specifically, we delve into ecological psychology and its intellectual roots, as it forms an important strand of RE (and CE) emanating from the Rasmussen's approach [9]. Ecological psychology has played a crucial role in human factors at large [13–15] and CE in particular, by providing a basis for human technology interaction. In the literature on CE, researchers have recognized and used the basic principles of ecological psychology. However, in this usage, the social basis of ecological psychology has been undermined and a more biological forefront has been emphasized by researchers [15, 16].

Typically, HRO theorists side-up with a "social" interpretivist basis of behavior; whereas, CE acquires a "naturalistic" understanding with a focus on the individual. The first step in reconciling the divide between RE and HRO is in terms of identifying pathways in which they share a common view of the individual as a social construct. Earlier, in Sect. 2, we had identified the necessity for treating this issue a central to the problem of safety in large sociotechnical systems. To enable a reconciliation, the psychologist James Gibson's attempt toward an anti-dualist stance is to be emphasized. Gibson's work is important as it forms the basis of one prominent strand of thought in CE [14]. Gibson throughout his career recognized the centrality of the social to cognition and more broadly, human life [17]. The individual for Gibson was inherently social. While, the social dimension of behavior lays at the central aspect of Gibson's thinking, largely, psychologists and engineers deriving from his approach have not addressed the social dimension of human behavior. Currently, CE emanating from Rasmussen's approach [14, 15] use the principles of ecological psychology but do not provide a recognition of the social basis of human behavior as a fundamentally social construct. Therefore, recognizing the social construction of the individual leads to the possibility of developing tools and frameworks that will enable safety in large-scale technical and sociotechnical systems. In addition, it presents a step toward considering the human as a social construct in RE and HRO and moving beyond the academic divide of social sciences on one hand and cognitive and engineering sciences on the other.

Apart from comprehending the social basis of the individual, the next step for developing a more socially situated CE and RE is to comprehend the social constitution of technology. Technologies, in various forms, such as devices and systems, among many others, are not only dictated by an internal logic but also embody values and social principles, consciously or unconsciously, that shape these technologies fundamentally. Historians of technology have identified several technologies and large-scale systems (e.g., electrical systems) that have been fundamentally shaped by choices of the various people and groups involved in various aspects of systems design and functioning. For example, the historian of technology, Hughes [18] notes that the development of large systems engineering projects such as the Boston Central Artery Project (a roadway project) involve various entities, interests, values, imaginations and visions of groups ranging from planners, engineers, politicians, citizens, among many others, who were involved in shaping the technological outcome. Hughes notes that the multiple negotiations and interactions with the groups led to the development of the "physical" infrastructure as "congealed politics [18]." The Boston Central Artery project is one of the several projects and systems that are socially constituted; historians and sociologists have significantly charted the social basis and evolution of technology and large technical systems to show that technologies are socially constituted. Toward this end, the RE/CE as well as HRO have both recognized the multitude of groups involved in risk management. Therefore, the use of social construction of technology perspective complements the already existing approach in safety science but also moves beyond it to recognize that technology is a social construct and should be designed and used with this view in mind.

Till now, we have highlighted that both people and technologies are social constructs. This dimension is necessary for safety science as a significant aspect of it deals with these two entities together. Once the socially sensitive nature of human performance is recognized, then there is a possibility to connect it more broadly to the design of artifacts and systems, which in themselves are social constructions. In addition, in order to bridge the gap between HRO and RE, the social basis of people and technologies from RE (Rasmussen's approach) will have to be extended to accommodate a more detailed understanding of human performance. Therefore, the challenge for the safety professional is to devise new concepts, frameworks and practical prescriptions, as well as tweak the existing ones for a comprehensive understanding of safety.

4 Epistemic Linkage 2: Perspectival View of Human Knowing and Acting in Technological Systems (Theme: Cognitive)

In developing the basis for safety, HRO adopts a view of interpretivism that has been quite prominent in the social sciences but less so in the psychological sciences and hence RE/CE. In contrast to this natural science orientation, a more humanistically oriented science was developed by social scientists. Scholars working in social sciences recognize that for a sustained and meaningful study of social life, there is the need for comprehending the ways in which people are constantly interpreting everyday meanings. This constant change in meanings and interpretive activity centered around these meanings is the proper basis for addressing human behavior. Social scientists deriving from the interpretive traditions recognize humans as moral entities that are socially situated in their milieus. People are constantly involved in interpreting, defining, negotiating, along with resisting the definitions and constructions of meaning. Therefore, the study of situations involving human life requires a coherent basis of understanding the various perspectives that are involved. Recognizing the value-laden multi-perspectival nature of human understanding is one of the key basis of comprehending the human. As a result, people in situations are not simply biological entities but are historical and societal-based constructs. They act from (dis-)vantage positions and adopt ethical and moral comprehension of any situation in which they are involved.

The interpretive basis of human behavior is best characterized in terms of roles and perspectives. Roles are parts that people play in various situations; whereas, "perspectives" involve a viewpoint with which the individual is engaged with the social milieu. Social scientists (Goffman, Blumer, Mead, Cooley, Schutz, Mead) from the interactionist sociological tradition have extensively addressed the need for a comprehension of human behavior in terms of multiple perspectives and roles that are shaping the situation as well as are shaped by it. In safety-critical systems, various roles exist which include managers, legislators, regulators, among many others. In terms of these modern day high-risk systems, along with interpretivism, the idea of roles and perspectives are already reflected in the discussions of social scientists and HRO [9]. While interpretation is broadly recognized as a foundational basis for social sciences, its inroads into cognitive and behavioral sciences have been quite limited. Therefore, a prominent challenge in linking the CE/RE to HRO is the need to show how both of them can be connected at a meta-theoretical level [7, 11, 12, 19]. One step in this linkage to show how CE deriving from the viewpoint of ecological psychology already adopts "perspectives" as a fundamental basis for theorizing cognitive activity.

In the last section above, we highlighted the social constitution of the individual and the need for building pathways between RE and HRO. In the following, we use an under-discussed theoretical feature of ecological psychology that can be used to build a linkage between CE and HRO in terms of perspectival nature of lived experience. Gibson's ecological psychology and specifically his approach to visual perception was inherently perspectival. His notion of perspective was quite literal and physical; it involved the eye as a nested station point within the body and the larger ecology. Gibson's ecological approach put forth in 1979 had been developing in his earlier books and papers (e.g., *The Senses Considered As Perceptual Systems*, 1966) as well as having its roots in his earlier World War II research involving pilots and aircrafts. In developing a theory of human perception, Gibson recognized the centrality of a perspectival basis, more literally, a physical and materially oriented point of observation. For Gibson, a pathway of locomotion involved a whole family of perspectives, based on an interconnected set of observation points. This shows that the notion of perspective is not simply a private mental concept where one perspective differs from another and is incommensurable. Rather, it is possible to adopt another person's perspective literally. In addition, the perspective is not just a physical description but involves a highly social basis. In explicating the social basis of knowledge emphasizing perspectival reality, Gibson (1979, p. 141) notes: "Only when each child perceives the values of things for others as well as for herself does she begin to be socialized [20]."

The important aspect for our purposes is not the actual mechanism of perception as Gibson describes it but the recognition that one strand of CE built on the ecological psychology takes into account the perspectival nature of reality. This perspectival nature from CE will have to be developed along with the perspectival nature of social interaction underlying the social sciences and hence HRO. Accompanying the perspectival nature of reality is the emphasis on the value and meanings that people have for their surrounds. In Gibson's viewpoint, this aspect is covered in detail in terms of the concept of affordance (note: Flach and Bennett [15] highlight this approach as meaning processing). The theory of affordances links the world of value and meaning with a physical basis of everyday experience from a psychological and hence RE/CE perspective. Similarly, the world of morality, value, meaning and power is also found in the social sciences (and hence HRO) and also a central concern for safety researchers. Thus to develop a foundational basis of models and methods for safety science requires that the cognitive basis of interaction with the everyday world be linked to the social basis of everyday interaction. This can be possible by comprehending one aspect of RE/CE in terms of ecological psychology and connecting it to a broader viewpoint of perspectivism and the interpretive basis of lived experience from the social sciences.

5 Epistemic Linkage 3: The Designed Dimension and Engineering Knowledge Involved in Sociotechnical Systems (Theme: Engineering-Based Philosophy)

The third theoretical linkage between RE/CE and HRO camps can be addressed by emphasizing the need for the designed dimension of safety systems as well as the practice-based approach required for safety in the field. Both these aspects are crucial because oftentimes they are under-emphasized in the formation of epistemic constructs, theories and frameworks. The designed dimension of systems requires that safety as a discipline should take into account, its epistemic basis in the form of correct functioning and malfunction. In a very fundamental sense, safety professionals take this epistemic uniqueness of technological systems into account by incident and accident investigations. However, the designed dimension of technical systems has remained slightly subdued in disciplinary discourses and needs a more reflexive engagement for addressing safety in terms of epistemic concepts and frameworks.

Toward this end, Rasmussen's approach from CE is not only cognizant of the engineered dimension of safety-critical systems but also develops its epistemic structures based on interpretive comprehension of everyday practice of operators in high-risk systems [21]. In deriving from everyday practice, this strand of CE emanating from Rasmussen shares an affinity to HRO, which has also placed a marked emphasis on everyday practice. Notwithstanding the practice approach, the CE strand emanating from Rasmussen also characterizes the designed dimension of technical systems. Drawing from the philosopher Michael Polanyi, Rasmussen notes that engineered systems cannot be completely addressed by scientific thinking. In particular, science as a discipline does not study the "logic of contrivance" and hence malfunctions. In contrast to scientific thinking, engineers produce machines and technologies that are designed to function correctly and avoid malfunction (engineers use science but engineering is not an application of science).

In this regard, consider Rasmussen's approach in terms of CE/RE that provides practical tools for everyday designers and engineers. Earlier, Rasmussen had started with the problem of technical reliability; over a period of time he recognized that the problem of reliability needed a formulation in terms of human systems reliability. In other words, the human had to be treated as a part of the technological system without being reduced to a mechanistic cog. To aid the synergy between the human and the technical system, Rasmussen devised new conceptual tools and frameworks that have become mainstream in CE in the present date [22]. Another crucial aspect of the AH and more broadly Rasmussen's approach is that even though it derives from scientific knowledge, it is not a direct application of the same, i.e., the engineering dimension shines through when we recognize the way in which the AH and other constructs straddle between the practitioner and the researcher. The AH and other related constructs are derived from everyday practice but constructed based on the viewpoint of the researcher. Epistemic structures such as the AH not only help in the design of complex technological systems but also toward supporting everyday practice of the people in those systems.

In HRO, the emphasis on the designed dimension of systems in terms of its constituent models and frameworks is not very prominent (one can only hypothesize because of its roots in social sciences). The main idea from the above discussion is that the notion of engineering knowledge and designed dimension of sociotechnical systems are important meta-theoretical aspects to be considered. The notion of engineering knowledge has to be reflexively explored by researchers and incorporated in their frameworks, tools and methodologies for linking CE to HRO and safety in general.

6 Conclusion

The aim of this article was to present reciprocities between RE/CE and HRO in light of broader foundational insights into the field of safety as well as in terms of an epistemological understanding. Specifically, three main aspects were discussed corresponding to a sociological, cognitive and engineering-based philosophical dimensions: social as a basic metric; perspectival nature of human knowing and acting; and the engineered dimension of sociotechnical systems. All these three aspects along with the broader issues of safety have to be taken together metatheoretically by researchers to find reciprocities between RE/CE and HRO. A considerable emphasis is placed by safety practitioners to ensure that systems and processes do not malfunction or fail due to internal or external circumstances and result in loss to life, society or the environment. Therefore, the epistemic constructs of safety should incorporate these aspects to arm safety experts with concepts that actually work in the field; thus, supporting the first group of "*anything goes as long as it works in practice*" and bringing it closer to the acceptability of the design. In addition, the intellectual scope of "*nothing goes, even if it works in practice*" can be broadened to recognize that other aspects of knowledge exist that goes beyond the disciplinary purviews of science. Consequently, in the search for foundations of safety science, our goal is to bridge the gap between academics and practitioners such that they can integrate various disparate strands of thought, such as HRO and RE/CE, and understand practice reflexively

References

1. A.W. Righi, T.A. Saurin, P. Wachs, A systematic literature review of resilience engineering: research areas and a research agenda proposal. Spec. Issue Resilience Eng. **141**, 142–152 (2015)
2. D.D. Woods, Four concepts for resilience and the implications for the future of resilience engineering. Spec. Issue Resilience Eng. **141**, 5–9 (2015)
3. M.D. Patterson, R.L. Wears, Resilience and precarious success. Spec. Issue Resilience Eng. **141**, 45–53 (2015)
4. J. Bergström, R. van Winsen, E. Henriqson, On the rationale of resilience in the domain of safety: a literature review. Spec. Issue Resilience Eng. **141**, 131–141 (2015)
5. K.A. Pettersen, P.R. Schulman, Drift, adaptation, resilience and reliability: toward an empirical clarification. Safety Science (2016)
6. R.L. Wears, K.H. Roberts, Special issue, Safety Science, High reliability organizations and resilience engineering. Safety Science (2018)
7. T.K. Haavik, S. Antonsen, R. Rosness, A. Hale, HRO and RE: a pragmatic perspective. Safety Science (2016)
8. J. Pariès, L. Macchi, C. Valot, S. Deharvengt, Comparing HROs and RE in the light of safety management systems. Safety Science (2018)
9. J.C. Le Coze, Vive la diversité! High reliability organisation (HRO) and resilience engineering (RE). Safety Science (2016)
10. P.E. Vermaas, *A Philosophy of Technology: from Technical Artefacts to Sociotechnical Systems*. Morgan & Claypool, San Rafael, California (2011)
11. J.C. Le Coze, K. Pettersen, T. Reiman, The foundations of safety science. Saf. Sci. **67**, 1–5 (2014)
12. T.K. Haavik, On the ontology of safety. Saf. Sci. **67**, 37–43 (2014)
13. K.J. Vicente, *The Human Factor: Revolutionizing the Way People Live with Technology*. Routledge (2013)
14. K.J. Vicente, *Cognitive Work Analysis: Toward Safe, Productive, and Healthy Computer-Based Work* (CRC-Press, Boca Raton, 2009)
15. K.B. Bennett, J.M. Flach, *Display and Interface Design Subtle Science, Exact Art* (CRC, Boca Ratón, 2011)
16. V. Kant, The sociotechnical constitution of cognitive work analysis: roles, affordances and malfunctions. Theor. Issues Ergon. Sci. **19**(2), 195–212 (2017)

17. H. Heft, *Ecological Psychology in Context: James Gibson, Roger Barker, and The Legacy of William James's Radical Empiricism* (L. Erlbaum, Mahwah, N.J., 2001)
18. T.P. Hughes, *Rescuing Prometheus: Four Monumental Projects that Changed Our World* (Pantheon, New York, 1998)
19. J.C. Le Coze, An essay: societal safety and the global. Safety Science (2017)
20. J.J. Gibson, *The Ecological Approach to Visual Perception* (Houghton Miffin, Boston, 1979)
21. J. Rasmussen, L.P. Goodstein, A.M. Pejtersen, Cognitive Systems Engineering. Wiley, New York (1994)
22. P. Waterson, J.C. Le Coze, H.B. Andersen, Recurring themes in the legacy of Jens Rasmussen. Appl. Ergon. **59**(Pt B), 471–482 (2017)

Improved Adaptive Response Surface Method for Reliability Analysis of Structures

Atin Roy and Subrata Chakraborty

Abstract An adaptive response surface method for improved reliability estimate of structures is explored in the present study. The approach primarily follows the iterative algorithm of obtaining improved converged centre point for the design of experiment (DOE) in which the bounds of random variables during iterations are set with a multiplier of the standard deviation of the corresponding random variables to obtain new DOE. However, there is no specific basis for such selection. An attempt has been made here to propose a strategy to select the bound for new DOE during each iteration. The boundary nearer to the new centre point is kept the same as the physical boundary for each random variable. The other boundary is set in such a way that both the upper limit and the lower limit of the corresponding random variable are equidistant from the new centre point of DOE. Further, a zone-based metamodelling strategy is explored to minimize such inaccuracy for approximating limit state function. The improved performance of the proposed approach with regard to the accuracy and efficiency of the proposed algorithm is demonstrated by considering two example problems.

Keywords Adaptive response surface method · Improved design of experiment scheme · Structural reliability · Zone-based strategy

1 Introduction

Structural reliability analysis (SRA) is a theoretical framework to consider the effect of uncertainty involved in parameters of an engineering structures [1, 2]. The first-order reliability method (FORM) and the second-order reliability method (SORM) are two analytical approaches for SRA. Monte Carlo simulation (MCS), a simulation-

A. Roy (✉) · S. Chakraborty
Indian Institute of Engineering Science and Technology, Shibpur, India
e-mail: atin.3222@yahoo.com

S. Chakraborty
e-mail: schak@civil.iiests.ac.in

© Springer Nature Singapore Pte Ltd. 2020
P. V. Varde et al. (eds.), *Reliability, Safety and Hazard Assessment for Risk-Based Technologies*, Lecture Notes in Mechanical Engineering,
https://doi.org/10.1007/978-981-13-9008-1_29

369

based approach is also widely used for SRA. However, such reliability analysis approaches require repetitive evolution of limit state function (LSF). The total number of LSF evaluations is not an important computational issue if the LSF is explicit in nature. But, if finite element (FE) analysis is necessary to define an implicit LSF for SRA, the computational demand for SRA of the large complex system becomes an important issue. In this regard, metamodelling is an efficient alternative technique for response approximation of complex structures by alleviating the computational burden without compromising with accuracy. Various metamodelling techniques have emerged as an effective solution to such problems and represent a convenient way to achieve a balance. Response surface method (RSM) is the most popular among various metamodelling techniques.

The application of RSM for SRA was first proposed by Faravelli [3] utilizing a polynomial expansion based on the least square method (LSM). Subsequent studies on the applications of LSM-based RSM are enormous [4–8]. The successful approximation by any metamodel predominantly depends on the proper selection of training points to construct the metamodel. Usually, the input–output data at a set of carefully selected points referred as the design of experiment (DOE) are evaluated by running FE analysis or numerical simulation. A sufficient number of points in the DOE is necessary for accuracy of a metamodel in approximating nonlinearity and local trends of the LSF. But, a small number of FE analysis is desirable for preparing input–output data to build metamodel for any complex structures. In this regard, it is of worth noting that the accuracy of metamodel-based SRA hinges on the proper selection of centre point of DOE to construct a metamodel. The centre point should be close to the most probable failure point (MPFP) so that the metamodel obtained ensures sufficient accuracy in most of the region near the boundary of the safe and unsafe domain. In fact, the early study by Bucher and Bourgund [4] to obtain an improved RSM based on this fact and subsequent developments on iterative metamodel construction based on adaptive DOE scheme is notable [5–8]. Besides the traditional RSM, the adaptive DOE scheme is also applied in conjunction with advanced metamodel like support vector regression [9], moving LSM [10], etc. It is observed from the mentioned studies [5–10] that DOEs are reconstructed adaptively centring around a point obtained based on MPFP evaluated by FORM/SORM with preceding response surface functions (RSF). These adaptive DOEs are made with a primary intention of achieving the MPFP and reliability index (β) accurately. Zhao and Qiu [11] proposed a control point as an alternative of MPFP to calculate the centre point of a new DOE, but they also utilized FORM to obtain the design point. On the contrary, Su et al. [12] searched the MCS point in the failure domain with the maximum joint pdf and considered it as MPFP. It is important to note that a series of studies suggested adding more samples to the previous design instead of shifting DOE based on a new centre point [13, 14]. These approaches do not involve the determination of a centre point to build a new DOE; rather, add points one by one to the previous DOE for achieving the desired accuracy of prediction. An arbitrary multiplication factor to standard deviation (SD) of each random variable is generally used in defining the domain of DOE. Most of the researchers suggested to reduce this multiplication factor empirically based on experience or engineering knowledge. However, Wong et al. [15] proposed a method to select this multiplier depending upon the coefficient of variation (COV) of the corresponding random variable. Alternatively, statistical resampling techniques to obtain the boundary for the next DOE is also utilized [14, 16].

In the present study, an adaptive RSM for improved reliability estimate is explored. The procedure primarily follows the iterative algorithm of Rajashekhar and Ellingwood [5], where the bounds of random variables during iterations are set with a reduced multiplier of SD of the corresponding random variables in the successive iterations till convergence. However, there is no specific basis or guideline for such selection. An attempt has been made to propose a strategy in the present study to select the bound for new DOE during each iteration. In the proposed algorithm, the boundary nearest to the new centre point is kept the same as the physical boundary for each random variable. The other boundary is set in such a way that both the upper limit and the lower limit of the corresponding random variable are equidistant from the new centre point of the DOE. In this regard, it is to be noted that a converged response surface obtained by adaptive DOE scheme will definitely yield satisfactory prediction of responses for samples within the boundary of the final DOE, but it may fail outside this boundary. Thus, a zone-based metamodelling strategy is further explored to minimize such inaccuracy for approximating LSF. It is to be noted here that a randomly simulated sample for reliability analysis may not be within the domain of the last DOE but may fall in any of the previous DOE domain(s). In case, the random sample point belongs to more than one domain, the response is evaluated using the response function obtained from the last iteration among those. The improved performance regarding the accuracy and efficiency of the proposed approach is illustrated considering two numerical examples.

2 Response Surface Method

RSM is a collection of mathematical and statistical techniques originally developed by Box and Wilson [17] for the empirical model building of a response (output variable) which is influenced by several independent variables (input variables). In SRA, RSM basically replaces the implicit relationship between different input parameters and desired output by a simple polynomial function. A complete second-order polynomial generally been avoided since it requires a greater number of DOE points. A reduced quadratic polynomial without cross-terms is usually used for efficient SRA [4, 5, 15]. The typical reduced quadratic polynomial RSF for approximating a response depended on n input variables can be expressed as follows:

$$\hat{y}(\mathbf{X}) = \beta_0 + \sum_{i=1}^{n} \beta_i x_i + \sum_{i=1}^{n} \beta_{ii} x_i^2 = \mathbf{f}(\mathbf{X})^{\mathrm{T}} \boldsymbol{\beta} \tag{1}$$

where $\mathbf{x}$ is the vector consisting of n input variables, $\hat{y}(\mathbf{X})$ is the predicted output response at $\mathbf{x}$, $\boldsymbol{\beta} = \{\beta_0, \ldots, \beta_i, \ldots, \beta_{ii}, \ldots\}^{\mathrm{T}}$ is the vector of unknown polynomial coefficients and $\mathbf{f}(\mathbf{X}) = \left\{1, \ldots, x_i, \ldots, x_i^2, \ldots\right\}^{\mathrm{T}}$ is a vector of the considered basis functions. The unknown coefficients β_0, β_i and β_{ii} are usually obtained by the LSM-based regression. Fitting procedure requires the evaluation of response values by the actual numerical or FE model at sample points selected as per the DOE. To estimate the unknown coefficients, the LSM basically minimizes the error norm defined for the prediction at all the DOE points with respect to the actual LSF, y as

$$e^2 = \sum_{l=1}^{p}\left(y^l - \beta_0 - \sum_{i=1}^{n}\beta_i x_i^l - \sum_{i=1}^{n}\beta_{ii}\left(x_{ii}^l\right)^2\right)^2 = (\mathbf{y} - \mathbf{F}\boldsymbol{\beta})^{\mathrm{T}}(\mathbf{y} - \mathbf{F}\boldsymbol{\beta}) \quad (2)$$

and the least squares estimate of the vector $\boldsymbol{\beta}$ is obtained as

$$\hat{\boldsymbol{\beta}} = \left(\mathbf{F}^{\mathrm{T}}\mathbf{F}\right)^{-1}\mathbf{F}^{\mathrm{T}} \quad (3)$$

where p is the total number of DOE points, $\mathbf{F}$ is the design matrix formed by p numbers of $\mathbf{f}(\mathbf{x})$ row vector corresponding to each DOE points and the square matrix $\mathbf{F}^T\mathbf{F}$ has the order of $(2n + 1)$. After obtaining $\boldsymbol{\beta}$ from Eq. (3), the approximate LSF for any set of input parameters can be found from Eq. (1).

For metamodel training, classical designs such as saturated design, factorial design, central composite design, etc. are suitable when outputs are obtained from physical or laboratory experiment, where outputs can be different if repeated for identical input. But, in the present study, outputs are generated by a computer analysis which are absolutely free from replication errors. Thus, computer experiments need separate design strategy. In this regard, space-filling design-based DOE can be employed for constructing an efficient metamodel. Therefore, the uniform design (UD) [18], which is a space-filling scheme for spreading samples uniformly in the entire design domain, is implemented in the present study. To obtain the DOE for the numerical study, the adopted UD tables of different dimensions are readily available [19].

3   A Simple Adaptive DOE Proposal

Each random variable involved in the LSF of a reliability analysis problem has a realistic boundary. All the simulation samples for MCS-based SRA should be generated within this domain only. The prediction accuracy of a metamodel should be such that it can successfully determine whether the sample is in the failure region or not. Hence, the accuracy in estimating the probability of failure is directly connected with the performance of metamodel for MCS samples close to the failure surface. With an aim of capturing the trends of LSF more accurately in the neighbourhood of the failure plane, an adaptive DOE procedure is proposed. In the proposed procedure, first, a zone of interest is decided where the accuracy of prediction is a major concern for SRA in the MCS framework. Subsequently, the DOE is built within this zone to update the metamodel.

The initial DOE is constructed over the entire physical domain with a space-filling design since simulation samples can be anywhere of the physical domain and minimum accuracy for prediction at all points is essential to decide whether failure or not. The vectors of the upper bound and the lower bound for all the random variables in the initial DOE have been denoted by $\mathbf{x}_{ub,0}$ and $\mathbf{x}_{lb,0}$, respectively. The metamodel obtained based on the initial DOE is used to estimate the P_f by MCS-based approach and associated reliability index, $\beta = -\Phi^{-1}(P_f)$ is also obtained and denoted hereafter as β_{MCS}. With this known β_{MCS}, the corresponding design point is obtained. Thereafter, the centre point of the next DOE is selected following Bucher and Bourgund [4]. Thus, after evaluating design point, $\mathbf{x}_{D,i}$ at i-th iteration, the new centre point, $\mathbf{x}_{C,i+1}$ for $i+1$-th iteration can be obtained on the straight line joining previous centre point, $\mathbf{x}_{C,i}$ and $\mathbf{x}_{D,i}$ [5]. Mathematically, the new centre point evaluation can be expressed as

$$\mathbf{x}_{C,i+1} = \mathbf{x}_{C,i} + \left(\mathbf{x}_{D,i} - \mathbf{x}_{C,i}\right) \frac{g\left(\mathbf{x}_{C,i}\right)}{g\left(\mathbf{x}_{C,i}\right) - g\left(\mathbf{x}_{D,i}\right)} \tag{4}$$

where $g(\mathbf{x}_{C,i})$ and $g(\mathbf{x}_{D,i})$ are the actual LSF evaluated at $\mathbf{x}_{C,i}$ and $\mathbf{x}_{D,i}$. It is to be noted that the bound for this new DOE need to be decided. The zone of interest is basically the region where the prediction accuracy of a metamodel to approximate LSF at simulation samples significantly influences the accuracy of estimating the P_f. Since, the MPFP evolution is based on the β_{MCS}, in both sides of failure surface the accuracy of simulation samples is important, otherwise the estimate may change significantly. Thus, both the upper limit $(x^j_{ub,i+1})$ and the lower limit $(x^j_{lb,i+1})$ for j-th random variable at $i+1$-th iteration should be equidistance from the new centre, $\mathbf{x}_{C,i+1}$. In the earlier studies [4–10], both the extents for each random variable in successive DOEs were decided by considering progressively smaller values of an arbitrary multiplying factor of corresponding SD from the new centre. A new strategy is proposed in the present study to select the bound for new DOE during each iteration. In doing so, for each random variable at $i+1$-th iteration, the initial boundary which is nearer to the new centre is kept same for a new boundary, and the opposite boundary is set such that both the new boundaries are equidistant from the new centre, $\mathbf{x}_{C,i+1}$. This can be expressed as

$$x^j_{ub,i+1} = x^j_{C,i+1} + \min\left(x^j_{ub,0} - x^j_{C,i+1}, x^j_{C,i+1} - x^j_{lb,0}\right),$$
$$x^j_{lb,i+1} = x^j_{C,i+1} - \min\left(x^j_{ub,0} - x^j_{C,i+1}, x^j_{C,i+1} - x^j_{lb,0}\right), \tag{5}$$

where $x^j_{C,i+1}$, $x^j_{ub,0}$ and $x^j_{lb,0}$ are representing the components of vectors $\mathbf{x}_{C,i+1}$, $\mathbf{x}_{ub,0}$ and $\mathbf{x}_{lb,0}$ corresponding to the j-th random variable. For a two-dimensional problem, Fig. 1a conceptually illustrates the proposed scheme of updating the centre point

(a)   **(b)**

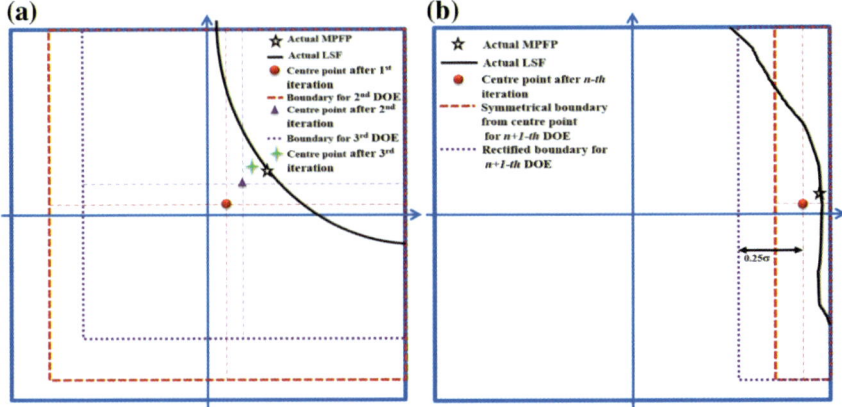

Fig. 1 **a** Representation of adaptive DOE for updating the centre point and **b** same to avoid numerical difficulty

and deciding the zone of interest for DOE for constructing next metamodel in the subsequent iteration.

However, it is to be noted that MPFP may be very close to the physical boundary for one or more random variables. In such cases, numerical problems like singularity, ill-conditioning may occur yielding erroneous results. To overcome this, a minimum spread of one quarter of SD of corresponding random variable from the updated centre point at opposite to the closest physical boundary at each iteration is maintained. Ensuring such can be mathematically expressed as

$$
x_{ub,i+1}^{j} = x_{C,i+1}^{j} + \max\left\{\min\left(x_{ub,0}^{j} - x_{C,i+1}^{j}, x_{C,i+1}^{j} - x_{lb,0}^{j}\right), 0.25\sigma^{j}\right\} \leq x_{ub,0}^{j},
$$
$$
x_{lb,i+1}^{j} = x_{C,i+1}^{j} - \max\left\{\min\left(x_{ub,0}^{j} - x_{C,i+1}^{j}, x_{C,i+1}^{j} - x_{lb,0}^{j}\right), 0.25\sigma^{j}\right\} \geq x_{lb,0}^{j},
$$
$$
(6)
$$

where σ^{j} represents the SD of the j-th random variable. The modification of the boundary to avoid numerical difficulty is illustrated in Fig. 1b for a two-dimensional problem.

4 Zone-Based Response Surface Application

Most of the previous studies on adaptive metamodelling-based SRA have considered the final metamodel obtained based on the DOE formed around the last converged centre point to approximate the response, discarding all the previous DOE. In this regard, it is to be noted that when the MPFP is obtained by FORM/SORM to evaluate β, the effect of prediction capability outside the range of the latest DOE does not affect accuracy. But, the accuracy of the FORM/SORM-based approach to evaluate β, particularly for highly nonlinear natures of LSF and higher level of randomness in the random variables involved in an LSF is questionable. Thus, the present study proposed to estimate the P_f value by MCS to ensure for better accuracy and the β value and MPFP is decided accordingly. The response approximation for each simulation point in such approach will be very much important to decide whether it is safe or in an unsafe region, as it directly influences P_f. Whenever a metamodel is build using the latest adaptive DOE, it undoubtedly gives satisfactory prediction if the sample falls within the boundary of the DOE formed around the last converged centre point, but the samples which are outside this boundary will definitely lack accuracy if the current metamodel is used. To minimize this inaccuracy, a zone-based metamodelling strategy is proposed here for response approximation. If a sample falls within the boundary of the present DOE, then the metamodel based on the current DOE will be used to predict the response for that sample. If the sample falls outside the boundary of the present DOE, then among the previous DOEs which include the said sample, the most updated one will be used to estimate the response for this sample. The strategy ensures that the response at any point can only be approximated by the metamodel constructed using the best possible updated DOE which is such that the point is within the boundary.

5 Numerical Study

To study the effectiveness of the proposed adaptive scheme, two numerical problems are taken up. The reliability analysis is performed by two adaptive schemes: i) Adaptive least square method (ALSM) where DOEs are updated according to the proposed procedure but uses only the final metamodel trained by the last updated DOE for predictions at MCS samples, and ii) Zone-based adaptive least square method (ZALSM) which is almost similar to the ALSM scheme, but it considers a zone-based training for approximating response for each MCS sample depending on the location of sample in the physical domain.

5.1 Example 1: A Five-Dimensional (5-D) Standard Test Problem

A five-dimensional (5-D) standard test problem used to study the response approximation capability in metamodelling application is taken first to demonstrate the effectiveness of the proposed scheme. The LSF is expressed as

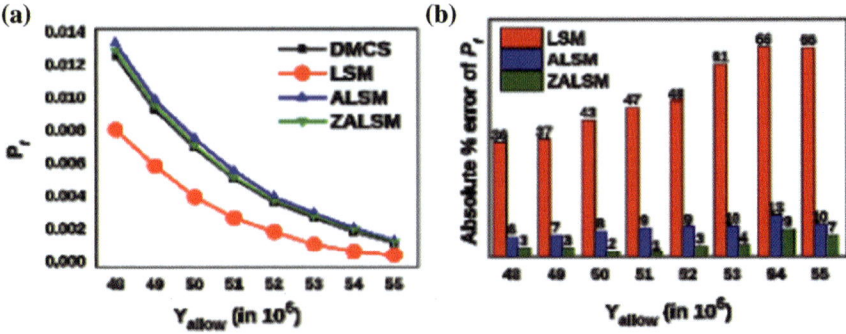

Fig. 2 **a** Comparison of P_f values and **b** absolute percentage (%) errors of P_f for different Y_{allow} values

$$g(\mathbf{X}) = Y_{\text{allow}} - \sum_{i=1}^{5} x_i^2 - \left(\sum_{i=1}^{5}\left(\frac{1}{2}\right)ix_i\right)^2 - \left(\sum_{i=1}^{5}\left(\frac{1}{2}\right)ix_i\right)^4, i = 1, 2, \ldots, 5.$$

$$(7)$$

where Y_{allow} is the allowable value of the function and all the five random input variables are considered to be lognormal with mean and SD of 10.0 and 1.0, respectively. The physical boundary for each random variable is considered as mean $\pm 3 \times$ SD. The DOE points are obtained in each iteration within the updated boundary according to the UD table, $U_{30}(30^5)$. The estimated P_f values obtained by the proposed two adaptive schemes, conventional LSM-based RSM (without considering the adaptive DOE scheme) and the most accurate direct MCS (DMCS) methods are shown in Fig. 2a. The absolute percentage errors in estimating P_f values by metamodels with respect to that obtained by DMCS ($P_{f,\text{DMCS}}$) are calculated as $100 \times (|P_{f,\text{DMCS}} - P_f| / P_{f,\text{DMCS}})$ and are shown in Fig. 2b.

As expected, the LSM-based RSM is observed to be far inferior than both of the adaptive schemes. Also, it may be noted that the ZALSM has better accuracy than the ALSM. It can be noted here that though the ALSM-based results are more conservative than other methods, its accuracy is far inferior than the proposed ZALSM approach.

5.2 Example 2: A Planar Ten-Bar Truss

The second problem considered is a planar ten-bar truss as detailed in Fig. 3. The material is considered to be homogenous without spatial variety. The reliability is evaluated with respect to the displacement at node 2, considering six random variables as detailed in Table 1. The physical bound for each random variable is assumed as

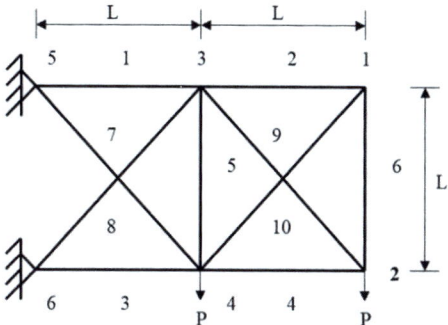

Fig. 3 Planar ten-bar truss

Table 1 Characteristics of the random variables of the ten-bar truss

Random variables (unit)	Base pdf	Mean	COV
Horizontal member's cross section, A_1 (m²)	Normal	7.5×10^{-3}	0.1
Vertical member's cross section, A_2 (m²)	Normal	1.5×10^{-3}	0.1
Diagonal member's cross section, A_3 (m²)	Normal	5.0×10^{-3}	0.1
Young's Modulus, E (N/m²)	Normal	7.0×10^{10}	0.05
Length, L (m)	Lognormal	9.0	0.05
Load, P (N)	Gumbel Max.	3.5×10^5	0.1

mean $\pm 3 \times$ SD. All the metamodels are constructed using UD table, $U_{30}(30^6)$. The LSF for the tip displacement at node 2 is given as follows [20]

$$g_{\text{disp}} = \frac{PL}{A_1 A_3 E D_T} \left[\begin{array}{l} 4\sqrt{2}A_1^3\left(24A_2^2 + A_3^2\right) + A_3^3\left(7A_1^2 + 26A_2^2\right) + \\ 4A_1 A_2 A_3\left\{\left(20A_1^2 + 76A_1 A_2 + 10A_3^2\right) + \sqrt{2}A_3(25A_1 + 29A_2)\right\} \end{array} \right]$$

$$- d_{\text{allow}} = 0$$

where, $D_T = 4A_2^2(8A_1^2 + A_3^2) + 4\sqrt{2}A_1 A_2 A_3(3A_1 + 4A_2) + A_1 A_3^2(A_1 + 6A_2)$ \hfill (8)

The P_f values obtained by the LSM, ALSM, ZALSM and direct MCS approaches are shown in Fig. 4a. Furthermore, to study the effectiveness of the adaptive schemes, the absolute percentage errors in estimating P_f values by metamodels are shown in Fig. 4b. It can be noted that ZALSM scheme is better than ALSM. The ALSM estimate P_f with less than 10% error; whereas, such error is around 1% in case of ZALSM approach.

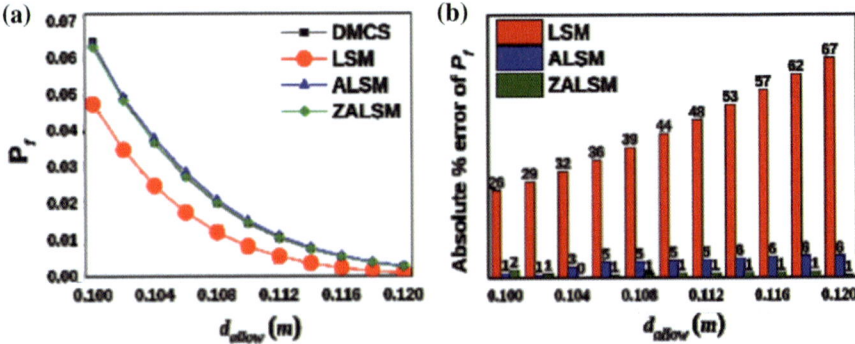

Fig. 4 **a** Comparison of P_f values and **b** absolute percentage (%) errors of P_f for different d_{allow} values

6 Summary and Conclusions

A strategy to select the bound for new DOE during each iteration for improved reliability estimate by adaptive RSM is proposed. Further, for improved accuracy in the approximation of LSF for SRA, a zone-based metamodelling strategy is adopted. The reliability analysis is performed by two adaptive schemes, i.e. ALSM in which DOEs are updated according to the proposed procedure and ZALSM which considers a zone-based training for metamodel approximating response for each MCS sample depending on the location of the sample in the physical domain. The effectiveness of the two schemes is elucidated numerically. It has been generally noted from the results of the numerical study that both the adaptive schemes, i.e. the ALSM and ZALSM outperform over the LSM-based RSM. The P_f values obtained by the LSM, ALSM, ZALSM and DMCS approaches show that the zone-based adaptive scheme, i.e. ZALSM is better than ALSM. Thus, the proposed ZALSM approach can be applied for improved SRA of the complex engineering system. However, the numerical study is made for two simple problems at this stage, and it needs to investigate the effectiveness of the proposed approach for high dimension and more complex engineering system involving implicit LSF.

References

1. O. Ditlevsen, H.O. Madsen, *Structural Reliability Methods* (John Wiley and Sons Ltd., Chichester, UK, 1996)
2. A. Haldar, S. Mahadevan, *Reliability Assessment Using Stochastic Finite Element Analysis* (John Wiley and Sons Inc., New York, USA, 2000)
3. L. Faravelli, Response surface approach for reliability analyses. J. Eng. Mech. ASCE **115**(2), 2763–2781 (1989)
4. C.G. Bucher, U. Bourgund, A fast and efficient response surface approach for structural reliability problems. Struct. Saf. **7**, 57–66 (1990)

5. M.R. Rajashekhar, B.R. Ellingwood, A new look at the response surface approach for reliability analysis. Struct. Saf. **12**, 205–220 (1993)
6. S.H. Kim, S.W. Na, Response surface method using vector projected sampling points. Struct. Saf. **19**, 3–19 (1997)
7. P.K. Das, Y. Zheng, Cumulative formation of response surface and its use in reliability analysis. Probab. Eng. Mech. **15**, 309–315 (2000)
8. F. Duprat, A. Sellier, Probabilistic approach to corrosion risk due to carbonation via an adaptive response surface method. Probab. Eng. Mech. **21**, 207–216 (2006)
9. B. Richard, C. Cremona, L. Adelaide, A response surface method based on support vector machines trained with an adaptive experimental design. Struct. Saf. **39**, 14–21 (2012)
10. S. Goswami, S. Ghosh, S. Chakraborty, Reliability analysis of structures by iterative improved response surface method. Struct. Saf. **60**, 56–66 (2016)
11. W. Zhao, Z. Qiu, An efficient response surface method and its application to structural reliability and reliability-based optimization. Finite Elem. Anal. Des. **67**, 34–42 (2013)
12. G. Su, L. Peng, L. Hu, A Gaussian process-based dynamic surrogate model for complex engineering structural reliability analysis. Struct. Saf. **68**, 97–109 (2017)
13. B. Echard, N. Gayton, M. Lemaire, AK-MCS: an active learning reliability method combining Kriging and Monte Carlo simulation. Struct. Saf. **33**(2), 145–154 (2011)
14. N. Roussouly, F. Petitjean, M. Salaun, A new adaptive response surface method for reliability analysis. Probab. Eng. Mech. **32**, 103–115 (2013)
15. S.M. Wong, R.E. Hobbs, C. Onof, An adaptive response surface method for reliability analysis of structures with multiple loading sequences. Struct. Saf. **27**, 287–308 (2005)
16. N. Gayton, J.M. Bourinet, M. Lemaire, CQ2RS: a new statistical approach to the response surface method for reliability analysis. Struct. Saf. **25**, 99–121 (2003)
17. G.E.P. Box, K.B. Wilson, On the Experimental Attainment of Optimum Conditions, *Journal of the Royal Statistical Society*. Series B (Methodological) **13**(1), 1–45 (1951)
18. K.T. Fang, Uniform design: application of number-theoretic methods in experimental design. Acta Math. Applicatae Sinica **3**, 363–372 (1980)
19. K.T. Fang, C.X. Ma, D. Maringer, Y. Tang, P. Winker, The Uniform Design, in *Website*, The Uniform Design Association of China, Hong Kong Baptist University. Retrieved from http://www.math.hkbu.edu.hk/UniformDesign/. Accessed Oct 2018
20. S.K. Choi, R. Grandhi, R.A. Canfield, *Reliability-Based Structural Design* (Springer Science & Business Media, 2006)

Adaptive Metamodel-Based Robust Design Optimization of RC Frame Under Stochastic Blast Load

G. Datta, A. K. Choudhary and S. Bhattacharjya

Abstract This paper presents robust design optimization (RDO) of a reinforced concrete (RC) frame subjected to stochastic blast-induced ground motion (BIGM). As a full simulation approach requires extensive computational time to solve such problem, the polynomial response surface method (RSM) has been applied in the present study to alleviate the associated computational burden. The least squares method (LSM) generally adopted in the conventional RSM is often reported to be a source of error in optimization. Hence, a moving least squares method (MLSM)-based adaptive RSM is explored in the RDO. Random blast load is modelled by spectral representation method. The record-to-record variation of BIGM time history is captured by applying dual RSM. The optimization problem has been developed as a cost-minimization problem subjected to the displacement constraints. The RDO is formulated by simultaneously optimizing the expected value and the variation of the performance function by using weighted sum method. The robustness in the constraint is ensured by limiting the probability of failure of limit state function. The results show that the proposed MLSM-based RDO strategy yields more accurate solutions than the conventional LSM-based RSM taking full simulation approach as reference. At the same time, the present procedure yields Pareto-front in significantly lesser computational time than the full-simulation-based RDO approach.

Keywords Blast-induced ground motion · Dual response surface method · Moving least squares method · Parameter uncertainty · Robust design optimization

G. Datta · A. K. Choudhary · S. Bhattacharjya (✉)
Indian Institute of Engineering Science and Technology, Shibpur, Howrah, India
e-mail: soumya@civil.iiests.ac.in

G. Datta
e-mail: gaurav.rs2015@civil.iiests.ac.in

A. K. Choudhary
e-mail: abkabk025@gmail.com

© Springer Nature Singapore Pte Ltd. 2020
P. V. Varde et al. (eds.), *Reliability, Safety and Hazard Assessment for Risk-Based Technologies*, Lecture Notes in Mechanical Engineering,
https://doi.org/10.1007/978-981-13-9008-1_30

381

1 Introduction

The optimal design of structure under blast load disregarding uncertainty may lead to an unsafe design, inviting catastrophic consequences [1]. Hao et al. [2] showed that neglecting the random fluctuations in blast-loading estimations leads to erroneous failure probability predictions. Conventionally, optimization under uncertainty is performed by reliability-based design optimization (RBDO). Altunc et al. [3] presented RBDO of blast-resistant composites. It has been realized that the RBDO brings specified target reliability of a design. But, the system may be still sensitive to the input parameter variation due to uncertainty. In this regard, the robust design optimization (RDO) approach is felt to be more elegant owing to its capability of assuring improved performance of structure by minimizing the variation of performance simultaneously ensuring necessary safety. The study on RDO of reinforced concrete (RC) structures subjected to stochastic seismic load is notable in this regard [4]. However, RDO of the structure under stochastic blast load is observed to be scant in literature.

Israel and Tovar [5] presented the robust and reliable design of blast-mitigating shell structures under stochastic blast loading by adopting univariate dimension reduction methodology. However, they mainly dealt with topology optimization of the plate structure of the lightweight ground vehicle. Marjanishvili and Katz [6] outlined a decision-based framework by which the magnitude of the consequences to blast-induced local damage can be calculated and used to assess structural resiliency. Stewart [7] predicted reliability-based design load factors considering air blast variability on a model of RC column in direct Monte Carlo Simulation (MCS) framework. Hadianfard et al. [8] conducted reliability analysis of steel columns under blast loading due to uncertainties associated with blast-loading and material properties. They have also used MCS to obtain the damage probability.

The full MCS approach is the most accurate way to obtain RDO solution under random dynamic loading [1, 6]. However, major limitation of this approach is the enormous computational time requirement. To overcome this problem, response surface method (RSM)-based metamodels are built to represent the limit state function explicitly in terms of random parameters, thereby evading repetitive time history analyses module inside the main optimization loop.

The application of RSM to deal complex mechanical model of structure under static or deterministic dynamic load is quite straightforward. But, there remains a critical issue for structural response approximation under stochastic BIGM, as the load-time history varies significantly even with the same value of load-related parameters. To take care of this aspect, the concept of dual RSM [9] is adopted in the present study. At first, the input variable space is separated into two groups: (i) the structural parameters and (ii) the stochastic sequences. Then, a suite of BIGM time histories is prepared considering randomness in the blast load. The structural responses are evaluated at each design of experiment (DOE) point for all the input loads in the suite. Thus, two RSM models are constructed; one for the mean of the random dynamic response and another for the standard deviation (SD) of the dynamic response. Using these RSM models, the RDO problem is formulated as a two-criteria equivalent optimization problem, where the expected value and SD of cost of structure are optimized ensuring constraint feasibility under input uncertainty. The concept of dual RSM has been successfully applied for fragility analysis and optimization under stochastic earthquake [10, 11]. However, dual RSM is not yet applied in RDO under blast load and builds the scope of this study.

It may be noted at this point that the conventional RSM is hinged on the concept of least squares method (LSM). However, it has been reported in the literature that LSM is a major source of error in the RSM [12]. Thus, in the present study, a moving least squares method (MLSM)-based adaptive RSM is explored for RDO of the structure under BIGM. The procedure is explained by optimizing an RC frame. The results by the present MLSM-based RDO approach is compared with the conventional LSM-based RSM, taking full-simulation-based RDO as reference.

2   Robust Design Optimization of Structure

The RDO is fundamentally concerned with minimizing the effect of uncertainty in the design variables (DVs), $\mathbf{x}$ and design parameters (DPs), $\mathbf{z}$. Let us consider two designs: the conventional optimal design $(\mathbf{x_{opt}})$ and a robust optimal design $(\mathbf{x_{rob}})$. When both the designs are perturbed with the same amount of input deviation Δx due to uncertainty, the output response probability density function (PDF) has considerably less deviation for RDO (Δf_{rob}) than the other case Δf_{opt}, i.e., $\Delta f_{rob} < \Delta f_{opt}$. The RDO captures comparatively a flatter insensitive region of the performance function. The RDO can be expressed as [13]:

$$\text{find } \mathbf{x}$$
$$\text{minimizing } \phi(\mathbf{x}, \mathbf{z}) = p_E E[f(\mathbf{x}, \mathbf{z})] + p_\sigma \sigma[f(\mathbf{x}, \mathbf{z})] + P_G(\mathbf{x}, \mathbf{z})$$
$$\text{subject to } \Gamma_j(\mathbf{x}, \mathbf{z}) = \text{CDF}_{G_j}^{-1}(1 - P_{f,j}^{acc}) \le 0, \forall j \in J$$
$$x_{min} \le x \le x_{max} \tag{1}$$

where $E[f(\mathbf{x}, \mathbf{z})]$ denotes the expected value of the objective function $f(.), \sigma[f(\mathbf{x}, \mathbf{z})]$ is its standard deviation, p_E and p_σ are suitable weighting coefficients. Function $P_G(\mathbf{x}, \mathbf{z})$ is a penalty that is added when the constraints (G_j) are violated. $P_{f,j}^{acc}$ is the accepted probability of constraint violation. Function $\Gamma_j(\mathbf{x}, \mathbf{z})$ is the value of the original constraint function obtained from the inverse cumulative distribution function (CDF) of $G_j(\mathbf{x}, \mathbf{z})$ for an user-specified acceptable probability of constraint feasibility, $(1 - P_{f,j}^{acc})$.

The implementation of the RDO by using Eq. (1) requires the construction of the CDF for $G_j(\mathbf{x}, \mathbf{z})$ by the MCS for each iteration of optimization. Construction of the CDF in turn needs several runs of finite element model for each update of DVs during optimization [13]. As a result, the computational cost for yielding RDO solutions in the full MCS framework becomes extremely high. Therefore, statistical techniques are used to build surrogate models, i.e., metamodels as partial substitutes for expensive computer codes to reduce enormous computational involvement of large-scale design optimization applications. Thus, in this paper, the MLSM-based RSM is used judiciously to approximate $G_j(\mathbf{x}, \mathbf{z})$ of the RDO problem.

3   The Dual RSM in RDO

In the dual RSM, the responses are evaluated at each DOE point for all the input blast load in the suite. Then, the mean y_μ and SD, y_σ of any desired response 'y' are computed at the considered design parameters. The response surface for mean and SD are obtained for the considered responses, i.e.,

$$y_\mu = g(\mathbf{x}, \mathbf{z}) \quad \text{and} \quad y_\sigma = h(\mathbf{x}, \mathbf{z}) \tag{2}$$

Finally, the CDF of the response is constructed by using y_μ and y_σ. For example, if the response is observed to follow Extreme Value type I distribution, then CDF of y is given as,

$$F_Y(Y) = \exp\left[-\exp\{-\alpha_n(Y - u_n)\}\right]$$
$$\text{with, } \alpha_n = 1 \Big/ \sqrt{6}(\pi / y_\sigma) \text{ and } u_n = y_\mu - 0.5772 / \alpha_n \tag{3}$$

In above, u_n and α_n are the parameters of distribution. The dual RSM can be implemented by both the LSM and the MLSM-based RSM. As already discussed, noting the drawbacks of the conventional LSM-based RSM for improved response approximation, the MLSM-based RSM is explored in the present study.

4   MLSM-Based RSM in RDO

The conventional LSM-based RSM is based on an average global prediction over the specified DOE. By the LSM, sum of the squares of error (L_y) is minimized to obtain polynomial coefficients (Ω) as,

$$\mathbf{L_y(x)} = \varepsilon^{\mathrm{T}}\varepsilon = (\mathbf{Y} - \mathbf{Q}\Omega)^{\mathrm{T}}(\mathbf{Y} - \mathbf{Q}\Omega); \quad \Omega = [\mathbf{Q}^{\mathrm{T}}\mathbf{Q}]^{-1}\mathbf{Q}^{\mathrm{T}}\mathbf{Y} \tag{4}$$

In above, $\mathbf{Y}, \mathbf{Q},$ and Ω are the response vector, the design matrix, and the coefficient vector, respectively. The MLSM-based RSM is a weighted LSM that has varying weight functions with respect to the position of approximation [14]. The weight function is defined around the prediction point $\mathbf{x}$ and its magnitude changes with $\mathbf{x}$.

The $\mathbf{L_y}$ and $\boldsymbol{\Omega}$, in this case, are dependent on $\mathbf{x}$ and can be obtained by modifying Eq. (4) as,

$$\mathbf{L_y}(\mathbf{x}) = \varepsilon^T \mathbf{W}(\mathbf{x})\varepsilon = (\mathbf{Y} - \mathbf{Q}\boldsymbol{\Omega})^T \ \mathbf{W}(\mathbf{x})(\mathbf{Y} - \mathbf{Q}\boldsymbol{\Omega});$$
$$\boldsymbol{\Omega}(\mathbf{x}) = [\mathbf{Q}^T \mathbf{W}(\mathbf{x})\mathbf{Q}]^{-1}\mathbf{Q}^T \mathbf{W}(\mathbf{x})\mathbf{Y} \tag{5}$$

In the above equation, $\mathbf{W}(\mathbf{x})$ is the diagonal matrix of the weight function. It depends on the location of the associated approximation point of interest $(\mathbf{x})$. $\mathbf{W}(\mathbf{x})$ is obtained by utilizing a weighting function as [14]:

$$w(\mathbf{x} - \mathbf{x_I}) = w(\mathbf{d}) = \left(e^{-(\frac{d}{cI})^{2k}} - e^{-(\frac{1}{c})^{2k}}\right) \bigg/ \left(1 - e^{-(\frac{1}{c})^{2k}}\right), \text{ if } \mathbf{d} < \Psi; \text{ else } 0 \tag{6}$$

where, Ψ defines the domain of the influence of point x_I; d is the Euclid distance between sampling point x_I and prediction point x; and c, k are the free parameters to be selected for better efficiency which is taken here as 0.4 and 1.0, respectively [14]. After evaluation of $\boldsymbol{\Omega}(\mathbf{x})$ the MLSM-based RSM expression for limit state function G_j is generated, which is used in Eq. (1).

5 Modelling of Underground Blast Loading

Artificial BIGM acceleration time histories compatible with prescribed power spectral density function (PSDF) have been generated by using MATLAB script. Based on field measured data and numerical simulation results, the effects of frequency content, amplitude, and duration of BIGM are approximately represented by the Clough-Penzien PSDF as [15];

$$S_g(\omega) = \frac{\omega^4}{\left(\omega_1^2 - \omega^2\right)^2 + 4\xi_1^2\omega_1^2\omega^2} \cdot \frac{1 + 4\xi_g^2\omega^2 / \omega_g^2}{\left(1 - \omega^2 / \omega_g^2\right)^2 + 4\xi_g^2\omega^2 / \omega_g^2} S_0 \tag{7}$$

where S_0 is the intensity of the white-noise random process and ω_1, ξ_1 are the central frequency and the damping ratio of the high-pass filter parameters of the ground. ω_g and ξ_g are the parameters of the low-pass filter as presented by Kanai-Tajimi PSDF [16, 17]. These low-pass filter parameters of Kanai-Tajimi PSDF can be thought of as foundation properties in a situation where a white-noise disturbance is applied at bedrock and the motion is transmitted to the ground surface through a soil layer. For any stationary process, the power spectral densities of velocity, $S_v(\omega)$ and displacement, $S_d(\omega)$ are related to the PSDF of the acceleration as below.

$$S_v(\omega) = \frac{1}{\omega^2} S_g(\omega) \text{ and } S_d(\omega) = \frac{1}{\omega^4} S_g(\omega) \tag{8}$$

For the present study, the values of the various parameters ($\omega_1 = 60$ Hz; $\xi_1 = 0.6$; $\omega_g = 300$ Hz; $\xi_g = 0.6$ and $S_0 = 6.6 \times 10^{-3}$ m^2/s^3) as considered in Hao and Wu [18] are taken up. The central or principal frequency of the input BIGMs varies from 2.5 to 300 Hz. The damping ratios ξ_1 and ξ_g are considered to be 0.6 for firm soil condition [19].

The stationary random acceleration time history, $\ddot{u}_g(t)$ is generated from the PSDF function [see Eq. (7)] as [20].

$$\ddot{u}_g(t) = \sum_{\overline{n}=1}^{\overline{N}} \sqrt{(S_g(\omega)\Delta\omega}(A_{\overline{n}}\cos(\omega_{\overline{n}}t) + B_{\overline{n}}\sin(\omega_{\overline{n}}t)) \qquad (9)$$

In Eq. (9), $A_{\overline{n}}$ and $B_{\overline{n}}$ are independent Gaussian random variables with zero mean and unit random variance. The frequency step, $\Delta\omega$ is given by ($\omega_u/\overline{N}$), where ω_u is the upper cut-off frequency, above which the values of the frequency spectrum are insignificant. The interval $[0, \omega_u]$ is divided into $\overline{N}$ equal parts. The non-stationary BIGM $\ddot{x}_g(t)$ is then obtained by multiplying a deterministic modulating function $\phi(t)$ to $\ddot{u}_g(t)$. Thereby, $\ddot{x}_g(t)$ is obtained as [21];

$$\ddot{x}_g(t) = \phi(t)\ddot{u}_g(t) \quad \phi(t) = e^{-\overline{c}(t-t_1)} \qquad (10)$$

In above, $\overline{c}$ is a constant and t_1 is the strong motion duration. A sample blast input acceleration PSDF and its corresponding acceleration time history are shown in Figs. 1 and 2, respectively.

Fig. 1 A typical PSDF input

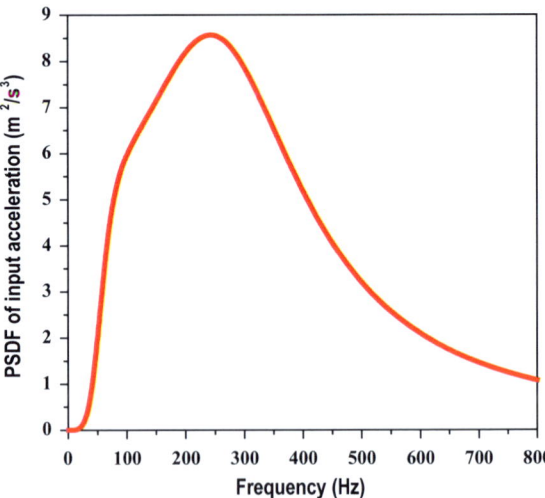

Fig. 2 A sample artificially generated input acceleration time history

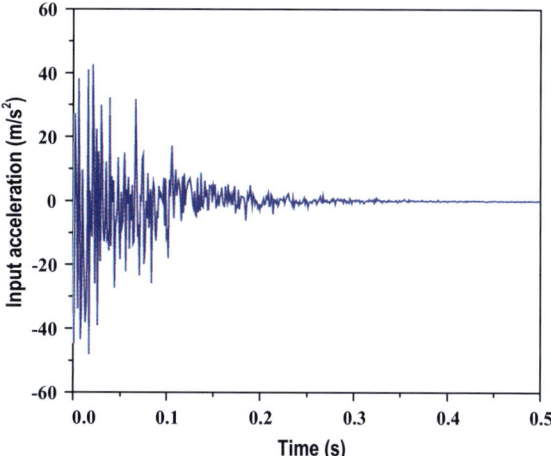

6 Numerical Study

The proposed RDO approach is elucidated by considering a twenty-storied RC-framed building (Fig. 3) considered to be located in the Kolkata city of India. The building consists of twelve bays of 4 m width in the longitudinal direction and five bays of 6 m width in the transverse direction. The total height of the building is 73 m with storey height 3.65 m each. The dead load consists of self-weight of structural and non-structural members as per Indian Standard code of practice. The grade of concrete is considered as M 35 (i.e., compressive strength of concrete is 35 N/mm^2). The reinforcing steel grade is taken as Fe500 (i.e., yield strength of steel is 500 N/mm^2). The finite element modelling is done in software SAP2000 using 3D frame elements for beams and columns and flat shell element for shear walls. The beams and columns are grouped to systematically design the building. Then, representative cross-sectional dimensions for each group are selected to formulate the optimization problem. The cross-section of the columns are kept uniform for consecutive five stories and then reduced by specific proportion. An initial design of column size performed as per Indian code suggests cross-sectional dimensions of (600 mm × 800 mm), (475 mm × 650 mm), (450 mm^2), and (400 mm^2) for ground to 5th floor, 6th to 10th floor, 11th to 17th floor, and 18th to 20th floor, respectively. The width (x_1) and depth (x_2) at the base section of column is chosen as the representative DVs of columns. Then, the cross-sectional dimensions of column for other stories are obtained by maintaining the same proportion as that obtained in the initial design. Similarly, width (x_3) and depth (x_4) of beam cross-section along the transverse direction of the building is selected as the representative DVs for beams. The thickness of the shear wall is also taken as a DV (x_5), which is kept uniform throughout the height of the building. Initially, the thickness of the shear wall is obtained as 200 mm after a deterministic design. Minimization of structural cost and

Fig. 3 A RC twenty-storied building

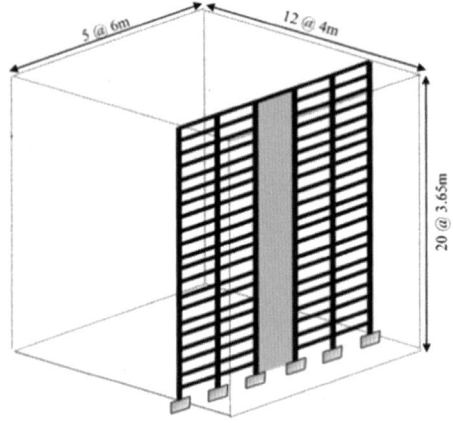

limiting maximum top displacement of the structure are taken as objective function and constraint, respectively. The uncertain DPs are the peak particle velocity (PPV) $(\dot{x}_g(t))$, the Young's Modulus of concrete (E_c), and unit cost of RC. The value of $\dot{x}_g(t)$ (assumed Extreme Value type I distributed), E_c (Lognormally distributed), and unit cost of RC (normally distributed) is taken as 10 cm/s, 29.58 GPa and Rs. 12,000/- per cu. m, respectively. The DOE is performed by uniform design scheme [22] with $U_{20}(20^8)$ scheme. A total of 400 time histories are generated to represent randomness in blast load. The deterministic design optimization (DDO) is formulated as:

Find **x**

to minimize $f(\mathbf{x}, \mathbf{z}) =$ Cost of the frame

subjected to: $G(\mathbf{x}, \mathbf{z}) = \delta(\mathbf{x}, \mathbf{z}) \leq \delta_{al}$ $\mathbf{x}^L \leq \mathbf{x} \leq \mathbf{x}^U$ (11)

where, $\delta(\mathbf{x}, \mathbf{z})$ and δ_{al} are the maximum top displacement and allowable displacement, respectively. δ_{al} is taken as $(H_s/350)$, where H_s is the height of the structure. $P_{f,j}^{acc}$ is taken as 0.1%.

The RDO problem is formulated as per Eq. (1) and solved by the Sequential Quadratic Programming routine available in MATLAB. The RDO is performed by the proposed dual MLSM-based RSM and the conventional LSM-based RSM approach. The results are then compared with the most accurate full MCS-based RDO. The results are presented for varying coefficient of variation (COV) of PPV. The optimal cost and the SD of optimal cost obtained by the RDO are presented in Figs. 4 and 5. It can be observed that the trend is similar to all the three approaches. It is also clear that the optimal cost and its SD by the dual MLSM-based RSM approach is in close conformity with the direct MCS-based approach. It has been further observed that the results by the conventional LSM-based RSM approach is significantly deviated from the direct MCS results. This warns the application of the LSM-based RSM in the RDO.

Fig. 4 Robust optimal cost with varying COV of peak particle velocity

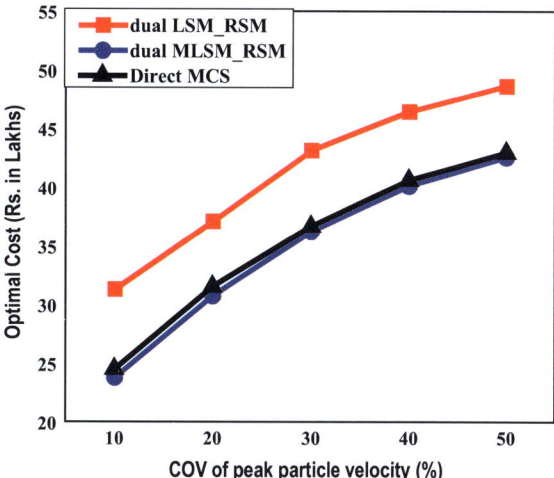

Fig. 5 SD of the robust optimal cost with varying COV of peak particle velocity

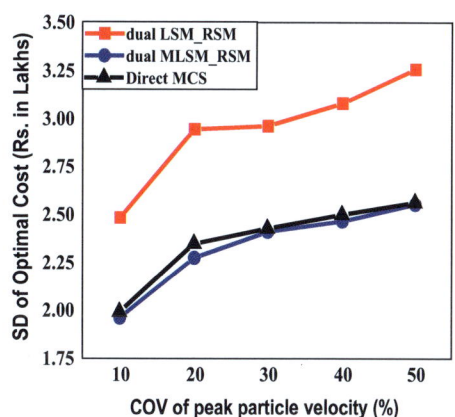

The Pareto-fronts obtained by the proposed and the conventional RDO approaches are plotted in Fig. 6. The Pareto-front obtained by the dual MLSM-based RSM are much closer to the direct MCS-based RDO results in comparison with that obtained by the LSM-based RSM. As can be observed from Figs. 5 and 6 that the maximum COV of optimal weight is 8.2% by the proposed RDO approach, even when input uncertainty is as high as 50%, which indicates the robustness of the structure with respect to the considered uncertainties. The MLSM-based RSM approach yields the Pareto-front in 43 min (including DOE construction), whereas the full MCS approach takes 11 h. This endorses the computational efficiency of the proposed approach. The LSM-based approach, though inaccurate, takes 39 min, which is less than the MLSM-based RSM. This is because one new RSM expression is generated at each iteration of RDO by the proposed dual MLSM-based RSM.

Fig. 6 Pareto-optimal curve

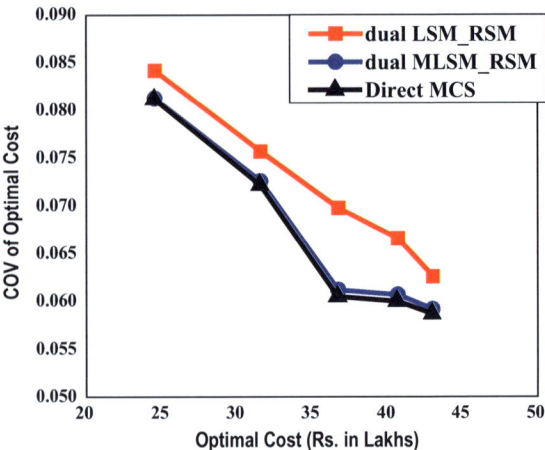

7   Summary and Conclusions

A computationally efficient RDO procedure of structure under random blast loading in the dual RSM framework is presented. The present procedure adopts MLSM-based adaptive RSM in place of conventional LSM-based RSM. The proposed approach is elucidated by optimizing a RC-framed structure. The results show that the RDO yields 12% higher optimal cost than that by the deterministic optimization. But, at the same time, COV of the objective function is observed to be reduced to maximum 8.2% by the proposed RDO approach. The proposed MLSM-based RDO approach is observed to be more accurate than the conventional LSM-based RDO approach when compared to full-simulation-based RDO as reference. Also, the proposed RDO approach is computationally efficient as it yields RDO results in substantially lesser time than the full-simulation-based RDO approach. The RDO procedure presented here is generic enough and can be easily extended to large complex structures, which is under consideration at this stage.

Acknowledgements The authors gratefully acknowledge the funding received from the CSIR to carry out this research work (Scheme No. 22(0779)/18/EMR-II dated 02/05/2018).

References

1. J. Xu, B.F. Spencer Jr., X. Lu, X. Chen, L. Lu, Optimization of structures subject to stochastic dynamic loading. Comput. Aided Civ. Inf. **32**, 657–673 (2017)
2. H. Hao, M.G. Stewart, Z.X. Li, Y. Shi, RC column failure probabilities to blast loads. Int. J. Protective Struct. **1**(4), 571–591 (2010)

3. A.B. Altunc, J.J. Kim, M. Al-Haik, M.M.R. Taha, Reliability-based design of blast-resistant composite laminates incorporating carbon nanotubes. Compos. Struct. **93**(8), 2042–2048 (2011)
4. S. Bhattacharjya, S. Chakraborty, Robust optimization of structures subjected to stochastic earthquake with limited information on system parameter uncertainty. Eng. Optimiz. **43**(12), 1311–1330 (2011)
5. J.J. Israel, A. Tovar, Investigation of plate structure design under stochastic blast loading, in *10th World Congress on Structural and Multidisciplinary Optimization*, USA (2013)
6. S.M. Marjanishvili, B. Katz, A framework for performance-based optimization of structural robustness, in *12th International Conference on Applications of Statistics and Probability in Civil Engineering, ICASP12*, Canada (2015)
7. M.G. Stewart, Reliability-based load factor design model for explosive blast loading. Struct. Saf. **71**, 13–23 (2018)
8. M.A. Hadianfard, S. Malekpour, M. Momeni, Reliability analysis of H-section steel columns under blast loading. Struct. Saf. **75**, 45–56 (2018)
9. D.K.J. Lin, W. Tu, Dual response surface optimization. J. Qual. Technol. **27**(1), 34–39 (1995)
10. S. Ghosh, S. Ghosh, S. Chakraborty, Seismic fragility analysis in the probabilistic performance-based earthquake engineering framework: an overview. Int. J. Adv. Eng. Sci. Appl. Math. (2017). https://doi.org/10.1007/s12572-017-0200-y
11. A.K. Rathi, A. Chakraborty, Reliability-based performance optimization of TMD for vibration control of structures with uncertainty in parameters and excitation. Struct. Control Health Monit. **24**(1), e1857 (2017). https://doi.org/10.1002/stc.1857
12. C. Kim, S. Wang, K.K. Choi, Efficient response surface modeling by using moving least squares method and sensitivity. AIAA J. **43**(11), 2404–2411 (2005)
13. I. Venanzi, A.L. Materazzi, L. Ierimonti, Robust and reliable optimization of wind-excited cable-stayed masts. J. Wind Eng. Ind. Aerodyn. **147**, 368–379 (2015)
14. A.A. Taflanidis, S.H. Cheung, Stochastic sampling using moving least squares response surface methodologies. Probab. Eng. Mech. **28**, 216–224 (2012)
15. R. Clough, J. Penzien, *Dynamics of Structures* (Computers & Structures Inc, Berkeley, 1995)
16. K. Kanai, Semi empirical formula for the seismic characteristics of the ground motion. Bull. Earthq. Res. I. Tokyo. **35**, 309–325 (1957)
17. H. Tajimi, A statistical method of determining the maximum response of a building structure during an earthquake, in *Proceedings of the 2nd World Conference on Earthquake Engineering*, Tokyo II, 781–798 (1960)
18. H. Hao, C. Wu, Numerical study of characteristics of underground blast induced surface ground motion and their effect on above-ground structures. Soil Dyn. Earthq. Eng. **25**(1), 39–53 (2005)
19. L.S. Katafygiotis, A. Zerva, A.A. Malyarenko, Simulation of homogeneous and partially isotropic random fields. J. Eng. Mech. **125**, 1180–1189 (1999)
20. S.O. Rice, Mathematical analysis of random noise. Bell Syst. Tech. J. **24**(1), 282–332 (1944)
21. P.C. Jennings, G.W. Housner, Simulated earthquake motion for design purpose, in *Proceedings of the 4th World Conference on Earthquake Engineering*, Santiago, A-1 145–160 (1969)
22. K.T. Fang, X. Lu, Y. Tang, J.X. Yin, Constructions of uniform designs by using resolvable packings and coverings. Discrete Math. **274**, 25–40 (2004)

Mesh Stiffness Variation Due to the Effect of Back-Side Contact of Gears

Jay Govind Verma, Shivdayal Patel and Pavan Kumar Kankar

Abstract The information regarding the source of vibration and noise is vital and depends on time-varying mesh stiffness of gear. The mesh stiffness can be estimated for the spur gear pair using both drive-side mesh stiffness and back-side mesh stiffness. Due to tight mesh or use of anti-backlash gears, gear back-side tooth contact takes place. Back-side mesh stiffness is calculated by the similar method as used for drive-side mesh stiffness. In this study, an analytical method is used for calculating the time-varying back-side mesh stiffness. The effect of addendum alteration and tip relief on back-side mesh stiffness is analyzed by knowing the association among mesh stiffness of drive-side and back-side of contacting gears. Also, the effect of gear's center distance modification, tooth thickness change on the backlash is calculated.

Keywords Anti-backlash · Addendum alteration · Back-Side contact · Gear mesh

1 Introduction

Gear knock or idle noise is due to meshing at back-side tooth contact (or impact). Back-side tooth contact denotes interaction of the tooth profiles those are not intended to convey power. Anti-backlash gears or tight mesh gears cause back-side gear tooth contact. Tight mesh gears are also caused by the change in backlash. The mounting errors of gears influence center distance, which created a change in backlash. Due to the unavoidable presence of gear backlash, the noise level of the gearbox increases

J. G. Verma (✉) · S. Patel
Discipline of Mechanical Engineering, PDPM IIITDM Jabalpur, Jabalpur, India
e-mail: royalgovindvrm@gmail.com

S. Patel
e-mail: shivdayal@iiitdmj.ac.in

P. K. Kankar
Discipline of Mechanical Engineering, IIT Indore, Indore, India
e-mail: pkankar@iiti.ac.in

© Springer Nature Singapore Pte Ltd. 2020
P. V. Varde et al. (eds.), *Reliability, Safety and Hazard Assessment for Risk-Based Technologies*, Lecture Notes in Mechanical Engineering, https://doi.org/10.1007/978-981-13-9008-1_31

393

either in a driving state (rattle sound) or in a neutral state (idle sound) [1, 2]. The application of tight mesh of gear can be observed in wind turbine gearboxes where the contact of drive-side along with back-side occurs simultaneously.

The value of mesh stiffness is considered as a parameter to measure the condition of paired gear. The deviation of mesh stiffness from the healthy gear pair and its influence on gear workings had been comprehensively examined [3–6]. The drive-side mesh stiffness (DSMS) was discussed by many researchers in the literature [3–8], and very few had reported about the back-side mesh stiffness (BSMS). The amount of reduction in DSMS can be used to measure the severity of damage in tooth [7]. An analytical model was established to estimate the mesh stiffness of pairing gear, and it was calculated by taking into account of bending deflection, Coulomb friction, axial compression, Hertzian contact, and fillet foundation deflection [7].

Further, shear energy was additionally introduced to calculate the mesh stiffness of spur gear [8]. The influence of time-varying mesh stiffness (TVMS) on the variation of center distance due to mounting error had been analytically modeled by Luo et al. [9]. They determined that the analytical method was capable of reflecting the influence of an actual number of contacting teeth with center distance variation. The reduction of mesh stiffness in the presence of tooth crack was calculated by analytical-finite element method [10]. The variation of mesh stiffness of pairing gear due to changes in web width and web hole was also calculated [10]. The dynamic behavior of a pierced gear pair due to the influence of crack propagation was studied by Ma et al. [11]. The severity of crack in gear also depends upon the direction in which it propagates. The crack propagation along the rim creates more influence on the sound and vibration than the crack propagating along with the tooth thickness [11]. The time-varying mesh stiffness for an asymmetric model was also studied by researchers [2], in which analytical equations were derived to compute the mesh stiffness at back-side. It was concluded that the model is applicable for standard spur gear and addendum alteration.

Gear teeth may have altered addendum with the purpose of avoiding undercut or, to adjust stresses due to bending in the gear pair or, to change the length of approach and recess. To obtain positive and negative addendum alteration, the height of the tooth's addendum is increased or decreased, respectively [12]. The influence of addendum alteration on BSMS was calculated [13]. It was found that addendum alteration influenced the dynamic transmission error and lateral vibration of gear.

Many research works are available for mesh stiffness calculation and considering its dynamic effect. Limited literature is available to calculate BSMS. In standard spur gear pair having a standard specification of gear parameters, the BSMS is presumed to be similar to the DSMS. While for a tight mesh case or anti-backlash case, the BSMS is not identical to the DSMS.

The primary objective of this study is to calculate the mesh stiffness at the back-side of the spur gear pair by knowing the mesh stiffness at drive-side. The general function for the BSMS is obtained which is used to calculate the significance of the addendum alteration on the gear. Further, this function is also used to calculate the effect of gear's center distance modification and tooth thickness changes.

2 Mesh Stiffness for the Spur Gear

In the direction of power transmission, the DSMS is a resultant mesh stiffness of teeth in contact. It varies with time during the meshing of the gear. The DSMS is influenced by the contact ratio, location of the applied load, tooth geometry, the material of gear, and error/faults in the tooth profile [7, 11]. It acts alongside the line of action. In contrast to this, the BSMS produced by contacting teeth acts alongside the line of action which is not involved in transmitting power. In the case of normal gear pair having standard gear's specifications, the BSMS is same as the DSMS. But, this is not correct for the case of tight mesh gear pair or anti-backlash gear. The different contact pair of the driver gear and the driven spur gear is represented in Fig. 1. The left and right portions of Fig. 1 are the driver gear and the driven gear, respectively. As individual tooth of the driving gear and the driven gear meshed at contact point L on the line of action, the tooth pairs in the onward and backward directions are two and one, respectively, and simultaneously as shown in Fig. 1a. But, when the individual tooth of the driving gear and the driven gear meshed at pitch point M, the tooth pairs in the onward and backward direction are one and two, respectively, and simultaneously as shown in Fig. 1b. Thus, the mesh stiffness is asymmetric in both directions which are well-known as the forward line of action and backward line of action. This asymmetric nature of mesh stiffness does not influence the dynamic behavior of gear if a significant backlash occurs. For the anti-backlash gears or tight mesh gears, the asymmetry in mesh stiffness is sound enough and cannot negligible.

2.1 Drive-Side Mesh Stiffness

The DSMS is the stiffness calculated along the power transmission [7]. The gear tooth is supposed as a cantilever beam of non-uniform thickness with effective length equal to tooth height for calculation of deflection. Also, the gear material assumes to be homogeneous and isotropic. The applied load is evenly scattered along the tooth width for the spur gear. The DSMS is the time-dependent factor which correlates

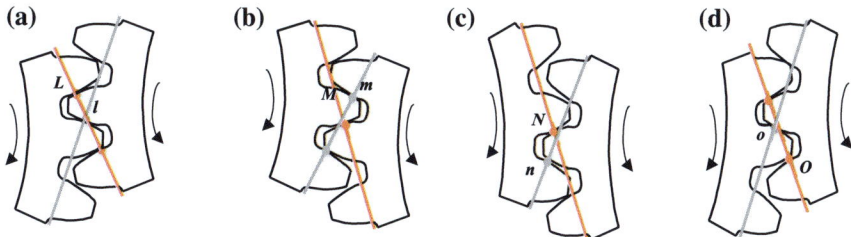

Fig. 1 Gear contact position of a mesh tooth pair **a** Initial point L, $t = 0$. **b** Pitch point M, $t = t_m$. **c** Symmetric point N, $t = t_n$. **d** Endpoint O, $t = t_o$

with contact ratio of gear pair also. The equivalent stiffness is a combined effect of the bending stiffness (K_b), shear stiffness (K_s), and compressive stiffness (K_a) for a tooth. This stiffness acts in the direction of load application. The stiffness due to the flexibility of the foundation is known as stiffness due to fillet foundation deflection (K_f). Hertzian contact stiffness (K_h) is the stiffness produced by local deformation due to contact in gear mesh and defined by Hertzian contact theory. With considering all these stiffness, the equivalent mesh stiffness (K_e) is calculated for a gear pair of the single tooth (contact ratio is one) as in Eq. (1) [7].

$$K_e = 1 \bigg/ \left(\frac{1}{K_{f1}} + \frac{1}{K_{b1}} + \frac{1}{K_{s1}} + \frac{1}{K_{a1}} + \frac{1}{K_h} + \frac{1}{K_{b2}} + \frac{1}{K_{s2}} + \frac{1}{K_{a2}} + \frac{1}{K_{f2}} \right) \quad (1)$$

where subscripts 1 and 2 represent pinion and gear, respectively.

For contact ratio lying from 1 to 2, the total mesh stiffness (K) along the line of action is calculated by addition of the known value of equivalent mesh stiffness as Eq. (2) [7].

$$K = K_e^1 + K_e^2 \quad (2)$$

where K_e^1 and K_e^2 are equivalent mesh stiffness for the first point of contact and second point of contact for a pair of a single tooth.

2.2 Back-Side Mesh Stiffness

The gear pair is mounted having a specific amount of center distance and backlash in normal condition. The gear pair is always having a small amount of backlash for lubrication and ease of engagement and disengagement. But, due to defects in manufacturing in gears or mounting errors creates tight mesh gear pair. In the case of gear pair with tight mesh or anti-backlash, the stiffness of the varying number of pairing teeth along the back-side is called BSMS. For standard spur gear pair without addendum alteration, as shown in Fig. 1, the individual tooth initiates to contact at the first mesh point at $t = 0$. If contact ratio varies from 1 to 2, at $t = 0$, the number of tooth pairs in the mesh are two. At $t = t_m$, the mesh point moves to the pitch point M. At $t = t_n$, the number of tooth pairs in contact along the drive-side is equal to the back-side. This implies that the BSMS function should be the same as the DSMS function at $t = t_n$. At $t = t_o$, the tooth pair leaves the contact. The mesh stiffness function for drive-side and back-side is represented by $K_d(t)$ and $K_b(t)$, respectively. A new time variable T is defined such that it is equal to zero at contact point N that is a symmetric point. Hence, the mesh stiffness functions are as follows [2]:

$$K_b(T) = K_d(-T) \quad (3)$$

Fig. 2 Different critical
mesh position of spur gear

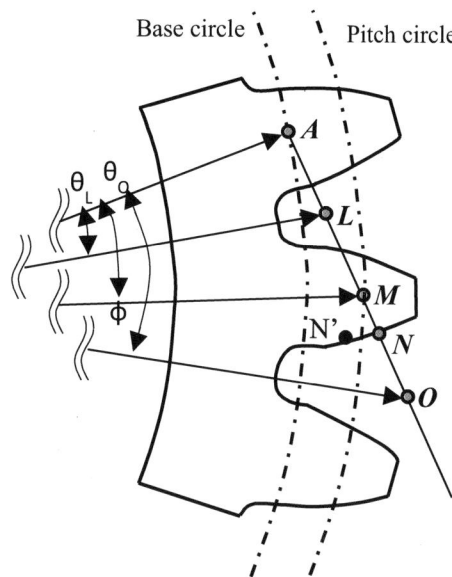

where $T = t - t_n$ therefore

$$K_b(t - t_n) = K_d(-t + t_n) \tag{4}$$

$$K_b(t) = K_d(-t + 2t_n) \tag{5}$$

The duration of time spent for tooth pair to move from mesh point L to the symmetric mesh point N along the forward line of action is given as [16]

$$t_n = \overline{LN} / (R_d \times \omega_d) \tag{6}$$

where $\overline{LN}$ is the distance of the point L from point N along the line of action of drive-side, R_d is the base radius of the driving gear, and ω_d is rotating speed of the driving gear.

From Fig. 2, $\overline{LN}$ is calculated as

$$\overline{LN} = \overline{AM} - \overline{LM} + \overline{MN} \tag{7}$$

$$\overline{AM} = R_d \times \tan\phi, \text{ and } \overline{LM} = R_d \times \theta_L \tag{8}$$

where ϕ is the pressure angle of the spur gear and θ_L is the initial mesh angle for contact at point L. The requisite time for movement of contacting tooth pair from mesh point L to pitch point M is,

Fig. 3 DSMS and BSMS of
ideal spur gear pair

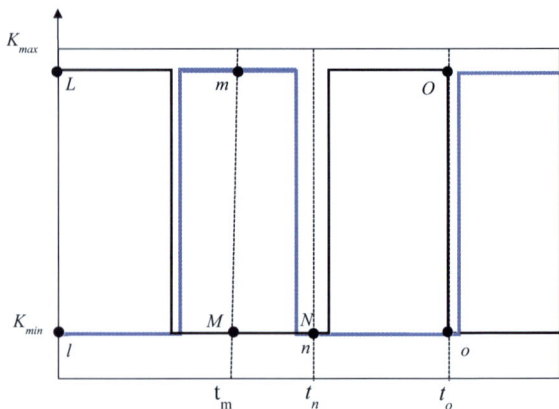

Fig. 3 DSMS and BSMS of ideal spur gear pair

$$t_m = (\tan \varphi - \theta_L) / \omega_d \tag{9}$$

The time required for the movement of tooth pair from the point of contact M that is pitch point to the point of contact N that is symmetric point along the line of action is equal to the rotation of driving gear from point M to point N' along the pitch circle.

$$t_{mn} = \overline{MN'} / (R_p \times \omega_d) \tag{10}$$

where R_p is the pitch radius of the driving gear and $\overline{MN'} = (\Pi R_p) / 2z_d$, z_d is the number of tooth of the driving gear. Thus,

$$t_n = t_m + t_{mn} = \frac{\tan \varphi - \theta_L}{\omega_d} + \frac{\Pi}{2 \times \omega_d \times z_d} \tag{11}$$

The ideal representation of mesh stiffness of drive-side and back-side spur gear pair is shown in Fig. 3. The amplitude of mesh stiffness varies with respect to time, or rotational speed of gear represents single tooth pair and double tooth pair. The DSMS at different contact points is represented by points L, M, N, and O. Similarly, the BSMS at corresponding points is represented by l, m, n, and o, respectively. The pattern of BSMS having position change with respect to DSMS by using Eq. (11). It is seen that the DSMS is equal to the BSMS at the point of contact N. The DSMS after the point of contact N is symmetrical to the BSMS before contact point N.

The specification of different parameters of gear is taken from Ref. [7]. According to Eqs. (1) and (2), the DSMS is calculated and represented in Fig. 5. Also, the BSMS of the spur gear for tight mesh gear or anti-backlash gear is calculated and depicted in Fig. 5.

According to Eq. (10), it is clear that the time span (t_{mn}) depends on chordal thickness at the gear. Therefore, addendum modification creates position change between BSMS function and DSMS function.

2.3 Altered Addendum Tooth with Tip Relief

Positive addendum modification resultant, higher load capacity and decreased contact ratio. Also, negative addendum modification resultant, higher contact ratio and reduced load capacity [13]. Figure 4 explains the tooth profile of driving gear with positive and negative addendum alteration, where x_{mc} is the alteration coefficient. Gears with different amount of addendum alteration having a corresponding value of chordal thickness. From Eq. (10), it is determined that addendum alteration reveals that changes of chordal thickness, due to which BSMS varies in terms of the position change with regard to equivalent DSMS. The change in tooth thickness due to addendum alteration is calculated by Ref. [2], which is equal to ($2 \times x_{mc} \times \tan \varphi$). Then the modified value,

$$\overline{MN'} = \frac{\Pi R_p}{2z_d} + (2 \times x_{mc} \times \tan \varphi) \tag{12}$$

Thus, the tooth thickness changes with addendum modification. Due to which, the DSMS function and BSMS function change for a different amount of addendum altered gear pair. With the addition of this, tip relief profile modification is introduced at the point of highest point of single tooth contact of spur gear (contact ratio lies between 1 and 2). To provide tip relief, a small amount of material along the tooth flank is removed from the involute profile of the gear as shown in Fig. 4. In this figure, the solid line and dashed line represent actual involute tooth profile and modified tooth

Fig. 4 Altered addendum tooth of spur gear with tip relief

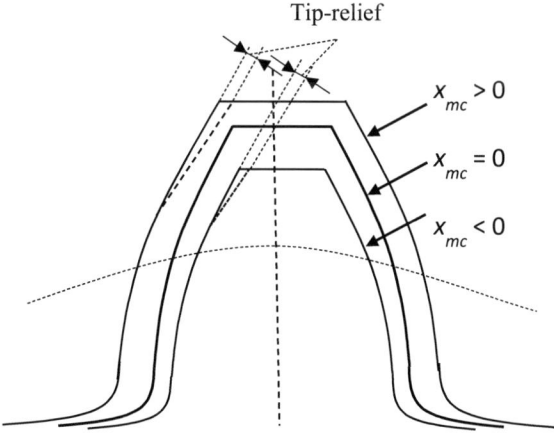

profile of tip relief, respectively. Tip relief profile modification of gear compensates the little amount of the sound created in anti-backlash gear. Due to tip relief, the time variable mesh stiffness of drive-side from Eqs. (1) and (2) as well as back-side from Eq. (11) is improved as shown in Fig. 6. The mating gear pair must have a small amount of clearance between the tip of the tooth and the mating gear's root for avoiding interference and smooth running. If positive addendum modification is within the clearance limit, then the amount of backlash change by using Eq. (12). Otherwise, it produces nasty sound and ruins the gear pair.

2.4 Operating Center Distance Modifications and Tooth Thickness Changes

There are some of the factors which affect the amount of backlash like errors in involute profile, tooth thickness, and center distance. A change in tooth thickness maintains the specific amount of backlash, required for smooth running of the gear pair. The amount of increase or decrease in operating center distance of gear pair cause an increase or decrease in backlash, respectively. The amount of backlash change due to operating center distance modifications and tooth thickness change [14] are calculated by Eqs. (13) and (14), respectively.

$$b_c = 2 \times \Delta_c \times \tan(\varphi) \tag{13}$$

$$b_t = t_i - t_a \tag{14}$$

where b_c is the amount of change in backlash due to operating center distance modifications, Δ_c is the difference between actual and theoretical operating center distance, b_t is a change in backlash due to chordal thickness changes, and t_i and t_a are theoretical and actual chordal thickness, respectively.

3 Conclusions

This study investigated the influence of the addendum variation on the BSMS of spur gear. The analytical method is derived from the correlation between DSMS and BSMS with different addendum alteration. The tooth thickness of spur gear at pitch circle is altered due to addendum alteration. It is seen that the position change occurred between DSMS and BSMS caused by the modified chordal thickness of gear. Also, it is understood from Figs. 5 and 6 that the DSMS and the BSMS influence on the amplitude of meshing stiffness with a change in addendum and modification of tip relief. The derived formulas for tooth contact in back-side are valuable for

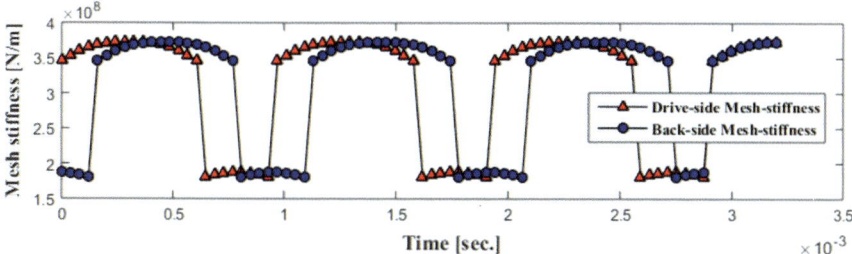

Fig. 5 DSMS and BSMS of standard spur gear pair

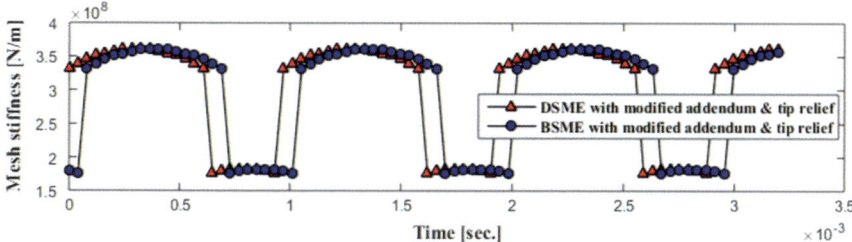

Fig. 6 DSMS and BSMS with modified addendum and tip relief

the static analysis as well as dynamic investigation of gear structures. So, the tooth contact in back-side may affect the vibration response significantly.

References

1. J.L. Dion, S. Le Moyne, G. Chevallier, H. Sebbah, Gear impacts and idle gear noise: experimental study and non-linear dynamic model. Mech. Syst. Signal Process. **23**, 2608–2628 (2009). https://doi.org/10.1016/j.ymssp.2009.05.007
2. Y. Guo, R.G. Parker, Analytical determination of back-side contact gear mesh stiffness. Mech. Mach. Theory **78**, 263–271 (2014). https://doi.org/10.1016/j.mechmachtheory.2014.03.011
3. Z. Chen, Y. Shao, Mesh stiffness calculation of a spur gear pair with tooth profile modification and tooth root crack. Mech. Mach. Theory **62**, 63–74 (2013). https://doi.org/10.1016/j.mechmachtheory.2012.10.012
4. Z. Chen, Y. Shao, Dynamic features of a planetary gear system with tooth crack under different sizes and inclination angles. J. Vib. Acoust. **135**, 31004 (2013). https://doi.org/10.1115/1.4023300
5. O.D. Mohammed, M. Rantatalo, J.O. Aidanpää, U. Kumar, Vibration signal analysis for gear fault diagnosis with various crack progression scenarios. Mech. Syst. Signal Process. **41**, 176–195 (2013). https://doi.org/10.1016/j.ymssp.2013.06.040
6. Z. Chen, W. Zhai, Y. Shao, K. Wang, G. Sun, Analytical model for mesh stiffness calculation of spur gear pair with non-uniformly distributed tooth root crack. Eng. Fail. Anal. **66**, 502–514 (2016). https://doi.org/10.1016/j.engfailanal.2016.05.006
7. O.D. Mohammed, M. Rantatalo, J.O. Aidanpää, Improving mesh stiffness calculation of cracked gears for the purpose of vibration-based fault analysis. Eng. Fail. Anal. **34**, 235–251 (2013). https://doi.org/10.1016/j.engfailanal.2013.08.008

8. Tian X, Dynamic simulation for system response of gearbox including localized gear faults. M.Sc. thesis, University of Alberta, Edmonton, Alberta, Canada (2004)

9. Luo Yang, Natalie Baddour, Ming Liang, Effects of gear center distance variation on time varying mesh stiffness of a spur gear pair. Eng. Fail. Anal. **75**, 37–53 (2017)

10. Q. Wang, K. Chen, B. Zao, H. Ma, X. Kong, An analytical-finite-element method for calculating mesh stiffness of spur gear pairs with complicated foundation and crack. Eng. Fail. Anal. **94**, 339–353 (2018)

11. H. Ma, X. Pang, J. Zeng, Q. Wang, B. Wen, Effects of gear crack propagation paths on vibration responses of the perforated gear system. Mech. Syst. Signal Process. **62**, 113–128 (2015). https://doi.org/10.1016/j.ymssp.2015.03.008

12. S. Baglioni, F. Cianetti, L. Landi, Influence of the addendum modification on spur gear efficiency. Mech. Mach. Theory **49**, 216–233 (2012)

13. Yu. Wennian, Chris K. Mechefske, Markus Timusk, Influence of the addendum modification on spur gear back-side mesh stiffness and dynamics. J. Sound Vib. **389**, 183–201 (2017)

14. *Backlash (due to operating center modifications).* Retrieved from https://www.fxsolver.com/browse/?like=2323

Estimation of Damage Due to Fatigue, Impact Loading in CFRP Laminates Through Multi-sampling Image Analysis Technique

Raghu V. Prakash, Mathew John and Michele Carboni

Abstract The carbon fiber-reinforced polymeric (CFRP) composite materials are the material of choice for the aircraft structures as the designers require lightweight structures with enhanced mechanical properties. These materials are susceptible to accidental impacts during service and maintenance, and the damage will progress under varying static or dynamic service load conditions leading to the ultimate failure of the component. Recent advancement in non-destructive techniques such as X-ray computed tomography provide excellent details about the presence of damages in 3-Dimension in a component, which is an useful input for failure prediction and remaining life estimation. However, the quality of X-ray CT imaging is dependent on the equipment used, its calibration and image settings which, in turn, may affect the reliability and repeatability of damage quantification, if damage analysis is done in a routine way using binarization algorithms. In this study, the defects as well as the damage present in the low-velocity impacted CFRP laminates subjected to fatigue loading conditions are quantified and analyzed by the analysis of CT scan images obtained from two different CT systems with images of different resolution and contrast. The results of the comparative study show that the damage analysis of polymer composites using X-ray CT depends largely on the image quality and the choice of right threshold level is important for accurate damage estimation.

Keywords CFRP · Low-velocity impact · Fatigue damage · X-ray computed tomography · Image analysis

R. V. Prakash (✉) · M. John
Indian Institute of Technology Madras, Chennai, India
e-mail: raghuprakash@iitm.ac.in

M. Carboni
Politecnico di Milano, Milan, Italy

© Springer Nature Singapore Pte Ltd. 2020
P. V. Varde et al. (eds.), *Reliability, Safety and Hazard Assessment for Risk-Based Technologies*, Lecture Notes in Mechanical Engineering,
https://doi.org/10.1007/978-981-13-9008-1_32

403

1 Introduction

Carbon fiber-reinforced polymeric (CFRP) composites are widely used in aircraft industry due to its high specific stiffness and strength. However, they are susceptible to low-velocity impact during manufacturing, service or maintenance which results in barely visible impact damage (BVID). When the low-velocity impacted structural components are subjected to cyclic service loads, the impact damage can trigger the fatigue failure mechanism causing damage progression; damage can be due to intra-laminar defects such as matrix cracking, fiber–matrix interfacial de-bonding, fiber breakage, and inter-laminar defect of delamination. Several non-destructive testing methods (NDT) such as infrared thermography, ultrasonic testing, and X-ray computed tomography (CT) are being widely employed for the qualitative as well as quantitative analysis of voids/defects formed during manufacturing or damage progression due to mechanical loadings mentioned above.

A non-destructive technique like X-ray CT recently finds wide application in damage analysis or characterization of various engineering materials, as, it provides complete information about the material as three-dimensional (3-D) images. X-ray CT is used in manufacturing sectors for the quality control of polymer products, as, it helps visualize the interior features within solid objects and gives it in the digital form in a three-dimensional manner [1]. But the reliability and repeatability of the defect/damage measurement are very important which depends on many factors. Pavana et al. have reported that the porosity measurement of polyamide-12 material produced by laser sintering [2] through the X-ray CT image is influenced by parameters such as selection of target material, influence of tube power, noise reduction algorithm, and influence of voxel size.

In this study, quasi-isotropic CFRP specimens taken from three different laminates are considered; they were subjected to different mechanical loading conditions (impact, fatigue, and impact followed by fatigue). The CT imaging of the specimens was done using two different systems—one at IIT Madras, India, and another at PoliMi, Italy. The defect/damage quantification of the specimens was done through digital image processing of the 2D X-ray CT images of chosen specimens. The main aim of this work is to understand the effect of X-ray CT system, specimen variables on the CT image quality, and its role on the subsequent damage estimation through the image processing.

2 Experimental Methodology

2.1 Materials and Specimen Fabrication

The CFRP laminates were prepared using carbon fiber (woven roving mat) of 480 gsm (g/m^2) as reinforcement and epoxy as the matrix with fiber–matrix ratio of 1:1. The mixing ratio of epoxy (Araldite® LY556—unmodified liquid epoxy resin

(a)

(b)

(c)

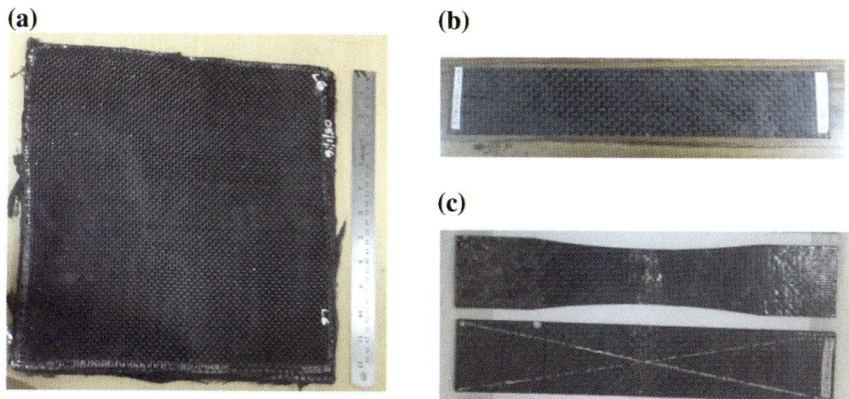

Fig. 1 **a** Quasi-isotropic CFRP laminate of 300 mm × 300 mm size; **b** The un-impacted fatigue specimen; **c** The 35 J impacted fatigue specimens of hourglass and flat shape

based on Bisphenol-A) to the curing agent Aradur® HY951 (unmodified aliphatic polyamine—try ethylene tetra amine) was 10:1 by weight. Hand-layup technique was used for making the laminates with quasi-isotropic (QI) stacking sequence of [0#90/±45/0#90/±45]s. The laminates of size 300 mm × 300 mm with 4.5 mm thickness were prepared; they were cured at 80 °C for 3 h in a compression molding machine. Tensile test specimens of size 250 mm × 25 mm × 4.5 mm and fatigue test specimens of size of 250 mm × 45 mm × 4.5 mm were cut from the laminates. Some of the fatigue specimens were machined to an hourglass shape with 35 mm width at the center to ensure damage concentration at the minimum width region. The specimens were cut using a CNC router from the three different laminates prepared (Refer Fig. 1) and subjected to different mechanical loadings conditions—impact, fatigue, impact + fatigue.

2.2 Mechanical Testing

The CFRP specimens were subjected to 35 J impact loading using drop weight impact testing machine with a drop weight impactor of mass 5.2 kg. It is ensured that the impactor did only single impact at the geometric center of the specimen. The load and corresponding displacement during impact for specimens from two different laminates were measured and plotted (Fig. 2). The average value of the actual energy absorbed by the specimens for the low-velocity impact with drop weight velocity of 3.66 m/s is found to be 2.75 J.

The tensile tests were conducted on the un-impacted and impacted specimens for determining the load range for the fatigue tests. The ultimate tensile strength of the specimens with quasi-isotropic stacking sequence is found to be 313 MPa. The constant amplitude (CA) fatigue tests on the specimens prepared were carried out at

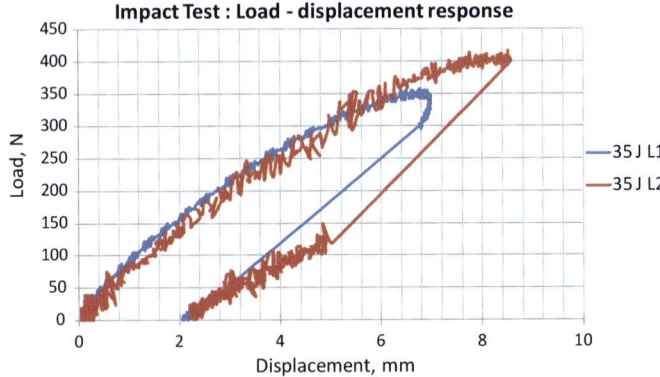

Fig. 2 Typical load–displacement curves for two CFRP specimens impacted with 35 J of energy

a stress ratio ($R = P_{min}/P_{max}$) of 0.1 and at a sinusoidal frequency of 5 Hz. The cyclic load range chosen for different specimens is as given as; (a) 0.882–8.82 kN for 35 J impacted hourglass specimen (b) 1.134–11.34 kN for 35 J impacted flat specimen and (c) 2.25–22.5 kN for the un-impacted flat specimen. The load and displacement data during the fatigue tests were continuously monitored for the purpose of stiffness estimation. The stiffness was estimated from the unloading segment of the load–displacement loop over a window of 50–90% of maximum load. The fatigue tests were stopped at half million cycles for an un-impacted specimen and at 0.1 million cycles for an impacted specimen, and the stiffness degradation due to fatigue loading for these specimens was estimated as 0.07 and 36.6%, respectively, for these specimens.

2.3 Computed X-Ray CT

3-D reconstructed X-ray CT images of damaged specimens that were extracted from three different laminates after different mechanical loading conditions were multiply sampled using two different CT imaging systems: (1) North Star Imaging 3D X-ray CT X25 tomographic system at PoliMi, Italy (referred to as XCT-A) and (2) GE make pheonix v-tomex-x s model tomographic system at CNDE lab, IIT Madras (referred to as XCT-B). The X-ray CT scanning of specimens A1, A2, A3, and A4 (refer Table 1 for specimen IDs and corresponding loading condition) are done with XCT-A; the specimens B1 and B2 with XCT-B. The scanning of specimens with XCT-A is done with a voltage of 76 kV and current 40 μA at the X-ray source and the scanning using XCT-B is done with a voltage of 50 kV and current of 100 μA at the X-ray source. The final voxel size (resolution) of the reconstructed images from machines XCT-A and XCT-B are 26 and 75 μm, respectively.

The 3-D reconstructed X-ray CT images of each specimen from these CT machines are sliced into multiple 2-D snapshots along the front (F) direction (planar

Table 1 Damage estimated for the specimens scanned with XCT-A and XCT-B with different threshold values

Specimen ID	Mechanical loading condition	X-ray CT machine used	Area considered Width × Height (mm × mm)	Damage volume percentage (Inter-modes algorithm)	Damage volume percentage (Threshold 100)	Void content percentage [$(\mu - 3\sigma)$ criteria]
A1	Un-impacted	XCT-A	44 × 34	6.47	22.05	0.89
A2	35 J Impact	XCT-A	44 × 34	3.54	10.55	–
A3	Fatigue	XCT-A	44 × 34	2.05	17.09	–
A4	35 J Impact, Fatigue	XCT-A	44 × 34	16.03	19.12	–
B1	Un-Impacted	XCT-B	24 × 44	0.86	1.14	0.97
B2	35 J Impact	XCT-B	34 × 44	1.59	1.47	–

sections). Such 272 numbers of orthographic snapshots (2-D slices) of the damaged region of the specimens (A1, A2, etc.) are taken along the F-direction from the 3-D computed tomography obtained from XCT-A. Similarly, the 3-D tomography obtained from XCT-B was sliced into 10 orthographic images for specimen B1 and 20 images for specimen B2. These images are converted to 8-bit grayscale images with pixel values in the range 0–255 by using the public domain image processing and analysis software ImageJ [3]. Typical 2-D CT slice images along the through thickness (F) direction of un-impacted and 35 J impacted specimens taken from the two X-ray CT systems are shown in Figs. 3 and 4 along with respective histogram of grayscale pixel intensities.

The 2-D X-ray CT image quality is evaluated from the histogram of the pixel intensity values of respective images. By comparing the images of un-impacted specimen taken from two different X-CT systems and image settings (Figs. 3 and 4), it is seen that the A1 specimen (from XCT-A) is of poor quality as its spread (or standard deviation) is more compared to specimen B1 from XCT-B as the former has more noise. Under such conditions, one needs to set different threshold levels for damage area measurement through digital image processing techniques. Thus, one may say that the repeatability of damage estimation through image processing is dependent on the right selection of threshold for binarization of the image.

In this study, the damage quantification of the specimens is done with digital image processing of the X-ray CT images with ImageJ software. The intra-laminar damage caused by impact, fatigue loading is estimated for its volume using the average area method which has been validated and reported earlier [4, 5]. The void

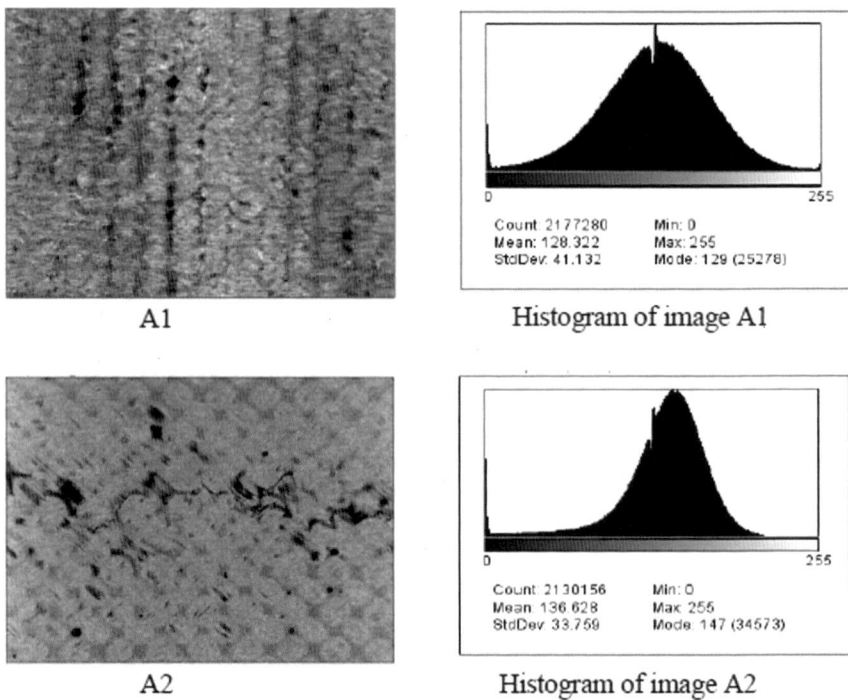

A1   Histogram of image A1

A2   Histogram of image A2

Fig. 3 CT images of Un-impacted (A1) and 35 J impacted (A2) specimens taken with XCT-A

and damage fraction of the specimens are estimated with the three different threshold values chosen; (a) using Intermodes algorithm (b) using Otsu algorithm [6] and (c) a fixed threshold value of 100. Here the threshold setting for (a) and (b) is based on the statistical parameters of the histogram of all the images as it ensures repeatable damage quantification.

3   Results and Discussion

The damage quantified through the image processing of the specimens for the different threshold values chosen are given in Table 1. The damage volume fraction estimated with a common threshold value of 100 overestimates the damage and is far from physical observations of damage. The quantification based on thresholding using Otsu algorithm for specimens scanned with XCT-A overestimate the damage, as, threshold value for all cases (A1, A2, A3, A4) are higher than those estimated using a threshold level of 100 and hence not shown in Table 1. In case of specimens (B1 and B2) scanned with XCT-B, the threshold based on Intermodes algorithm estimates relatively lesser damage. But the Otsu algorithm gives reasonably good results

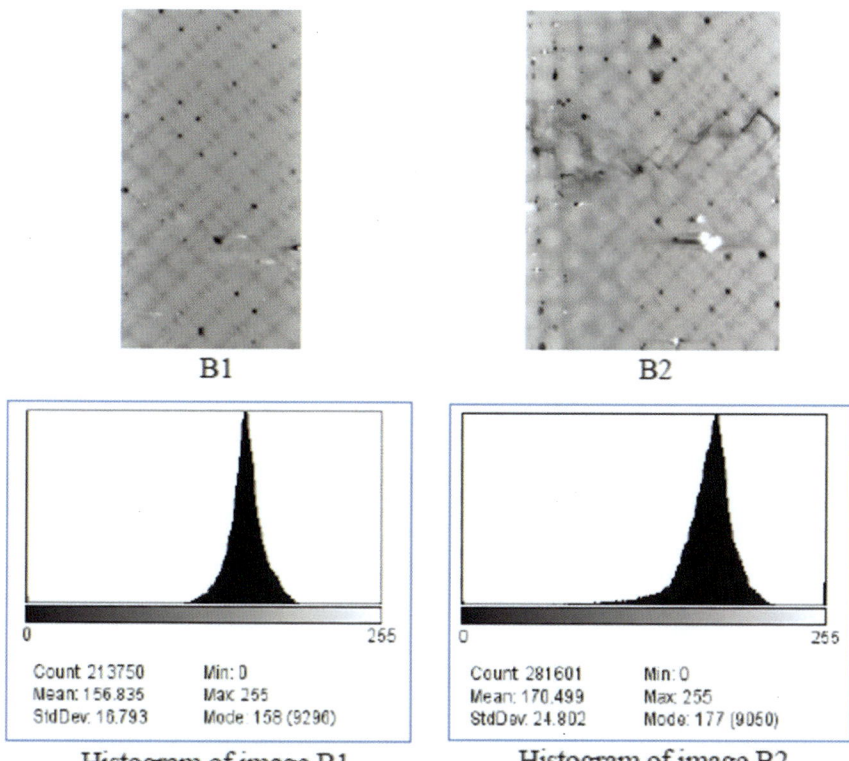

Histogram of image B1   Histogram of image B2

Count 213750	Min: 0
Mean: 156.835	Max: 255
StdDev: 16.793	Mode: 158 (9296)

Count 281601	Min: 0
Mean: 170.499	Max: 255
StdDev: 24.802	Mode: 177 (9050)

Fig. 4 CT images of Un-impacted (B1) and 35 J impacted (B2) specimens taken with XCT-B

for B2 specimen as it gives damage fraction of 4% which is a reasonable agreement. Hence, the choice of thresholding algorithm depends on the respective histogram and demands manual interpretation for reliable estimation of damage. However, the void contents in the two laminates (L1 and L2) are estimated by choosing the right threshold through a statistical analysis [7] from the mean (μ) and standard deviation (σ) of the pixel intensity value of the X-ray CT binary images of pristine specimens and the same is shown in Table 1. It is seen that the void content measured for the laminates through the binary images from machines XCT-A and XCT-B through this threshold criteria is comparable and is within the specified limits of 1.5%—where the mechanical properties are not affected by the presence of voids [8].

It is to be noted that the mechanical loading can impart both intra-laminar and inter-laminar damages. These two modes can be clearly distinguished from the histograms of the X-ray CT binary images [7]. In this study, the intra-laminar damage caused to the specimens by impact, fatigue loading is quantified by digital image processing of the CT images through the average area method. Since the damage quantification is very much sensitive to image characteristics, thresholding has been done by going through images one after another with auto-thresholding and by manu-

(a)   **(b)**   **(c)**

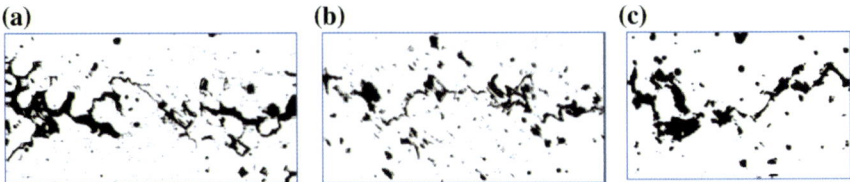

Fig. 5 Binary image of **a** 35 J Impact, Fatigue specimen XCT-A, **b** 35 J Impact only XCT-A, (c) 35 J Impact only specimen XCT-B

Table 2 Intra-laminar damage due to impact, fatigue loading quantified through digital image processing

Specimen ID	Mechanical loading condition	X-ray CT machine used	Area considered Width × Height (mm × mm)	Damage volume estimated (mm³)	Damage volume fraction (%)
A2	35 J Impact	XCT-A	44 × 22	186.13	4.27
A4	35 J Impact, Fatigue	XCT-A	44 × 22	228.94	5.25
B2	35 J Impact	XCT-B	34 × 22	141.07	4.10

ally checking the respective histograms for its correctness to avoid any overestimation of intra-laminar damage due to the inclusion of explicit inter-laminar delamination damage captured in the images. The digital image processing of planar section 2-D CT images gives intra-laminar damage volume which is the main focus of this study; it appears to be correlated with the stiffness degradation due to fatigue loading after specimen impact [9]. The damage area is measured from the binary image based on the right threshold which is set by evaluating the histogram of each image. For this purpose, a 44 mm × 22 mm window for flat specimens (A2 and A4) and 34 × 22 mm for hourglass specimen (B2) is chosen at the impacted region where the damage is mostly concentrated. The damage area identified through binarization at approximately 0.45 mm depth from the impacted surface for the specimens is as shown in Fig. 5. The damage thus quantified through the image processing is presented in Table 2.

The comparison of damage quantified through the image processing for 35 J post-impacted fatigue specimens shows that the manual thresholding instead of auto-thresholding using any algorithm may be preferred even though the images are from different machines with different image settings. It is seen that the damage quantified for A2 and B2 for 35 J impacted specimens are more or less equal (Table 2) even though the images are from two different CT machines.

4 Conclusions

The quantification of damage through multiple sampling (different samples with varying physical conditions scanned with CT scan machines of different make with different parameter settings) is carried out. It is understood that the area measurement through image processing is highly sensitive to the threshold levels chosen for the measurement; use of a common threshold for specimens scanned with the same system (say XCT-A, Threshold 100) does not provide reliable damage estimation. This is because the 2-D CT image quality may vary depending on machine settings, slicing, and 3-D reconstruction settings, number of slices from the CT reconstructed image, brightness/contrast settings of 2-D CT image settings, physical conditions of the specimen, etc. Hence if the images are from the same machine, the auto-thresholding using suitable algorithm can be employed for damage analysis purpose by way of intra-laminar impact, fatigue damage (in the absence of delamination defect) provided the image setting parameters are consistent. But when the CT images are from different machines that have different machine parameter settings, proper choice of threshold is to be selected manually on a case-to-case basis (depending on the aim of the analysis) for the purpose of meaningful comparative damage analysis. Use of such an approach resulted in similar damage parameter quantification for 35 J impacted specimens scanned using two different X-Ray CT systems.

Acknowledgements The authors would like to thank Prof. Krishnan Balasubramanian, CNDE, Department of Mechanical Engineering, IIT Madras for providing X-ray CT facility for scanning the specimens.

References

1. M. Pavana, T. Craeghs, R. Verhelst, O. Ducatteeuw, J.P. Kruth, W. Dewulf, CT-based quality control of laser sintering of polymers. Case Stud. Nondestr. Test. Eval. **6**, 62–68 (2016)
2. M. Pavana, T. Craeghs, J.P. Kruth, W. Dewulf, Investigating the influence of X-ray CT parameters on porosity measurement of laser sintered PA12 parts using a design-of-experiment approach. Polym. Testing **66**, 203–212 (2018)
3. ImageJ, Open source Java image processing programme. https://imagej.nih.gov/ij/index.html
4. M. John, R.V. Prakash, Quantification of fatigue damage in carbon fiber composite laminates through image processing. Mater. Today Proc. **5**, 16995–17005 (2018)
5. M. John, R.V. Prakash, Void content measurement in fiber reinforced plastic composites by X-ray computed tomography. Mater. Sci. Forum **928**, 38–44 (2018)
6. N. Otsu, A threshold selection method from gray-level histograms. IEEE Trans. Syst. Man Cybern. **9**, 62–66 (1979). https://doi.org/10.1109/tsmc.1979.4310076
7. R.V. Prakash, M. John, M. Carboni, A multiple-loading single-sample exploratory method of estimating damage in polymer composite materials through analysis of X-ray tomography images, in *Proceedings of the ASME 2018 Pressure Vessels and Piping Conference (PVP-2018)*, Prague, Czech Republic, 15–20 July 2018

8. C. Dong, Effects of process-induced voids on the properties of fibre reinforced composites. J. Mater. Sci. Technol. **32**, 597–604 (2016)
9. R.V. Prakash, M. John, D. Sudevan, A. Gianneo, M. Carboni, Fatigue studies on impacted and unimpacted CFRP laminates, in *ASTM STP Fatigue and Fracture Test Planning, Test Data Acquisitions and Analysis,* ASTM STP1598, pp. 94–118. http://dx.doi.org/,10.1520/STP1598201600946

Size Effect in Bimodular Flexural Cylindrical Specimens

Awani Bhushan and S. K. Panda

Abstract The present work deals with the strength scaling to characterize size effect of the cylindrical ceramic specimen for bimodular materials. The analysis is based on Weibull statistical theory. The derived semi-analytical expressions for effective volume and effective surface area for characterizing volume and surface flaws, respectively, have been validated with the experimental based numerical model. The flexural testing on cylindrical specimens has been done. The characteristic unimodular and bimodular strength has been estimated with different estimators like linear regression and maximum likelihood estimators for two different sized flexural specimens. The strength distribution analysis is controlled by 90% of the confidence interval, and coefficient of determination has been tabulated for goodness of the fit of Weibull distribution curve. The numerical based unimodular and bimodular models have been developed and also its post-processing code has been developed to evaluate the effective volume and surface area numerically. The comparison has been made for different fracture criteria such as principal of independent action (PIA) and normal stress averaging (NSA).

Keywords Bimodularity · Effective surface area · Effective volume · Flexural cylindrical specimen · Size effect

1 Introduction

Ceramic materials are desirable in many applications such as nuclear, aerospace, electronics fields, etc. and as a machine part in various industries, because of their ability to withstand strength at elevated temperature, high thermal conductivity, low thermal expansion, and outstanding wear resistance. Despite some good characteristic

A. Bhushan (✉) · S. K. Panda
Indian Institute of Technology (BHU) Varanasi, Varanasi, India
e-mail: awanibhu@gmail.com

S. K. Panda
e-mail: skpanda.mec@itbhu.ac.in

© Springer Nature Singapore Pte Ltd. 2020
P. V. Varde et al. (eds.), *Reliability, Safety and Hazard Assessment for Risk-Based Technologies*, Lecture Notes in Mechanical Engineering, https://doi.org/10.1007/978-981-13-9008-1_33

413

properties, structural ceramics are generally brittle and are associated with low strain tolerance and low fracture toughness. That means that it is very susceptible to the cracks and flaws. The tensile strength of ceramics is much lower than the compressive strength. So its design for flexural application becomes critical to neglect bimodularity. The phenomenon of showing different mechanical properties in tension and compression is called bimodularity. It is commonly found in ceramics and composites. The bimodularity is first reported by Saint-Venant in 1864 [1]. The phenomenon did not get much attention for a long time by the researchers. Later on, the concept of a bimodularity was rediscovered by Timoshenko in 1941 [2] while considering the flexural stress in such a material undergoing pure bending. The effective modulus for stiffness of such a beam in pure bending was given by Marin [3] in 1962. The bimodulus concept was extended to two-dimensional materials by Ambartsumyan in various research articles [4–7]. Within the last few decades, several attempts have been taken to establish constitutive relationships for such materials, Bert et al. [8–11]; Green and Mkrtichian [12]; Isabekian and Khachatryan [13]; Jones [14, 15].

The flexural strengths of similar specimens of ceramic materials vary considerably and also bimodularity plays a significant role, and therefore, failure predictions for ceramics are performed using statistics and reliability analysis. The most widely accepted statistical method for characterizing the strength behavior of brittle materials is Weibull analysis, which is based on the weakest link theory [16]. Furthermore, volume or surface dominated flaws are characterized by Weibull effective volume or effective surface area. Effective volume is the volume of a hypothetical tensile specimen, which, when subjected to the maximum stress, has the same probability of fracture as the flexure specimen stressed at same maximum stress [17, 18]. The Weibull weakest link model [19–21] leads to a strength dependency on component size:

$$\frac{\sigma_1}{\sigma_2} = \left(\frac{V_{E2}}{V_{E1}}\right)^{1/m_V} \tag{1}$$

where the volume-based Weibull analysis is used when the failure initiating flaws exist in the bulk of the material.

$$\frac{\sigma_1}{\sigma_2} = \left(\frac{A_{E2}}{A_{E1}}\right)^{1/m_A} \tag{2}$$

when failure mainly occurs due to surface flaws, then the area Weibull analysis can better describe the strength distribution. σ_1 and σ_2 are the mean strengths of the specimen type 1 and 2; V_{E1}, V_{E2} are effective volumes of the specimen type 1 and 2; A_{E1}, A_{E2} are effective surfaces of the specimen type 1 and 2, respectively.

Thus, calculation of effective volume and effective surface is a key step in estimating size effect in the ceramic component. Sufficient test specimens have been prepared and tested under ASTM condition to estimate the Weibull parameters of ceramic materials. The testing of actual large size components subjected to complex stress fields at high temperature is both expensive and difficult to test. Therefore, the size effect analysis is very much important to establish the correlation between the actual strength of component used and strength of test specimen tested in the laboratory. However, estimation of the effective volumes and effective surfaces area of bimodular material and thereby the more accurate strength prediction considering size effect remain really a challenging issue. In the present work, the derived integral expressions of Weibull effective volume and effective surface area for the cylindrical beam specimen having bimodularity [16] is used to predict the size effect. Dependence of bimodularity on effective volume and effective area was estimated with C++ program for analytical integral form and observed variation of effective volume/surface area against deviation of the Weibull modulus. The experiment-based unimodular and bimodular FE (finite element) model is used to evaluate the effective volume and effective surface area. And the comparison has been made for both analytical and numerical results.

2 Weibull Effective Volume and Effective Surface Area

In order to describe the effective volume and correlated with probability of failure as well as strength-size-scaling. We started the derivation from standard Weibull distribution. Weibull postulated that for a stressed volume in the failure strength, σ has a set of failure statistics described by the three-parameter equation

$$P_f = 1 - \exp\left[-\left\{\frac{\sigma - \sigma_u}{\sigma_0}\right\}^m\right] \quad \sigma > \sigma_u \tag{3}$$

$$P_f = 0 \quad \sigma \leq \sigma_u \tag{4}$$

where
P_f = Probability of failure or the cumulative distribution function for a three-parameter Weibull distribution σ = A random variable representing failure strength
σ_u = A threshold stress parameter for which P_f is finite, often assumed zero
σ_o = Weibull scale parameter
m = Weibull modulus.
    The derivative of the cumulative distribution function is given by

$$f(\sigma) = \left\{\frac{m}{\sigma_0}\right\}\left\{\frac{\sigma - \sigma_u}{\sigma_0}\right\}^{m-1} \exp\left[-\left\{\frac{\sigma - \sigma_u}{\sigma_0}\right\}^m\right] \quad \sigma > \sigma_u \tag{5}$$

$$f(\sigma) = 0 \quad \sigma \leq \sigma_u \tag{6}$$

in which $f(\sigma)$ is the frequency or the probability density function for the cumulative distribution function in Eqs. (3) and (4). However, consideration of threshold parameter to be zero results in a two-parameter Weibull formulation, which has wide applicability in component design. Hence, the cumulative distribution function [Eqs. (3) and (4)] for a two-parameter Weibull distribution is given by

$$P_f = 1 - \exp\left[-\left\{\frac{\sigma}{\sigma_0}\right\}^m\right] \quad \sigma > 0 \tag{7}$$

$$P_f = 0 \quad \sigma \leq 0 \tag{8}$$

and from Eqs. (5) and (6), the expression for the probability density function simplifies to

$$f(\sigma) = \left\{\frac{m}{\sigma_0}\right\}\left\{\frac{\sigma}{\sigma_0}\right\}^{m-1} \exp\left[-\left\{\frac{\sigma}{\sigma_0}\right\}^m\right] \quad \sigma > 0 \tag{9}$$

$$f(\sigma) = 0 \quad \sigma \leq 0 \tag{10}$$

with reference to Eq. (7), the cumulative distribution function can be reformulated to account for the size and volume effect. The Weibull characteristic strength depends on the geometry of the test specimen and also has units of stress. From a modeling point of view, σ_0 and m are the first and second Weibull distribution fitting parameters for a set of data. The characteristic strength is representative of the test specimen or component and is dependent on the flexural test specimen. When there is a change in specimen geometry, the magnitude of the characteristic distribution parameters strength also changes. This change in magnitude is directly related to the size effect. As the size of a component or test specimen geometry is increased, then, on average, the tensile strength of the component decreases. The reason for this is that as the volume (or surface area) of the component is increased, the likelihood of encountering a critical flaw with deleterious orientations to the load applied increases. For a spatial flaw distribution over the volume of the test specimen, $(\sigma_o)_V$ and m_V represent Weibull characteristics strength and Weibull modulus, respectively. Similarly, the corresponding Weibull parameters for flaw distribution over the surface area have been defined as $(\sigma_o)_A$ and m_A for all the test specimens. If flaw density and distribution are edge dominant, Weibull parameters for strength and modulus are designated as $(\sigma_o)_L$ and m_L. The subscripts V, A, and L, respectively, denote the volume, area, and edge dependent terms. Weibull theory correlates the strength distribution parameters based on specimen geometry to a strength distribution parameter based on the material property. The probability of failure expression for a two-parameter Weibull Stochastic model for the volume scattered critical flaws can be written as

$$P_f = 1 - \exp\left[-\int_V \left\{\frac{\sigma}{(\sigma_0)_V}\right\}^m dV\right] \quad \sigma > 0 \tag{11}$$

$$P_f = 0 \quad \sigma \leq 0 \tag{12}$$

The above integral has to be carried out for all tensile regions of the specimen volume. From the viewpoint of fracture, the function inside the integral is defined as the crack density function and the total parameter within exponential expression is known as risk of rupture. The integration is sometimes performed over an effective gauge section instead of over the total volume of the test specimen. Accurate evaluation of Weibull scale (strength) parameter $(\sigma_o)_V$ and the shape parameter (Weibull modulus) m_V enables specific limit for the designed characteristic strength with a certain degree of strength of dispersion. These are evaluated from a large set of flexure specimen fracture history data of test specimens. Let us define the so obtained Weibull statistical parameters from failure data as follows

$$m_V = m \tag{13}$$

$$(\sigma_0)_V = \sigma^* \tag{14}$$

Equation (11) can be expressed as

$$P_f = 1 - \exp\left[-\int_V \left\{\frac{\sigma}{\sigma^*}\right\}^m dV\right] \tag{15}$$

Rearranging the terms

$$P_f = 1 - \exp\left[-\left\{\frac{\sigma_{max}}{\sigma^*}\right\}^m \int_V \left\{\frac{\sigma}{\sigma_{max}}\right\}^m dV\right] \tag{16}$$

where σ_{max} is the maximum stress at failure. Now, substituting

$$k_V V = \int_V \left\{\frac{\sigma}{\sigma_{max}}\right\}^m dV = V_E \tag{17}$$

V_E is Effective Volume and we obtain,

$$P_f = 1 - \exp\left[-V_E\left\{\frac{\sigma_{max}}{\sigma^*}\right\}^m\right] \tag{18}$$

where $k_V V$ is the effective volume that accounts for specimen geometry and stress gradients and k_V is considered a coefficient based on the geometry of test specimen and typically considered one for uniaxial tensile test specimen. The term $k_V V$ is a function of Weibull modulus and being less than or equal to gauge volume V of the test specimen. This can be considered as the size of an equivalent uniaxial tensile specimen having the same probability of failure as the test specimen. The Weibull's weakest-link theory effectively demonstrates the strength dependency on component size. The method thereby delineates scaling of failure strength from test component size to prototype or from model loading configuration to actual application. However,

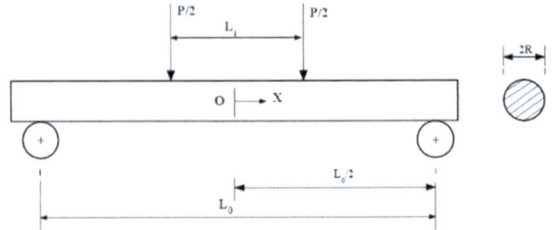

Fig. 1 Four-point flexure specimen

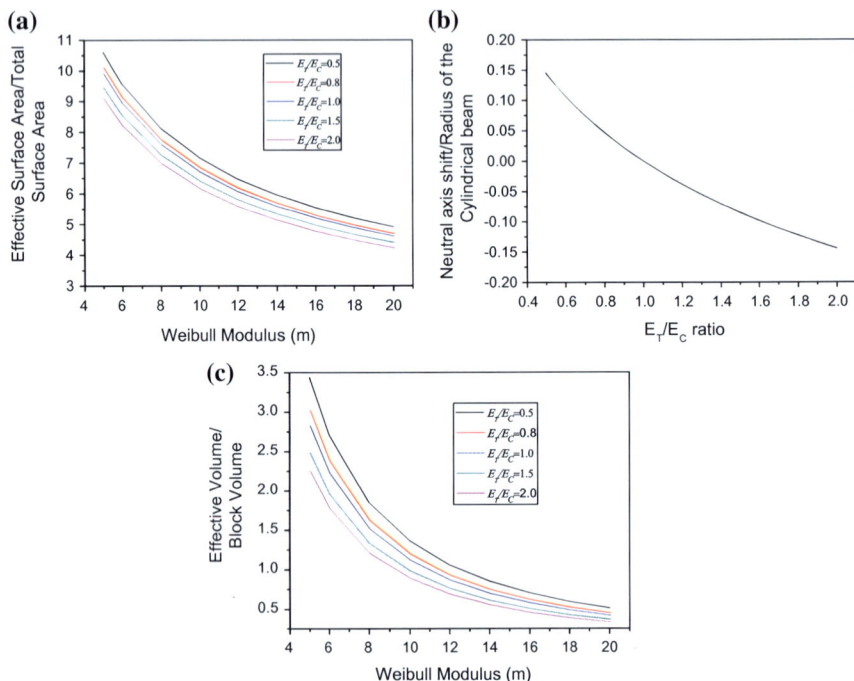

Fig. 2 **a** The neutral axis shift from the centroidal axis is plotted against the range of bimodular ratio (E_T/E_C). **b** Effective volume plot against Weibull modulus for a range of bimodular ratio. **c** Effect of bimodularity on effective surface in four-point-¼ point flexure

in the present analysis, the specimen cross-section is circular, the cylindrical beam represented in Fig. 1.

In order to derive the effective volume and effective surface area, the neutral surface shift has been evaluated through equating and balancing the force between the tensile and compressive region. After evaluation of the neutral axis, it is plotted against the range of bimodular ratio in Fig. 2a. The shift of neutral axis moves downward with increase of bimodular ratio. The effective volume has been formulated in Table 1. The detailed derivation is available in [16].

Table 1 Integral expressions of the effective volume and effective surface area for cylindrical beam loaded in flexure

Effective volume	Four-point bend test	$V_E =$ $\frac{4}{(R-\delta)^m}\left(1+m\frac{L_i}{L_0}\right)\left(\frac{L_0}{2(m+1)}\right)\int_{r=\delta}^{R}(r-\delta)^m\sqrt{(R^2-r^2)}dr$
Effective surface area	Four-point bend test	$A_E =$ $\frac{2R}{(R-\delta)^m}\left(1+m\frac{L_i}{L_0}\right)\left(\frac{L_0}{2(m+1)}\right)\int_{\sin^{-1}\frac{\delta}{R}}^{\frac{\pi}{2}}(R\sin\phi-\delta)^m d\phi$

The above integral for effective volume and surface area has been used to evaluate the effective volume and plotted against the Weibull modulus in Fig. 2b, c. The integral value has been calculated through the C++ program using Newton-Raphson method. The nature of the plot of effective volume or effective surface area against Weibull modulus is similar to existing literature [16, 22]. The increase of bimodular ratio reduced the tensile volume due to neutral axis shift in downward direction, and therefore, effective volume is also reduced. A similar trend is observed in the effective surface area.

3 Experimental and Numerical Study

Bimodularity is an inherent property of many ceramics. To explore this phenomenon, one should get to estimate the modulus of elasticity in tension and compression. The ratio of modulus of elasticity in tension and compression gives the bimodular ratio. In this manuscript, we have tested the twenty tensile specimens and twenty compressive specimens made up of highly purified graphite to evaluate the bimodular ratio as per ASTM standards ASTM C-749 and ASTM C-695. The experimental average value of bimodular ratio for highly purified graphite is found to be 1.51. The sixty small (diameter 10 mm; length 64 mm; support span 50 mm; load span 25 mm) and sixty large (diameter 20 mm; length 128 mm; support span 10 mm; load span 50 mm) cylindrical specimens have been tested as per ASTM C-651. The fixture arrangement for flexural specimen testing is shown in Fig. 3.

(a)   (b)

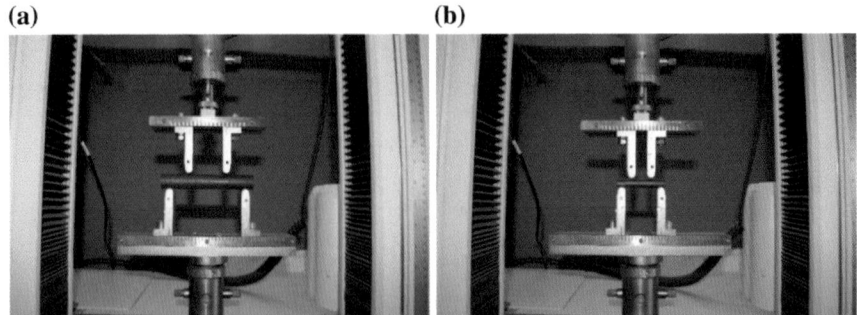

Fig. 3 Four-point ¼ loading flexure assembly for **a** large, **b** small bimodular cylindrical graphite specimens

The tested flexural specimens' results have been plotted against the cumulative probability of failure in Fig. 4a for small flexural specimen and Fig. 4b for large flexural specimens. The probability of failure has been calculated by following ranking scheme:

$$P_f = \frac{n - 0.5}{N} \qquad (19)$$

The best data-fitting lines have been drawn using linear regression (LIN 2) and maximum likelihood with unbiased (MLE2-U) and biased estimators (MLE2-B). The Weibull parameters and coefficient of determination have been tabulated in the Table 2. The coefficient of determination shows that the fracture data with 90% confidence interval fitted the regression lines more than 95.5%. That means the fracture data follows the Weibull distribution.

The effective volume has been calculated trough mesh convergent bimodular FE model and compared with the unimodular model. The both unimodular and bimodular Weibull models have been with PIA and NSA fracture criterion. The results are found to be very close to each other. The effective volume and effective surface area for bimodular graphite calculated through the bimodular FE model are found to be more close to the bimodular analytical results as tabulated in Table 3. Also, the comparison of effective volume and effective surface area has been made for unimodular analytical results developed by Quinn [18] with unimodular FE model followed with WeibPar and CARES life program [23]. The characteristic strength for large flexural specimens is found to be close to 67 MPa with different estimators, whereas the characteristic strength for large flexural specimens is found be near about 77 MPa. The specimen size effect is clearly seen here into two different sizes of the flexural specimens. The characteristics strength for smaller specimens is higher than the larger specimens. This is due to increase in number of flaws inside the gauge volume. This is also proved that the weakest link approach is working on estimating size effect on ceramic materials.

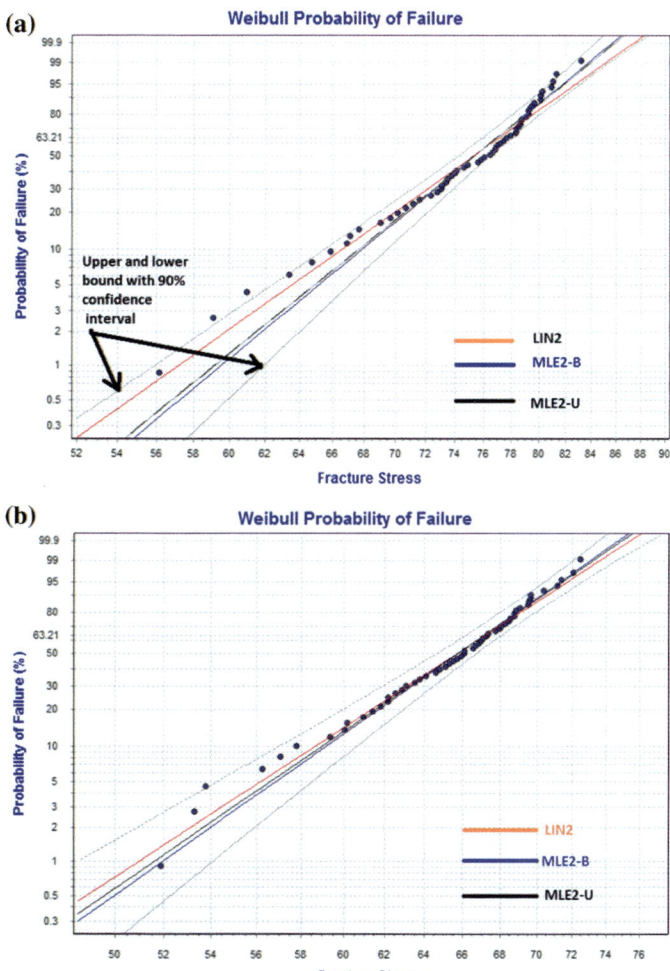

Fig. 4 **a** Weibull probability plot for small and **b** large four-point flexural ¼ loading specimen

Table 2 The characteristic strength (strength at 63.21% of probability of failure) and Weibull Modulus for large and small flexural specimen having corresponding coefficient of determination for different loading configurations

	Size	Weibull Modulus (m)			Characteristic strength (σ_θ)			R^2-value		
		LIN2	MLE2-B	MLE2-U	LIN2	MLE2-B	MLE2-U	LIN2	MLE2-B	MLE2-U
Four-point bend 1/4 loading	Small	15.38	17.92	17.48	77.06	76.91	76.91	0.986	0.980	0.983
	Large	16.65	17.95	17.50	67.13	67.09	67.09	0.977	0.955	0.961

Table 3 Effective volume tabulation for analytical and experimental analysis

Four-point bend 1/4 loading	Size	Bimodular integral	Bimodular FE model	Difference (%)	Unimodular integral [18]	Unimodular FE model	Difference (%)
Effective volume	Small	17.51	17.21	1.71	19.98	18.03	9.57
	Large	139.99	138.01	1.14	159.65	145.31	8.92
Effective surface area	Small	74.48	74.98	0.67	77.87	76.12	2.23
	Large	298.75	300.02	0.42	311.31	304.57	2.29

4 Conclusions

For estimating the size effect in any ceramic bimodular material, the semi-analytical approach is used to derive the expressions for Weibull effective volume and surface area for the beam with circular cross-section. The effect of bimodularity has been reported in the Weibull effective volume and effective surface plots against the Weibull modulus. The bimodular ratio (E_T/E_C) has a significant effect on the asymptomatic variation of Weibull volume/surface vs Weibull modulus variation. Those bimodular materials characterized with lower Weibull modulus have more significant effect of the bimodularity on effective volume. The Weibull model has been developed using concept of weakest link for experimental validation of proposed integrals for effective volume/surface area. The Weibull characteristic strength and Weibull modulus have been estimated through experimentation for four-point bend specimen made up of nuclear grade graphite with a 90% confidence bound level using WeibPar and CARES life estimating software package. A post-processing program has been written in FORTRAN language to quantify the effect of bimodularity which take the input from the FE model developed. The bimodular experimental based numerical solution for effective volume is very close to semi-analytically derived results. The characteristic strength for small and large flexural specimens is significantly different, which shows the size effect on the specimen size. It also reveals that the larger specimens increase the numbers of flaws in the gauge volume leading to lower failure strength.

Acknowledgements Financial support from the Board of Research on Nuclear Sciences, Department of Atomic Energy, India (Project Sanction No. 2011/36/62-BRNS) is greatly acknowledged.

References

1. B. Saint-Venant, *Notes to Navier's Resume des lecons dela resistance des corps solids*, 3rd edn. (Paris, 1864)
2. S. Timoshenko, *Strength of Materials, Part 2, Advanced Theory and Problems*, 2nd edn. (Van Nostrand, Princeton, NJ, 1941)

3. J. Marin, *Mechanical Behavior of Engineering Materials* (Prentice-Hall, Englewood Cliffs, NJ, 1962)
4. A.A. Ambartsumyan, S.A. Khachatryan, Elasticity for Materials with Different Resistance to Tension and Compression, Mekhanika Tverdogo Tela (in Russian), vol. 2 (1966)
5. S.A. Ambartsumyan, The axisymmetric problem of a circular cylindrical shell made of material with different strength in tension and compression. Izvestia. Mekhanika, vol. 4 (1965), pp. 1055–1067. http://oai.dtic.mil/oai/oai?verb=getRecord&metadataPrefix=html&identifier=AD0675312
6. S.A. Ambartsumyan, A.A. Khachatryan, A Multi-modulus elasticity theory. Mekhanika Tverdogo Tela **6**, 64–67 (1966)
7. S.A. Ambartsumyan, Equations for a plane problem of the multi-modulus theory of elasticity. Isvestia An Armyaskoy SSR, Mekhanika, vol. 2 (1966)
8. C.W. Bert, Model for fibrous composites with different properties in tension and compression. J. Eng. Mater. Technol. ASME **99**, 344–349 (1977)
9. C.W. Bert, F. Gordaninejad, Deflection of thick beams of multimodular materials. Int. J. Numer. Meth. Eng. **20**, 479–503 (1984)
10. C.W. Bert, J.N. Reddy, V.S. Reddy, W.C. Chao, Bending of thick rectangular plates laminated of bimodulus composite materials. AIAA J. **19**, 1342–1349 (1981). https://doi.org/10.2514/3.60068
11. C.W. Bert, J.N. Reddy, Mechanics of bimodular composite structures, in *Mechanics of Composite Materials, Recent Advances*, pp. 323–337 (1983)
12. A.E. Green, J.Z. Mkrtichian, Elastic solids with different moduli in tension and compression. J. Elast. **7**, 369–386 (1977). https://doi.org/10.1007/BF00041729
13. N.G. Isabekian, A.A. Khachatryan, On the multimodulus theory of elasticity of anisotropic bodies in plane stress state. Ivestiya Akademii Nauk Armianskoi SSR, Mekhanika **22**, 25–34 (1969)
14. R.M. Jones, Stress-strain relations for materials with different moduli and compression. AIAA **15**, 16–23 (1977)
15. R.M. Jones, Buckling of stiffened multilayered circular cylindrical shells with different orthotropic moduli in tension and compression. AIAA **9**, 917–923 (1971)
16. A. Bhushan, S.K. Panda, D. Khan, A. Ojha, K. Chattopadhyay, H.S. Kushwaha, I.A. Khan, Weibull effective volumes, surfaces, and strength scaling for cylindrical flexure specimens having bi-modularity. J. Test. Eval. ASTM **44**, 1977–1997 (2016). https://doi.org/10.1520/JTE20150301
17. G.D. Quinn, Weibull strength scaling for standardized rectangular flexure specimens. J. Am. Ceram. Soc. **10**, 508–510 (2003)
18. G.D. Quinn, Weibull effective volumes and surfaces for cylindrical rods loaded in flexure. J. Am. Ceram. Soc. **86**, 475–479 (2003). https://doi.org/10.1111/j.1151-2916.2003.tb03324.x
19. F.T. Peirce, Tensile tests for cotton Yarns v.—"The Weakest Link" theorems on the strength of long and of composite specimens. J. Text. Inst. Trans. **17**, T355–T368 (1926). https://doi.org/10.1080/19447027.1926.10599953
20. W. Weibull, The phenomenon of rupture in solids. Proc. R. Swed. Inst. Eng. Res. **153**, 1–55 (1939)
21. W. Weibull, A statistical distribution function of wide applicability. J. Appl. Mech. **18**, 293–297 (1951). citeulike-article-id:8491543
22. D.L. Shelleman, O.M. Jadaan, J.C. Conway, J.J. Mecholsky, Prediction of the strength of ceramic tubular components: part I—analysis. J. Test. Eval. **19**, 192–200 (1991). https://doi.org/10.1520/JTE12556J
23. WeibPar V-4.3 and CARES V-9.3, WeibPar (Weibull Distribution Parameter Estimation) V-4.3 and CARES (Ceramics Analysis and Reliability Evaluation of Structures) V-9.3, Developed by Life Prediction Banch NASA Glenn Research Center, Procured by Connecticut Reserve Technologies; Inc., (n.d.)

Bayesian Model Calibration and Model Selection of Non-destructive Testing Instruments

Sharvil A. Faroz, Devang B. Lad and Siddhartha Ghosh

Abstract Concrete acquires a major share of infrastructure and building stock. However, degradation and deterioration of reinforced concrete (RC) structures have been a major concern for the construction industry in recent years. Evaluating the current health of the structure is important for taking a decision regarding future action on the structure. Structural health monitoring (SHM) becomes an essential step for an engineer to gain knowledge about the health of the structure. SHM faces various challenges due to site conditions as well as the limitations present in the NDT tool itself. SHM compromises of collecting health data using NDT tools and analyzing it with the physical or empirical model. These models are fitted using various techniques to develop the relationship between the NDT readings and the corresponding actual quantity of interest. Before the NDT tool is taken on site for actual investigation, the model should be calibrated in the laboratory. The model calibration of the NDT tool is prone to measurement uncertainties which are not properly incorporated in the commonly adopted regression method of calibration. This paper focuses on the model calibration and selection of the best model using Bayesian inference. Bayesian inference helps to quantify the measurement uncertainties. For this, a measurement error model (MEM) is adopted to relate the NDT readings to the property being estimated. An illustration of the calibration and selection process is demonstrated for the proposed approach. For the demonstration, we adopt rebound hammer which is one of the most common NDT tools used to evaluate the present strength of concrete by relating the NDT readings to the crushing strength values obtained in the laboratory. As multiple models are available in the literature for both the cases, Bayesian model selection method is used for selecting the most plausible

S. A. Faroz
Infrastructural Risk Management, Mumbai, India
e-mail: iamsharvil.faroz@gmail.com

D. B. Lad · S. Ghosh (✉)
Structural Safety, Risk and Reliability Lab, Department of
Civil Engineering, Indian Institute of Technology Bombay, Mumbai, India
e-mail: sghosh@civil.iitb.ac.in

D. B. Lad
e-mail: devangladb@gmail.com

© Springer Nature Singapore Pte Ltd. 2020
P. V. Varde et al. (eds.), *Reliability, Safety and Hazard Assessment for Risk-Based Technologies*, Lecture Notes in Mechanical Engineering,
https://doi.org/10.1007/978-981-13-9008-1_34

model from all available models. This will help us to identify which model represents the NDT instrument in the best possible way.

Keywords Bayesian inference · Measurement uncertainty · Model calibration · Model selection · NDT · RC structures · Rebound hammer · SHM

1 Introduction

With the commencement of the twentieth century, there was a surge in the use of the reinforced concrete structures. It was eventually when concrete structures started failing and deteriorating in an aggressive environment, engineers realized that concrete durability concerns were far more complex [1, 2]. Concrete degradation over an extended period with an increased rate of aggressive environments has challenged engineers for its maintainability and safe functioning. Infrastructure location exposes concrete to environment highly susceptible to degradation. Thus evaluation of its present state to closely accurate results is subject that to be dealt with caution. These factors lead to either repair, retrofit, rehabilitate, and replace the structure based on the vulnerability of the structure [3]. Before it is repaired or retrofitted, it is necessary that the condition of the present structure is evaluated. Several methods like destructive, semi-destructive, and non-destructive techniques are available for evaluation of concrete properties [1]. Spatial variation and instrument precision affect the engineering judgments. Various standards viz., [4–7] etc. are available which provides guidelines for the assessing concrete properties using a Non-Destructive Testing (NDT) method.

NDT inspection involves the measurement recorded on the structure, evaluation of concrete strength/property and analyzing the results obtained. This measurement is prone to various errors and contains inherent deficiency due to various factors involved in the inspection process [8]. The quality of measurement recorded depends on the calibration of the instrument before it is actually taken on the site. For example, rebound hammer before used on the actual structure, correlation graph is provided on the instrument that is calibrated using the laboratory data of concrete cubes [9]. Thus, it becomes necessary to calibrate the instrument correctly to incorporate the uncertainties involved in the process. Various techniques are available for calibrating the NDT instruments such as regression analysis, curve fitting, artificial neural networks (ANN), and Bayesian updating. A huge amount of work is carried out by researchers for calibration of NDT instruments and providing conversion models and calibration curves. For e.g., Ploix et al. [10] carried out bilinear regression between UPV versus water saturation and porosity rate where they successfully combined providing a surface plot, which shows that the properties of the concrete can be correlated while Sbartaï et al. [11] correlated UPV, RN, and compressive strength of concrete cubes using 3-D plots. EN 13791: 2007 [12] suggest the calibration of NDT instrument before using it for evaluation of structure. In case of corrosion, various models for Linear Polarisation Resistance (LPR) are developed based on time-varying nature of the corrosion rate which is based on power law, Faraday's law, Fick's law, etc. [13]. These model parameters are to be first calibrated using experimental data to represent the actual phenomena in the most accurate way. The process of evaluating these parameters is called as model calibration. Once calibration of models is achieved, it is necessary to check the appropriateness of each model and select the best model which fits the calibration data with least error, and it is called as model selection [14, 15]. In this paper, a demonstration of the model calibration and model selection are discussed using Bayesian updating. This paper focuses on calibration through Bayesian updating of the instrument using a probabilistic measurement error model (MEM), in order to capture the measurement uncertainty and selecting the best model [8]. Major advantages of the Bayesian model calibration are that it allows to incorporate the variation in the calibration data and the parameters can be continuously updated with new incoming data. But however it does not take into effect the variation (or uncertainty) in the structure itself and does not involve special cases like carbonated concrete, corroded, etc.

2 Bayesian Updating

Bayes' theorem is given by Eq. (1),

$$f\left(\theta | D^c\right) = \frac{f(D^c | \theta) \times f(\theta)}{\int_\Theta f(D^c | \theta) f(\theta) \mathrm{d}\theta} \tag{1}$$

where $f(\theta | D^c)$ is posterior probability, $f(D^c | \theta)$ is the likelihood obtained from observed/measured/calibration data of a system, and $f(\theta)$ is the prior distribution of the model parameters [8]. The prior and the likelihood are the pillars of Bayesian inference. The prior distribution of the parameters is obtained from the literature which is known as informative prior while if this is unknown a (non-informative) suitable distribution can be also be assumed. Likelihood function provides with the necessary input to update the past knowledge (prior) which is incorporated using experimental data. Generally, for n observed/measured statistically independent data, the likelihood function is given by [16],

$$f(D^c | \theta_l) = L(\theta_l) = \prod_{i=1}^{n} L_i(\theta_l) \tag{2}$$

Here, the subscript is added for θ_l which corresponds to the specific model parameters. There are several limitations in the computation of posterior distribution for various reasons viz. (1) multiple number of parameters are involved, (2) large number of samples are required for simple Monte Carlo Simulation (MCS), (3) posterior is not of a standard form, etc. [8]. Hence in this study, we adopt Markov Chain Monte Carlo (MCMC) simulation where correlated random samples are generated enabling the user with a less computational intensive for posterior distribution. Several MCMC techniques are available for posterior computation, this study adopts Metropolis–Hastings algorithm for Bayesian updating [17].

3 Bayesian Model Calibration

The calibration problem can be summarized as one relating the NDT measurement recorded to the quantity of interest defined using a mathematical model under some specific condition. The parameters of the mathematical model are to be evaluated and the quality of the fit obtained depends on the scatter of the data, calibration procedure adopted, number of training data (calibration data) available, etc. This may include uncertainties due to the measurement process (instrument) and model itself. These uncertainties present in the instruments can be accounted for by formulating a probabilistic measurement error model (MEM) to relate NDT readings to the true values. From the available deterministic models (Sect. 3.2), we may apply probabilistic context taking measurement uncertainty into account by adding a random variable $E \sim N(\mu = 0, \sigma = e_0)$. E is a correcting random variable where e_0 is the random error/noise which is independent and identically distributed (i.i.d.) which represents measurement uncertainty [8]. The model now can be written as,

$$Y = M(B; X) + E \tag{3}$$

On the basis of the probabilistic analysis, here B of $M(\cdot)$ are the parameters of the models that are needed to be evaluated. Such regression models involving inspection or measurement are MEM models.

Calibration is an inverse problem where the aim is to evaluate the values of parameters with the known value of Y and X (measurements which are known as training data). In Bayesian calibration, these parameters are modeled as random variables quantified by their respective probability distribution function (PDFs). The data obtained through repeated experiments are combined with the prior distribution of the parameters to obtain an updated posterior distribution. The prior can be represented by the joint PDF assuming that no correlation exists between each of the parameters, $f(\theta_l | M_l)$ [18]. The likelihood function $f(D^c | \theta_l, M_l)$ can be formulated as given in Eq. (2) where D^c is the training data and θ_l are the model parameters for the particular model class, M_l. Hence from the Eq. (1), posterior is the product of prior and likelihood, given by [8],

$$f(\theta_l | D^c, M_l) = \frac{f(D^c | \theta_l, M_l) f(\theta_l | M_l)}{f(D^c | M_l)} \tag{4}$$

Below two sections present the calibration illustration for case of corrosion rate measurement and crushing strength (f_c) estimation from rebound number values, respectively.

3.1 Rebound Hammer Calibration

Rebound hammer testing is an important step in structural auditing. The crushing strength (f_c) is obtained from the graph given in the instrument manual or calculating from model available from the literature. The quality and accuracy of f_c estimation rely on the precision of the calibration technique and calibration data (D_n^c). Literature studies show that a great amount of work is being carried our correlating rebound number (RN) and crushing strength (f_c) of concrete [19, 20]. Linear law, bilinear law, power law, double power law, etc. are being used to relate RN and f_c [1]. Codes and standards also suggest calibration of rebound hammer instrument using suitable technique and also provide calibration curves [7, 12, 21]. In this section, Bayesian updating is demonstrated for calibration of rebound hammer instrument. In this study, linear law, power law, and exponential law are adopted as there exist more than 20 models calibrated by different researchers [19]. Equation for each model is given in Table 1.

In the above-given models, a and b are model parameters, and E is the random error discussed in above section. The training data is sourced from Qasrawi, 2000 [22], using the WebPlotDigitizer [8], 118 points set of pairs of the corresponding f_c and RN were extracted. These were bifurcated into "Training Data" (D_n^c) and "Validation Data" (D_n^v) as show in Fig. 1. Validation Data is used for model validation which is beyond the scope of this paper. The prior parameters (a, b, E) have to be defined in such a way that f_c obtained should not be negative. For e.g., only in linear model, a can take a negative value. In a linear model, the parameters b and σ can take only positive values. To keep them positive, we rewrite them as: $b = e^{b_1}$, $\sigma = e^s$. Thus, the parameters in case of linear law are [a b_1 s]. Similarly for power and exponential model, the parameters are [$a_1 b_1 s$]. The prior parameters in this study considered as non-informative and a standard normal distribution $N(\mu = 0, \sigma = 1)$ are assumed [8]. The joint prior distribution function (JPDF) is taken as the product of the densities of individual parameters, by assuming each parameter to be independent of others. Effect of prior in the presence of large number of training data is negligible; however, calibration results are influenced for small number of training data. Likelihood is formulated according to Eq. (2). Posterior was evaluated using Eq. (4) and samples set of parameters were obtained. In the MCMC simulation, a Gaussian proposal density is chosen: $q(\Theta^+|\theta_m) \sim N(\theta_m, C)$. Here, C is an appropriate diagonal covariance matrix, chosen to ensure sufficient mixing of the Markov Chain [23]. Covariance matrix has to be so tuned that the acceptance rate should lie between 0.20 and 0.35 for three parameter [24]. As it was difficult to tune the covariance matrix manually, DRAM was used to obtain C by running it for an initial run of $N = 2000$ or

Table 1 Models relating rebound number (R) and crushing strength of concrete	Linear Law	$f_c = a + b.\text{RN} + E$
	Power Law	$f_c = a.\text{RN}^b + E$
	Exponential Law	$f_c = a.\exp(b.\text{RN}) + E$

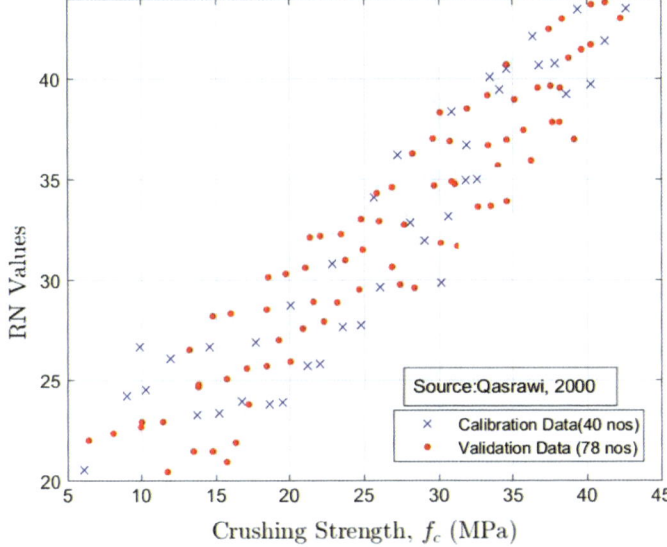

Fig. 1 Relationship of rebound number (R) and crushing strength of concrete (f_c) [22]

$N = 5000$ simulations. The covariance matrix obtained from DRAM was then used in Metropolis–Hastings (MH) algorithm for generating samples of $N = 10,000$ of each parameter [8]. Now to check diagrammatically the bounds and the fitting of the calibration ("training") data, we carry out a forward problem for prediction of crushing strength f_c versus the same calibration data. Figures 2, 3 and 4 show the prediction of each model using the calibrated data with 95% bounds. To carry out this procedure, we substitute all N samples of for particular R_i measurement reading, thus we get $N = 10,000$ f_c values for one NDT reading. Mean is evaluated and two times the standard deviation is plotted for 95% bounds.

4 Bayesian Model Selection

As discussed in Sect. 3.1, three models were selected for Bayesian calibration, and thus there can be large number of varied types of model available. Models with less number of parameters are called "simple" while with more number of parameters is called "complex." It is necessary to select a model which is a compromise between model complexity and goodness-of-fit, and fits best to the calibration data. The appropriateness model can be selected based on criteria such as: checking the difference of error between the measured value and the predicted value of the model or curve fitting. In this study, model selection is carried out using Bayes' theorem.

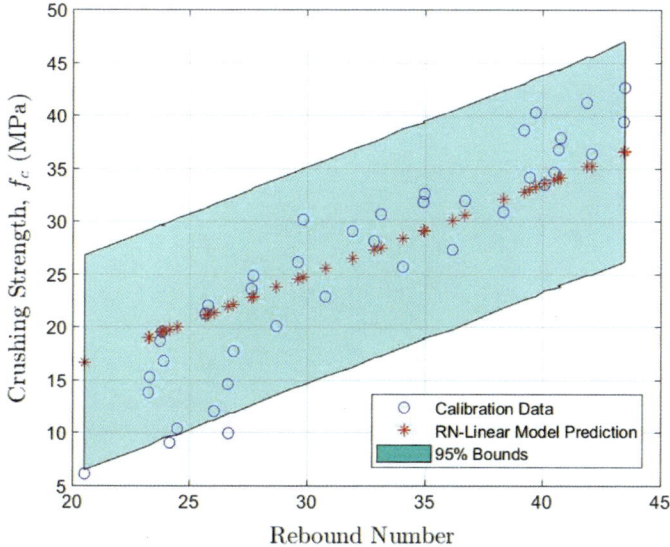

Fig. 2 Model Prediction of f_c versus the calibration data (D_n^c) for linear model

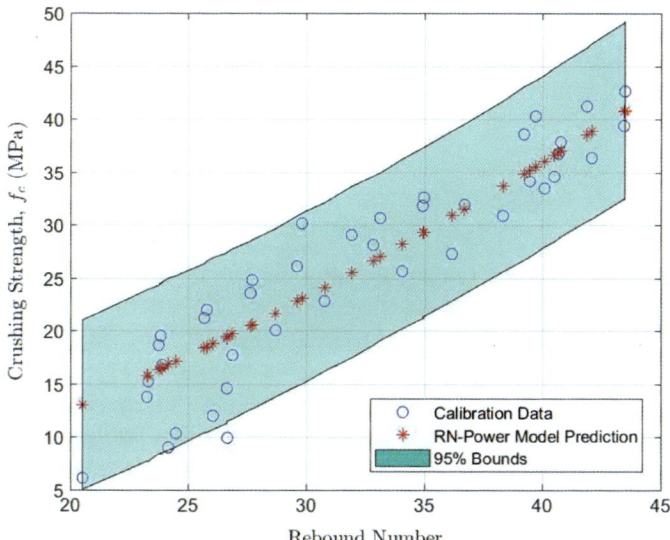

Fig. 3 Model Prediction of f_c versus the calibration data (D_n^c) for power model

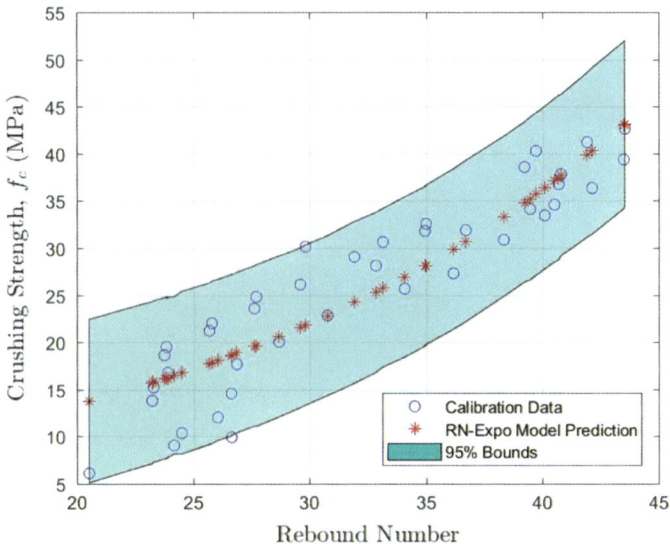

Fig. 4 Model Prediction of f_c versus the calibration data (D_n^c) for exponential Model

In Sect. 3.2, we have three a priori models (M_l), the probability of a particular model M_l, conditioned on **M** and the data D_n^c, can be obtained by Bayes' theorem [14],

$$p(M_l | D_n^c, \mathbf{M}) = \frac{f(D_n^c | M_l) \cdot p(M_l | \mathbf{M})}{\sum_{l=1}^{L} f(D_n^c | M_l) \cdot p(M_l | \mathbf{M})} \tag{5}$$

where $p(M_l | D_n^c, \mathbf{M})$ is the prior probability of each model M_l, which relates the plausibility of each model, such that $\sum p(M_l | \mathbf{M}) = 1$. The term $f(D_n^c | M_l)$ is called as the evidence, and its calculation is generalized by Cheung and Beck, 2010 [25]. The best model maximizes the posterior model plausibility $p(M_l | D_n^c, \mathbf{M})$, compared to other models. This can be extended that the numerator of Eq. (5) should be maximized. The number of models available for a particular law was random in number and taking into consideration that these laws can be infinitely fitted, the prior probability is considered to be equal: $p(M_l | \mathbf{M}) = 1/3$. This can also be calculated based on the ratio of total number of each model available, but this is very uncertain in this case hence we take it equal.

Application of Bayes' theorem enforces the belief that it is better to avoid undesired extra effort which can be done in few steps. This belief is carried out when the evidence is written as [14],

$$\ln f(D_n^c | M_l) = \mathrm{E}\left[\ln f(D_n^c | \theta, M_l] - \mathrm{E}\left[\ln \frac{f(\theta | D_n^c, M_l)}{f(\theta | M_l)}\right]\right. \tag{6}$$

Table 2 Model selection results

Models from Sect. 3.2	Prior probability	Rank-based prior probability	Log evidence	Expected information gain
M_l	$p(M_l\|D_n^c, \mathbf{M})$		$\ln f(D_n^c\|M_l)$	$\mathrm{E}\left[\ln \frac{f(\theta\|D_n^c, M_l)}{f(\theta\|M_l)}\right]$
Linear Law	0.03681	2	-128.48	6.81
Power Law	0.9675	1	-122.92	11.02
Exponential Law	0.02881	3	-126.43	12.29

where $\mathbf{E}[\cdot]$ is the expectation with respect to the posterior $f(\theta|D_n^c, M_l)$. The first term on the right-hand side of Eq. (6) is a measure of the data-fit and the second term represents the Kullback–Leibler information gain from the data. The information gain is higher for complex models as they over-fit the data, thereby penalizing the evidence of a complex model [8]. Thus, the evidence is a trade-off between how well the model fits the data versus its complexity. Based on the posterior model probability Eq. (5), the models are ranked in Table 2 and provide the values of the log evidence and expected information gain.

Thus from Table 2, it can be concluded that power law strikes the best figure for the most plausible model and expected information gain with respect to other models. Thus using Bayesian model selection technique, comparing two parameters, the best model can be inferred.

5 Conclusions

NDT instrument calibration is a vital step for good quality assessment of the concrete properties. This study provides the framework of probabilistic approach adopting Bayesian model calibration which is superior to deterministic method as it takes into account the uncertainty involved. The key conclusions are summarized below:

1. Bayesian calibration results in sample values of parameters rather than single point value evaluated using deterministic analysis. Hence a PDF is obtained for the final estimate which gives better reliability on the prediction/estimation of concrete properties.
2. Calibration parameters can be continuously updated with new incoming data which will reduce the uncertainty involved in the measurement phenomenon.
3. Concrete properties of old RC structures having high spatial variation, probabilistic technique provides a technique to integrate the structural properties as a whole.
4. Bayesian model selection provides with an approach which evaluates the best model which is simple and fits the calibration data.

6 Future Scope

1. In this work, only three model laws were selected for calibration. More number of complex models can be incorporated in the calibration and model selection process. Also, bivariate models can be taken into consideration.
2. Model validation by challenging the calibrated model with a different set of data.
3. Data fusion of two different NDT instrument using Bayesian inference.

References

1. V.M. Malhotra, N.J. Carino, *Handbook on Nondestructive Testing of Concrete*, 2nd edn. (CRC Press, Boca Raton, FL, 2003)
2. M. Shetty, *Concrete Technology* (Chand Co. LTD, 2005), pp. 420–453
3. S.K. Verma, S.S. Bhadauria, S. Akhtar, In-situ condition monitoring of reinforced concrete structures. Front. Struct. Civ. Eng. **10**(4), 420–437 (2016)
4. BS:1881-202, *Recommendations for surface hardness testing by rebound hammer*. BSI Publiscationa, London (1986)
5. BS:1881-203, *Recommendations for measurement of ultrasonic pulse for concrete*. BSI Publiscationa, London (1986)
6. BIS 13311, *Non-Destructive Testing of concrete—Methods of Test : Part 1—Ultrasonic Pulse Velocity*. Bureau of Indian Standards, New Delhi, India (1992)
7. BIS 13331, *Non-Destructive Testing of concrete—Methods of Test : Part 2—Rebound Hammer*. Bureau of Indian Standards, New Delhi, India (1992)
8. S.A. Faroz, *Assessment and Prognosis of Corroding Reinforced Concrete Structures through Bayesian Inference*. Ph.D. Thesis, Indian Institute of Technology Bombay, Mumbai, India (2017)
9. P.K. Mehta, Durability of concrete–fifty years of progress? Spec. Publ. **126**, 1–32 (1991)
10. M. Ploix, V. Garnier, D. Breysse, J. Moysan, NDE data fusion to improve the evaluation of concrete structures. NDT E Int. **44**(5), 442–448 (2011)
11. Z. Sbartaï, D. Breysse, M. Larget, J. Balayssac, Combining NDT techniques for improved evaluation of concrete properties. Cem. Concr. Compos. **34**(6), 725–733 (2012)
12. B. EN:13791, *Assessment of Insitu Compressive Strength in Structures and Precast Concrete Components* (2007)
13. C. Lu, W. Jin, R. Liu, Reinforcement corrosion-induced cover cracking and its time prediction for reinforced concrete structures. Corros. Sci. **53**(4), 1337–1347 (2011)
14. M. Muto, J.L. Beck, Bayesian updating and model class selection for hysteretic structural models using stochastic simulation. J. Vib. Control **14**(1–2), 7–34 (2008)
15. J.L. Beck, K.-V. Yuen, Model selection using response measurements: Bayesian probabilistic approach. J. Eng. Mech. **130**(2), 192–203 (2004)
16. S.A. Faroz, N.N. Pujari, R. Rastogi, S. Ghosh, Risk analysis of structural engineering systems using Bayesian inference, in *Modeling and Simulation Techniques in Structural Engineering* (IGI Global, 2017), pp. 390–424
17. S.A. Faroz, N.N. Pujari, S. Ghosh, A Bayesian Markov Chain Monte Carlo approach for the estimation of corrosion in reinforced concrete structures, in *Proceedings of the Twelfth International Conference on Computational Structures Technology*, vol. 10, Stirlingshire, UK (2014)
18. A.H.-S. Ang, W.H. Tang, *Probability Concepts in Engineering Planning and Design* (1984)
19. D. Breysse, Nondestructive evaluation of concrete strength: An historical review and a new perspective by combining NDT methods. Constr. Build. Mater. **33**, 139–163 (2012)

20. E. Arioğlu, O. Koyluoglu, Discussion of prediction of concrete strength by destructive and nondestructive methods by Ramyar and Kol. Cement and Concrete World (1996)
21. ASTM:C805/C805 M-13a, *Test for Rebound Number of Hardened Concrete*. ASTM International, West Conshohocken, PA (2013)
22. H.Y. Qasrawi, Concrete strength by combined nondestructive methods simply and reliably predicted. Cem. Concr. Res. **30**(5), 739–746 (2000)
23. S. Chib, E. Greenberg, W.R. Gilks, S. Richardson, D. Spiegelhalter, Markov chain Monte Carlo in practice. Am. Stat. **49**(4), 327–335 (1995)
24. G. Roberts, J.S. Rosenthal et al., Optimal scaling for various Metropolis-Hastings algorithms. Stat. Sci. **16**(4), 351–367 (2001)
25. S.H. Cheung, J.L. Beck, Calculation of posterior probabilities for Bayesian model class assessment and averaging from posterior samples based on dynamic system data. Comput. Aided Civ. Infrastruct. Eng. **25**(5), 304–321 (2010)

Fusion of Rebound Number and Ultrasonic Pulse Velocity Data for Evaluating the Concrete Strength Using Bayesian Updating

Devang B. Lad, Sharvil A. Faroz and Siddhartha Ghosh

Abstract Reinforced concrete (RC) is one of the most widely used construction material. However, failure in old RC structures is primarily due to degradation and deterioration of concrete, while failure in new structures is due to various other reasons. Most of the current RC structure stock is reaching its service life limit and many are still being used beyond their anticipated life span. Structural health monitoring (SHM) becomes an essential step to decide whether the structure is to be repaired, retrofitted, replaced, or allowed to continue without any action. The concrete strength assessment at any stage of the RC structure is necessary for SHM. Non-destructive testing (NDT) methods are widely accepted techniques and commonly used tools for SHM. Different NDT tools are used to understand different properties of the structure such as homogeneity, strength, presence of crack, and carbonation. Two different NDTs may deliver complementary information or conflicting results on the same parameter. Combining the measurements of these different techniques will incorporate various effects of the property on concrete strength resulting in better assessment. This paper focuses on using Bayesian updating for fusing data of rebound number (RN) and ultrasonic pulse velocity (UPV), which are two most commonly used NDT tools to evaluate the strength of concrete. Bayesian inference helps to quantify the epistemic uncertainty (in measurements) and integrate both UPV and RN measurements into a combined estimate. To account for the uncertainty, the measurement error model (MEM) is calibrated through Bayesian updating using the Markov Chain Monte Carlo (MCMC) algorithm. A framework of data fusion is discussed in detail considering three prior cases (non-informative, partially informative and fully informative), which are adopted based on IS 456:2000 specifications. A

D. B. Lad · S. Ghosh (✉)
Structural Safety, Risk and Reliability Lab, Department of Civil Engineering,
Indian Institute of Technology Bombay, Mumbai, India
e-mail: sghosh@civil.iitb.ac.in

D. B. Lad
e-mail: devangladb@gmail.com

S. A. Faroz
Infrastructure Risk Management, Mumbai, India
e-mail: iamsharvil.faroz@gmail.com

© Springer Nature Singapore Pte Ltd. 2020
P. V. Varde et al. (eds.), *Reliability, Safety and Hazard Assessment
for Risk-Based Technologies*, Lecture Notes in Mechanical Engineering,
https://doi.org/10.1007/978-981-13-9008-1_35

robust estimation of the estimated strength density functions (PDFs) is obtained, which becomes unconditioned on the MEM parameters obtained from the Bayesian calibration of a MEM. Robust estimates from two different NDT tools are then fused to a single (probabilistic) estimate of the crushing strength of concrete.

Keywords Bayesian updating · NDT · Data fusion · MCMC · Rebound hammer · UPV · Robust estimation

1 Introduction

Reinforced concrete structures were acknowledged for their long life span, fire resistant and high durability. However, in the last few decades, deterioration and degradation of concrete have emerged as a major challenge. This phenomenon appears to be far more complex with increasing age of the concrete and varied exposure conditions. Hence it is necessary to evaluate the current conditions of concrete structures to investigate the residual life. Various destructive, semi-destructive and non-destructive techniques (NDT) are available for investigation purpose [1]. For the past few decades, extensive research and experimental investigations have been carried out to discover the NDT techniques for reliable assessment which has led to a new branch called as structural health monitoring (SHM). Since the NDT techniques measure the different physical property of the concrete, the results of each NDT may be complementary or conflicting to the other depending on the properties of the current concrete [2]. Combining different techniques for strength or property assessment of concrete structure is a necessity before arriving to an engineering judgement. Different techniques evaluate varied properties which lead to more reliable assessment and in the similar context we can find a huge amount of work done to combine different NDTs using statistical or deterministic analysis [2–4].

Rebound hammer (RN) and ultrasonic pulse velocity (UPV) test are the most common NDT techniques used for structural auditing. This paper aims at combining these two techniques using the probabilistic approach to assess the current crushing strength (f_c) of the concrete structure. Bayesian updating is adopted to incorporate the measurement uncertainty in the testing process [5]. Bayesian updating leads to the formulation of a framework of integration of both different techniques with continuous incoming data. On the other hand, deterministic analysis leads to the point value of the estimate of strength to which the estimated value is dependent on the calibration technique [6]. This paper provides the framework of the fusion of the two NDTs using Bayesian inference and inferring result from it.

2 Literature Review

NDT reading is affected by noise "error" either due to variability in the material or inaccuracy in the measurement instrument itself. Thus relying on the single instrument for the overall assessment is not a good approach [2]. Indian standard [7, 8] suggests that the rebound hammer test results should be complemented with ultrasonic velocity test results. American code [9, 10] and British codes [11, 12] suggest combing these two NDT using regression analysis. EN:13791, 2007 [13] propose two approaches using 3-14 nos. core samples values to fit a curve for the two techniques before the actual assessment. There are various nomograms available provided by various authors giving a relationship between RN versus UPV versus f_c [6, 14]. RILEM technical committee (TC): 207 recommend a "SonReb" approach for the combination of the two NDT and relating to strength based on regression analysis and further research is carried for enhancement of the technique [15, 16]. Researchers have tried to combine the techniques using bilinear law, double power law, exponential law, etc. using various techniques like regression, artificial neural networks (ANN), response surface method (RSM), statistical analysis, etc. [2–4]. Incorporation of the uncertainty in the evaluation process is the important factor that lacks in the literature. The number of models available in the literature is considerable which itself means the lack of proper understanding of the phenomenon. Bayesian updating for NDT data fusion of the rock strength was formulated by Ramos et al. [17] and Bayesian data fusion from Linear Polarisation Resistance (LPR) and Ground Penetration Radar (GPR) is formulated by Faroz [5]. This paper adopts Markov Chain Monte Carlo (MCMC) technique for the combination of Rebound number (RN) and UPV which is based on Bayes' theorem.

3 Bayesian Model Formulation

To account for the uncertainty, a probabilistic measurement error model (MEM) is adopted to relate NDT readings to the true values. This MEM needs to be calibrated before carrying out the data fusion. The calibration and model selection process is summarized in the paper titled "Bayesian Model Calibration and Model Selection of Non-Destructive Testing Instruments" written by authors of the current paper. After the calibration process and model selection, power law ($f_c = a.\mathrm{RN}^b + E$) for rebound hammer and exponential law ($f_c = a. \exp(b.\mathrm{UPV}) + E$) were concluded as the best model for correlating the crushing strength of concrete to the NDT reading. Sample values of the parameters, $\theta = [abE]$, are obtained after calibration that will further be used during data fusion. Bayesian updating is the process of generating posterior samples of f_c using prior and the likelihood distribution. Let $\beta = (\mu = e^m, \sigma = e^s)$ be the samples that is need to be evaluated; hence the parameters of f_c are $[ms]$. As mentioned above Metropolis–Hastings algorithm of MCMC is adopted for generating the posterior samples. Below gives the formulation of the prior and likelihood for carrying out data fusion.

3.1 Prior ($f'(\beta)$) of Crushing Strength (f_c)

As the aim is to evaluate the strength of concrete, it is very difficult to have some prior knowledge of concrete strength unless the concrete in newly constructed or records is still available. Hence we consider three cases:

- **Non-informative Prior**—Here, we have no idea about concrete strength. This situation arises while monitoring old structures where the drawings are not available.
- **Partial Informative Prior**—Table 1 of IS 10262:2009 [18] has given a standard deviation for concrete grade from 10 N/mm² (M10)—$\sigma = 5$–55 N/mm² (M50) with a maximum value being $\sigma = 6$ N/mm². Assuming σ to follow lognormal distribution (cannot be negative) and taking these values as 5 and 95 percentile value we calculate the parameter values of prior as $(\sigma = e^s) = N(1.522; 0.1375)$.
- **Informative Prior**—This is the case where we know the strength of concrete through available drawings or any other specifications. For example, suppose the concrete structure is made of M10, here 10 N/mm² is the 5 percentile value as per IS 456:2000 [19] using the formula of target mean strength we can calculate the prior parameter for only mean value of M10 as $\beta(\mu = e^m) = N(2.625, 0.195)$ considering that the prior parameters as lognormally distributed. Distribution of standard deviation of M10 is calculated as in case of partial informative step. Refer Fig. 1 for clear understanding.

3.2 Likelihood ($L(\beta)$) of Crushing Strength (f_c)

The parameter (β) condition estimative distribution of f_c for power model of rebound number can be written as,

$$f(f_c|\mu, \sigma) = f(f_c|\beta) = \frac{1}{E \cdot \sqrt{2\pi}} \exp\left[-\frac{(f_c - \mu)^2}{2\sigma^2}\right] \tag{1}$$

Table 1 Parameters distribution of $\beta = [\mu\sigma]$ after data fusion

	μ (mean) of f_c	σ (standard deviation) of f_c
Fully informative prior	$N(37.84, 3.96)$	$LN(4.57, 1.14)$
Partial informative prior	$N(41.15, 4.35)$	$LN(4.55, 1.14)$
Non-informative prior	t – location $(41.47, 3.13, 8.69)$	Inverse Gaussian $(2.60, 3.39)$

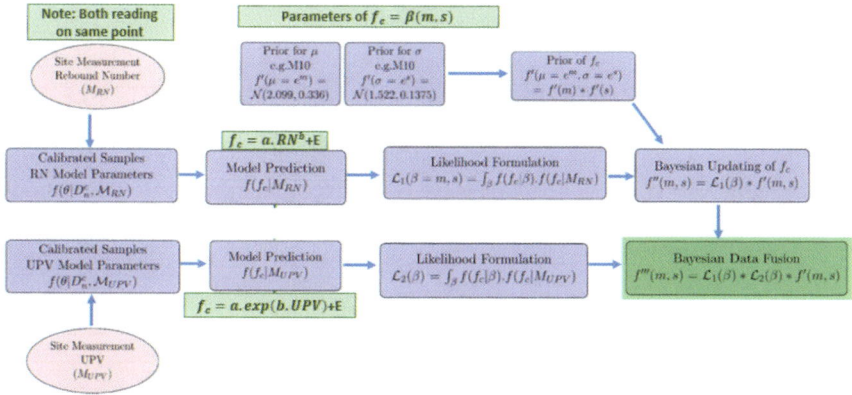

Fig. 1 Formulation for Bayesian data fusion considering fully informative prior

where $\beta = [\mu = e^m \sigma = e^s]$ are the parameters of f_c. This can be similarly written for exponential model of UPV. From calibration, we have N sample values of model parameters θ, hence f_c value obtained is not in the form of point value; rather, it is a distribution for one single measurement reading M_{RN_i} or M_{UPV_i}. This distribution can be written as $f(f_c|M_i)$ where M_i is the measurement of NDT tool. The likelihood function from the given measurement can be integrated over the β values and written as,

$$L_j(\beta) = \int_\beta f(f_c|\beta) \cdot f\left(f_c|M_{RN_i}, D^c\right) d f_c \tag{2}$$

where Eq. (2) is written for the case power model of rebound number and similar can be formulated for exponential model of UPV. Here D_n^c (calibration data) term is dropped as it denotes the estimate distribution using the calibrated samples. Equation (2) can be approximated using the f_{c_i} of $f(f_c|M_{RN_i})$ and obtained using principle of robust estimation [5]. This is discussed in Sect. 3.3. Using samples from robust estimation, the likelihood function can be approximated as:

$$L_j(\beta) \approx \frac{1}{N} \sum_{i=1}^{N} f\left(f_{c_i}|\beta\right) \tag{3}$$

For m NDT tool reading and assuming that each reading is statistically independent, overall likelihood can be written as:

$$L(\beta) = \prod_{j=1}^{m} L_j(\beta) \tag{4}$$

3.3 Robust Estimation Based on Parameter Uncertainty

Samples values ($N = 10{,}000$) of model parameters generated from MCMC simulation have to be supplied to that particular model to obtain the PDF $f(\hat{f}_c|\hat{x}, \theta, M_l)$ of the crushing strength f_c for one value of NDT reported value, where $\hat{x}$ is NDT tool reading, here $\hat{f}_c = f_{c_i}$ as in Eq. (3). Hence because of multiple sample values available for the model parameters, a family of distributions is available for the quantity of interest. The PDF $f(\hat{f}|\hat{x}, \theta, M_l)$ can be approximated by using N samples $\theta^{(k)}$ from $(\theta|D_n^c, M_l)$ [5],

$$f\left(\hat{f}_c|\hat{x}, \theta, M_l\right) \approx \frac{1}{N} \sum_{k=1}^{N} f\left(\hat{f}_c|\hat{x}, \theta^{(k)}, M_l\right) \tag{5}$$

The above expression contains summation of N PDFs. The form of the PDF $f\left(\hat{f}_c|\hat{x}, \theta, M_l\right)$ may not be similar to that of each component PDF in the summation or may not be even of any standard form.

4 Data Fusion Illustration

Fusion is defined as the part where input from two different NDT tool measuring two different characteristics of concrete will yield one single result of strength. To fuse the data, we have two options, first update using RN_i and then with UPV_i or either way. It is important to note that both the reading should be taken at the same location. Prior is obtained considering the case as non-informative, partially know or fully know discussed in Sect. 3.1. Likelihood has to be obtained using the Eq. (2) and approximated using Eqs. (3) and (4). So now suppose we have prior formulated as $f'(\beta)$ as given in Sect. 3.1. Now using MCMC posterior generation, first update can be written as,

$$f''(\beta) = f'(\beta) \times L_1(\beta) \tag{6}$$

Again using the likelihood function from another NDT instrument $L_2(\beta)$, second update is obtained as,

$$f'''(\beta) = f''(\beta) \times L_2(\beta) = f'(\beta) \times L_1(\beta) \times L_2(\beta) \tag{7}$$

For three different formulation of the given problem of data, fusion is diagrammatically shown in Figs. 1, 2, and 3. The discussion of fusion for each prior case is as follows:

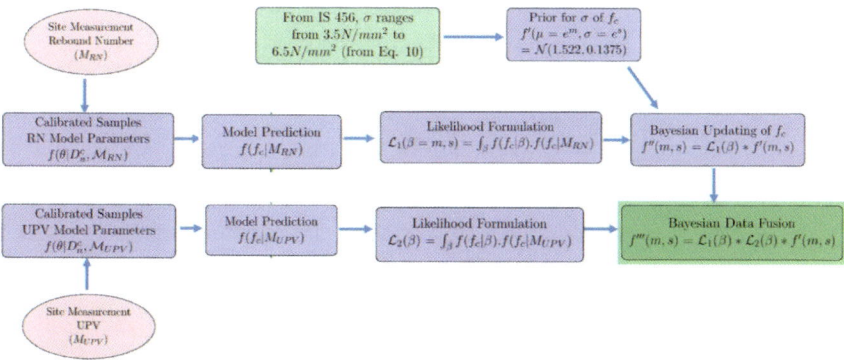

Fig. 2 Formulation for Bayesian data fusion considering partial informative prior

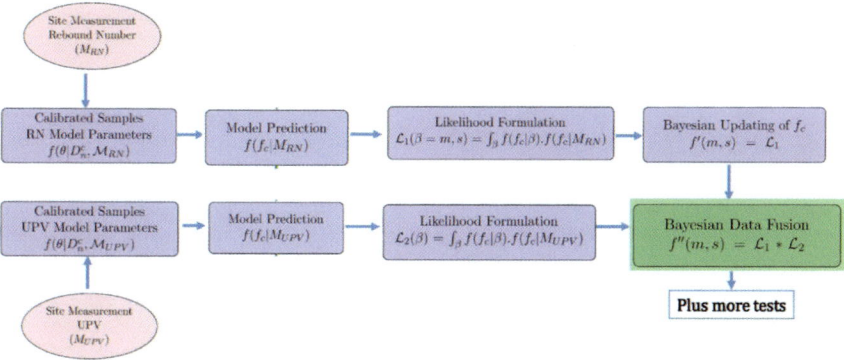

Fig. 3 Formulation for Bayesian data fusion considering non-informative prior

Case 1—Fully Informative prior (IP)

As discussed in the above Sect. 3.1, we assume the prior strength of concrete to be M10 (for experimental basis) and take $RN_i = 41$ and $UPV_i = 4.866$ kmph [20]. For validation first using RN_i and subsequently UPV_i and also vice versa, the samples generated were similar which concludes that it does not matter on which updating is done first. Unconditioned PDF (PDF averaging) is shown in Fig. 4 calculated using the Eq. (5), when compared to other unconditioned PDF, fully informative prior gives better results. f_{ck} value at the test location is reported in the same figure given by Eq. (8) for the unconditioned PDF.

$$0.05 = \int_{-\infty}^{f_{ck}} f(f_c)\mathrm{d}f_c \qquad (8)$$

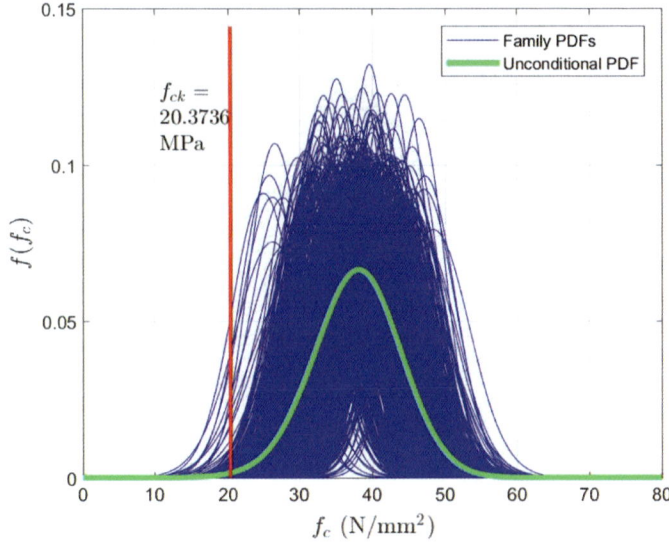

Fig. 4 Family of PDFs and unconditional PDF after data fusion of fully informative prior case

Parameters $\beta = (\mu = e^m, \sigma = e^s)$ follows normal and lognormal distribution, respectively, and verified through Kolmogorov–Smirnov (K–S) test [21]. With multiple sets of measurement at the various location, Eq. (8) can be used to evaluate the f_{ck} of the whole structure.

Case 2—Partial Informative Prior (PIP)

The values of μ and σ of f_c are almost same as that to case 1. Family of PDFs shown in Fig. 5 is more spread than of case 1 (Fig. 4). Hence it means that the standard deviation increase which we can infer that with less knowledge about prior causes less convergence of the results.

Case 3—Non-informative Prior (NIP)

Trace Plot of MCMC sample generated was not stabilized. Family of PDFs and unconditional PDF are generated using the samples whose standard deviation fall in the range 1–8 N/mm² as shown in Fig. 6. It was observed that non-informative case results are more inclining on the model. Results are not as per expectation and sometimes difficult to interpret. However, this can be used for getting some idea of the evaluated structure and make a judgement in the case where there is no data available.

Comparing all the cases, the results are as expected, that is with an increase in information of prior the results are stabilized. From Fig. 7, it can be seen that CDF plot is grown steeper with more information.

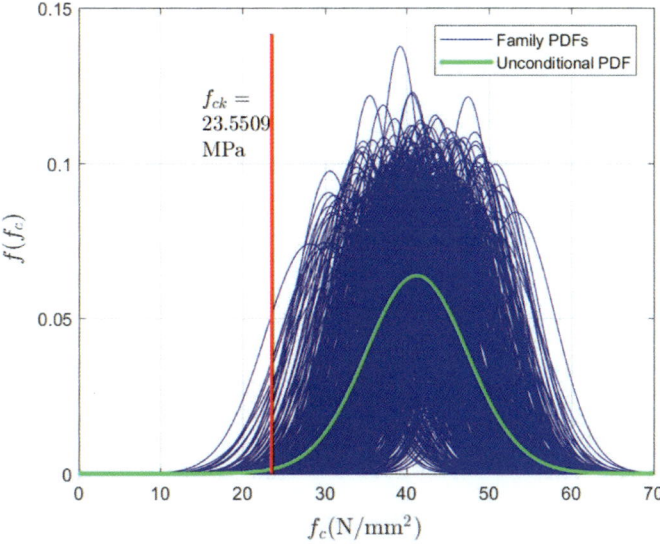

Fig. 5 Family of PDFs and unconditional PDF after data fusion of partial informative prior case

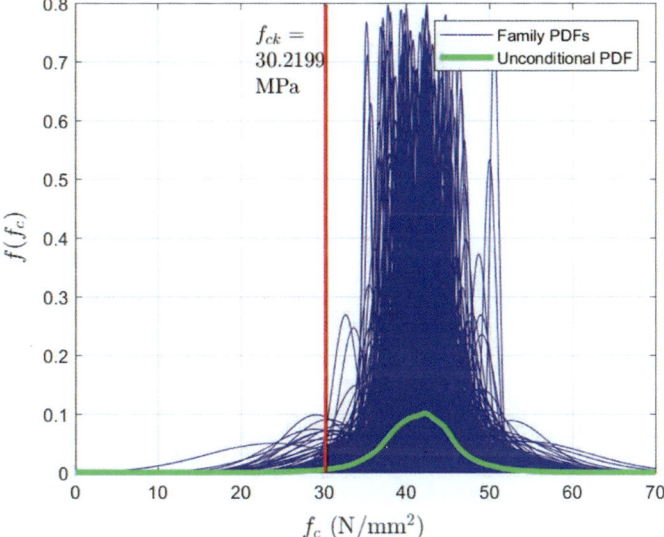

Fig. 6 Family of PDFs and unconditional PDF after data fusion of non-informative prior case

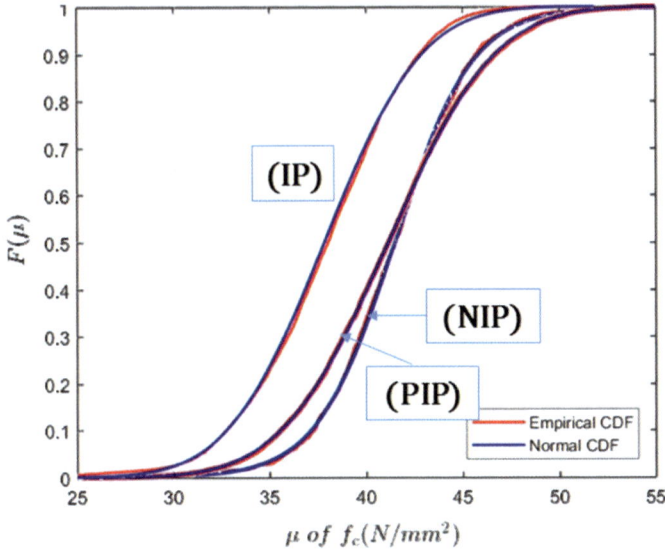

Fig. 7 CDF plots for μ of f_c obtain after data fusion from MCMC simulation

5 Results

Table 1 gives the parameter distribution fit and verified through K-S test for each prior case after fusion.

6 Conclusions

Following are the conclusion summarized below:

1. In the present study, a probabilistic framework is developed for fusing two NDT tools, namely the rebound hammer and ultrasonic pulse velocity. A more detailed safety assessment of the present condition of the structure can be done using Bayesian inference by quantifying the uncertainty using measurement error model (MEM).
2. Three techniques of data fusion (in the form of prior consideration) were formulated which is generalized and can be used in any circumstances. Formulation flow charts for these techniques are prepared. Also, it is seen that the models are behaving as per the engineering judgement. With more availability of the prior knowledge, better is the data fusion achieved.
3. As the prior was assumed to be M10, in table one we can see that the case 1 gave the least value of μ of f_c. This indicates that prior value should be carefully formulated for better accuracy as posterior is directly dependent on prior. Hence,

if the prior adopted is lower than the actual value, posterior strength evaluated will also be on a lower side.

4. Robust estimation using the principle of PDF averaging after data fusion helps us to obtain a single distribution of f_c to be independent to the model parameters which aids in inferring the results unconditional of the parameter variations.

5. This framework enables the engineer to keep on updating the results with increasing incoming data in the form of measurement or using a different technique.

7 Future Scope

1. Uncertainty in the material or structure itself can also be considered in the fusion process.

2. Only two NDT tools were fused together in this study which can be extended to additional tools. Semi-destructive and destructive test results can be fused with these models.

References

1. S.K. Verma, S.S. Bhadauria, S. Akhtar, In-situ condition monitoring of reinforced concrete structures. Front. Struct. Civ. Eng. **10**(4), 420–437 (2016)
2. Z. Sbartaï, D. Breysse, M. Larget, J. Balayssac, Combining NDT techniques for improved evaluation of concrete properties. Cem. Concr. Compos. **34**(6), 725–733 (2012)
3. D. Breysse, G. Klysz, X. Dérobert, C. Sirieix, J. Lataste, How to combine several non-destructive techniques for a better assessment of concrete structures. Cem. Concr. Res. **38**(6), 783–793 (2008)
4. D. Breysse, Nondestructive evaluation of concrete strength: An historical review and a new perspective by combining NDT methods. Constr. Build. Mater. **33**, 139–163 (2012)
5. S.A. Faroz, *Assessment and Prognosis of Corroding Reinforced Concrete Structures through Bayesian Inference*. Ph.D. Thesis, Indian Institute of Technology Bombay, Mumbai, India (2017)
6. E. Arioğlu, O. Koyluoglu, Discussion of prediction of concrete strength by destructive and nondestructive methods by Ramyar and Kol. Cement and Concrete World (1996)
7. BIS 13311, *Non-Destructive Testing of Concrete—Methods of Test : Part 1—Ultrasonic Pulse Velocity*. Bureau of Indian Standards, New Delhi, India (1992)
8. BIS 13331, *Non-Destructive Testing of Concrete—Methods of Test : Part 2—Rebound Hammer*. Bureau of Indian Standards, New Delhi, India (1992)
9. ASTM:C805/C805 M-13a, *Test for Rebound Number of Hardened Concrete*. ASTM International, West Conshohocken, PA (2013)
10. ASTM:C597-02, *Standard Test Method for Pulse Velocity Through Concrete*. ASTM International, West Conshohocken, PA (2003)
11. BS:1881-202, *Recommendations for Surface Hardness Testing by Rebound Hammer*. BSI Publicationa, London (1986)
12. BS:1881-203, *Recommendations for Measurement of Ultrasonic Pulse for Concrete*. BSI Publicationa, London (1986)
13. B. EN:13791, *Assessment of Insitu Compressive Strength in Structures and Precast Concrete Components* (2007)

14. H.Y. Qasrawi, Concrete strength by combined nondestructive methods simply and reliably predicted. Cem. Concr. Res. **30**(5), 739–746 (2000)
15. D. Breysse et al., Non destructive assessment of in situ concrete strength: comparison of approaches through an international benchmark. Mater. Struct. **50**(2), 133 (2017)
16. M. Alwash, D. Breysse, Z.M. Sbartaï, Using Monte-Carlo simulations to evaluate the efficiency of different strategies for nondestructive assessment of concrete strength. Mater. Struct. **50**(1), 90 (2017)
17. L.F. Ramos, T. Miranda, M. Mishra, F.M. Fernandes, E. Manning, A Bayesian approach for NDT data fusion: the Saint Torcato church case study. Eng. Struct. **84**, 120–129 (2015)
18. BIS, *IS 10262 Concrete Mix Proportioning—Guideline*. Bureau of Indian Standards, New Delhi, India (2009)
19. BIS, *IS 456 Plain and Reinforced Concrete—Code of Practice*. Bureau of Indian Standards, New Delhi, India (2000)
20. R. Domingo, S. Hirose, Correlation between concrete strength and combined nondestructive tests for concrete using high-early strength cement, in *The Sixth Regional Symposium on Infrastructure Development* (2009), pp. 12–13
21. A.H.-S. Ang, W.H. Tang, *Probability Concepts in Engineering Planning and Design* (1984)

Stochastic Free Vibration Analysis of Sandwich Plates: A Radial Basis Function Approach

R. R. Kumar, K. M. Pandey and S. Dey

Abstract In this paper, stochastic free vibration analysis of sandwich plate using a radial basis function (RBF) approach is carried out. Due to uncertain material anisotropy and inherent manufacturing inaccuracies, such types of structures are subject to variability. It is therefore required to consider the randomness in input parameters to assess the stochastic free vibration behaviour of sandwich plates which have significant computational challenges. The mathematical formulation is developed based on the C^0 stochastic finite element method (SFEM) coupled with higher-order zigzag theory (HOZT). Natural frequency analysis is conducted for stochasticity in ply orientation angle, face sheet thickness, core thickness, face sheet material properties, core material properties and skew angle. The cost-effective and computationally efficient RBF model is utilized as the surrogate to obtain the uncertain first five natural frequencies. In a single array, the global stiffness matrix is stored by using skyline techniques, while it is solved by subspace iteration.

Keywords Natural frequency · Radial basis function · Randomness · Sandwich plate

1 Introduction

Sandwich structure is widely used for small components such as thermostat, oil filter, automotive rim in addition to large structure such as wind turbine blade, spacecraft, aircraft, marine structure (either military or civilian) due to its high safety measures

R. R. Kumar (✉) · K. M. Pandey · S. Dey
Mechanical Engineering Department, National Institute
of Technology Silchar, Silchar, India
e-mail: ravinits2014@gmail.com

K. M. Pandey
e-mail: kmpandey2001@yahoo.com

S. Dey
e-mail: infodrsudip@gmail.com

© Springer Nature Singapore Pte Ltd. 2020
P. V. Varde et al. (eds.), *Reliability, Safety and Hazard Assessment
for Risk-Based Technologies*, Lecture Notes in Mechanical Engineering,
https://doi.org/10.1007/978-981-13-9008-1_36

449

.t lower price which is achieved due to its high specific strength, stiffness and weight sensitive nature. Every structure requires these properties (lightweight, high specific strength and stiffness) in order to fulfil specific application requirement as well as to save natural resources such as material and fuel, but the production cost is one of the parameters which restricts its use in general purpose structure. The sandwich plate consists of two face sheets and core in between them. An adhesive is used to attach these three components together to get the sandwich plate. The manufacturing of sandwich plate is challenging and costly due to its inherent complexity in honeycomb core, but with the evolution of technology, it is made cost sensitive by employing soft core sandwich plate, while complexity is still a measure issue. Laminated soft core sandwich plate is very difficult to manufacture accurately as per the design specification, resulting in undesirable uncertainties [1]. Randomness in geometric and material properties causes stochastic nature of mass and stiffness matrices. Source of uncertainties in the sandwich composite is produced from uncertainties in fabrication parameters and randomness in material properties. Therefore, it is necessary to consider the effect of stochasticity in free vibration analysis to quantify the uncertainties. Surrogate-based uncertainty quantification approach [2–13] is adopted by various researchers in case of problems where an efficient solution is not available. In general, the randomized output frequency of large number of input parameter is generated by using the Monte Carlo simulation (MCS) techniques. Although MCS can be used to quantify the uncertainties observed in material and geometric properties, it is found to be inefficient. High-dimensional input–output relationship can be mapped very efficiently on using radial basis function as a surrogate model. A uniformly selected random sample over the entire range ensures the good predictability of RBF for the whole design space. The aim of the present work is to develop an algorithm for stochastic free vibration analysis of simply supported soft core laminated sandwich plate using computationally efficient RBF approach. Flow diagram corresponding to the algorithm is illustrated in Fig. 1.

2 Theoretical Formulation

2.1 Governing Equation for Natural Frequency

For structural deformation, strain–displacement relation [14] can be expressed as

$$\{\bar{\varepsilon}(\varpi)\} = \left[\frac{\partial U(\varpi)}{\partial x} \frac{\partial V(\varpi)}{\partial y} \frac{\partial W(\varpi)}{\partial z} \frac{\partial U(\varpi)}{\partial x} + \frac{\partial V(\varpi)}{\partial y} \frac{\partial U(\varpi)}{\partial z} + \frac{\partial W(\varpi)}{\partial x} \frac{\partial V(\varpi)}{\partial z} + \frac{\partial W(\varpi)}{\partial x} \right]$$

(1)

i.e. $\{\bar{\varepsilon}(\varpi)\} = [A(\varpi)]\{\varepsilon(\varpi)\}$

where $[A]$ is the unit step function. In general, at any design point the displacement vector can be written as

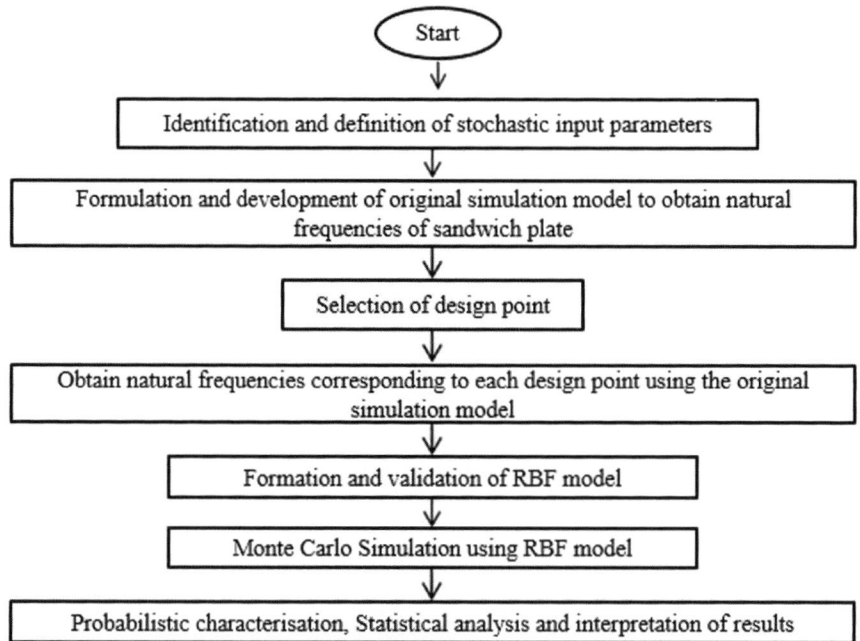

Fig. 1 Flow diagram for stochastic free vibration analysis using RBF model

$$\{S(\varpi)\} = \sum_{k=1}^{n} \zeta_i(\varpi) S_i(\varpi) \tag{2}$$

where $\{s\} = \{u_0 v_0 w_0 \theta_x \theta_y u_u v_u w_u u_l v_l w_l\}^T$ By using Eq. (1), strain vector can be expressed as

$$\{\varepsilon(\varpi)\} = [A(\varpi)]\{S(\varpi)\} \tag{3}$$

where $[A]$ is a strain–displacement matrix. In case of free vibration analysis, the dynamic equilibrium equation by using Hamilton's principle can be written as

$$[R(\varpi)]\{\overline{S}\} = \lambda^2 [M(\varpi)]\{\overline{S}\} \tag{4}$$

where $[R(\varpi)]$ is the stochastic free vibration frequency, and $[M(\varpi)]$ is a global mass matrix which can be expressed as

$$[M(\varpi)] = \sum_{k=1}^{n_u+n_l} \iiint \rho_k(\varpi)[N]^T[J]^T[N][J]\mathrm{d}x\mathrm{d}y\mathrm{d}z = \iint [N]^T[K(\varpi)][N]\mathrm{d}x\mathrm{d}y \tag{5}$$

where the random mass density of kth order is represented as $\rho_k(\varpi)$, and $[J]$ contain Z terms and other constants, which is of the order of 3×11 and shape function matrix is indicated by $[N]$. The stiffness matrix $[K(\varpi)]$ equation can be written as

$$[K(\varpi)] = \sum_{k_l}^{n_u+n_l} \rho_k(\varpi)[J]^T[J]dz \tag{6}$$

2.2 Radial Basis Function

The Euclidean distances of linear combination presented in surrogate-based model are represented as [15]

$$\hat{Y}(x) = \sum_{k=1}^{n} w_k \varphi_k(X, x_k) \tag{7}$$

Weight determined by using the least squares method is represented by w_k, number of sampling points by n, while kth basis function determined at the sampling point x_k is described by $\varphi_k(X, x_k)$.

RBF model is represented by using the radial function, which is expressed as,

$$F(x) = \frac{1}{\sqrt{1 + \frac{(x-c)^T(x-c)}{r^2}}} \quad \text{(For inverse multi-quadratic)} \tag{8}$$

$$F(x) = \exp\left(-\frac{(x-c)^T(x-c)}{r^2}\right) \quad \text{(For Gaussian)} \tag{9}$$

$$F(x) = \frac{1}{1 + \frac{(x-c)^T(x-c)}{r^2}} \quad \text{(For Cauchy)} \tag{10}$$

$$F(x) = \sqrt{1 + \frac{(x-c)^T(x-c)}{r^2}} \quad \text{(For multi-quadratic)} \tag{11}$$

Here, $r^2 = 1$ is assumed to be fixed for the application of Gaussian basis function. Since the function value from approximate function equals to that of true function, it exactly passes through all the sampling point. It acts in a similar way to that of the brain. It is having versatile problem-solving ability like pattern recognition, prediction, optimization, associative memory and control tool which is modelled as per our biological brain, which made it of keen interest to the researcher.

Table 1 Material properties

	Face sheet	Core
E_1 (GPa)	131	0.00689
E_2, E_3 (GPa)	10.34	0.00689
G_{12}, G_{13} (GPa)	6.895	0.00345
G_{23} (GPa)	6.205	0.00345
v_{12}, v_{13}, v_{21}, v_{31}	0.22	0
v_{23}, v_{32}	0.49	0
ρ kg/m^3	1627	97

3 Results and Discussion

In this paper, simply supported soft core skewed sandwich plate, having length $(l) =$ 10 cm, width $(b) = 10$ cm, thickness $(t) = 1$ cm having core to face sheet thickness ratio of ten, skew angle $(\varphi) = 30°$, four cross-ply (in both face sheet) is considered for free vibration analysis until otherwise mentioned. Degree of stochasticity (Δ) is taken as ±10 for ply orientation angle, ±10% for face sheet thickness, core thickness, face sheet material properties and core material properties, whereas ±5 for the skew angle. The deterministic finite element code used for the present analysis is validated by the author in the previous study [14]. RBF is used as a surrogate model which is verified by probability density function (PDF) and scatter plot. It is lucid clear from PDF plot that the result obtained from 256 samples size (N) by using RBF model is very close to results obtained from 10,000 sample size by direct MCS technique as depicted in Fig. 2a. Although RBF uses only 256 samples input data, it gives output equivalent to that of 10,000 samples in small time. For further confirmation of the RBF model as the surrogate to the original FE model, the scatter plot is presented in Fig. 2b. Least variation of a scatter plot from the diagonal line shows the predictability and applicability of the RBF surrogate model. Material properties [16] considered for the analysis are given in Table 1. Here, the surrogate based MCS approach is employed for probabilistic description of natural frequencies in order to achieve computational efficiency without affecting the accuracy [17–21]. Figure 3a indicates the effect of different degree of stochasticity on natural frequencies. Here, stochasticity is taken as ±5, ±10 and ±15 in skew angle, whereas it is taken as ±10%, ±15%, and ±20% in all other parameters. The sparsity of the PDF is observed to be increasing with an increase in the degree of stochasticity. A common range is observed for third and fourth natural frequency even in case of ±10% degree of stochasticity. Hence, it is not recommended to design the structure in that range to avoid resonance phenomenon. It is reported in Fig. 3b that there is an insignificant effect of ply orientation angle on overall response of the structure irrespective of their mode. Figure 4 presents the effect of plate thickness on natural frequencies of a sandwich plate. It shows that natural frequency increases with an increase in plate thickness and least increment are observed in case of fundamental natural frequency. Effect of plate length and width

(a)   **(b)**

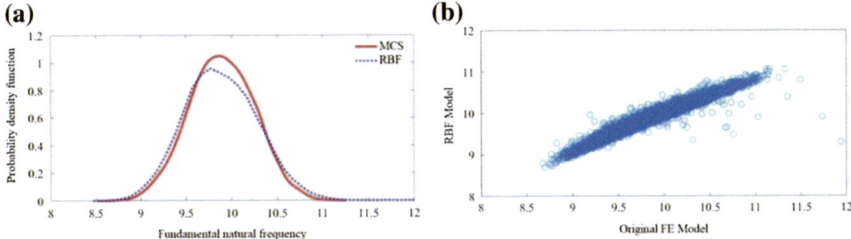

Fig. 2 a Probability density function (PDF) plot of fundamental natural frequency using MCS and RBF model, **b** scatter plot of original finite element (FE) model to that of RBF model for simply supported sandwich plate, considering variation in all input parameters

(a)   **(b)**

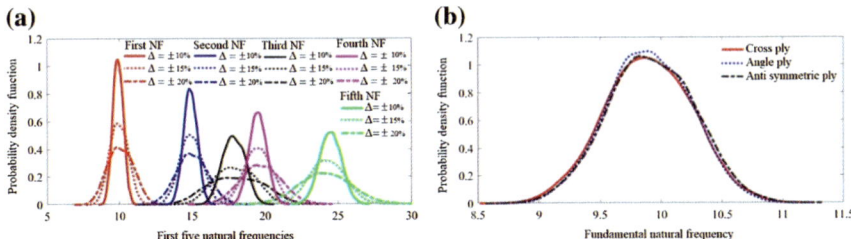

Fig. 3 a PDF plot of first five natural frequencies for ±10%, ±15% and ±20% degree of stochasticity, **b** probability density function (PDF) plot of fundamental natural frequency using cross-ply, angle ply and anti-symmetric ply laminate for simply supported sandwich plate, considering variation in all input parameters

on overall response of the structure is presented in Fig. 5. It is observed that natural frequency decreases with an increase in either length or width or both, whereas it is found to be the same by interchanging the length and width.

4 Conclusion

This paper presents the surrogate-based bottom-up stochasticity propagation approach for frequency analysis of the sandwich plate. The radial basis function is used as a surrogate model and observed to be highly efficient with respect to conventional Monte Carlo simulation techniques. The natural frequency analysis of sandwich plate by using the RBF approach considering uncertainties in input parameters are the novelty of the present study. In addition to free vibration analysis for a different degree of stochasticity, it is also analysed for different length–width ratio, different ply orientation angle, different plate thickness keeping core thickness to face sheet thickness ratio constant considering a combined variation of all input parameters. Stochastic first five natural frequencies as discussed in the previous section shows a significant deviation of result from their deterministic value insists design

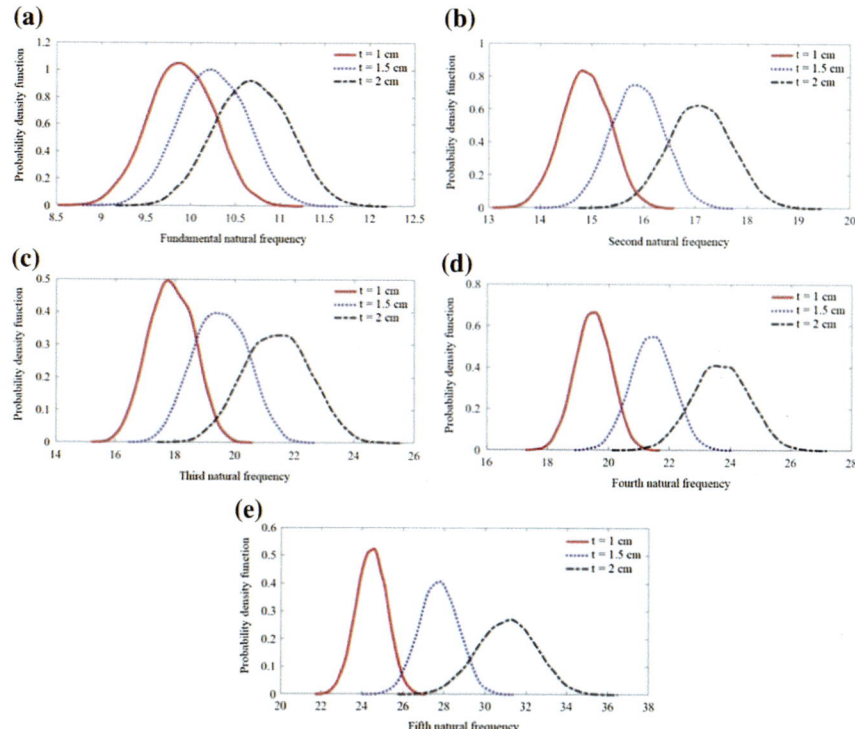

Fig. 4 **a** Fundamental natural frequency, **b** second natural frequency, **c** third natural frequency, **d** fourth natural frequency, and **e** fifth natural frequency for simply supported sandwich plate of thickness $(t) = 1$ cm, 1.5 cm and 2 cm, considering variation in all input parameters

engineers consider the randomness in input parameters resulting due to source uncertainties. Hence, for reliable and safe design it is required to consider stochasticity in the design and analysis of the sandwich structure, and use of sandwich structure will avoid hazardous environment impact by reducing material and fuel consumption. This analysis can be further extended for the analysis of more complex and real-life structure.

Acknowledgements The first author would like to acknowledge the financial support received by the Ministry of Human Resource and Development (MHRD), Government of India, during this research work.

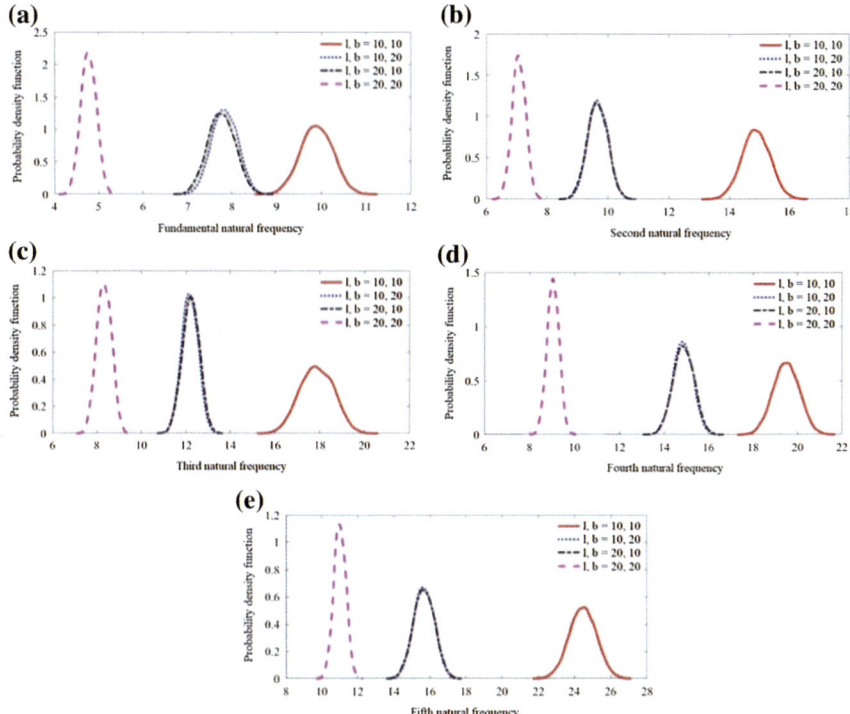

Fig. 5 a Fundamental natural frequency, **b** second natural frequency, **c** third natural frequency, **d** fourth natural frequency, and **e** fifth natural frequency for simply supported sandwich plate of length (*l*) width (*b*) ratio = 10,10; 10,20; 20,10; 20,20, considering variation in all input parameters

References

1. S. Dey, T. Mukhopadhyay, S. Adhikari, Stochastic free vibration analysis of angle-ply composite plates–A RS-HDMR approach. Compos. Struct. **1**(122), 526–536 (2015)
2. T. Mukhopadhyay, R. Chowdhury, A. Chakrabarti, Structural damage identification: A random sampling-high dimensional model representation approach. Adv. Struct. Eng. **19**(6), 908–927 (2016)
3. T.K. Dey, T. Mukhopadhyay, A. Chakrabarti, U.K. Sharma, Efficient lightweight design of FRP bridge deck. Proc. Inst. Civil Eng. Struct. Buildings **168**(10), 697–707 (2015)
4. A. Mahata, T. Mukhopadhyay, S. Adhikari, A polynomial chaos expansion based molecular dynamics study for probabilistic strength analysis of nano-twinned copper. Mater. Res. Express **3**, 036501 (2016)
5. T. Mukhopadhyay, A. Mahata, S. Dey, S. Adhikari, Probabilistic analysis and design of HCP nanowires: an efficient surrogate based molecular dynamics simulation approach. J. Mater. Sci. Technol. **32**(12), 1345–1351 (2016)
6. S. Metya, T. Mukhopadhyay, S. Adhikari, G. Bhattacharya, System reliability analysis of soil slopes with general slip surfaces using multivariate adaptive regression splines. Comput. Geotech. **87**, 212–228 (2017)
7. S. Dey, T. Mukhopadhyay, S.K. Sahu, S. Adhikari, Stochastic dynamic stability analysis of composite curved panels subjected to non-uniform partial edge loading. Eur. J. Mech./A Solids **67**, 108–122 (2018)
8. S. Dey, T. Mukhopadhyay, H.H. Khodaparast, S. Adhikari, A response surface modelling approach for resonance driven reliability based optimization of composite shells. Periodica Polytech., Civil Eng. **60**(1), 103–111 (2016)
9. P.K. Karsh, T. Mukhopadhyay, S. Dey, Spatial vulnerability analysis for the first ply failure strength of composite laminates including effect of delamination. Compos. Struct. **184**, 554–567 (2018)
10. S. Naskar, T. Mukhopadhyay, S. Sriramula, Probabilistic micromechanical spatial variability quantification in laminated composites. Compos. B Eng. **151**, 291–325 (2018)
11. K. Maharshi, T. Mukhopadhyay, B. Roy, L. Roy, S. Dey, Stochastic dynamic behaviour of hydrodynamic journal bearings including the effect of surface roughness. Int. J. Mech. Sci. **142–143**, 370–383 (2018)
12. S. Naskar, T. Mukhopadhyay, S. Sriramula, S. Adhikari, Stochastic natural frequency analysis of damaged thin-walled laminated composite beams with uncertainty in micromechanical properties. Compos. Struct. **160**, 312–334 (2017)
13. T. Mukhopadhyay, A multivariate adaptive regression splines based damage identification methodology for web core composite bridges including the effect of noise. J. Sandwich Struct. Mater. **20**(7), 885–903 (2018). https://doi.org/10.1177/1099636216682533
14. S. Dey, T. Mukhopadhyay, S. Naskar, T.K. Dey, H.D. Chalak, S. Adhikari, Probabilistic characterisation for dynamics and stability of laminated soft core sandwich plates. J. Sandwich Struct. Mater. 01–32 (2017)
15. S. Dey, T. Mukhopadhyay, S. Adhikari, Metamodel based high-fidelity stochastic analysis of composite laminates: a concise review with critical comparative assessment. Compos. Struct. **171**, 227–250 (2017)
16. H.D. Chalak, A. Chakrabarti, M.A. Iqbal, A.H. Sheikh, Free vibration analysis of laminated soft core sandwich plates. J. Vib. Acoust. **135**(1), 011013 (2013)
17. R.R. Kumar, T. Mukhopadhyay, K.M. Pandey, S. Dey, Stochastic buckling analysis of sandwich plates: the importance of higher order modes. Int. J. Mech. Sci. **152**, 630–643 (2019)
18. P.K. Karsh, T. Mukhopadhyay, S. Dey, Stochastic dynamic analysis of twisted functionally graded plates. Compos. Part B: Eng. **147**, 259–278 (2018)
19. T. Mukhopadhyay, S. Naskar, P.K. Karsh, S. Dey, Z. You, Effect of delamination on the stochastic natural frequencies of composite laminates. Compos. Part B: Eng. **154**, 242–256 (2018)

20. P.K. Karsh, T. Mukhopadhyay, S. Dey, Stochastic low-velocity impact on functionally graded plates: probabilistic and non-probabilistic uncertainty quantification. Compos. Part B: Eng. **159**, 461–480 (2019)
21. P.K. Karsh, T. Mukhopadhyay, S. Dey, Stochastic investigation of natural frequency for functionally graded plates. IOP Conf. Ser.: Mater. Sci. Eng. **326**, 012003 (2018)

Risk Analysis of Piping System of NRNF

Sachin Singh, M. Hari Prasad and P. K. Sharma

Abstract A preliminary risk analysis of piping system of non-reactor nuclear facility (NRNF) has been carried out for radiological hazards. In order to estimate the risk, one needs to estimate the frequency of occurrence of piping failure and consequences due to that failure. The piping system is subjected to both static and dynamic loads. Hence, in estimating the failure frequency of piping system, seismic hazard curves have been utilised besides its fragility curves. In the present case, these hazard curves have been taken from the available literature. In general, fragility curves are developed by using conventional factor of safety approach. However, in the present study, generation of fragility curves based on response surface methodology by using finite element modelling of piping system has been proposed. Based on the limit state functions developed, failure probability of a component has been obtained by using structural reliability method. The piping failure frequency has been estimated by convoluting both seismic hazard curves and the fragilities of the piping system. Finally, consequences have been estimated by using point kernel technique and thereby estimated the risk. It is found that the risk is in acceptable limits.

Keywords Finite element modelling · Fragility curves · Hazard curves · Limit state function · NRNF · Response surface methodology · Risk

1 Introduction

NRNF such as solvent extraction facilities is generally designed to extract heavy metals from the irradiated fuels. The irradiated fuel is processed using solvent extraction technique. The nuclear materials in NRNFs are present mostly in solution form and operate at ambient conditions. The treatment processes use large quantities of haz-

S. Singh (✉) · M. H. Prasad · P. K. Sharma
Bhabha Atomic Research Centre, Trombay, Mumbai, India
e-mail: ssingh@barctara.gov.in

S. Singh
Homi Bhabha National Institute, Trombay, Mumbai, India

© Springer Nature Singapore Pte Ltd. 2020
P. V. Varde et al. (eds.), *Reliability, Safety and Hazard Assessment for Risk-Based Technologies*, Lecture Notes in Mechanical Engineering,
https://doi.org/10.1007/978-981-13-9008-1_37

ardous chemicals, which can be toxic, corrosive and combustible. Hence, inadvertent release of hazardous chemical or radioactive substances needs to be prevented. The range of hazards can include radiation hazard, conventional chemical reactions, fire and explosion, etc., which can occur in different locations and in association with different operations. As these facilities involve large number of piping systems, failure of the piping system can lead to severe consequences in terms of release of radioactivity and effect on operating personnel. Hence, in the present study, risk analysis of piping system has been considered. The present analysis consists of following steps:

1. Selection of piping system.
2. Finite element modelling of piping system.
3. Generation of response surface based on finite element analysis.
4. Definition of limit state function.
5. Fragility analysis of piping system by using structural reliability methods.
6. Consequence estimation of piping failure and thereby estimating risk of the piping system.

In order to estimate the risk, one needs to estimate the frequency of occurrence of piping failure and consequences due to that failure. The piping system is subjected to both static and dynamic loads. Hence, one needs to consider seismic hazard curves while estimating the failure frequency of piping system besides its fragility curves. In the present case, these hazard curves have been taken from the available literature. In general, fragility curves are being developed by using conventional factor of safety approach [1]. However, in generating the required data, lot of assumptions is being made, and in some cases, generic data are being used for generating the fragility curves [2]. Hence, in this project, generation of fragility curves based on response surface methodology by using finite element modelling of piping system has been proposed.

2 Modelling of a Piping System

There are two kinds of methods of stress analysis for nuclear piping: simplified calculation method and numerical simulation method. Simplified calculation method based on the empirical formula, which is suitable for small piping, could not accurately describe the mechanical behaviour of the piping system. Considering the high requirements for security, large piping systems with complex mechanical properties must resort to numerical simulation method with the aid of computer program. With the development and application of computer technology, stress analysis of larger piping system can be achieved by finite element analysis.

In the present analysis, the piping system considered is class 2 nuclear piping which is composed of three lines: line 1, line 2 and line 3 as shown in Fig. 1. The material of the piping is shown in Tables 1 and 2. Software simulation of piping system arrangement is shown in Fig. 2. This piping system has 13 pipe supports, one

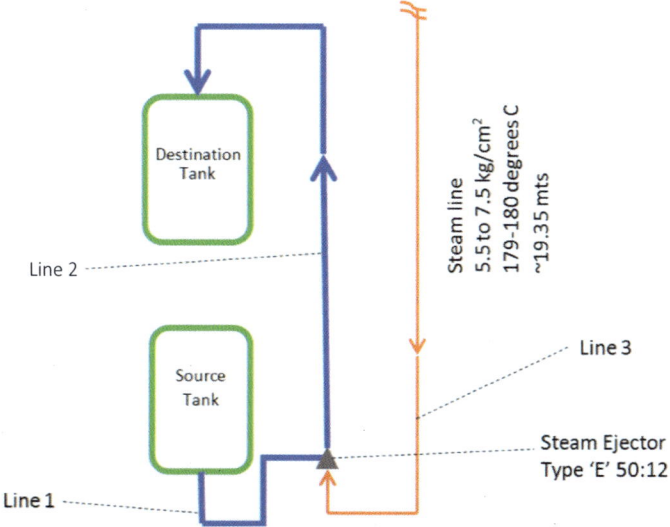

Fig. 1 General layout of the piping system

Table 1 Operational and geometric parameters of the piping system

Line number	Operating parameters			Geometrical parameters		
	Pressure (bar)	Temperature (°C)	Specific gravity	Outer diameter (mm)	Wall thickness (mm)	Piping class
Line 1	0.26–0.39	40–45	1.17	48.26	3.683	Class 2
Line 2	1.2	55–60	1.17	48.26	3.683	Class 2
Line 3	7	175–180	0.003599	33.401	3.3782	Class 2

steam ejector, bends and three anchor points in the piping system connecting lines 1, 2 and 3.

Once the piping system is modelled, static and dynamic analysis has been carried out. All operational inputs are utilised to find out the stress at nodal points for modelled piping system. The stresses obtained in the piping system are shown in Table 3.

3 Response Surface Methodology

In this section, the methodology to develop the response surface has been discussed. In most cases, the exact relationship between a response and a set of input variables that influence the response is implicit and the response has to be computed by running a "black box" complex computer code. In many circumstances, however,

Table 2 Other operational and material parameters in the piping system

S. No.	Parameters	Values
1	Reference temperature	28 °C
2	Density of MOC (SS304L)	8027 kg/m3
3	Young's modulus (E)	193.95 GPa
4	Temperature coefficient of expansion	15.4×10^{-6} mm/mm/°C
5	Ejector type	E (50:12) 50 indicates flow in lpm and 12 indicates head in metres
6	Mill tolerance	12.5%
7	Coefficient of friction	0.5

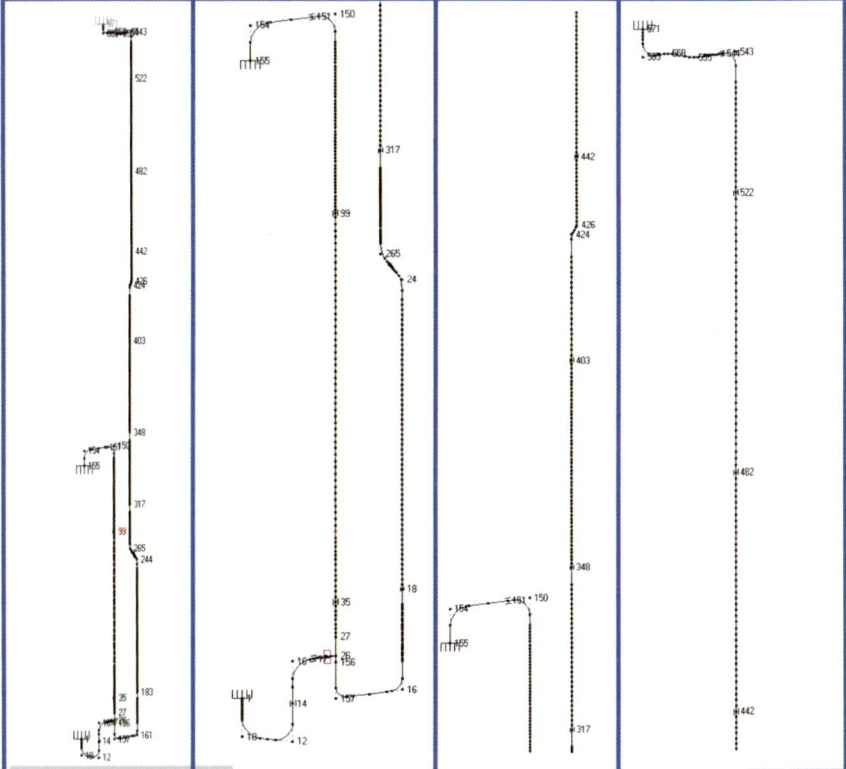

Fig. 2 FE modelling of the piping system

Table 3 Maximum nodal stress at different lines

Maximum nodal stress (in MPa)	Line 1	Line 2	Line 3
Sustained	44.62(169.5)	30.35(162.8)	27.79(127)
Expansion	37.96(172.2)	21.78(171.1)	112.5(165.1)
Sustained + expansion	64.33(285.2)	20.93(279.6)	131.8(249.8)
Occasional (smoothened spectrum)	104.8(254.2)	105.1(244.2)	101.5(190.5)
Occasional (raw spectrum)	92.95	91.33	81.07

computational expense of running computer analysis codes may become prohibitive when a large number of models are involved.

One of the most widely used models is the RSM [3]. The origin of the RSM can be traced back to the works done by several researchers in the early 1930s or even earlier. The experimental design systematically defines an efficient set of experimental sampling points at which the responses must be computed or observed. Design-of-experiment (DOE) techniques provide the needed basis for this critical step in the methodology. There are many types of experimental design that can be used for this purpose, but the most common ones are a full factorial design (FFD) and a central composite design (CCD). The response function considering second-order polynomial model is as follows:

$$y = \beta_0 + \sum_{i=1}^{k} \beta_i x_i + \sum_{i=1}^{k} \beta_{ii} x_i^2 + \sum_{i=1}^{k-1} \sum_{j>i}^{k} \beta_{ij} x_i x_j + \varepsilon \tag{1}$$

where
y = Response or dependent variable
x_i, x_j = The coded input or independent variables
$\beta_0, \beta_i, \beta_{ii}, \beta_{ij}$ = Unknown coefficients to be estimated
ε = Bias or lack of fit error term
k = Number of input variables

As discussed above, the response function depends on several input variables; in the present case, it depends on the pressure, diameter, thickness, dead weight, damping ratio and peak ground acceleration (PGA). Here, response of the system is defined in terms of maximum principal stress observed. Before conducting the design of experiments, all the above-mentioned variables have been studied from sensitivity as well as uncertainty point of view in order to find out the key parameters that will have more impact on the response of the system. Key parameters come out as pressure and damping ratio. Based on the design of experiments, finite element analysis has been carried out for several combinations of input variables, and regression analysis has been carried out to generate the response surfaces [4]. The response surfaces so obtained are given for three different lines in the piping system as follows:

$$RS_1 = 3.2758P + 0.237\xi^2 - 4.5234\xi + 103.11 + R1$$

$$RS_2 = 3.2758P + 0.3309\xi^2 - 6.0187\xi + 104.57 + R2$$
$$RS_3 = 2.4718P + 0.1174\xi^2 - 3.9549\xi + 90.49 + R3 \tag{2}$$

where
ξ Damping ratio
$P =$ Internal pressure
$R =$ Aleatory uncertainty in the input ground motion which has mean value as 0 and standard deviation as β
R is calculated by generating multiple artificial time histories for corresponding PGA and damping ratio values.

4   Fragility Analysis

The seismic fragility of a structure is mathematically defined as the probability of failure of the structure conditional on specific ground motion intensity. Let I_j be a specific value of the hazard intensity at the jth level. Let D be the load effect due to this random event on the global response of the structure. Let C_i be the structural capacity to withstand this load effect corresponding to the ith limit state. Accordingly, the seismic fragility is calculated as:

$$\mathrm{PF}_{ij} = P\big[D \geq C_i | I_j\big] \tag{3}$$

where PF_{ij} denotes the probability of failure with respect to the ith limit state at the jth hazard intensity level. The fragility curve for a limit state i can be constructed by evaluating PF_{ij} at different levels of hazard intensity (j) as shown in Fig. 3.

As it is explained for generating the fragility curves, one needs to establish the limit state function. In the present analysis, the following limit state function has been considered:

$$\mathrm{LS} = \text{Allowable Stress } (S_a) - \text{Applied Stress} \tag{4}$$

Here, the applied stress is the summation of pressure stress, dead weight stress and dynamic stress which can be obtained from the response surface. The piping system which is considered in the analysis consists of three lines that are subjected to different temperature, pressure, dead weight and dynamic stresses. As the allowable stresses depend on the temperature, all the three lines will have three different allowable stresses. By combining the response surfaces generated for calculating the applied stresses, the limit state equations can be written as:

$$LS_1 = S_{a1} - \big(3.2758P + 0.237\xi^2 - 4.5234\xi + 103.11 + R1\big)$$
$$LS_2 = S_{a2} - \big(3.2758P + 0.3309\xi^2 - 6.0187\xi + 104.57 + R2\big)$$

Table 4 Random variables and their distribution parameters

Parameters	Distribution	Mean value			Standard deviation		
		Line 1	Line 2	Line 3	Line 1	Line 2	Line 3
Sa (MPa)	Normal	254.4	244.2	190.5	12.72	12.21	9.52
P (MPa)	Normal	0.039	0.12	0.7	0.00042	0.001667	0.001667
$\xi(\%)$	Normal	4	4	4	0.4	0.4	0.4
R	Normal	0	0	0	5.59	8.17	3.55

$$LS_3 = S_{a3} - \left(2.4718P + 0.1174\xi^2 - 3.9549\xi + 90.49 + R3\right) \qquad (5)$$

Similar types of equations have been developed for various PGA values. Based on the limit state functions developed, failure probability of a component can be obtained from Eq. (3). Here, the failure of the component takes place if the applied stress is more than the allowable stress. The random variables and their corresponding distribution parameters that are considered in the present analysis are shown in Table 4.

In order to find out the failure probability from the limit state function, one can utilise structural reliability methods such as simulation methods or numerical methods. In the present case, numerical methods such as first-order reliability method (FORM) have been used. Finally, the failure probability of the system has been estimated by combining the individual failure probabilities of each line as the failure of the system takes place if any one of the lines fails. The failure probability of the system can be written as:

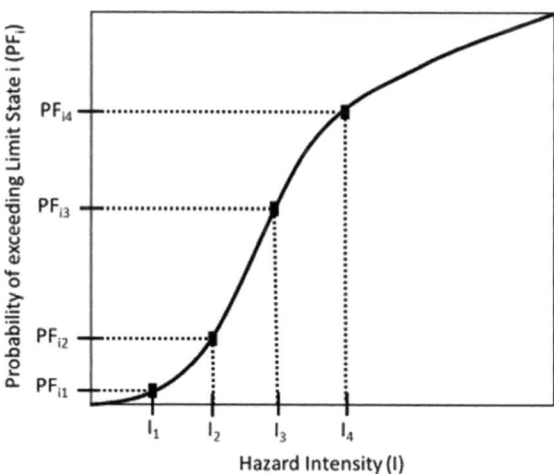

Fig. 3 Graphical representation of fragility curve for limit state i

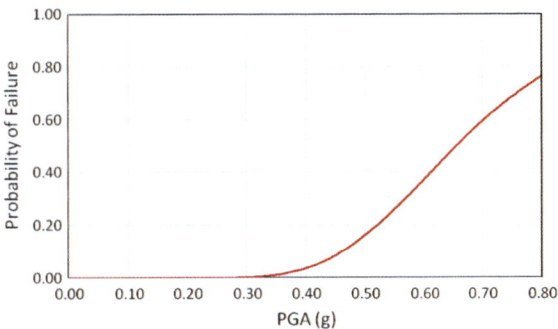

Fig. 4 Fragility curve of the piping system under consideration

$$PF_{System} = PF_1 + PF_2 + PF_3 \tag{6}$$

The failure probability so estimated is for a particular value of PGA. In order to find out the failure probabilities at different PGA values, the similar exercise has been carried out such as the development of limit state functions by using various response surfaces generated for various PGA values and calculated the probability of failure by using FORM. The failure probabilities so obtained are plotted against PGA values to generate the fragility curve which is shown in Fig. 4.

5 Risk Estimation

As per the definition of the risk, it is defined as likelihood of occurrence of an undesirable event and the consequences due to that event. In the present analysis, undesirable event is considered as failure of piping system and the consequences considered as release of radioactivity into the working environment. It is clear from the definition of the risk that there are two terms in it: one is the frequency of occurrence and other one is the consequences. In order to estimate the risk, one needs to estimate both frequency of occurrence of failure of piping system and consequences of piping failure. As the piping system is subjected to both static and dynamic loads, one needs to consider frequency of occurrence of earthquakes while estimating the failure frequency of piping system. This can be mathematically written as:

Failure Frequency = Earthquake Occurence Frequency × Failure Probability|Earthquake

One can obtain the earthquake occurrence frequency from seismic hazard curves. Hazard curve is a graph drawn between seismic ground motion parameter such as PGA and frequency of exceedance of particular level of PGA. In the present case, these hazard curves have been taken from the available literature for a typical site under consideration as shown in Fig. 5.

The second term in the frequency estimation is the estimation of failure probability which can be obtained from the fragility curves. Now, the piping failure frequency can be calculated by convoluting both seismic hazard curves and the fragilities of the piping system as given in Eq. (7) and is also illustrated in Fig. 6. Based on the present analysis, the failure frequency of piping system is obtained as 2.6294E−06/yr.

$$P_F = \int_0^\infty \left(\frac{dH}{da}\right) p_f da \tag{7}$$

where
H = Hazard curve
A = PGA level
p_f = Conditional failure probability at a given PGA level
P_F = Total failure frequency.

In this analysis, consequence due to failure of piping system has been estimated. In general, plant areas have been divided into red, amber, green and white zones. The radiological condition of the plant is continuously monitored on centralised radiation protection console (CRPC) located in main control room (MCR) and HP room. Adequate shielding provision has been made to ensure that radiation level is maintained ≤ 1 μSv/hr (0.1 mR/hr) in all continuously occupied areas (Fig. 7).

Radiation level outside shield can be calculated using proven gamma shielding codes which use point kernel technique to calculate the dose at desired point. In case of spillage, it has been found that source is spread over larger floor area due to which

Fig. 5 Hazard curve for a typical site

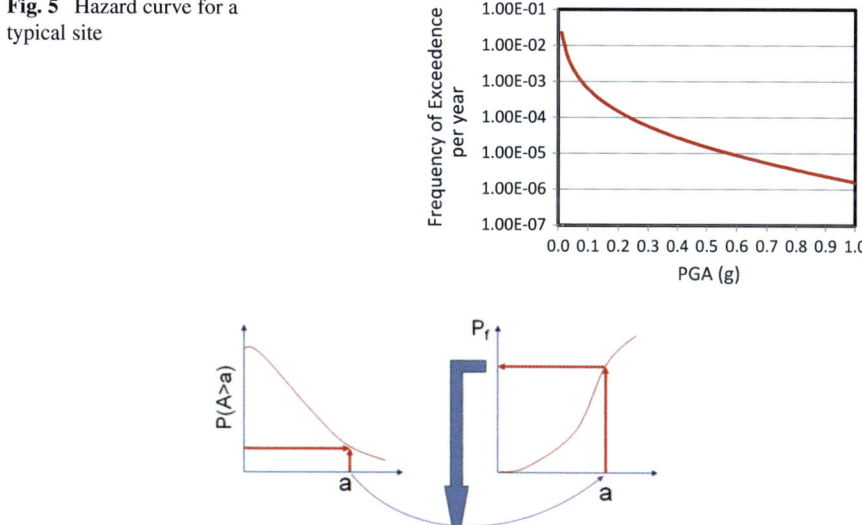

Failure Frequency = Hazard X Fragility

Fig. 6 Convolution of seismic hazard and seismic fragility curves

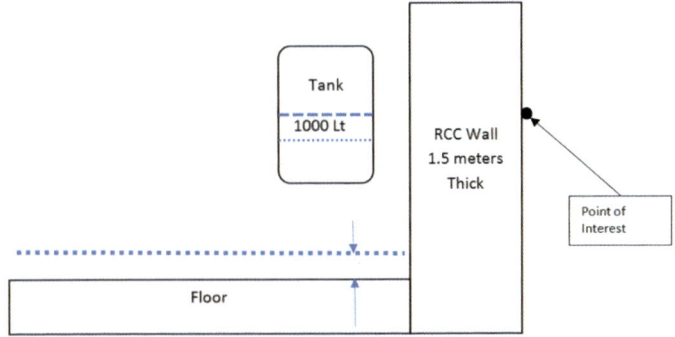

Fig. 7 Schematic for area representing radiation facility

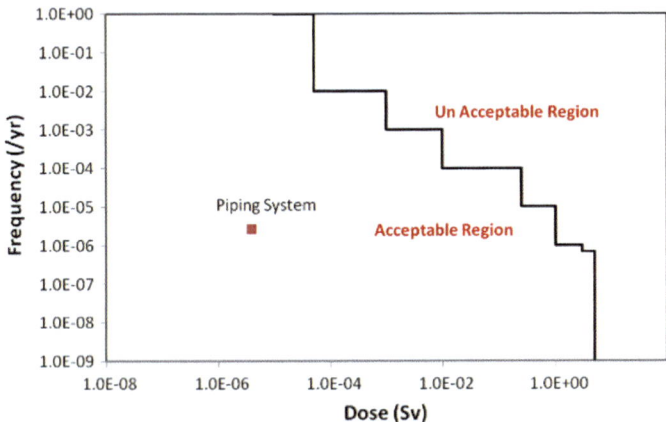

Fig. 8 F-C curve showing the failure frequency and dose for current piping system

the contribution of dose inside cell increases to a greater extent. But contribution of dose outside cell wall becomes lesser than the 1 μSv/hr. As per designer experience, a typical value of 0.5 μSv/hr is considered as the dose value outside the cell wall during the spillage. For personnel exposure, the time for exposure has been taken conservatively considered as one shift, which is around 8 h and the dose is estimated as 4 μSv. From this analysis, risk is estimated as 1.05E−05 μSv/yr. The results obtained are plotted on F-C curve which is as per NUREG-1860 [5] as shown in Fig. 8. It can be observed that risk is within acceptable region for the present study.

6 Conclusion

A preliminary risk analysis of piping system of NRNF has been carried out for radiological hazards. The piping system is subjected to both static and dynamic loads.

Hence, in estimating the failure frequency of piping system, seismic hazard curves have been utilised besides its fragility curves. In the present case, these hazard curves have been taken from the available literature. Fragility curves have been developed by response surface methodology by using finite element modelling of piping system. Based on the limit state functions developed, failure probability of a component has been obtained by using FORM. The piping failure frequency has been estimated by convoluting both seismic hazard curves and the fragilities of the piping system. Finally, consequences have been estimated by using point kernel technique. The risk due to the failure of piping system has been estimated as 1.05E−05 μSv/yr and is found to be in acceptable limits.

Acknowledgements The authors would like to thank Shri. K.V Ravi, Chief Executive, Shri. Sanjay Pradhan, Deputy Chief Executive, Shri. K.K.Singh, Project Director, and Shri. B.K Sinha for giving opportunity to execute this project work.

References

1. I. Zentner, A. Nadjarian N. Humbert, E. Viallet, Numerical calculation of fragility curves for probabilistic seismic risk assessment, in *The 14thWorld Conference on Earthquake Engineering* (2008)
2. IAEA TECDOC-1267, Procedures for conducting PSA of NRNF
3. G.E.P. Box, K.B. Wilson, On the experimental attainment of optimum conditions. J. Roy. Stat. Soc. B **13**(1), 1–45 (1951)
4. T.W. Simpson, J.D. Peplinski, P.N. Koch, J.K. Allen, Metamodels for computer-based engineering design: survey and recommendations. Eng. Comput. **17**(2), 129–150 (2001)
5. NUREG-1860, *Feasibility Study for a Risk-Informed and Performance-Based Regulatory Structure for Future Plant Licensing*. U.S. Nuclear Regulatory Commission, Dec 2007

Stress Analysis and Qualification of Non-standard Gasket Joint of Heavy Water Heat Exchanger

Nitin Joshi, Mayank Agarwal, R. C. Sharma, U. D. Patil, Umesh Kumar, K. Rama Varma and O. P. Ullas

Abstract Dhruva is a 100 MWth Research Reactor having heavy water as primary coolant and demineralized water as secondary coolant. There are three shell and tube type vertical heat exchangers in the primary coolant system. The heat exchanger is having single-pass construction on both sides, with heavy water on the tube side and demineralized light water on the shell side. Due to identifying a minor leak from tube side to shell side, one of the heat exchangers had to be taken out for repairs, which would eventually require the replacement of cover joint gaskets. The dished end covers of the heat exchanger are bolted to the shell with a double tongue and groove joint having one inner ring gasket and an outer ring gasket to provide the sealing. The gaskets used in these grooves are spiral wound metallic gasket with graphite as the filler medium. Because of the inherent design of the joint, there is a limitation in tightening torque of the flanged bolts. The inner and outer gaskets are compressed by the 60 numbers of flange bolts. The allowable torque on the bolts is limited due to the configuration. Excess tightening will result in flange rotation and induce undue stresses as well as leakage from the inner gasket. Hence to optimize the bolt loading, a stress analysis of the joint was carried out. To validate the results, a mock set-up simulating the site condition and configuration of the joint was fabricated. New gaskets were procured and tested in the mock set-up for various bolt load conditions and the results were found to be in agreement with the results of analysis. Subsequently, the heat exchanger was removed and after repair and leak rectification, the dished end covers were assembled with new gaskets as per the test results. The heat exchanger is performing satisfactorily without any leaks.

Keywords Bolt torque · Heat exchanger · Qualification · Spiral wound gasket · Stress analysis

N. Joshi · M. Agarwal (✉) · R. C. Sharma · U. D. Patil · U. Kumar · K. R. Varma · O. P. Ullas
Bhabha Atomic Research Centre, Mumbai, India
e-mail: aamayank@gmail.com

© Springer Nature Singapore Pte Ltd. 2020
P. V. Varde et al. (eds.), *Reliability, Safety and Hazard Assessment for Risk-Based Technologies*, Lecture Notes in Mechanical Engineering,
https://doi.org/10.1007/978-981-13-9008-1_38

1 Introduction

Dhruva is a research reactor with a rated power of 100 MWth. The reactor is vertical tank type and uses metallic natural uranium as a fuel with heavy water as coolant, moderator and reflector. The main coolant heavy water is further cooled by secondary coolant which is demineralized light water. For this, there are three vertical shell and tube type heat exchangers installed in separate shielded rooms.

On one occasion of routine sampling of secondary coolant water, traces of tritium were observed. The same was confirmed through repeated sampling for a few days, as the level of tritium was very minimal. This indicated a leak of the primary coolant to the secondary coolant system. After detailed checking by isolating suspected locations one by one, it was concluded that the leak was in the main coolant heat exchanger#3. Heavy water is in the tube side, single-pass top-to-bottom flow and secondary water in the shell side, single-pass bottom-to-top flow. The heat exchanger has two dished end covers—one at the top and other at the bottom. The material of construction of shell is SA 515 Gr-70 and the dish end covers, tubes, and tube sheets are SS-304L (SA-240-TP-304L). The repair of the heat exchanger is called for its dismantling.

2 Construction of Heat Exchanger

The main body or shell is cylindrical, with top and bottom flanged tube sheet weld joined to the shell. The top and bottom dished end covers are also flanged. The sealing is provided by double tongue and groove (two concentric ones) in both top and bottom cover joints, the tongue being on the shell side flanges and groove in the cover flanges.

Also, a stop collar is provided at the tongue location to limit the compression of the gasket. Provision to detect any leak in the inter-gasket space is also provided. Sealing gaskets for both the grooves are spiral wound types, SS316 with Graphite filler, with no inner and outer rings. Since the assembly was done by OEM and no maintenance has so far been carried out, it was required to analyse the gasket joints to understand the behaviour of the joint and establish parameters for effective sealing, before an attempt was made for the repair of the heat exchanger (Fig. 1) [1].

3 Details of the Joint

In a standard spiral wound gasket, there is an inner or outer ring or both. These rings assist in locating the gasket on the flange faces and prevents over-compression of the spiral. According to OEM of spiral wound gaskets, over-compression of the gaskets will lead to gasket collapse. Hence, the rings serve as a stopper also. However, the

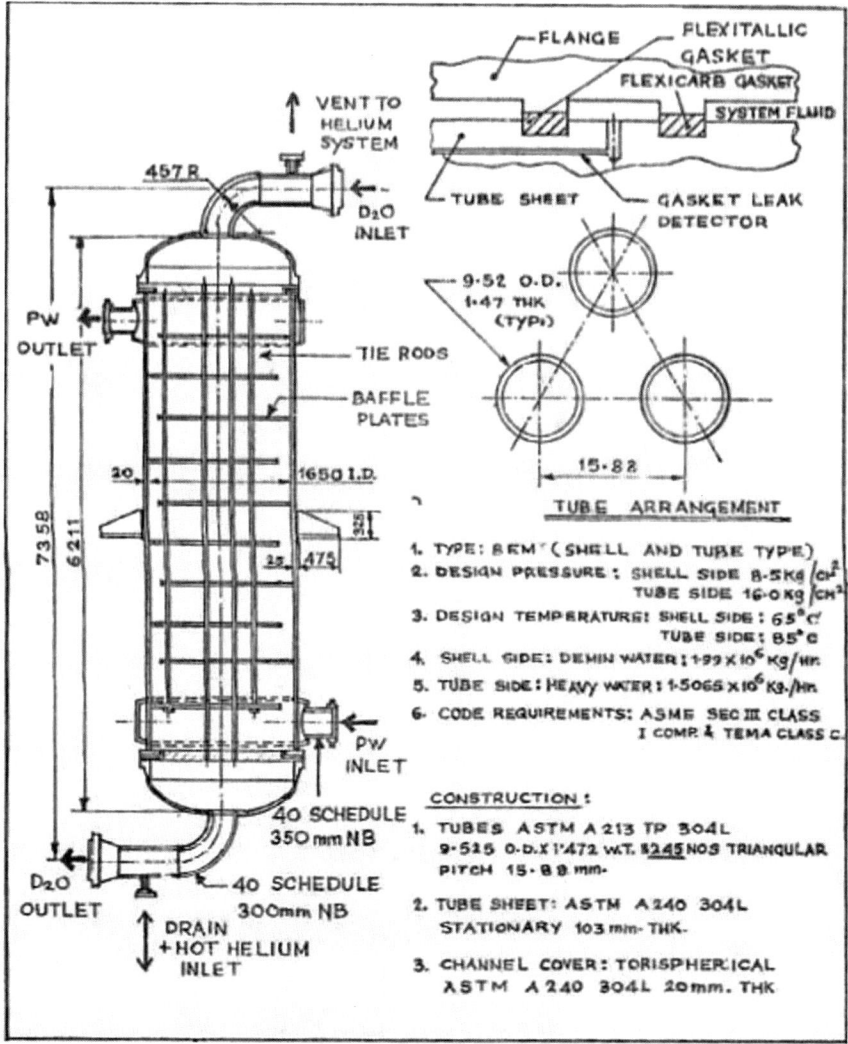

Fig. 1 Schematic of heat exchanger

gasket used in the heat exchanger does not contain inner and outer rings and the compression of the sealing element is controlled by the stopper collar provided in the outer tongue location. The detailed dimensional drawing of the flange is given in Fig. 2. The thickness of each gasket is 7.2 mm. The width of each gasket is 15 mm. The fasteners are M24, ASTM F568 Class 10.9-60 no.

The primary side experiences a maximum pressure of 10 kg/cm^2 with a design pressure of 16 kg/cm^2. So, the gasket has to seal against this pressure. A detailed

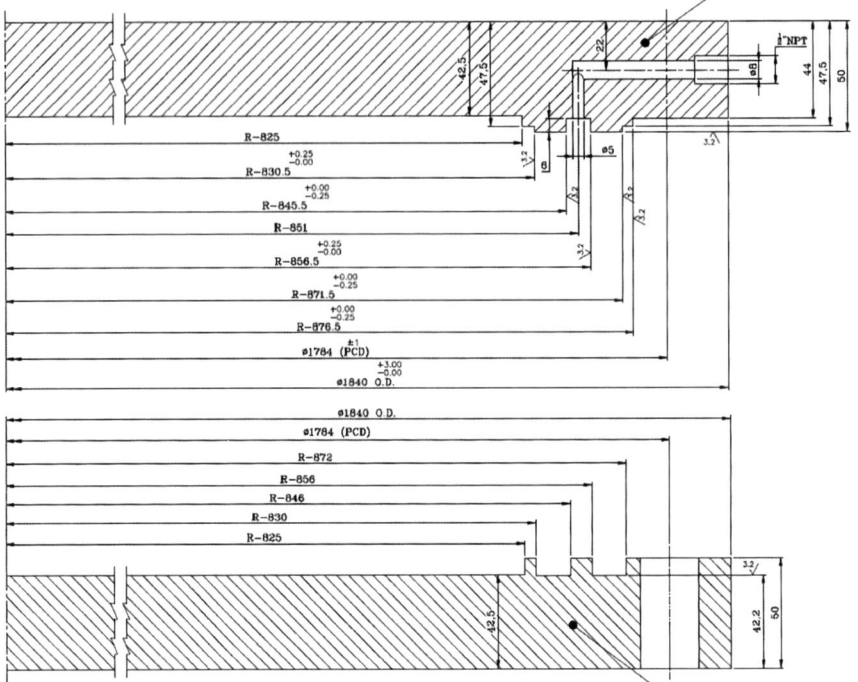

Fig. 2 Joint details with dimensions

FEM analysis was carried out to understand the joint behaviour at various amount of bolt torque.

4 Analysis of the Joint

The analysis was required to establish the variation of gasket seating pressure, gasket closure and the flange closure with change in bolt-tightening torque and subsequent application of internal pressure. For the analysis, the yield strength of CS (SA 515 Gr-70) was taken as 250 MPa and Poisson ratio as 0.3, whereas the yield strength of SA-240-304L was taken as 170 MPa and Poisson ratio 0.29 [2]. The material properties of the gasket are shown in Fig. 3 for the old gasket as well as the new gasket. The analysis was to be vetted by a mock set-up, simulating the parameters of the original heat exchanger joint. Accordingly, a mock set-up was designed and given for fabrication. The mock-up facility was designed so as to simulate the actual heat exchanger, for qualifying the new gaskets of graphite filled spiral wound type, of the same dimensions as will be used in actual heat exchanger.

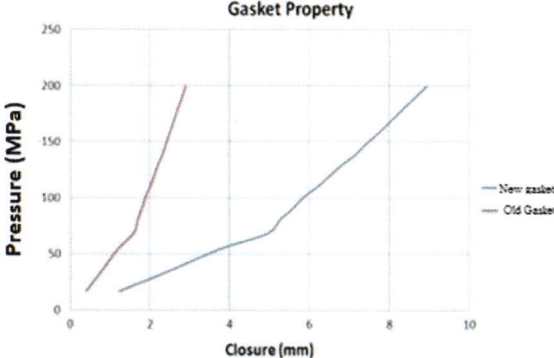

Fig. 3 Gasket properties of old and new gasket

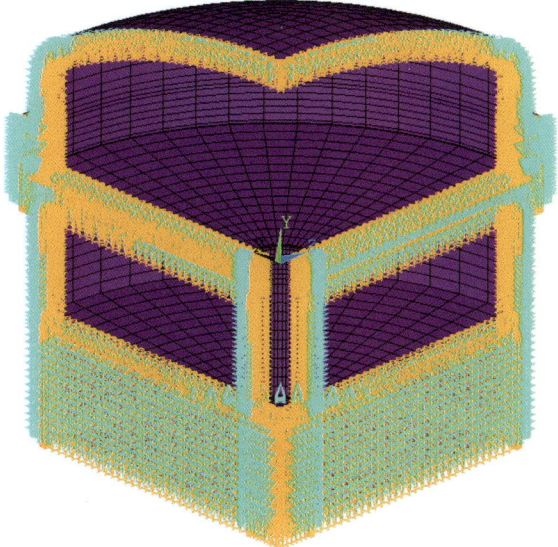

Fig. 4 Mock-up facility FE model

A quarter model of the mock-up facility was made for the analysis. Solid, shell gasket elements were used (Fig. 4).

The lower end of the shell was fixed in the vertical direction, and symmetric boundary conditions were applied in the circumferential direction. The internal pressure was applied on the projected areas, and bolt load was applied as pressure load on the PCD of the bolt. The bolt torque was converted into axial load using the following equation [3].

$$T = KFd \qquad (1)$$

T is the bolt torque, F is the axial bolt load, d is the size of the bolt, K is the coefficient depending on material in contact, thread angle and the lubricant used between nut and mating surfaces.

Here, $K = 0.2$ (assuming no use of lubricant and metal to metal contact is there).

The bolt torque was varied from 34.5 to 51 kgm. The internal pressure was applied to the exposed area of shell and gasket. The maximum internal pressure given was 16 kg/cm². The perforated plate (tube sheet) was modelled as a solid plate with modified Poisson's ratio and elasticity properties as per ASME Section III Appendices A-8000 Figure A-8131 [4]. The mock-up facility was analysed at different internal pressures, i.e. 0, 12 kg/cm² (pump shut off pressure) and 16 kg/cm² (hydro test pressure) and varied bolt torque. It was observed that while keeping the bolt torque constant, with an increase in internal pressure, the flange moment increases due to which the gasket pressure decreases significantly.

The effect of operating temperature can result in bolt elongation, creep relaxation of gasket material or thermal degradation resulting in the reduction of flange load. However, the effect of temperature has not been taken in this work. The flanges must be sufficiently rigid to prevent the unloading of the gasket due to the flange rotation when the internal pressure is introduced. The properties of the old and the new gaskets were different and hence the gasket seating stress and closure were different than the original one (Figs. 5, 6 and Table 1).

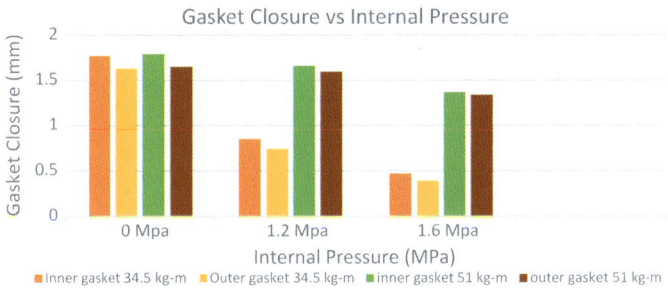

Fig. 5 Gasket closure with change in bolt torque and internal pressure

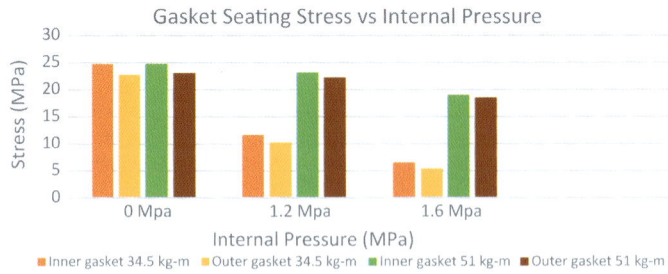

Fig. 6 Gasket seating stress with change in torque and internal pressure

Table 1 Gasket stress and closure for different torque and load values

Torque	Load	Gasket stress (MPa)						Gasket closure (mm)							Flange edge closure
kgm	Mpa	Inside of- inner gasket	Outside of- inner gasket	Avg inner gasket	Inside of- outer gasket	Outside of- outer gasket	Avg outer gasket	Inside of- inner gasket	Outside of- inner gasket	Avg inner gasket	Inside of- outer gasket	Outside of- outer gasket	Avg outer gasket	
51.00	0	24.17	25.35	24.76	22.75	23.35	23.05	1.77	1.81	1.79	1.63	1.67	1.65	2.09
51.00	1.2	22.56	23.74	23.15	21.61	22.84	22.23	1.62	1.70	1.66	1.55	1.64	1.60	2.17
51.00	1.6	18.29	19.72	19.01	17.79	19.23	18.51	1.32	1.42	1.37	1.29	1.39	1.34	1.98
34.50	0	24.45	25.05	24.75	22.42	23.04	22.73	1.75	1.79	1.77	1.61	1.65	1.63	2.05
34.50	1.2	11.23	12.07	11.65	9.82	10.67	10.25	0.82	0.88	0.85	0.71	0.77	0.74	1.21
34.50	1.6	5.97	6.95	6.46	4.85	5.88	5.37	0.43	0.51	0.47	0.35	0.43	0.39	0.92

From the analysis, it was observed that for internal pressure of 16 kg/cm^2 and 51 kgm torque sufficient gasket seating stress is obtained. Taking an internal pressure of 1.2 MPa (upper limit), the gasket factor in each case is greater than 3, which is the gasket factor as per OEM of the gasket, thus assuring the leak tightness of the joint. The change in the gap on the application of internal pressure in different bolt torque conditions is due to prestressing of the bolts, since the load in the bolt increases once the compression stopper meets the upper flange. On analysing different cases, it was concluded that the above-proposed mock-up model presented satisfactory results. Hence, this mock-up facility can be used to replicate the actual conditions in heat exchanger. The bolts in the mock-up facility are to be tightened by 51 kgm bolt torque condition to achieve a satisfactory result with sufficient margin.

5 Validation in Mock Set-up

Based on the above analysis, the mock-up facility was fabricated to test and validate the joint. The mock-up facility was designed to simulate the actual heat exchanger for qualifying the new gaskets. In the existing heat exchanger, the shell and head thickness is 20 mm, tube-sheet thickness is 103 mm. There are 60 numbers of M24 bolts at 1784 mm pitch circle diameter (PCD) used to join the flanges. The head of the mock-up facility is designed with the same geometry as in the actual heat exchanger without the inlet nozzle The head is fabricated by forming a 20-mm-thick plate and 100-mm-thick flange welded with it to form existing geometry. In the mock-up, the tube sheet was replaced with a solid plate with a 200-NB pipe welded at the middle to stiffen solid plate and 60-mm-thick plate. The bottom end of the set-up is closed by a 60-mm-thick plate. This plate was stiffened by eight stiffeners on outer side and a 200-NB pipe was connected to tube sheet on inner side (Fig. 7).

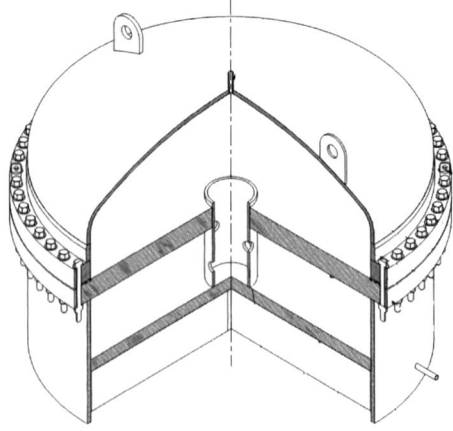

Fig. 7 Mock-up facility for qualification of gasket

Fig. 8 Hydrotest set-up for metallic gasket

Table 2 Flange gap for different torque values at different bolt locations

Bolt no.	Torque values (kgm)						Final gasket compression
	0	10	20	30	40	51	
1–2	5.41	4.84	4.54	4.07	3.57	3.4	2.01
16–17	5.42	4.77	4.46	4.03	3.53	3.38	2.04
31–32	5.42	4.82	4.51	4.14	3.58	3.48	1.94
46–47	5.42	4.85	4.51	4.09	3.52	3.43	1.99

The bolt torque was given up to 51 kgm in steps. The bolt tightening was done in a pre-defined sequence as per the HX manufacturer (Fig. 8).

Based on the geometry of the joint, the maximum deflection of gasket that can be achieved is 1.9 mm. From Table 2, it is observed that after the final tightening, the maximum compression is nearly 1.9 mm. This indicates the gasket has been fully compressed and stopper collar has touched the face. The set-up was hydro-tested at 12 kg/cm^2 and checked for any leakage from the inter-gasket positions. No leakage was observed from any of the inter-gasket leak collection point, indicating that joint is holding. Further torquing was not attempted since it can result in flange rotation and thereby reducing the gasket stress.

6 Conclusion

The heat exchanger was being taken under maintenance, for the first time, after commissioning of the reactor. Even after testing and qualification, during service, very minor inter-gasket leakage was continuing. Hence, it was required to analyze the joint before attempting the removal of heat exchanger so as to carry out maintenance in minimum time possible and commission back, to reduce the reactor downtime. Through the stress analysis and subsequent successful mock set-up testing of the joint, the procedure of actual heat exchanger assembly could be established. The testing was not carried out at the test pressure deliberately as it can overcompress the gasket and there was a chance that its resilience may be affected due to flange rotation.

The heat exchanger was subsequently repaired, assembled, and commissioned and is functioning satisfactorily without any inter-gasket leakage.

Acknowledgements The authors are highly thankful to Shri. C. G. Karhadkar, (Head ROD), Shri. P. Sumanth (Head RRMD) and Shri Sujay Bhattacharya (Director, RPG) for their guidance throughout.

References

1. Heavy water design manual of Dhruva
2. S. Kumar et al., Internal report on analysis of the heavy water heat exchanger
3. Gasket manual of M/s Flexitallic
4. ASME Section III Appendices A-8000

Reliability-Based Design Optimization of a Large-Scale Truss Structure Using Polynomial Chaos Expansion Metamodel

Subhrajit Dutta and Chandrasekhar Putcha

Abstract Optimal design of engineering systems considering uncertainties is broadly dealt in the area of reliability-based design optimization (RBDO). These RBDO problems involve a costly inner loop reliability estimate involving a computation-intensive true model solver. The current work is aimed at reducing this cost of computation in RBDO. To this end, a polynomial chaos expansion (PCE) metamodel is used. This PCE metamodel is later used to substitute the actual expensive true model in the reliability computation phase. A stochastic optimizer—particle swarm optimization (PSO) is used for the outer optimization loop. The effectiveness of PCE metamodel combined with PSO is demonstrated for a large-scale truss structural system in reducing the overall computation cost involved in RBDO.

Keywords Reliability-based design optimization · Metamodel · Polynomial chaos expansion · Structural optimization

1 Introduction

Structural optimization is a challenging research domain both for academicians and practicing engineers. Design optimization of real structural systems, like bridges, towers, buildings, etc., is complex in nature, primarily due to the heavy computation effort required to reach a desired accuracy in solution. During the analysis and design stage of a structure, the input variables must be adjusted properly to ensure that the structure performs optimally in the presence of external forces. However, predicting optimal input (design) variables becomes difficult due to the uncertainty associated

S. Dutta (✉)
Department of Civil Engineering,
National Institute of Technology Silchar, Silchar, India
e-mail: subhrajit.nits@gmail.com

C. Putcha
Department of Civil and Environmental Engineering,
California State University Fullerton, Fullerton, CA, USA
e-mail: cputcha@fullerton.edu

© Springer Nature Singapore Pte Ltd. 2020
P. V. Varde et al. (eds.), *Reliability, Safety and Hazard Assessment for Risk-Based Technologies*, Lecture Notes in Mechanical Engineering,
https://doi.org/10.1007/978-981-13-9008-1_39

481

with the environmental loads that act on a structure, the material properties, geometry and boundary conditions. In particular, the inherent uncertainty in environmental loads must be considered during the design phase of a structure in order to avoid instabilities. Therefore, the overall aim is to search for optimum values of design variables considering uncertainty. Such an optimal solution can be dealt with in the broad area of reliability-based design optimization (RBDO) [1, 7, 14].

The solution scheme of an RBDO problem has two major parts: the optimization part and the probability of failure computation or reliability analysis part [10]. For an RBDO problem, the reliability computation may appear either in objective function(s) or in constraint(s). Selection of both optimization algorithm and reliability analysis method determines the solution accuracy and the computation time required. For an RBDO problem, the reliability-based constraints involve a significant amount of computation. The primary objective of an efficient reliability analysis is to compute the desired solution, while significantly reducing the computation cost. Probability of failure computation using explicit limit-state functions are lighter in simulation. However, due to the nature of most real engineering problems, the reality demands the use of high-fidelity solvers to evaluate limit-state functions. Hence, the reliability computation becomes intractable with most of the available computational resources, in general. The computation even shoots up when an optimization at the outer loop needs to be executed in RBDO. Moreover, depending on the problem at hand, a more rigorous simulation may be needed to ensure that a global optimum is reached. The present work is aimed at reducing this computation cost of reliability analysis, which in turn reduces the overall cost of RBDO, particularly when a high-fidelity solver is involved during the computation of various limit states. To this end, a *metamodel*, also termed as a *surrogate model*, is used. Metamodels are computationally inexpensive mathematical models that approximate the actual physical (high-fidelity) model. A metamodel is constructed using the model output/responses at support points and following a learning algorithm to obtain its parameters. In general, a metamodel is created to approximate the responses over a range of input (design) variables. As the dimensionality of these variables increases, improvement of the metamodel is required. Thus, there is a trade-off between the cost of the metamodel and the desired accuracy.

From the aforementioned discussion, the contributions of this work can be pointed out as:

- development of a stochastic metamodel for RBDO in order to obtain optimal solutions within manageable computation time, without compromising the accuracy level.
- applying this computationally lighter metamodel-based approach to the RBDO of a large-scale truss structure.

2 Reliability-Based Design Optimization: A Polynomial Chaos-Based Approach

Reliability-based design optimization (RBDO) provides a robust optimal solution to practical design problems. In general, a typical RBDO problem is solved using a double-loop approach [1, 14], wherein the inner loop reliability analysis using numerical solver is used to compute the mathematical functions in optimization. In the present work, to reduce this computation time, the computation-intensive solver involved in reliability computation is substituted by a PCE-based metamodel.

2.1 Polynomial Chaos Expansion

A square-integrable random variable, random vector, or random process can be expressed as a mean square convergent series using random orthonormal polynomial bases, known as polynomial chaos expansion (PCE) for Hermite bases, and generalized polynomial chaos expansion (gPCE) for other bases [8]. The Hermite bases are used for normal/Gaussian random variables. However, if a random variable is non-normal, then isoprobabilistic transformation is required, if Hermite polynomials are to be used as bases. Suppose a vector of random variables, $\boldsymbol{\xi}$ with the support, $\mathscr{D}_{\boldsymbol{\xi}}$, be described by the joint probability density function (PDF), $f_{\boldsymbol{\xi}}$. Considering a finite variance of the physical model $Y = \mathscr{M}(\boldsymbol{\xi})$, such that

$$\mathbb{E}[Y^2] = \int_{\mathscr{D}_{\boldsymbol{\xi}}} \mathscr{M}^2(\boldsymbol{\xi}) \, f_{\boldsymbol{\xi}} \, d\boldsymbol{\xi} \tag{1}$$

the polynomial chaos expansion of $\mathscr{M}(\boldsymbol{\xi})$ becomes [13]

$$Y = \mathscr{M}(\boldsymbol{\xi}) = \sum_{\alpha} y_{\alpha} \, \Psi_{\alpha}(\boldsymbol{\xi}) \tag{2}$$

where Ψ_{α} are multivariate orthonormal polynomials and $\boldsymbol{\alpha}$ are the multi-indices that maps the multivariate Ψ_{α} to their corresponding bases, where the base coordinates are denoted by the PCE coefficients y_{α}. For real applications, the expansion of Y must be truncated by retaining significant number of PCE terms. The full N-order PCE extending to M random variables is defined as the set of all multidimensional Hermite polynomials (Ψ), whose degree does not exceed N [8]

$$Y(\xi_1, \xi_2 \ldots \xi_M) \approx \sum_{\alpha=0}^{P-1} y_{\alpha} \, \Psi_{\alpha}(\boldsymbol{\xi}) \tag{3}$$

Here, P is the number of terms in the truncated PCE given as

$$P = \sum_{k=0}^{N} C_{M+k+1}^{k} = \frac{(M+N)!}{M!N!} \tag{4}$$

The primary computation in the chaos expansion lies in the determination of its significant coefficients. These can be computed by an intrusive Galerkin method, by non-intrusive spectral projections, such as Gauss quadrature and Smolyak's coarse tensorization, or by collocation [5, 13]. Recent studies on sparse PCE—a computationally lighter expansion of the "full" PCE, where only the low-ranked polynomials are included—proposed a least-angle regression (LARS) method for determining the PCE coefficients [5]. Determination of these coefficients using LARS is computationally efficient for RBDO problems, as demonstrated in Dutta et al. [6]. In the present study, LARS is used for computing PCE coefficients.

2.2 Development of PCE-Based Metamodel for RBDO

In this work, a double-loop RBDO approach is adopted. For reliability computation, PCE-based metamodeling is used, which substitutes the multiple evaluations of a costly numerical solver. This saves the computation cost significantly. In general, for RBDO problems the physical model responses depend on both deterministic and random variables. However, the responses formulated using PCE, as described above, consider only the random input variables. Hence, to make the design variables dependent on random responses Y in a PCE framework, they must be defined in a random space. To define the design variables in a random space, the working principle of a stochastic optimization technique is utilized in this case. For the present work, the outer loop corresponds to the particle swarm optimization (PSO) technique. PSO characterizes the design variables as particles in a random field [9]. However, it must be noted that even though the optimum search process in PSO begins with random initialization (in random space) of to-be-optimized design variables, the final outcome of optimization is deterministic optimum values only. The use of random design variables in building PCE models has already been performed in robust design optimization problems [12].

Next, to build the proposed PCE-based metamodel for RBDO, the following steps are adopted:

1. Generate n samples of design variables: $\{x_1, \ldots, x_n\}$ following a probabilistic distribution in the random search domain. The design variables are assumed to be characterized as PSO particles in design space.
2. Once the (random) samples of design variable are generated, same number (n) of samples for an intrinsic random variable are generated following its probabilistic distribution: $\{\xi_1, \ldots, \xi_n\}$.

3. Evaluate the "true" model responses with the costly numerical solver at support points using a design of experiment scheme. Note that the true model is evaluated only once for the input random samples.
4. Build the N-order PCE metamodel from the true model responses using a non-intrusive LARS scheme as outlined in the previous Sect. 2.1.

This PCE-based metamodel construction can be decoupled from the outer optimization loop. Once the metamodel is built, the "true" model solver involved in reliability computation is replaced by it for the further optimization process.

3 RBDO of a Steel Tower

A three-dimensional truss structure (see Fig. 1) is considered for testing the efficiency of the proposed RBDO methodology following the work of Papadrakakis et al. [11]. The height of the tower is 16 m, while its base is an equilateral triangle of side 6.93 m as shown in Fig. 1. A vertical load, $V = 2$ kN, is applied to all the nodes, while an uncertain horizontal load, H, with mean value 8 kN is applied to the top nodes in the global X direction.

The optimization problem solved here involves minimization of the tower weight (W), with the design variables being the dimensions of the truss members. Circular hollow sections (CHS) are adopted to design this tower following the standard design practice. For each design variable, the dimension variables are the outer diameter, d_0, and the thickness, t, of the circular hollow section. Three types of constraints have been considered—maximum allowable stress to avoid yielding, compressive force for buckling, and displacement constraints. The allowable values for these

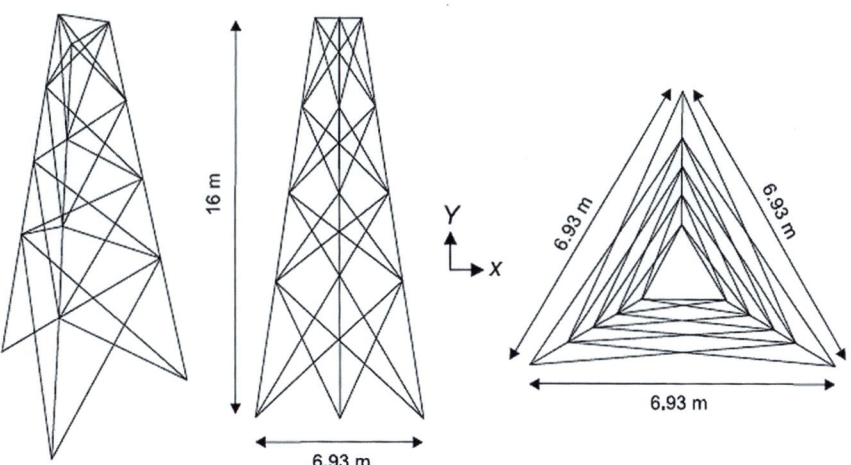

Fig. 1 A 3D truss structure with its elevation and plan

Table 1 Probabilistic characterization of input/design random variables

Random variables (X)	Probability distribution	Mean (μ_X)	CoV[a] (V_X) (%)
Modulus of elasticity (E)	Normal	2×10^8 kN/m²	7
Yield stress (f_y)	Normal	3.5×10^5 kN/m²	10
Horizontal load (H)	Normal	8 kN	37.5
CHS outer diameter (d_0)	Normal	150 mm	10
CHS thickness (t)	Normal	5 mm	5

[a]Coefficient of variation

constraint parameters are obtained from Indian Standard (IS) code of practice [4]. A finite element (FE) analysis is carried out using stiffness method to obtain the model responses [3]. Fixed boundary conditions are applied at the base of the tower. The probabilistic characterization of input/design variables is given in Table 1, considering the design variables (d_0 and t) to be random in order to do robust design optimization. The stress and buckling constraints are taken to be probabilistic in nature, while the displacement constraint is a deterministic one. The present RBDO problem is formulated as

$$\begin{aligned} \underset{\mu_{d_0},\, \mu_t}{\text{minimize}} \quad & F = \mu_W \\ \text{subject to:} \quad & \mathbb{P}\left(p_{1_{\max}} > p_{\text{alw}}\right) < 10^{-4} \quad \text{probabilistic constraint} \\ & \mathbb{P}\left(P_{c_{\max}} > P_e\right) < 10^{-4} \quad \text{probabilistic constraint} \\ & d_{\max} < d_{\text{alw}} \qquad\qquad\quad \text{deterministic constraint} \end{aligned} \tag{5}$$

where μ_W is the mean weight of the tower, which is a function of mean values of the design variables, μ_{d_0} and μ_t. $p_{1_{\max}}$ is the maximum principal axial stress considering all the members and for all loading cases, p_{alw} is the allowable axial stress. $P_{c_{\max}}$ is the maximum axial compressive force for all loading cases, P_e is the critical Euler buckling force in compression. The maximum value of nodal displacement, $d_{\max}$ is constrained with an allowable deterministic value, $d_{\text{alw}} = 200$ mm. It must be noted that the optimization parameters ($p_{1_{\max}}$, $P_{c_{\max}}$ and $d_{\max}$) are obtained using computation-intensive FE analysis. A target failure probability of 10^{-4}, that is typical for civil structures, is chosen here [7].

For the present case, the objective function for the RBDO problem is the mean weight of the tower. The PCE metamodel contains input/design variables as listed in Table 1 as the random variables. For construction of the PCE metamodel, 1200 Latin hypercube samples of input/design variables are used. A fifth-order PCE ($N = 5$) is constructed for the response parameters. The chaos coefficients y_α are generated using the LARS method as mentioned earlier. The probability density function (PDF) of W estimated using the fifth-order PCE is plotted in Fig. 2a and is compared with the "true" FE model PDF estimated using 10^4 crude Monte Carlo simulation (MCS). Since the target probability of failure for optimization problem was set as 10^{-4}, a total

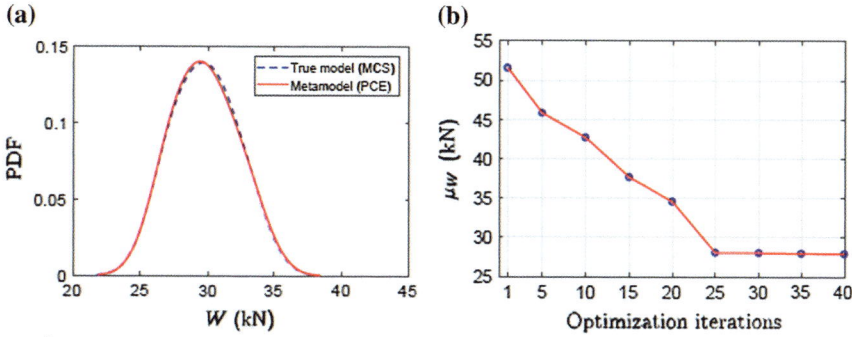

Fig. 2 **a** Comparison of response PDFs; **b** Convergence of objective function

Table 2 Comparison of optimization results with normalized CPU time

	PCE metamodel	MCS
No. of analyses	1200	10,000
CPU time	1	129
Minimized μ_W (kN)	27.88	27.83

of 10^4 realizations of random variable are used for MCS [2]. The comparison shows a good match, which indicates that the PCE can provide a better approximation of the "true" FE model for the chosen problem. To verify the goodness of fit, leave-one-out error (ϵ_{LOO}) for W is obtained as 2.635%.

The RBDO problem is next solved using the formulation given in Eq. 5 in order to verify the accuracy of the proposed method. Based on the PCE-based metamodel, the optimum CHS dimension values for the RBDO are: $d_0 = 165$ mm and $t = 4.5$ mm (the objective function for these optimum values is, $\mu_W = 27.88$). The convergence history of μ_W is shown in Fig. 2b. The PCE metamodel-based results are close to those obtained using MCS scheme. However, it must be noted that the normalized CPU time required by MCS is almost 129 times more than that by PCE. This time saving makes a high-dimensional optimization under uncertainty problem tractable. These results are reported in Table 2.

4   Concluding Remarks

In this work, a non-intrusive procedure has been proposed to construct a PCE-based metamodel to solve RBDO problems. The RBDO using PCE-based metamodel is demonstrated for the optimization of a large-scale truss structure. The optimization results with PCE-based RBDO have been validated with MCS-based RBDO for the same structure. Furthermore, the results are obtained with a significant reduction in computation cost. From this study, it can be concluded that the proposed PCE-based metamodeling technique can be successfully applied for the robust optimal design of large structural systems, and also it is expected to perform better for other design under uncertainty problems, in general.

References

1. Y. Aoues, A. Chateauneuf, Benchmark study of numerical methods for reliabilitybased design optimization. Struct. Multidisc. Optim. **41**(2), 277–294 (2009)
2. S.-K. Au, J.L. Beck, Estimation of small failure probabilities in high dimensions by subset simulation. Probab. Eng. Mech. **16**(4), 263–277 (2001)
3. K.J. Bathe, *Finite Element Procedures* (PHI Learning Private Limited, India, 2010)
4. BIS, Indian standard for general construction in steel: code of practice. IS:800 (Bureau of Indian Standards, 2007)
5. G. Blatman, B. Sudret, Adaptive sparse polynomial chaos expansion based on least angle regression. J. Comput. Phys. **230**(6), 2345–2367 (2011)
6. S. Dutta, S. Ghosh, M.M. Inamdar, Optimisation of tensile membrane structures under uncertain wind loads using PCE and kriging based metamodels. Struct. Multidisc. Optim. **57**(3), 1149–1161 (2018)
7. S. Dutta, S. Ghosh, M.M. Inamdar, Reliability-based design optimisation of framesupported tensile membrane structures. ASCE-ASME J. Risk Uncertainty Eng. Syst. Part A Civ. Eng. **3**(2), G4016001 (2017)
8. R. Ghanem, P.D. Spanos, *Stochastic Finite Elements: A Spectral Approach* (Springer, Berlin, 1991)
9. J. Kennedy, R.C. Eberhart, *Swarm Intelligence* (Morgan Kaufmann Publishers Inc., San Francisco, 2001)
10. A.S. Nowak, K.R. Collins, *Reliability of Structures*, 2nd edn. (CRC Press, Boca Raton, 2013)
11. M. Papadrakakis, N.D. Lagaros, V. Plevris, Design optimization of steel structures considering uncertainties. Eng. Struct. **27**(9), 1408–1418 (2005)
12. S. Poles, A. Lovison, A polynomial chaos approach to robust multiobjective optimization, in *Hybrid and Robust Approaches to Multiobjective Optimization—Dagstuhl Seminar Proceedings* (2009)
13. C. Soize, R. Ghanem, Physical systems with random uncertainties: chaos representations with arbitrary probability measure. SIAM J. Sci. Comput. **26**(2), 395–410 (2005)
14. M.A. Valdebenito, G.I. Schuëller, A survey on approaches for reliability-based optimization. Struct. Multidisc. Optim. **42**(5), 645–663 (2010)

Influence of System Compliance on Contact Stresses in Fretting

Pankaj Dhaka and Raghu V. Prakash

Abstract System compliance is one of the critical parameters which need to be accounted for precise measurement of loads and displacements from a fretting test rig, irrespective of whether it is fretting fatigue or fretting wear. The system compliance in a fretting setup consists of compliance due to driving element, driven element, and other intermediate subassemblies which are part of the load train. The issue becomes more pertinent in fretting applications which involve loads (of the order of a few Newtons) together with displacement amplitudes (of a few microns), where the stiffness of the system can affect the contact variables significantly. In general, the experimental setup is calibrated at the beginning and requisite corrections are made to account for the system rigidity, but there could be a gradual stiffness degradation over the time which could influence the accuracy of the results. In the present study, two-dimensional finite element analysis has been carried out for a representative fretting test setup which consists of a flat specimen in contact with a cylindrical pad and subjected to normal load and tangential displacement. The loading elements have been represented through the elastic springs whose stiffness can be varied in both tangential and normal direction, to understand the implications of system compliance on both normal and shear tractions. The results of the finite element model are validated with the analytical solutions proposed by Mindlin (J Appl Mech 16:259–268, 1949 [1]).

Keywords Contact stresses · Fretting fatigue · Fretting wear · System compliance

1 Introduction

Fretting is a contact phenomenon which is generally observed when two bodies, in contact, are subjected to small amplitude of vibration or cyclic loading. The fretting behavior of contact has been studied using both numerical and experimental meth-

P. Dhaka · R. V. Prakash (✉)
Indian Institute of Technology Madras, Chennai, India
e-mail: raghuprakash@iitm.ac.in

© Springer Nature Singapore Pte Ltd. 2020
P. V. Varde et al. (eds.), *Reliability, Safety and Hazard Assessment for Risk-Based Technologies*, Lecture Notes in Mechanical Engineering,
https://doi.org/10.1007/978-981-13-9008-1_40

489

ods, with each method having its own advantages and limitations. System rigidity is one of the crucial parameters while analyzing fretting behavior using experimental methods. The influence of system compliance becomes even more prominent for the applications which involve load or displacements of a few newtons or microns, respectively. A typical fretting test setup involves various elements like fixtures to hold the specimen, load train, etc. All these elements together constitute the compliance of fretting setup. In general, the effect of system compliance is accounted for by calibrating the fretting setup at the beginning of the experimental investigation and obtaining the rigidity coefficients [2–4].

Many researchers have carried out finite element studies to integrate the effect of system compliance by modeling the specimen and fretting pad along with supporting fixtures [5, 6]. The objective of these studies has been primarily to incorporate the influence of rigidity of the supporting test-rig elements (as observed in the experiments) on the contact behavior rather than on understanding the effect of system compliance exclusively. Specifically, only the influence of normal compliance on fretting fatigue life has been studied by modeling an extended compliant section along with pad geometry. Madge [7] studied the flat with rounded edge-on-flat configuration and found a sigmoidal relationship between modulus of the compliant section and fretting fatigue life, with negligible influence at lower (about 1/10th of elastic modulus of fretting pad) and higher (approximately equal to elastic modulus of fretting pad) values of modulus of compliant section. Later, Mutoh et al. [8] also made a similar observation for bridge type pad and found a decrease in fretting fatigue strength with increase in rigidity of contact pad; though in their work, the contact pad rigidity was relatively lower and varied approximately between 1/20th and 1/80th of elastic modulus of the fretting pad.

In the present work, finite element analysis has been carried out for cylinder-on-plate configuration to understand the influence of both normal and tangential stiffness for loading conditions prevailing in the case of fretting wear.

2 Methodology

A two-dimensional finite element analysis was carried out for a cylinder-on-plate configuration. The geometric dimensions of the model were taken from Tsai and Mall [6]. The fretting pad and specimen were modeled with first-order, 2D plane strain elements with reduced integration (CPE4R). First-order elements are generally preferred over second-order elements which could lead to oscillations in pressure distribution for frictional contact problems [9]. Sufficiently refined mesh was used in the contact zone with an element size of approximately $10 \times 30 \, \mu m$, while regions away from the contact zone were modeled with coarser mesh to minimize computational effort. Both fretting pad and specimen were modeled with elastic–plastic properties of Ti-6Al-4V [10] which is one of the commonly used Ti alloys in fretting applications. The Ramberg–Osgood deformation plasticity model available in ABAQUS® 6.12 was used to define elastic–plastic material behavior and does not

Table 1 Material properties of Ti-6Al-4V

Young's modulus (GPa)	Poisson's ratio	Yield strength (MPa)	α	n
121	0.29	677.6	0.43	10

require a definition for elastic and plastic behavior separately [11]. The corresponding material constants were obtained using Eq. (1) by fitting a power law relationship in plastic strain versus true stress response, normalized with respect to corresponding yield point values as described in [12]:

$$\frac{\varepsilon_p}{\varepsilon_0} = \alpha \left(\frac{\sigma}{\sigma_0} \right)^n \tag{1}$$

where ε_p is plastic strain, $(\varepsilon_0, \sigma_0)$ are yield point values for true strain and true stress, and (α, n) are Ramberg–Osgood coefficient and exponent, respectively (Table 1).

Coulomb's friction model was used to define the frictional interaction between the cylindrical pad and plate with a constant friction coefficient of 0.75, a value typically observed for Ti-6Al-4V/Ti-6Al-4V mating pair [13]. A combined penalty and augmented Lagrange approach was used to model tangential and normal interaction, respectively. The combined approach has been found stable and computationally less expensive compared to a pure penalty and pure Lagrange algorithm for fretting contact problems [14]. The top end of the fretting pad and bottom face of the plate were restrained to move in the Y-direction, while left end of the fretting pad and side faces of the specimen were restrained in the X-direction. A normal load (P) was applied at the interface between the fretting pad and the top compliant section in negative Y-direction, while a sinusoidally varying tangential load (Q) was applied (in the X-direction) at the interface between left extended section and the fretting pad. Similar boundary conditions have been used earlier by Tsai and Mall [6]. A counteracting couple (M) was applied on the fretting pad to avoid its rotation by nullifying the moment which would be generated due to applied tangential load [14]. To understand the influence of tangential (Kt) and normal stiffness (Kn) of loading fixtures, an extended section was modeled with a cylindrical pad in both normal and tangential direction as shown in Fig. 1. The elastic modulus of the extended section was varied to simulate the effect of tangential and normal stiffness.

The validation of the finite element model was carried out by comparing the contact pressure and contact shear distribution obtained from elastic finite element analysis with that of analytical solutions proposed in the literature [1, 15–17]. Finite element analysis results were compared for two different loading conditions, (a) Hertzian case, in which a normal load of 570 N was applied, and (b) Mindlin case, in which both normal load and tangential load were applied along with counteracting couple. For validation, both fretting pad and plate were modeled with elastic properties of aluminum, viz. elastic modulus and Poisson's ratio of 75 GPa and 0.3, respectively.

(a) **(b)**

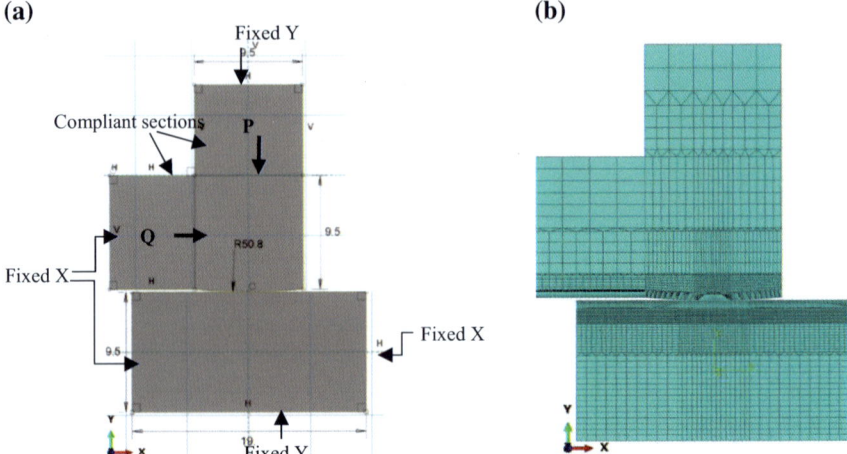

Fig. 1 a Geometry and schematic of loading configuration, and **b** meshed model. All dimensions are in mm

The effect of system stiffness was studied for loading condition similar to that of fretting wear. All the analyses were carried out with a tangential load of about 75% of the maximum sliding friction force. An initial iteration with equal normal and tangential stiffness (elastic modulus of compliant sections being 1/100th of the modulus of elasticity of fretting pad/specimen) was taken as a reference condition and results from all the analysis cases were normalized with respect to it. To understand the influence of system compliance, the stiffness of normal and tangential compliant sections was varied independently by ±60% in incremental steps of 20%. All the analysis cases are given in Table 2.

3 Results and Discussion

3.1 Finite Element Validation

The contact pressure and contact shear distribution were extracted from the plate surface and compared with the analytical solution proposed by Hertz (from Ref. [16]) and Mindlin [1]. From Fig. 2a, it can be seen that a good agreement is found between the contact pressure distribution (normalized with respect to maximum contact pressure) obtained using elastic finite element analysis and analytical solution given by Hertz. A deviation of about 0.16 and 5.6% was obtained for maximum contact pressure and contact zone size, respectively. Further, contact shear stress distribution was extracted for combined normal and tangential loading, and the results were compared with analytical formulation given by Mindlin [1]. It was found that

Table 2 Analysis cases

Analysis case	S. No.	P (N)	Q (N)	M (N-m)
Hertz	LC1	570	–	–
Mindlin	LC2	570	95	451.25

	S.No.	Effect of K_t at constant $K_n = E/100$			
		K_t/K_n	P (N)	Q (N)	M (N-m)
Effect of system stiffness	LC3	1.6	570	332.5	1496.25
	LC4	1.4			
	LC5	1.2			
	LC6	1.0	570	332.5	1496.25
	LC7	0.8	570	332.5	1496.25
	LC8	0.6			
	LC9	0.4			

S. No.	Effect of K_n at constant $K_t = E/100$			
	K_n/K_t	P (N)	Q (N)	M (N-m)
LC10	1.6	570	332.5	1496.25
LC11	1.4			
LC12	1.2			
Reference case		570	332.5	1496.25
LC13	0.8	570	332.5	1496.25
LC14	0.6			
LC15	0.4			

LC load case, *Kn* normal stiffness, *Kt* tangential stiffness, and *E* modulus of elasticity of pad/specimen

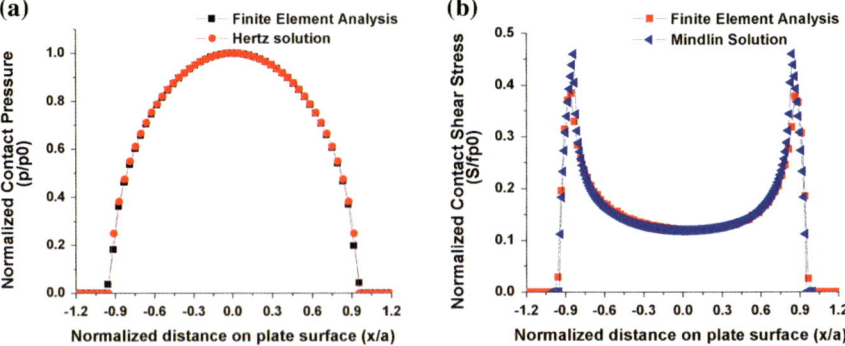

Fig. 2 Comparison between finite element analysis results and analytical solution given by **a** Hertz, and **b** Mindlin

finite element analysis results for maximum contact pressure, contact zone size, and stick zone size correlate well with analytical results and the error is only about 0.18, 4.23, and 1.8%, respectively, but the peak contact shear stress at the stick–slip boundary shows a deviation and a more accurate value can be obtained by further refinement of mesh.

3.2 Effect of Tangential Stiffness

The effect of tangential stiffness was studied by varying the elastic modulus of the compliant section which is attached horizontally to the fretting pad. The results were extracted for the contact parameters which are critical in the context of fretting, viz. contact pressure, contact zone size, stick zone size, contact slip, and normal stress on plate surface in the tangential direction. While contact pressure and contact slip have a direct implication on the quantification of fretting wear volume as given by Archard's law [18], the normal stress on the plate surface in the tangential direction has generally been considered to identify possible location of fretting fatigue crack initiation [5]. The identification of stick zone is particularly relevant because no wear takes place inside the stick zone. Further, the location of peak tensile stress in tangential direction also coincides with the stick–slip interface.

From Fig. 3a, it can be seen that loss or increase in tangential stiffness has no effect on contact zone size, but rise and fall of about 6% was observed in the stick zone size with 60% gain or loss in tangential stiffness, respectively. Further, there was no influence of tangential stiffness on the maximum contact pressure at the interface, as shown in Fig. 3b. This is because, from the Hertz formula (Eq. 2), the contact zone size and contact pressure are a function of normal load, contact geometry, and material properties and has no influence of tangential load unless there is a mismatch in the material properties of mating pair. On the other hand, from the analytical formulation

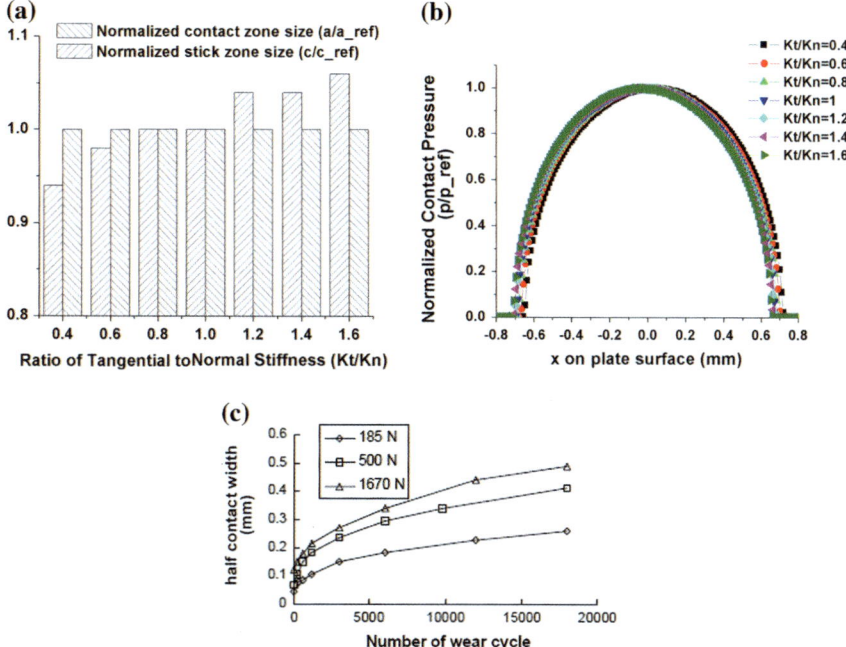

Fig. 3 Effect of tangential stiffness on **a** contact zone and stick zone size, **b** contact pressure distribution, and **c** evolution of half-contact width with number of cycles [19]

proposed by Mindlin (Eq. 5), the stick zone size is dependent on the tangential load and decreases with an increase in tangential load. As tangential stiffness acts as a restraint to the tangential load, an increase in the tangential stiffness will lead to a decrease in tangential load and hence an increase in stick zone size. This implies that, if during the course of experiments, there is loss of tangential stiffness due to slackness, and it may lead to slip in the regions which were in stick condition earlier. This could lead to an increase in the fretting wear volume because a portion of the contact area which was not contributing to fretting wear earlier (due to prevailing stick conditions) will also start wearing out due to slip.

$$a = \left(\frac{4PR}{\pi E^*}\right)^{0.5} \tag{2}$$

where a is half-contact zone size (mm), P is normal load (N), and R is relative radius of curvature (mm) given by

$$\frac{1}{R} = \frac{1}{R_1} + \frac{1}{R_2} \tag{3}$$

and E^* is the composite modulus of elasticity (MPa) given by

$$\frac{1}{E^*} = \frac{1 - v_1^2}{E_1} + \frac{1 - v_2^2}{E_2} \tag{4}$$

where R_i, E_i, v_i represent the radius of curvature, elastic modulus, and Poisson's ratio for cylinder and plate

$$\frac{c}{a} = \left(1 - \frac{Q}{\mu P}\right)^{0.5} \tag{5}$$

where c is half-stick zone size (mm), Q is applied tangential load (N), μ is the coefficient of friction, and P is the applied normal load (N).

Further, a lateral shift of about 10% of contact zone size (toward positive X-direction) was observed in the contact pressure distribution with 60% loss in Kt which could be because of relaxation in lateral constraint with loss in tangential stiffness. This is critical because it will lead to a shift in the location where wear is taking place and an increase in the wear scar width might occur. For example, considering the evolution of half-contact width (with a number of cycles) for a cylinder-on-plate configuration (as shown in Fig. 3c), it can be seen that initially, there is a rapid increase in the contact zone width but as the wear progresses, the contact zone size grows at a slower pace. But, if there is a lateral shift in the contact zone size, it could lead to contact between the unworn surfaces and hence an increase in contact width. This could aggravate fretting wear, but further work is required to ascertain this. No shift was observed with an increase in lateral stiffness beyond Kt/Kn = 1.

Figure 4a shows the distribution of normal stress (in X-direction) on the plate surface, normalized with respect to maximum contact pressure for reference condition, i.e., Kt/Kn = 1. It can be seen that peak tensile stress at the stick–slip boundary had the negligible influence of variation in tangential stiffness, but its location shows a lateral shift of about 20 μm (in negative X-direction) and 40 μm (in positive X-direction) with an increase and decrease of 60% in Kt, respectively. This could be misleading for experimentalists because the location of peak tensile stress is considered to be a possible fretting crack initiation site, and a shift in its location with time can delay the crack initiation. Establishing fidelity of this statement would require more work. Figure 4b shows the contact slip distribution in the half-contact zone. It represents a typical stick–slip distribution with a central stick zone and partial slip at the outer edges. It can be observed that contact slip distribution shows a shift in negative X-direction with an increase in tangential stiffness. Further, the central stick zone size also increases with an increase in tangential stiffness. Also, as can be seen from Fig. 4c, initially at Kt/K = 1, both leading edge and trailing edge have same contact slip but with loss in tangential stiffness, the contact slip at trailing edge increases by about 10%, while a drop of about 5% takes place at a leading edge. This could be because of partial release in lateral constraint due to a loss in tangential stiffness. Further, the asymmetry in contact slip could lead to the asymmetry in the wear scar. Reverse trend was observed for the increase in tangential stiffness, with larger slip being at the leading edge than the trailing edge.

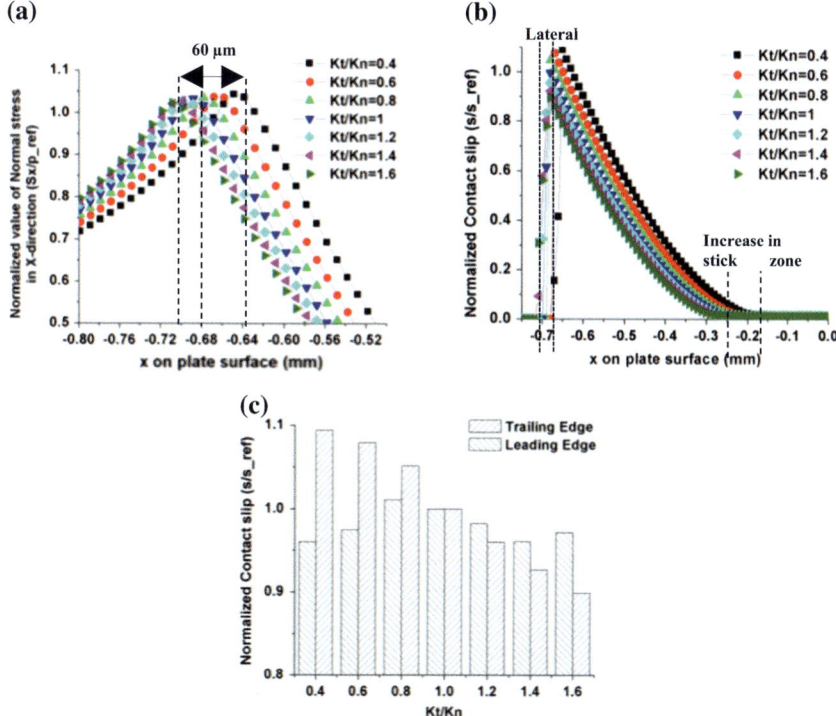

Fig. 4 Effect of tangential stiffness on **a** normal stress (in X-direction) distribution on the plate surface, **b** contact slip distribution in the contact zone, and **c** maximum contact slip at the leading and trailing stick–slip boundary

3.3 Effect of Normal Stiffness

The effect of normal compliance of test rig was simulated by modeling a compliant section on the top of the fretting pad, and its elastic modulus was varied between Kn/Kt = 0.4 to Kn/Kt = 1.6. From Fig. 5a, it can be seen that an increase of about 10% is observed in the stick zone size with 60% loss in normal stiffness, whereas 60% increase in Kn leads to a reduction of 13% in stick zone size. The contact zone size was found to have significantly lesser variation (about 1.5%) with 60% loss or gain in Kn. Further, the ratio of stick zone to contact zone size shows that relative proportion of stick zone in contact zone increases by about 3% with 60% loss in Kn, while a relatively higher drop (about 4.5%) in the stick zone proportion was observed with increase in Kn. Though the stick zone size is not just dependent on the effective normal load getting transferred to the contact zone and is also affected by other loading conditions, but the variation of stick zone size with variation in Kn can be partially explained by the variation in normal load. For this, the normal force acting in the contact zone was extracted, and its variation with normal stiffness is shown

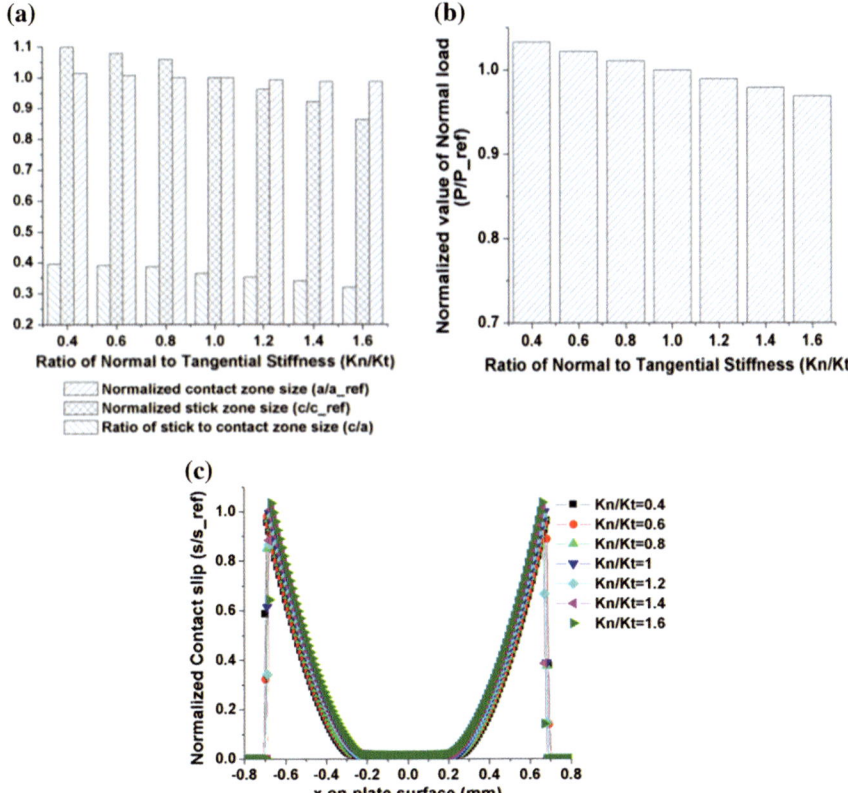

Fig. 5 Effect of normal stiffness on **a** contact zone and stick zone size, **b** normal load in the contact zone, and **c** contact slip

in Fig. 5b. It can be seen that a 60% increase or decrease in normal stiffness leads to about 3% decrease and increase in the normal load respectively. This variation in normal load with normal stiffness could have possibly contributed to the variation in the stick zone size. Further, an increase in contact slip (about 3.5%) was observed in the slip region due to an increase in normal stiffness by 60% and vice versa as shown in Fig. 5c, but no asymmetry was observed in this case in contrast to the earlier case of variation in tangential stiffness. The normal stress on the plate surface (in X-direction) had the negligible influence of variation in normal stiffness.

4 Conclusions

A two-dimensional finite element analysis was carried out in the present work to study the effect of variation in system stiffness on the contact behavior in fretting wear. From the present study, it is found that variation in tangential and normal stiffness has a significant influence on the stick zone size and contact slip distribution. Further, the variation in tangential stiffness also leads to a lateral shift in the contact pressure and normal stress distribution. Also, comparatively tangential stiffness has more impact on fretting behavior than normal stiffness. Further, loss in stiffness has quantitatively more influence than the increase in stiffness for tangential stiffness while vice versa is true for normal stiffness. This is of practical relevance because generally there is a higher possibility of loss in system stiffness rather than gain in stiffness, due to a slackening of joints in assemblies over the course of experiments. The present study could be useful from the viewpoint of incorporating the effect of variation in system compliance during the course of an experiment on the contact parameters.

References

1. R.D. Mindlin, Compliance of elastic bodies in contact. J. Appl. Mech. **16**, 259–268 (1949)
2. S. Fouvry, Ph Kapsa, L. Vincent, Analysis of sliding behavior for fretting loadings: determination of transition criteria. Wear **185**, 35–46 (1995)
3. E. Marui, H. Endo, N. Hasegawa, H. Mizuno, Prototype fretting-wear testing machine and some experimental results. Wear **214**, 221–230 (1998)
4. O. Jin, S. Mall, Effect of slip on fretting behavior: experiments and analyses. Wear **256**, 671–684 (2004)
5. P.A. McVeigh, T.N. Farris, Finite element analysis of fretting stresses. J. Tribol. **119**, 797–801 (1997)
6. C.T. Tsai, S. Mall, Elasto-plastic finite element analysis of fretting stresses in pre-stressed strip in contact with cylindrical pad. Finite Elem. Anal. Des. **36**, 171–187 (2000)
7. J.J. Madge, *Numerical Modelling of Effect of Fretting Wear on Fretting Fatigue* (University of Nottingham, 2008)
8. Y. Mutoh, M. Jayaprakash, K. Asai, K. Ichikawa, Effect of contact pad rigidity on fretting fatigue design curve. Trans. Indian Inst. Metals **63**(2–3), 181–186 (2010)
9. J. Ding, S.B. Leen, I.R. McColl, The effect of slip regime on fretting-wear induced stress evolution. Int. J. Fatigue **26**, 521–531 (2004)
10. J.M. Ambrico, M.R. Begley, Plasticity in fretting contact. J. Mech. Phys. Solids **48**, 2391–2417 (2000)
11. Abaqus 6.12-1 documentation
12. L.A. James, Ramberg-Osgood strain hardening characterization of an ASTM A302-B steel. J. Press. Vessel Technol. **117**, 341–345 (1995)
13. S. Fouvry, C. Paulin, S. Deyber, Impact of contact size and complex gross-partial slip conditions on Ti-6Al-4V/Ti-6Al-4V fretting wear. Tribol. Int. **42**, 461–474 (2009)
14. K. Anandavel, R.V. Prakash, Effect of three-dimensional loading on macroscopic fretting aspects of an aero-engine blade-disc dovetail interface. Tribol. Int. **44**, 1544–1555 (2011)
15. D. Nowell, D.A. Hills, Mechanics of fretting fatigue tests. Int. J. Mech. Sci. **29**(5), 355–365 (1987)
16. K.L. Johnson, *Contact Mechanics* (Cambridge University Press, Cambridge, UK, 1985)
17. D.A. Hills, D. Nowell, *Mechanics of Fretting Fatigue* (Kluwer Academic Publishers, Dordrecht, Netherlands, 1994)
18. J.F. Archard, Contact and rubbing of flat surfaces. J. Appl. Phys. **24**, 981–987 (1953)
19. I.R. McColl, J. Ding, S.B. Leen, Finite element simulation and experimental validation of fretting wear. Wear **256**, 1114–1127 (2004)

Development of Methodology for Seismic Qualification of Dhruva Reactor Components

Vinayak A. Modi, Manish Bhadauria, Arihant Jain, N. S. Joshi
and P. V. Varde

Abstract Safety critical systems and components in a nuclear plant are to be qualified for expected seismic activity at the site. This ensures safety in case of any expected seismic event. IEEE STD 344 allows for seismic qualification by analysis and testing. Testing is carried out by mounting the device on a shaker to simulate earthquake ground motion. This paper discusses the methodology for seismic qualification of components using the electrodynamic shaker. Power spectral density is first obtained from site response spectrum using analytical methods. The PSD profile is used to generate the ground response in the form of random vibrations. Device response to the applied ground motion is recorded during experimentation. The comparison between the performance of the component before, during and after seismic testing is used for qualification purpose.

Keywords Seismic testing · Seismic qualification · Random vibration · Power spectral density

V. A. Modi
Vellore Institute of Technology, Vellore, India
e-mail: vinayaka.modi2015@vit.ac.in

M. Bhadauria
Homi Bhabha National Institute, Mumbai, India
e-mail: manish550000@gmail.com

A. Jain (✉) · N. S. Joshi · P. V. Varde
Bhabha Atomic Research Center, Mumbai, India
e-mail: arihantj@barc.gov.in

N. S. Joshi
e-mail: nsjoshi@barc.gov.in

P. V. Varde
e-mail: varde@barc.gov.in

© Springer Nature Singapore Pte Ltd. 2020
P. V. Varde et al. (eds.), *Reliability, Safety and Hazard Assessment for Risk-Based Technologies*, Lecture Notes in Mechanical Engineering,
https://doi.org/10.1007/978-981-13-9008-1_41

501

1   Introduction

Reliability of nuclear power plant operation is very critical because failure can result in fatal effects in the nearby areas. Seismic disturbances pose a viable threat to the nuclear power plant which can result in both structural and functional failure of the NPP. One notable disaster occurred due to earthquake inducing a tsunami in Japan which resulted in Fukushima Daiichi accident. The local effects of which resulted in cancer deaths from contamination due to more than 1 curie per square kilometre (density of contamination) are likely to scale correspondingly—on the order of 1000 [1]. The global effects were estimated as an additional 130 (15–1100) cancer-related mortalities and 180 (24–1800) cancer-related morbidities. Around ∼600 mortalities have been reported due to non-radiological causes [2]. It is clearly apparent that the nuclear power plant should be designed and qualified in a conservative way to avoid such disasters. Dhruva is a 100 MW high neutron flux research reactor located at Bhabha Atomic Research Centre (BARC). It is located in Mumbai which is a seismic zone III according to the IS:1893-2002 (BIS standard, 2002), implying that it can be affected up to intensity VII earthquake according to MSK64 Intensity Scale [5]. In such regions, seismic qualification for components becomes relatively more important. Standards are there for seismic qualification of equipment for nuclear power plants like IEEE Std 344-2013 [3] and IEC 60980 [4] which were referred while developing the methodology for seismic qualification of components for Dhruva reactor.

Earthquakes for seismic qualifications were agreed to be classified into two types, namely Safe Shutdown Earthquake (SSE) and Operating Basis Earthquake (OBE), and Safe Shutdown Earthquake (SSE) is having high intensity even though it has a low probability of occurrence, i.e. once in 10,000 years and the test component is designed to maintain its structural integrity and functional operability during such an event. The systems are usually designed to remain operational for an Operating Basis Earthquake (OBE) having an intensity which is lower than SSE but can be expected at the site once in 100 years.

2   Objective

The goal of the project is to compare the standards for seismic testing of structural and non-structural components and hence develop a methodology suitable for test as well as testing equipment. The standard referred here was "*IEEE Std 344-2013 (Revision of IEEE Std 344-2004)—IEEE Standard for Seismic Qualification of equipment for Nuclear Power Generating Stations*" [3].

The developed methodology was tested and verified for a sample test component to know the scope of the methodology and implications.

3 Measures for Qualifying Equipment According to IEEE 344 [3]

For seismic qualification of Category I equipment, the equipment should be able to perform its required safety function(s) during and/or after the time it is subjected to the forces resulting from a Safe Shutdown Earthquake (SSE). In addition, the equipment shall withstand the effects of a number of Operating Basis Earthquakes (minimum of five OBEs) prior to the application of an SSE. The methods for seismic qualification in IEEE standard 344 are grouped into four general categories as:

(a) Prediction of test equipment by analysis.
(b) Testing of test component under simulated seismic conditions.
(c) Qualification of test component by combined testing and analysis.
(d) Using experience data to qualify the test equipment.

3.1 Qualification by Testing

Test methodologies are of three categories, namely proof testing, generic testing and fragility testing. The seismic environment is categorised into two categories, namely single-frequency testing and multi-frequency testing. The method is dependent on the vibration environment and test equipment.

The proof or generic test seismic simulation waveforms should produce a Test Response Spectrum (TRS) that closely envelops the Required Response Spectrum (RRS) using single- or multiple-frequency input. The peak acceleration should be equal to or greater than the RRS zero period acceleration (ZPA), and the frequency content should not be above RRS ZPA asymptote.

3.2 Qualification of Candidate Equipment

The requirements for qualifying a candidate equipment item using earthquake experience data are as follows:

(a) The RRS should be enveloped by the Earthquake Experience Spectrum (EES) for the reference equipment class over the frequency range of interest (1 Hz to the cut-off frequency of the response spectra). Failure of the EES to envelop the RRS shall be justified.
(b) The minimum RRS should be 5% critically damped which is used for comparison with the EES. This RRS should be derived from the SSE.

4   Test Component Set-up

Methodology for seismic qualification of components for both structural and functional testing is developed for single-degree-of-freedom (or rectilinear) testing with the help of electrodynamic shaker. The electrodynamic shaker is designed to give excitation in one direction. The component set-up has to be adjusted for excitation in three directions. For mounting, the component suitable fixture has to be developed which induces the same excitement in the test component as it occurs in the event of seismic activity. For this, the stiffness of the fixture should be high.

4.1   Mounting System for Test Component and Accelerator Positions

A global coordinate system was assumed on the shaker table according to which x-direction was taken along the excitation direction when the shaker table was attached in a horizontal position. A local coordinate system is attached to the test component (Fig. 1). For qualification of any component, nine tests (3 * 3) are required which is due to excitation along one axis and its respective effect on the test component which is measured by accelerometers. The number of tests can be reduced by a factor of 3 if three accelerometers are mounted simultaneously.

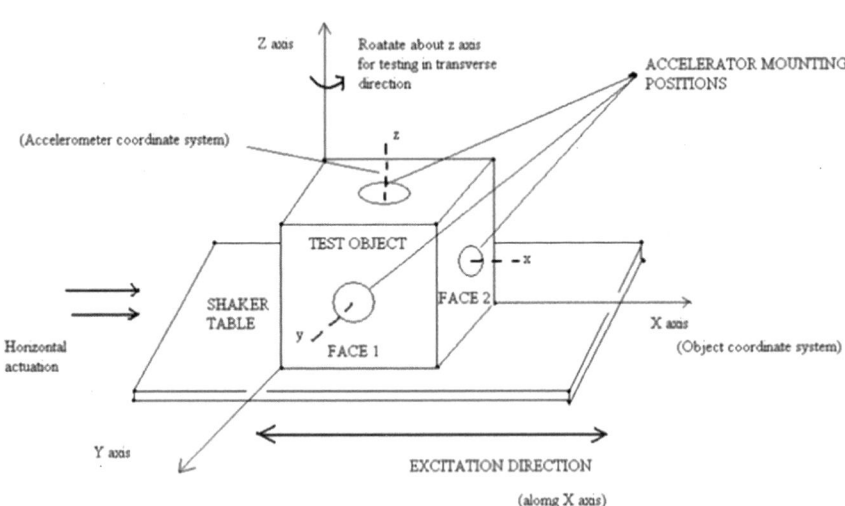

Fig. 1 Mounting system for seismic testing of component in x- and y-direction

5 Seismic Qualification Methodologies

Two methodologies were developed for seismic qualification of the component which was based on single-frequency test and multi-frequency test, respectively. Both the tests can be used for a single component or an individual test can be used for qualification of the test equipment depending on the requirement.

5.1 Single-Frequency Test Methodology

For a single-frequency test, firstly the functionality, mounting methods, location, etc. is determined for the test component which provides the information about the mounting fixtures so that the mounting on shake table is identical to the real-time environment. A suitable fixture has to be developed for every test component such that the fixture does not affect the vibrations for the test component. Modal analysis should be performed for the fixture to ensure that it does not affect test results. For this, the minimum modal frequency obtained through analysis should be significantly greater than the test frequency range (0–50 Hz). Fabrication of a suitable fixture is done as per the design requirements.

In this methodology, the sine sweep signal is given to component in the frequency range typical of seismic events. The component resonance frequency is obtained via such testing. The component is then tested for the prolonged duration at the resonant frequency for qualification. For testing, the component is mounted in vertical and horizontal directions, and the suitable excitation is provided (Figs. 2 and 3).

5.2 Multiple-Frequency Test Methodology

Multiple-frequency test may be performed by applying to the equipment a specified time history that has been synthesised to simulate the seismic input. It shall be demonstrated that the actual motion of the shake table is as severe as, or more severe

Fig. 2 Test component in vertical direction

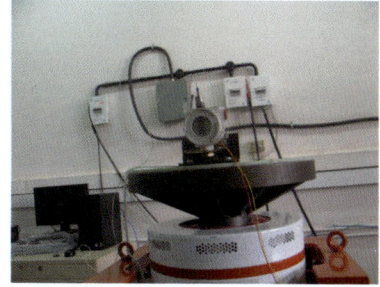

Fig. 3 Test component in
horizontal direction

Fig. 4 Required response
spectra of Dhruva reactor site

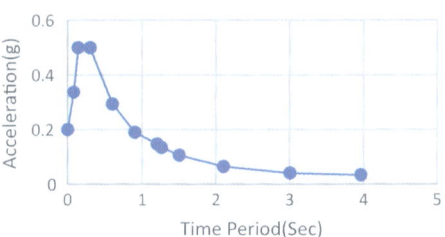

than, the required motion. This can be accomplished by a direct comparison of the
table motion time history with the specified motion by means of an accelerograph
reading. Comparison can further be made through the use of response spectra of the
required motion and the table motion. In the latter method, a response spectrum of
the specified motion (RRS) is developed for the appropriate damping given in Clause
6 of IEEE standard 344. Subsequently, a table motion is developed so that its TRS
envelops the RRS.

The shake table takes the data in the form of power spectral density versus frequency. So, the spectrum compatible time history generated from RRS is converted
into power spectral density (PSD).

5.2.1 Response Spectrum

From the data available at the test site, Required Response Spectrum (RRS) was
generated as shown in Fig. 4.

It is observed that slight variation in natural frequency of component causes a
significant change in spectral acceleration of RRS, and broadening of RRS (Fig. 5)
was done to cover the uncertainty in component frequency in accordance with IEEE
standard. This also has an effect of making the RRS artificially conservative.

The RRS is usually specified at several levels of damping. The RRS with damping
of 5% was chosen for doing shake table experiment since it is the recommended value
given in IEEE standard irrespective of component damping value.

Fig. 5 Broadened RRS

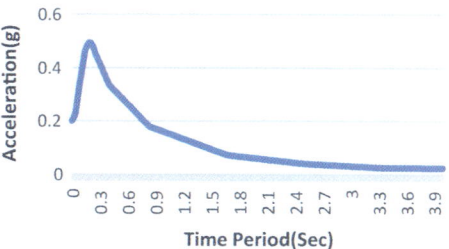

5.2.2 Time History

The Required Response Spectrum (RRS) is the only available data for the location. Corresponding time history is required for PSD generation, and hence, synthetic acceleration method [6] is used for conversion of response spectrum to time history. After having defined the required response spectrum and synthetic accelerogram, spectrum compatible accelerogram is generated using real acceleration adjustment method.

In the real acceleration adjustment method, the artificial accelerogram is defined starting from a real one and adapting its frequency content to match the target spectrum using the Fourier transformation method. The correction of the random process [7] is carried on at every iteration using Eq. 1:

$$F(f)_{i+1} = F(f)_i \left[\frac{\text{SRT}(f)}{\text{SR}(f)i} \right] \tag{1}$$

SRT(f) is the value of the target spectrum, and SR(f)i is the value of response spectrum corresponding to the accelerogram of the current iteration for frequency f. $F(f)_{i+1}$ and $F(f)_i$ are the values of the accelerogram for the current and previous iteration in the frequency domain, respectively (Fig. 6). This process is iterated till the convergence is achieved with the target spectrum. In every iteration, Fourier transformation is used to move from the time domain to frequency domain. The correction is performed in the frequency domain. After this, an inverse Fourier transform is done to verify for convergence with target spectrum. If convergence is achieved time history is obtained if not then the process is iterated until convergence is achieved (Fig. 7).

The acceleration time history (after several iterations) generated as shown in the figure.

PSD plot is generated from time history as shown in Fig. 8.

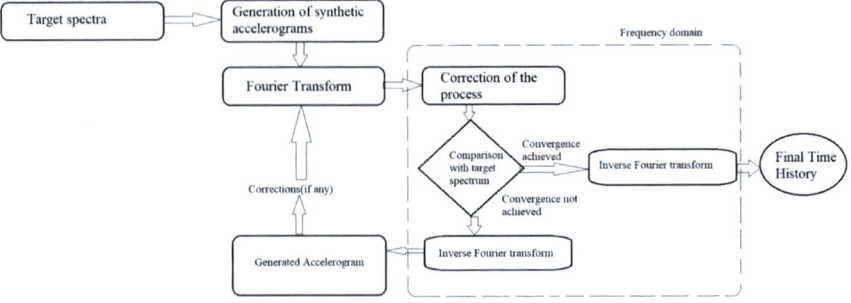

Fig. 6 Flow process for conversion of response spectra to corresponding time history [6]

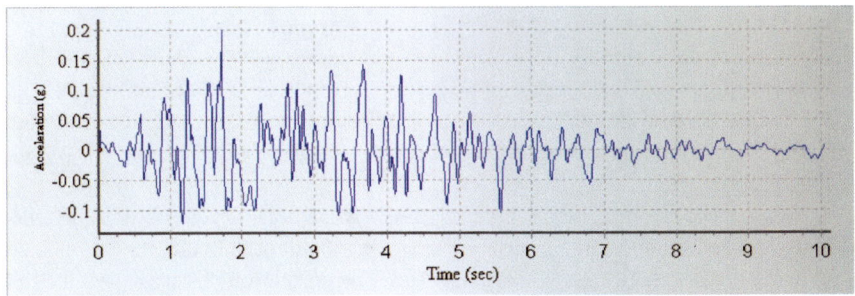

Fig. 7 Final acceleration time history

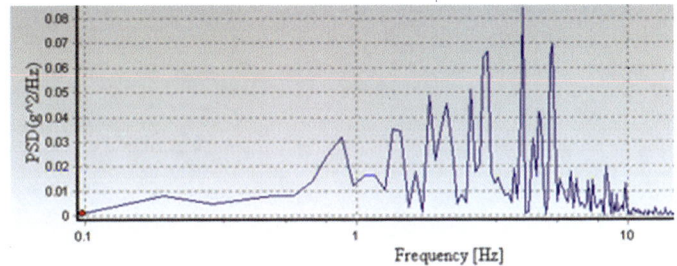

Fig. 8 Power spectral density plot

6 Results

From the shake table experiment (Fig. 9), it was found that the Test Response Spectra (TRS) enveloped the Required Response Spectra (RRS) in the higher-frequency range greater than 3 Hz. Below 3 Hz frequency, it can be seen that the Test Response Spectra (TRS) does not envelop Required Response Spectrum (RRS), and after 3 Hz, the TRS envelops satisfactorily. Hence, the test component can be qualified for a frequency range of 3 Hz and above.

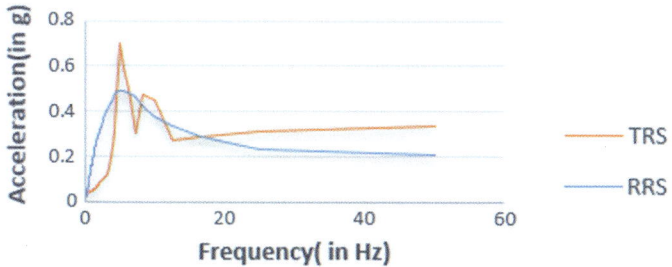

Fig. 9 Target response spectrum versus required response spectrum

6.1 Failure and Acceptance Criteria

For functional testing, after suitable envelope of TRS over RRS for a specified frequency range, it is verified that the component works suitably under seismic excitations and is able to do its intended function throughout the period. This type of acceptance criteria is accepted for electrical or functional components.

For structural testing, after suitable envelop of TRS over RRS for specified frequency range, the test component is verified for structural integrity. This type of acceptance criteria is accepted for mechanical structures.

Reason of improper response for lower frequencies than 3 Hz

The electrodynamic shaker used for qualification process cannot provide excitation below a frequency of 3 Hz. Hence, there is no suitable excitation below 3 Hz; hence, Test Response Spectrum does not envelop Required Response Spectrum, and the equipment cannot be qualified for low-frequency excitation (<3 Hz).

7 Conclusions

The following conclusions were drawn for the testing equipment used for the qualification process.

- The test methodology was developed for one degree-of-freedom electrodynamic shaker but can be extended to similar equipment.
- The equipment can provide suitable excitation for single-frequency testing. The test component after prolonged testing at resonance frequency can be conservatively qualified for seismic activity.

- As the major cause of failure occurs due to resonance, for smaller and stiff components resonance condition occurs at higher-frequency levels. So smaller electrical and mechanical components can be successfully qualified with multi-frequency testing.

References

1. F.N. Von Hippel, The radiological and psychological consequences of the Fukushima Daiichi accident. Bull. At. Sci. **67**(5), 27–36 (2011)
2. J.E. Ten Hoeve, M.Z. Jacobson, Worldwide health effects of the Fukushima Daiichi nuclear accident. Energy Environ. Sci. **5**(9), 8743–8757 (2012)
3. IEEE Std 344-2013 (Revision of IEEE Std 344-2004), IEEE standard for seismic qualification of equipment for nuclear power generating stations
4. ASME QME-1-2017, Qualification of active mechanical equipment used in nuclear facilities
5. R. Sinha et al., An earthquake risk management master plan for Mumbai: risk assessment and its mitigation, in *Proceedings of 15WCEE: World Conference on Earthquake Engineering, Lisbon. Portugal* (2012)
6. B. Halldorsson, A.S. Papageorgiou, Calibration of the specific barrier model to earthquakes of different tectonic regions. Bull. Seismol. Soc. Am. **95**(4), 1276–1300 (2005)
7. A. Masi, M. Vona, M. Mucciarelli, Selection of natural and synthetic accelerograms for seismic vulnerability studies on reinforced concrete frames. J. Struct. Eng. **137**(3), 367–378 (2010)

Experimental Modal Analysis of Polypropylene Fiber Reinforced Concrete

Radhika Sridhar and Ravi Prasad

Abstract This paper experimentally investigates the static and dynamic properties of polypropylene fiber reinforced concrete (PPFRC). And also, the main objective of this present study is to examine the dynamic behavior of damaged and undamaged PPFRC beams in free-free constraints. The static properties have been carried out by conducting compression, splitting tensile and flexural strength. In addition to this, dynamic properties such as damping ratio, mode shape and fundamental frequency of PPFRC beams were also determined experimentally. In this paper, the influence of 0.1, 0.2, 0.3 and 0.4% variation of fiber contents by volume fraction of concrete with constant water-cementitious ratio of 0.31 was investigated for M50 grade of concrete. In order to determine the effect of polypropylene fibers in concrete and to compare, the properties of plain concrete have also been studied. The test results emphasize that the addition of fibers reduces the fundamental frequency and increase the damping ratio of concrete. Damping of PPFRC beams increases with the increase of damage while the fundamental natural frequency decreases with an increase of damage. The changes of fundamental natural frequency and damping values have been correlated to the damage degree of the prismatic beam experimentally.

Keywords Damping · Natural frequency · Mode shape · Flexure · Polypropylene fiber · Damage

1 Introduction

Historically, concrete is the most commonly used construction material because of its versatility and availability all over the world [1, 2]. Due to the environmental service conditions and complexity of structures such as high-rise buildings, dams, bridges,

R. Sridhar (✉) · R. Prasad
National Institute of Technology, Warangal, India
e-mail: radhika.sridhar2210@gmail.com

R. Prasad
e-mail: drprasadravi@gmail.com

© Springer Nature Singapore Pte Ltd. 2020
P. V. Varde et al. (eds.), *Reliability, Safety and Hazard Assessment for Risk-Based Technologies*, Lecture Notes in Mechanical Engineering,
https://doi.org/10.1007/978-981-13-9008-1_42

511

hydraulic structures and offshore structures, any normal grade or ordinary concrete could no longer meet the requirements in generality [3]. It is well known, nevertheless, the ratio between compressive strength and flexural strength or split tensile strength of ordinary concrete or high-strength concrete will be inevitably brittle in nature [4]. In order to overcome this drawback, researches have been focused on various necessary treatments such as adding fibers to concrete, use of supplementary mineral admixtures and externally strengthening the members with fiber reinforced polymers [5]. Among these, addition of fibers like steel, polypropylene, polyvinyl alcohol, natural and synthetic fibers into concrete has been accepted widely because of its improved mechanical properties such as compressive strength, split tensile strength, flexural strength and flexural toughness [6]. In recent years, mainly steel fibers have been used by numerous researchers, especially in concrete industry. By adding steel fibers to concrete significantly increases the flexural strength, toughness and its post-cracking behavior [7, 8]. Most of the synthetic fibers have a low modulus of elasticity such as polypropylene, polyethylene, polyester and nylon. The main advantages of using synthetic fibers are its alkaline resistance, high melting point and low cost of raw materials [9]. While steel fibers improve the hardened concrete properties, the synthetic fibers provide benefits even when the concrete is in plastic state. Besides, in recent few decades, inclusion of non-metallic polypropylene fiber has also become one of the most extensively used fibers in concrete which in turn enhances the dynamic as well as durability properties of the composite matrix [10]. Generally, the concrete structures will always be subjected to vibration forces such as impact loading on dynamic shock of moving vehicles. Depending on the type of structure and impact or dynamic load, harmonic excitation exists through the external force of a certain frequency applied to a system for given amplitude. Resonance can also occur when the external excitation has the same frequency as the natural frequency of the system. In this regard, it is necessary to study the dynamic property of any structure at typical modes which will be helpful in reducing resonance and attenuating vibration [11, 12]. The essential concept concealed with vibration monitoring technique for any concrete structure damage analysis would be based on dynamic characteristics and does not rely on the geometry of the structure; as a result, changes will occur in the dynamic response behavior [13]. The dynamic behavior of concrete is one of the important parameters which can be determined by its dynamic properties, such as dynamic modulus of rigidity, dynamic modulus of elasticity and dynamic Poisson's ratio which present different values compared to their static or mechanical properties [13]. Some of the aforementioned research work has been studied about the static properties of different composite matrices and there is a dearth of knowledge available on the study of dynamic properties of fibrous concrete. Therefore, the main aim of the present experimental work is to study the effect of polypropylene fiber on mechanical and dynamic properties and also the effect under damaged and undamaged condition of polypropylene fiber reinforced concretes (PPFRCs).

2 Experimental Program

In this present investigation, fiber dosages of 0.1, 0.2, 0.3 and 0.4% by volume fraction were considered to investigate the static properties including compressive, splitting tensile and flexural strength and dynamic properties of polypropylene fiber reinforced concrete.

2.1 Materials

Ordinary Portland cement (OPC) of 53 grade of cement according to bureau of Indian standard code has been used in this present investigation. Good-quality river sand was used as fine aggregate. The coarse aggregate used in this work was of 20-mm-sized crushed angular shape. The coarse and fine aggregates are free from dust before used in concrete. The differences and usage of fiber depend on the requirement of behavior and property for a concrete, allowing the increase in explicit effects and mechanical properties. Figure 1 displays the typical view of PP fiber used in this study. And the mechanical and physical properties of PP fiber are illustrated in Table 1.

Fig. 1 Typical view of polypropylene fiber

Table 1 Physical and mechanical properties of polypropylene fiber

Fiber type	Shape of fibers	Length (mm)	Diameter (mm)	Aspect ratio	Tensile strength (MPa)	Density (kg/m^3)
PP	Straight	12	0.038	315	420	990

2.2 Mixing and Casting Procedure

The determination of mix design for this research to obtain the desired strength of M50 grade of concrete was performed based on IS 10262:2009. Five mixtures of PPFRCs were included in this experimental work to investigate the mechanical, dynamic properties and damage assessment of PPFRCs. Non-fibrous specimens have also been cast to ascertain the effect of the polypropylene fiber reinforced concrete. Mix proportion of M50 grade of concrete is shown in Table 2. In this experimental work, a portable mixer has been used throughout the study. Firstly, the fine and coarse aggregates were added into the mixture machine and allowed to mix for 55 s to 1 min and the required amount of cement was added into it. Then, the water was added gradually to the mixture and it has been continued in order to get the proper workability. Then, the fibers were added gradually and it was mixed for another 2 min to get better performance in mechanical properties. In generality, for the determination of mechanical and dynamic properties of PPFRC, three specimens of standard cubes (150 mm × 150 mm × 150 mm), cylinders (150 mm diameter and 300 mm height) and prisms (cross section 100 mm × 100 mm and 500 mm length) were cast for each percentage variation of PP fibers as well as plain concrete (reference concrete) for comparison.

2.3 Test Procedure of Mechanical and Dynamic Properties

To examine the composite performance of fibrous reinforced concrete, three cubes and cylinders of standard size (150 mm × 150 mm × 150 mm) and (150 mm diameter and 300 mm height), respectively, for each mixture were cast and tested in UTM (universal testing machine) with the capacity of 3000 kN after 28 days of its curing period in accordance with Bureau of Indian Standards code. All prismatic beams were tested in a dynamic universal testing machine of 1000 kN capacity under four-point bending test to acquire modulus of rupture which is shown in Fig. 2.

Table 2 Mix proportions of plain and PPFRCs

Mix No.	Designation	Fiber V_f (%)	Cement (kg/m^3)	Sand (kg/m^3)	Coarse aggregate (kg/m^3)	Water (kg/m^3)	SP (%)
1	M1	0	1	1.56	2.94	0.31	0.3
2	M2	0.1	1	1.56	2.94	0.31	0.3
3	M3	0.2	1	1.56	2.94	0.31	0.3
4	M4	0.3	1	1.56	2.94	0.31	0.3
5	M5	0.4	1	1.56	2.94	0.31	0.3

Fig. 2 Typical view of four-point bending test setup

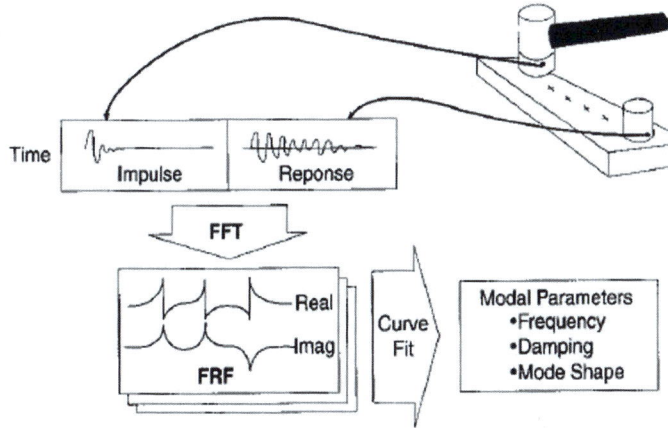

Fig. 3 Schematic view of modal test setup

The prismatic specimens of aforementioned size were cast for each percentage variation of PPFRCs to examine their natural frequency in free-free condition according to ASTM C-215. A total of 15 prisms were cast and tested in order to evaluate the dynamic properties and vibration-based damage assessment of PPFRCs. All specimens were tested using dynamic analyzer and impact hammer technique which consists of forced transducer used for excitation and accelerometer to collect the response from the analyzer which is used to carry out the frequency response function. Fundamental resonant frequency between 1 and 4000 Hz has been considered in this present study. Figure 3 displays the schematic view of dynamic test setup.

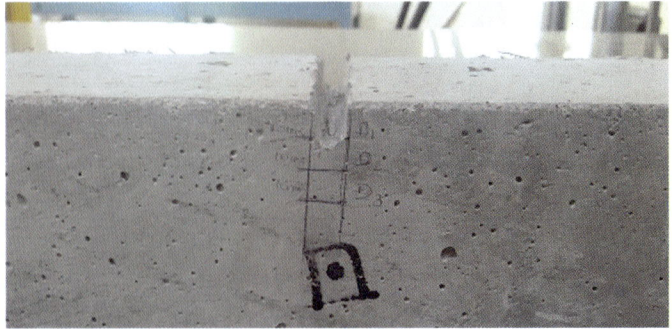

Fig. 4 Typical view of specimen at first damage level

2.4 Test Procedure of Damage Evaluation

Prismatic beams of aforementioned size were cast and tested at the age of 28 days for all mixtures to evaluate the damage characteristics of PPFRCs in free-free condition. At the mid-span of the specimen, a small damage has been induced with the help of concrete cutter machine. Before inducing damage to the prismatic beams, all specimens were tested for its undamaged condition (free-free supports) in order to compare and assess the damage in concretes with polypropylene fibers inclusion. Figure 4 displays the specimen at first damage level. Totally, four damage stages were considered in this present study:

(1) Undamaged beam (D_0),
(2) Damage induced at the mid-span of width 10 mm and depth 10 mm (D_1),
(3) Damage induced at the mid-span of width 10 mm and depth 20 mm (D_2),
(4) Damage induced at the mid-span of width 10 mm and depth 30 mm (D_3).

3 Results and Discussions

The results acquired through experimental tests are discussed in this chapter. Static result shows the behavior of plain and fibrous concrete composites, and the dynamic result shows the dynamic response behavior of the beams in both damaged and undamaged states under free-free constraints.

3.1 Mechanical Property Tests

The mechanical property tests have been carried out at the age of 28 days, and the acquired results are shown in Table 3. The compressive strength of concrete

Table 3 Mechanical property results of plain and PPFRCs

Mix No.	Designation	Compressive strength (MPa)	Splitting tensile strength (MPa)	Flexural strength (MPa)
1	M1	57.53	4.36	5.34
2	M2	58.93	4.52	5.61
3	M3	59.68	4.77	6.12
4	M4	60.43	5.36	7.13
5	M5	58.91	4.93	6.23

is the most important strength parameter, and it is a qualitative measure for other properties of hardened concrete. When the concrete fails under compressive load, it is a mixture of crushing and shear failure. Compressive strength results emphasize that the increase in percentage volume fraction of PP fibers led to 2.4–5% increases in compressive strength for M2 and M4 specimens, respectively, when compared to plain concrete. The splitting tensile strength was carried out on a concrete; cylindrical specimens were cast for four sets of PP fibers and one set of plain concrete at the age of 28 days. From the results, the maximum load which was taken by the cylinder was considered for splitting tensile strength evaluation, and the specimens were broken into two halves at the same load for plain concrete, whereas for PPFRC, cylindrical specimens were held together after cracks and even when the test was continued up to more than the maximum load.

3.2 Dynamic Property Tests

Damping ratio of PPFRC prismatic beams increases with the addition of fiber content. Fundamental transverse frequencies have been carried for all mixtures according to ASTM C215 [14]. The test results emphasize that the frequency for each and every damage level as well as undamaged level, non-fibrous concrete has attained the highest natural frequency when compared to that of PPFRCs. It can be also observed that the resonant transverse frequencies decrease with the increase of fiber content. And also, it was observed that the fundamental transverse frequency decreased was about 2.5% for M4 mixture when compared to the frequency of plain concrete in undamaged condition.

3.3 Damage Evaluation of Polypropylene Fiber Reinforced Concrete

Fundamental transverse frequency and damping ratio values obtained for undamaged and three different damage levels of PPFRCs are illustrated in Tables 4 and 5,

Table 4 Natural frequency (ω) of plain and PPFRCs

Mix No.	Designation	D_0		D_1		D_2		D_3	
		Mode 1	Mode 2	Mode 1	Mode 2	Mode 1	Mode 2	Mode 1	Mode 2
1	M1	1558	3728	1536	3722	1531	3716	1512	3705
2	M2	1547	3712	1523	3703	1502	3682	1486	3671
3	M3	1531	3701	1506	3684	1484	3661	1442	3642
4	M4	1519	3693	1491	3669	1463	3652	1426	3634
5	M5	1503	3688	1475	3653	1442	3632	1411	3625

Table 5 Damping ratio (ζ) of plain and PPFRCs

Mix No.	Designation	D_0		D_1		D_2		D_3	
		Mode 1	Mode 2	Mode 1	Mode 2	Mode 1	Mode 2	Mode 1	Mode 2
1	M1	0.56	0.42	0.58	0.45	0.61	0.47	0.64	0.51
2	M2	0.58	0.45	0.59	0.47	0.62	0.48	0.66	0.53
3	M3	0.60	0.47	0.62	0.48	0.64	0.50	0.69	0.55
4	M4	0.62	0.48	0.64	0.50	0.65	0.52	0.71	0.56
5	M5	0.65	0.51	0.66	0.52	0.67	0.54	0.72	0.58

respectively, in free-free condition. The values of damping ratio have been decreased gradually for second mode of vibration when compared to first mode because of the increment in natural frequency.

The polypropylene fiber reinforced concrete beam with 0.4% fiber content has the highest damping value for both first and second vibration modes. From the table, it can be observed that the fundamental frequency decreases with increasing of fiber content as well as with an increase of structural damage, whereas damping ratio increases with the increase of PP fiber content as well as with an increase of damage, when compared to the properties of undamaged stage. Percentage decrease of fundamental transverse frequency values for three damage levels are shown in Figs. 5 and 6 of PPFRCs with the comparison of undamaged specimens for first and second vibration modes, respectively. From these figures, it can be observed that the percentage decrease of transverse frequency in first mode is lower than that of the percentage decrease of frequency in second mode. The comparison of damping values of PPFRCs for three different damage levels are displayed in Figs. 7 and 8, respectively. The damping ratio value of first mode for all the mixtures was higher than that of the damping ratios in second mode, because the fundamental frequency of the second mode is always higher than the first mode.

Fig. 5 Comparison of natural frequency for PPFRC—Mode 1

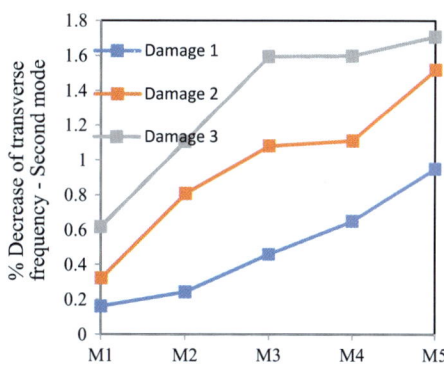

Fig. 6 Comparison of natural frequency for PPFRC—Mode 2

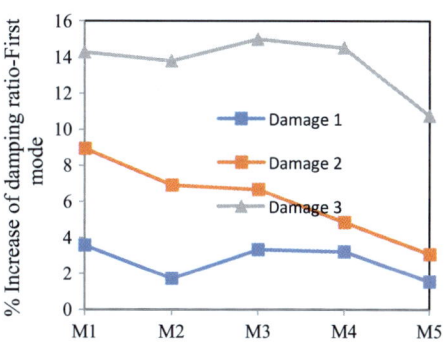

Fig. 7 Comparison of damping ratios for PPFRC—Mode 1

Fig. 8 Comparison of damping ratio for PPFRC—Mode 2

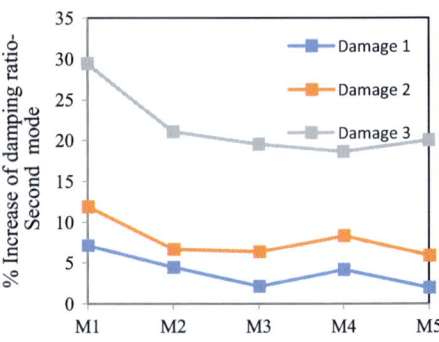

4   Conclusions

Based on the experimental tests, specific conclusions can be drawn as follows:

1. The static results emphasize that the addition of PP fibers in concrete significantly improves the flexural strength considerably when compared to compression and splitting tensile strength.
2. Addition of PP fibers to concrete resulted in an inherent increase of compressive strength, splitting tensile strength and flexural strength of PPFRCs. The result showed that 2.5–5% increase of compressive strength, 3.7–22.9% increase of splitting tensile strength and 5.1–33.5% increase of flexural strength for M2 and M4 mixtures than that of plain concrete.
3. The fundamental transverse frequency of concrete specimens tends to slightly change with environmental effect such as wind, humidity, temperature.
4. The damping ratio increased with the addition of PP fiber content and also with an increase of damage.
5. The natural frequency of plain and PPFRCs decreases with an increase of fiber content and also decreases with an increase of structural damage.

References

1. A. Sivakumar, N. Santhanam, A qualitative study on plastic shrinkage cracking in high strength hybrid fiber reinforced concrete. Cem. Concr. Compos. **29**, 575–581 (2007)
2. A. Senthil Kumar, S. Natesan, Effect of polypropylene addition on restrained shrinkage cracking of cement composites. J. Inst. Eng. Div. 100–106 (2004)
3. T. Harun, Statistical analysis for mechanical properties of polypropylene fiber reinforced light weight concrete containing silica fume exposed to high temperature. J. Mater. Des. **30**, 3252–3258 (2009)
4. A. Badr, A.F. Ashour, A.K. Platten, Statistical variations in impact resistance of polypropylene fiber reinforced concrete. Int. J. Impact Eng. **32**, 1907–1920 (2006)
5. M. Nili, V. Afroughsabet, The effects of silica fume and polypropylene fibers on the impact resistance and mechanical properties of concrete. Int. J. Impact Eng. **24**, 927–933 (2010)

6. Y. Mohammed, R. Azad, S. Singh, S. Kaushik, Impact resistance of steel fibrous concrete containing fibers of mixed aspect ratio. Constr. Build. Mater. **23**, 183–189 (2009)
7. M. Hsie, C. Tu, P.S. Song, Mechanical properties of polypropylene fiber reinforced concrete. Constr. Build. Mater. **492**, 153–157 (2008)
8. T. Gupta, R. Sharma, S. Chaudar, Impact resistance of concrete containing waste rubber and silica fume. Int. J. Impact Eng **83**, 76–87 (2015)
9. P.S. Song, H. Hwang, B.C. Sheu, Strength properties of nylon and polypropylene fiber reinforced concrete. Cem. Concr. Compos. **35**, 1546–1550 (2005)
10. O. Karahan, C.D. Atis, Durability properties of polypropylene fiber reinforced fly ash concrete. J. Mater. Des. **32**, 1044–1049 (2011)
11. H. Mohammadhosseini, A.S.M.A. Awal, Influence of palm oil fuel ash on fresh and mechanical properties of self-compacting concrete. Sadhana **40**, 1–11 (2015)
12. S. Orak, Investigation of vibration damping on polymer concrete with polyester resin. Cem. Concr. Res. **30**, 171–174 (2000)
13. L. Zheng, X.S. Huo, Y. Yuan, Experimental investigation on dynamic properties of rubberized concrete. Constr. Build. Mater. **22**, 939–947 (2008)
14. ASTM C 215, standard test method for fundamental transverse, longitudinal and torsional resonant frequencies of concrete specimens. American Society for Testing Materials (2008)

Design and Health Monitoring of Tensegrity Structures: An Overview

Neha Aswal and Subhamoy Sen

Abstract Tensegrity structures can be defined as structural mechanisms having separate tension and compression members, where compression members are discontinuous and float in a network of tension members. Before incorporating tensegrity into major construction works, stability and safety of tensegrity as a structure has to be studied and scrutinized properly. Being a recent design philosophy, there are not many elaborate literature reviews available on tensegrity structures pertaining to the field of civil engineering. This paper aims to bring together most of the research works done in design as well as health monitoring of tensegrity structures. So far, they have found practical application in stadium roofs, domes and bridges having demand of large column-free spaces. The studies relevant to design procedures involving form finding, structural stability and load analysis have been discussed. Very few researches, focusing on health monitoring of tensegrities, are available, which have also been discussed, thereafter highlighting the need of more research work in this field.

Keywords Form finding · Load analysis · Overview · Structural health monitoring · Structural stability · Tensegrity structures

1 Introduction

The formal origin of tensegrity structures dates back as early as 1973 when Snelson [1] introduced this particular structure type in the form of sculptures. The term "Tensegrity," a combination of two words—tension and integrity—was first coined by Buckminster Fuller in 1960s. As the name suggests, tensegrities are the structure types that derive their integrity from the balance of tension members. Fuller [2] defined tensegrity as: "An assemblage of tension and compression components

N. Aswal (✉) · S. Sen
Indian Institute of Technology Mandi, Mandi, India
e-mail: nehaaswal96@gmail.com

S. Sen
e-mail: subhamoy@iitmandi.ac.in

© Springer Nature Singapore Pte Ltd. 2020
P. V. Varde et al. (eds.), *Reliability, Safety and Hazard Assessment for Risk-Based Technologies*, Lecture Notes in Mechanical Engineering,
https://doi.org/10.1007/978-981-13-9008-1_43

arranged in a discontinuous compression system." Thus, these structures are basically made of two types of components: bars as compression members and cables or tendons as tension members, which together ensure a stable volume in the space and subsequently the stability and integrity of the structure through pre-stressing. Tensegrities are lightweight, self-stressed, pin-jointed bar framework with pure compression and tension members with discontinuous compression members. Though a relatively new research field, tensegrities have already found application in various fields, namely, sculpting, architecture, biomechanics, space [3] and civil structures.

Concept of tensegrity in construction has been in existence since long. Compared to their widespread applications at present date, earlier applications were mostly confined to mast, tent or dome constructions. With the improvement in design philosophy with tensegrity arrangements, these structure types became very popular due to their ease of construction, deployability and excellent relative stiffness. Within the domain of civil construction, suspension bridges can be categorized as typical tensegrities in which the tension is carried by cables and the compression by the posts. Tensegrity construction has also seen the application in roof construction requiring large column-free spaces, and thus has found a practical application in stadium roofs. The option of deployability has also been exploited in making the deployable roof, robotic arms, space telescopes, etc. Gilewski et al. [4] list various applications of tensegrities in civil engineering.

To name a few iconic constructions that used the concept of tensegrity structural system, there are Needle Tower I and II, located outside of the Hirshhorn Museum in Washington, D.C. and Kröller-Müller Museum, Netherlands, respectively, that are examples of public artwork designed by American sculptor, Kenneth Snelson. The much known, Munich Olympic Stadium (1972) designed by the architects Gunther Behnisch and Frei Otto closely resemble typical tensegrity structures. The tension in these lightweight structures was nullified by a network of compression members and tendons. The Seoul Olympic Gymnastics Hall (1988) and Georgia Dome (1996) are examples of the application of tensegrity structures, to make lightweight roofs of large covered structures. The Millennium Dome, made by Richard Rodgers Partnership (1996–1999), is another example, in which a membrane roof is kept at a place by a system of compression struts and cables. Other applications in tensegrity grid structures in civil engineering can be found in [5, 6]. Kurilpa Bridge, a bicycle and pedestrian bridge over the Brisbane River, designed by Cox Rayner Architects and Arup Engineers is one of the largest tensegrity bridges in the world.

Tensegrities are broadly classified into Class I (compression members are entirely discontinuous) and Class II (compression members are not discontinuous to their entirety and may originate/ terminate from/ to the same node) types. Tensegrities may be regular (faces are regular polygon) or irregular (faces are irregular polygon) and are geometrically nonlinear [7]. Designing a stable tensegrity structure involves majorly three steps: choosing the suitable pre-stress, form finding under that pre-stress and finally analyzing the structure under external loads [8]. When subjected to different external loads, the pre-stress, and hence the form of tensegrities, changes, thereby, creating a need for structural health monitoring (SHM) of tensegrities. There are several studies on the design concepts, form finding approaches and stability analysis of tensegrities, while an insignificant number of studies are focused on the SHM studies of tensegrities. To the authors' knowledge, there is no article that summarizes all the existing studies on tensegrities. This paper thus attempts to review the existing literature on tensegrities in order to identify the research gaps and the scope for further research.

2 Literature Review

Though the first tensegrity structure was built in 1948 [9], it was not till 1981 [10] that the design aspects of this type of structure were studied. Three major steps are identified in designing a tensegrity—form finding, structural stability and load analysis [11]. Form finding [3] and structural stability steps [12–15] of designing go hand-in-hand. Every time the external loading changes significantly, the form changes and hence the response of the structure, but for better understanding all topics, though very much interconnected, are discussed separately.

2.1 Form Finding

Form finding of a structure is the science and art of finding a suitable configuration of a structure conditioned on a particular loading arrangement under which the structure satisfies all its operational requirements perfectly. Hence, the very first step in the design of a tensegrity structure is to find its stable form. The first method introduced for form finding was a modified existent force density method that was well suited for tensile reticulated systems [16]. The result obtained from this method gives multiple solutions (geometries) for a given set of self-stresses. Various methods for form finding of tensegrities have been developed since then.

Form finding methods are broadly classified into kinematic and static methods. Kinematic methods imitate how the tensegrities are formed practically, i.e., by keeping the lengths of cables constant while increasing (or decreasing) the strut lengths until a maximum (or minimum) is reached. Under the umbrella of kinematic methods, there exist several contributions from different researchers: analytical solutions, nonlinear programing and dynamic relaxation [17].

For analytical solutions of form finding, it is considered that the tensegrity structure is globally rigid, i.e., there exists no other exactly similar configuration that could satisfy the cable and strut constraints [18]. Analytical solutions are, however, found to be suitable only for small, regular and symmetric tensegrities [17]. Nonlinear programing method introduced by Ohsaki and Zhang [19] is a force-deformation problem that can be solved by employing nonlinear optimization algorithms. However, the computational demand for such a method is reported to be huge [19]. Dynamic relaxation method, introduced by Motro [20], utilizes the dynamics of the tensegrities for form finding. It has been efficiently used for form finding of non-regular tensegrity structures [21] as well. This method, however, becomes cumbersome if form finding is done for several different ratios of strut and cable lengths [17]. Mathematical model for this method is tedious as it involves solving PDEs which are difficult to solve both analytically and numerically [22]. It has been reported that kinematic methods cannot be used when huge numbers of unknowns are present and they tend to become difficult to solve as the complexity of structure increases [17].

Over the time, due to the reported limitations of kinematic methods, static methods are given preference over them [17]. Static methods are characterized as methods in which a relationship is set up between equilibrium configurations of a structure with given topology and the forces in its members. Static methods generally include analytical solutions, force density method, energy method and reduced coordinates method [17].

Existent force density method had been modified to well suit the tensile reticulated systems and was applied to basic tensegrities [16]. Force density method is easier to compute as it makes the initial equilibrium equations linear by using tension coefficients. While it is best suited for searching new configurations, there is a difficulty in controlling the shape of the structure [10]. Force density method with a slight modification has been used by most of the researchers [23, 24] to handle cases with initially unknown element lengths. Variations include incorporation of shape constraints [23], assumption on structural symmetry [24], etc. Theoretical basis for numerical optimization tools used for large-scale problems is provided by simplifying problems using symmetric analysis.

Energy method uses the concept that a structure is stable when it is at its minimum potential energy [25, 26]. Force density and potential energy methods are however equivalent [17]. The reduced coordinate method introduced by Sultan [22] offers a greater control on the shape of the structure but requires extensive symbolic manipulations. This method was used for tensegrity towers and pre-stress condition are provided using the virtual work principle [27].

Many numerical methods have also been introduced to ease form finding of larger complex tensegrities [28–30]. Genetic algorithms have also been successfully applied in form finding of irregular structures and are found to be capable of giving rise to new unknown configurations [30, 31]. Stiffness matrix-based form finding method has been introduced by Zhang et al. [28] that utilizes stiffness and total potential energy to direct the rapid convergence of the structural configuration to the self-equilibrated and stable state from an arbitrary initial state. Numerical method independent of the material properties of the structures has also been developed in this attempt [32]. The method can be accurately applied to large structures and requires only initial stable topology of a tensegrity structure as input [32]. Form finding using FEM has been developed by Pagitz and Tur [33] that can be used to find configurations for topologies that are statically indeterminate, determinate or even kinematic. For this, a computational framework has been proposed by Koohestani [34] and design of tensegrity structures with/without super stability using unconstrained minimization. This method requires connectivity data and a random set of force densities for the initialization, enabling the formation of a wide variety of tensegrities with different geometrical and mechanical characteristics [34].

2.2 Structural Stability

Many propositions are given regarding stability analysis of tensegrity structures. One of the important proved [35] conjectures [10] is that a tensegrity framework with a strict proper self-stress is the second-order rigid if and only if the equivalent bar framework is the second-order rigid. This implies that stability of tensegrity depends on its degree of rigidity which is in turn dependent on the pre/self-stress of the tensegrity. Degree of rigidity is identified by finding the second-order rigidity of an equivalent bar framework [10, 35]. A plane tensegrity framework, in which the vertices and bars form a strictly convex polygon with additional cables across the interior, is rigid if and only if it is the first-order rigid [26]. In context to tensegrities, a stronger pre-stress stability is known as super stability [36] but a very high pre-stress level may lead to zero stiffness condition, i.e., it may lead to instability in an already stable structure [37].

Two principles are found to be true regarding tensegrities—*pretension vs. stiffness principle and small control energy principle* [38]. These are briefly summarized as follows: unless a string slackens, the stiffness of tensegrity structure does not decrease on increasing the applied load. Greater load will be required for slackening the strings having higher pretension. The shape of the structure can be changed with very small control energy as shape changes are achieved by changing the equilibrium of the structure.

If the force density matrix of a d-dimensional tensegrity is positive semi-definite (Hermitian matrix with positive eigenvalues) and has a minimum rank deficiency of $d + 1$ with the rank of geometry matrix being $d(d + 1)/2$, then the d-dimensional tensegrity structure is stable, irrespective of selection of materials and level of self-stresses [39]. Stability of dihedral symmetric prismatic tensegrity structures is not only related directly to the connectivity of members, but also sensitive to their geometry (height/radius ratio) and the level of self-stress and stiffness of members [40]. Equilibrium of a tensegrity is preserved under affine node position transformations [23]. It has been observed that the stability of the original tensegrity structure is not necessarily maintained after actuating into a new configuration as mechanisms may be developed; hence, stability conditions for clustered actuated tensegrity structures have been identified [41] in a later article.

2.3 Load Analysis

Mechanical behavior of self-stressed reticulated spatial systems is nonlinear due to their flexibility. Numerical method to study the behavior of tensegrities when acted upon by external actions (traction, compression, flexion and torsion actions) has been developed by Kebiche et al. [7]. It is based on the geometrical nonlinear analysis. Investigation of static stability of a three bar-six tendon tensegrity structure under external loads has been done by Lazopoulos [42] conforming to delay convention for stability (system state remains in a stable or meta-stable equilibrium state until that state disappears), considering two kinds of instabilities—global (overall) and local (strut buckling). Although force–displacement relationship is nonlinear, an analytically verified energy dissipation efficiency of a linear dynamic model of a three bar-six tendon tensegrity structure has been developed by Oppenheim and Williams [43].

For dynamic analysis of tensegrity structures, Lagrangian approach is typically used [44–47]. Linear dynamic model using harmonic modal analysis has been investigated for a three bar-six module tensegrity structure and a six-stage tensegrity beam [44]. Analysis of static and dynamic responses requires computation of infinitesimal mechanisms of tensegrity module and interpretation global deformation modes in terms of infinitesimal mechanism modes [45]. Member buckling and cable slacking scenario were considered to estimate the critical load [45].

Stiffness of a tensegrity mechanism increases with the application of an external tensile load and decreases with an application of compression load [46]. Under large displacements, the response due to compression loading on tensegrity prism changes from stiffening to softening [48]. Numerical method based on dynamic relaxation method for form finding and load analysis, particularly for tensegrity spline beam (Class II tensegrity) and grid shell structures, has also been developed [5]. Nonlinear elastoplastic analysis of tensegrity systems under static loads, using an updated Lagrangian formulation and a modified Newton–Raphson iterative scheme, has been developed by Kahla et al. [49] and Tran et al. [8], which considers both geometric and material nonlinearities. Member forces and deflections in double layer tensegrity grids (bars are confined between two parallel layers of the cables) are strongly affected by the span, structural depth and level of pre-stress [50].

2.4 Health Monitoring of Tensegrities

From analysis point of view, it is safe to say that there exists sufficient literature that deals with different structural aspects of a tensegrity structure. However, from the SHM perspective of such structures, not much studies have been done. With the advent of cheap sensor technology and computational facilities, vibration-based SHM has come up as an excellent approach to assess structural health. Typically, the health of any structure reflects on its dynamic properties. Thus, an anomaly in structure due to damage can always be sensed through analyzing its vibrational response [51–58].

Due to its unique approach to achieve stability and integrity, the stiffness of tensegrities largely depend on the pre-stressing in the tendons and the static and dynamic buckling properties of the compression members. For example, slacking in the cable may cause reduction in stiffness and also may potentially hamper the stability of the structure as a whole. Safety of tensegrity structures can be qualitatively measured by two factors: (a) global stiffness and stability of the structure and (b) the residual stiffness that may come useful in the occurrence of abnormal loading conditions

[59]. To prevent buckling in the compression members, the flexural rigidity can be increased in the expense of more material which in turn increases the dead weight. On the other hand, increasing pre-stressing in the cables while may ensure better stiffness can actually reduce the residual stiffness in the system which makes the structure more vulnerable to sudden failure.

While improved design concept can avert failure of a structure that is yet to be constructed, for the in-service structures regular monitoring, maintenance and retrofitting is the only way. SHM is one such approach that can be employed to continually record the structural response and identify any anomaly that has occurred. Typically, vibrational properties of tensegrities are found to be well correlated with the actual health of the structure and its components [15]. Any change in tension in the cable, buckling in the bars or damage in any of the components should thus ideally be possible to be identified through proper analyzing the vibrational signatures masked in the recorded measurements such as acceleration.

In the context of dynamic property-based designing of tensegrity structures, several researchers have studied the relation between tensegrity design parameters to its corresponding modal parameters employing a forward approach [60, 61]. Ashwear et al. [15] discussed the effect of pre-stress level on the natural frequencies of tensegrity structures. In a later research, Ashwear et al. [59] investigated the impact of temperature on the vibrational properties of tensegrity structure. In a recent article, Ashwear et al. [62] demonstrate the usage of vibrational SHM techniques for tensegrity arrangements. Sultan [60] attempted natural frequency separation technique for designing tensegrity structure. Faroughi et al. [61] studied an inverse approach to investigate the impact of vibrational properties on the structural design. They further introduced an algorithm for designing tensegrities that takes the dynamic properties into account. This approach although is not targeted to SHM of tensegrity systems but is relevant because of the inverse approach that has been considered to take dynamic behavior into consideration for deign purpose. Typical SHM techniques follow a similar line of inverse approach. One of the relevant studies of SHM of tensegrities can be found in Panigrahi et al. [63] and Bhalla et al. [64] in which experiments are performed on a single module tensegrity structure in order to employ SHM techniques to detect damage in it.

However, apart from this, studies that discuss the identification of vibrational properties of any tensegrity structure and use them inversely to detect any anomaly in the structure are not abandon. Raja [65] investigated control aspects of tensegrities under random excitation. Prof. Suresh Bhalla and his group contributed significant numbers of research articles which are more pertinent to structural design research. Their studies are related to behavior study [66], system identification and damage detection [63], design of deployable structures [67, 68], etc. Panigrahi et al. [69] experimented on a tensegrity structure in their effort to employ a low-cost electro-mechanical impedance-based technique for SHM. Their research also investigated the application of neural network in tensegrity structure design [70] and form finding [71].

Besides, the dynamic property of tensegrities gets affected by temperature variations. Detection of cable slacking, member buckling or damage through modal comparison thus may often lead to a false prediction. Moreover, tensegrities are perceived to be affected the most due to these variations because of their lightweight construction philosophy [59]. However, there is only one instance of studies that discussed this issue for tensegrities [59].

3   Conclusions

This paper briefly summarizes all the existing research on tensegrity structures relevant to form finding, stability and/or load analysis and SHM. It has been observed that unlike form finding or load analysis, the research in SHM on tensegrities is significantly limited. However, the plausibility of developing a reliable SHM tool for tensegrities lies in the fact that a proper forward modeling approach should be available, which in turn depends on a clear understanding about the form finding and load analysis steps. Thus, SHM for tensegrities cannot be isolated from the other aspects described in this paper which proves the relevance as well as the requirement of discussion on all these aspects in a single premise.

References

1. K. Snelson, *Tensegrity Masts* (Shelter Publication, Bolinas, CA, 1973)
2. R.B. Fuller, Tensile-integrity structures, U.S. Patent 3,063,521, issued November 13, 1962
3. G. Tibert, Deployable tensegrity structures for space applications, Ph.D. dissertation, KTH (2002)
4. W. Gilewski, J. Kłosowska, P. Obara, Applications of tensegrity structures in civil engineering. Proc. Eng. **111**, 242–248 (2015)
5. S.M.L. Adriaenssens, M.R. Barnes, Tensegrity spline beam and grid shell structures. Eng. Struct. **23**(1), 29–36 (2001)
6. J. Quirant, M.N. Kazi-Aoual, R. Motro, Designing tensegrity systems: the case of a double layer grid. Eng. Struct. **25**(9), 1121–1130 (2003)
7. K. Kebiche, M.N. Kazi-Aoual, R. Motro, Geometrical non-linear analysis of tensegrity systems. Eng. Struct. **21**(9), 864–876 (1999)
8. H.C. Tran, J. Lee, Geometric and material nonlinear analysis of tensegrity structures. Acta. Mech. Sin. **27**(6), 938–949 (2011)
9. K. Snelson, Snelson on the tensegrity invention. Int. J. Space Struct. **11**(1–2), 43–48 (1996)
10. B. Roth, W. Whiteley, Tensegrity frameworks. Trans. Am. Mat. Soc. **265**(2), 419–446 (1981)
11. M. Schenk, Statically balanced tensegrity mechanisms-A literature review, *Department of Bio-Mechanical Engineering, Delft University of Technology* (2005)
12. H. Furuya, Concept of deployable tensegrity structures in space application. Int. J. Space Struct. **7**(2), 143–151 (1992)
13. C. Sultan, M. Corless, R.E. Skelton, Linear dynamics of tensegrity structures. Eng. Struct. **24**(6), 671–685 (2002)
14. B. Moussa, N.B. Kahla, J.C. Pons, Evolution of natural frequencies in tensegrity systems: a case study. Int. J. Space Struct. **16**(1), 57–73 (2001)
15. N. Ashwear, A. Eriksson, Natural frequencies describe the pre-stress in tensegrity structures. Comput. Struct. **138**, 162–171 (2014)

16. N. Vassart, R. Motro, Multi-parametered form finding method: application to tensegrity systems. Int. J. Space Struct. **14**(2), 147–154 (1999)
17. A.G. Tibert, S. Pellegrino, Review of form-finding methods for tensegrity structures. Int. J. Space Struct. **26**(3), 241–255 (2011)
18. R. Connelly, M. Terrell, Globally rigid symmetric tensegrities. Struct. Topol. *1995 núm 21* (1995)
19. M. Ohsaki, J.Y. Zhang, Nonlinear programming approach to form-finding and folding analysis of tensegrity structures using fictitious material properties. Int. J. Solids Struct. **69**, 1–10 (2015)
20. R. Motro, Tensegrity systems and geodesic domes. Int. J. Space Struct. **5**(3–4), 341–351 (1990)
21. L. Zhang, B. Maurin, R. Motro, Form-finding of nonregular tensegrity systems. J. Struct. Eng. **132**(9), 1435–1440 (2006)
22. C. Sultan, Modeling, design, and control of tensegrity structures with applications (1999)
23. M. Masic, R.E. Skelton, P.E. Gil, Algebraic tensegrity form-finding. Int. J. Solids Struct. **42**(16–17), 4833–4858 (2005)
24. R.P. Raj, S.D. Guest, Using symmetry for tensegrity form-finding. J. Int. Assoc. Shell Spatial Struct. **47**(3), 245–252 (2006)
25. R. Connelly, Rigidity and energy. Inventiones Mathematicae **66**(1), 11–33 (1982)
26. R. Connelly, Rigidity, in *Handbook of Convex Geometry, Part A* (1993), pp. 223–271
27. C. Sultan, M. Corless, R.E. Skelton, Reduced prestressability conditions for tensegrity structures, in *40th Structures, Structural Dynamics, and Materials Conference and Exhibit* (1999), p. 1478
28. L.Y. Zhang, Y. Li, Y.P. Cao, X.Q. Feng, Stiffness matrix based form-finding method of tensegrity structures. Eng. Struct. **58**, 36–48 (2014)
29. H.C. Tran, J. Lee, Advanced form-finding of tensegrity structures. Comput. Struct. **88**(3–4), 237–246 (2010)
30. X. Xu, Y. Luo, Form-finding of non regular tensegrities using a genetic algorithm. Mech. Res. Commun. **37**(1), 85–91 (2010)
31. C. Paul, H. Lipson, F.J.V. Cuevas, Evolutionary form-finding of tensegrity structures, in *Proceedings of the 7th Annual Conference on Genetic and Evolutionary Computation* (ACM, New York, 2005), pp. 3–10
32. A. Micheletti, W. Williams, A marching procedure for form-finding for tensegrity structures. J. Mech. Mater. Struct. **2**(5), 857–882 (2007)
33. M. Pagitz, J.M.M. Tur, Finite element based form-finding algorithm for tensegrity structures. Int. J. Solids Struct. **46**(17), 3235–3240 (2009)
34. K. Koohestani, A computational framework for the form-finding and design of tensegrity structures. Mech. Res. Commun. **54**, 41–49 (2013)
35. R. Connelly, W. Whiteley, Second-order rigidity and prestress stability for tensegrity frameworks. SIAM J. Discrete Math. **9**(3), 453–491 (1996)
36. R. Connelly, Tensegrity structures: why are they stable? in *Rigidity Theory and Applications* (Springer, Boston, 2002), pp. 47–54
37. S.D. Guest, The stiffness of tensegrity structures. IMA J. Appl. Math. **76**(1), 57–66 (2010)
38. R.E. Skelton, R. Adhikari, J.P. Pinaud, W. Chan, J.W. Helton, An introduction to the mechanics of tensegrity structures, in *Decision and Control, 2001. Proceedings of the 40th IEEE Conference, vol. (5)* (2001), pp. 4254–4259
39. J.Y. Zhang, M. Ohsaki, Stability conditions for tensegrity structures. Int. J. Solids Struct. **44**(11–12), 3875–3886 (2007)
40. J.Y. Zhang, S.D. Guest, M. Ohsaki, Symmetric prismatic tensegrity structures: Part I. Configuration and stability. Int. J. Solids Struct. **46**(1), 1–14 (2009)
41. K.W. Moored, H. Bart-Smith, Investigation of clustered actuation in tensegrity structures. Int. J. Solids Struct. **46**(17), 3272–3281 (2009)
42. K.A. Lazopoulos, Stability of an elastic tensegrity structure. Acta Mech. **179**(1–2), 1–10 (2005)
43. I.J. Oppenheim, W.O. Williams, Geometric effects in an elastic tensegrity structure. J. Elasticity Phys. Sci. solids **59**(1–3), 51–65 (2000)

44. H. Murakami, Static and dynamic analyses of tensegrity structures. Part 1. Nonlinear equations of motion. Int. J. Solids Struct. **38**(20), 3599–3613 (2001)
45. H. Murakami, Static and dynamic analyses of tensegrity structures. Part II. Quasi-static analysis. Int. J. Solids Struct. **38**(20), 3615–3629 (2001)
46. M. Arsenault, C.M. Gosselin, Kinematic, static, and dynamic analysis of a planar one-degree-of-freedom tensegrity mechanism. J. Mech. Des. **127**(6), 1152–1160 (2005)
47. M. Arsenault, C.M. Gosselin, Kinematic, static, and dynamic analysis of a spatial three-degree-of-freedom tensegrity mechanism. J. Mech. Des. **128**(5), 1061–1069 (2006)
48. A. Amendola, G. Carpentieri, M. De Oliveira, R.E. Skelton, F. Fraternali, Experimental investigation of the softening–stiffening response of tensegrity prisms under compressive loading. Compos. Struct. **117**, 234–243 (2014)
49. N.B. Kahla, K. Kebiche, Nonlinear elastoplastic analysis of tensegrity systems. Eng. Struct. **22**(11), 1552–1566 (2000)
50. A. Hanaor, M.K. Liao, Double-layer tensegrity grids: static load response. Part I: analytical study. J. Struct. Eng. **117**(6), 1660–1674 (1991)
51. S.W. Doebling, C.R. Farrar, M.B. Prime, D.W. Shevitz, Damage identification and health monitoring of structural and mechanical systems from changes in their vibration characteristics: a literature review (1996)
52. H. Sohn, A review of structural health monitoring literature: 1996–2001, in *LANL Report* (2004)
53. E.P. Carden, P. Fanning, Vibration based condition monitoring: a review. Struct. Health Monit. **3**(4), 355–377 (2004)
54. S.W. Doebling, C.R. Farrar, M.B. Prime, A summary review of vibration-based damage identification methods. Shock Vibr. Digest **30**(2), 91–105 (1998)
55. Y.J. Yan, L. Cheng, Z.Y. Wu, L.H. Yam, Development in vibration-based structural damage detection technique. Mech. Syst. Signal Process. **21**(5), 2198–2211 (2007)
56. W. Fan, P. Qiao, Vibration-based damage identification methods: a review and comparative study. Struct. Health Monit. **10**(2), 83–111 (2011)
57. C.R. Farrar, S.W. Doebling, P.J. Cornwell, E.G. Straser, *Variability of modal parameters measured on the Alamosa Canyon Bridge*. No. LA-UR-96-3953; CONF-970233-7. Los Alamos National Lab., NM (United States), 1996
58. O.S. Salawu, Detection of structural damage through changes in frequency: a review. Eng. Struct. **19**(9), 718–723 (1997)
59. N. Ashwear, A. Eriksson, Influence of temperature on the vibration properties of tensegrity structures. Int. J. Mech. Sci. **99**, 237–250 (2015)
60. C. Sultan, Designing structures for dynamical properties via natural frequencies separation: Application to tensegrity structures design. Mech. Syst. Signal Process. **23**(4), 1112–1122 (2009)
61. S. Faroughi, J.M.M. Tur, Vibration properties in the design of tensegrity structure. J. Vib. Control **21**(3), 611–624 (2015)
62. N. Ashwear, A. Eriksson, Vibration health monitoring for tensegrity structures. Mech. Syst. Signal Process. **85**, 625–637 (2017)
63. R. Panigrahi, A. Gupta, S. Bhalla, Damage assessment of tensegrity structures using piezo transducers, in *ASME 2008 Conference on Smart Materials, Adaptive Structures and Intelligent Systems* (2008), pp. 21–25
64. S. Bhalla, R. Panigrahi, A. Gupta, Damage assessment of tensegrity structures using piezo transducers. Meccanica **48**(6), 1465–1478 (2013)
65. M.G. Raja, S. Narayanan, Active control of tensegrity structures under random excitation. Smart Mater. Struct. **16**(3), 809 (2007)
66. A. Gupta, S. Bhalla, R. Panigrahi, Behaviour of foldable tensegrity structure, in *Keynote paper, 3rd Specialty Conference on the Conceptual Approach to Structural Design* (2005), pp. 9–16
67. R. Panigrahi, A. Gupta, S. Bhalla, Dismountable steel tensegrity grids as alternate roof structures. Steel Compos. Struct. **9**(3), 239–253 (2009)

68. R. Panigrahi, S. Bhalla, A. Gupta, Development and analysis of a prototype dismountable tensegrity structures for shelter purposes. Int. J. Earth Sci. Eng. **3**(4), 561–578 (2010)
69. R. Panigrahi, S. Bhalla, A. Gupta, A low-cost variant of electro-mechanical impedance (EMI) technique for structural health monitoring. Exp. Tech. **34**(2), 25–29 (2010)
70. S.N. Panigrahi, C.S. Jog, M.L. Munjal, Multi-focus design of underwater noise control linings based on finite element analysis. Appl. Acoust. **69**(12), 1141–1153 (2008)
71. R. Panigrahi, A. Gupta, S. Bhalla, K. Arora, Application of artificial neural network for form finding of tensegrity structures, in *IICAI* (2005), pp. 1950–1962

Performance Prediction of Rolling Element Bearing with Utilization of Support Vector Regression

Shivani Chauhan, Pradip Yadav, Prashant Tiwari, S. H. Upadhyay and Niraj Mishra

Abstract Bearings are customary and significant components in any rotating machinery. For an ideal operational rotating system, bearing assumes an indispensable part. Once a catastrophe happens, it will cause gigantic budgetary incidents and even safety hazards. Therefore, it is required to execute a performance assessment and state prediction adequately. Roused from the ongoing works and progress outline, we propose a one-step-ahead prediction strategy in light of the support vector regression analysis over a degradation indicator. From the obtained signal, time-domain feature was computed with the usage of ensemble empirical mode decomposition (EEMD) strategy and further categorized through k-means. The categorization procedure creates the clusters and healthy data cluster has been taken as the reference for degradation indicator calculation. The same has been anticipated utilizing support vector regression (SVR) with one-step-ahead strategy. The outcomes showed the proficiency is at standard and impressively higher than existing strategies.

Keywords Bearings · Performance assessment · Prediction strategy · Support vector regression

S. Chauhan · P. Yadav · P. Tiwari (✉) · N. Mishra
National Institute of Technology Uttarakhand, Srinagar Garhwal, Uttarakhand, India
e-mail: prashant.tiwari.om@gmail.com

S. Chauhan
e-mail: shivanichauhan4148@gmail.com

P. Yadav
e-mail: pradipy51@gmail.com

N. Mishra
e-mail: nkm@nituk.ac.in

P. Tiwari · S. H. Upadhyay
Indian Institute of Technology Roorkee, Roorkee, Uttarakhand, India
e-mail: shumefme@iitr.ac.in

© Springer Nature Singapore Pte Ltd. 2020
P. V. Varde et al. (eds.), *Reliability, Safety and Hazard Assessment for Risk-Based Technologies*, Lecture Notes in Mechanical Engineering,
https://doi.org/10.1007/978-981-13-9008-1_44

1 Introduction

Bearings are conventional as well as crucial components in any rotating machinery. For the perfect operational rotating system, bearing plays a vital role. As per statistics, 30% of the failures of rotating apparatus are caused by local harm or deformities of rolling bearing [1]. Once a failure occurs, it will cause enormous financial misfortunes and even safety hazards. Several kinds of literature reported the importance and progress of the prognostic approach for assessment of bearing conditions. Degradation starts as the life progresses. Whatever be the reason of degradation of performance, the product can no longer meet the required level of service. Consequently, failure detection and degradation assessment become highly crucial. To forecast the impending failure, changes in the level of performance is very convenient. With this, prognostics come into the picture. With prognostics, usage of equipment can be increased and maintenance due to sudden failure can be avoided [2–9].

Conclusions can be derived from the above exploration which can be established as fundamental for the proposed bearing health assessment method. Needless to say, the vibration signals of bearings contain countless information which can change the working states of machinery [3]. A health indicator is an essential part of the assessment modal which clearly confirms the current working state of the component at an interval of the defined time step. Prediction of the health-state indicator is the last stage in evaluation modal which may be performed step-by-step to carry out the decisions on scheduling the maintenance. Selection of the correct regression method and classification modal governs the efficiency and productivity of method.

Inspired by the recent works and trend, a one-step-ahead prediction method is proposed based on the support vector regression analysis over degradation indicator. From the acquired signal, a time-domain feature was calculated with the implementation of ensemble empirical mode decomposition (EEMD) strategy and further classified through k-means. The classification process generates the clusters, and healthy data cluster has been taken as the reference for degradation indicator calculation. The same has been predicted using SVR with one-step-ahead strategy. The results demonstrated the efficiency is at par and considerably higher than existing methods.

2 Assessment Procedure and Prediction Model

2.1 Review of EEMD

EMD is an efficient tool for non-stationary signal analysis and proven as better over fast Fourier transform (FFT), short-time Fourier transform (STFT), and wavelet packet transform (WPD). However, the decomposition of a signal into monocomponents, i.e., intrinsic mode function (IMF), by EMD suffers from end-effect complications and mode mixing [7]. Mode mixing, the phenomenon of retaining a signal belonging to a particular scale over many IMFs, results undistinguished erratic bearing signal. To ease this disadvantage, Wu and Huang proposed another noise-assisted information examination strategy named ensemble EMD (EEMD), which fundamentally diminishes the possibility of undue mode blending, strikingly improves the soundness of calculation, and reflects the dyadic property of the deterioration for any information. It can shortly present the mathematical algorithm of EEMD as follows [8]:

1. Add a white noise, $w(t)$, to target signal, named $t_1(t)$; therefore,

$$t_2(t) = t_1(t) + w(t) \tag{1}$$

2. Use EMD algorithm to decompose the acquired signal $t_2(t)$.

    i. Recognize all the extremes, i.e., maxima and minima of the signal $x(t)$.

    ii. Find out the mean function of upper and lower envelopes, $m(t)$ and taking the difference,

$$D(t) = x(t) - m(t). \tag{2}$$

    iii. The iteration will stop if the difference signal will become the zero mean process and $D(t)$ will be considered as IMF_1, $C_1(t)$. If it is not the case, then replace $x(t)$ with $D(t)$ in step 1.

    iv. Determine the residue signal

$$E(t) = x(t) - C_1(t). \tag{3}$$

$$x(t) = \sum_{i=1}^{n} c_i(t) + E(t) \tag{4}$$

3. Repeat the above-stated steps, taking different white noise series of same power, every time. The new IMF combination $F_{ij}(t)$ is obtained

where $i = $ iteration number, $j = $ IMF scale.
Determine the mean of finally obtained IMF of decompositions as desired output:

$$EEMD = \sum_{i=1}^{ni} F_{ij}(t) \tag{5}$$

where ni are white noise trial numbers.

2.2 Support Vector Regression (SVR)

The basic idea of SVR [9] to transform the low-dimensional data L to a high-dimensional space M by utilizing nonlinear mapping function Ω is used in SVR. This is followed by linear regression in the high-dimensional space. For provided data set $(x_i, y_i)(i = 1, 2, \ldots, k)$ where y_i is the desired value. Regression can be understood by the following equations [10]:

$$y = f(x) = (\omega, \Omega(x)) + v \tag{6}$$

where v is offset value, $\Omega : R^n \rightarrow M, \omega \rightarrow M$

$$\min y = 1/2\|\omega^2\| + T \sum_{j=1}^{k}(\beta_i^* + \beta_i) \tag{7}$$

$$y_i - \omega_i, \Omega(x_i) - v \ll \varepsilon + \beta_i \tag{8}$$

$$(\omega, \Omega(x)) + v - y \ll \varepsilon + \beta_i^*$$

where T is a penalty parameter β_i^* and β_i are relaxation considered for linear inseparable boundary. By introducing the kernel function equation becomes:

$$w(\alpha_i, \alpha_i^*) = -1/2 \sum_{i=1}^{k}((\alpha_i - \alpha_i^*)(\alpha_j - \alpha_j^*)(\Omega(x_i).\Omega(x_j)))$$

$$+ \sum(\alpha_i + \alpha_i^*)(j_i - \varepsilon) \tag{9}$$

The kernel function $k(x_i, x) = (\Omega(x_i), \Omega(x))$ is utilized to figure out the inner product when the support vector regression is nonlinear.

2.3 Layout of Proposed Work

The figure demonstrates the step-by-step procedure followed in the prediction of performance of the rolling element bearing. After collection of considered signals, IMFs were calculated; clustering is done using k-means followed by health assessment and prediction (Fig. 1).

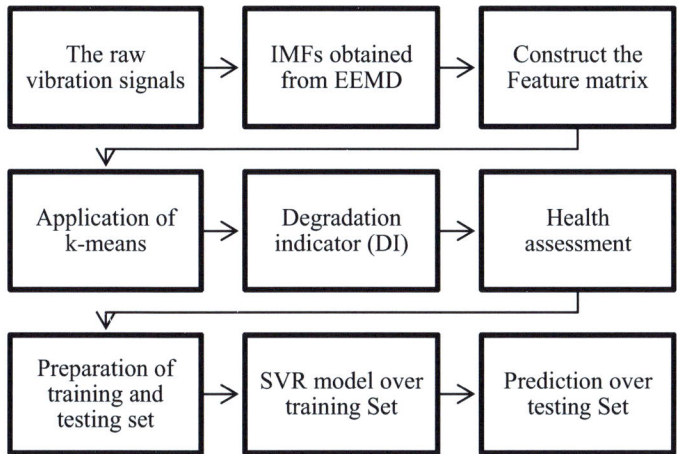

Fig. 1 Flowchart of the proposed health assessment approach

3 Validation

For validation of proposed method, the vibration signatures considered for computation purpose are taken from the experiment runs in the Industry/University Cooperative Research Centers Program (I/UCRC) for Intelligent Maintenance Systems. An alternating motor is used to rotate the bearing at 2000 RPM with load of 6000 lbs, bearing signal is acquired using national instrument data acquisition (NI DAQ) card 6062E, the accelerometer utilized is PCB 353B33 and the sampling rate was 20 kHz with data length were taken 20,480 points [11].

4 Result and Discussion

4.1 Considered Signal for Validation

In this section, the proposed prognostic approach is tested on the bearing experimental signals to show its adequacy in bearing flaw prognostics. The aftereffects of the recommended approach are evaluated and talked about in detail. The prime intent of PDA approach is to arrange the bearing degradation over the entire lifetime of the bearing into four phases, particularly normal, slight, severe, and failure. The time-domain vibration signal of the bearing for the whole life span is shown in Fig. 2. It tends to be understood that no proper conclusion can be made by observing the vibration signals. Hence, a correct and proper health assessment has to be conducted for the same.

Fig. 2 Bearing vibration
signal for complete lifetime

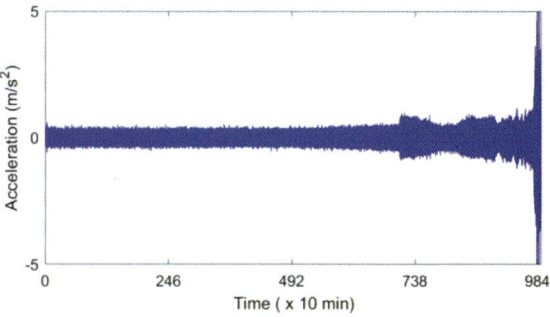

4.2 Heath Assessment

After obtaining the signals, steps mentioned in Sect. 2.3 are executed to acquire the
degradation indicators for the bearing. The signal samples gathered at progressive
time moments are first decomposed through the EEMD approach. The next stage
includes the extraction of features from the bearing signals. The IMF energy for the
bearing has been taken as features for characterization of condition. Figure 3 portrays
the degradation index of the bearing signal in the whole life and in the expanded
view, respectively. It is stable up to a time step of 534. At 534 inspection point,
the DI curve drops to a lower value of 0.9941 indicating that the slight degradation
in bearing has begun. These variations become more and more pronounced as the
bearing defect advances. After which, the DI curve keeps on fluctuating, which is
due to the smoothening effect of the crack edges. While the fault propagates, the
crack surfaces are repeatedly smoothened and sharpened as the bearing roller meets
them. Beyond the 800 steps, the HI curve continues to fall and the bearing progresses
rapidly from severe degradation phase to failure.

4.3 Prediction

The effectiveness of the degradation indices and its efficiency in prediction making
has been cross-verified utilizing the one-step-ahead prediction model discussed in
Sect. 2.2. The training set constructed according to illustration provided in Table 1
contains the healthy and faulty DI (400 to 550) to construct the SVR model. The
same model is then utilized to predict the next 50 steps with one-step-ahead strategy.
Figure 4 clearly indicates that the predicted value follows the similar trend as the
original DI. It may be concluded that the proposed DI is well groomed to indicate
the condition without perusing the flaws in trend ability and monotonicity. Selected
parameters for the preparation of support vector regression are mentioned in Table 2.

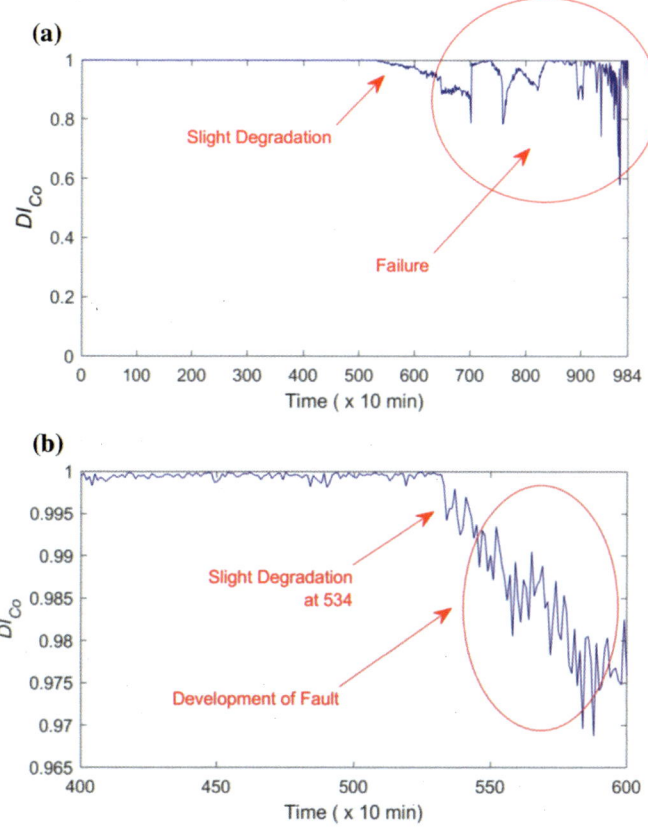

Fig. 3 **a** Degradation indicator for the health assessment of bearing corresponds to whole life, **b** zoomed view of (**a**)

Table 1 Training set for direct prediction technique

Input $X = [X_1, X_2, X_3, \ldots, X_d]$	Output $Y = [Y_1, Y_2, Y_3, \ldots, Y_d]$
$[x_1, x_2, x_3, \ldots, x_d]$	$[x_{d+1}]$
$[x_2, x_3, x_4, \ldots, x_{d+1}]$	$[x_{d+2}]$
$\vdots$	$\vdots$
$[x_{t-h-d+1}, x_{t-h-d+2}, x_{t-h-d+3}, \ldots, x_{t-h}]$	$[x_{t-h+1}]$

Table 2 Selected parameters for the preparation of support vector regression

Parameters	Values
Kernel function	Gaussian
Kernel offset	0
Kernel scale factor	0.1

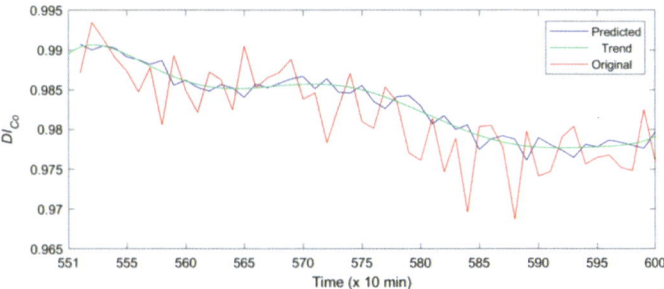

Fig. 4 Predicted and original values of DI on SVR-based model

5 Conclusion

This section summarizes the work done in this paper. Based on the above discussions, the conclusions drawn from this present work are as follows:

1. Bearing prognostics is crucial to the uninterrupted operation of rotating machines. Degradation tracking and remaining useful life (RUL) estimation are the two milestones in prognostics and health management (PHM) of bearings.
2. The fault feature extraction plays an important role in predicting the bearing degradation phases. The combination of EEMD, k-means, and degradation indicator is capable to obtain an effective index for tracking the evolution of degradation in bearings.
3. Direct prediction technique based on SVR is validated on the generated DI which is found to be effective and has ability to forecast the degradation state with minimum error. It can track the performance and degradation behavior of bearing with the ability of judgment toward opening and flattening of cracks. This ability of model is crucial for the advance knowledge of RUL of bearings.

References

1. X. Zhao, M. Jia, Fault diagnosis of rolling bearing based on feature reduction with global-local margin fisher analysis. Neurocomputing (2018)
2. H.E. Kim, A.C. Tan, J. Mathew, E.Y. Kim, B.K. Choi, Prognosis of bearing failure based on health state estimation, in *Engineering Asset Lifecycle Management* (Springer, London, 2010), pp. 603–613
3. S. Janjarasjitt, H. Ocak, K.A. Loparo, Bearing condition diagnosis and prognosis using applied nonlinear dynamical analysis of machine vibration signal. J. Sound Vib. **317**(1–2), 112–126 (2008)
4. P. Boškoski, M. Gašperin, D. Petelin, Đ. Juričić, Bearing fault prognostics using Rényi entropy based features and Gaussian process models. Mech. Syst. Signal Process. **52**, 327–337 (2015)
5. H.E. Kim, A.C. Tan, J. Mathew, B.K. Choi, Bearing fault prognosis based on health state probability estimation. Expert Syst. Appl. **39**(5), 5200–5213 (2012)

6. B.S. Yang, A.C.C. Tan, Multi-step ahead direct prediction for the machine condition prognosis using regression trees and neuro-fuzzy systems. Expert Syst. Appl. **36**(5), 9378–9387 (2009)
7. X. Zhang, Y. Liang, J. Zhou, A novel bearing fault diagnosis model integrated permutation entropy, ensemble empirical mode decomposition and optimized SVM. Measurement **69**, 164–179 (2015)
8. K.M. Chang, Arrhythmia ECG noise reduction by ensemble empirical mode decomposition. Sensors **10**(6), 6063–6080 (2010)
9. R. Golafshan, K.Y. Sanliturk, SVD and Hankel matrix based de-noising approach for ball bearing fault detection and its assessment using artificial faults. Mech. Syst. Signal Process. **70**, 36–50 (2016)
10. X. Zhang, J. Zhou, Multi-fault diagnosis for rolling element bearings based on ensemble empirical mode decomposition and optimized support vector machines. Mech. Syst. Signal Process. **41**(1–2), 127–140 (2013)
11. J. Lee, H. Qiu, G. Yu, J. Lin, Rexnord Technical Services 'Bearing data set', IMS, University of Cincinnati. NASA Ames prognostics data repository (2007)

Statistical Variation in Creep of Concrete—An Experimental Investigation

G. S. Vijaya Bhaskara, K. Balaji Rao and M. B. Anoop

Abstract Design of creep-sensitive concrete structures requires not only prediction of mean values of creep strains but also bounds to account for deviations. Hence, there is a need to understand how mean and deviations of the creep strains from its mean values evolve in time, to develop an appropriate stochastic model. Such studies are scant since it requires high-speed data acquisition. Results of experimental investigations carried out at CSIR-SERC, on creep and shrinkage of OPC concrete for the loading duration of 505 days after loading at 28 days, under both sealed and unsealed conditions, are presented in this paper. The rates of variation in basic and total creep suggest that basic creep rates are higher at ages beyond ten days and also noted that percentage contribution of basic creep towards total creep is higher at later ages due to moisture presence in sealed specimen; hence, consideration of basic creep is imperative in the design of closed web structures (viz. box girders) and mass concrete structures (viz. dams). From a statistical analysis of creep strains, it is noted that the histograms of square of instantaneous creep strain deviations from its mean value ε_{dev}^2 are unimodal up to 0.93 days after loading, bi-modal from 0.93 to 40.56 days after loading and tri-modal after 40.56 days after loading. From the variations of mean and mean-square deviations of creep strains with respect to time, it is noted the creep evolution is a non-stationary stochastic process. Moisture presence and local load variations in specimen in the initial stages after loading seem to set the initial conditions for variation in the evolution of creep strains with time. More experimental investigations need to be carried out in this direction.

Keywords Creep · OPC concrete · Statistical variation · Experimental investigation

---

G. S. Vijaya Bhaskara (✉) · K. Balaji Rao · M. B. Anoop
CSIR-Structural Engineering Research Centre, Chennai, India
e-mail: vbhaskara@serc.res.in

K. Balaji Rao
e-mail: balaji@serc.res.in

M. B. Anoop
e-mail: anoop@serc.res.in

545

P. V. Varde et al. (eds.), *Reliability, Safety and Hazard Assessment for Risk-Based Technologies*, Lecture Notes in Mechanical Engineering,
https://doi.org/10.1007/978-981-13-9008-1_45

1 Introduction

A survey on several concrete bridges monitored over a period of more than 20 years has shown that creep deflections far exceed the predictions made based on recommendations given in several national codes, giving rise to a wake-up call for better understanding and characterization of long-term creep of concrete [1, 2]. Creep is a complex phenomenon involving several possible mechanisms of material motion. Design of creep-sensitive concrete structures requires not only prediction of mean values of creep strains but also bounds to account for deviations. Hence, there is a need to understand how deviations of the creep strains from its mean values evolve with time that helps in the development of an appropriate stochastic model. Such studies are scanty in the literature. From the review of the literature on experimental investigations, it is noted that a number of measured creep strain data points over time are very less and in most of the studies initial reading after loading is recorded after several hours. Due to limited measured creep strain data from each test, it is difficult to estimate deviations of creep strains at different instants of time and hence predict the evolution of mean and deviations. To understand the variation of concrete creep strain deviations with time, an experimental investigation is carried out at CSIR-SERC, Chennai. The evolution of strains is monitored at closer time intervals in concrete cylinders made of the same concrete mix, simultaneously loaded in the same testing machine, exposed to the same exposure conditions, to avoid the deviations due to external factors, such as environment and loading. A statistical analysis of experimental data is carried out to see how creep strains vary with time. The details and the results of experimental investigations carried out at CSIR-SERC on creep and shrinkage (measured from load-free specimens) are given in the next section.

2 Experimental Investigations at CSIR-SERC

With an aim to understand the time evolution of mean and deviation of creep strains, ordinary Portland cement concrete specimens for creep and shrinkage and for compressive strength of concrete are cast. Twelve 150-mm dia. and 300-mm Ht. cylinders are cast. Six specimens are loaded, in which three specimens are sealed and three specimens kept unsealed to obtain both basic and total creep components. Remaining six specimens are used for measuring shrinkage strains, in which three specimens are sealed and remaining three specimens are kept unsealed to obtain both autogenous and total shrinkage strains. Eighteen companion concrete cubes intended to test for compressive strength at the concrete age of 7, 28, 56, 90, 180 and 365 days are also cast. Specimens intended for creep are loaded at 28 days. The evolution of strains and internal temperature with time are being monitored in loaded (under uniaxial compressive stress of about 8.5 MPa) and load-free cylindrical specimens. While early age strains and internal temperature in concrete cylinders are measured from instant of casting using embedment vibrating wire strain gauge; in this study, only relevant details of creep tests are given.

2.1 Characterization of Materials

Cement: Ordinary Portland cement of grade 53, confirming to IS 12269-2004 [3], was used in this study. The soundness and specific gravity of cement are, respectively, 0.2% and 3.145. The chemical composition of cement is given in Table 1.

Fine aggregate: Natural river sand was used for concrete mixtures. Sieve analysis was carried out using mechanical sieve shaker. The fine aggregate used was conforming to Zone II as per IS 383-1970 [4]. The physical properties are given in Table 2.

Coarse aggregate: Crushed granite aggregates of nominal size 10 and 20 mm were combined in 40:60 ratio, respectively, by weight, such that the grading of

Table 1 Chemical composition of cement

Composition	Cement (%)
SiO_2	16.16
Al_2O_3	3.67
Fe_2O_3	4.76
CaO	70.46
MgO	0.73
SO_3	2.7
Na_2O	0.1
K_2O	0.67

Table 2 Characteristics of aggregates

Description	Fine aggregate	Coarse aggregate	
		10 mm	20 mm
Grading zone conforming to IS: 383-1970	Zone II	–	–
Specific gravity	2.67	2.8	2.85
Water absorption (%)	0.8	0.8	0.2
Fineness modulus	3.81	4.6	5.46
600 micron passing (%)	36.18	–	–
Dry rodded density (kg/m^3)	–	1508	1517

Table 3 Details of concrete mix proportions

Ingredients	OPC
The free water–cement ratio	0.49
Cement content (kg/m^3)	367.3
Superplasticizer (% of the binder content)	1
Water content (kg/m^3)	180
Fine aggregate content (kg/m^3)	871.6
Coarse aggregate content (kg/m^3)	1065

combined aggregate was conforming to IS 383–1970 [4]. The characteristics of coarse aggregates are presented in Table 2.

2.2 Mix Proportioning of Concrete

The mix proportions were determined for normal concrete, to achieve specified 28-day average compressive strength of about 48 MPa and slump of 75–100 mm, using the method proposed by the Department of Environment (DoE), UK [5]. The mix proportioning details are given in Table 3. Slump of fresh concrete was measured for all batches before casting, and the average value of slump was 89 mm.

2.3 Casting and Preparation of Test Specimens

The concrete cylinders (of size 150-mm diameter × 300-mm height) were cast and were moist-cured under water-saturated burlap for 28 days. The companion speci mens, namely cubes, cylinders and prisms, were also cast. After the curing period, the specimens for basic creep and autogenous shrinkage were wrapped with self-adhesive aluminium foil tape and polythene sheet, to protect against hygral exchange with the surrounding environment [6]. Ends of all shrinkage specimens were protected against moisture desiccation by covering with self-adhering aluminium foil tape just before the loading of creep specimens.

2.4 Testing of Concrete Specimens

After surface preparation, both sealed and unsealed specimens intended for measurement of creep are placed concentrically in creep testing machine ECH-4c/1200 kN. Two creep testing machines were chosen for a given mix. In the first machine, three unsealed cylinders and in second machine three sealed cylinders are stacked one over

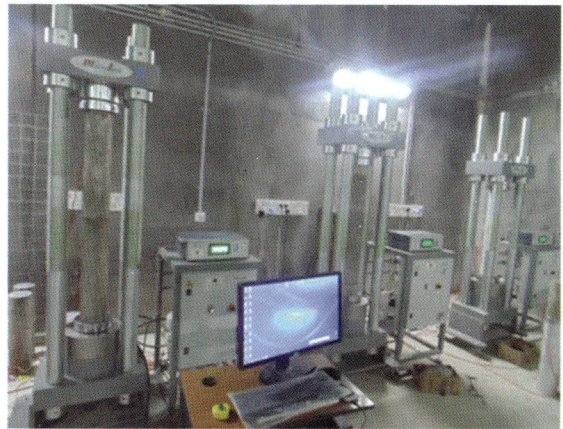

Fig. 1 Creep testing of concrete at CSIR-SERC

the other and placed centrally (see Fig. 1). Acquisition of load, strain and internal temperature started just before loading. Initially, a preload of magnitude, corresponding to a stress level not exceeding 1 MPa, is applied, to establish contact between specimens and upper plates. Though average compressive strength of cubes is 48 MPa, average compressive strength of cylinders at the time of loading was found to be 28.34 MPa. Hence, the concrete cylinders for creep were loaded to the stress value of about 8.5 MPa, which corresponds approximately to 30% of average compressive strength of these cylinders at the time of loading (28 days) [6, 7]. After reaching target load, the load is maintained constantly within ±1%, by the machine, using its integrated automatic closed-loop servo-controlled system. The creep and shrinkage specimens are monitored for strains and internal temperature using embedded vibrating wire strain gauges having 150-mm gauge length. To obtain the amount of water loss from specimens for estimating ultimate shrinkage, initial weights of specimens are recorded immediately after demoulding the specimens and range from 13.172 to 13.213 kg. Specimens will be again weighed at the end of the test. The concrete specimens intended for compressive strength determination were tested at the concrete age of 7, 28, 56, 90, 180 and 365 days as per [8].

2.5 Data Acquisition

Measuring creep strains at the core of the concrete specimens gives more realistic values of creep of a given concrete mix than at the concrete surface. Embedment-type vibrating wire strain gauge is used in this study due to its long-term stability, accuracy and integral thermistor. To carry out statistical analysis of creep strains, in the present study creep strains are measured at closer intervals of time by using

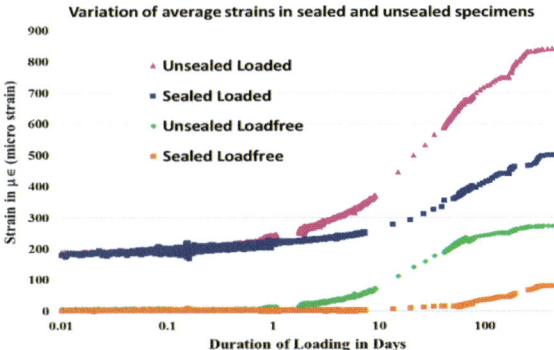

Fig. 2 Variation of average strains in sealed and unsealed specimens

automatic dataTaker. Acquisition interval at which data is acquired is 30 s for first 24 h, 1 min for next 3 days, later 15 min till 320 days and weekly till 505 days of loading duration.

Temperature correction for the measured strains is carried out as follows

$$(T_{\text{current}} - T_{\text{initial}})(C_1 - C_2) \tag{1}$$

where C_1 and C_2 are coefficients of thermal expansion of concrete and vibrating wire, respectively.

3   Results and Discussion

The evolution of average strains with time, measured in both sealed and unsealed specimens for the loading duration of 505 days after loading at 28 days, under loaded and load-free conditions is shown in Fig. 2. These results are based on the average of strains measured in three cylinders in a given creep testing machine for the concrete mix considered. The measured strains will also be useful for recalibration/updating coefficients of creep models, since most of available creep models are semi-empirical. Development of strains with time in individual loaded and load-free concrete specimens is typically given for unsealed specimens in Fig. 3. Sparsely spaced data points in the plots correspond to the time period over which the memory card in dataTaker was not functioning. Hence, only intermittent strain data could be retrieved. Creep strains are obtained by deducting elastic strains and average shrinkage strains from measured strains in loaded specimens. From the measured strains, the basic creep and drying creep are determined. Figure 4 shows the evolution of creep strains with time in unsealed individual concrete specimens. The basic creep and total creep rates are calculated as per ASTM [7] for concrete mix considered and are given in Figs. 5 and 6.

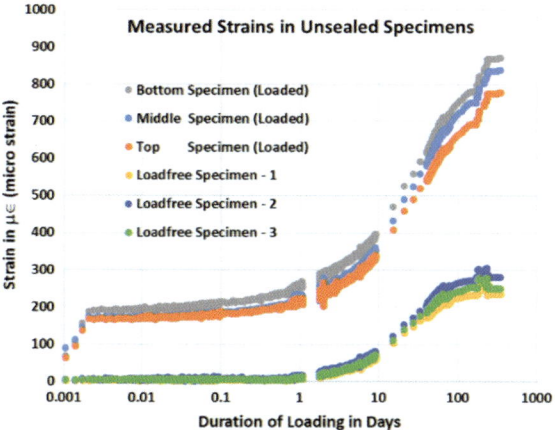

Fig. 3 Evolution strains with time in individual unsealed loaded and load-free concrete specimens

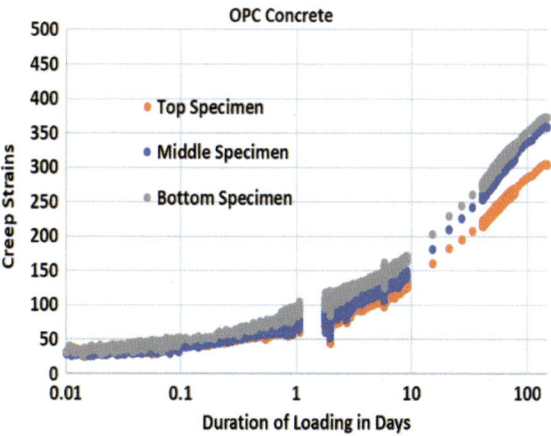

Fig. 4 Evolution of creep strains with time in unsealed individual concrete specimens

It is observed from Fig. 2 that strains are increasing with time gradually and strains in sealed specimens are less compared to strains in unsealed specimens. It may be primarily due to no moisture exchange between ambient environment and specimens in sealed specimens. As can be noted from Fig. 3 that the strain evolution in individual unsealed specimens for loaded and load-free conditions, though concrete mix composition and exposure conditions are same and subjected to same loading conditions, there is variations in strains and evolves with time. The average basic creep and total creep strains are obtained as 261.3 and 406.34 μm/m, respectively, at 505 days after loading. During the same time period, the average shrinkage strain development in sealed and unsealed specimens is 82.3 and 273.5 micro-strain, respectively. It is noted that percentage contribution of basic creep towards total

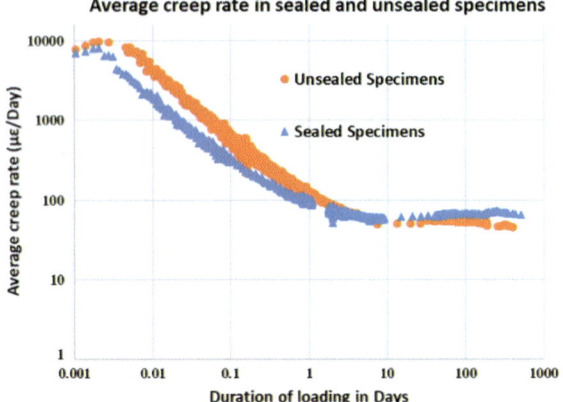

Fig. 5 Variation of average creep rates in unsealed and sealed specimens

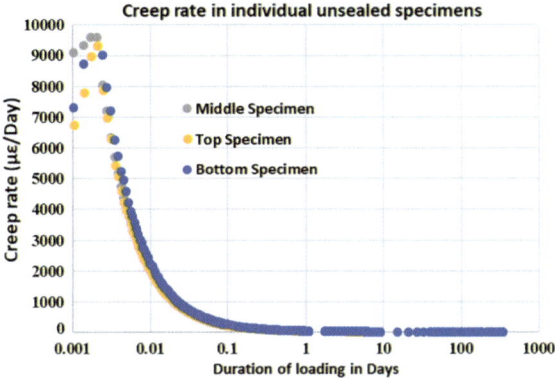

Fig. 6 Variation of creep rates with time in individual unsealed concrete specimens

creep is higher at later ages due to the moisture presence in sealed specimen. It is observed from Fig. 5 that creep rate is higher during initial days after loading and reduces with time in both sealed and unsealed specimens. Short-term creep rate (during initial ages after loading, i.e. ≤ 10 days) is lesser in sealed specimens compared to unsealed specimens. However, long-term creep rate (≥ 10 days) is higher in sealed specimens than that of unsealed specimens. This is primarily due to the presence of higher moisture content in the sealed specimens than that of unsealed specimens at later ages after loading, i.e. beyond ten days. A higher rate of long-term basic creep than rate of long-term total creep shows the importance of consideration of basic creep in design of structures, especially in design of mass concrete structures, such as dams and closed web structures (viz. box girders), wherein the loss of moisture with the ambient atmosphere is less. Also, for given environmental conditions, basic creep is higher in locations of structure (such as junctions of girder and abutment;

cross-girder/end beam joins main girders with the deck slab of bridge), where loss of moisture is less due to effect of local environment.

It is observed from Fig. 4 that creep strains in the bottom specimen are higher than that of the middle and top specimens. Creep strains are lesser in top specimen than that of the middle and bottom specimen. This may be due to higher stress on CSH gel units caused by additional weight of the middle and top specimens on the bottom specimen. Further, as specimen creeps, there is reduction in applied load due to relaxation, and creep machine applies load to maintain the target load, using its integrated automatic closed-loop servo-controlled system. During this process, relaxation of stress in top specimen is faster than the bottom specimen, as weights of the top and middle specimens acting on the bottom specimen. This relaxation due to creep and restoring load by machine is frequent during initial days, as creep rate is higher during initial days after loading. Creep in concrete specimens starts immediately after loading, and creep rate is increasing in first few minutes and then starts decreasing with time (see Figs. 5 and 6). From the review of the literature on creep tests, Hubler et al. [9] noted that in most of the creep tests initial reading was recorded after several hours. Hence, increasing creep rate during first few minutes was not observed in the literature. Observing increasing creep rate during initial few minutes after loading was possible due to faster acquisition of data (with time intervals of 15 s) immediately after loading using embedded vibrating strain gauge and high-speed data acquisition system. Variation of creep rate is higher in initial ages in individual specimens (see Fig. 6) and may be responsible for setting initial conditions for variation in creep strains with time. This mechanism may be a reason for variation of creep curves of individual specimens with time.

4　Statistical Analysis

To understand the time evolution of mean and deviation of creep strains, a statistical analysis of total creep strains (measured using unsealed specimens) is carried out. From measured strains in unsealed concrete specimens, corresponding elastic strains are deducted. The average of shrinkage strains is measured in three concrete cylinders deducted from the strains measured in each of three loaded cylinders, and creep strain curves are obtained. Based on the slope of average creep strains and keeping sample size of each segment same, creep curves are divided into 24 segments. Variation of instantaneous mean creep strains with time is shown typically for segments 1, 3, 13 and 17 in Fig. 7. At each point of time, mean strains are deducted from creep strains, to obtain instantaneous creep strain deviations from its mean value (ε_{dev}). Histogram of squares of ε_{dev}, i.e. ε_{dev}^2, is plotted for each segment and shown typically for segments 1, 3, 13 and 17 in Fig. 8.

From the variations in instantaneous mean values of creep strain with time, it is noted that creep phenomenon is non-stationary stochastic process (see Fig. 7). It is also noted from Fig. 8 that the histograms of ε_{dev}^2 are unimodal up to 0.93 day after loading (Seg. 1–2), bi-modal from 0.93 to 40.56 days after loading (Seg. 3–14) and

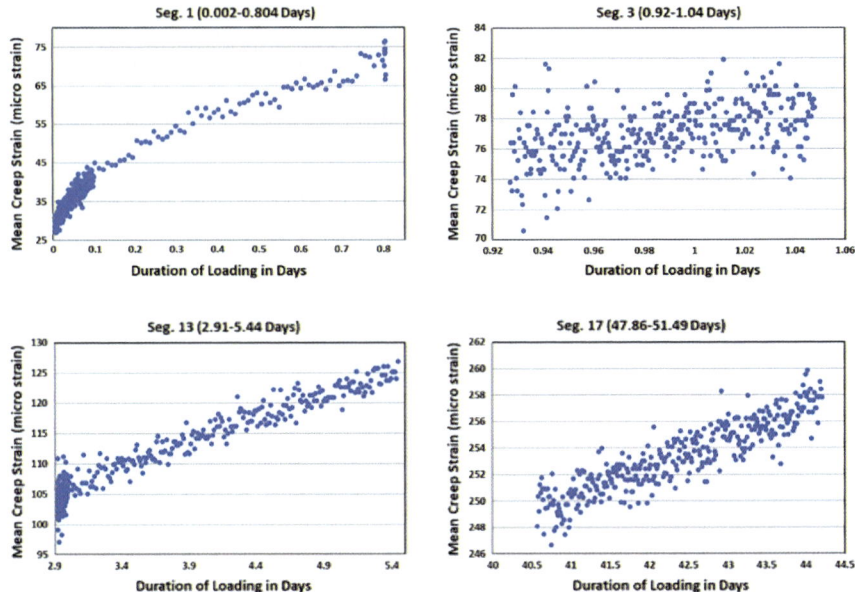

Fig. 7 Variation of instantaneous mean creep strains with time (typically for segments 1, 3, 13 and 17)

tri-modal after 40.56 days after loading (Seg. 15–25). This may be due to variation of creep curves of individual specimens with time (see Fig. 4). This perhaps could be attributed to different moisture movements under different stresses in individual specimens, i.e. faster reduction of stress in top specimen than that of the bottom specimen during process of relaxation and restoring load by creep machine using its integrated automatic closed-loop servo-controlled system. Slow relaxation of stress in the bottom specimen may be due to higher stress on CSH gel units caused by additional weight of the middle and top specimens on the bottom specimen. Thus, the moisture presence and local load variations in specimen in the initial stages after loading seem to set the initial conditions for variation in the evolution of creep strains with time [10, 11]. This also can be seen from Fig. 6 that variation of creep rate is higher in initial ages in individual specimens. However, more experimental investigations need to be carried out at different stress levels and different ages at loading to ascertain the above hypothesis.

5 Conclusions

From statistical analysis of the results of creep tests carried out on OPC concrete at CSIR-SERC, the following conclusions have been drawn:

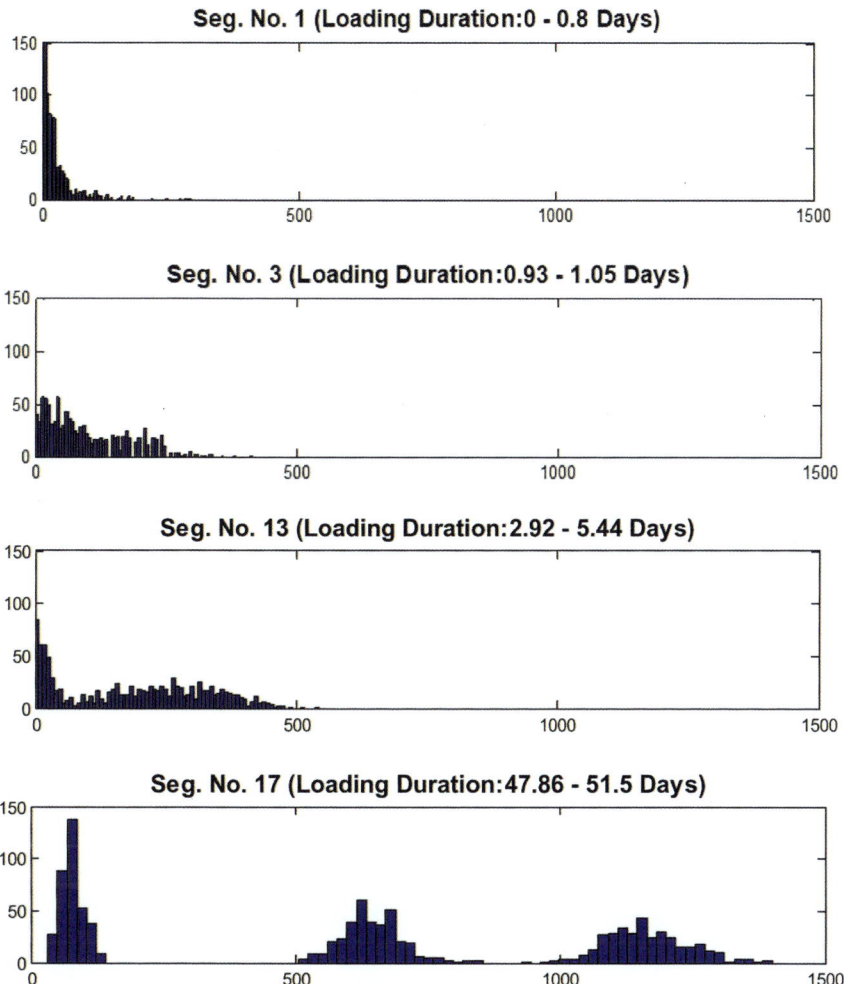

Fig. 8 Histograms of square of instantaneous creep strain deviations from its mean value (*x*-axis: square of deviation around mean value of creep strain in $\mu\epsilon^2$ (ϵ^2_{dev}); *y*-axis: frequency)

- From the variations of mean and mean-square deviations of creep strains with respect to time (Figs. 7 and 8), it is noted the creep evolution is a non-stationary process.
- Increasing creep rate during initial few minutes after loading is observed. This observation was possible due to faster acquisition of data (with time intervals of 15 s) immediately after loading using embedded vibrating strain gauge.
- The rates of variation in basic and total creep suggest that basic creep rate is higher at ages beyond ten days. This observation is significant with respect to the design

of closed web structures (viz. box girders) and effect of local environment on creep deformations of even open web structures.

- Based on the tests, though limited, due to high-speed data acquisition (at small intervals of time), increase in creep rate during initial few minutes after loading is observed and also it was possible to infer that the moisture presence and local load variations in specimen in the initial stages after loading (≤ 3 days) seem to set initial conditions for variation in the evolution of creep strains with time [10, 11]. More experimental investigations need to be carried out in this direction to ascertain this hypothesis.

References

1. Z.P. Bažant, M.H. Hubler, Q. Yu, Excessive creep deflections: an awakening. Concr. Int. **33**(8), 44–46 (2011)
2. Z.P. Bazant, Q. Yu, M. Hubler, V. Kristek, Z. Bittnar, Wake-up call for creep, myth about size effect and black holes in safety: what to improve in fib model code draft, in *Concrete Engineering for Excellence and Efficiency* (2011), pp. 731–746
3. Indian Standard, Specification for 53 grade ordinary Portland cement, IS 12269, in *Bureau of Indian Standards, New Delhi* (2004)
4. Indian Standard, Specification for coarse and fine aggregates from natural sources for concrete, IS 383, in *Bureau of Indian Standards, New Delhi* (1970)
5. D.C. Teychenné, R.E. Franklin, H.C. Erntroy, *Design of Normal Concrete Mixes* (Building Research Establishment, Garston, CRC, 1997)
6. P. Acker, Z.P. Besant, C. Chern, C. Huet, F.H. Wittmann, RILEM recommendation on measurement of time-dependent strains of concrete. Mater. Struct. **31**(212), 507–512 (1998)
7. C. ASTM-512, Test method for creep of concrete in compression, in *American Society of Testing and Materials* (West Conshohocken, Pennsylvania, 1994)
8. Indian Standard, Method of test for strength of concrete, IS 516, in *Bureau of Indian Standards, New Delhi* (1959)
9. M.H. Hubler, R. Wendner, Z.P. Bazant, Comprehensive database for concrete creep and shrinkage: analysis and recommendations for testing and recording. ACI Mater. J. **112**(4), 547–558 (2015)
10. F. Ulm, F. Le Maou, C. Boulay, Creep and shrinkage of concrete-kinetics approach. ACI Spec. Publ. **194**, 135–154 (2000)
11. Z.P. Bazant, S. Baweja, Creep and shrinkage prediction model for analysis and design of concrete structures: model B3. ACI Spec. Publ. **194**, 1–84 (2000)

Evaluation of Fracture Parameters of Cracks in Compressor Blade Root Using Digital Photoelasticity

Muktai Thomre and K. Ramesh

Abstract This study focuses on the failure of blades which is one of the critical parts of the compressor and turbine. Experiments are performed on the aero-engine compressor blade model of dovetail joint using the technique of photoelasticity. First, the high-stress concentration zone is found. This is along the contact surface of the blade and disc. Two different configurations of cracks located in this zone are studied further. The stress intensity factors for mixed mode loading are evaluated using over-deterministic nonlinear least squares approach.

Keywords Dovetail joint · Image processing · Least squares analysis · Photoelasticity · Stress intensity factor

1 Introduction

Aero-engine compressor blade and disc assemblies are subjected to high thermo-mechanical loads due to high-speed rotation and the fluid flow over the blades. Aero-engine compressor assembly can be divided into three critical regions for which inspection is required: the dovetail joint used to fasten blade and disc, the assembly holes or weld areas and the hub region [1]. Because of varying thermo-mechanical loads, a complex stress distribution is developed near the blade–disc joint. In the majority of cases, cracks are initiated at the dovetail joint region due to the fretting action at the blade/disc interface [2]. Rajasekaran et al. [3] conducted the experiments and identified the same kind of failure in blades.

With the new advancement in the aero-engines, several researchers have paid their attention in the area to understand the stress distribution of the dovetail joint region. Durelli et al. [4] performed series of experimental techniques like brittle coatings,

M. Thomre · K. Ramesh (✉)
Indian Institute of Technology Madras, Chennai, India
e-mail: kramesh@iitm.ac.in

M. Thomre
e-mail: muktaithomre88@gmail.com

© Springer Nature Singapore Pte Ltd. 2020
P. V. Varde et al. (eds.), *Reliability, Safety and Hazard Assessment for Risk-Based Technologies*, Lecture Notes in Mechanical Engineering,
https://doi.org/10.1007/978-981-13-9008-1_46

557

electrical resistance strain gages and two-dimensional photoelasticity to get the solution of the problems in turbine and compressor blade attachments, viz. transmission of forces from the airfoil into the dovetail joint, optimization of the dovetail fillets. Durelli et al. considered that as the dovetail geometry is the same for any transverse cross section the dovetail stresses could be taken as a two-dimensional problem. They also concluded that the most important forms of loading are the radial centrifugal force due to blade rotation and the bending of the blade due to the gas pressure. Naumann [5] concluded the same loading conditions for the blade. Plesiutschnig et al. [6] made the simplified cantilever model for blade–disc arrangement under these loads.

In the current study, attention is devoted to find the high-stress concentration area when the aero-engine compressor blade and disc model is subjected to the combined loading of tension and bending. After finding the highly stressed zone, cracks with two different orientations are studied to evaluate their fracture parameters to understand the criticality of the cracks. The experimental technique of two-dimensional photoelasticity is used to get the information about the stress field in the presence of the crack. Using digital photoelasticity [7], the fringe order data of the experimental images is obtained with the help of in-house developed software DigiTFP® [8] which is further used to find stress intensity factors (SIFs) using over-deterministic nonlinear least squares approach.

2 Experiments Using Two-Dimensional Photoelasticity and Evaluation of SIFs Using Least Squares Approach

2.1 Experimental Model

A scaled down aero-engine compressor blade and disc dovetail joint arrangement is considered for the experiments. The specimen is made from 6 mm thick epoxy sheet cast by mixing Araldite CY 230 epoxy resin and HY 951 hardener in the ratio of 10:1 by weight. The mixture is mixed carefully to avoid any air bubble formation. The mixture is then allowed to cure in the mould for 24 hours at room temperature. The specimens of the required size as shown in Fig. 1 are then machined by a high-speed router with precautions to avoid excessive heating and formation of residual stresses. The material stress fringe value for the epoxy specimens is found out using circular disc of 50 mm diameter which comes out to be 11.90 N/mm/fringe for the wavelength 546.1 nm.

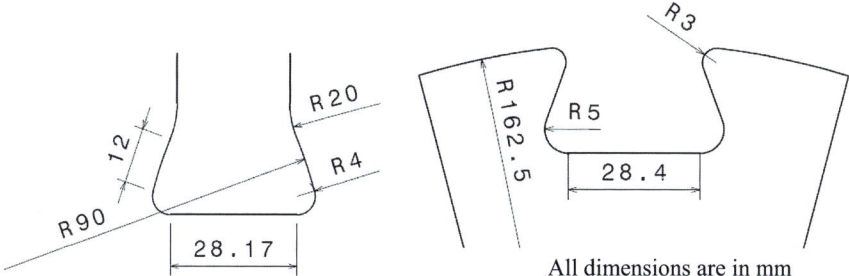

Fig. 1 Blade and disc geometry

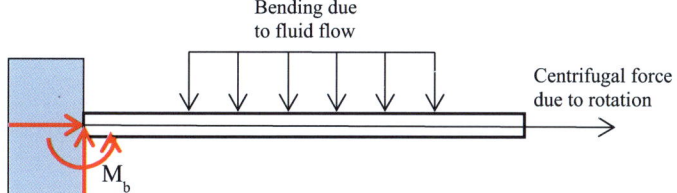

Fig. 2 Simplified blade and disc model subjected to loading

2.2 Forces Experienced by the Blade

The blade is subjected to the centrifugal forces, centrifugal bending and unsteady centrifugal forces due to lateral shaft vibration and alternating bending [5] along with thermal stresses. These loads are the effect of the blade rotation and fluid flowing over the blades. In these loads, the two most significant ones are radially outward centrifugal force and bending of the blade and hence mainly considered for design purpose. The bending is produced by the action of the gas pressure on the airfoil and by the tangential component of the centrifugal forces due to tilting of the airfoil [4]. With this consideration, the dovetail assembly is subjected to two elementary forms of loading, namely tensile and bending, as represented in Fig. 2.

From the experimental observations done by Forshaw et al. [9], the ratio between the maximum stress produced by tensile loading and the maximum stress produced by bending loading, at the root of the blade is calculated for the model used by Forshaw et al. considering speed of 6000 rpm, and material of the blade as titanium alloy with density 4.5 g/cm^3. Further, maintaining the same ratio, the force ratio is calculated for the current model considered for the study, which comes out to be 27.216:1. For simplicity, the load ratio is kept constant as 25:1 for all the tests performed. To apply combined loading, a rope-pulley arrangement is made to the existing body jig.

(a)   **(b)**

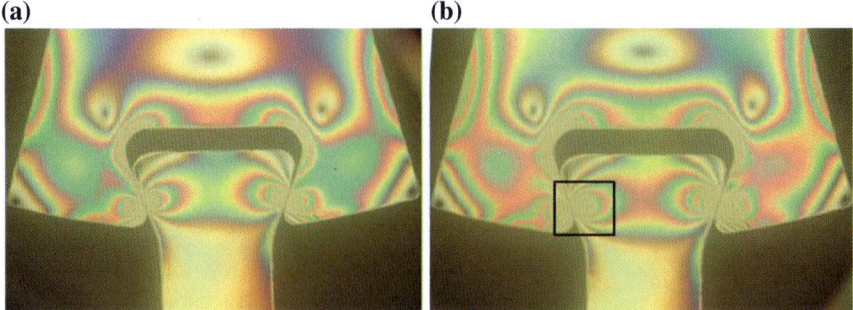

Fig. 3 Dark field isochromatic fringe pattern of joint under. **a** Tensile loading. **b** Combined loading

Fig. 4 Orientation of crack
with respect to contact
surface

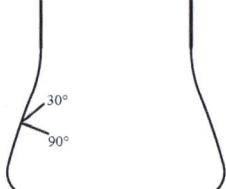

2.3 Stress Distribution at the Joint

From the simplified model, it is concluded that the maximum stress is at the region
where the blade is attached to the disc. Hence, in all the tests, the root of the blade is
focused. Using the technique of photoelasticity, the distribution of stress is obtained
for the calculated loads. Figure 3a gives the stress distribution only under tension
whereas Fig. 3b indicates stress distribution under combined loading.

The experimental study done by Rajasekaran and Nowell [3] on dovetail joints
showed that highlighted region in Fig. 3b is more susceptible to crack growth. Hence,
the square of area 25 sq. mm near the contact region is taken as the region of interest
for all the experiments.

2.4 Experimental Analysis

Once the area of interest is identified, edge cracks of 5 mm length each are cut on
the blade model using a saw of 0.1 mm thickness creating 0.05 mm crack tip at two
different orientations with respect to contact surface as represented in Fig. 4.

Figure 5 shows the experimental arrangement of transmission photoelastic setup
used in this study. The images are recorded using Canon 1200D DSLR camera with
a circular polariscope arrangement. Canon 50–300 mm zoom lens is used along

Fig. 5 Transmission photoelasticity experimental setup

with extenders to get the magnification as high as 30 pixel/mm. After deciding the distance between the object and the camera to get a suitable magnification factor, the same setup was maintained throughout the test. Each time, blade and disc assembly is loaded first to get the exact location of the contact and then the cracks are cut. Combined loading in the 25:1 ratio is applied for each specimen and the images are captured.

2.5 Fringe Order Estimation

With reference to the work proposed by Ramesh et al. [10], the evaluation of the fringe order data using a single-colour image is done using the twelve fringe photoelasticity (TFP) methodology. For this, 0–6 fringe order calibration table is made using the channel-shaped specimen for the same camera settings and is used further in the evaluation process. The fringe order data corresponding to each positional coordinate is found for all the two specimens.

2.6 Evaluation of SIF Using Least Square Analysis

The fracture parameters are calculated for each crack tip by processing the fringe order data obtained from twelve fringe photoelasticity (TFP) methodology using digital photoelastic along with an over-deterministic nonlinear least squares approach. The in-house developed software PSIF uses a multi-parameter crack tip stress field equation proposed by Atluri and Kobayashi, which was later corrected by Ramesh et al. [11]. The corrected multi-parameter stress field equation proposed by Atluri and Kobayashi in terms of the positional coordinates (r, θ) is given as

$$
\begin{Bmatrix} \sigma_x \\ \sigma_y \\ \tau_{xy} \end{Bmatrix} = \sum_{n=1}^{\infty} \frac{n}{2} A_{In} r^{n-2/2} \begin{Bmatrix} \left(2 + \frac{n}{2} + (-1)^n\right)\cos\left(\frac{n}{2}-1\right)\theta - \left(\frac{n}{2}-1\right)\cos\left(\frac{n}{2}-3\right)\theta \\ \left(2 - \frac{n}{2} - (-1)^n\right)\cos\left(\frac{n}{2}-1\right)\theta + \left(\frac{n}{2}-1\right)\cos\left(\frac{n}{2}-3\right) \\ \left(\frac{n}{2}-1\right)\sin\left(\frac{n}{2}-3\right)\theta - \left\{\frac{n}{2} + (-1)^n\right\}\sin\left(\frac{n}{2}-1\right)\theta \end{Bmatrix}
$$
$$
- \sum_{n=1}^{\infty} \frac{n}{2} A_{IIn} r^{n-2/2} \begin{Bmatrix} \left(2 + \frac{n}{2} - (-1)^n\right)\sin\left(\frac{n}{2}-1\right)\theta - \left(\frac{n}{2}-1\right)\sin\left(\frac{n}{2}-3\right)\theta \\ \left(2 - \frac{n}{2} + (-1)^n\right)\sin\left(\frac{n}{2}-1\right)\theta + \left(\frac{n}{2}-1\right)\sin\left(\frac{n}{2}-3\right)\theta \\ -\left(\frac{n}{2}-1\right)\cos\left(\frac{n}{2}-3\right)\theta - \left\{-\frac{n}{2} + (-1)^n\right\}\cos\left(\frac{n}{2}-1\right)\theta \end{Bmatrix}
$$

$$(1)$$

where $A_{I1}, A_{I2}, \ldots$ and $A_{II1}, A_{II2}, \ldots$ are the unknown mode-I and mode-II parameters, respectively. The SIFs are computed from the coefficients as $K_I = A_{I1}\sqrt{2\pi}$ and $K_{II} = -A_{II1}\sqrt{2\pi}$.

For a plane stress conditions, the stress components σ_x, σ_y and τ_{xy} are related to the principal stresses as

$$
\sigma_1, \sigma_2 = \frac{\sigma_x + \sigma_y}{2} \pm \sqrt{\frac{(\sigma_x - \sigma_y)^2}{4} + \tau_{xy}^2} \tag{2}
$$

The fringe order N and the in-plane principal stresses σ_1 and σ_2 are related using the stress-optic law as

$$
\frac{N F_\sigma}{t} = \sigma_1 - \sigma_2 \tag{3}
$$

where F_σ is the material stress fringe value which is calculated experimentally and t is the thickness of the specimen.

Substituting Eq. (2) in (3) the error function g for the mth data point can be obtained as:

$$
g_m = \left\{\frac{\sigma_x - \sigma_y}{2}\right\}^2_m + (\tau_{xy})^2_m - \left\{\frac{N_m F_\sigma}{2t}\right\}^2 \tag{4}
$$

From Eq. (4), it is observed that error function g is nonlinear in terms of the unknown parameters $A_{I1}, A_{I2}, \ldots$ and $A_{II1}, A_{II2}, \ldots$. If the Eq. (4) is solved with an initial estimate for these unknown parameters, there is a possibility that the error will not be zero due to inaccuracy in the estimate. Hence, iterative process based on Taylor series expansion of g_m is used to get corrected values of parameters. This simplifies to a matrix problem represented as follows [11].

$$\{g\}_i = -[b]_i \{\Delta A\}_i \tag{5}$$

where $\{g\}_i$, $[b]_i$ and $\{\Delta A\}_i$ are given by following matrices, respectively.

$$\{g\}_i = \begin{Bmatrix} g_1 \\ g_2 \\ \vdots \\ g_m \\ \vdots \\ g_M \end{Bmatrix}_i \tag{6}$$

$$[b]_i = \begin{bmatrix} \frac{\partial g_1}{\partial A_{I1}} & \frac{\partial g_1}{\partial A_{I2}} & \cdots & \frac{\partial g_1}{\partial A_{Ik}} & \frac{\partial g_1}{\partial A_{II1}} & \frac{\partial g_1}{\partial A_{II2}} & \cdots & \frac{\partial g_1}{\partial A_{IIl}} \\ \frac{\partial g_2}{\partial A_{I1}} & \frac{\partial g_2}{\partial A_{I2}} & \cdots & \frac{\partial g_2}{\partial A_{Ik}} & \frac{\partial g_2}{\partial A_{II1}} & \frac{\partial g_2}{\partial A_{II2}} & \cdots & \frac{\partial g_2}{\partial A_{IIl}} \\ \vdots & \vdots & & \vdots & \vdots & \vdots & & \vdots \\ \frac{\partial g_M}{\partial A_{I1}} & \frac{\partial g_M}{\partial A_{I2}} & \cdots & \frac{\partial g_M}{\partial A_{Ik}} & \frac{\partial g_M}{\partial A_{II1}} & \frac{\partial g_M}{\partial A_{II2}} & \cdots & \frac{\partial g_M}{\partial A_{IIl}} \end{bmatrix} \tag{7}$$

$$\{\Delta A\}_i = \begin{Bmatrix} \Delta A_{I1} \\ \Delta A_{I2} \\ \vdots \\ \Delta A_{Ik} \\ \Delta A_{II1} \\ \Delta A_{II2} \\ \vdots \\ \Delta A_{IIl} \end{Bmatrix}_i \tag{8}$$

In the above equations, i represents the ith iteration and M is the total number of data points. The number of mode-I and mode-II parameters is represented by k and l, respectively. From Eq. (5), we get the following set of equations

$$\{\Delta A\}_i = -[c]_i^{-1}\{d\}_i \tag{9}$$

where

$$[c]_i = [b]_i^T [b]_i \text{ and } \{d\}_i = [b]_i^T \{g\}_i$$
$$\{A\}_{i+\infty} = \{A\}_i + \{\Delta A\}_i \tag{10}$$

The value evaluated from Eq. (9) is used to find the $\{A\}_{i+\infty}$ for the next iteration as given in Eq. (10). For limiting the number of iterations, either of the two convergence criteria can be used. One is the parameter error minimization and other is the fringe order error minimization. The fringe order error minimization criterion is used here as it gives better results [11]. The convergence criterion is satisfied if

$$\frac{\sum |N_{\text{theory}} - N_{\text{exp}}|}{\text{total no. of data points}} \leq \text{convergence error} \tag{11}$$

where N_{theory} is the theoretical fringe order obtained from multi-parameter solution and N_{exp} is the actual fringe order observed in the experiment. The convergence error considered in this study is 0.1 as below this value the data points echoed well on the theoretically generated image. The parameters thus obtained are used to calculate the SIFs as per the relation given in Eq. (1).

2.7 Result and Discussion

The theoretical reconstruction is accepted after satisfying the convergence criteria. A number of data sets are processed to get the optimal reconstruction for each crack orientation. Figure 6 shows the experimental image, fringe order plot with selected data and reconstructed image with echoed data, for the two cracks.

After satisfying the convergence criteria and proper reconstruction, the SIF values for all the crack orientations are obtained as given in Table 1.

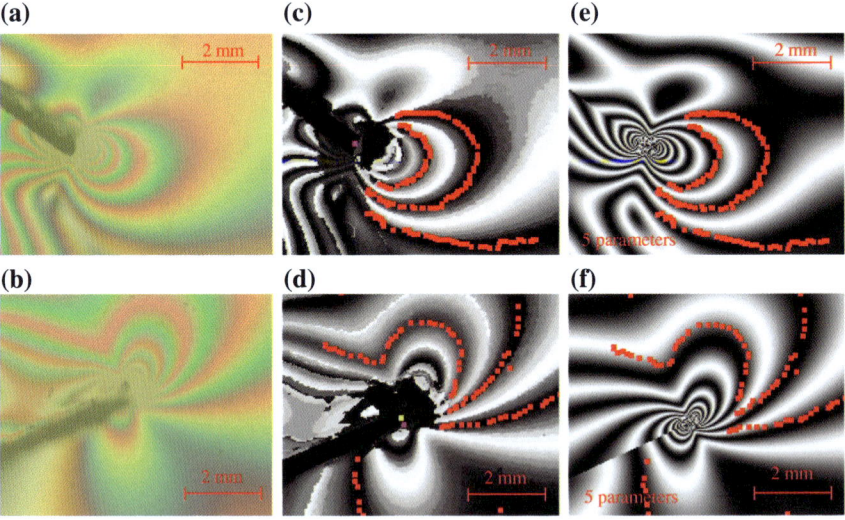

Fig. 6 Near the crack tip isochromatic view obtained under combined loading conditions. Experimental Image at **a** 30°, **b** 90° crack orientation. Whole field fringe order data evaluated using TFP at **c** 30°, **d** 90° crack orientation. Theoretically reconstructed image with echoed data at **e** 30°, **f** 90° crack orientation

Table 1 SIFs values at different orientations of crack

Sr. No	Angle (deg)	Magnification factor (pixel/mm)	K_I (MPa√m)	K_{II} (MPa√m)	No. of parameters	Convergence error
1	30	30	0.2358	0.3421	5	0.05
2	90	30	0.0112	0.3158	5	0.042

From these results, it is observed for 90° mode-II is dominant. For angle 30°, both the modes are significant. As the stress field is very complex in the zone of interest, even for the small change in the crack orientation, there is a considerable change in the SIFs values. Also, while collecting the data points from the resolved fringe order plot for the least squares analysis, it is collected manually for each fringe from the annular zone. Because of the manual selection, it is observed that in some regions the data points are not continuous which will have its effect on the correctness of the SIFs values to some extent. To overcome this, the algorithm for automatic selection of data points from any required region is under development.

3 Conclusions

In this work, the dovetail joint used in the blade and disc assembly in the aero-engine compressor is studied. The assumption of plane stress conditions is made. Out of different forces, namely centrifugal force, bending, vibrational forces, thermal force, the two major forces, (1) centrifugal force and (2) bending due to fluid flowing over the airfoil, are considered in the simplified model. The dovetail joint assembly is replaced by the cantilever beam model. The ratio of stress due to centrifugal force and the bending stress is then calculated for the given geometry and is maintained for all the tests performed.

Using the circular polariscope arrangement, the maximum stress location is found to be near the contact surface at the root of the blade. Two different orientations of cracks with reference to the contact surface are put. An over-deterministic least squares approach is used to evaluate the fracture parameters at this location at two different orientations. It has been observed that even for a small change in the orientation of the crack, there is a significant change in the SIFs values. The highest mode-I SIF value is observed for 30°. Mode-II SIF value comes out to be the highest for 90°. Still, the accuracy of SIF estimation could be improved by improving automatic data collection facility for continuous collection of data points. Currently, authors are working on that. Also, the study of cracks for other orientations is under consideration along with variation in loading conditions. The results obtained from the experimentations can be used for the validation of FE results. Obtaining the data for other orientations, on which the authors currently working on, the empirical relations can be obtained between SIF value with the angle and crack length as variables.

References

1. P. Papanikos, S.A. Meguid, Z. Stjepanovic, Three-dimensional nonlinear finite element analysis of dovetail joints in aero-engine discs. Finite Elem. Anal. Des. **29**, 173–186 (1998)
2. P. Papanikos, S.A. Meguid, Theoretical and experimental studies of fretting initiated fatigue failure of aero-engine compressor discs. Fatigue Fract. Eng. Mater. Struct. **17**(5), 539–550 (1994)
3. R. Rajasekaran, D. Nowell, Fretting fatigue in dovetail blade roots: experiment and analysis. Tribol. Int. **39**, 1277–1285 (2006)
4. A.J. Durelli, J.W. Dally, W.F. Riley, Stress and strength studies on turbine blade attachment. SESA Proc. **XVI**(1), 171–186 (1957)
5. H.G. Naumann, Steam turbine blade design options: how to specify or upgrade, *proceedings of the eleventh turbomachinery symposium*, 29–50 (1982)
6. Ernst Plesiutschnig, Patrick Fritzl, Norbert Enzinger, Christof Sommitsch, Fracture analysis of a low pressure steam turbine blade. Case Stud. Eng. Fail. Anal. **5–6**, 39–50 (2016)
7. K. Ramesh, *Digital Photoelasticity-Advanced Techniques and Applications*, Springer, Berlin, Heiderlberg, New York (2000)
8. DigiTFP® Digital Twelve Fringe Photoelasticity Software developed in Digital Photomechanics Lab, IIT Madras (2017). https://home.iitm.ac.in/kramesh/dtfp.html
9. T.R. Forshaw, H. Taylor, C. Chaplin, Alternating Pressures and Stresses in an Axial Flow Compressor, aeronautical research council report, 11–25 (1956)
10. K. Ramesh, V. Ramakrishnan, C. Ramya, New initiatives in single-colour image-based fringe order estimation in digital photoelasticity. J. Strain Anal. **50**(7), 488–504 (2015)
11. K. Ramesh, S. Gupta, A.A. Kelkar, Evaluation of stress field parameters in fracture mechanics by photoelasicity-revisited. Eng. Fract. Mech. **56**(1), 25–45 (1997)

Hybrid Response Surface Function-Based Metamodeling of Response Approximation for Reliability Analysis

Sounak Kabasi and Subrata Chakraborty

Abstract Response surface method (RSM) is mostly adopted to overcome the computational challenge of Monte Carlo simulation (MCS)-based reliability assessment of system comprising implicit limit state function (LSF). In the present study, a hybrid response surface function (HRSF) based on exponential approximation and second-order polynomial estimation is investigated for improved response approximation for reliability analysis. The method calibrates each random variable according to the response function using an exponential function having a varying parameter that attempts to accommodate the nonlinearity of each variable in the LSF. The exponential function is further regressed using conventional quadratic polynomial model. The effectiveness of the proposed HRSF-based approach is elucidated numerically by considering several saturated designs and uniform design schemes. The performance of the proposed procedure is demonstrated by comparing the proposed response approximation and the usual polynomial RSM-based approximation with that of obtained by the most accurate direct MCS technique.

Keywords Hybrid response surface function · Monte Carlo simulation · Reliability analysis · Response surface method

S. Kabasi (✉) · S. Chakraborty
Indian Institute of Engineering Science and Technology, Shibpur, India
e-mail: sounak.kabasi001@gmail.com

S. Chakraborty
e-mail: schak@civil.iiests.ac.in

© Springer Nature Singapore Pte Ltd. 2020
P. V. Varde et al. (eds.), *Reliability, Safety and Hazard Assessment for Risk-Based Technologies*, Lecture Notes in Mechanical Engineering,
https://doi.org/10.1007/978-981-13-9008-1_47

567

1 Introduction

The primary task of structural reliability analysis is to obtain the probability of failure (P_f) by solving a multidimensional integral, i.e.

$$P_f = \int_{g(\mathbf{X}) \leq 0} f_{\mathbf{X}}(\mathbf{x}) d\mathbf{X} \tag{1}$$

In the above, X represents a vector of dimension n characterizing the associated limit state function (LSF), $g(X)$ whereas $f_X(\mathbf{x})$ is the joint probability density function (pdf) of the random variables. The exact computation of the above is often computationally demanding. Several approximations based on simulation and analytical approaches are usually adopted in order to calculate the probability of failure. The second moment-based methods are widely used due to their straightforwardness and low-computational involvement. Such approach has, however, several known disadvantages with regard to the assumption about the curvature of the failure surface and approximation accuracy. The most direct, conceptually straightforward and accurate means of reliability analysis is based on the direct Monte Carlo Simulation (MCS) technique. However, the direct MCS-based approach involves repetitive evaluations of LSF. For a sufficiently reliable estimate of probability of failure, several thousands of such simulations might be required [1]. If the LSF is known explicitly, evaluating the LSF several times is not at all difficult. However, if there is an involvement of a finite element (FE) model with highly nonlinear behaviour, then each such performance function evaluation might become computationally taxing [2]. In order to assess the reliability of a complex structure and to model its mechanical behaviour, a trade-off between the computational involvement of the numerical method and the accuracy of the reliability algorithms is a necessity. Response surface method (RSM)-based metamodels provide a reasonable solution to such problems [3]. The choice of initial input data is an important criterion for obtaining a fairly accurate solution of response approximation in a reasonable time. The common practice is to adopt an adaptive scheme in order to derive a response surface function (RSF) in a region where the contribution of probability of failure is maximum. Bucher [4] proposed a modified RSM where the response surface (RS) model is obtained by using the design of experiment (DOE) points centred around the mean points of the random variables and then updating the centre point further by using a linear approximation of the LSF. Rajashekhar [5] further proposed an iterative scheme to construct improved response surface where the centre point of DOE scheme is updated till convergence. Subsequently, a good number of studies to obtain DOEs adaptively centring around a point obtained based on the most probable point of failure evaluated by first order reliability method (FORM) to construct RSF are notable [6–11]. However, such adaptive schemes need several iterations before convergence depending on the complexity and nonlinearity of the associated LSF of a reliability analysis problem. The application of hybrid response surface function (HRSF) which utilizes fewer training samples to obtain response approximating functions as no iteration is involved is noted to be useful in this regard [12, 13]. An HRSF based on exponential approximation and second-order polynomial estimation is investigated in the present study for an improved approximation of response for reliability analysis. The method calibrates each random variable according to the response function using an exponential form having a varying parameter that attempts to accommodate the nonlinearity of each variable in the LSF. The exponential form is expected to tackle problems of varying nonlinearity in adaptive manner. Basically, such a calibration can be obtained by regressing the variables to any suitable function whose functional range covers all possible real numbers in order to accommodate any possible exponent. The exponential function is further regressed using conventional quadratic polynomial model. The proposed HRSF-based RSM is elucidated numerically to study its effectiveness with respect to the usual polynomial RSM by comparing the results obtained from the most accurate direct MCS technique.

2 Structural Reliability Analysis by RSM

The RSM primarily uncovers complicated relationships that exist between the several input variables and the corresponding responses through the use of empirical models. The RS is usually represented by a simple algebraic relation that is obtained by using cautiously chosen training data points termed as a DOE. This RS serves as a metamodel of the response as obtained by FE analysis of a complex mechanical model.

For n number of output values y_i conforming to n input data, x_{ij} (i.e. ith sample for input x_j in a DOE), the input–output relation can be expressed as:

$$\mathbf{y} = \mathbf{X}\boldsymbol{\beta} + \varepsilon_y \tag{2}$$

where $\mathbf{X}$ represents the design matrix, $\mathbf{y}$ is output vector; $\boldsymbol{\beta}$ and ε_y are the unknown coefficient and the error vectors, respectively. In most of the cases, Eq. (2) is represented by a quadratic polynomial form as:

$$\mathbf{y} = \beta_0 + \sum_{i=1}^{k} \beta_i x_i + \sum_{i=1}^{k} \sum_{j \geq i}^{k} \beta_{ij} x_i x_j \tag{3}$$

The least squares method (LSM) is used to determine the unknown coefficients to minimize the error norm as:

$$e(\boldsymbol{\beta}) = \sum_{i=1}^{n} \left(y_i - \beta_0 - \sum_{i=1}^{k} \beta_i x_i - \sum_{i=1}^{k} \sum_{j \geq i}^{k} \beta_{ij} x_i x_j \right)^2 = (\mathbf{y} - \mathbf{X}\boldsymbol{\beta})^{\mathrm{T}} (\mathbf{y} - \mathbf{X}\boldsymbol{\beta}) \tag{4}$$

And the least squares estimate of $\boldsymbol{\beta}$ is obtained as,

$$\beta = \left[\mathbf{X}^\mathsf{T}\mathbf{X}\right]^{-1}\left\{\mathbf{X}^\mathsf{T}\mathbf{y}\right\} \tag{5}$$

The implicit response y can be easily estimated for any desired input parameters, once Eq. (5) is solved.

3 The Hybrid Response Surface Function

The HRSF based on exponential approximation and second-order polynomial estimation to improve the accuracies for modelling monthly pan evaporations was studied by Keshtegar and Kisi [12] where data are normalized based on normal distribution assumption. Following the similar basic principle, each variable in the present study is calibrated according to the response function using an exponential function by introducing a varying parameter P. This parameter is expected to improve the response approximation as it adaptively accommodates the nonlinearity of each variable in the LSF. The HRSF-based response approximation involves three layers to predict an event using hybrid polynomial and exponential functions. The underlying procedures are presented in the following subsection.

3.1 The HRSF Model

The proposed model of the hybrid function can be presented in three layers as follows:

Layer 1: The values of the random variables X_i along with their mean and standard deviation (sd) are the input data. The random variables are normalized by the following relation:

$$U_i^k = \frac{x_i^k - \mu_{x_i}}{P_i \sigma_{x_i}} \tag{6}$$

where P_i is the adjusting parameter to take into account the nonlinearity of the ith random variable in the LSF, U_i^k is the ith normalized input variable of the kth observation, μ_{x_i} is the mean and σ_{x_i} is the sd of the ith random variable ($\mathbf{X_i}$).

Layer 2: The response output is calibrated using each normalized variable by an exponential basic function as:

$$Z_i^k = b_i + c_i e^{U_i^k} \tag{7}$$

where b_i and c_i are the unknown coefficients of the predicted exponential function based on the kth observation of the ith normalized variable U_i^k and Z_i^k is the corresponding output response. The coefficient P_i is an unknown coefficient which can be effectively varied according to the nonlinearity of the response output. It may be noted that the parameter P_i takes into account the nature of nonlinearity of the ith random variable in the LSF.

In this layer, there are $3n$ numbers of total unknown coefficients. In order to find these coefficients, an axial saturated design (SD) can be used. The least squares approximation can be performed conveniently by taking logarithm of Eq. (7), i.e.

$$\log(Z_i^k - b_i) = \log c_i + U_i^k \tag{8}$$

The regression is done at this stage to obtain the unknown coefficient c_i (Eq. 7) and parameter P_i (Eq. 6). To start with, the values of all b_i are assumed to be equal for all the random variables and are taken as the response values at the design points. To be specific, the initial values of all the b_i are taken slightly less than the minimum value of the responses obtained at the design point (as per the SD scheme) such that the regression can be performed conveniently in the logarithmic domain. This will enable to move to the next layer (layer 3) of response approximation. However, once the HRSF is obtained in layer 3 for response approximation, the parameter b_i can be optimized which is explained in Sect. 3.2.

Layer 3: Once the regression is done in layer 2, the response function is regressed by a second-order polynomial function. The approximated function of HRSF with second-order polynomial function and normalized exponential function as given by Eq. (3) can be obtained as,

$$Y = a_0 + \sum_{i=1}^{n} a_i Z_i + \sum_{i=1}^{n} a_{ii} Z_i^2 + \sum_{i=1}^{n} \sum_{j=i+1}^{n} a_{ij} Z_j Z_j \tag{9}$$

In this layer, a more improved DOE scheme like uniform design (UD) scheme can be used in order to find the unknown coefficients a_{ij}. In addition to the data points obtained as per the UD scheme, the training data points already obtained as per the adopted axial SD scheme in the previous layer may be also used at this layer in order to avoid unnecessary wastage of data points. It may be noted that the proposed HRSF needs to calibrate $(n + 1) * (n + 2)/2$ coefficients in layer 3.

3.2 Optimum Choice of b_i

As mentioned earlier in layer 2, the initial values of all b_i were fixed at the minimum value of the responses at the corresponding design points so that the HRSF can be constructed for necessary response approximation. However, in order to find the optimum value of b_i, the leave-one-out cross-validation scheme is used to obtain the optimum value of b_i. In this method, one row is left out of the design matrix, and the HRSF is built using the other points of the design matrix. The HRSF thus generated is used to predict the response of the combination that was left out. This process is carried out sequentially for every row of the design matrix, leaving out one row at a time. The coefficient of determination (R^2) is then evaluated using the predicted responses and the actual responses for a particular value of b_i. The b_i for which the maximum value of R^2 is obtained is used to generate the final HRSF. Maximizing R^2 is a good scheme for efficient response approximation. Hence, it has been used to find the optimum values of b_i.

4 Numerical Study

The effectiveness of the proposed HRSF algorithm is elucidated numerically by considering three examples. The HRSF scheme was tested considering three different SD schemes, i.e. $2n + 1$, $4n + 1$ and $6n + 1$ number of axial points, respectively, and different UD schemes. The various statistical norms, i.e. the R^2 value, mean absolute error (MAE) and root-mean-square error (RMSE) usually used to study the capability of metamodels [14, 15], are obtained for each scheme, and the results are compared with the conventional LSM-based polynomial RSM model. The expressions of these norms are

$$\text{RMSE} = \sqrt{\frac{1}{m} \sum_{i=1}^{m} (y_i - \hat{y}_i)^2}; \text{MAE} = \frac{\sum_{i=1}^{m} |y_i - \hat{y}_i|}{m};$$

$$R^2 = 1 - \sum_{i=1}^{m} \left(y_i - \hat{y}_i \right)^2 / \sum_{i=1}^{m} (y_i - \bar{y}_i)^2 \tag{10}$$

where y_i is the actual response, $\hat{y}_i$ is the predicted response by the metamodel and m is the total number of observations.

4.1 Example 1: Response of a Sphere Subjected to Internal Pressure

The equivalent von Mises stress (σ_{eq}) of a sphere under internal pressure as shown in Fig. 1 with material properties described as homogenous without spatial variety is given by Goswami et al. [10],

$$\sigma_{eq}(r = r_0) = \sqrt{(\sigma_r^2 + \sigma_\theta^2 - 2\sigma_r\sigma_\theta)} = (|\sigma_r - \sigma_\theta|) = \frac{3p_0 r_1^3}{2(r_1^3 - r_0^3)} \tag{11}$$

where r_0 and r_1 are the internal and external radii, respectively, p_0 is the internal pressure, σ_r and σ_θ are the radial and circumferential stress, respectively. p_0, r_0 and r_1 are assumed to follow lognormal distribution having mean values 1200 MPa, 50

Table 1 Results of statistical norms obtained by the proposed HRSF method with different DOE schemes

Statistical norms	Number of UD runs	
	21	30
$6n + 1$ SD in layer 3		
R^2	0.9799	0.9811
RMSE	0.28	3.73E−01
MAE	38.8235	5.33E + 01
$4n + 1$ SD in layer 3		
R^2	0.9755	0.9792
RMSE	3.58E−01	3.74E−01
MAE	4.55E + 01	5.28E + 01
$2n + 1$ SD in layer 3		
R^2	0.9775	0.9816
RMSE	2.99E−01	3.73E−01
MAE	4.16E + 01	5.53E + 01

and 100 mm. In general, the COV varies between 5 and 10%. Hence, as a preliminary test for validating the method's capabilities, the COV has been assumed to be 7.5%.

The statistical norms as obtained by adopting different SD schemes applied at Level 2 and UD schemes at Level 3 are shown in Table 1. The response approximation was further performed by the usual quadratic polynomial RSM. The R^2, RMSE and MAE values obtained by the usual RSMs (using 58 UD points thus keeping the total number of data points same for both the schemes) are 0.9783, 0.405 and 58.4, respectively. Furthermore, it can be observed that 41 number of data points used in HRSF for $2n + 1$ SD and 30 UD runs gives sufficiently better results than the usual RSM that used a total of 58 data points. These results clearly indicate the improved response approximation capability of the proposed approach. The results of the statistical norm indicate that a $2n + 1$ saturated design scheme is economical and yields sufficiently better results than conventional LSM-based RSM model.

Fig. 1 Sphere under internal pressure

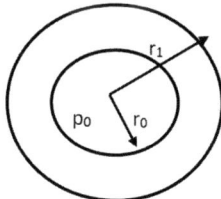

4.2 Example 2: An Implicit Response Function

In structural reliability analysis, the response of the structure is usually found out by numerical method. In such cases, the explicit performance is unknown. Hence, this example presents a problem where the responses are obtained by solving a matrix equation as following,

$$[K]\{X\}=\{F\} \tag{12}$$

where $\{X\}$ is the unknown response vector, the known deterministic vector, $F = [10,11,12,13,14,15]^{T}$ and

$$
K = \begin{bmatrix}
s_1 s_2 & \frac{s_1 s_2^{2.2}}{s_3^4} & \frac{s_5 s_1}{s_2 s_4^6} & \frac{s_8 s_1}{s_4^3} & \frac{s_7 s_3}{s_8} & \frac{s_6 s_2^{4.1}}{s_8} \\[2mm]
\frac{s_1 s_6}{s_3} & s_1 s_3^{0.2} & s_4 s_1^{0.1} & \frac{s_1 s_5}{s_2 s_4^6} + \frac{s_1 s_8}{s_4^3} & \frac{s_7 s_8}{s_5^3} & 1 \\[2mm]
s_3 s_6^2 & s_3 + s_5^{0.6} s_7 & s_2 s_4 + \frac{s_8 s_6}{s_5} & \frac{s_8 s_6^3}{s_3^4 s_5^6} & s_1 + s_1 s_2 & \frac{s_1 s_8^4}{s_4} \\[2mm]
\frac{s_8^4 s_2}{s_3} & s_7 & 0 & \frac{s_7 s_4}{s_5} + \frac{s_4 s_3}{s_5 s_6} & s_4 s_8 s_7 s_6 & \frac{s_4}{s_2^4} \\[2mm]
\frac{s_4 s_7}{s_1} & s_2^2 & s_1^2 s_7^{0.1} & \frac{s_1^2 s_8^3}{s_7 + s_1 + s_2} & s_1 s_4 s_5 & \frac{s_2 s_4}{s_5 + s_6} + \frac{s_1 s_7}{s_6} \\[2mm]
9 & \frac{s_4 s_1}{s_8} + \frac{s_2^5 s_1}{s_5} & s_1 + s_2 + s_3 + s_4 & \frac{1}{s_5 + s_2 + s_1 + s_8} & \frac{s_1 s_4 s_2}{s_7} & s_1^2 s_8
\end{bmatrix}
$$

where $s_1, s_2, s_3, s_4, s_5, s_6, s_7, s_8$ are the lognormal random variables having mean of 5,10,1,0.01,0.5,0.02,2 and 4, respectively, and sd of 0.5,1,0.2,0.002,0.075,0.002,0.2 and 0.4, respectively. The LSF is considered as, $g(X) = X_1 - X_0$, where X_1 is the response obtained by solving the matrix equation for the first degree of freedom and X_0 is the allowable response. It is to be noted that the LSF is now implicit and needs to be approximated by RSM.

The HRSF was constructed by using $2n + 1$ SD and 30 UD. A total of 47 UD points were used for generating the polynomial-based RSM, keeping the total number of data points same for both the schemes. The probability of failures for varying allowable response X_0 is obtained by the proposed HRSF, conventional LSM-based RSM and also by the most accurate direct MCS techniques by using 10^6 sample size based on convergence study. The results are shown in Fig. 2. It can be clearly noted from this plot that the proposed HRSF can estimate the probability of failures far better than the usual LSM-RSM-based approach for all the allowable value of the responses, X_0.

4.3 Example 3: A Two Degree of Freedom Dynamic System

A primary–secondary spring mass system (Fig. 3) is now considered. The LSF can be expressed as [16],

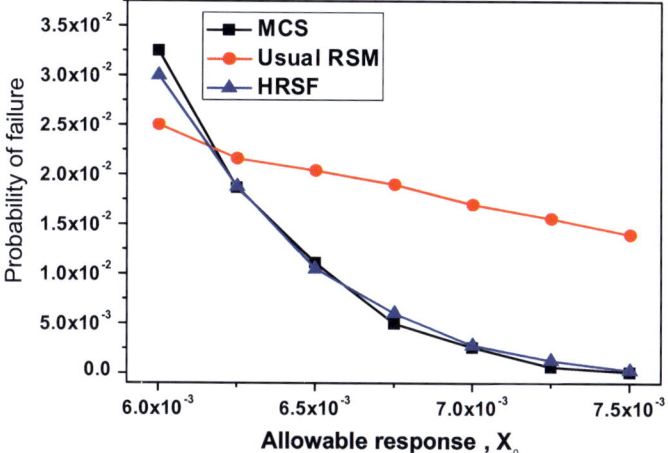

Fig. 2 Comparison of performance of the HRSF and the usual RSM in estimating probability of failure

Fig. 3 Two degree of freedom dynamic system

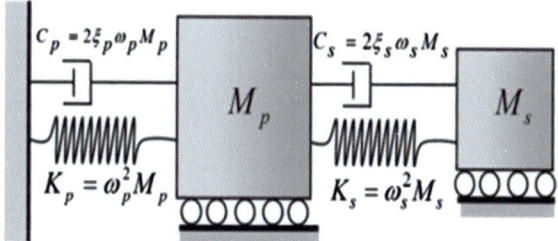

$$g = F_s - K_s \times P(E[x_s^2])^{0.5} \tag{14}$$

In the above, F_s is the capacity of the secondary spring, P is the peak factor having a constant value of 3 and $E[x_s^2]$ is the mean-square relative displacement response of the secondary spring given by,

$$E[x_s^2] = \frac{\pi S_0}{4\xi_s \omega_s} \left[\frac{\xi_a \xi_s}{\xi_p \xi_s (4\xi_a^2 + \theta^2) + \gamma \xi_a^2} \times \frac{(\xi_p \omega_p^3 + \xi_s \omega_s^3)\omega_p}{4\xi_a \omega_a^4} \right] \tag{15}$$

where $\gamma = M_s/M_p$ is the mass ratio, $\omega_a = (\omega_p + \omega_s)/2$ and $\xi_a = (\xi_p + \xi_s)/2$ are the average frequency and damping ratio of the two systems, $\theta = (\omega_p - \omega_s)/\omega_a$ is a tuning parameter and S_0 is the intensity of white noise. The means and sd of the eight independent random variables with lognormal distribution are listed in Table 2.

The variation of probability of failure for different means of M_p obtained by the proposed HRSF and the usual polynomial RSM is shown in Fig. 4. The exact probability of failure is obtained using 10^6 MCS samples. The HRSF was constructed

Table 2 Statistical properties of the random variables

Random variable	M_p	M_S	K_P	K_S	ξ_p	ξ_s	F_s	S_0
Mean	Variable	0.01	1	0.01	0.05	0.02	100	15
Standard deviation	$0.1 * \mu$	0.001	0.2	0.001	0.02	0.01	10	1.5

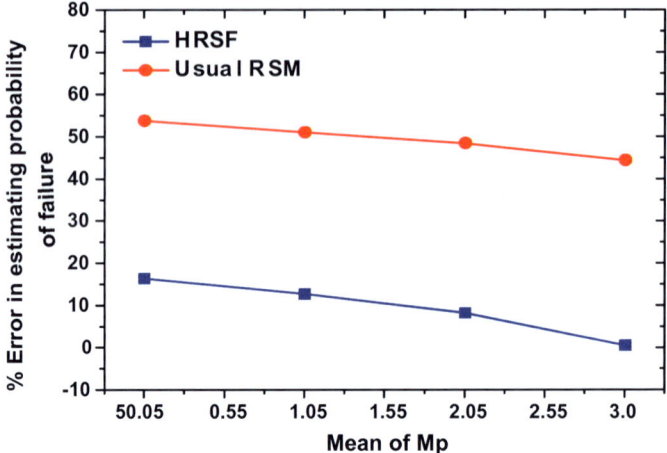

Fig. 4 Comparison of proposed HRSF and the usual polynomial RSM

based on $2n + 1$ number of SD points and 30 numbers of UD points. It can be noted from the plot that the performance of HRSF is much better than that obtained by the usual polynomial RSM generated using 47 number of UD points.

5 Summary and Conclusions

A hybrid response surface function to predict the response of structure by calibrating each random variable using an exponential function is presented for reliability analysis. In doing so, a varying parameter is introduced in the calibration process to accommodate the nonlinearity of each variable in the LSF. The effectiveness of the proposed HRSF-based approach is studied considering three numerical examples. The statistical norms are obtained by adopting different SD schemes at layer 2 and also different UD schemes at layer 3. The comparison of various statistical norms obtained by the proposed HRSF approach and by the usual polynomial RSM clearly indicates improved response approximation capability of the proposed approach. The reliability analysis results also confirm the improved capability of the proposed approach. The results indicate that the exponential basis function can be used to trace the trend of the nonlinear curves much better than the conventional LSM. The approach is flexible with respect to the nonlinearity of the response function as it has the capability of self-calibration based on the chosen problem. The exponential calibration in layer 2 allows to calibrate a function of even higher degree using a fewer number of data points. If the least square approximation has to be used for fitting higher-degree polynomials, the number of data points will increase drastically. Hence, higher accuracy using less number of data points can be achieved using HRSF. It may be noted that the method's capability does not directly depend on the nature of the distribution of the random variables, but depend on the DOE scheme. The nature of distribution does play an important role in deciding the most probable point of failure, and hence, if the DOE points are lying far away from the LSF, then HRSF's accuracy for reliability analysis like any other metamodel will definitely suffer. To address this issue, an adaptive scheme for updating the design point can be attempted which needs further study.

References

1. N.R. Mann, R.E. Schafer, N.D. Singpurwalla, *Methods for Statistical Analysis of Reliability and Life Data* (Wiley, New York, 1974)
2. B. Richard, C. Cremona, L. Adelaide, A response surface method based on support vector machines trained with an adaptive experimental design. Struct. Saf. **39**, 14–21 (2012)
3. L. Faravelli, Response surface approach for reliability analyses. J. Eng. Mech. **115**(2), 2763–2781 (1989)
4. C.G. Bucher, U. Bourgund, A fast and efficient response surface approach for structural reliability problems. Struct. Saf. **7**(1), 57–66 (1990)
5. M.R. Rajashekhar, B.R. Ellingwood, A new look at the response surface approach for reliability analysis. Struct. Saf. **12**(3), 205–220 (1993)
6. S.H. Kim, S.W. Na, Response surface method using vector projected sampling points. Struct. Safety **19**, 3–19 (1997)
7. I. Kaymaz, C.A. McMahon, A response surface method based on weighted regression for structural reliability analysis. Probab. Eng. Mech. **20**, 11–17 (2005)
8. D.L. Allaix, V.I. Carbone, An improvement of the response surface method. Struct. Saf. **33**, 165–172 (2011)
9. W. Zhao, Z. Qiu, An efficient response surface method and its application to structural reliability and reliability-based optimization. Finite Elem. Anal. Des. **67**, 34–42 (2013)
10. S. Goswami, S. Ghosh, S. Chakraborty, Reliability analysis of structures by iterative improved response surface method. Struct. Saf. **60**, 56–66 (2016)
11. G. Su, L. Peng, L. Hu, A Gaussian process-based dynamic surrogate model for complex engineering structural reliability analysis. Struct. Saf. **68**, 97–109 (2017)
12. B. Keshtegar, O. Kisi, Modified response-surface method: new approach for modeling pan evaporation. J. Hydrol. Eng. **22**(10) (2017). https://doi.org/10.1061/(ASCE)HE.1943-5584.0001541
13. B. Keshtegar, S. Heddam, Modelling daily dissolved oxygen concentration using modified response surface method and artificial neural network: a comparative study. Neural Comput. Appl. 1–12 (2017)
14. N.R. Draper, H. Smith, *Applied Regression Analysis* (Wiley-Interscience, 1998). ISBN 0-471-17082-8
15. C.J. Willmott, K. Matsuura, Advantages of the mean absolute error (MAE) over the root mean square error (RMSE) in assessing average model performance. Clim. Res. **30**, 79–82 (2005)
16. A.D. Kiureghian, M. De Stefano, Efficient algorithm for second order reliability analysis. J. Eng. Mech. **117** (1991)

Durability-Based Service Life Design of RC Structures—Chloride-Induced Corrosion

N. J. Yogalakshmi, K. Balaji Rao and M. B. Anoop

Abstract In this present study, chloride ion diffusion into cover concrete is modelled as a sub-diffusion process. A time-fractional partial differential equation is used to describe the sub-diffusion process and the numerical solution of this equation is obtained using explicit finite difference scheme. The parameters C_s, α and D_α involved in the solution are estimated by data fitting performed using genetic algorithm-based optimization technique in MATLAB. This study is carried out for predicting the chloride profiles of OPC concrete that are available in the literature, exposed to marine environment for a long time. The collected profiles correspond to both laboratory samples exposed to field exposure and sample drawn from a real existing structure. The initiation time of corrosion is then determined numerically by finding the time at which chloride concentration at rebar reaches the critical chloride concentration. The results of the sub-diffusion model are also compared with *fib* model, whose parameters are also fitted using optimization technique. Based on field observations reported in the literature, it is found that the initiation times of corrosion predicted by sub-diffusion model are less conservative and more realistic compared to the predictions of *fib* model.

Keywords Chloride-induced corrosion · Fick's law · Sub-diffusion · Initiation time

N. J. Yogalakshmi (✉)
AcSIR, CSIR-Structural Engineering Research Center, Chennai, India
e-mail: yoga.engg10@gmail.com

K. B. Rao · M. B. Anoop
CSIR-Structural Engineering Research Center, Chennai, India
e-mail: balaji@serc.res.in

M. B. Anoop
e-mail: anoop@serc.res.in

© Springer Nature Singapore Pte Ltd. 2020
P. V. Varde et al. (eds.), *Reliability, Safety and Hazard Assessment for Risk-Based Technologies*, Lecture Notes in Mechanical Engineering,
https://doi.org/10.1007/978-981-13-9008-1_48

579

1 Introduction

A major problem concerning the durability of reinforced concrete structures in marine and coastal environment is the corrosion of steel reinforcement due to the presence of chloride ions. Corrosion of reinforcement leads to a series of structural problems such as reduction in cross-sectional area of steel, and cracking and spalling of cover concrete, loss of bond between concrete and steel and also reduction in material strength. Therefore, in most of the cases, initiation time of corrosion limits the service life time of the structure. Modelling the initiation time of corrosion, which in turn depends on the chloride ion ingress into cover concrete is an important step in durability-based service life design of RC structures.

Collepardi et al. [1] were the first to use Crank's solution [2] of Fick's second law equation to model the chloride ingress into the concrete. The surface chloride concentration, C_s, and the apparent chloride diffusion coefficient, D_a, were the significant parameters and were derived by fitting the error function equation to the measured chloride profiles. This equation was based on the assumptions that: (i) the concrete is a homogenous material; (ii) initial chloride content inside concrete is zero; (iii) the surface chloride concentration and diffusion coefficient are constant; and (iv) the chloride binding is linear. Despite these inappropriate assumptions, this equation is the widely used model for service life design till date. From their experiments, Takewaka and Matusumoto [3] showed that the diffusion coefficient is not constant but varies with time. The process of diffusion retards with time, due to the development of denser concrete pore structure by hydration products and chloride binding and this gets reflected in diffusion coefficient which reduces with time. After this observation, a number of time-dependent chloride ingress models were proposed by assuming a power law relationship between $D(t)$ and t [4]. The expression for $D(t)$ was either directly inserted into Crank's solution [4] or integrated over the time and then inserted into Crank's solution [5] in different chloride ingress models. Also, the reference time that is involved in $D(t)$ calculation varied from 1 s to 1 year depending on the type of experiments conducted by the researcher proposing the model. The diffusion coefficient at reference time and power law exponent mostly depended on degree of hydration and water-to-cement ratio. Later, correction factors for temperature and humidity were also included in the power law expression of $D(t)$ [6]. It is noted that surface chloride concentration also increases with time depending on the exposure condition [7] and several models also exist considering time-dependent models for $D(t)$ and $C_s(t)$ in Fickian framework.

One thing that is clear from the above review is that the requirement of a constant D for using Crank's solution to model the process of chloride ingress has been violated by all the above-mentioned models. Also, treating of concrete as a homogenous material for using such models is not acceptable. Therefore, the models based on the solution of Fick's second law are more of empirical type than of phenomenological nature. Wei et al. [8] modelled the time-dependent chloride ingress into the heterogeneous concrete matrix as a sub-diffusion process using time-fractional diffusion equation. This model overcomes some of the limitations caused by the assumptions mentioned above.

Tateishi et al. [9] in their work brought out the changes in waiting time distribution, mean square displacement and probability distribution of fractional diffusion equation when different fractional operators are used. The use of Caputo differential operator models the mean square displacement as that of normal diffusion process at smaller times and transforms into confined diffusion process at larger times. On the other hand, the use of Riemann–Liouville operator models the mean square displacement as power law function of time, and waiting time as a power law distribution [9]. The modelling by Riemann–Liouville operator is more representative of the actual diffusion phenomenon in concrete compared to Caputo operator, where the diffusion is at steady state at larger times. It is noted that the solution of fractional diffusion equation using Riemann–Liouville and Grunwald–Letnikov operators coincides when the function is continuous and is integrable within a given interval [10]. Also, the fractional derivatives can be numerically estimated in simple and efficient way by Grunwald–Letnikov operator [10].

In this paper, the initiation time of corrosion is estimated using sub-diffusion model and compared with the prediction of Fickian-based *fib* model. Grunwald–Letnikov operator is used in numerical solution of fractional diffusion equation which is different from the Caputo operator used by Wei et al. [8]. C_s is assumed to be constant in the present study, although it varies with different exposure condition.

2 Model for Corrosion Initiation Time

The time at which the chloride concentration at the level of reinforcement exceeds the critical chloride concentration is the corrosion initiation time. At this time, the passive layer, which is a protective layer, formed around steel reinforcement due to the alkaline concrete surrounding gets destroyed in the presence of chloride ions. The value of critical chloride content, C_{cr}, depends on the steel–concrete interface, pH level, electrochemical potential and composition of the steel, binder type, water–binder ratio (w/b), moisture and oxygen availability at the steel surface, temperature and so on [11]. In this paper, the concentration of chloride inside the concrete is estimated by modelling the chloride ingress process as a sub-diffusion process as given in Sect. 2.1.

2.1 Sub-diffusion Process to Model Chloride Ingress

The chloride ingress into concrete is mostly viewed as a Fickian diffusion process. While it is known that simple diffusion coefficient needs to be modified, as time-dependent diffusion coefficient, to account for modification in pore structure due to hydration as a first step, there is a need to develop models that take into account the geometrical form of pore structure. Many investigations showed that the pore structure of cement-based material has fractal characteristics at some length scales [12]. Also the colloidal-based model for CSH gel [13], a product of hydration, showed that the CSH clusters grow into diffuse fractal structures. Based on the results of experimental studies on pore structure of concrete, Konkol and Prokopski [14] and Jin et al. [15] proposed a fractal structure for pores in concrete. From these studies, it is inferred that pores in concrete can be modelled using a fractal geometry. Janett [16] explained the diffusion process on such complex fractal-state space by timescaling. It is important to note that the fractal space is related to scaling of time, not space, because of the most fundamental principle of thermodynamics. The entropy of any existing physical system increases by becoming increasingly disordered, diluted or randomized. Thereby, the arrow of time can only move forward, but speed of occurrence can be relative. In this problem, as the diffusion retards with time, the anomalous exponent of time lies between 0 and 1, and hence treated as a sub-diffusion process. Similar to the analogy that exists between diffusion process and Brownian motion process, sub-diffusion process is analogous to continuous time random walk accompanied by long rests [17]. The long rests can also be seen as the outcome of sinks that are distributed inside concrete as explained by Bazant [18]. Sinks are places where hydration products block the connectivity of pores and make the path tortuous; as a result, the diffusing ions get entrapped. Hence, diffusion process retards with time due to refinement of pore structures, and the process needs to be modelled as an anomalous diffusion process and not as a normal Fickian diffusion process.

A time-fractional partial differential equation is used, in this paper, as the tool to describe the process. The same is given by,

$$\frac{\partial C}{\partial t} = D_\alpha \mathbb{D}_t^{1-\alpha} \frac{\partial^2 C}{\partial x^2} \tag{1}$$

where α is fractional order of time $0 < \alpha < 1$; D_α is the chloride sub-diffusion coefficient; $\mathbb{D}_t^{1-\alpha}$ is the fraction derivative term. Metzler and Klafter [17] report approximate analytical solutions of time-fractional diffusion equation using Fox function. Such functions are complex and are not easy to handle. Also, these analytical solutions are specific to a given boundary condition and are not applicable for boundary conditions that represent the chloride ingress into the cover concrete (i.e. $(0, t) = C_s$ and $C(x \gg \tilde{x}, t) = 0$). Therefore, as an alternative to analytical solution, numerical solution is used [8].

2.2 Numerical Solution for Time-Fractional Partial Differential Equation

In this study, explicit finite difference schemes are used to provide a numerical solution [10]. The fractional derivative term is defined by Grunwald–Letnikov definition.

It is noted that there also exist many other definitions, and Grunwald–Letnikov definition is the one considered by [10] in their numerical solution. A similar study performed by Wei et al. [8] used Caputo definition for fractional derivative and implicit finite difference scheme to provide the numerical solution. The numerical representation of fractional derivative defined by Grunwald–Letnikov is given as [10],

$$\mathbb{D}_t^{1-\alpha} f(t) = \lim_{h \to 0} \frac{1}{h^{1-\alpha}} \sum_{k=0}^{m} w_k^{1-\alpha} f(t - kh) \tag{2}$$

where h is the time interval, m is number of time steps and $w_k^{1-\alpha} = \left(\frac{k-2+\alpha}{k}\right) w_{k-1}^{1-\alpha}$.

A forward time and centred space scheme (FTCS) is used and the resulting numerical solution for chloride concentration is given by,

$$C_j^{m+1} = C_j^m + S_a \sum_{k=0}^{m} w_k^{1-\alpha} \left[C_{j-1}^{m-k} - 2C_j^{m-k} + C_{j+1}^{m-k} \right] \tag{3}$$

where C_j^m denotes the chloride concentration at mth time step $(m.\Delta t)$ and jth step length $(j.\Delta x)$ and $S_a = \frac{D_\alpha (\Delta t)^\alpha}{(\Delta x)^2}$.

The procedure adopted, in the present study, for estimating the initiation time of corrosion using the numerical solution of fractional diffusion equation is as follows:

i. The time step, Δt, and jump length, Δx, are taken as 0.04 years and 0.005 m, respectively. The value of Δt and Δx is fixed such that $0 \leq S_a \leq 1/2^{(2-\alpha)}$, for the numerical solution to be stable.
ii. The initial condition is set as $C(0, 0) = C_s$.
iii. The boundary conditions are set as $C(0, t) = C_s$ and $C(.5, t) = 0$.
iv. The concentration values at every other time step and jump length are calculated using Eq. (3). A matrix containing chloride concentration values with the rows representing jump length and columns representing time step is thus generated.
v. The row corresponding to the location of reinforcement is scanned to find the time at which the chloride concentration value equals the critical chloride concentration. This defines the initiation time of corrosion.

3 Model Performance

The performance of the proposed model is checked by comparing the model predictions with the chloride profiles of concrete samples exposed to field conditions. The parameters C_s, α and D_α involved in Eq. (1) are found by fitting the chloride concentration values obtained from the numerical solution with the field data. Genetic algorithm-based optimization technique is used to perform the least square fitting in

MATLAB. The least square error (Eq. 4) is given as the objective function for optimization. Iterations are carried out till the average change in the fitness value of the
objective function is less than 1.0e−20. It is noted that objective function contains
the numerical solution of fractional diffusion equation as given in Sect. 2.2.

$$\text{LSE} = \sum_{i=1}^{n} \left[\frac{C_i^d(t)}{C^n(x_i, t)} - 1 \right]^2 \tag{4}$$

where C_i^d is the measured chloride concentration from laboratory samples exposed
to field conditions; $C^n(x_i, t)$ is the chloride concentration estimated using considered
model at x_i; n is the number of data points.

The chloride profiles of OPC (ordinary Portland cement) concrete specimens
obtained from field exposure trials of Chalee et al. [19] and Bamforth [20] are used.
The specimens of Chalee et al. had water–cement ratios of 0.65 and 0.45. These
specimens were exposed to tidal environment in the Gulf of Thailand for periods
of 2, 3, 4, 5 and 7 years. The temperature at site varied around 25–35 °C. The
parameters involved in the models are fitted using chloride profiles at 2, 3, 4 and
5 years collectively and the estimated values of parameters are used for predicting the
chloride profile measured at 7 years. The specimens of Bamforth had water–cement
ratio of 0.66 and were exposed in the splash zone at Folkestone on the southeast
coast of England. The parameters are fitted using chloride profile at 0.5, 1, 2, 3 and
6 years collectively and the estimated values of parameters are used for predicting
the chloride profile measured at 8 years.

The chloride profiles predicted using the proposed model are also compared with
those estimated using equation in *fib* model code 2006 [5]. The equation (Eq. 5) in *fib*
model code is based on DuraCrete model and uses a time-variant diffusion coefficient
and constant surface chloride concentration. The expression for diffusion coefficient
is based on a power law relationship between $D(t)$ and t with correction factor for
temperature. The chloride concentration increases with increase in temperature. The
model for chloride ingress recommended by *fib* model code is,

$$C(x, t) = C_s \left[1 - \text{erf} \left(\frac{x}{2\sqrt{D_0 \left(\frac{t_0}{t}\right)^a \exp\left(b\left[\frac{1}{T_{\text{ref}}} - \frac{1}{T}\right]\right)t}} \right) \right] \tag{5}$$

The parameters C_s, a and D_0 are found by fitting the chloride concentration
values obtained from the Eq. (5) with the field data. The value of T_{ref} is 293 K and
b is 4800 K. In this study, the convection zone, which is the zone up to which the
transport mechanism is not by diffusion, is neglected and also the initial chloride
concentration inside concrete is assumed to be zero. To estimate the parameters, C_s,
a and D_0, in Eq. (5), Genetic algorithm-based optimization procedure is used.

Table 1 Values of parameters involved in prediction models and its least square error

Investigator	Sub-diffusion process				Fickian diffusion process (*fib* model)			
	C_s(% by weight of cement)	α	D_α (m^2/s^α)	LSE (Eq. 4)	C_s(% by weight of cement)	a	D_0 (m^2/s)	LSE (Eq. 4)
Chalee—w/b = 0.65	5.74	0.94	2.08×10^{-11}	0.063	5.13	0.22	1.14×10^{-11}	0.056
Chalee—w/b = 0.45	5	0.82	5.08×10^{-11}	0.66	4.47	0.26	3.70×10^{-11}	0.57
Bamforth— w/b = 0.66	3.77	0.88	4.64×10^{-11}	0.221	3.39	0.13	9.67×10^{-11}	0.136

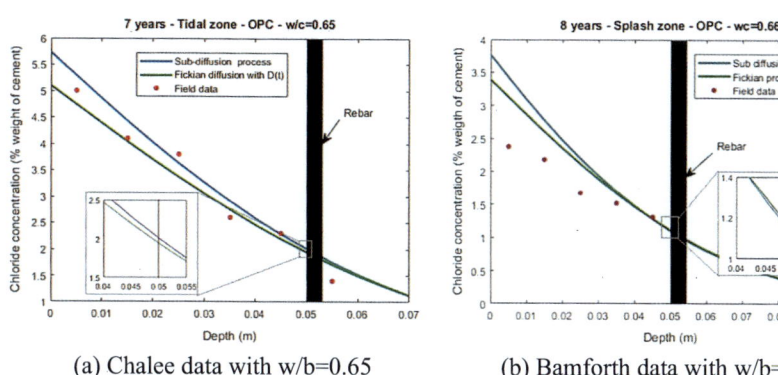

(a) Chalee data with w/b=0.65   (b) Bamforth data with w/b=0.66

Fig. 1 Comparison of chloride profiles predicted by the models

4 Results and Discussion

The values of the parameters obtained from optimization studies are given in Table 1. The chloride profiles predicted for a higher age (reported in a given investigation) using the parameters fitted to chloride profiles of lower ages are shown in Fig. 1.

4.1 Predictive Capability of Models on Concrete Specimens Exposed to Field Condition

The data used for prediction correspond to two different exposure conditions: (i) tidal zone and (ii) splash zone. The chloride profiles predicted by sub-diffusion model perform well in the case of Chalee data which represent tidal zone exposure (blue line in Fig. 1a), while deviates from the field data near the surface in the case of Bamforth data which represent splash zone exposure (blue line in Fig. 1b). The concentration values are overestimated up to a couple of centimetres from the concrete surface (Fig. 1b). This may be because the model neglects the occurrence of convection zone for partially saturated concrete in which the dominant transport mechanism is by capillary suction. This effect is not felt in Chalee data, since it is exposed to tidal zone where the samples get wetted every now and then without subsequent time for drying. It is also noted here that there is a possibility of chlorides being washed away during chemical analysis, thereby giving lesser concentration value [21]. There is no difference observed in the predictive capability of the model with respect to the two different water–cement ratios.

The chloride profiles predicted using both *fib* model and sub-diffusion model follow similar trend and represent reasonably well the measured profiles of the concrete samples considered in this study (Fig. 1). This is also true for chloride profile at lower ages using which the parameters are fitted. (Figures are not presented here.) One thing that can be observed from Fig. 1 is that even though the surface chloride concentration is less in the case of *fib* model, the chloride concentration near the reinforcement is same as that of sub-diffusion model. This means Fickian-based *fib* equation models the ingress at a slightly higher rate than sub-diffusion model. This observation can have an effect on the initiation time of corrosion especially when the structure has prolonged initiation time of corrosion. Therefore, the predictive capability of both models is examined, in the next section, for corrosion initiation time of actual OPC concrete structures exposed to marine environment for several years.

4.2 Predictive Capability of Models on Actual Structures

The chloride profiles of OPC core samples from real structures reported by Pack et al. [22], Liam et al. [23], Oslakovic et al. [24] and Michael [25] are considered. Liam et al. and Michael considered concrete cores from 24-year-old reinforced concrete jetty structure in Woodlands, Singapore, and 54-year-old reinforced concrete Lloyd Roberts jetty structure in Shute Harbour, Australia, respectively. On the other hand, Pack et al. and Oslakovic et al. considered concrete samples from 22.5-year-old bridge pier at South Korea and 25-year-old KrK Arch Bridge columns located in Croatia, respectively. The average temperature ranged from 24 to 32 °C in Woodlands, 13.5 °C near KrK Arch Bridge and 13.8 °C in South Korea. The model parameters that are fitted to the profiles and its LSE of prediction are given in Table 2. The initiation time of corrosion estimated using both the models is given in Table 3. The cover depth and critical chloride concentration used for estimating the initiation time of corrosion are taken from the respective literature in order to compare the results of the initiation time with the field observations reported in the literature.

It can be seen from Table 3 that in the case of Liam data and Michael data, the prediction of sub-diffusion model is more realistic as the results of initiation time

Table 2 Values of parameters and its least square error for predicting chloride profiles of actual structures

Investigator	Sub-diffusion process				Fickian diffusion process (*fib* model)			
	C_s (% by weight of cement)	α	D_α (m^2/s^α)	LSE	C_s (% by weight of cement)	a	D_0 (m^2/s)	LSE
Pack—bridge pier	2.65	0.96	4.42×10^{-12}	0.273	2.64	0.21	9.69×10^{-12}	0.275
Liam—jetty pile	1.2	0.88	3.02×10^{-11}	0.21	1.17	0.13	3.55×10^{-12}	0.23
Oslakovic—bridge column	2.57	0.76	5.72×10^{-11}	0.036	2.4	0.2	1.98×10^{-12}	0.058
Michael—jetty deck	1.54	0.725	9.82×10^{-11}	0.12	1.35	0.37	4.47×10^{-12}	0.29

Table 3 Estimated initiation times of corrosion of actual structures using both the models

Investigator	Exposure	Cover depth (mm)	C_{cr} (% by weight of cement)	Initiation time (years)		Remarks related to field observation
				Sub-diffusion process	Fickian diffusion process (*fib* model)	
Pack—bridge pier w/c = 0.47	Tidal zone	50	0.4	9.34	6.81	Initiation time reported in the study is around 5 years. This is not compared with the corrosion state of the structure. State of structure not known
Liam—jetty pile w/c = 0.5	Mean tidal zone	70	0.4	34.34	32.13	No delamination observed at 24 years in mean tidal zone of piles. Initiation time reported in study is 35 years
Oslakovic—bridge column w/c = 0.36	Splash zone and wind blown	35	0.6	35.85	39.64	Average initiation time of all the columns and arch is between 9 and 28 years for splash zone (age factor not considered). It is also reported that the repair works have been carried out after 15 years of construction due to corrosion cracks
Michael—jetty deck w/c = N/A	Splash zone	50	0.2	57.2	44.87	No delamination observed at 54 years

N/A not available

are comparable with field observation. For Oslakovic data, it is highly difficult to compare the results because the reported initiation times are average of all samples collected from different structural components at different heights, while the chloride profile considered in this study is from columns at a height 40 m above mean sea level. However, when compared to *fib* model, the initiation time is closer to the range reported by Oslakovic et al.

Therefore, it is concluded that although *fib* model is a good representation of the chloride ingress process (Fig. 1), it may be over-conservative when used for estimating initiation time of corrosion of real structure, especially when initiation time of corrosion is long (Table 3). The longer initiation time can be seen as an attribute of good-quality workmanship since the exposure condition in all the cases is severe. Also, as the accuracy of any empirical model becomes questionable outside the range of data used in its fitting, it is better to use a phenomenological model such as sub-diffusion for modelling chloride ingress into cover concrete.

5 Conclusions

The ingress of chlorides into the cover concrete is often modelled using Fick's second law of diffusion. Constant and time-varying diffusion coefficients are used in the literature for this purpose. From the brief review of literature presented in Sect. 2.1, it is inferred that the diffusion model should take into account the development of geometry of the pore structure with age. Based on the theoretical investigations of Janett [16], a sub-diffusion model is proposed in this paper. The predictions made by this model are compared with: (i) actual chloride profiles of OPC concrete samples exposed to marine environment and (ii) observed chloride profiles from actual structures. Also, the predictions of the proposed model are compared with those of *fib* model code. It is found that the predictions by the proposed model, estimated in terms of corrosion initiation time, can be significantly different from those of the *fib* model, especially when the quality of workmanship is good.

References

1. M. Collepardi, A. Marcialis, R. Turriziani, Penetration of chloride ions into cement pastes and concrete. J. Am. Ceram. Soc. **55**(10), 534–535 (1972)
2. J. Crank, *The Mathematics of Diffusion*, 2nd edn. (Clarendon Press, Bristol, 1975)
3. K. Takewaka, S. Mastumoto, Quality and cover thickness of concrete based on the estimation of chloride penetration in marine environments. ACI Spec. Publ. **109**, 381–400 (1988)
4. M. Maage, S. Helland, J.E. Carlsen, Practical non-steady state chloride transport as a part of a model for predicting the initiation period, *Chloride Penetration into Concrete, RILEM PRO*, 2 , ed. L.-O. Nilsson, J. Ollivier (1995), pp. 398–406
5. L. Tang, L.-O. Nilsson, Chloride diffusivity in high strength concrete at different ages. Nord. Concr. Res. **11**, 162–171 (1992)
6. *fib* Model Code, Bulletin No. 34, Service Life Design (2006)

7. A. Costa, J. Appleton, Chloride penetration into concrete in marine environments—part II: prediction of long term chloride penetration. Mater. Struct. **32**, 354–359 (1999)
8. S. Wei, W. Chen, J. Zhang, Time-fractional derivative model for chloride ions sub-diffusion in reinforced concrete. Eur. J. Environ. Civil Eng. **21**(3), 319–331 (2017)
9. A.A. Tateishi, H.V. Ribeiro, E.K. Lenzi, The role of fractional time- derivative operators on anomalous diffusion. Front. Phys. **5**, 52 (2017)
10. S.B. Yuste, L. Acedo, An explicit finite difference method and a new von Neumann-type stability analysis for fractional diffusion equations. SIAM J. Numer. Anal. **42**(5), 1862–1874 (2005)
11. U. Angst, B. Elsener, C.K. Larsen, O. Vennesland, Critical chloride content in reinforced concrete—a review. Cem. Concr. Res. **39**(12), 1122–1138 (2009)
12. S.W. Tang, E. Chen, Z.J. Li, Fractal model for pore structure in cement-based materials, in *2nd International conference on Microstructural-related Durability of Cementitious Composites*, *RILEM Publications SARL* (2012), pp. 997–1004
13. J.J. Thomas, H.M. Jennings, A colloidal interpretation of chemical aging of the CSH gel and its effects on the properties of cement paste. Cem. Concr. Res. **36**(1), 30–38 (2006)
14. J. Konkol, G. Prokopski, The effect of concrete mix composition on the character of fracture of set concrete. J. Geogr. Geol. **6**(4), 29–41 (2014)
15. X. Jin, B. Li, Y. Tain, N. Jin, A. Duan, Study on fractal characteristics of cracks and pore structure of concrete based on digital image technology. Res. J. Appl. Sci. Eng. Technol. **5**(11), 3165–3171 (2013)
16. P. Janett, Diffusion on fractals and space-fractional diffusion equations. Technische Universiẗat Chemnitz, Fakulẗat fur Naturwissenschaften, Dissertation (2010)
17. R. Metzler, J. Klafter, The random walk's guide to anomalous diffusion: a fractional dynamics approach. Phys. Rep. **339**(1), 1–77 (2000)
18. Z.P. Bazant, Physical model for steel corrosion in concrete sea structures-theory. ASCE J. Struct. Div. **105**(6), 1137–1153 (1979)
19. W. Chalee, C.A. Jaturapitakkul, P. Chindaprasirt, Predicting the chloride penetration of fly ash concrete in seawater. Mar. Struct. **22**(3), 341–353 (2009)
20. P.B. Bamforth, The derivation of input data for modelling chloride ingress from eight-year UK coastal exposure trials. Mag. Concr. Res. **51**(2), 87–96 (1999)
21. K.Y. Ann, J.H. Ahn, J.S. Ryou, The importance of chloride content at the concrete surface in assessing the time to corrosion of steel in concrete structures. Constr. Build. Mater. **23**(1), 239–245 (2009)
22. S.W. Pack, M.S. Jung, H.W. Song, S.H. Kim, K.Y. Ann, Prediction of time dependent chloride transport in concrete structures exposed to a marine environment. Cement Concr. Res. **40**(2), 302–312 (2010)
23. K.C. Liam, S.K. Roy, D.O. Northwood, Chloride ingress measurements and corrosion potential mapping study of a 24-year-old reinforced concrete jetty structure in a tropical marine environment. Mag. Concr. Res. **44**(160), 205–215 (1992)
24. I.S. Oslakovic, D. Bjegovic, D. Mikulic, Evaluation of service life design models on concrete structures exposed to marine environment. Mater. Struct. **43**(10), 1397–1412 (2010)
25. M. Large, Physical testing: chloride ingress, in *prepared for Cardno NSW Water and Environment*, ed. by L.R. Jetty, S. Harbour (Australia, 2016)

Artificial-Neural-Network-Based Uncertain Material Removal Rate by Turning

Subhankar Saha, Saikat Ranjan Maity and Sudip Dey

Abstract The present paper deals with the uncertain material removal rate (UMRR) of cylindrical turning of AISI 52100 steel (as workpiece) being turned by cubic boron nitride (CBN) (as single-point cutting tool insert). During machining operations such as cylindrical turning, assessment of material removal rate is often inharmonious, non-uniform and unpredictable due to variabilities in rotational speed of workpiece, feed and depth of cut provided by the cutting tool. It occurs due to unforeseen operational and manufacturing uncertainties. The present work is aimed to develop a computational model in conjunction with artificial neural network (ANN) approach. The constructed computational model is validated with the previous published experimental results. The traditional Monte Carlo simulation (MCS) is employed to compare the efficacy and accuracy of the constructed artificial neural network (ANN)-based surrogate model. The effect of both individual and combined variations of input parameters such as cutting speed, feed and depth of cut on the material removal rate is portrayed. The surrogate model is validated with the original Monte Carlo simulation (MCS), and the intensity of variation of output quantity of interest (QoI) is presented by the probability density function plots. The statistical analyses are carried out based on parametric studies, and the subsequent results are illustrated. The effect of depth of cut is observed to be maximum sensitive to influence the uncertain material removal rate, followed by feed and cutting speed.

Keywords Artificial neural network · Turning · Monte Carlo simulation · Material removal rate

S. Saha (✉) · S. R. Maity · S. Dey
National Institute of Technology Silchar, Silchar, India
e-mail: sahamech90@gmail.com

S. R. Maity
e-mail: saikat.jumtech@gmail.com

S. Dey
e-mail: infodrsudip@gmail.com

© Springer Nature Singapore Pte Ltd. 2020
P. V. Varde et al. (eds.), *Reliability, Safety and Hazard Assessment for Risk-Based Technologies*, Lecture Notes in Mechanical Engineering,
https://doi.org/10.1007/978-981-13-9008-1_49

591

1 Introduction

Over the last few decades, researchers have rigorously attempted to investigate the role of deterministic process parameters like the cutting speed, feed and depth of cut on several performance measures of the machining operation in both conventional and automated environments. Among them, the material removal rate (MRR) is considered to be the most important parameter in the literature. A study based on Taguchi method is accomplished to optimize the process parameters such as the cutting speed, feed and depth of cut so as to maximize the material removal rate (MRR) of SAE 1020 steel in CNC lathe. The individual effect of the concerned process parameters is also portrayed [1]. Response surface methodology (RSM) is employed by some researchers to analyze the influence of four process parameters such as the cutting speed, feed, depth of cut and tool nose radius on surface roughness and MRR of EN-24 alloy steel turned by cemented carbide tool [2]. An L-9 orthogonal array, analysis of variance (ANOVA) and signal-to-noise ratio (S/N) are used to investigate the role of spindle speed, feed rate and depth of cut on MRR and surface roughness. The process parameters are also optimized [3]. But on the basis of exhaustive literature survey, few work is reported in the past which addresses the issue of unforeseen operational as well as manufacturing uncertainties in the context of turning operation which actually leads to variability in the machining characteristics. The random fluctuation in process conditions in the machining zone, uncertainties in handling, uncertainties in tool wear, random fluctuation in machine tool vibration, and geometric error in machine tool are some of the types of operational uncertainties, whereas manufacturing uncertainties include the variability in material deformation behavior of the workpiece, the inherent material defects and a band of geometric tolerance allowed for cutting tool. In this paper, uncertainty analysis of the material removal rate (MRR) is performed for turning operation taking into account the effect of both individual and combined variations of the uncertain input parameters, namely feed, cutting speed and depth of cut using artificial neural network (ANN) approach.

2 Mathematical Formulation

2.1 Material Removal Rate (MRR)

The material removal rate in turning is the volume of a single layer of material removed during each revolution of the workpiece. It is expressed in mm^3/min. The material removal rate (MRR) is expressed mathematically as

$$MRR = 1000 \times V \times f \times d \tag{1}$$

where V is the cutting velocity in m/min, f is the feed in mm/rev, and d is the depth of cut in mm.

2.2 Artificial Neural Network (ANN)

Machining processes are no doubt stochastic in nature with high degree of nonlinearity and considered to be time-dependent. They also vary with material properties and machining conditions. ANN is a mathematical technique, which with the aid of proper learning and training can handle such phenomena. Here in this paper, a feed-forward (ANN) is employed for computational modeling [4] and the network architecture is of the form that it consists of a single input layer followed by a hidden layer and then the output layer. Hyperbolic tangent sigmoid transfer function (tansig) is used for the input and hidden layer nodes. Quick propagation (QP) algorithm is executed to train the network. Previously published literature ensures that this algorithm can be employed for the training of all ANN models [5]. The performance of the model is statistically measured by root-mean-squared error (RMSE).

$$\text{RMSE} = \sqrt{\frac{1}{N} \sum_{i=1}^{N} (X_i - X_{id})^2} \qquad (2)$$

where N is the number of points, X_i is the predicted value, and X_{id} is the actual value.

2.3 Uncertain Material Removal Rate (UMRR) Using Artificial Neural Network (ANN) Model

The uncertainty analysis of MRR is carried out considering the role of both individual and combined variations of the uncertain input parameters. To make it possible, the distribution of randomness is allowed to lie within a tolerance limit of 10% of the deterministic value of the three concerned input parameters. The flowchart of the proposed analysis using ANN model is shown in Fig. 1.

3 Results and Discussion

The purpose of the earlier published literature [6] which is considered here for reference was to arrest the effect of process parameters such as the cutting speed, feed and depth of cut on MRR, cutting power and MRR per cutting power during turning of AISI 52100 steel using polycrystalline cubic boron nitride (CBN) insert in the deterministic framework. But the intention of the present work is to assess the effect of the same parameters but considered herein as random parameters to predict their influence on the uncertain material removal rate (UMRR) using artificial neural network (ANN) approach. It is found that the results obtained matched with the deterministic MRR values found from the experimental results. Moreover, the artificial

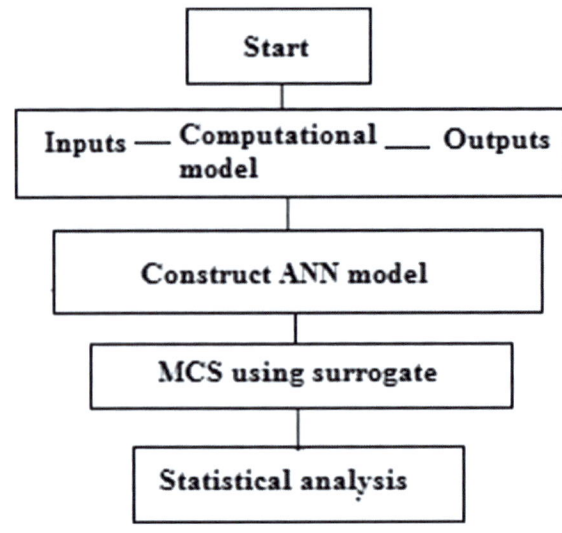

Fig. 1 Flowchart of the proposed analysis

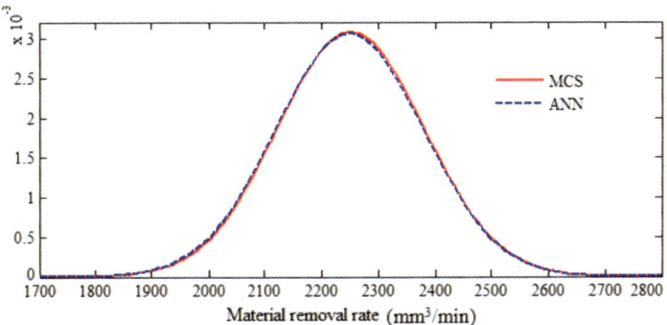

Fig. 2 Probability density function plot of UMRR by MCS and ANN for the combined variation of input parameters

neural network (ANN) model is being found to be well converged with the Monte Carlo simulation (MCS) analysis for the combined variation of input parameters as shown by the probability density function plot in Fig. 2 corroborating the accuracy of the surrogate model.

In view of the above, study on the individual effect of process parameters on MRR is carried out by varying the input parameter of interest keeping the other input parameters fixed and both the deterministic MRR and the uncertain mean MRR based on ANN approach is presented in Table 1 and the subsequent probability density function plots of the random response used for statistical analyses are also illustrated in Fig. 3. It is witnessed that with incorporating marginal uncertainties, which is likely to be inevitable in the input parameters like the cutting speed, feed

Table 1 Effect of process parameters variation on MRR and UMRR

Process parameters	Cutting speed (m/min)	Feed (mm/rev)	Depth of cut (mm)	Deterministic MRR (mm³/min)	Uncertain mean MRR (mm³/min)
Cutting speed	250	0.03	0.1	750	750.65
	300	0.03	0.1	900	899.61
	350	0.03	0.1	1050	1050.4
Feed	250	0.03	0.1	750	750.34
	250	0.04	0.1	1000	1000.1
	250	0.05	0.1	1250	1250.1
Depth of cut	250	0.03	0.1	750	750.29
	250	0.03	0.2	1500	1501.3
	250	0.03	0.3	2250	2249

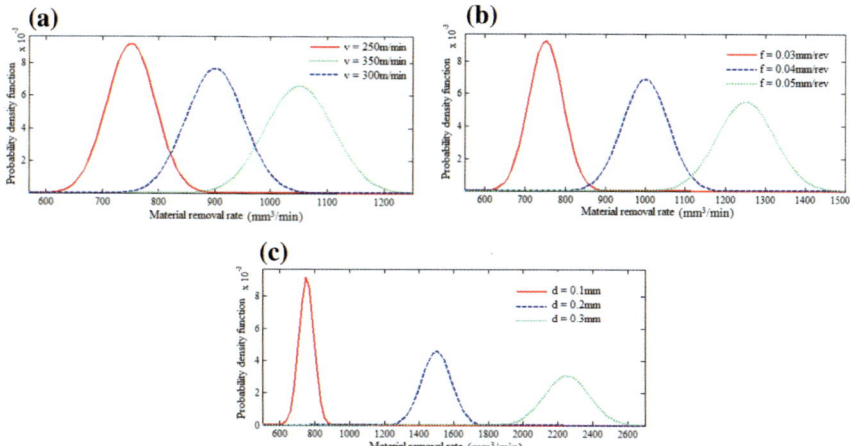

Fig. 3 a–c Effect of individual variation of process parameters on UMRR

and depth of cut, respectively, to analyze the individual effect on MRR, it is found that the MRR is no more deterministic as found in the earlier published literature but a completely random response is obtained with considerable sparsity from the mean as shown by the probability density function plot in Fig. 3. This depicts the credibility of ANN-based surrogate model which is found to capture the uncertainty in the output quantity of interest, namely MRR. Beyond that, it is observed both in Table 1 and the subsequent plots in Fig. 3 that the depth of cut is the most sensitive input parameter affecting the UMRR followed by feed and cutting speed.

Figure 4 depicts the combined effect of the variation in the input parameters at different degrees of uncertainty (P), i.e., 10, 20 and 30% on the material removal rate (MRR). It is found that with the increase in the degree of uncertainty, the frequency

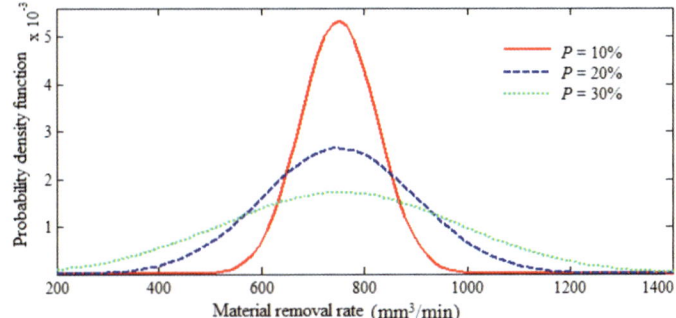

Fig. 4 Combined effect of the variation in the input parameters on UMRR at different degree of uncertainty

band of the response spread over a larger window indicating increase in the degree of uncertainty on the response

4 Conclusions

It is found that with incorporating the artificial neural network (ANN) model, the uncertain material removal rate (UMRR) is captured effectively with only a sample size of 512 which is computationally cost-effective and time-efficient than the original Monte Carlo simulation (MCS) which uses sample size of 1000. The ANN-based approach to uncertainty quantification can be considered to deal with more complex machining phenomena in near future.

References

1. S. Mukherjee, A. Kamal, K. Kumar, Optimization of material removal rate during turning of SAE 1020 material in CNC lathe using Taguchi technique. Procedia Eng. **97**, 29–35 (2014)
2. S. Dinesh, K. Rajaguru, V. Vijayan, Investigation and prediction of material removal rate and surface roughness in CNC turning of EN24 alloy steel. Mech. Mech. Eng. **20**(4), 451–466 (2016)
3. A. Abdallah, B. Rajamony, A. Embark, Optimization of cutting parameters for surface roughness in CNC turning machine with aluminium alloy 6061 material. IOSR J. Eng. **4**(10), 01–10 (2014)
4. A.W. Minns, M.J. Hall, Artificial neural networks as rainfall-runoff models. Hydrol. Sci. J. **41**(3), 399–417 (1996)
5. A. Ghaffari, H. Abdollahi, M.R. Khoshayand, I.S. Bozchaloo, A. Dadgar, M. Rafiee-Tehrani, Performance comparison of neural networks. Environ. Sci. Technol. **42**(21), 7970–7975 (2008)
6. A.D. Bagawade, P.G. Ramdasi, Effect of cutting parameters on material removal rate and cutting power during hard turning of AISI 52100 steel. Int. J. Eng. Res. Technol. **3**(1), 1018–1025 (2014)

RAMS

Failure Analysis and Process Capability Study of Transmission System at the End of Assembly Line

Aman Panse and Prasad V. Kane

Abstract This paper discusses the particular issues at the end of assembly line inspection of gearbox. This paper discusses various faults that incur during newly manufactured and assembled transmission system. Existing manual diagnostic technique that is applied for the end of assembly line inspection of transmission is also discussed. The paper presents the study of vibration-based approach to evaluate the process capability of gearbox inspection. The study is useful to propose the benchmark for accepting or rejecting the gearbox.

Keywords Failure analysis · Process capability · Online inspection · Vibration-based approach

1 Introduction

This paper discusses the issue of failure analysis and process capability of newly assembled transmission system. At the end of assembly line of gearbox, manual inspection is carried out to assess the quality of gearbox which is based on the judgement of operator. The various faults observed at the end of assembly line are gear shaft assembly error, profile errors of gear teeth, burrs and dents on teeth and missing component [1].

At the end of assembly line, the gearbox is operated by electric motor on a test bench. Based on the sound emitted, operator takes decision on the quality of gearbox with his past experience and judgement [2], which is not a scientific method and hence efforts are made in this paper to collect the data of vibration signal emitted by gearbox and obtain the process capability indices and control limit values. Based on

A. Panse (✉) · P. V. Kane
Visvesvaraya National Institute of Technology, Nagpur, India
e-mail: panse21@gmail.com

P. V. Kane
e-mail: prasadkane20@gmail.com

© Springer Nature Singapore Pte Ltd. 2020
P. V. Varde et al. (eds.), *Reliability, Safety and Hazard Assessment for Risk-Based Technologies*, Lecture Notes in Mechanical Engineering,
https://doi.org/10.1007/978-981-13-9008-1_50

599

Fig. 1 Flow chart of
methodology

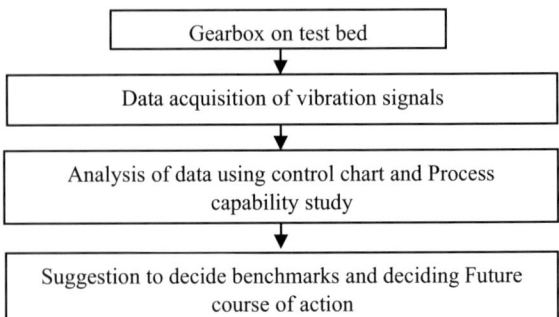

experience and questionnaire survey from the automobile manufacturer of central India, rejection rate of gearbox based on the existing manual inspection is 2–4%.

In this study, the vibration signals of tractor transmission are acquired from the gearbox at the end of assembly line. The signals are processed to obtain signal statistical feature like kurtosis, root mean square (rms), mean, skewness [3]. It is observed that kurtosis is the parameter, which is sensitive to the error causing deviations from normal operating condition [4]. This parameter is selected to obtain C_P and C_{Pk} values of the process. The control limit of the kurtosis of vibration signal is proposed as quality index to decide acceptance or rejection of gearbox instead of perception/opinion of operator based on sound emitted by gearbox.

2 Methodology

The methodology followed in this work is shown in (Fig. 1) the following flow chart.

3 Experimental Set-Up

The induction motor of 5hp is used to operate sliding mesh gearbox of tractor, which is shown in Fig. 2. To acquire signal IEPE accelerometer is used which is mounted on the bearing housing of gearbox. The accelerometer is attached to data acquisition system [DAQ card] of National Instruments [N19234]. The output of DAQ card is given to a laptop with LabVIEW as interfacing software for data acquisition. The programme with block diagram is developed in LabVIEW to acquire vibration signal and extracts signal statistical features. 100 vibration signals have been taken which is divided into 10 subgroups of equal size 10 for the analysis (Table 1).

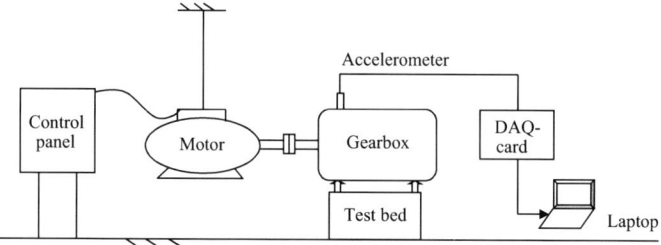

Fig. 2 Experimental set-up

Table 1 Kurtosis values of vibration signal

size (n)	Subgroup (K)									
	K1	K2	K3	K4	K5	K6	K7	K8	K9	K10
1	3.154	3.780	3.134	2.950	2.897	2.898	2.807	3.094	3.484	2.858
2	3.798	3.050	2.766	2.983	4.602	2.926	3.195	2.947	3.011	2.812
3	4.513	3.431	2.884	2.903	2.619	3.043	2.731	2.947	3.084	3.048
4	4.273	3.712	2.942	3.055	4.383	2.959	3.301	2.947	3.096	3.142
5	3.637	4.833	3.040	3.095	2.916	2.868	3.043	3.013	2.996	3.195
6	3.781	3.326	2.783	3.001	3.407	3.314	3.228	3.093	3.085	3.217
7	4.075	3.526	2.870	5.270	3.771	3.182	3.017	3.219	3.099	3.353
8	3.381	3.238	2.715	3.004	3.011	2.969	2.933	3.027	3.495	3.477
9	3.465	3.282	2.994	3.031	2.909	2.759	3.021	3.241	3.084	3.126
10	3.689	3.383	3.010	3.152	2.954	3.103	2.948	3.186	3.021	3.343

4   Failure Analysis and Process Capability Study

4.1   Failure Analysis

Failure of gearbox to get accepted at the end of assembly line is a serious matter for industry. It would be more serious for industry to have annoyed customer for automobile industry, due to irritation high-frequency noise generated by gearbox in driver cabin, which is not masked by engine noise. Hence, failure analysis at the end of assembly line is very important task which needs to be carried out. The sound emitted from gearbox is categorized as rattle sound and whining noise. The various reasons for it found after failure analysis are misalignment of gear shaft, profile error on gear tooth, burrs and dents on gear tooth occurrences were observed. The issue of missing components has also been identified during inspection.

4.2   Process Capability Studies

The main objective of process capability study is to identify the consistency and variation observed in any process. For this, mean and standard deviation are the parameters used for measurement of central tendency and variation in the process. The control chart using mean, standard deviation and range, i.e. X-bar, S and R chart can be plotted to monitor central tendency and spread in the process. The process capability indices C_P and C_{Pk} can also be used as a metric to evaluate these parameters of the process [5].

X-bar chart is the average of sample of subgroup. It measures the central tendency of response variable over time. Both S chart and R chart are used to analyse its spread or dispersion. S chart uses the standard deviation to represent spread in the data, and R chart uses the range. This chart is frequently used for monitoring variable data. These charts also introduces some process capability concept by comparing it with specifications limits [6].

The average of each subgroup is calculated by summing the individual measurements in the subgroup and dividing by the number of measurement. The central line for the X-bar chart is found by summing the averages and dividing by the number of subgroups. Where n is the subgroup size and k the number of subgroups, the formulas for these calculations are as follows:

$$\bar{x}_i = \sum_{j=i}^{n} \frac{x_{ij}}{n} \tag{1}$$

$$\bar{\bar{x}} = \sum_{i=1}^{k} \frac{\bar{x}}{k} \tag{2}$$

$$s_i = \sqrt{\frac{\sum_{j=i}^{n} \left(x_{ij} - \bar{x}_i\right)^2}{n-1}} \tag{3}$$

$$\bar{s} = \frac{\sum_{i=1}^{k} s_i}{k} \tag{4}$$

where

$\bar{x}_i$ ith subgroup mean
$\bar{\bar{x}}$ grand mean, i.e. X-bar chart central line
s_i subgroup standard deviation
$\bar{s}$ mean standard deviation of subgroups
n subgroup size
k number of subgroup

The control limits for $\bar{x}$ and s charts are given as

$$\text{UCL}_x = \bar{\bar{x}} + A_2 \bar{R} \tag{5}$$

$$LCL_x = \overline{\overline{x}} - A_2\overline{R} \tag{6}$$

$$UCL_s = \overline{s} + 3\sigma\sqrt{(1 - C_4)^2} \tag{7}$$

$$LCL_s = \overline{s} - 3\sigma\sqrt{(1 - C_4)^2} \tag{8}$$

The factors A_2, C_4 for determining control limits are taken from factors table for $\overline{X}$ and S charts.

In this work, Kurtosis of the acquired vibration signals are used to characterise as the output of the process. The process capability indices provide a common metric to evaluate and predict the performance of process. C_P tells about the overall variability of the process.

A potential process capability ratio (Cp) is commonly used for quality control. Small value of C_p indicates the low process capability. C_p value more than 1 is always acceptable and less than 1 will never be acceptable.

$$C_p = \frac{USL - LSL}{6\sigma} \tag{9}$$

Process capability index (C_{pk}) is used for computing process capability index (C_{pk}) of each experiment

$$\text{Capability index } C_{pk} = \frac{\min(\mu - LSL, USL - \mu)}{3\sigma} \tag{10}$$

where,

USL   Upper specification kurtosis of acceleration signal of gearbox
LSL   Lower specification kurtosis of acceleration signal of gearbox

4.3 Results and Discussions

The input data collected from gearbox are used to plot X-bar and S control charts as shown in Fig. 3.

The $\overline{X}$ chart shows the deviation from central tendency and value of $\overline{X} = 3.224$ is obtained for kurtosis of collected vibration signals. The Lower Control Limit (UCL) and Upper Control Limit (LCL) of the X-bar chart are found to be 2.847 and 3.6 respectively with the assumption that the process is following a Normal Distribution.

In S chart, the mean value of standard deviation obtained is 0.3863. The LCL and UCL values for S chart are 0.663 and 0.1096, respectively. The chart shows that the variation within the subgroup is more. The variation in subgroup 4 and 5 is more (Fig. 4).

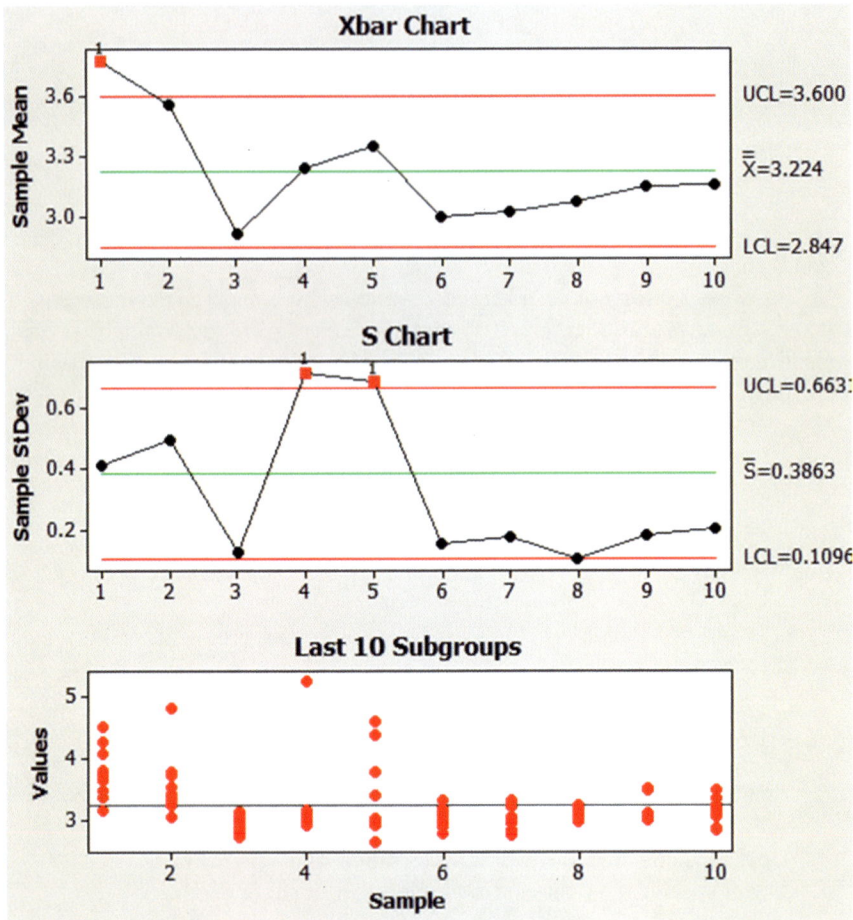

Fig. 3 $\overline{X}$, S control charts

The histogram of data, normal probability plot and capability plot are also obtained of kurtosis. The keen observation of histogram of kurtosis reveals that the data are skewed towards left with the mode value of 3. This distribution fitting can also be observed from normal probability plot. In normality plot, the P value is less than 0.05 shows that alpha error is more dominant.

The specification level values of 0 and 4 are assigned for calculating the C_P and C_{Pk} indices. The general thumb rule says that the gearbox is faulty with the kurtosis value more than 3. Hence, if it is assumed that the gearboxes to be acceptable up to the value of 4, the C_P and C_{Pk} values obtained are 1.68 and 0.65. C_{pk} accounts the variability within the subgroup whereas P_{pk} accounts for the overall variation of all the measurements.

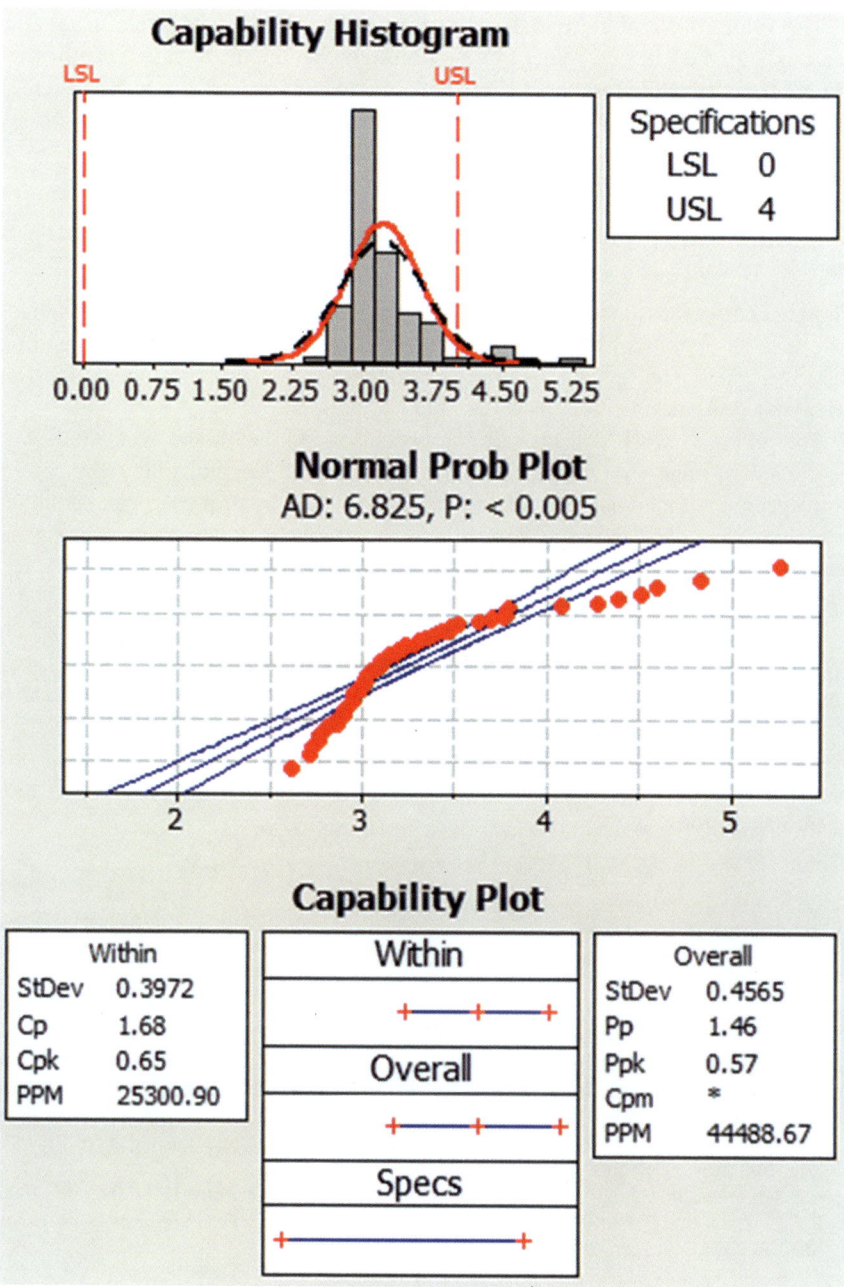

Fig. 4 Capability histogram, normal probability plot and capability plot

The value of C_P more than 1 indicates that the process is capable to produce gearbox with the Kurtosis values of vibration signal within the selected specification limits. The C_{Pk} value less than 1 indicates the readings are deviated from central value.

5 Conclusion

This paper discusses the issue of quality assessment of gearbox at the end of assembly line. It is observed that no scientific method is adapted for inspection of gearbox. Based on sound manifested by gearbox, operator decides the acceptance or rejection criteria of gearbox based on his perception and opinion.

The proposed work will provide the benchmark to decide the acceptable level in terms of C_P and C_{Pk} values. If a particular gearbox does not follow the control limit specified, then it can be sent back for rework. The control chart can be plotted to monitor the consistency and spread involved during each inspection process and corrective action can be taken. The C_p value will be more if the rejections are less and the deviation in obtained data is less.

It was observed that the burrs and dents are more frequently occurring faults from the past data in the industry. Hence, kurtosis which effectively quantifies the spikiness involved in signal has been selected as a variable to plot the control charts and to obtain C_P and C_{Pk} values.

The reference or benchmark obtained from this study would be a useful proposal for industry to quantitatively take decision regarding acceptance or rejection of similar gearbox.

References

1. W. Fan et al., Sparse representation of transients in wavelet basis and its application in gearbox fault feature extraction. Mech. Syst. Sig. Process. **56**, 230–245 (2015)
2. P. Kane, A. Andhare, Application of psychoacoustics for gear fault diagnosis using artificial neural network. J. Low Freq. Noise Vibr. Active Control **35**(3), 207–220 (2016)
3. A.R. Mohanty, *Machinery Condition Monitoring: Principles and Practices* (CRC Press, 2014)
4. V. Sharma, A. Parey, A review of gear fault diagnosis using various condition indicators. Procedia Eng. **144**, 253–263 (2016)
5. R.S. Leavenworth, E.L. Grant, *Statistical Quality Control* (Tata McGraw-Hill Education, 2000)
6. R.C.H. Chua, J.A. DeFeo, *Juran's Quality Planning and Analysis: For Enterprise Quality* (Tata McGraw-Hill Education, 2006)

Ball Bearing Fault Diagnosis Using Mutual Information and Walsh–Hadamard Transform

Vipul K. Dave, Vinay Vakharia and Sukhjeet Singh

Abstract Bearing failure may result in the breakdown of machinery or possibly damage the human being operating the machinery. It is therefore necessary to diagnose bearing faults at early stage. Vibration signals are contaminated with noise. To minimize the effect of noise, Walsh–Hadamard transform is used for feature extraction. Statistical features from coefficients are calculated to form feature vector. Mutual information a type of feature ranking method is used to select most informative feature and subsequently to reduce size of feature vector. Coefficients from healthy and faulty bearings at different rotational speeds were calculated from signals. Statistical features are calculated from the acquired signals with different speeds. Two different classifiers like support vector machine and artificial neural network have been used for finding the accuracy. Result shows the methodology adopted is effective to diagnose various bearing faults.

Keywords Mutual information · SVM · ANN · Walsh–Hadamard transform · Bearing

1 Introduction

Mechanical equipment plays core roles in the industries. With the rapid development of industry, the mechanical equipment is more and more required in their reliability, safety, economy, and intelligence [1]. Bearings are the crucial parts of industries, which provide support to relative motion between parts. It also concerns

V. K. Dave (✉) · V. Vakharia
Department of Mechanical Engineering, PDPU, Gandhinagar 382007, India
e-mail: vipul.dphd16@sot.pdpu.ac.in

V. Vakharia
e-mail: vinay.vakharia@sot.pdpu.ac.in

S. Singh
Department of Mechanical Engineering, GNDU Amritsar, Punjab 143005, India
e-mail: sukhjeets@iitrpr.ac.in

© Springer Nature Singapore Pte Ltd. 2020
P. V. Varde et al. (eds.), *Reliability, Safety and Hazard Assessment for Risk-Based Technologies*, Lecture Notes in Mechanical Engineering,
https://doi.org/10.1007/978-981-13-9008-1_51

with reliability and safety. Majority of failure are concern with bearing. High-speed rotation and wear are the main causes of bearing failure. With the passage of time, it will become severe and catastrophic failure of the components [1]. Earlier fault detection may avoid dangerous failure of bearing. Various methods like vibration measurement, lubricant analysis, infrared thermography, acoustic measurement are frequently used to detect and diagnose the faults associated with bearing. It is difficult for conventional techniques to identify faults at early stage, because of noise and higher vibration of bearing signal. It is possible to obtain the necessary diagnostic information from vibration signal of faulty bearings after applying signal processing techniques. In non-stationary vibration signals, due to change in operating and loading conditions with respect to time makes vibration signal to possess nonlinear characteristics. Signal processing methods such as fast Fourier transform and wavelet are used to identify faults by various authors. A machine fault identification methodology essentially consists of two main steps. In the initial stage, the useful information is extracted after applying signal processing techniques, which is useful for formation of feature vector, and in the later stage, the state of various faults is identified using various artificial intelligence techniques. Kankar et al. [2] presented fault diagnosis methodology using CWT after comparison of different mother wavelets. Shannon entropy and maximum energy to Shannon entropy criteria were used for wavelet selection. Mohanty et al. [3] have used variational mode decomposition (VMD) and FFT for fault diagnosis and discussed its effects for fault detection on bearing. The fault in the bearing is observed clearly. Vakharia et al. [4] have used minimum multiscale permutation entropy criteria for the scale selection for wavelet transform. Features were extracted and fed into supervised and unsupervised learning algorithms support vector machine (SVM) and self-organizing map (SOM). Recently Walsh–Hadamard transform (WHT) a signal processing method is developed which will be useful for classification of faults. The WHT decomposes a signal into a set of orthogonal, rectangular wave from Walsh function. The advantage of WHT is fast computation of Walsh–Hadamard coefficients and requires less storage space and fast signal construction.

Research work of this paper is arranged by following way: Sect. 2 presents introductions of Walsh–Hadamard transform (WHT) and its applications in a signal processing technique. Section 3 discussed about mutual information as feature ranking method. Section 4 represents SVM and ANN classifier. Section 5 describes experimental procedure. Section 6 talks about results and discussion and finally conclusion part of the paper (Figs. 1 and 2).

2 Walsh–Hadamard Transform

Fourier transform is a mathematical function which can describe or represent given function in series of sinusoidal functions. The basics about Walsh functions depend on it, where various operations like orthogonal, symmetric, involute are performed. WHT is 2-size discrete Fourier transform matrix which depends upon Walsh func-

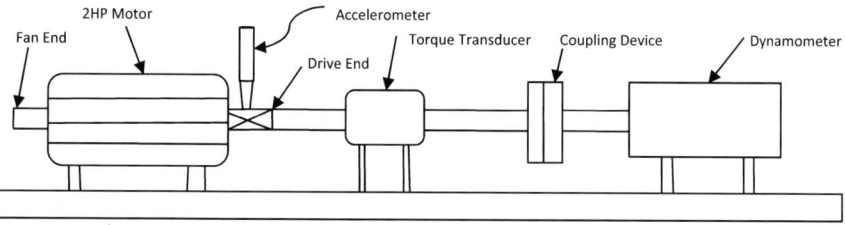

Fig. 1 Schematic diagram of experimental process

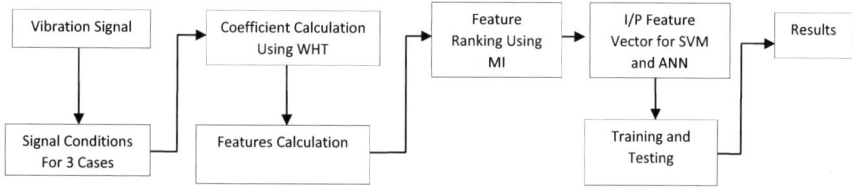

Fig. 2 Flow chart of the process

tions. The values in the Hadamard matrix may be $+1$ or -1. Mathematical operations such as additions and subtractions are possible in the calculation of Walsh–Hadamard transform [5]. Using discrete Fourier transform (DFT), the information about approximate finite set of frequencies can be revealed. Aliasing, i.e., distortion and leakage are the two main problem with DFT which can be overcome with use of WHT. It is observed that the WHT does not possess the shift-invariant and conjugate symmetry properties of the discrete Fourier transform. Another important feature of the WHT is, it can copy the concept of frequency where the sign of a function changes (e.g., from positive to negative) [6]. The authors are successful in a variety of applications in image and video coding, speech processing, data compression, communications, etc. to apply WHT as a signal processing method.

The Hadamard transform can be defined in two ways: recursively, or by using the binary (base-2) representation of the indices n and k. Recursively, we define the 1×1 Hadamard transform H_0 by the identity $H_0 = 1$, and then define H_m for $m > 0$ by [7]:

$$H_1 = \frac{1}{\sqrt{2}} \begin{pmatrix} H_{m-1} & H_{m-1} \\ H_{m-1} & -H_{m-1} \end{pmatrix} \tag{1}$$

where the $1/\sqrt{2}$ is a normalization factor which can be omitted based on applications. Variants of Hadamard matrices can be represented as follows.

$$H_0 = +1 \tag{2}$$

$$H_1 = \frac{1}{\sqrt{2}} \begin{pmatrix} 1 & 1 \\ 1 & -1 \end{pmatrix} \tag{3}$$

$$(H_n)_{ij} = \frac{1}{2^{\frac{n}{2}}} (-1)^{i*j} \tag{4}$$

Here $i * j$ is the bitwise dot product of the binary representations of the numbers i and j.

3 Mutual Information

Various feature ranking methods are available which can be applied to variety of applications. Selection of features, which describe the condition of the bearings, depends on the mathematical calculations and effective use of signal processing method. For better accuracy, one feature is not enough to describe the condition of it so, multiple features are required for identification of fault condition. It is observed that the information and the interaction between these features are important for efficient fault diagnosis [8–10]. To measure the relationship between two random variables which are sampled simultaneously, mutual information (MI) is one of the best methods. In particular, it measures how much information is communicated, on average, in one random variable about another. The advantage of Mutual Information it is independent of the type of classifier and provides optimal feature set from different domain together [8]. The mutual information between two random variables X and Y whose joint distribution is defined by $P(X, Y)$ is given by [9]

$$I(X; Y) = \sum_{y \in Y} \sum_{x \in X} p(x, y) \log \frac{p(x, y)}{p(x)p(y)} \tag{5}$$

In the above formula, $P(X)$ and $P(Y)$ are the marginal distributions of X and Y obtained through the marginalization.

4 Classifiers

4.1 Support Vector Machine (SVM)

SVM is basically a concept derived from statistical learning theory developed by Vapkin [11]. In SVM can be represented through a line or hyper plane is formed between two sets of data which is useful for differentiation between the data sets. When feature vector is supplied to the SVM algorithm, then it tries to orientate the boundary in such a way that the distance between the boundary and the nearest data point in each class is maximal [12, 13]. By using various types of kernels,

SVM can also be used for nonlinear classification. The penalty parameter which is represented as C is important to avoid misclassification for each training example. For higher value of C, smaller margin hyperplane will be selected for better classification correctly. In the current study, the value of C chosen is 10 and PUK is selected as a kernel function for analysis. The reason behind to choose PUK as a kernel is better variable performance depending on dimensionality, and efficient for low dimensional data.

4.2 Artificial Neural Network (ANN)

When features are obtained from nonlinear signals such as machine condition monitoring, then ANN can map them easily with signal conditions and this makes them ideal for applications where scattered data are generalized [13]. In ANN, neuron is a key element which is present in the biological brains and signals can be transmitted from one neuron to another. The connections between artificial neurons are called 'edges.' The combinations of artificial neurons and edges have a weight which adjusts as the learning proceeds. One input layer, hidden layer and output layer are present in feed-forward neural network [14]. The inputs to the network correspond to the attributes measured for each training point. The input information is passing through the first layer which then weighted and fed simultaneously to a second layer known as a hidden layer [15]. The outputs of the hidden layer units can be input to another hidden layer, and the process repeated so on. The number of hidden layers is arbitrary, although in practice, only one is used (Fig. 3).

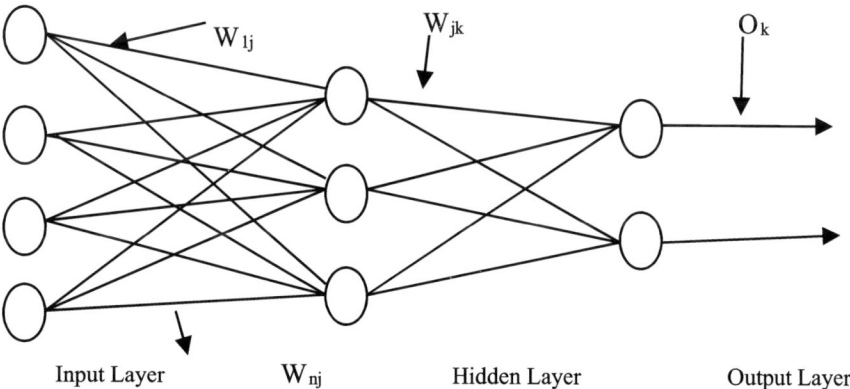

Fig. 3 A multilayer feed-forward neural network

5 Experimental Procedure

The experiments were conducted on 6205 bearing with given specifications in Table 1. Vibration signals are captured from drive end of bearing with varying speed of shaft. For various bearing conditions, raw vibration signals are used for features extractions using WHT. Three different bearing conditions are considered (1) Healthy bearing (2) Inner race fault (3) Outer race fault. 0.2 mm fault size has been considered. Schematic diagram shown in Fig. 1 consists of 2HP three phase induction motor which is coupled with dynamometer and torque transducer device. Accelerometer is used for measurement of vibration signals. The sampling frequency was set to 12.841 kHz.

Following features are calculated using mutual information feature ranking method for time-frequency domain. Some of them are explained below:

1. Kurtosis: It gives information whether the data are peaked or flat relative to a normal distribution.
2. Average: It represents the general significance of a set of unequal values.
3. Median: It is a value from midpoint of the result.
4. Geometric mean: It is defined the central number in a geometric progression.
5. Peak Value: It represents the maximum value of the signal during cycle.
6. Skewness: It reflects the symmetry of the data.
7. Variance: It measures the dispersion of a signal about its mean value.
8. Standard deviation: It is a value that defines the difference from average value of system or parameters.
9. Root Mean Square: It indicates the severity of a bearing defect by the signal intensity.
10. Crest Factor: It is defined as the ratio of maximum positive peak value of the signal to its RMS value.

$$CF = \frac{X_{0-pk}}{\mathrm{rms}_x} \qquad (6)$$

11. Shape Factor: It is used to represent the tim e series distribution of the signal in the time domain.
12. Form factor: It is a value obtained after dividing RMS value to average value of signal.

Table 1 Bearing data

Bearing type	6205 (SKF)
Outer race diameter (mm)	52
Inner race diameter (mm)	25
Ball diameter (mm)	7.94
Ball number	8
Contact angle	0°

13. Variance: It represents the squared deviation of a random variable from its mean value.

$$\mathrm{Var}\left(s^2\right) = \frac{\sum(Xi - X)^2}{n - 1} \tag{7}$$

14. Energy Ratio: It is defined as the RMS of the difference signal s_d divided by the RMS of the signal
   Containing only the regular meshing components, y_d and is given by

$$\mathrm{ER} = \frac{\mathrm{RMS}_d}{\mathrm{RMS}_{y_d}} \tag{8}$$

15. Shannon Entropy: It is an important statistical parameter which describes the degree of disorder or randomness in the signal. The uncertainty of signal coefficients is measured by Shannon entropy which is defined by

$$H_e = -\sum_{j=1}^{J-1} p_j \log p_j \tag{9}$$

Other than the above features remaining are able to understand well like sure entropy, norm entropy, impulse factor, clearance factor, minimum and maximum values.

6 Results and Discussions

In the present paper, authors evaluated the accuracy of various bearing faults using SVM and ANN classifier. Three cases of bearings are considered, i.e., ORD, IRD, and HB respectively. In total, 36 instances and 23 features are used for feature vector. Feature set matrix is formed and fed into the different classifiers. Class-wise fault identification accuracy can be considered using confusion matrix. Table 2 is the result of feature rank after applying mutual information algorithm. The highest value of MI is 5.3081 which corresponds to Shannon entropy. Remaining features are arranged in descending order with respect to the magnitude of mutual information. Tables 3 and 4 show the confusion matrix which summarizes the performance of SVM and ANN algorithm. Training efficiency of both classifiers is 97.22 and 94.22%, respectively. Testing accuracy is 100% for both classifiers.

Table 2 Feature ranking using MI

Sr. No	Feature name	Value	Sr. No	Feature name	Value
1	Shannon entropy	5.3081	12	Peak to peak	4.7889
2	Max. energy to entropy ratio	5.2872	13	Crest factor	4.7889
3	Peak to RMS	5.2179	14	Shape factor	4.7369
4	Median	5.1521	15	Kurtosis	4.7259
5	Maximum value	5.1418	16	Minimum value	4.7059
6	Average deviation	5.0744	17	RMS	4.6767
7	Variance	5.0527	18	Deviation square	4.6569
8	Skewness	5.0527	19	Relative wavelet energy	4.4830
9	Sure entropy	5.0517	20	Norm entropy	3.9252
10	Mean	4.9734	21	Log energy	3.5081
11	Root sum of square	4.8572	22	Impulse factor	3.1481
			23	Clearance factor	2.7349

Table 3 Confusion matrix with SVM for training and testing

Types of accuracy	HB	ORD	IRD	Identification results	Results (%)
Training	12	0	0	HB	97.22
	0	12	0	ORD	
	1	0	11	IRD	
Testing	0	0	0	HB	100
	0	1	0	ORD	
	0	0	3	IRD	

Table 4 Confusion matrix with ANN for training and testing

Types of accuracy	HB	ORD	IRD	Identification results	Results (%)
Training	12	0	0	HB	94.44
	0	12	0	ORD	
	2	0	11	IRD	
Testing	0	0	0	HB	100
	0	1	0	ORD	
	0	0	3	IRD	

7 Conclusion

In this paper, we proposed WHT for bearing fault identification algorithm with the help of mutual information. Twenty-three statistical features are calculated for three bearing fault cases. Fault size 0.2 mm is considered for analysis. SVM and ANN are used as classifiers for training and testing of feature vector. Results of SVM and ANN for training accuracy are 97.22 and 94.22%, respectively. Testing accuracy is 100% for both classifiers.

Acknowledgements The authors would like to acknowledge the support of PDPU, Gandhinagar and GNDU, Amritsar for providing the infrastructure required for carrying out the study.

References

1. V. Vakharia, V.K. Gupta, P.K. Kankar, Efficient fault diagnosis of ball bearing using Relief and random forest classifier. J. Braz. Soc. Mech. Sci. Eng. **39**(8), 2969–2982 (2017)
2. P.K. Kankar, S. Sharma, S.P. Harsha, Fault diagnosis of ball bearings using continuous wavelet transforms. Appl. Soft Comput. **11**, 2300–2312 (2011)
3. Mohanty, K.K. Gupta, K.S. Raju, Bearing fault analysis using variational mode decomposition. J. Instrum. Technol. Innov. **4**(2), 1–6 (2014)
4. V. Vakharia, V.K. Gupta, P.K. Kankar, Ball bearing fault diagnosis using supervised and unsupervised machine learning methods. Int. J. Acoust. Vibr. **20**(4), 244–250 (2015)
5. A. Ashrafi, Walsh–Hadamard transforms a review. Department of Electrical and Computer Engineering, San Diego State University, San Diego, CA, USA, vol. 201 (2017). ISSN 1076-5670
6. M. Irfan, S.Y. Shin, Robust Walsh-Hadamard transform-based spatial modulation. Digit. Sig. Process **S1051–2004**(17), 30023–30024 (2017)
7. R.C. Roy, The relationship between WHT and CIT. Second Int. Symp. Comput. Vis. Internet Procedia Comput. Sci. **58**, 321–332 (2015)
8. K. Kappaganthu, C. Nataraj, Feature selection for fault detection in rolling element bearings using mutual information. Contributed by the Technical Committee on Vibration and Sound of ASME for publication in the Journal of Vibration and acoustics, vol. 133, December (2011), vol. 133/061001-1
9. S. Verron, T. Tiplica, A. Kobi, Fault detection and identification with a new feature selection based on mutual information. J. Process Control **18**, 479–490 (2004)
10. B. Li, P. Zhang, S. Liang, G. Ren, Feature extraction and selection for fault diagnosis of gear using wavelet entropy and mutual information. IEEE (2008). 978-1-4244-2179-4
11. V.N. Vapnik, *the Nature of Statistical Learning Theory* (Springer, New York, 1995)
12. A. Widod, B.-S. Yang, Support vector machine in machine condition monitoring and fault diagnosis. Mech. Syst. Sig. Process. **21**, 2560–2574 (2007)
13. V. Vakharia, V.K. Gupta, P.K. Kankar, Bearing fault diagnosis using feature ranking methods an fault identification algorithms. Procedia Eng. **144**, 343–350 (2016)
14. J.B. Ali, N. Fnaiech, L. Saidi, B. Chebel-Morello, F. Fnaiech, Application of empirical mode decomposition and artificial neural network for automatic bearing fault diagnosis based on vibration signals. Appl. Acoust. **89**, 16–27 (2015)

15. B.A. Paya, I.I. East, M.N.M. Badi, Artificial neural network based fault diagnosis of rotating machinery using wavelet transform as a processor. Mech. Syst. Sig. Process. **11**(5), 751–765 (1997)

Storage Service Reliability and Availability Predictions of Hadoop Distributed File System

Durbadal Chattaraj, Sumit Bhagat and Monalisa Sarma

Abstract Hadoop is a de facto standard for Big Data storage and provides a complete arrangement of components for Big Data elaboration. Hadoop Distributed File System (HDFS), the fundamental module of Hadoop, has been evolved to deliver fault-tolerant data storage services in cloud. This work proposes a precise mathematical model of HDFS and estimates its data storage service availability and reliability. In this connection, a stochastic Petri net (SPN)-based dependability modelling strategy is adopted. In addition, a structural decomposition technique has been advocated to address the state space complexity of the said model. The proposed model is useful to measure crucial quality of service parameters, namely storage service reliability and availability for emerging distributed data storage systems in the context of cloud.

Keywords Big data · Cloud storage · Dependability · Stochastic process · Stochastic petri nets · Quality of service

1 Introduction

Hadoop Distributed File System, the primary module of Hadoop, is an "Apache Software Foundation venture" and a sub-project of the Apache Hadoop project [1]. Hadoop is ideal for storing and processing a massive amount of unstructured data (i.e. terabytes to petabytes of data) and utilizes HDFS as its underlying storage system [2, 3]. HDFS links desktop computers (or nodes), which are commodity hardware

D. Chattaraj (✉) · M. Sarma
Subir Chowdhury School of Quality and Reliability,
Indian Institute of Technology Kharagpur, Kharagpur, India
e-mail: dchattaraj@iitkgp.ac.in

M. Sarma
e-mail: monalisa@iitkgp.ac.in

S. Bhagat
Tata Motors, Pune, India
e-mail: sumitbhagat1076@gmail.com

© Springer Nature Singapore Pte Ltd. 2020
P. V. Varde et al. (eds.), *Reliability, Safety and Hazard Assessment for Risk-Based Technologies*, Lecture Notes in Mechanical Engineering, https://doi.org/10.1007/978-981-13-9008-1_52

617

restricted within clusters over which data files are disseminated. The data files can then be accessed and stored as one seamless file system. The detailed description about HDFS can be found in [3, 4]. M. K. McKusick and S. Quinlan have mentioned in "GFS: Evolution on fastforward" [5] about various components (i.e. Namenodes, Datanodes) failure and suggests different strategies for recovering the system from these critical failures [3]. Shvachko [6] studies the performance overhead of HDFS's fault-tolerant mechanism and highlights the dependencies among user's tasks versus number of Datanodes. Moreover, in the recent state of the art [7–15], various data storage availability models have been proposed in the context of cloud. In this synergy, authors were adopted different platforms such as IaaS cloud, virtualized (machine-inside-machine) system and cloud-assisted data storage cluster. However, these state-of-the-art mechanisms do not explicitly address the performance modelling issue of Big Data-assisted distributed file system vis-a-vis HDFS modelling. Li et al. [10] proposed a SPN-based model for HDFS, and from the same model, they formulate three dependability metrics. Moreover, the authors were adopted the old version of HDFS [3]. Hence, this work proposes a mathematical model for HDFS federation architecture (recent release) [3] to estimate its dependability metrics, namely storage service availability and reliability. To achieve this, we adopt Generalized Stochastic Petri Net (GSPN) [16, 17] based modelling technique. After designing the GSPN model in Platform-Independent Petri net Editor (PIPE) tool version 3.3,[1] we simulate the model and obtain an equivalent Continuous Time Markov Chain (CTMC) [17]. Finally, a Markovian state-based analysis [17] has been carried out to estimate the storage service reliability and availability.

2 Proposed Methodology

2.1 Model Design

We map the federated HDFS data storage system into GSPN. Note that, to find the details about the federated HDFS data storage system and mathematical preliminaries of GSPN, see [3, 17]. The proposed GSPN-based failure-repair model of federated HDFS data storage system is shown in Fig. 1a. From Fig. 1a, it is easy to observe that the HDFS federated data storage system consists of four crucial sub-systems, namely Namenode (a set of N Namenodes), Datanode (collection of m Datanodes), HDFS-client (application software to access Namenodes and Datanodes) and network module (collections of switch, router, communication media, etc.). The Namenodes store the meta-data ("fsimage" and "editlogs" information) [2] wherein Datanodes keep client's Big Data files in terms of raw data blocks with replicas of each block (default replication factor (RF) = 3). In order to analyse the storage service reliability and

---

[1]Platform-Independent Petri net Editor—http://pipe2.sourceforge.net.

availability, utilizing Markovian state-based analysis for such a complex system, in this study, we consider three strict assumptions as follows:

A. The client-side application software (HDFS-client) and network module are reliable and available during mission time.
B. The Namenode sub-system consists of two Namenodes (i.e. $n = 2$), and they are operating in hot standby mode. Note that the failure distribution of both Namenodes is independent and identically distributed.
C. The Datanode sub-system (see module B in Fig. 1a) consists of m numbers of identical Datanodes, and their failure distributions are independent and identically distributed. Keeping the eye on [10], we structurally decompose the sub-system into a single module say Datanode module. Further, depending on data blocks distribution and replication into different Datanodes, we consider two different scenarios of a typical HDFS data storage system. In the first one, a client's file (i.e. an unstructured file) is stored into distinct Datanodes. Here, we consider an idealistic strategy where the client file divides into η number of blocks and η blocks store in ξ different Datanodes provided $\eta = \xi$. In another case, we appraise a practical scenario of HDFS storage strategy where η blocks store in ξ different Datanodes such that $\xi < \eta$. The detail descriptions about both the cases are presented as follows.

- **Scenario 1**:
  Assumptions: The following assumptions have been made while carrying out the analysis on the Datanode module.
  1. Files to be stored are divided into several blocks and stored in different Datanodes.
  2. The replication factor is three, which is the default of HDFS configuration. This means that the content of each Datanode is replicated again in two different Datanodes.
  3. The contents of a Datanode are limited to only one block.
  4. Failure in retrieval of the data from the Datanode is considered as failure of the Datanode. This may happen owing to several failures and crashes that the machine might suffer during the data retrieval process.
  5. Failures and repairs of the Datanode are presumed to follow exponential distribution by means of failure rate λ and repair rate μ.
  6. The connections in the network topology are considered to be perfect, and the only concern for the analysis is storage and retrieval of data.

- **Scenario 2**:
  Assumptions: The following assumptions have been made while carrying out the analysis on the Datanode module.[2]
  1. Files to be stored are divided into several blocks and kept in different Datanodes.

---

[2]A small part of this work has been accepted in IEMIS-2018, Kolkata, India. https://doi.org/10.1007/978-981-13-1498-8_42.

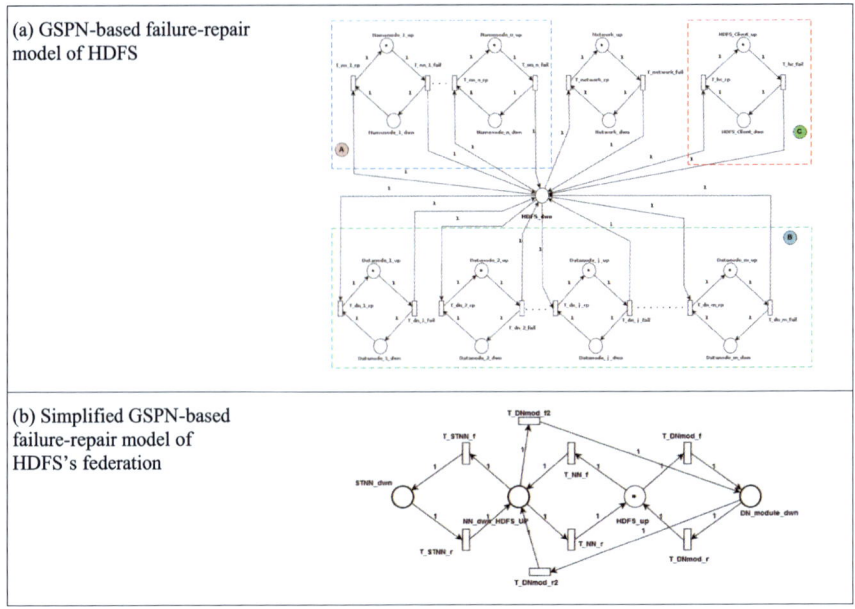

(a) GSPN-based failure-repair model of HDFS

(b) Simplified GSPN-based failure-repair model of HDFS's federation

Fig. 1 GSPN-based failure-repair model of federated HDFS data storage system

2. The RF is three, which is the default according to the HDFS configuration. This means that the content of each Datanode is replicated again in two different Datanodes.
3. The contents of a Datanode are not limited to only one block. This means that a Datanode stores multiple blocks depending upon the storage capacity of the node.
4. The connections in the network topology are considered to be perfect, and the only concern for the analysis is storage and retrieval of data. Inaccessibility of the data is considered as failure of the Datanode.
5. Failures of the Datanode are assumed to follow exponential distribution.
6. Failures of the Datanode module are assumed to follow a constant failure rate.
7. The failure rate of DN module is being analysed at a presumed time of one year.

After considering all the aforesaid assumptions, we construct an equivalent GSPN model of federated HDFS data storage system and it is shown in Fig. 1b.

2.2 Model Analysis

In order to do the analysis of the proposed GSPN model (refer Fig. 1b), we formulate the failure rate of the Datanode module for both the cases (as mentioned above) as follows:

Case I: According to the assumption of scenario 1, the mean time to failure (MTTF) of one block with replication factor one is reciprocal of the failure rate of the block, which is $MTTF_{(j=1)} = \lambda$. The state transition diagram for one block with RF $= 2$ is shown in Fig. 2a. Utilizing this state transition diagram, we obtain $MTTF_{(j=2)}$ is $\frac{\mu + 2\lambda}{\lambda^2}$. Similarly, we obtain the $MTTF_{(j=3)}$ for one block with RF $= 3$ is $\frac{\mu^2 + 3\lambda^2 + 2\mu\lambda}{\lambda^3}$. Therefore, the $MTTF_{(j=n)}$ for replication factor n of one block can be generalized as follows:

$$MTTF_{(j=n)} = \left(n \cdot \lambda^{(n-1)} \cdot \mu^{(n-n)} + (n-1) \cdot \lambda^{(n-2)} \cdot \mu^{n-(n-1)} + (n-2)\right.$$
$$\left. \cdot \lambda^{(n-3)} \cdot \mu^{n-(n-2)} + \cdots + (n - (n-1)) \cdot \lambda^{(n-n)} \cdot \mu^{n-(n-(n-1))}\right)/(\lambda_n)$$

The failure rate of one block says the ith block with replication factor n, denoted by λ_i is $\lambda_i = \frac{1}{MTTF_{(j=n)}}$. Let the size of the client file (Big Data) that has to be stored in the HDFS be X MB. The block size (typically 64 or 128 MB) into which the file has to be broken down and stored into the Datanodes is X_{block}. Thus, the total number of blocks per file is represented as $m = \frac{X}{X_{block}}$. Therefore, the failure rate of the Datanode module denoted by λ_m comprising of m blocks, and it is represented as

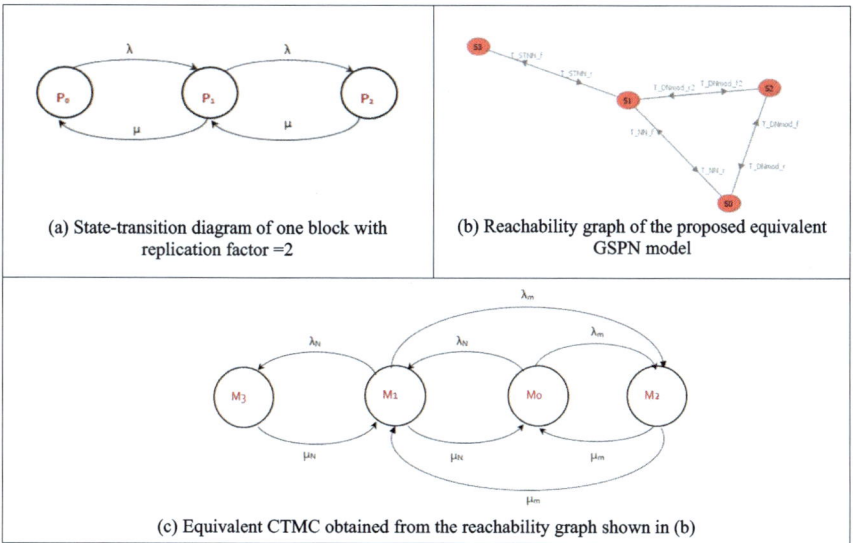

(a) State-transition diagram of one block with replication factor =2

(b) Reachability graph of the proposed equivalent GSPN model

(c) Equivalent CTMC obtained from the reachability graph shown in (b)

Fig. 2 Markovian state-based analysis [3]

$$\lambda_m = \sum_{i=1}^{m} \lambda_i. \tag{1}$$

Note that, the failure rate of DN module is solely influenced by the RF, and the number of blocks.

Case II: Scenario 2 is quite admissible and comparable with actual data storage mechanism of HDFS. In this synergy, the end user's file is divided into multiple data blocks and each data blocks are stored randomly into multiple DNs. In fact, it is not permissible for a single Datanode to store the same data blocks twice, but it could store multiple data blocks of different files [3]. To support the aforesaid claim, in [3], it is specified that "no two replicas of the identical data blocks need to be stored into same Datanode". This mechanism is fundamentally used in HDFS to fulfil the fault-tolerant property of the system. Therefore, the failure probability for the Datanode module (collection of k *DNs*) is represented as $F_{DM} =$ (3 DNs fail) $\cup$ (4 DNs fail) $\cup$ (5 DNs fail) $\cup \dots \cup$ (k DNs fail). It is assumed here that the DN module will fail to access the data blocks due to the failure of three or greater DNs at the same time instances [3]. For brevity, the above condition is valid as we assume the replication factor (RF) is equal to three in scenario 2. The failure of 3 DNs or greater for a particular time instance t follows a binomial distribution with parameters $k \in N$ and $p, q \in [0, 1]$. Thus, the cumulative distribution function (CDF) of DN module is represented as $F_{DM} =$ Prob. $[Y \geq 3] = (1 -$ Prob. $[Y \leq 2])$ where Y denotes the failure of individual DN. So, $F_{DM} = 1 - \left(\sum_{r=0}^{2} \binom{k}{r} p^r q^{k-r} \right)$. Since each node is assumed to follow exponential distribution with constant failure rate λ, so we can write

$$F_{DM} = 1 - \left(\sum_{r=0}^{2} \binom{k}{r} \right) (e^{-\lambda \cdot t})^r \cdot (1 - e^{-\lambda \cdot t})^{k-r}.$$

The Datanode module follows the exponential distribution with constant failure rate λ_m, for the mission time of one year. So, $R_{DM} = 1 - F_{DM} = e^{-\lambda t}$. Therefore, the failure rate of the Datanode module is obtained as

$$\lambda_m = -\left(ln \left[(1 - e^{-\lambda t})^k + k \cdot e^{-\lambda t} \cdot (1 - e^{-\lambda t})^{k-1} + \frac{k \cdot (k-1)}{2} \cdot e^{-2\lambda t} \cdot (1 - e^{-\lambda t})^{k-2} \right] \right) / t. \tag{2}$$

Here, the failure rate of DN module is being analysed at a presumed instantaneous time of one year or 8760 h [3]. Once the failure rate of the Datanode module has been obtained for both the cases (refer Eqs. 1 and 2), we perform a Markovian state-based analysis of the equivalent GSPN model (refer Fig. 1b). Before this operation, we generate a reachability graph vis-a-vis reachability marking of the same model by simulating it using PIPE tool. The reachability graph and the reachability markings are shown in Fig. 2b and Table 1, respectively.

Table 1 Reachability markings obtained from proposed GSPN-based failure-repair model [3]

Reachability markings	System states			
	HDFS_up	DN_module_dwn	NN_dwn_HDFS_up	STNN_dwn
M_0	1	0	0	0
M_1	0	0	1	0
M_2	0	1	0	0
M_3	0	0	0	1

From the reachability graph, we obtain an equivalent CTMC (refer Fig. 2c) and apply the Markovian state-based analysis technique to estimate the storage service reliability (SSR) and storage service availability (SSA) as follows.

From Fig. 2c, M_0 and M_1 are the success states of the HDFS data storage system. So, the SSR (i.e. transient reliability or $R(t)$) and the steady-state availability (A_S) of storage services are estimated as

$$R(t) = M_0(t) + M_1(t) = e^{-(\lambda_N + \lambda_m) \cdot t} \cdot [1 + \lambda_N \cdot t] \tag{3}$$

and

$$\text{SSA} = A_S = \frac{\left(1 + \frac{2\lambda_N + \lambda_m}{2\mu_N + \lambda_m}\right)}{1 + \left(\frac{2\lambda_N + \lambda_m}{2\mu_N + \lambda_m}\right) + \frac{\lambda_N}{\mu_N} \cdot \left(\frac{2\lambda_N + \lambda_m}{2\mu_N + \lambda_m}\right) + \frac{\lambda_m}{2\mu_m} \cdot \left(1 + \frac{2\lambda_N + \lambda_m}{2\mu_N + \lambda_m}\right)} \tag{4}$$

where λ_N, μ_N, λ_m and μ_m represents failure rate of Namenode, repair rate of Namenode, failure rate of Datanode module and repair rate of Datanode module, respectively. Note that, in SSR (i.e. transient reliability) estimation, we assume that the states M_2 and M_3 are absorbing states.

3 Case Study

We consider a virtual Hadoop cluster which consists of two Namenode servers (each having 4 tera bytes storage capacity) and two hundred Datanode servers (each having 5 tera bytes storage capacity). That means, we can store roughly 1 petabytes of data (Big Data) into the virtual cluster. In this study, we take the transition firing rates for T_NN_f and T_STNN_f is equal to 0.001, and T_NN_r = T_STNN_r = 1 from [10], and assume MTTR and MTTF of each Datanode are 1 and 1000-unit time, respectively. After taking all these parameters into consideration, we code Eqs. (1), (2), (3) and (4), respectively, using Java programming language and collect the output from the same code for plotting the results. The following (refer Fig. 3) are few observations as the realization of the proposed model:

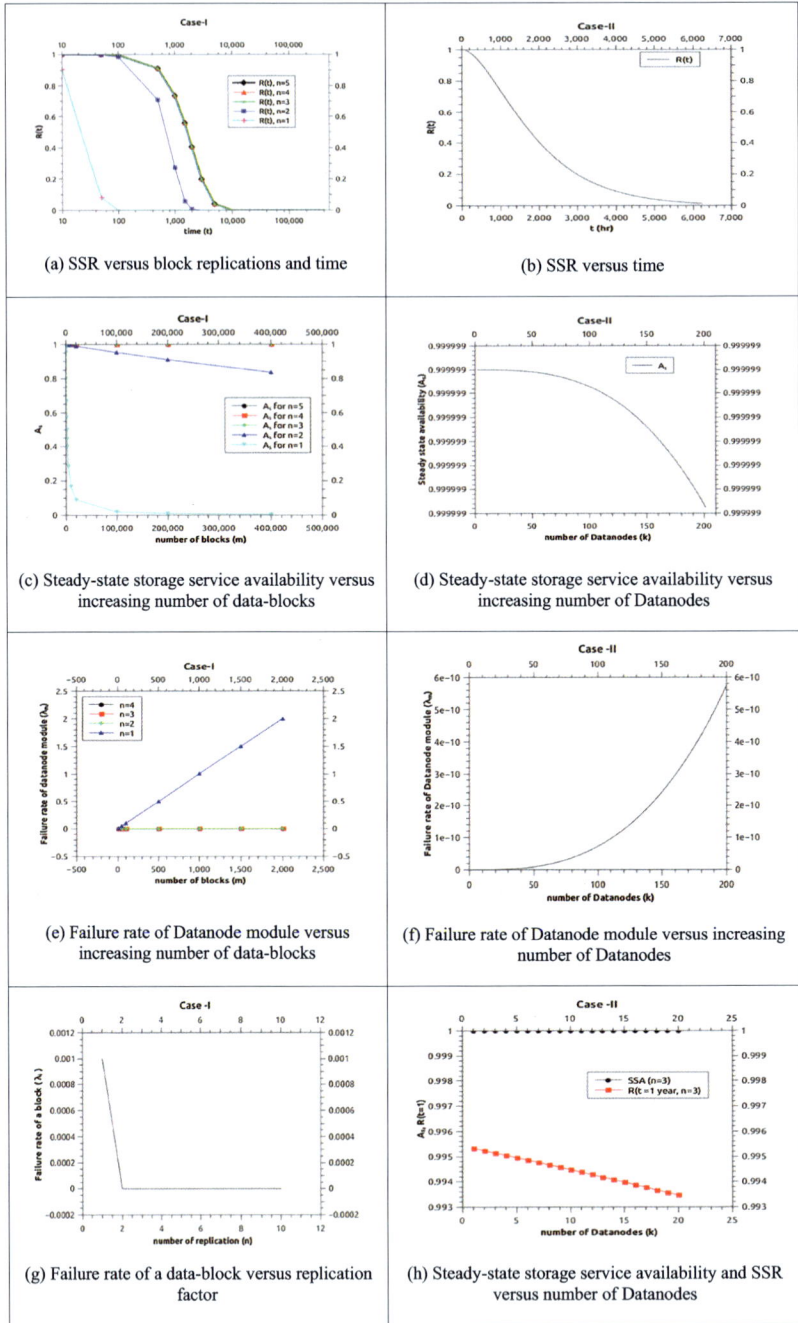

(a) SSR versus block replications and time

(b) SSR versus time

(c) Steady-state storage service availability versus increasing number of data-blocks

(d) Steady-state storage service availability versus increasing number of Datanodes

(e) Failure rate of Datanode module versus increasing number of data-blocks

(f) Failure rate of Datanode module versus increasing number of Datanodes

(g) Failure rate of a data-block versus replication factor

(h) Steady-state storage service availability and SSR versus number of Datanodes

Fig. 3 Summary of observations after analysing the proposed GSPN model

Case I (a) SSR will improve with respect to time (t) when we increase the RF (from $n = 1$ to $n = 5$), (c) storage service availability reaches its steady state when RF $\geq$ 3, (e) failure rate of the Datanode module becomes constant when RF ≥ 1, and (g) failure rate of a data block remains constant if we increase the replication factor (i.e. RF ≥ 2).

Case II (b) As time progresses, new components will be added with the Hadoop cluster (HC). So, the SSR will decrease, (d) the data storage service availability reaches its steady state after adding certain numbers of Datanode into the HC, and afterwards, it does not depend on the Datanode's scale-up or scale-down phenomenon, (f) more Datanodes induction into the HC leads to more failure of the Datanode module, hence reduce SSR, and (h) Datanode scale-up leads to SSR drop but achieve better storage service availability.

Remark 1 In recent time, the evolution of Redundant Array of Independent Disks (RAID level 0 to RAID level 10) with Solid State Device (SSD) technology improvizes data storage dependability a lot [18]. But, in this case study, we use traditional commodity-hardware-based storage platform to model the basic characteristics of HDFS. Moreover, due to the page constraint, presently it is not feasible to incorporate the effects of aforesaid technology in the proposed dependability modelling of HDFS. Hence, we are planning to address this issue in the extended version of the proposed study.

4 Conclusion

In this work, a stochastic Petri net-based model is proposed for HDFS federation architecture. Based on the proposed model, two crucial dependability attributes, namely the storage service reliability and storage service availability, are estimated. To make HDFS fault-tolerant and attain better storage service reliability and availability, it is advocated to set the replication factor as three (i.e. RF $= 3$). Our case study substantiates the efficacy of the aforesaid claim. In addition to this, we highlight the impact of scale-up or scale-down Datanode servers and block replications on the performance enhancement of the Hadoop cluster. Moreover, the proposed model analysis technique is scalable in nature, so it could be utilized for analysing the performance of other distributed storage systems. In the future, we plan to validate our proposed GSPN model with the real-time cluster settings considering real-time failure datasets.

References

1. Apache Hadoop, http://hadoop.apache.org/. Last access: Aug. 2018

2. K. Shvachko, H. Kuang, S. Radia, R. Chansler, The Hadoop distributed file system, in *Proceedings of 26th symposium on Mass Storage, Systems and Technologies*, IEEE (2010), pp. 1–10
3. D. Chattaraj, M. Sarma, D. Samanta, Stochastic petri net based modeling for analyzing dependability of big data storage system, in *Proceeding of 1st International Conference on Emerging Technologies in Data Mining and Information Security (IEMIS'18)*, Advances in Intelligent Systems and Computing, vol. 813 (2019). https://doi.org/10.1007/978-981-13-1498-8_42
4. HDFS Architecture, https://hadoop.apache.org/docs/current/hadoop-project-dist/hadoop-hdfs/HdfsDesign.html. Last access: Aug. 2018
5. K. McKusick, S. Quinlan, GFS: evolution on fast-forward. Commun. ACM **53**(3), 42–49 (2010)
6. K.V. Shvachko, H.D.F.S. Scalability, The limits to growth. Mag. USENIX & SAGE **35**(2), 6–16 (2010)
7. D. Bruneo, F. Longo, D. Hadas, H. Kolodner, Availability assessment of a vision cloud storage cluster, in *Proceedings of European Conference on Service-Oriented and Cloud Computing* (Springer, 2013), pp. 71–82
8. R. Ghosh, F. Longo, F. Frattini, S. Russo, K.S. Trivedi, Scalable analytics for IaaS cloud availability. Trans. Cloud Comput. IEEE **2**(1), 57–70 (2014)
9. F. Longo, R. Ghosh, V.K. Naik, K.S. Trivedi, A scalable availability model for infrastructure-as-a-service cloud, in *Proceedings of 41st International Conference on Dependable Systems & Networks (DSN)*, IEEE (2011), pp. 335–346
10. H. Li, Z. Zhao, L. He, Model and analysis of cloud storage service reliability based on stochastic petri nets. J. Inf. Comput. Sci. **11**(7), 2341–2354 (2014)
11. D. Bruneo, F. Longo, D. Hadas, E.K. Kolodner, Analytical investigation of availability in a vision cloud storage cluster. Scalable Comput.: Pract. Experience **14**(4), 279–290 (2014)
12. J. Shafer, S. Rixner, A.L. Cox, The Hadoop distributed file system: balancing portability and performance, in *International Symposium on Performance Analysis of Systems & Software (ISPASS-10)*. IEEE (2010), pp. 122–133
13. D. Bruneo, A stochastic model to investigate data center performance and QOS in IaaS cloud computing systems. Trans. Parallel Distrib. Syst. IEEE **25**(3), 560–569 (2014)
14. J. Dantas, R. Matos, J. Araujo, P. Maciel, Models for dependability analysis of cloud computing architectures for eucalyptus platform. Int. Trans. Syst. Sci. Appl. **8**, 13–25 (2012)
15. X. Wu, Y. Liu and I. Gorton, Exploring performance models of hadoop applications on cloud architecture, in *Proceedings of the 11th International ACM SIGSOFT Conference on Quality of Software Architectures (QoSA)*, IEEE, (2015), pp. 93–101
16. J. Wang, Petri nets for dynamic event-driven system modeling, in *Handbook of Dynamic System Modeling*, edited by P. Fishwick (CRC Press, 2007)
17. R. Zeng, Y. Jiang, C. Lin, X. Shen, Dependability analysis of control center networks in smart grid using stochastic petri nets. Trans. Parallel Distrib. Syst. IEEE **23**(9), 1721–1730 (2012)
18. S.A. Chamazcoti, S.G. Miremadi, Hybrid RAID: a solution for enhancing the reliability of SSD-based RAIDs. Trans. Multi-Scale Comput. Syst. IEEE **3**(3), 181–192 (2015)

Quality Assurance in Sustaining Reliability of Nuclear Reactor Safety Systems

Vivek Mishra and P. V. Varde

Abstract In nuclear reactors, a tremendous amount of thermal energy and radioactivity is generated due to fission reactions. Safety is in general an important aspect for any industry; however, the emission of significant nuclear radiation into the environment during an accident causing harm to humans both at the reactor site and off-site is the main safety concern. The three major nuclear accidents releasing radiation to the environment were–Three Mile Island, Chernobyl and Fukushima. To avoid these accidents in future, the demand for safety in nuclear installations has become more important. Nuclear industry from these accidents has learned and taken stringent steps toward ensuring a better reliability of safety systems that ensure cooling of reactor and fuel storage pond, shutting down of reactor and containing the radioactivity. Plant experience indicates that quality assurance during design, procurement, operation, maintenance, surveillance and human resource management have a direct bearing on the reliability of these safety systems. Based on a philosophy that effective quality assurance is inseparably linked to the fostering of an excellent safety culture and ensuring reliability, the implementation of quality management system based on IAEA standard is needed to achieve the desired safety and reliability objective. This paper presents the detailed activities carried out in quality assurance programme in a nuclear reactor to enhance the safety and reliability of nuclear reactor safety systems.

Keywords QA · Safety function · CDF · PIE · IAEA · ISO

1 Introduction

Plant and personnel safety during all states of reactor operation is an important criterion for nuclear reactors. Nuclear reactors generate thermal energy due to fission of fissile material, e.g., Uranium-235 during reactor operation. The fission process in the nuclear reactor also produces a large quantity of radioactive material. The

V. Mishra (✉) · P. V. Varde
Research Reactor Services Division, Bhabha Atomic Research Centre, Mumbai, India
e-mail: vmishra@barc.gov.in

© Springer Nature Singapore Pte Ltd. 2020
P. V. Varde et al. (eds.), *Reliability, Safety and Hazard Assessment for Risk-Based Technologies*, Lecture Notes in Mechanical Engineering,
https://doi.org/10.1007/978-981-13-9008-1_53

presence of these radioactive materials even after the reactor shutdown is responsible for heat production in the reactor. This heat is called decay heat. The decay heat is the main safety concern for a nuclear reactor from core (nuclear fuel) cooling point. In case the fuel melts uncontrolled radiation into the environment can be released causing irreversible damage to human and environment both at reactor site and off-site. Since, the first recorded one in 1952 at Chalk River in Ontario nuclear industry has witnessed around 33 serious incidents and accidents at nuclear power stations. A special International Nuclear and Radiological Events Scale (INES) has been developed by International Atomic Energy Authority to assess the severity of these accidents and events [1].

2 Major Accidents and INES Level

Some of the major accidents/incidents and their INES level are presented in Table 1.

In the INES scale, "incidents" falls in levels 1–3 levels 4–7 are termed as "accidents." In the INES scale, the severity of an event is about ten times greater for

Table 1 Major nuclear events

Level	Event	Country year	IAEA description
7	Major accident	Chernobyl Russia 1986	Significant on-site and off-site health and environmental effects due to external release of radioactivity
6	Serious accident	Kyshtym Russia 1957	Significant release of radioactivity from radioactive waste from explosion of a high activity waste tank
5	Accident with wider consequences	Three Mile Island 1979	Beyond damage to reactor core
4	Accident with local consequences	Fukushima 2011	Reactor shutdown after a Tsunami and earthquake, followed by failure of emergency cooling, hydrogen explosion
3	Serious incident	Sellafield, UK 2005	Release of large quantity of radioactivity within the installation
2	Incident	Atucha Argentina 2005	Exposure of a worker exceeding annual limit
1	Anomaly		

each increase in level. Deviations which are classified below scale/level 0 are events without safety significance.

The data in Table 1 indicates that any activity release from the reactor core has been categorized as serious incident and further into accidents. The fundamental objective of nuclear safety is the prevention of such release into the environment which is met by safety functions.

3 Safety Functions and Defense-in-Depth

Since the design stage of nuclear reactor, three fundamental goals of the safety functions are distinguished to ensure safety of nuclear reactor during all states of reactor operation [2].

- The nuclear chain reaction will be controlled and a safe shutdown will be ensured whenever needed; (*This function is achieved by reactor shutdown system.*).
- The reactor shall always be cooled as long as fuel is present in the reactor; (*This function is achieved by reactor emergency cooling system.*).
- In case of accident radioactive materials shall not be released into the environment; (*This function is achieved by containment isolation system.*).

The system, structure and components that implement these safety functions are called safety systems.

Implementation of these safety functions in nuclear power plants is achieved by applying the principle of defense-in-depth.

Defense-in-depth is carried out at various levels. The objective is to prevent a particular level to progress to the next level. In nuclear reactor, this means having safety functions designed for preventing postulated initiating events (PIEs) and their further progression. Following are five levels of defense-in-depth:

- Prevention of abnormal, normal operating conditions, malfunctions and failures;
- Detection of abnormal operating conditions and prevention of anticipated operational occurrences from becoming design basis accidents;
- Effective control of design basis accidents;
- Prevention of accident progression and mitigation of severe accidents consequences;
- Mitigation of radiological consequences in the case of a significant release of radioactive materials.

4 Reliability Targets for Safety Systems

Each accident and incidents are a lesson to learn. On the basis of the experience gained during the operation of reactors and during incidents and accidents that have occurred, nuclear power plants safety standards are continuously increasing. Nuclear industry from these accidents has learned and taken stringent steps toward ensuring a better reliability of safety systems that ensure cooling of reactor and fuel storage pond, shutting down of reactor and containing the radioactivity. An increased reliability of safety system ensures that safety system will meet their functionality requirement in any design basis accident.

The safety standard in nuclear industry is also characterized, among others with core damage frequency (CDF). CDF is the probability of occurrence of significant physical damage to (e.g., melting of) the core of a nuclear power plant that contains a large quantity of radioactive material. This damage is the outcome of an accident and malfunction of safety systems. CDF is a numerical value obtained through probabilistic safety analysis (PSA) of a nuclear reactor using fault tree and event tree analysis approach.

To ensure adequate safety of new nuclear plant and protect against possible core damage, the regulation stipulates that the applicant who intends to build a new nuclear power plant shall prove through analysis that the core damage frequency is below $1E-5$ per year [3]. To achieve this value of CDF, design of a reliable safety system and maintaining its functionality throughout lifetime of a nuclear reactor is a prerequisite. As per guide [4], reliability targets for the safety systems should be designed to meet the following values.

- The probability of failure/demand for each shutdown system should be $<10^{-3}$
- The probability of failure/demand for emergency core cooling system should be $<10^{-3}$
- The probability of failure/demand for containment isolation system should be $<10^{-3}$.

Quality assurance plays a major role in ensuring target reliability of these safety systems from design stage till life of a nuclear reactor.

5 Quality Assurance for Reliability

Quality assurance in general refers to those activities and systematic actions that provide adequate confidence for intended functional requirements of system, structure and components whenever needed.

In the nuclear industry, quality assurance applies to all activities and organizations which influence nuclear reactors safety (such as design, component manufacture, construction and on-site installation, start-up qualification tests, plant commissioning and operation).

Quality assurance programmes in nuclear reactor are governed by the following.

5.1 Quality Policy

The applicant/licensee and other organizations who affect the safety of the nuclear power plant shall prepare documents mentioning the objectives of their quality policy and their commitment to safety culture and high quality. The applicant will ensure that the quality assurance is correctly understood and implemented in all the activities that affect the safety [5].

5.2 Quality Assurance Requirements

As per IAEA safety guide [6], the safety of a nuclear reactor requirement for the planning and quality assurance implementation contains following items:

- An advanced safety culture shall be maintained during design, construction and operation of a nuclear power reactor. This safety culture is based on the safety-oriented attitude of the senior level management of the facilities involved with the upcoming nuclear plant. Such practice ensures a well-organized working environment and an interactive working atmosphere as well as the encouragement of deep association with activities to detect and eliminate causative factors which endanger safety.
- All activities which affect safety and relate to the design, construction and operation of a nuclear reactor shall have well-structured quality assurance programmes.
- A well proven or otherwise carefully evaluated high-quality technology shall be used to prevent operational transients and accidents (preventive measures) in all activities starting from design to operation of nuclear reactor.
- All activities of design, fabrication, installation and operation shall be carried out so that their quality level and the in-service inspections and tests required for verification of their quality level are adequate to ensure item's safety significance.
- A systematic and structured approach shall be applied for ensuring the plant operators' continuous awareness of the state of the plant's SSCs while carrying out any activity that affects the operation of a nuclear reactor and its sub-systems.

5.3 Establishment of Quality Assurance Programmes

The responsibility to establish quality assurance programmes for reactor design, fabrication of components, construction at site and operation lies with applicant/licensee. In addition, the applicant/licensee shall also ensure that:

- Vendors and organizations with the prerequisites and will for high-quality performance and with well-laid procedures that ensure quality assurance may only participate in the activities that affect the safety of the nuclear reactor.

Table 2 Severity levels

Severity	Consequence
High	Leading to a release of radioactive material that exceeds the limits accepted by the regulatory body for design basis accidents; or Cause the values of key physical parameters to exceed acceptance criteria for design basis accidents
Medium	Leading to a release of radioactive material that exceeds limits established for anticipated operational occurrences; or Cause the values of key physical parameters to exceed the design limits for anticipated operational occurrences
Low	Leading to doses to workers above authorized working limits

- The main suppliers to the facility supplying items affecting reactor safety shall submit a high-quality assurance programme covering various aspects such as material and chemical testing, NDT methods and codal requirement.

5.4 Implementation of Quality Assurance

The quality assurance programme submitted by licensees shall mention the procedures by which the implementation and viability of the quality assurance activities of the organizations referred in Sect. 5.3 are ensured.

5.4.1 QA During Design Stage

The implementation of quality assurance programmes starts right from the design stage of nuclear components. Nuclear components are categorized in various safety classes as per their functional requirement during a postulated initiating event (PIE). Failure of these functions leads to the consequence of different severity. The three level of severity as defined by IAEA based on failure of safety function is presented in Table 2 [7].

The severity of consequences of failures of SSCs is the direct basis of safety classification of the SSC. Each class of the component is designed as per applicable code.

Table 3 gives brief design details of different class of nuclear components and applicable design code.

Subsequent to several reviews of the design and improvement incorporated with respect to functionality and redundancy, the design for safety systems is frozen and procurement/fabrication of these components is initiated.

Table 3 Nuclear component design

Component	Severity level	Design code
Safety class I	High severity	ASME Section III, Div. 1-NB
Safety class II	Medium severity	ASME Section III, Div. 1-NC
Safety class III	Low severity	ASME Section III, Div. 1-ND

Fig. 1 Application of IAEA 50-C-Q and ISO 9001:2000

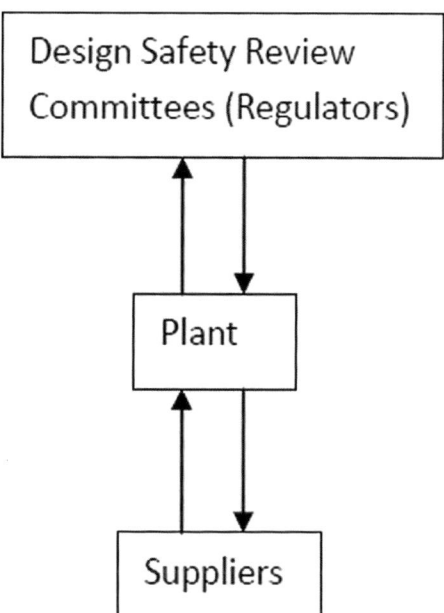

5.4.2 QA During Procurement/Fabrication

A nuclear utility interacts with two different agencies where the role of the quality management system is applied. On the one hand, the utility interacts with the regulator, who expects (in some Member States it is specified as a licensing requirement) that the utility will develop and implement a quality management system which satisfies the IAEA 50-C-Q requirements toward reactor safety.

The suppliers of items for the reactor forms are the second agency. The utility emphasize that suppliers should develop and implement a quality management system that complies with ISO 9001:2000 for the supply and delivery of items and services that satisfies the requirements of IAEA 50-C-Q [8]. The interaction with these agencies is presented in Fig. 1.

To initiate procurement, prospective suppliers/vendors are identified and evaluated based on their capabilities to supply items or services as per the requirements specified in the procurement documents. The procurement document is sent to would be suppliers, and it is ensured that the information given to the suppliers meets requirement as specified in applicable codes, concise and unambiguous, fully describes the items, warranty period and all services required.

In general, the procurement documents include objective scope of assignment, technical requirements, material and chemical testing, visual checks, NDT and dimensional checks, access to supplier facilities, identification of quality assurance standard, delivery schedules, acceptance criteria and documentation. This is followed by the evaluation of suppliers and awarding the contract of supply to a suitable supplier.

5.4.3 QA During Commissioning Activities

Plant commissioning is a process by which installed plant equipments and systems are brought into active service and are tested in detail to ensure that their performance is matching with the design intent. After commissioning stage, the ownership of the systems/plant is handed over from the construction organization to the operating organization/facility owners.

As a part of commissioning programme, a detailed commissioning document is prepared by involved agencies and reviewed by plant management/commissioning in-charge. The approved document contains objective of the testing, information about executing agency, design parameters, limiting conditions, associated hazard with safety precautions to be followed, important data recording, acceptance criteria and time schedule.

Plant management ensures that personnel involved in commissioning activities are well aware about their duties. This is accomplished by carrying out lecture programmes, briefings and exposure to simulation modules of plant systems.

Post commissioning of all the systems and results of commissioning activities are reviewed by the commissioning in-charge. It is ensured that commissioning tests/procedures have demonstrated that equipment is functionally acceptable. If required, changes to the design as a result of commissioning activities are reviewed, approved and carried out. Commissioning tests are repeated till design objective is met.

5.4.4 QA During Operation and Maintenance

High standards of safety culture shall be maintained in the nuclear power plants operation to ensure plant and personnel safety. To achieve this goal, an application for a license to operate the plant is submitted to regulatory authority along with various technical documents such as detailed deterministic and probabilistic safety analysis report, design basis report, design and operating manuals for various systems, standard and emergency operating procedures [9].

A technical specification containing the following information in detail is also submitted:

- Process and nuclear parameters (safety limits) that will never be exceeded during any state of reactor operation.
- Normal operating parameters for all the systems, accepted deviation and action levels.
- Radiation dose limits to occupational personnel and action level.
- Limits for discharge of radioactive gaseous, liquid and solid waste.
- Organization's chart and the respective duties, areas of responsibility and authorities of various units (Operation, maintenance, radiation safety and other technical support services).
- Surveillance schedules for equipments of safety systems.
- Reporting criteria for events and significant events.
- Qualification and licensing criteria for plant personnel engaged in operation and maintenance activities.

Based on these documents, license to operate the plant for a period is granted by regulatory authorities.

Operation of the reactor is entrusted to the competent personnel (engineering graduates) qualified and licensed for the duties assigned to them. Qualification process starts with familiarization of the plant through checklists, lecture programmes, plant visits and discussion through seniors. This is followed by examination and interview by plant management. On successful completion of qualification, plant personnel are licensed for reactor operation for a period. At the end of this period, their knowledge is assessed by a committee and relicensing is done.

To assure the quality and reliability of various system structures and components important to safety, in-service inspection and preventive maintenance programmes are prepared, and these activities are carried out periodically and records are maintained. A work permit system is implemented for carrying out any maintenance activity in the plant. To improve the system performance, design review, modifications and replacement of the components are also carried out.

Repaired or modified equipment is reinducted into service only after it has been tested for all operating conditions thoroughly and its operating parameters are within limits.

Periodic test for the plant systems is carried out to ascertain their performance. These tests are witnessed/approved by the persons other than those performing the tests.

Design, operating manuals and system drawings are periodically updated based on changes carried out in the system.

6 Regulatory Review to Ensure QA

It is the plant management who owes the primary responsibility of ensuring safety of nuclear reactor during its design, construction and operation. Regulatory authorities, i.e., various safety committees lay down additional necessary safety requirements and their enforcement during all stages of nuclear utility.

Regulators are independent authorities other than plant management who ensure that safety evaluations carried out for the plants and activities ensure that an adequate level of safety has been achieved, and that the objectives and criteria for safety established by the designers and the regulatory authorities have been met [10]. Regulators accomplish this task by reviewing the implementation of the quality assurance process at each stage of nuclear reactor.

Subsequent to design, construction and commissioning, a reactor remains in service for around 30–40 years. A multitier regulatory process throughout the service life of the reactor ensures plant and personnel safety.

At plant level, internal regulatory inspections are carried out by the plant personnel, deficiency in the operating practices, surveillance schedule, repeated component failure and human error induced incidents are brought out and responsible agencies are informed. An action taken report is prepared and discussed in the plant level safety committee having members from the same unit. Plant level committee is also entrusted to review the events associated with the plant and any safety-related modifications, etc. To minimize the events of repeated nature or of safety significance, root cause analysis is carried out and reviewed by plant level safety committee.

Items of higher safety significance are submitted to unit level safety committee after review by plant level safety committee, recommendations of the committees are implemented and compliance reports are prepared and submitted.

Technical audit reports and safety performance reports prepared by the plant are periodically reviewed by regulators.

Above all, a periodic safety analysis report (PSR) [11] containing various safety factors, e.g., five- or ten-year operational data, performance/maintenance records of safety systems, radiological safety data, events history of safety system malfunction, radiological waste disposal data and human factors is prepared and submitted to regulatory authorities for relicensing of the plant. A thorough review of this document is carried out by regulators. The regulators after ensuring that plant has operated safely and has all measures to maintain safe operation grant the license to operate the plant for a given period.

7   Conclusion

Nuclear reactors consider postulated initiating events during the design stage of the plant. Plant provisions under the umbrella of defense-in-depth capture the precursors to these events and activate the safety systems which stop the propagation of the initiating events before they result into incident/accident and mitigate if the events occur. A reliable performance of these safety systems can be ensured by applying a well-laid quality assurance programme though all stages of a nuclear utility. The role of regulators in ensuring compliance to quality assurance programme further strengthens safety.

References

1. International Atomic Energy Agency, INES, The International Nuclear and Radiological Event Scale User's Manual, IAEA, Vienna (2008)
2. International Atomic Energy Agency, Safety of Research Reactors, SSR-3, IAEA, Vienna (2016)
3. Nuclear Energy Agency, Probabilistic Risk Criteria and Safety Goals, NEA/CSNI/R (2009)
4. Atomic Energy Regulatory Board, Safety Systems for Pressurised Heavy Water Reactors, AERB/NPP-PHWR/SG/D-10 (2005)
5. Finnish Centre for Radiation and Nuclear Safety, Quality Assurance for Nuclear Power Plants (1991)
6. International Atomic Energy Agency, Quality Assurance for Safety in Nuclear Power Plants and Other Nuclear Installations, Code and Safety Guides Q1-Q14, IAEA, Vienna (1996)
7. International Atomic Energy Agency, Safety Classification of Structures, Systems and Components in Nuclear Power Plants, SSG-30, IAEA, Vienna (2014)
8. International Atomic Energy Agency, Quality Standards Comparison between IAEA50-C/SG-Q and ISO 9001-2000, Safety Report Series 22, IAEA, Vienna (2002)
9. International Atomic Energy Agency, Licensing Process for Nuclear Installations, SSG 12, IAEA, Vienna (2010)
10. International Atomic Energy Agency, Organization and Staffing of the Regulatory Body for Nuclear Facilities, Safety Guide GS-G-1.1, IAEA, Vienna (2002)
11. Bhabha Atomic Research Center, Procedure for Renewal of Authorisation for Operation of Radiation Installations, BSC/SG/2015/2 Rev.-1, BARC, India (2015)

Safety and Security of Process Plants: A Fuzzy-Based Bayesian Network Approach

Priscilla Grace George and V. R. Renjith

Abstract Chemical process industries handling hazardous chemicals are potential targets for deliberate actions by criminals and disgruntled employees. It is therefore imperative to have efficient risk management programs to identify prospective threats and to develop suitable mitigation strategy. In this study, a probabilistic security risk assessment approach based on dedicated Bayesian networks is presented. Bayesian networks are capable of capturing interdependencies between different factors involved in site security and are found to be very effective in performing analysis even in conditions of uncertainty due to unavailable data by making use of expert opinion. The human subjectivity associated with expert judgment is overcome by incorporating concepts of fuzzy logic. The effectiveness of this approach is demonstrated in an illustrative case study. The site-specific Bayesian network developed for the case study is found to be effective in analyzing the security status of the facility and the performance of security systems installed.

Keywords Bayesian networks · Fuzzy logic · Hazardous · Security risk

1 Introduction

Chemical industries handling hazardous substances at elevated temperatures and pressure are always a major safety concern. Accidents may occur in such units due to design flaws, system failures, operational errors, incompetence of workers, lack of proper maintenance, violation of safety procedures, etc. Such safety incidents may lead to serious consequences affecting people, environment, and property. Process industries therefore have a responsibility to reduce the occurrence of such events. This has lead to increased focus on risk assessment methodologies and safety evalu-

P. G. George (✉) · V. R. Renjith
Cochin University of Science and Technology, Cochin, Kerala, India
e-mail: priscillagrace.mec@gmail.com

V. R. Renjith
e-mail: renjithvr75@gmail.com

© Springer Nature Singapore Pte Ltd. 2020
P. V. Varde et al. (eds.), *Reliability, Safety and Hazard Assessment for Risk-Based Technologies*, Lecture Notes in Mechanical Engineering,
https://doi.org/10.1007/978-981-13-9008-1_54

ation programs. There is always scope for improvement in the safety culture of such facilities [1].

Risk assessment for process industries did not include deliberate actions from outsiders before September 11, 2001. Such ill-use of Hazardous chemicals involved in process industries can lead to huge fatalities, economic loss, and environmental damage. Disgruntled employees also pose serious threats to process industries [2]. Security focuses on protection against such intentional harm. Both internal and external threats must be considered while assessing the security risk of a facility to find out whether existing security measures are sufficient. It is therefore imperative to have efficient risk management programs to identify prospective threats and to develop suitable mitigation strategy. A basic security risk assessment involves threat analysis and vulnerability analysis. Threat analysis is used to identify sources, types of threats, and their likelihood. Vulnerability analysis describes ways in which the identified threats could be realized. Likelihood of attacks and severity is evaluated to assess the risk and suitable countermeasures are suggested [3].

Bayesian networks can be used to aid probabilistic risk and vulnerability assessment. They are acyclic graphs which enable qualitative and quantitative estimation of vulnerability. The system variables are represented by nodes with links among them showing conditional dependencies. The network structure forms the qualitative part and the conditional probability table associated with nodes form the quantitative part [4].

The objective of this study was to employ Bayesian network to portray the security aspects of a chemical plant handling hazardous chemicals. The vulnerability analysis is performed under conditions of uncertainty by incorporating fuzzy logic along with expert opinions to obtain unavailable failure data. This approach improves both qualitative and quantitative analysis as it takes into consideration the conditional dependencies among various events.

The paper is organized as follows: In Sect. 2, the framework for security risk assessment, Bayesian network methodology, and Fuzzy theory is discussed; Sect. 3 presents an illustrative case study; in Sect. 4, the results obtained by quantification of Bayesian network approach is discussed in detail; and Sect. 5 describes the main conclusions drawn.

2 Material and Methodology

2.1 Overview

The most accepted framework for security risk combines likelihood of an attack, attractiveness of an asset, vulnerability, and consequences. Probabilistic terms are used to deal with the uncertainties involved in the evolution of an attack.

Security issues arise due to deliberate actions from various categories of threat agents. The threat agents could be individuals like former disgruntled employees seeking a suitable opportunity to show their revenge or frustration by inflicting limited damage on the facility. Small groups such as political parties or activists may also target chemical facilities to cause major damage for political radicalism or financial advantages. Another class of threat agents is terrorists attacking the facility to induce maximum possible damage with no consideration for human life. The likelihood for attacks by various threat agents can be estimated by analyzing site-specific details such as threat history, political backgrounds, previous instances of terrorists' attacks, and information from intelligence agencies [5].

The attractiveness of an asset from the security perspective is analyzed in terms of the consequences of successful attack. The type of chemical handled, quantity, and conditions under which it is stored and its impact are important aspects to be considered. Terrorists are more likely to target assets which are capable of maximum possible damage whereas individuals mostly target assets with low damage potential.

Several attack modes are possible such as use of explosives, shooting, vehicle bombs, and deliberate tampering with equipment. The intrusion of threat agents into the facility depends on the attack mode. Deliberate tampering with equipment within facility requires intrusion of agents whereas shooting and vehicle bombs installed outside the plant perimeter does not require intrusion.

The effectiveness of security system determines the success level of attacks. Physical security includes perimeter fencing, proper lighting, CCTV installation, assignment of security guards, alarm system, access control, and involvement of law enforcement agencies.

2.2   Bayesian Networks

In this study, Bayesian network is used to quantify the probabilistic assessment of security risk in chemical facilities. Various factors involved in the evolution and realization of an attack are represented as nodes with links showing conditional dependencies [6]. The conditional probability tables associated with each node are filled using failure data available from standard data sources. Expert opinion is incorporated to deal with the uncertainty of unavailable failure data. Bayesian networks can be used to update prior probabilities with given new evidence by making use of Bayes' Theorem.

Updated probability [6],

$$P(U/E) = \frac{P(U, E)}{P(E)} \tag{1}$$

HUGIN EXPERT software is used in this study to quantify Bayesian network for security risk assessment of chemical facilities.

2.3 Expert Elicitation and Fuzzy Logic

Expert opinion can be used to generate the probabilities associated with nodes in Bayesian network in instances where there is no record of failure data available. Experts are selected based on their experience, knowledge, and expertize in the relevant field of study. This method is prone to human subjectivity as different experts may give different scores based on their knowledge. Fuzzy logic can be used to overcome this subjectivity [7].

Experts are asked to express their opinion of failure possibilities in linguistic terms. These linguistic terms are converted into corresponding fuzzy numbers by making use of trapezoidal membership function. Sum-production algorithm [8] is used to aggregate the fuzzy numbers corresponding to different experts. Weighting of experts is done based on their experience, age, and knowledge. Fuzzy numbers are defuzzified and converted into fuzzy possibility scores (FPS) by performing max–min aggregation method [8]. Referring to Fig. 1, the left and right scores for fuzzy number Z are calculated as,

$$\mu_L(Z) = \frac{(1-a)}{1 + (b-a)} \tag{2}$$

$$\mu_R(Z) = \frac{d}{1 + (d-c)} \tag{3}$$

The total fuzzy possibility score (FPS) is calculated as,

$$FPS = [\mu_R(Z) + (1 - \mu_L(Z))]/2 \tag{4}$$

Failure probabilities are calculated from FPS based on the following equations [8]:

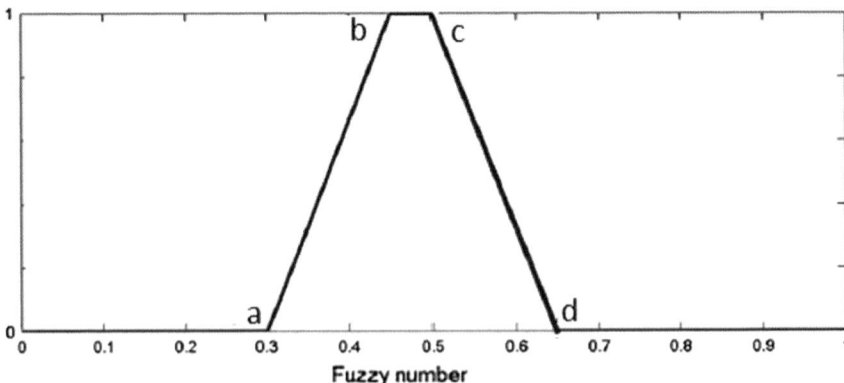

Fig. 1 Computation of left and right score for fuzzy number

$$\text{Failure Probability} = \begin{cases} 1/10^k, & \text{FPS} \neq 0 \\ 0, & \text{FPS} = 0 \end{cases} \tag{5}$$

$$k = 2.301 \times [(1 - \text{FPS})/\text{FPS}]^{1/3} \tag{6}$$

3 Case Study

The probabilistic security risk assessment method described above is applied on a fertilizer plant (X) which produces Ammonia from Naphtha and is finally converted into Urea. This is an illustrative case study which was utilized for security risk assessment by developing security risk factor table proposed by Bajpai et al. [9]. Sketch of plant X is shown in Fig. 2.

Plant X is located at a distance of about 25 km from a major city. Ammonia is stored in two storage tanks of 20 m diameter and 7500 MT capacity. Naphtha is stored in six tanks of 5000 MT capacity each. The plant maintains a good safety record and there have not been previous instances of terrorist activities. A few minor security incidents have been reported though. Proper perimeter fencing and lighting

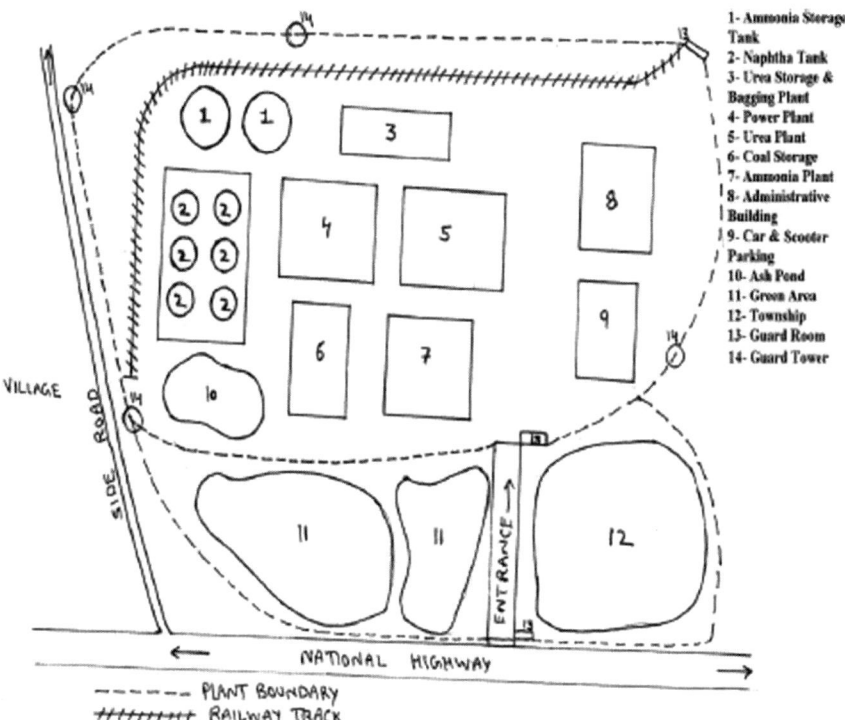

1- Ammonia Storage Tank
2- Naphtha Tank
3- Urea Storage & Bagging Plant
4- Power Plant
5- Urea Plant
6- Coal Storage
7- Ammonia Plant
8- Administrative Building
9- Car & Scooter Parking
10- Ash Pond
11- Green Area
12- Township
13- Guard Room
14- Guard Tower

Fig. 2 Layout for Plant X. (*Courtesy* Bajpai and Gupta [9])

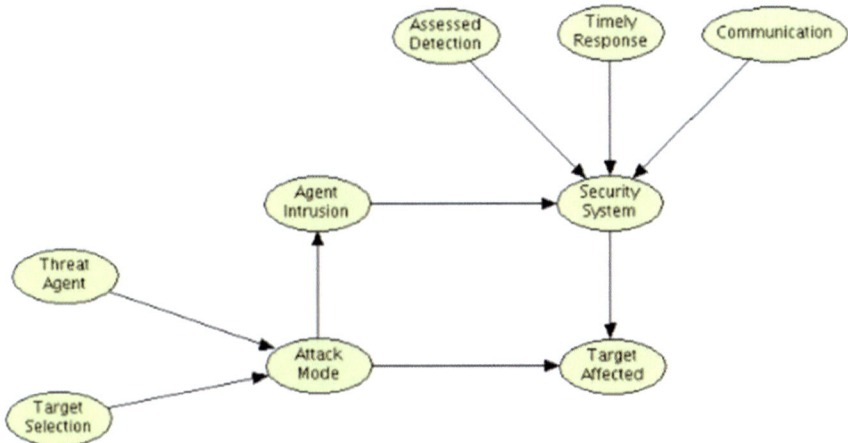

Fig. 3 Bayesian network for plant X

are done. CCTVs are installed to monitor the activities in the plant. Trained security guards are also deployed at various stations within the plant [9].

The schematic of the Bayesian network developed for plant X is shown in Fig. 3. Nodes are assigned for target selection, threat agents, attack modes, threat agent intrusion path, security system performance, and attack consequences.

Ammonia tank and Naphtha tank are the targets under consideration for this study. The marginal probabilities for target selection are assigned based on the severity of consequences and hazard potential of the chemicals involved. Here probabilities are assigned under the assumption that target with most severe consequences is selected. Comparing the hazard potential of Ammonia and Naphtha, the probabilities assigned for their selection are 0.35 and 0.65, respectively. The threat history and information from law enforcement agencies concerning plant X are analyzed to obtain probability values for type of threat agents involved. The values assigned for disgruntled individuals, activists, and terrorists are 0.75, 0.24, and 0.01, respectively.

Attack modes considered for this study are use of explosives, shooting from distance, vehicle bombs installed outside plant X, and deliberate tampering with equipment inside plant. Use of explosives and deliberate tampering requires intrusion of threat agents into the plant, so the effectiveness of security system determines the success probability for attack.

The conditional probability table (CPT) for the node representing attack mode is quantified by judging the preferred mode chosen by different threat agents based on their capability and intention. Table 1 shows the CPT for attack mode (the values are assumed to be same for Ammonia tank and Naphtha tank).

The performance of security system depends on successful detection of intrusion, successful timely response to intrusion by taking suitable measures to prevent threat agents from reaching target, and successful communication to responsive force [5]. These factors are assigned as parent nodes to security system. The CPT for these

Table 1 CPT for node representing attack mode

Attack mode	Threat agent type		
	Individuals	Activists	Terrorists
Explosives	0	0.4	0.5
Shooting	0.1	0.2	0.2
Vehicle bomb	0	0	0.2
Deliberate tampering	0.9	0.4	0.1

Table 2 Failure probabilities for factors affecting performance of security system

Factor	Failure probability
Assessed detection	0.000143
Timely response	0.000555
Communication to responsive force	0.000163

parent nodes is obtained by expert elicitation and fuzzy logic. In this study, five experts in the relevant field of study are chosen. The experts are assigned weights depending on their designation, education level, service time and age. The weights calculated for the five experts are 0.20, 0.23, 0.22, 0.18, and 0.15, respectively. Expert opinions are obtained in linguistic terms scaled at five levels: very low, low, medium, high, and very high. Trapezoidal membership function is used to convert expert opinion in linguistic terms to fuzzy numbers. The aggregation of expert opinion is done using sum-production algorithm. Fuzzy possibility scores are obtained by performing defuzzification based on max–min aggregation method [8]. The failure probabilities thus obtained for the factors affecting performance of security system are listed in Table 2.

The effectiveness of security system also depends on the intrusion path followed by threat agents. The security system elements for plant X are installed within its perimeter, so its success probability becomes zero for attack modes which does not require intrusion into the facility. The node named security system is modeled as an AND gate among its parent nodes.

The node named target affected represents the success probability of attack and it relates performance of security system with different attack modes. This node has the following states: Ammonia release, Ammonia theft, Naphtha fire, Naphtha contamination, and attack prevented (to account for attack modes requiring intrusion where the security system has been successful in preventing attack).

4 Results and Discussion

The Bayesian network specific to plant X as described above was quantified using HUGIN EXPERT software, and the probability distributions obtained are shown in Fig. 4.

Fig. 4 Probability distributions of Bayesian network for plant X

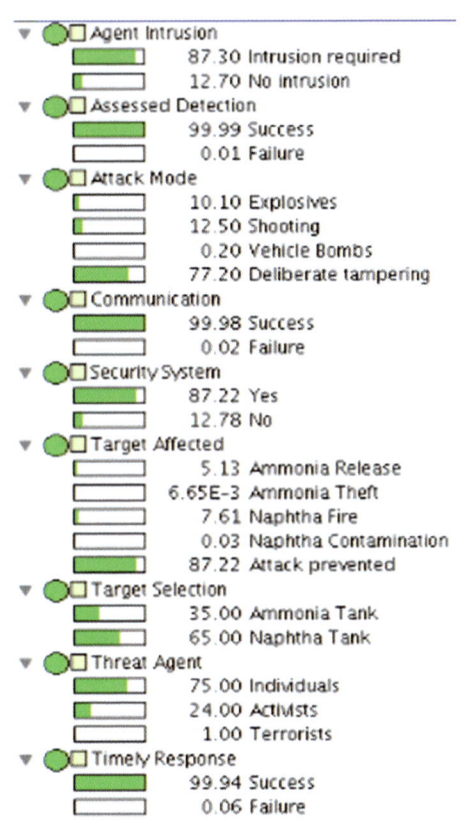

Using this software, evidences can be given and the in-built sum-propagation algorithm can be performed to obtain the corresponding updated posterior probabilities. Figure 5a shows the updated probabilities for the evidence no intrusion. It is observed that the probabilities for shooting mode and vehicle bomb mode have increased as they do not require threat agent intrusion. The probabilities for use of explosives and deliberate tampering are null. Also the security system has failed, since they are installed within the plant perimeter and hence is not effective in preventing attacks which do not require intrusion.

Figure 5b shows the updated probabilities for the evidence that deliberate tampering has occurred. It is observed that the probability for attack prevention has increased to 0.99 accounting for the high success probability of security system. The security system employed for plant X is efficient in preventing attacks which requires intrusion of threat agent. This shows the significance of an effective system in chemical facilities handling hazardous chemicals. The success probability for security system can be improved by enhancing the security further to enable fast detection of intrusion and reducing the delay in response after detection.

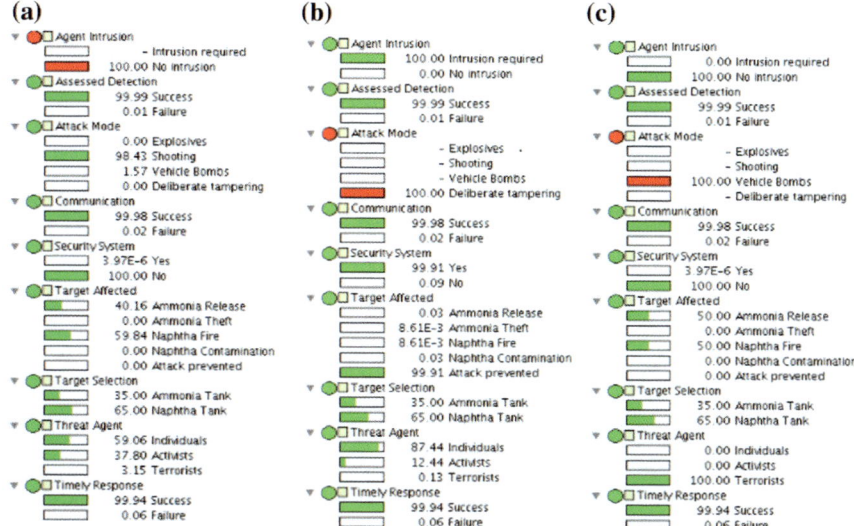

Fig. 5 Updated probabilities for evidences: **a** no intrusion **b** deliberate tampering, **c** vehicle bomb

Figure 5c shows updated probabilities for the evidence that vehicle bomb attack has occurred. The updated probabilities indicate that the security system has failed as vehicle bomb was installed outside plant perimeter thereby requiring no intrusion. Also the probability that the threat agent is a terrorist has tremendously increased. As discussed, various evidences can be given and hence site-specific detailed security risk analysis can be performed using Bayesian network approach.

5   Conclusions

This study makes use of Bayesian network approach to perform probabilistic security risk assessment. Bayesian networks can be developed for specific chemical sites to qualitatively and quantitatively analyze the security aspects of the site. Bayesian networks take into consideration the conditional interdependencies between various factors involved in security risk assessment. This approach has the advantage that effective analysis can be performed even in conditions of uncertainty and unavailable data by making use of expert judgments. The human subjectivity associated with expert opinion is overcome by incorporating fuzzy logic. This methodology is illustrated for a specific case study and the results obtained show the performance of security systems and its effectiveness in preventing attacks. The updated posterior probabilities can be obtained for different given evidences to perform detailed analysis.

Acknowledgements This work is supported by the Department of Safety and Fire Engineering, CUSAT, India.

References

1. B. Knegtering, H.J. Pasman, Safety of the process industries in the 21st century: a changing need of process safety management for a changing industry. J. Loss Prev. Process Ind. **22**, 162–168 (2009)
2. P. Baybutt, V. Reddy, Strategies for protecting process plants against terrorism, sabotage and other criminal acts (Primatech Inc., 2003)
3. P. Baybutt, Assessing risks from threats to process plants: threats and vulnerability analysis. Process Saf. Prog. **21**(4), 269–275 (2002)
4. J. Bhandari, R. Abbassi, V. Garaniya, F. Khan, Risk analysis of deepwater drilling operations using Bayesian network. J. Loss Prev. Process Ind. **38**, 11–23 (2015)
5. G. Landucci, F. Argenti, V. Cozzani, G. Reniers, Assessment of attack likelihood to support security risk assessment studies for chemical facilities. Process Saf. Environ. Prot. **110**, 102–114 (2017)
6. F. Argenti, G. Landucci, G. Reniers, V. Cozzani, Vulnerability assessment of chemical facilities to intentional attacks based on Bayesian Network. Reliab. Eng. Syst. Saf. **169**, 515–530 (2018)
7. M. Yazdi, S. Kabir, A fuzzy Bayesian Network approach for risk analysis in process industries. Process Saf. Environ. Prot. **111**, 507–519 (2017)
8. V.R. Renjith, G. Madhu, V. Lakshmana Gomathi Nayagam, A.B. Bhasi, Two-dimensional fuzzy fault tree analysis for chlorine release from a chlor-alkali industry using expert elicitation. J. Hazard. Mater. **183**, 103–110 (2010)
9. S. Bajpai, J.B. Gupta, Site security for chemical process industries. J. Loss Prev. Process Ind. **18**, 301–309 (2005)

Determination of Prony Series Coefficient of Polymer Based on Experimental Data in Time Domain Using Optimization Technique

Neha Singh and Mitesh Lalwala

Abstract Polymers due to their versatile properties and characteristics are very promising material for the future. But the mechanical behavior of the polymer depends on loading type, temperature and time. These behaviors of the polymer can be mathematically modeled using the phenomenological Prony series models. But the challenge is to find the parameters τ_q and E_q to provide a good conformance with the experimental data. In the present work methodology for inverse identification in the time domain for experimental data using mixed optimization techniques (Genetic Algorithm and Nonlinear Programing) is used to determine Prony series coefficients. The results are compared with the experimental data to validate the method.

Keywords Polymer · Wiechert model · Prony series coefficients · Optimization

1 Introduction

Polymers are widely used in modern days due to their versatile properties and characteristics. However, the behavior of the polymer when subjected to different kinds of loading is still under study due to their complex molecular structure, which molds mechanical properties that change according to time, temperature and loading.

The mechanical behavior of polymer can be modeled using the spring and damper as seen in Maxwell and Kelvin phenomenological models. Where spring represents instantaneous elastic behavior and damper represents delayed elastic behavior of the polymer. These models are very useful in understanding the physical relation between stress and strain that occurs in polymers and other viscoelastic material.

N. Singh · M. Lalwala (✉)
Department of Mechanical Engineering, Indian Institute of Technology Delhi,
Hauz Khas, New Delhi 110016, India
e-mail: lalwalamitesh@gmail.com

N. Singh
e-mail: ineha.m27@gmail.com

© Springer Nature Singapore Pte Ltd. 2020
P. V. Varde et al. (eds.), *Reliability, Safety and Hazard Assessment
for Risk-Based Technologies*, Lecture Notes in Mechanical Engineering,
https://doi.org/10.1007/978-981-13-9008-1_55

But since the polymers contain molecules of various chain length/molecular weight and due to various degree of polymerization above-mentioned simple models cannot predict the exact behavior of the polymers. A better approximation of a phenomenological model with the actual polymer can be obtained by various combinations (parallel or series) of above simple models known as the generalized model such as generalized Maxwell and generalized Kelvin model. Usually as many as 5–15 or more elements are used to represent the complete behavior of polymer. However, generalized Maxwell model can only be used to represent thermoplastic if all of the damper terms μ_i are nonzero. In order to represent a thermoset, a free spring is included in the model which results in Wiechert model, also known as Prony series model.

But the challenge is to find the relaxation times τ_i and elastic terms E_i, to provide a good fit over all time of the experimental data. This problem has been addressed by a number of methods in the literature, including procedure X (due to Tobolsky and Murakami and discussed by Tschoegl 1989), the collocation method by Schapery (1962), the multidata method (Cost and Becker 1970) and the windowing method (Emri and Tschoegl 1993; [1]).

In the present work, we are trying to determine the Prony series coefficient of polymer using inverse identification process in time domain using experimental data and mixed optimization technique (Genetic algorithm and Nonlinear programing) as suggested by Pacheco (2013) using MATLAB [2].

2 Theoretical Concept and Mathematical Formulation

2.1 Prony Series Model of Polymer

Wiechert model is obtained by adding one more free spring parallel to the generalized Maxwell model as shown in Fig. 1. Relaxation modulus $E(t)$ of the Wiechert model is defined as

Fig. 1 Wiechert Model

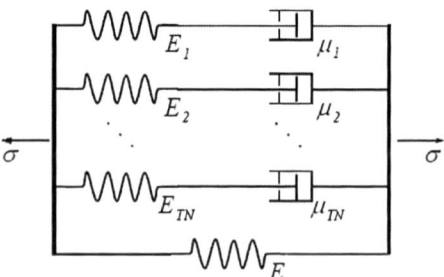

$$E(t) = E_\infty + \sum_{q=1}^{TN} \left(E_q e^{\frac{-t}{\tau_p}} \right)$$

Here, E_∞ is equilibrium modulus, E_q and τ_q are the elastic component and relaxation time associated with the qth component. Now using Boltzmann superposition theorem and hereditary integral, stress $\sigma(t)$ developed in given model for any variable strain input $\varepsilon(t)$ can be expressed as

$$\sigma(t) = E_\infty \varepsilon(0) + \int_0^t E(t - \tau) \frac{d\varepsilon(\tau)}{d\tau} d\tau$$

where $\varepsilon(0)$ is accumulated strain up to the initial instant.

2.2 Identification Process Formulation in Time Domain

We are using experimental data obtained through uniaxial traction test performed on the material, according to norm ISO 527/1B (ISO 2012) with constant strain rate,

$$\frac{d\varepsilon}{dt} = \dot{\varepsilon} = \text{constant for all } t$$

Using this strain, input stress developed in Wiechert model can be simplified as

$$\sigma_{\text{prony}} = E_\infty \dot{\varepsilon} t_k + \sum_{q=1}^{TN} E_q \dot{\varepsilon} \tau_q \left(1 - e^{\frac{-t}{\tau_p}} \right)$$

Here, we are assuming that no previous strain history has been applied to the model that is $\varepsilon(0) = 0$. We are considering a set of experimental data in which for each temperature T_j, traction test was performed at a constant strain rate $\dot{\varepsilon}_i$.

We are considering the total number of temperature NTemp, total number of strain rate NSr for the total no of sample points NPt. Absolute error D_{ijk} in stress associated with k-th point t_k ($1 \leq k \leq$ NPt), at j-th temperature T_j ($1 \leq j \leq$ NTemp) and i-th strain rate $\dot{\varepsilon}_i$ ($1 \leq i \leq$ NSr) can be expressed as

$$D_{ijk} = \left(\sigma_{\text{prony}} \left(E_\infty, \dot{\varepsilon}_i, t_k, T_j, E_q \right) - \sigma_{\text{exp}} \left(\dot{\varepsilon}_i, t_k, T_j \right) \right)$$

Here, σ_{prony} stress is evaluated according to Prony series model by the hereditary integral and σ_{exp} is the measured experimental stress, both obtained at k-th point at j-th temperature and i-th strain rate.

A measurement of total error associated with the model can be obtained by total quadratic error D_T^2, given by

$$D_T^2 = \sum_{i=1}^{\text{NSr}} \sum_{j=1}^{\text{NTemp}} \sum_{k=1}^{\text{NPt}} \left(\sigma_{\text{prony}}\left(E_\infty, \dot{\varepsilon}_i, t_k, T_j, E_q \right) - \sigma_{\text{exp}}\left(\dot{\varepsilon}_i, t_k, T_j \right) \right)^2$$

Or we can use average quadratic error

$$\overline{D_T} = \frac{D_T^2}{\text{Total Number of points}}$$

This scalar function will serve as an objective function in the optimization process for identification of the Prony series coefficients.

We note that, at this stage, we are not considering the effect of temperature in the determination of Prony series coefficient. We are using experimental data for various strain rates but at constant temperature.

2.3 Influence of Temperature on Behavior of Polymer

Effect of temperature on mechanical behavior of thermorheologically simple polymer can be expressed using WLF model (Williams et al. 1955) as

$$\log_{10} \alpha_T = \log_{10} \frac{\tau(T)}{\tau(T_o)} = \frac{-C_1(T - T_o)}{C_2 + (T - T_o)}$$

where T is the temperature at which response is measured, T_o is the reference temperature, C_1 and C_2 are material property constants and α_T is shift factor for all relaxation times at temperature T. By considering shift factor, the stress value at time t_k is given by

$$\sigma_{\text{prony}} = E_\infty \dot{\varepsilon}_j t_k + \sum_{q=1}^{TN} E_q \dot{\varepsilon}_j \alpha_T(T_i) \tau_q \left(1 - e^{\frac{-t}{\alpha_T(T_i)\tau_p}} \right)$$

By using the above equation for the stress, we can incorporate the effect of temperature in the identification process.

2.4 Optimization Formulation

Using above-mentioned theories, the identification process can be formulated in terms of optimization problem as below:

$$\text{Minimize } f(x) = \overline{D_T}$$

$$\text{where } x = \{E_\infty, E_i, C_1, C_2, T_o\} \quad (i = 1, \ldots, TN)$$

$$\text{Constraints :} \begin{cases} 0 \leq E_\infty \leq E_\infty^{\text{upp}} \\ 0 \leq E_i \leq E_i^{\text{upp}} \\ C_1^{\text{low}} \leq C_1 \leq C_1^{\text{upp}} \\ C_2^{\text{low}} \leq C_2 \leq C_2^{\text{upp}} \\ T_o^{\text{low}} \leq T_o \leq T_o^{\text{upp}} \end{cases}$$

where $E_\infty^{\text{upp}}, E_i^{\text{upp}}, C_1^{\text{low}}, C_2^{\text{low}}, T_o^{\text{low}}, C_1^{\text{upp}}, C_2^{\text{upp}}, T_o^{\text{upp}}$ are arbitrated values which represent the upper and lower limits of design variables x. In this standard optimization problem, goal is to minimize error $\overline{D_T}$.

This formulation is for the instance when we assume fixed relaxation times τ_i as suggested in the literature [2]. Better conformation with the experimental data can be obtained if we optimize τ_i also in the optimization algorithm. Under this condition, optimization formulation will be as follows:

$$\text{Minimize } f(x) = \overline{D_T}$$

$$\text{where } x = \{E_\infty, E_i, C_1, C_2, T_o, \tau_i\} \quad (i = 1, \ldots, TN)$$

$$\text{Constraints :} \begin{cases} 0 \leq E_\infty \leq E_\infty^{\text{upp}} \\ 0 \leq E_i \leq E_i^{\text{upp}} \\ C_1^{\text{low}} \leq C_1 \leq C_1^{\text{upp}} \\ C_2^{\text{low}} \leq C_2 \leq C_2^{\text{upp}} \\ T_o^{\text{low}} \leq T_o \leq T_o^{\text{upp}} \\ \tau_i \text{ are assumed to be uniformly} \\ \text{distributed on log scale.} \end{cases}$$

Flow chart of the process adopted is indicated in Fig. 2.

3 Results and Discussion

We are using experimental stress-strain data, for Stamax 30YM240 (Polypropylene-based Thermoplastic) with 30% concentration of long fiber glass, in accordance with norm ISO 527/1B (ISO 2012). The material was subjected to three sets of tests of pure traction, each at a different but constant temperature (−35, 23 and 80 °C) and at each temperature four different strain rates, constant throughout each particular test (of 0.0001 (mm/mm)/s, 0.01 (mm/mm)/s, 0.1 (mm/mm)/s, 1 (mm/mm)/s) [2]. These data have been used as entry file in identification process.

Fig. 2 Identification process
flow chart

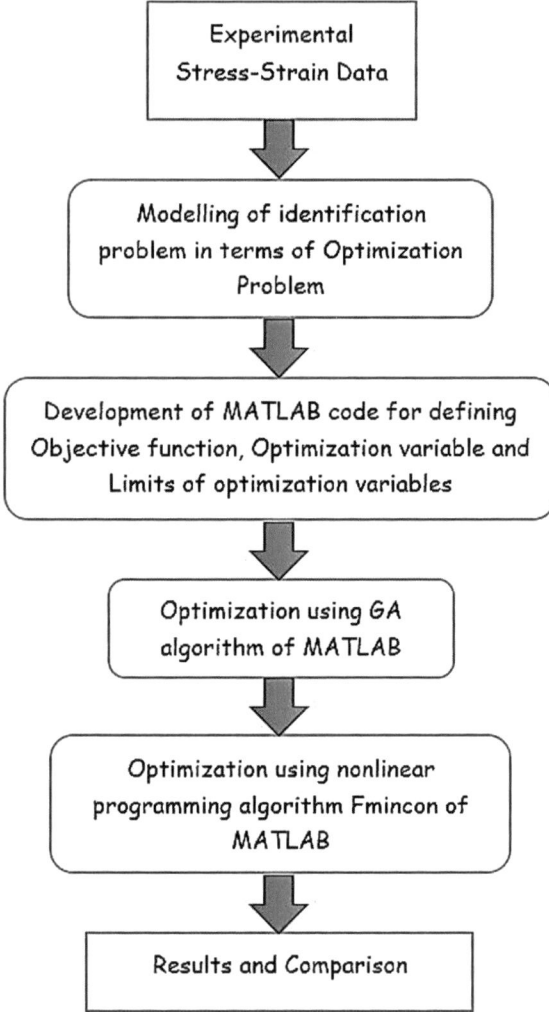

3.1 Parameters, Limits and Constraints of Identification Process

Options and limits implemented for optimization algorithm in MATLAB are shown in Table 1.

For constant relaxation time case, we are using values of τ_1 represented in Table 2.

Table 1 Parameters, limits and constraints of identification process

GA algorithm	
GA-population	200
GA-generations	200
GA-function tolerance	1e−6
GA-mutation	@mutationadaptfeasible, 0.02
NLP algorithm	
Algorithm	Fmincon
NLP method	Active set
Constraints and limits	
No of elements in Prony series, TN	8
Equilibrium modulus, E_∞	0 MPa $\leq E_\infty \leq$ 10,000 MPa
Prony series components, E_i $(i = 1...TN)$	0 MPa $\leq E_i \leq$ 5000 MPa
Material constant, C_1	$-10 \leq C_1 \leq 10$
Material constant, C_2	$-200\ °C \leq C_2 \leq 200\ °C$
Reference temperature, T_o	$-100\ °C \leq T_o \leq 100\ °C$
Relaxation times, τ_i $(i = 1...TN)$	Uniformly distributed between 0 to 10^4 on log scale

Table 2 Value of constant τ_i for identification process

$\tau_1 = 0.01$ s	$\tau_5 = 26.825$ s
$\tau_2 = 0.07196$ s	$\tau_6 = 193.06$ s
$\tau_3 = 0.5179$ s	$\tau_7 = 1389.5$ s
$\tau_4 = 3.7243$ s	$\tau_8 = 10,000$ s

3.2 Numerical Data

(a) Constant Temperature and all strain rates

This section presents the results of identification process for constant temperature and all strain rates of 0.0001 (mm/mm)/s, 0.01 (mm/mm)/s, 0.1 (mm/mm)/s, 1 (mm/mm)/s. Tables 3 and 4 represent the results of the identification process. From these tables, it becomes clear that Prony series coefficients depend on the temperature.

From Tables 3 and 4, it can be observed that the identification problem yields good results in determining Prony series coefficient with average error of 1.5 MPa. Also, average error in results obtained by considering τ constant and τ variable are almost same.

(b) All Temperatures and constant strain rate

This section presents the result of constant strain rate and temperature -35, 23 and 80 °C. Figure 3 represents comparison of experimental stress-strain data and data

Table 3 Identification process results for constant temperature using constant τ

τ_i (s)	$T = -35\,°C$	$T = 23\,°C$	$T = 80\,°C$
	E_i(MPa)	E_i(MPa)	E_i(MPa)
0.01	1532.3	1078.9	669
0.07196	349.57	359.32	470.89
0.5179	100.37	209.14	180.79
3.7273	253.76	0	0
26.825	1	0	0
193.06	469.31	263.84	0
1389.5	214.88	22.032	715.17
10,000	3226.4	858.69	1328.4
	$T = -35\,°C$	$T = 23\,°C$	$T = 80\,°C$
E_∞(MPa)	1768.395	2093.469	19.4197
GA Min (MPa)^2	2.47783	11.2512	7.11847
NLP Min (MPa)^2	2.40046	10.3755	5.0564
Mean error (MPa)	0.9266	2.3688	1.3593
Mean % error	4.2999	10.1818	8.7436

Table 4 Identification process results for constant temperature using variable τ

$T = -35\,°C$		$T = 23\,°C$		$T = 80\,°C$	
τ(s)	E_i(MPa)	τ(s)	E_i(MPa)	τ(s)	E_i(MPa)
0.01	1314.3	0.01	1432.5	0.01	797.46
0.1	437.63	0.13371	316.73	0.12612	552.52
6.3578	0	1.3165	0	9.9028	0
100	459.23	33.74	0	12	0
620.65	491.23	691.79	865.51	659.27	0
850.33	432.7	735.51	629.45	102.52	0
2198.1	507.02	5323.9	233.2	1078	1214.8
1033.8	3492.3	1099	477.98	3214.5	4.5503
		$T = -35\,°C$	$T = 23\,°C$	$T = 80\,°C$	
E_∞(MPa)		516.2256	1120.53	872.8524	
GA Min (MPa)^2		2.8223	10.8687	5.6174	
NLP Min (MPa)^2		2.37495	10.3902	5.0651	
Mean error (MPa)		0.8912	2.2877	1.3508	
Mean % error		4.13	9.5258	8.766	

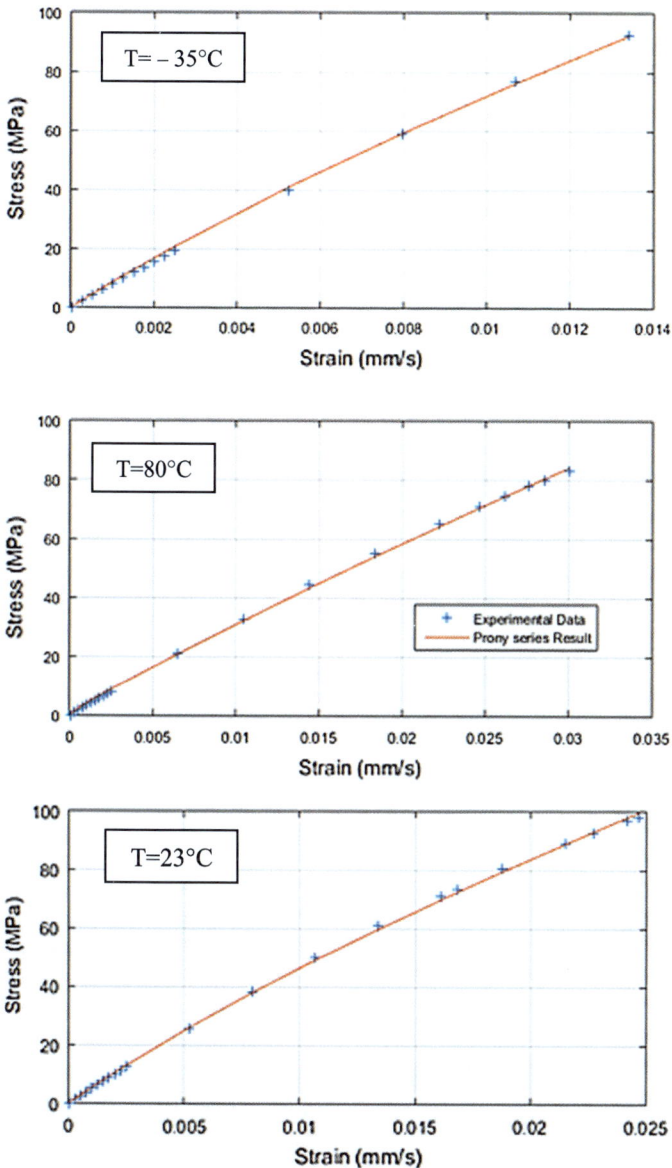

Fig. 3 Stress–Strain data comparison of experimental and Prony series result for strain rate of 1 (mm/mm)/s at various temperature using variable τ

obtained by Prony series for constant strain rate of 1(mm/mm)/s. Tables 5 and 6 represent results of identification process. From tables, it is clear that Prony series coefficient depends on strain rate applied.

Table 5 Identification process results for constant strain rate using constant τ

$\tau(s)$	$\dot{\varepsilon} = 1$ (mm/mm)/s	$\dot{\varepsilon} = 0.1$ (mm/mm)/s	$\dot{\varepsilon} = 0.01$ (mm/mm)/s	$\dot{\varepsilon} = 0.0001$ (mm/mm)/s
	E_i(MPa)	E_i(MPa)	E_i(MPa)	E_i(MPa)
0.01	1954.5	2475.3	3802.8	939.52
0.07196	1146.3	3004	2378.8	2515.4
0.5179	1183.2	2818.1	1023.3	1208.7
3.7273	2083.6	891.26	0.71985	467.11
26.825	0	489.3	1106.4	2398.5
193.06	1907.3	2368.2	2421.9	1421.6
1339.5	451.99	390.9	1054.4	784.76
10,000	790.63	1586,1	2493.9	2472.1

	$\dot{\varepsilon} = 1$ (mm/mm)/s	$\dot{\varepsilon} = 0.1$ (mm/mm)/s	$\dot{\varepsilon} = 0.01$ (mm/mm)/s	$\dot{\varepsilon} = 0.0001$ (mm/mm)/s
T_o(°C)	−75.2872	−61.5289	−80.0308	−18.3069
C_1	10	5.8091	5.4765	4.4794
C_2(°C)	133.3752	58.0661	69.3511	200
E_∞ (MPa)	2101.289	531.944	748.9165	0
GA Min (MPa)^2	1.99003	9.86412	2.26563	18.7795
NLP Min (MPa)^2	0.522586	1.40348	1.533	1.57974
Mean error (MPa)	0.562	0.9619	0.9342	0.8559
Mean % error	7.4016	11.4263	11.3418	9.81

From Tables 5 and 6, it can be observed that identification problem yields give results in determining Prony series coefficient with average error of 0.75 MPa. Also, results obtained by considering τ variable are better than that obtained by considering τ constant.

(c) *All temperatures and all strain rates*

This section presents the identification process results considering all temperature and all the strain rate data. Figure 4 represents the relaxation modulus resulting from identification process using constant and variable τ at their respective reference temperatures. Table 7 represents the results of the identification problem.

From Table 7, it can be seen that the identification process gives comparable results in conformance with experimental data. However, average error obtained here are comparatively large of the order of 2.5 MPa. There is not much difference in error in results obtained by constant and variable τ approach.

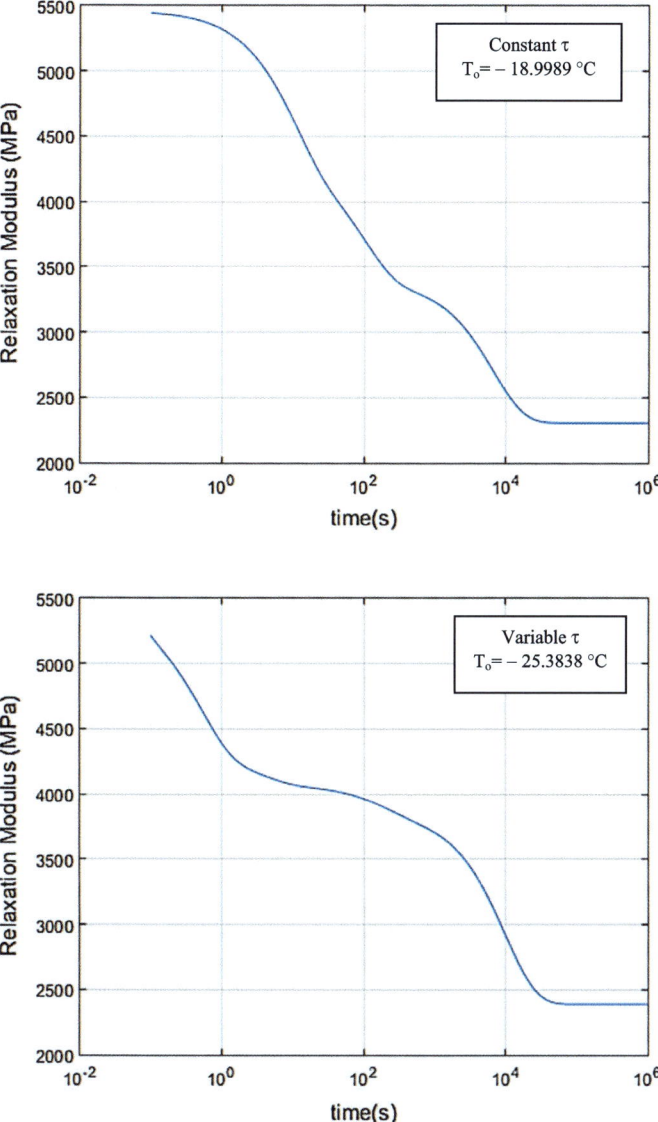

Fig. 4 Relaxation modulus obtained from identification process at all temperatures and all strain rates using constant τ and variable τ at their respective reference temperatures

Table 6 Identification process results for constant strain rate using variable τ

$\dot{\varepsilon} = 1$ (mm/mm)/s		$\dot{\varepsilon} = 0.1$ (mm/mm)/s		$\dot{\varepsilon} = 0.01$ (mm/mm)/s		$\dot{\varepsilon} = 0.0001$ (mm/mm)/s	
$\tau_i(s)$	E_i(MPa)	$\tau_i(s)$	E_i(MPa)	$\tau_i(s)$	E_i(MPa)	$\tau_i(s)$	E_i(MPa)
0.001	19.022	0.001	4988.9	0.001	520.84	0.001	4599
0.001	517.63	0.001	2288.2	0.001	66.433	0.001	302.43
0.03148	0	0.01	1809.2	0.01	3914.2	0.01	2137.3
0.1	3490.9	0.10712	4133.5	0.32927	3526.9	0.1	274.42
10	286.94	8.0306	786.98	10	565.35	1	1159.8
28.08	2092.5	10	1409.8	15.284	1898.8	16.58	2697.9
615.85	88.098	480.77	2102.3	1000	1582.6	504.57	2377.2
3684.1	1682.9	8895.1	2475.7	4914.1	2625.7	4376.5	1054

	$\dot{\varepsilon} = 1$ (mm/mm)/s	$\dot{\varepsilon} = 0.1$ (mm/mm)/s	$\dot{\varepsilon} = 0.01$ (mm/mm)/s	$\dot{\varepsilon} = 0.0001$ (mm/mm)/s
T_o (°C)	−56.5558	−58.3903	−46.5733	−16.5582
C_1	10	5.9699	4.1979	3.9042
C_2 (°C)	158.3113	45.576	30.6706	163.7036
E_∞ (MPa)	1303.362	594.3954	301.32	1253.9741
GA Min (MPa)^2	0.520216	1.71084	1.87016	1.9156
NLP Min (MPa)^2	0.461711	1.14238	1.33295	1.2231
Mean error (MPa)	0.5271	0.7651	0.7966	0.8242
Mean % error	6.1404	8.387	8.1096	10.1927

4　Conclusion

From the above results, following conclusions can be drawn:

- Prony series coefficient, to represent actual behavior of polymer, can be effectively approximated using inverse identification process in time domain for experimental data using mixed optimization technique.
- Stress–strain results obtained from the Prony series data show good conformance with the actual experimental data.
- Inverse identification process for constant strain rate data yields better conformance with experimental data as compared to constant temperature data. It is observed that with constant temperature data, average error between experimental and Prony series data is around 1.5 MPa, while in case of constant strain rate data it is around 0.75 MPa. It indicates that strain rate have significant influence on mechanical behavior of polymer, and further study is required to relate effect of strain rate on stress behavior.
- On considering stress-strain data for all temperature and all strain rate in inverse identification process, it is observed that average error is comparatively higher of

Table 7 Identification process results for all temperature and all strain rate data using constant τ and variable τ

Constant τ		Variable τ	
$\tau_i(s)$	$E_i(MPa)$	$\tau_i(s)$	$E_i(MPa)$
0.01	1240.9	0.001	0
0.07196	261.36	0.001	394.21
0.5179	1049.7	0.01	2220.7
3.7273	220.41	1	44.264
26.825	0	10	495.14
193.06	228.86	10	606.52
1389.5	0	102.43	935.88
10,000	1454.4	6804.1	1075

	Constant τ	Variable τ
$T_o(°C)$	−18.9989	−25.3838
C_1	10	10
$C_2(°C)$	67.337	91.3081
$E_\infty(MPa)$	2388.171	2304.417
GA Min $(MPa)^2$	37.04932	28.509
NLP Min $(MPa)^2$	20.7189	20.611
Mean error (MPa)	2.5041	2.4954
Mean % error	10.2298	9.6598

the order of 2.5 MPa as compared to the previous cases. But still results are in good conformation with experimental data as this error accounts to only 10% average error in each plot.

- Identification process yields better or approximately same result for variable τ method as compared to constant τ method.
- Advantage of using variable τ method is that the optimization process shows faster convergence as compared to constant τ method. It is observed that there is less difference in GA best value and NLP best value using this method.

References

1. H.F. Brinson, L.C. Brinson, *Polymer Engineering Science and Viscoelasticity-An Introduction* (Springer Publication, New York City, 2008)
2. J.E.L. Pacheco et al., Viscoelastic materials characterization with Prony series application. Latin Am. J. Solid Struct. **12**, 420–425 (2015)

Mode Selection in Variational Mode Decomposition and Its Application in Fault Diagnosis of Rolling Element Bearing

Pradip Yadav, Shivani Chauhan, Prashant Tiwari, S. H. Upadhyay
and Pawan Kumar Rakesh

Abstract Bearings are standard and furthermore enormous parts in any rotating machinery. For the perfect operational rotating system, bearing assumes a key responsibility. Once a calamity happens, it will cause huge budgetary episodes and even security threats. Proper fault diagnosis of the rolling element bearings will put a check on all such issues. In the past, several fault diagnosis strategies have been implemented and found effective due to the involvement of signal decomposition. A proper choice of decomposed modes may extract vital information and evidence on fault type, location, and condition. Inspired by the same fact, the work proposes statistical criteria for mode selection in the vibrational mode decomposition method utilizing entropy and covariance parameters. The criteria have been validated on a test signal sample gathered over bearing with inner race failure. The illustration proves the promising nature of the method which makes the method superior till date.

Keywords Bearings · Fault diagnosis · Variational mode decomposition · Spectrum analysis

P. Yadav · S. Chauhan · P. Tiwari (✉) · P. K. Rakesh
National Institute of Technology Uttarakhand, Srinagar Garhwal, Uttarakhand, India
e-mail: prashant.tiwari.om@gmail.com

P. Yadav
e-mail: pradipy51@gmail.com

S. Chauhan
e-mail: shivanichauhan4148@gmail.com

P. K. Rakesh
e-mail: pawanrakesh@nituk.ac.in

P. Tiwari · S. H. Upadhyay
Indian Institute of Technology Roorkee, Roorkee, Uttarakhand, India
e-mail: shumefme@iitr.ac.in

© Springer Nature Singapore Pte Ltd. 2020
P. V. Varde et al. (eds.), *Reliability, Safety and Hazard Assessment for Risk-Based Technologies*, Lecture Notes in Mechanical Engineering,
https://doi.org/10.1007/978-981-13-9008-1_56

1 Introduction

Bearing plays an essential role throughout the operation in all rotating machinery. Once a failure occurs in the bearing, it will cause financial misfortunes and safety hazards. According to the statistics, bearing failure leads to the sudden breakdown of the gearbox motor and wind turbine. Vibration-based conditioning assessment and fault diagnosis more important at regular interval and is the effective way to prevent losses. The vibration signal acquired by the bearing setup has lots of information but it also contains superfluous which creates mislead in assessment, the proper strategy will help in the extraction of vital information and supersedes deceives. The exploration of several methods has been conducted; the methods, namely spectral kurtosis and kurtogram [1], cyclostationary methods [2], wavelet-dependent techniques [3], empirical mode decomposition (EMD), ensemble empirical mode decomposition (EEMD), local mean decomposition (LMD), morphological signal processing technique [4], matching pursuit [5], order tracking [6], data mining tools in signal processing [7], have been reported and found efficient; out of these methods, signal decomposition has proven itself an efficient fault diagnosis and failure pattern recognition. Previous work on signal decomposition techniques EMD, LMD, and EEMD demonstrates that modes after decomposition give the essential information regarding fault in the bearing components. Signal decomposition decomposes the complex signal into the basic modes which can be simply studied in the time and frequency domains [8–11]. In 2013, Dragomiretskiy et al. suggested new adaptive method, variational mode decomposition (VMD) [12]; it is a non-stationary signal processing method which decomposes the signal into a predefined number of modes; these modes are non-recursive modes. After decomposition, VMD rationally manages the condition of convergence; therefore, it can eliminate the problem of mode mixing.

In this work, the variational mode decomposition (VMD) method is used for signal decomposition to gather the mode functions. The novelty of the paper is in selection criteria of the mode function which has been performed by the covariance and entropy parameters. This one is proven and promising criterion for the selection of an effectual and worthy mode for fault diagnosis. The process ends up in the recognition of fault using the envelope spectrum. Section 2 explains the methods and procedures to be followed in the entire prognostic model; Sect. 3 explains the experimentation over which the validation of the proposed model has been performed; and also describes the outcomes; and Sect. 4 concludes the work.

2 Methodology and Procedure

2.1 Review and Algorithm of VMD

VMD decomposes a given signal (real-valued) f into a number of discrete sub-signals or modes (u_k) which have some specific sparsity properties, and these modes are band-limited mode function. It is expected that individual mode k is to be mostly compact around a center pulsation (ω_k). For each mode u_k, construct the analytical signal by means of Hilbert transform to obtain a unilateral frequency spectrum. Then, shift the mode's frequency spectrum to "baseband" by mixing with an exponential tuned to the respective estimated center frequency and the bandwidth of the signal is calculating using H^1 Gaussian smoothness of the demodulated signal, i.e., the squared L^2-norm of the gradient. The goal of optimization is to minimize the sum of the spectral widths of all the IMFs as follows [12]:

$$\min_{\{u_k\},\{\omega_k\}} \left\{ \sum_k \left\| \partial_t \left[\left(\delta(t) + \frac{j}{\pi t} \right) * u_k(t) e^{-j\omega_k t} \right] \right\|_2^2 \right\} \tag{1}$$

Subject to $\sum_k u_k = f$

where $\{u_k\} : \{u_1, u_2, \ldots, u_k\}$ and $\{\omega_k\} : \{\omega_1, \omega_2, \ldots, \omega_k\}$ are shorthand notations for the set of all modes and their center frequencies, respectively. To solve the above-constrained optimization problem, a quadratic penalty term and the Lagrangian multiplier are introduced to convert it into an unconstrained problem as follows:

$$L(\{u_k\}, \{\omega_k\}, \lambda) := \alpha \sum_k \left\| \partial_t \left[\left(\delta(t) + \frac{j}{\pi t} \right) * u_k(t) \right] e^{-j\omega_k t} \right\|_2^2$$

$$+ \left\| f(t) - \sum_k u_k(t) \right\|_2^2 + \left\langle \lambda(t), f(t) - \sum_k u_k(t) \right\rangle \tag{2}$$

where α and λ are penalty parameter and Lagrangian multiplier, respectively. VMD uses the alternating direction method of multipliers (ADMM) to solve the above equation iteratively. Eventually, the signal is disintegrated into K IMFs. The algorithm of VMD is summarized as follows:

Initialize $\{u_k^1\}, \{\omega_k^1\}, \lambda^1, n \leftarrow 0$
Repeat
$n \leftarrow n + 1$
For $k = 1 : K$ do
Update u_k:

$$u_k^{n+1} \leftarrow \underset{u_k}{\arg \min} \; L(\{u_{i<k}^{n+1}\}, \{u_{i\geq k}^n\}, \{\omega_i^n\}, \lambda^n) \tag{3}$$

end for
for $k = 1 : K$ do
update ω_k:

$$\omega_k^{n+1} \leftarrow \underset{\omega_k}{\arg \min} \; L(\{u_i^{n+1}\}, \{\omega_{i<k}^{n+1}\}, \{\omega_{i\geq k}^n\}, \lambda^n) \tag{4}$$

end for
dual ascent:

$$\lambda^{n+1} \leftarrow \lambda^n + \tau \left(f - \sum_k u_k^{n+1} \right) \tag{5}$$

Until convergence: $\sum_k \left\| u_k^{n+1} - u_k^n \right\|_2^2 / \left\| u_k^n \right\|_2^2 < \varepsilon$

2.2 Mode Selection Parameter

VMD decomposes the signal into various modes; the main mode which gives the information of faulty location is selected by comparing the covariance and energy entropy of each mode. Mode having the highest value of covariance and energy entropy is selected as a principal or vital mode. Finally, the defect in the bearing can be identified by the examination of the principal mode's envelope spectrum. The covariance (cov) and energy entropy (H_i) are expressed as follows:

$$\text{cov} = \frac{\sum_{i=1}^n (x_i - \bar{x})(y_i - \bar{y})}{n - 1} \tag{6}$$

$$H_i = -p_i \times \log p_i; \; p_i = \frac{E_i}{E}; \; E_i = \sum_{i=1}^k |c_i(t)|^2; \; E = \sum_{i=1}^n E_i \tag{7}$$

where H_i is energy entropy of ith IMF, E_i is energy of ith IMF, E is sum of energy of n IMFs, and p_i is the percentage of energy of ith IMF in the whole signal energy E.

2.3 Procedure of Fault Diagnosis

The sequence of the process followed in the proposed model of fault diagnosis is illustrated in Fig. 1.

3 Experimentation, Result, and Discussion

Dataset exploited for the investigation and validation of fault diagnosis methodology has been taken from experimentation performed by Society for Machine Failure Prevention Technology [13]. The tests were conducted on bearing test rig equipped with the NICE bearing. The aforementioned dataset is utilized for the illustration of the implementation of the proposed method. However, a signal sample from real machine may also be utilized and the result so generated has a minute difference in frequency analysis. Therefore instead of including unnecessary haziness and for having effective observation bearing with preinstalled fault in a test rig is utilized.

The data has been acquired with 48 kHz of frequency and for the duration of 3 s. In this section, an illustration has been delivered in support of the anticipated diagnostic approach as presented in Fig. 1. Figure 2 helps in visualizing the data sample considered for validation. The signal sample is first decomposed into six IMFs utilizing VMD algorithm as enlightened in Sect. 2.1, and the gathered IMFs are demonstrated in Fig. 3. In this progression, the next step involves the selection of vital mode which provides the essential information in the fault diagnosis process. The selection criteria have been explained in Sect. 2.2 which involves the calculation of covariance and entropy. In support of this, the calculated values of the mentioned parameters are provided in Table 1. Among the six IMFs, the third one is found to be the most appropriate and further analysis based on envelope spectrum has been executed. The basic requirements of this study are the fault frequencies which depend on the operational condition and geometry of bearing; Table 2 shows the same.

Further, the last step mentioned in Fig. 1 is executed on the selected principal mode. Figure 4 demonstrates the envelope spectrum calculated by elementary fast Fourier transform (FFT) over the principal mode, which clearly indicates the presence of inner race fault frequencies. That is, the multiples of the value 118.87 appear in the one-sided spectrum. Hence, the conclusion can be gathered that the inner race catastrophe has occurred in the bearing. Such inferences may be conducted with the proposed method for various types of defects present in the bearing and other rotating components.

Fig. 1 Procedure of the proposed approach

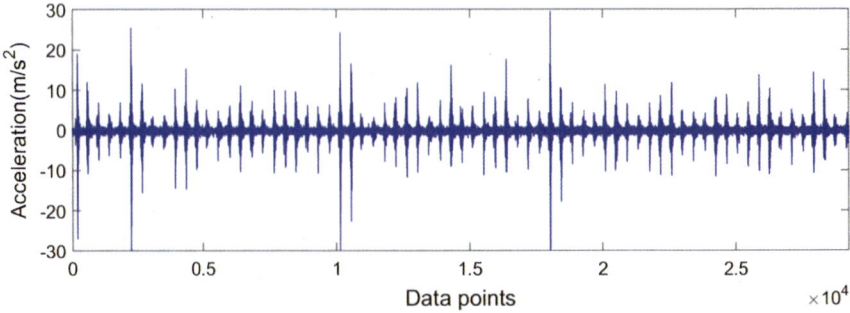

Fig. 2 Sample of bearing vibration signal with inner race defect

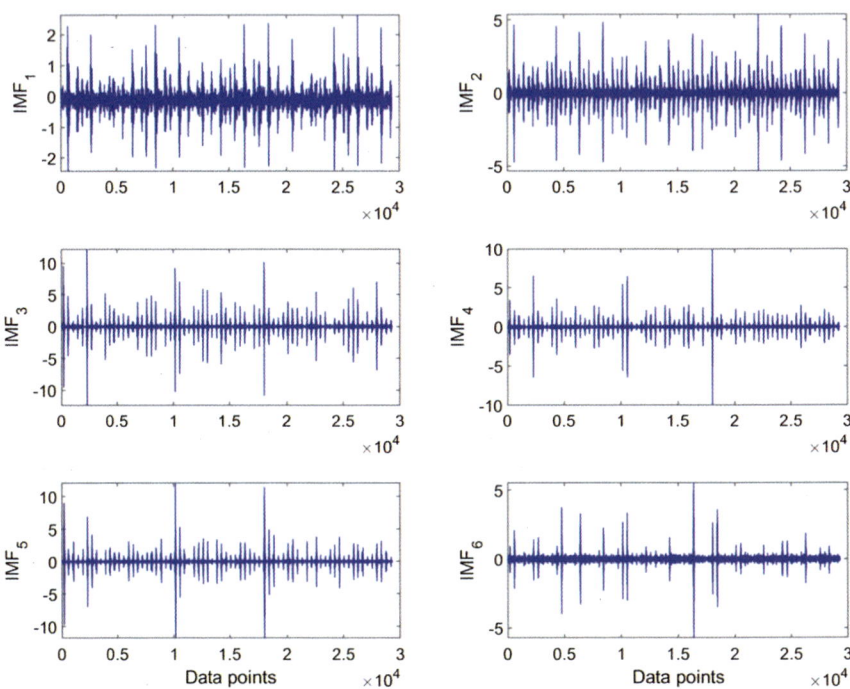

Fig. 3 Extracted IMFs of vibration signal sample envisaged in Fig. 4.1

4 Conclusion

This section summarizes the work done in this paper. Based on the results obtained in Sect. 4, several conclusions are drawn. Finally, the work to be carried out in the future is summarized. The work proposes a bearing diagnosis approach based on a combination of VMD, statistical parameter, and energy entropy of the signal; these parameters give effective result than the correlation and energy ratio. VMD

Table 1 Entropy and covariance calculation corresponding to the considered signal and their IMFs

Product functions derived from VMD	Faulty bearing		Parameter		Selected IMF
	Entropy	Covariance	Entropy	Covariance	
IMF1	0.1885	0.1738			
IMF2	0.3108	0.4489			
IMF3	**0.3635**	**0.8014**	**0.3635**	**0.8014**	**IMF3**
IMF4	0.2753	0.4289			
IMF5	0.3475	0.6393			
IMF6	0.1309	0.1200			

Table 2 Ball pass frequencies for considered experiment

BPFI	BPFO	BSF	FFT
118.87	81.12	63.91	14.837

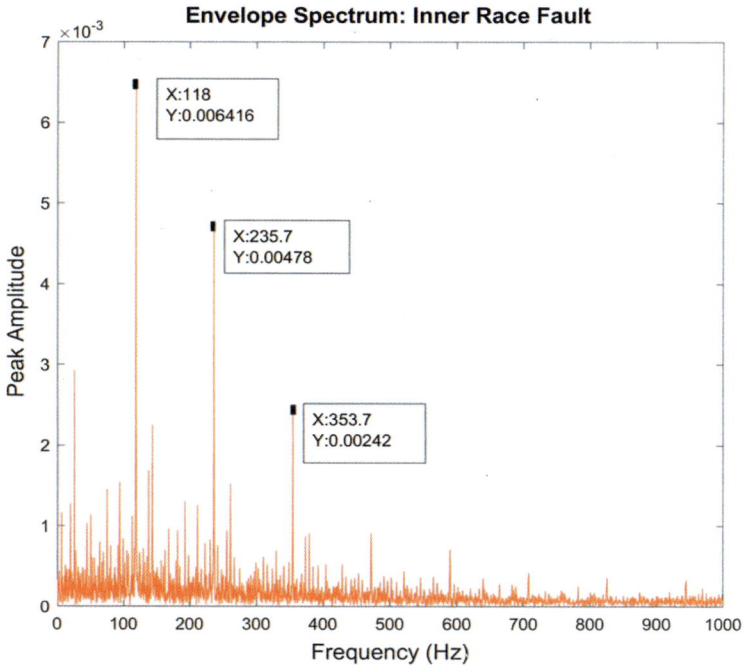

Fig. 4 One-sided envelope spectrum of the considered sample

extracts the modes at faulty location effectively and shows sound result than other decomposition methods like LMD, EMD. It remedied the problem of mode mixing since it considered individual mode, and it gives the precise result since predefined some parameter from the beginning. Moreover, the recommended method is also effective for fault diagnosis in gears and other rotating components.

References

1. Y. Lei, J. Lin, Z. He, Y. Zi, Application of an improved kurtogram method for fault diagnosis of rolling element bearings. Mech. Syst. Sig. Process. **25**(5), 1738–1749 (2011)
2. Y. Ming, J. Chen, G. Dong, Weak fault feature extraction of rolling bearing based on cyclic Wiener filter and envelope spectrum. Mech. Syst. Sig. Process. **25**(5), 1773–1785 (2011)
3. X. Wang, Y. Zi, Z. He, Multiwavelet denoising with improved neighboring coefficients for application on rolling bearing fault diagnosis. Mech. Syst. Sig. Process. **25**(1), 285–304 (2011)
4. Y. Dong, M. Liao, X. Zhang, F. Wang, Faults diagnosis of rolling element bearings based on modified morphological method. Mech. Syst. Sig. Process. **25**(4), 1276–1286 (2011)
5. L. Cui, J. Wang, S. Lee, Matching pursuit of an adaptive impulse dictionary for bearing fault diagnosis. J. Sound Vib. **333**(10), 2840–2862 (2014)
6. D. Siegel, H. Al-Atat, V. Shauche, L. Liao, J. Snyder, J. Lee, Novel method for rolling element bearing health assessment—a tachometer-less synchronously averaged envelope feature extraction technique. Mech. Syst. Sig. Process. **29**, 362–376 (2012)
7. B. Li, P.L. Zhang, D.S. Liu, S.S. Mi, G.Q. Ren, H. Tian, Feature extraction for rolling element bearing fault diagnosis utilizing generalized S transform and two-dimensional non-negative matrix factorization. J. Sound Vib. **330**(10), 2388–2399 (2011)
8. N.E. Huang, Z. Shen, S.R. Long, M.C. Wu, H.H. Shih, Q. Zheng, H.H. Liu, The empirical mode decomposition and the Hilbert spectrum for nonlinear and non-stationary time series analysis. Proc. R. Soc. Lond. A: Math. Phys. Eng. Sci. **454**(1971), 903–995 (1998)
9. Z. Wu, N.E. Huang, Ensemble empirical mode decomposition: a noise-assisted data analysis method. Adv. Adapt. Data Anal. **1**(01), 1–41 (2009)
10. J.S. Smith, The local mean decomposition and its application to EEG perception data. J. R. Soc. Interface **2**(5), 443–454 (2005)
11. Y. Wang, Z. He, Y. Zi, A demodulation method based on improved local mean decomposition and its application in rub-impact fault diagnosis. Meas. Sci. Technol. **20**(2), 025704 (2009)
12. K. Dragomiretskiy, D. Zosso, Variational mode decomposition. IEEE Trans. Sig. Process. **62**(3), 531–544 (2014)
13. E. Bechhoefer, Condition based maintenance fault database for testing of diagnostic and prognostics algorithms (2013)

Performance Enhancement of Reactor Building Containment Isolation System by Use of Direct-Acting Solenoid Valves

Jigar V. Patel, Kaustubh Gadgil, C. Sengupta and P. Sumanth

Abstract Containment isolation system is an important safety system of nuclear reactors, designed to reduce the radiological consequences and risk to the public from various postulated design basis accident conditions. Reliability and performance of containment isolation system to assure containment integrity during power operation and other stated activities are ascertained by routine functional checks on system components. The work presented here focuses on performance and reliability enhancement of containment isolation system of Dhruva Research Reactor, BARC, Mumbai. The enhancement was achieved by replacing pilot-operated solenoid valves (SVs) with direct-acting SVs in the logic circuit. On several occasions, during routine functional checks of the logic circuit of isolation damper actuators, isolation dampers were observed to be partial/incompletely closed, leading to uncertainty of containment integrity on actual demand conditions. Faults in the logic circuit of isolation damper actuators were analyzed, and the root cause was identified to be lack of minimum operating differential pressure (MOPD) available across the pilot-operated SVs during routine checks. Based on manufacturer's product literature [1], direct-acting SVs were found to be most suitable for use in the concerned application. The direct-acting SVs were qualified for use in the logic circuit of isolation damper actuators, after a series of performance and environmental tests. The performance of the logic circuit of isolation damper actuators with direct-acting SVs was improved along with no reported failure occurrence. A fault tree analysis [2] was carried out based on the failure data of pilot-operated and direct-acting SVs for a period of 5 years, to compare the failure probability of the logic circuit of isolation damper

J. V. Patel (✉) · K. Gadgil · C. Sengupta · P. Sumanth
Bhabha Atomic Research Center (BARC), Mumbai, India
e-mail: jigar@barc.gov.in

K. Gadgil
e-mail: kaustubh@barc.gov.in

C. Sengupta
e-mail: csen@barc.gov.in

P. Sumanth
e-mail: spanyam@barc.gov.in

© Springer Nature Singapore Pte Ltd. 2020
P. V. Varde et al. (eds.), *Reliability, Safety and Hazard Assessment for Risk-Based Technologies*, Lecture Notes in Mechanical Engineering,
https://doi.org/10.1007/978-981-13-9008-1_57

actuators. The failure probability of the logic circuit of isolation damper actuators after installation of direct-acting SVs was found to be improved by a decade.

Keywords Containment isolation system · Dampers · Dhruva Research Reactor · Direct-acting · Fault tree analysis · Failure probability · Fault tree analysis · Isolation dampers · Minimum operating differential pressure (MOPD) · Nuclear reactor · Pilot-operated · Pneumatic ladder circuit · Solenoid valves (SVs)

1 Introduction

A/C and ventilation system of typical nuclear reactor supplies fresh, clean, and conditioned air to the reactor building. This is achieved by a set of supply and exhaust fans, which maintain a continuous air flow through reactor building. Supply fans take air from the atmosphere, and exhaust fans continuously remove air from the reactor building to stack through a bank of high-efficiency particulate air (HEPA) filters. Containment isolation system is an important part of ventilation system of nuclear reactors, designed to achieve positive isolation for reducing the radiological consequences and risk to the public from various postulated design basis accident conditions. Reliability and performance of containment isolation system to assure containment integrity during power operation and other stated activities are ascertained by routine functional checks on system components.

Containment isolation system of Dhruva Research Reactor, BARC, Mumbai, consists of two butterfly dampers provided in the supply duct and exhaust duct of the reactor building ventilation system. The functioning of containment isolation system can be broadly segregated into three stages, viz. (1) trip generation: instruments which continuously monitor critical parameters like reactor building pressure abnormally high or low and abnormally high exhaust air gamma activity and send initiating signal to the logic circuit of final actuation devices; (2) logic circuit: pneumatic ladder circuit consisting of SVs based on 2/3 coincidence logic to operate final actuation device; and (3) final actuation devices: pneumatic cylinder actuator type butterfly dampers (air to close and dead weight to open) in the supply side and exhaust side of the reactor building. The study presented here focuses on problems persisting in the logic circuit of isolation damper actuators which consisted of two-way pilot-operated SVs and modifications implemented in the logic to improve the reliability of the system.

Figure 1 shows a schematic of the logic circuit consisting of 12 numbers of SVs arranged in a pneumatic ladder to facilitate the closing of isolation dampers in 2/3 coincidence sequence. Six numbers of two-way normally closed (NC) SVs form the supply line, and six numbers of two-way normally open (NO) SVs form the exhaust line of the logic circuit. The SVs at supply side are energized to open and at exhaust side energized to close. During normal operation of the system, i.e., isolation dampers in open position, all the SVs are energized. These SVs are de-energized on registration of any two channels of the following conditions: (i) reactor building high

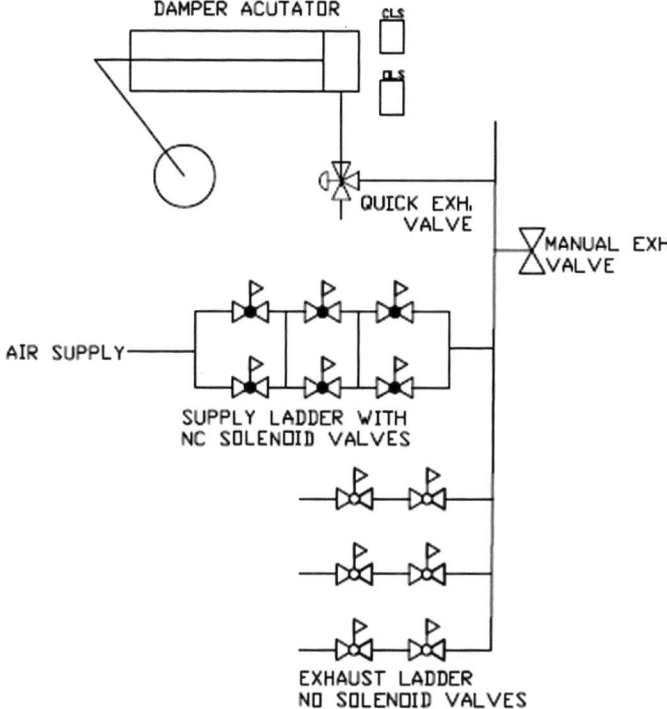

Fig. 1 Pneumatic damper actuator circuit consisting of ladder of solenoid valves

pressure, (ii) reactor building low pressure, and (iii) high exhaust air gamma activity. When SVs are de-energized, the supply line is isolated and the exhaust line is open to the atmosphere causing the isolation damper to close by its dead weight. Isolation dampers fail to close or close partially on the following faults: (i) incomplete isolation of air supply from any one supply path due to fault in SVs on the supply line or (ii) block in two or all three exhaust lines due to the failure in SVs on the exhaust line.

Routine functional checks on the logic circuit of isolation damper actuators are carried out to ascertain reliability and performance of containment isolation system and thereby assure the integrity of containment on actual demand conditions. On several occasions, during routine functional checks on the logic circuit of isolation damper actuators, isolation dampers were observed to be partial/incompletely closed. Analysis of faults, identification of the root cause, and modifications applied in the logic circuit of isolation damper actuators are described in the proceeding sections. The comparison of the failure probabilities in the logic circuit consisting of pilot-operated SVs and direct-acting SVs using fault tree analysis [2] is also depicted in the trailing section.

2 Fault Analysis, Identification, and Rectification in the Logic Circuit of Isolation Damper Actuators

2.1 Classification of Solenoid Valve: Pilot-Operated and Direct-Acting [3]

SVs are basically electromechanical ON/OFF devices in which a valve plunger is moved by an electromagnetic field of the solenoid coil, thereby opening/closing the valve plug. Higher flow rate of fluid through the valves demands larger orifice size of the valves; hence, the size of valve plunger and solenoid coil increases. Predominantly, pilot-operated SVs were designed to meet the criteria of larger flow rates at reduced coil size by introducing a pilot stage in the valve. Typical schematic of pilot-operated SVs (two-way normally closed) is shown in Fig. 2. When the solenoid coil is energized, a pilot plunger is moved by the magnetic field of the coil. The movement of pilot plunger creates a differential pressure across the main plunger forcing it to operate, and hence, the valve is opened. When the coil is de-energized, differential force created due to pilot plunger is removed; pressure at inlet port assists the spring force to restore the normal condition of the valve by moving the main plunger back to the closed position. Thus with reduced coil size (low power consumption), higher orifice valves can be designed. The main drawback of these types of valves is the minimum operating differential pressure (MOPD) required across the main plunger. Typically, the MOPD required for proper functioning of the valves is around 2 bar [2]. If the pressure at inlet port is less, the MOPD available across main valve plunger would be insufficient and the spring force alone would not close the valve completely, leading to continuously air leak from the valve.

Fig. 2 Typical schematic of two-way normally closed (NC) pilot-operated solenoid valve [1]

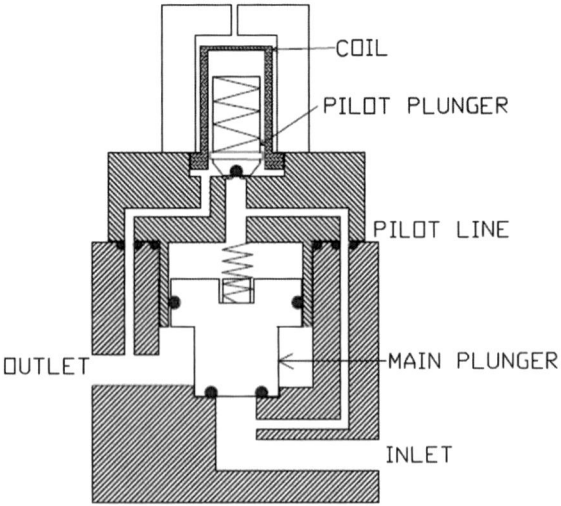

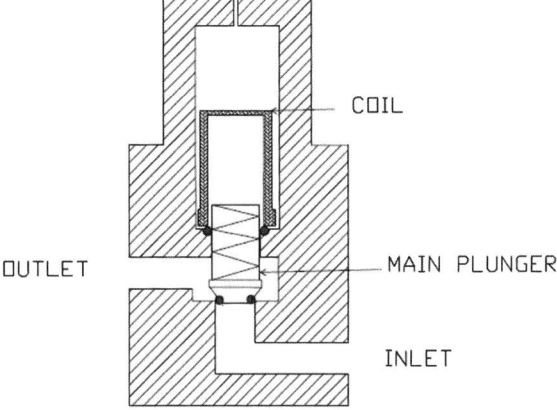

Fig. 3 Typical schematic of two-way normally closed (NC) direct-acting solenoid valve [1]

Like NC-type pilot-operated valves, NO valves also require the MOPD across main valve plunger for proper functioning. In case of NO-type pilot-operated SVs, the inlet pressure assists the spring force to open the main plunger when the valve is de-energized. If inlet pressure is less, the MOPD available across main valve plunger would be insufficient and the spring force alone would not open the valve completely, leading to partial/complete closure of the valve. Hence, pilot-operated SVs should be used in applications where minimum air supply at the inlet port is maintained above the MOPD value at all conditions.

Direct-acting solenoid valves do not require any MOPD for operating the main valve plunger. Typical schematic of direct-acting valves (two-way normally closed) is shown in Fig. 3. Unlike the pilot-operated type, direct-acting valves do not have any pilot plunger. The solenoid coil creates a magnetic field when energized, which forces the main plunger to open position. As soon as the coil is de-energized, the main plunger is moved back to its normal position by the spring action and the valve is closed. Thus, only the spring force is required to move the main plunger of valves. Hence, these types of valves could be used in applications where air supply to the inlet port is zero or minimal. But as discussed earlier, the size limitation of direct-acting valve limits their use for high flow requirements where pilot-operated valves hold the supremacy.

2.2 Fault Analysis, Identification, and Rectification

The logic circuit of isolation damper actuators used in reactor containment system of Dhruva Reactor consisted of pilot-operated SVs. During routine checks of the system, on several occasions, it was observed that isolation dampers failed to close when a trip command was simulated in two trip channels. On one occasion, all the SVs of the logic circuit were refurbished with new O-ring seal kits and installed

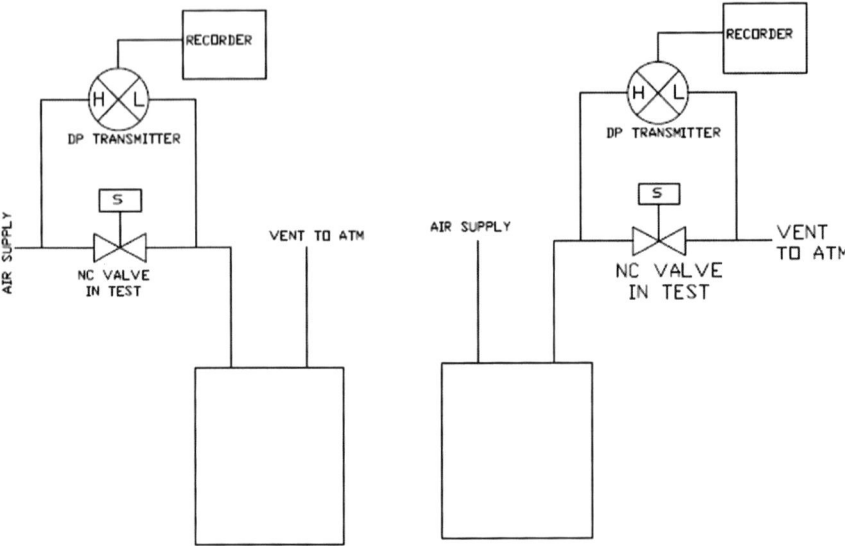

Fig. 4 Test setup for testing solenoid valves

back to check the system functioning. However, the damper was found to be not moving from its open position when the close command was simulated. Fault in the logic circuit was analyzed by monitoring pressure decay of pneumatic actuators of isolation dampers during routine functional checks. It revealed that, when SVs are de-energized, the actuator was exhausted to the atmosphere and air pressure at the inlet port of the SVs gradually reduces to zero. As soon as the inlet pressure reduces to less than the MOPD value, the exhaust line SVs were closed partially and supply line SVs were leak past, leading to the delayed/partial closing of the isolation dampers.

To study this aspect, trials were conducted on a test setup as shown in Fig. 4. The setup consisted of a cylinder, differential pressure (DP) transmitter, recorder, and two numbers of SVs (NC and NO). Differential pressure across the SV under test was recorded for the analysis. For testing NC-type SV, the cylinder was pressurized by energizing the SV and pressure curve was monitored for calculating the time required to completely pressurize the cylinder. When cylinder pressurizes completely, DP across the SV becomes zero. For testing NC-type SV, the cylinder was first pressurized with SV in the energized state (i.e., DP across SV was maximum), and thereafter, SV was de-energized to de-pressurize the cylinder.

Both pilot-operated SVs and direct-acting SVs were tested for performance on the test setup. Figures 5 and 6 show pressurization and de-pressurization graphs recorded for pilot-operated SVs, whereas Figs. 7 and 8 show similar graphs for direct-acting SVs.

Observations revealed that NC-type pilot-operated SVs required around 370 s to fully pressurize the cylinder. During de-pressurization through NO-type pilot-

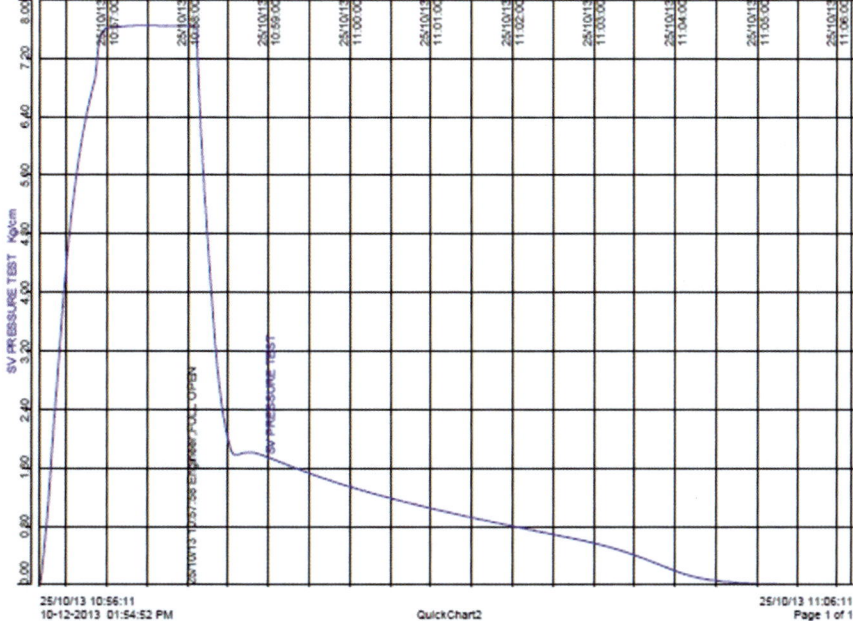

Fig. 5 Pressurization of cylinder with pilot-operated valve

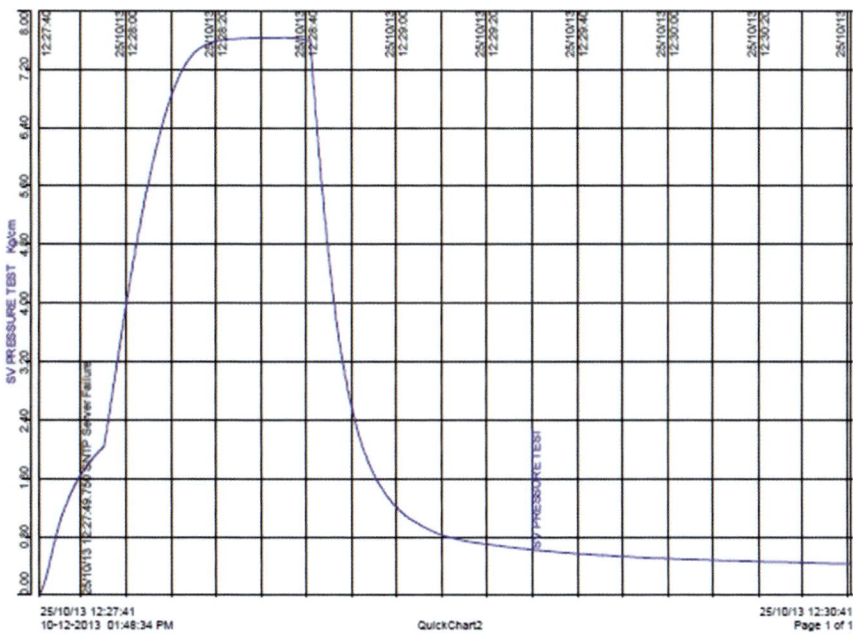

Fig. 6 De-pressurization of cylinder with pilot-operated valve

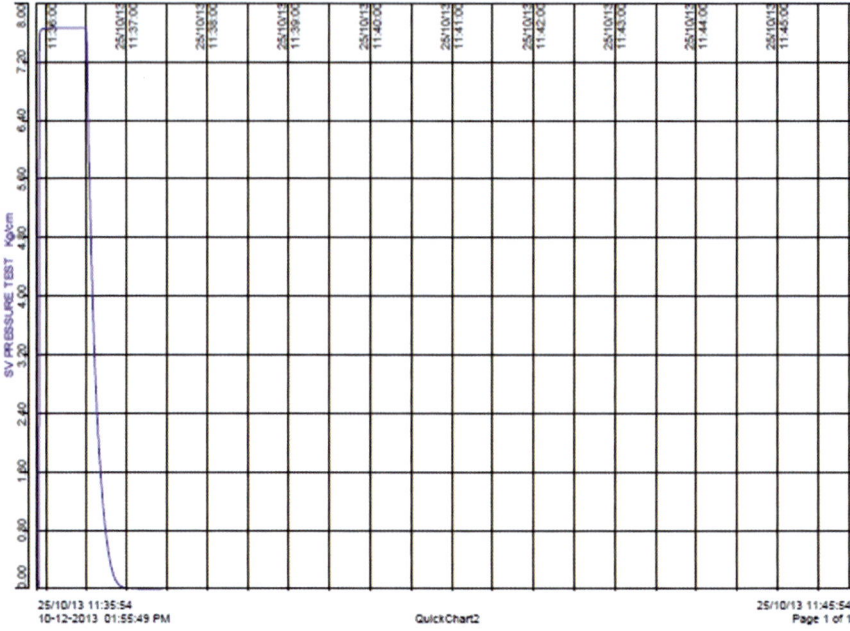

Fig. 7 Pressurization of cylinder with direct-acting valve

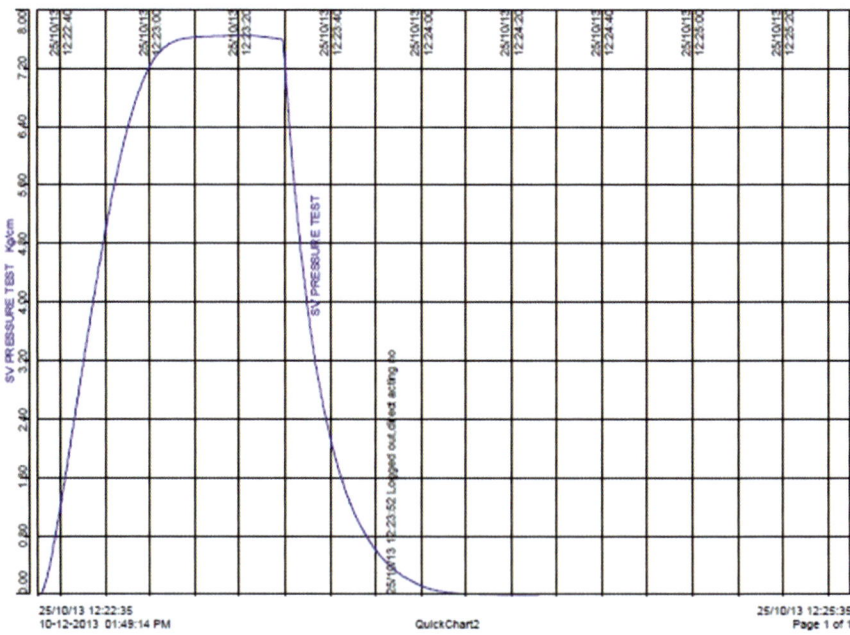

Fig. 8 De-pressurization of cylinder with direct-acting valve

operated SVs, the cylinder pressure quickly decayed to 1.8 kg/cm^2; thereafter, pressure decayed gradually and cylinder could not be de-pressurized completely due to partial/complete closure of the exhaust line SV. The residual pressure in the cylinder even after 7 min was around 0.3 kg/cm^2. The results clearly indicate improper functioning of pilot-operated SVs when the inlet pressure falls below the MOPD value required.

The results as shown in Figs. 7 and 8 revealed that direct-acting SV took just 25 s to fully pressurize the same cylinder, and thereafter, the cylinder could be completely de-pressurized in 55 s. The results revealed that the use of direct-acting SVs was most appropriate for use in the logic circuit of isolation damper actuators, as the pressure at the inlet port of the SVs is expected to be zero or minimal on close demand. Also, based on manufacturer's product literature [1], direct-acting SVs were found to be most suitable for use in the concerned application.

Further, a test was conducted to check the operability of the coil, O-rings, and the solenoid valve at elevated temperature of around 50 °C for one NC type and one NO type of direct-acting SV in an environmental chamber [4]. The functioning of both the valves was satisfactory with no deterioration in the condition of O-rings and solenoid coil. Based on the above tests, direct-acting SVs were qualified for actual site installation and the logic circuit of both the isolation damper actuators was modified. Thereafter, the performance of the containment isolation system of Dhruva Reactor was observed by recording the actual closing and opening times of the isolation dampers during routine functional checks. Damper opening times and closing times improved with direct-acting SVs (refer Figs. 9 and 10), i.e., 9 s opening and 8 s closing, as compared to that with pilot-operated SVs, i.e., 18 s to open and 9 s to close (refer Figs. 11 and 12).

The performance of the direct-acting SVs was monitored for around 3 years, and no failure of these valves was found. The damper opening and closing times were also consistent which enhanced reliability and performance of the containment isolation system.

3 Reliability Enhancement in Ladder Circuit by Use of Direct-Acting Solenoid Valves

A fault tree analysis was done on the logic circuit of isolation damper actuators prior to and after installation of direct-acting SVs [2, 5]. Partial/delayed closing of isolation damper was considered as a failure event considering the safety aspects of Dhruva Reactor. A leak in any one supply line or block in two or all three exhaust lines of the ladder circuit on demand was considered as the initiating event of failure as it would lead to incomplete or delayed closing of isolation damper. Other components of the logic circuit of isolation damper actuator, viz. quick exhaust valve, pressure regulating valve (PRV), air filter regulator (AFR), and associated system piping, had no reported failures; hence, the failure of the isolation dampers was entirely

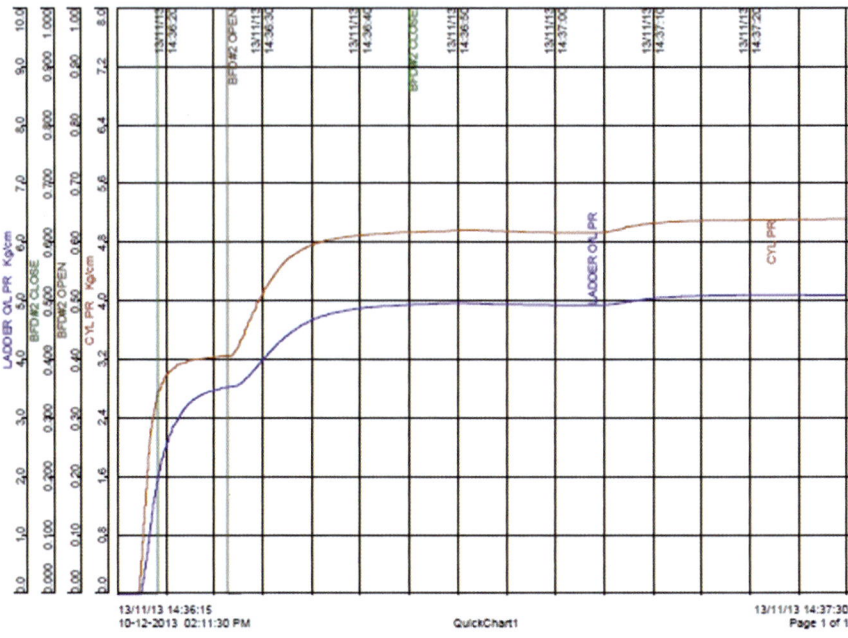

Fig. 9 Isolation damper opening with direct-acting valve

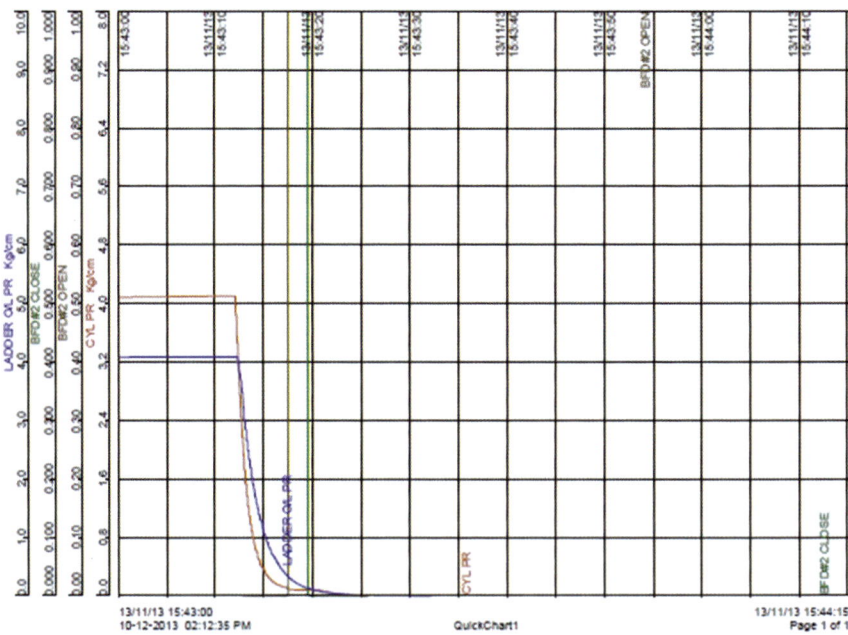

Fig. 10 Isolation damper closing with direct-acting valve

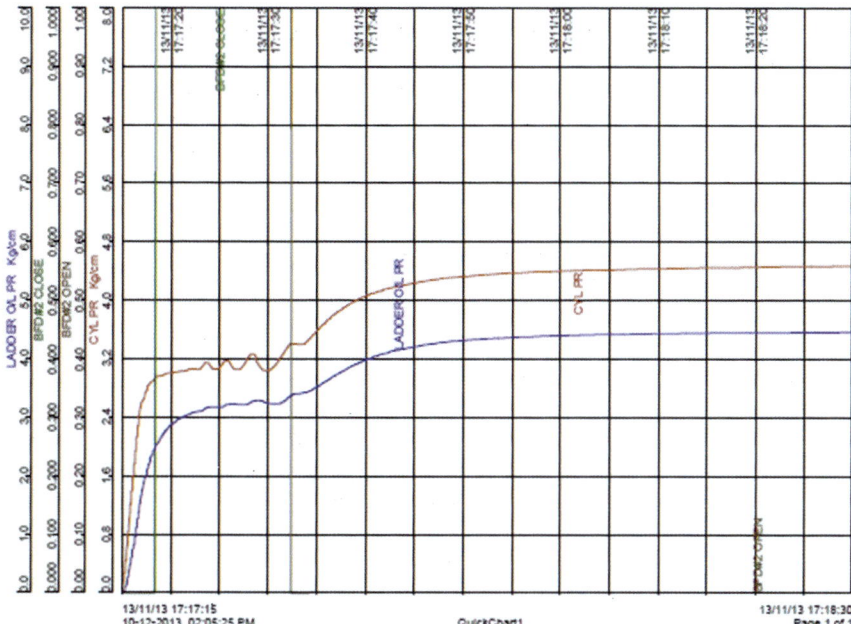

Fig. 11 Isolation damper opening with pilot-operated valve

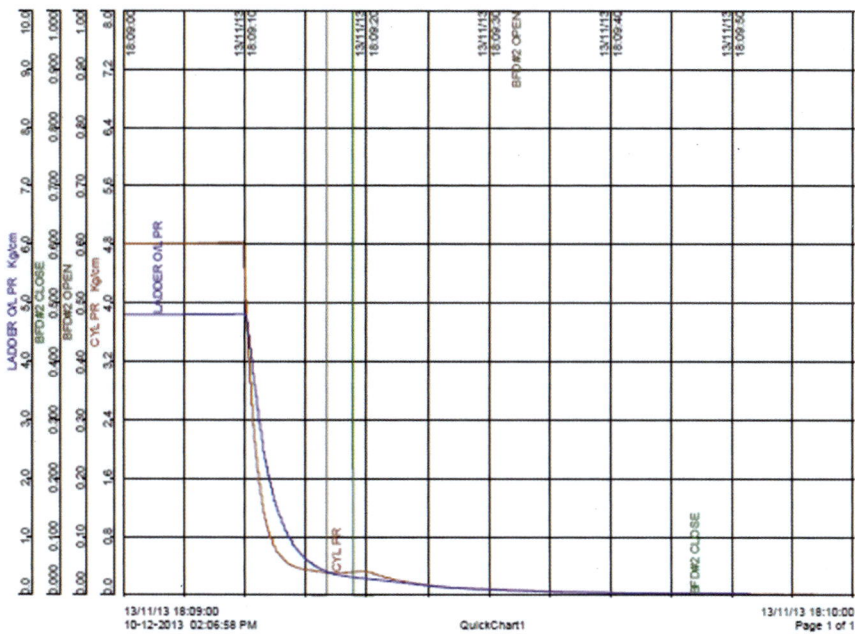

Fig. 12 Isolation damper closing with pilot-operated valve

Table 1 Failure paths in ladder circuit of isolation dampers

Description of path	No. of paths	No. of SVs required to fail in each path	Types of failure in SV
Supply line	8	3	Leak pastness
Exhaust line	4	1	Blockage

Table 2 Failure probability of pilot-operated SVs and direct-acting SVs

Types of SV	Location	Demands/year	No. of failures for 5 years	Failure/demand
Pilot-operated	Supply line	72	9	$1.04e^{-3}$
Pilot-operated	Exhaust line	72	3	$3.47e^{-4}$
Direct-acting	Supply line	72	0	$1.16e^{-4}$
Direct-acting	Supply line	72	0	$1.16e^{-4}$

dependent on faults in the logic circuit. Blockage in supply line SVs, leak pastness in exhaust line SVs, power supply failure, and air supply failure are not considered for failure analysis, as such failures would not lead to unsafe failure of the isolation dampers. A total of 24 numbers of SVs used in the logic circuit of both supply side and exhaust side isolation damper actuators were considered for reliability analysis. Table 1 gives failure paths' ladder circuit which would lead to partial/delayed closing of the isolation dampers.

The SVs in the logic circuit face around 72 demands per year considering routine functional checks and trips on containment isolation system. Plant data for pilot-operated SVs and direct-acting SVs for a period of 5 years was considered for calculating the overall failure probability of the ladder logic. A total of nine pilot-operated SVs in the supply line and three pilot-operated SVs in the exhaust line failed during this period during routine functional checks. Failure data direct-acting SVs was collected for a period of 5 years after modification in the logic circuit, and no failure was reported. Table 2 lists the failure probability of pilot-operated SVs and direct-acting SVs used in the ladder circuit which was utilized for arriving the overall failure probability of the logic circuit of isolation damper actuators.

Figures 13 and 14 show the fault tree of the logic circuit with failure data of pilot-operated SVs. The same fault tree was utilized for calculating the failure probability of the logic circuit after modification with direct-acting SVs. With pilot-operated SVs, the failure probability of the logic circuit was found to be $1.454e^{-6}$ which improved 10 times, i.e., $1.705e^{-7}$ after installation of direct-acting SVs.

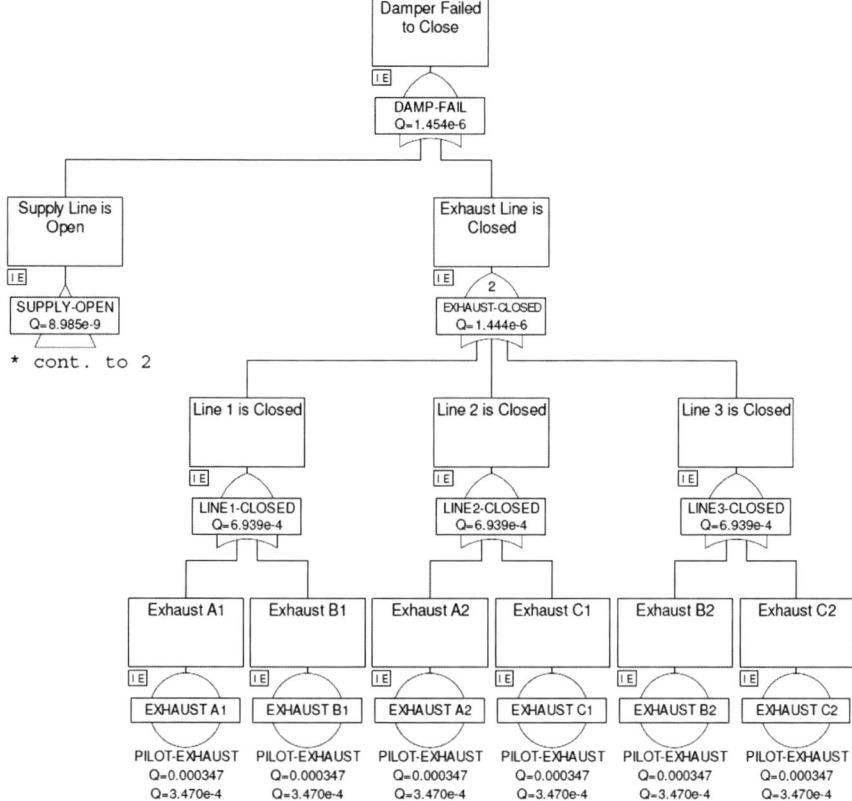

Fig. 13 Fault tree of ladder circuit for isolation dampers (Part 1)

3.1 Conclusions

As pilot-operated valves require minimum operating differential pressure (MOPD) across them for proper functioning, earlier ladder logic of the containment isolation system had inferior performance as pilot-operated SVS failed several times during routine functional checks. By proper selection of SVs, i.e., direct-acting valves, the failure probability of the ladder logic improved by a decade and no failures were reported. Thus, the selection of valves has played an important role in enhancing the performance of the ladder logic and thereby increasing the overall reliability of containment isolation system of Dhruva Reactor.

Acknowledgements This work is supported by Research Reactor Maintenance Division and Research Reactor Services Division of Bhabha Atomic Research Centre (BARC), India.

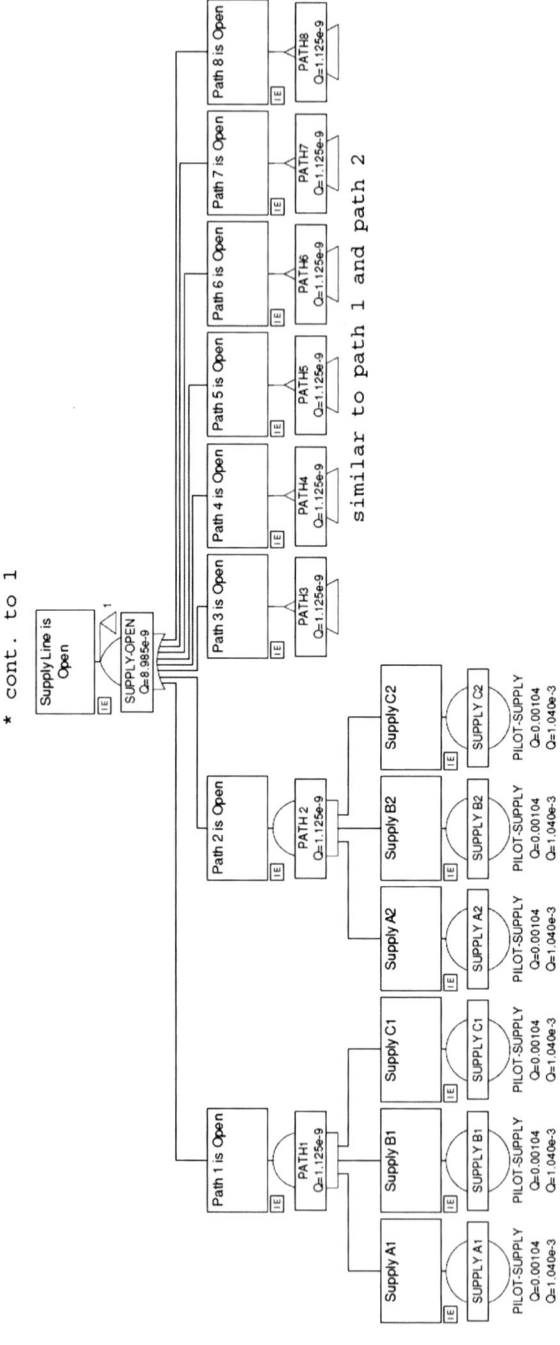

Fig. 14 Fault tree of ladder circuit for isolation dampers (Part 2)

References

1. Product Catalogue by Manufacturer
2. Isograph: Reliability Analysis Software tool
3. NUREG/CR-4819 (Volume: 1): Aging and Service Wear of Solenoid Operated Valves used in Safety Systems of Nuclear Power Plants
4. IS:8935-1985: Standard for Specification for Electric Solenoid Operated Actuators
5. Level 1+ PSA Report of Dhruva Reactor

Surveillance on Nuclear Instrumentation of Dhruva Research Reactor

N. V. Patel, Parag A. Punekar, M. N. Korgaonkar, Sparsh Sharma, N. Ramkumar and P. Sumanth

Abstract Dhruva is a 100 MWth research reactor utilized for the production of radioisotopes and offers irradiation facilities for testing various materials such as nuclear fuel and structural materials, cooled by heavy water, moderated by heavy water, and fueled by natural metallic uranium. The Nuclear Instrumentation (NI) monitors the neutron flux in the reactor using neutron detectors which are located out of core and feeds appropriate signals to the Reactor Regulating System (RRS) and Reactor Protection System (RPS). NI plays an important role in the safety of the reactor. The log power (log P), linear power (lin P), log rate (log R), i.e., the inverse of reactor period and linear rate (Lin-R) are generated using neutron detector signals based on which reactor (power) regulation and protection are achieved. NI also generates signals for post-accident analysis. For covering fourteen decades of in-core neutron flux, triplicated nuclear channels consisting of fission chambers (FCs) and pulse processing electronics, triplicated log and linear safety channels consisting of B-10 lined uncompensated ion chambers (UICs) and DC amplifiers, multi-range DC channels (MRDC) based on UICs for main control room (MCR) and supplementary control panel (SCP) are used. Additionally, for post-accident analysis, high range channels based on UICs are provided. The neutron detectors are located in triplicated basket arrangement around radial beam holes at the nearly same elevation and spread

N. V. Patel (✉) · P. A. Punekar · M. N. Korgaonkar · S. Sharma · N. Ramkumar · P. Sumanth
Bhabha Atomic Research Centre (BARC), Mumbai, India
e-mail: patelnv@barc.gov.in

P. A. Punekar
e-mail: punekar@barc.go.in

M. N. Korgaonkar
e-mail: mnk@barc.gov.in

S. Sharma
e-mail: sparsh@barc.gov.in

N. Ramkumar
e-mail: nram@barc.gov.in

P. Sumanth
e-mail: spanyam@barc.gov.in

© Springer Nature Singapore Pte Ltd. 2020
P. V. Varde et al. (eds.), *Reliability, Safety and Hazard Assessment for Risk-Based Technologies*, Lecture Notes in Mechanical Engineering,
https://doi.org/10.1007/978-981-13-9008-1_58

687

at 120° w.r.t. each other, in out of core region. Considering the importance of NI in reactor regulation and protection and with the fact that reactor design lifetime is much more than the NI design lifetime, commensurate surveillance including preventive maintenance (PM) as per plant technical specification is followed. The surveillance mechanisms, which have evolved over a period of time, based on operational data, help in ensuring intended performance of NI and thereby assuring overall safety. The objective of this paper is to present a detailed account of surveillance of nuclear instrumentation of Dhruva which includes various diagnostic features available and maintenance methodology formulated.

Keywords Diagnostics · Nuclear instrumentation · Preventive maintenance · Surveillance

1 Introduction

Neutron flux measurement is an important safety function in nuclear reactors. Based on neutron flux signals from nuclear detectors, reactor regulating system (RRS) and reactor protection system (RPS) initiate control/protection action through important safety devices for controlling reactor power or shutting down the reactor. To cover 14 decades of in-core flux, 13 detectors, viz. three fission counters (FCs) and 10 B-10 lined uncompensated ion chambers (UICs) are used in Dhruva. FCs are used as detectors for reactor regulation and protection. Using FCs in pulse mode as well as in Campbell mode has helped in covering entire flux range using the same detectors. Pulse mode operation of FCs covers lower flux range and Campbell mode operation covers higher flux range with over two decades of overlapping in range. FCs feed signal to triplicated Campbell channels as well as triplicated pulse channels. Campbell technique, which is a statistical method for avoiding conventional DC channels and its associated problems like drift, low current measurement etc., is deployed in Dhruva which also offers the additional benefit of better gamma discrimination, detector saving, and diverse design. Each Campbell channel generates four signals, viz. linear power, linear rate, log power, and log rate. These four signals from all three Campbell channels are fed to all three channels of RRS. Comparators, averaging units, function generator, auto–manual valve controller forms the basic blocks of the (triplicated) RRS processing circuit that generates valve control signals for the (three) control valves. As Dhruva uses moderator level control for power regulation, detectors for nuclear instrumentation (NI) are installed below moderator level at which criticality is possible. UICs are being used as detectors for triplicated linear safety channels and triplicated log rate safety channels, (single) high range linear safety channel and (single) high range log rate safety channel. One nuclear detector is connected to one nuclear channel except for FC, where one FC feed to one Campbell channel and one pulse channel. Considering the advantage of logarithmic representation of power over six decades using log-amplifiers, it is implemented in log rate and high range log rate safety channels. Linear amplifier is used for processing detector current in case of linear safety and high range linear safety channels. Two multi-range DC (MRDC) channels are used to depict reactor power from 10 W to 100 MW linearly in each decade. In the case of main control room, MRDC channel uses log–antilog design where range changing is accomplished by switching the reference voltages. As a diverse design, supplementary control panel MRDC uses an amplifier with high Meg feedback resistances along with range changing band switches.

In order to meet the requirement of reliable nuclear Instrumentation, in Dhruva, a combination of diversity, redundancy, coincidence, testability, and failsafe features are deployed. Further, to ensure high reliability and availability of NI, during the operational phase, surveillance based on design recommendations and operating experiences and as mentioned in the technical specification of reactor operation is strictly followed. Apart from these, additional surveillance such as taking weekly readings of nuclear channels and RRS, monitoring power supplies, taking periodic characteristics of nuclear detectors is also followed. Observations/data obtained during surveillance are scrutinized and analyzed for identifying any discrepancy or onset development of faults/failures, e.g., degradation in insulation resistance, gas leakage in detectors.

2 Surveillance on Nuclear Instrumentation and Analysis of Surveillance Data

Nuclear instrumentation consists of neutron detectors and associated electronics. Surveillance methodology and requirements differ for neutron detectors and electronics. Surveillances are carried out individually on detectors and electronics as well as detectors and electronics together.

2.1 Surveillance on Nuclear Detectors

10 nos. of UICs and 3 nos. of FCs are used for measurement of neutron flux in Dhruva. For evaluating the performance of these nuclear detectors and to identify any developing faults, surveillance schedules are in place. Insulation resistance (IR) between HV electrode and signal electrode is an essential requirement of gas-filled detectors, particularly UICs used for NI, wherein currents of the order of one pico-ampere have to be measured. Degradation of insulation due to several mechanisms may affect detector performance. Though signal and HV cables have been observed to be unaffected over the operational period of more than three decades, it is understood that it is necessary that periodic measurement of IR of these cables is carried out and documented for reference. HV saturation characteristics of the UICs are taken at maximum permissible power once in two years. Shutdown current measurement

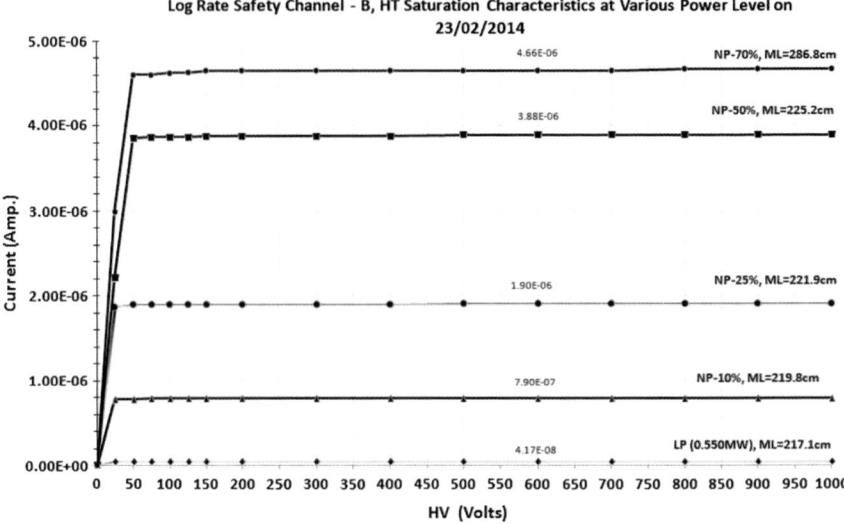

Fig. 1 HV saturation characteristic of UIC of log rate safety channel-B

of UICs is also carried out once in two years. HV saturation characteristics of UICs are taken at various power levels once in five years or as and when required. Characteristics data are analyzed for determining the health of the nuclear detectors. HV saturation characteristics at various power levels of UIC of log rate safety channel-B of Dhruva taken in year 2014 are shown in Fig. 1. Linearity characteristics of UIC of log rate safety channel-B taken in year 2014 is shown in Fig. 2.

Figure 1 shows HV saturation characteristics of UIC of log rate safety channel-B at low power (550 KW), 10, 25, 50, and 70 MW. HV saturation characteristics are taken at a power level by measuring UIC current using electrometer at different HV starting from 0 V up to 1000 V. Figure 2 shows linearity characteristics of log rate safety channel-B which shows UIC current at different power level.

HV saturation characteristics and discrimination bias characteristics of FCs is taken under reactor shutdown conditions once in five year schedule. Figures 3 and 4 show HV saturation characteristics and discrimination bias characteristics of FC of Campbell channel-A.

2.2 Analysis of Surveillance Data of Nuclear Detectors

For reviewing the observations of surveillance, typical guidelines have been framed for establishing healthiness of the UICs so that their usage can be continued. Accordingly, guidelines given in Table 1 are followed.

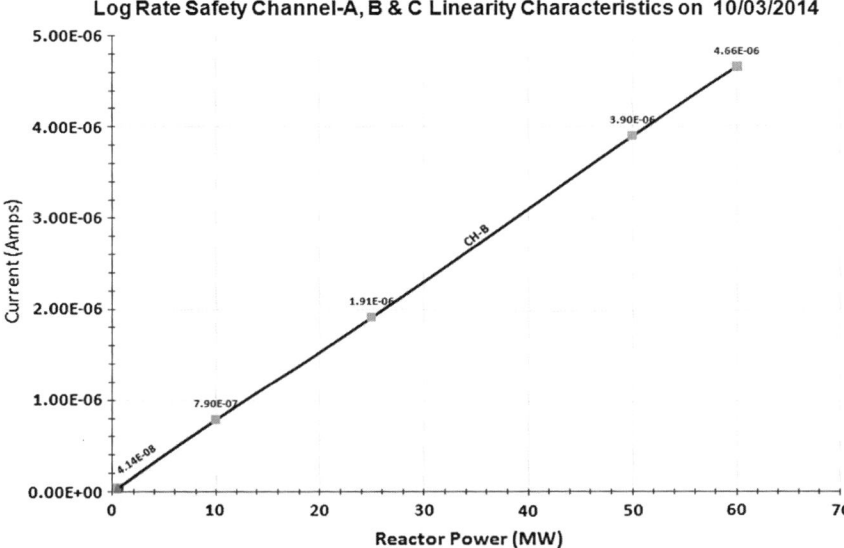

Fig. 2 Linearity characteristic of UIC of log rate safety channel-B

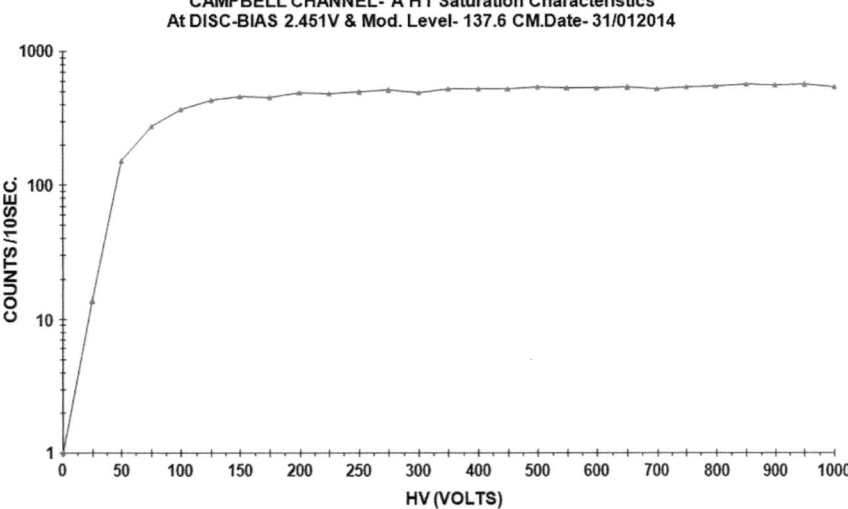

Fig. 3 HV saturation characteristics of FC of Campbell channel-A

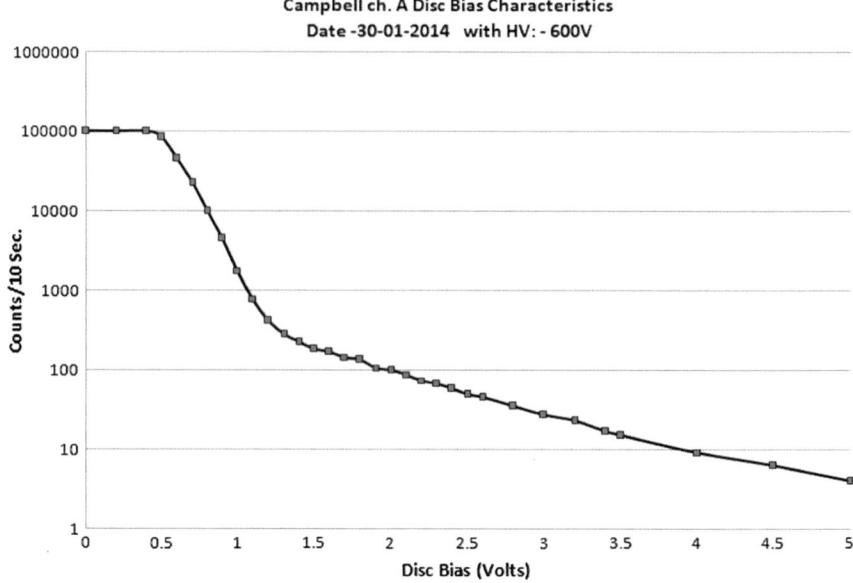

Fig. 4 Discrimination bias characteristics of FC of Campbell channel-A

Table 1 Typical guidelines for continuing usage of the UICs in Dhruva [1]

S. No.	Parameters	Criterion
1	Neutron sensitivity	<50% if other factors are okay
2	90% saturation voltage	<200 V
	Operating voltage	<700 V
3	Errors of expected current versus actual current (i) Minimum range (ii) Maximum range	<50% if expected current is <1 nA <20% if expected current is >1 nA
4	Leakage current	<50% of current at minimum of range if minimum current <1 nA

Based on the criterion mentioned in Table 1, observation data collected though HV saturation characteristics are analyzed and decisions are made whether the UIC is suitable for continuing its usage or needs replacement. Figure 5 shows HV saturation characteristic of UIC of log rate safety channel-C taken at 50 MW in year 2003. The characteristics show that 90% saturation voltage has gone beyond 200 V and at around 700 V HV, there are variations in current. These observations indicate toward loss of integrity of UIC. The UIC was replaced with the spare after observing these characteristics.

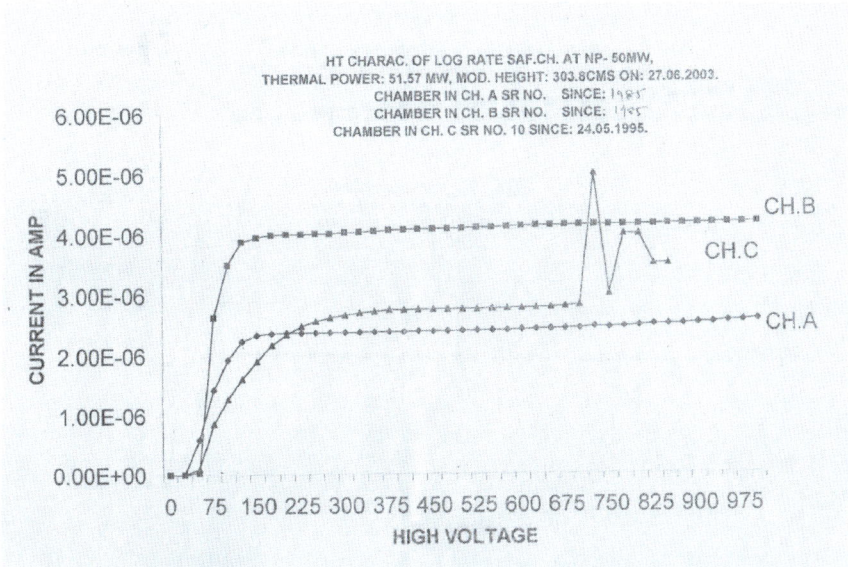

Fig. 5 HV saturation characteristic of UIC of log rate safety channels

2.3 Surveillance on Electronics and Preventive Maintenance

Weekly functional checks and quarterly detailed calibration checks are performed on electronics part of nuclear instrumentation. Using in-built test facilities, functional checking of trips and alarms, generated by nuclear channels, are checked online during reactor operation on a weekly basis. These functional checks are important for enhancing confidence in the ability of nuclear instruments to bring the reactor to the safe state when demanded upon. During quarterly detailed calibration checks, calibration of processing electronics is checked for entire range and calibration is ensured to be within the defined tolerance band. Set-points for trips and alarms are also checked and ensured to be as desired. Circuits required for generating test signals are in-built to the electronics. Special test circuits are in-built to Campbell channels, which process signals from fission chambers in Campbell mode, which is also known as mean square mode. 10 and 100 kHz signals, one at a time, are fed though HV cable and received at channel end through signal cable. This test confirms proper connectivity between detector and channel which are more than 100 m apart. Provision for in-built test circuit is also there for calibrating Campbell channels using 10 and 100 kHz signal for log module and linear module, respectively.

The RRS varies the reactivity of the reactor in such a way as to maintain the average output of the fission chambers in the three channels at a value corresponding to the desired power level. If any one of the fission chambers or Campbell channels is faulty, the mean signal delivered to the next component in the channel may be in error and the reactor would then be controlled at a wrong power level. To avoid the consequences of such failures, comparators are incorporated at various stages in the RRS. To benefit by the advantages of a triplicated system and yet to have the three final control signals of the same magnitude, comparators are used to average the signals from the three channels. Functional checking of comparators by feeding test signal using in-built test facilities is carried out on weekly basis as part of trips and alarms checks of RRS where in disagreement in a signal of a channel is simulated with respect to other two channel. During quarterly checks, calibration of all comparator cards is carried out and set-points for disagreement levels are verified.

During preventive maintenance of NI which have their input as low current, its important characteristics like high input impedance, low input offset or leakage current, low drift for offset voltage, low operating temperature, and low noise are verified. This is achieved by proper servicing of NI during PM and ensuring cold, clean, dry operating conditions, and tight/proper connectivity. Apart from these, power supply filter capacitors (solid Aluminum Electrolyte) in nuclear channels are replaced on preventive basis at every two years. During calibration checking, cable connectors are ensured to be firmly in place.

To achieve higher availability without sacrificing safety, grouped local coincidence logic is implemented in RPS. For this, related absolute trip parameters of NI are put in a group and total seven groups have been formed for NI parameters. For trip parameters within a group, coincidence logic in RPS is global. But for trips parameters of different groups, coincidence logic in RPS is local. For example, if one parameter in a group crosses trip set-point in one channel (e.g., Group-1 in Channel-A) and some other parameter in some other group in other channel (e.g., Group-2 in Channel-B), then there will be no reactor trip. If one parameter in a group crosses trip set-point in one channel (e.g., Group-1 in Channel-A) and some other parameter in the same group in other channel (e.g., Group-1 in Channel-B), then two channels will give channel trip and the reactor will trip. This coincidence logic has resulted in a reduced probability of spurious reactor trips and it has helped in efficient maintenance management.

2.4 Online Diagnostic Features in NI

For linear safety channels and log rate safety channels, a diagnostic signal of 4 kHz AC is continuously superimposed on HV and received through signal cable at channel after it has made its way through detector due to detector capacitance. This 4 kHz test signal is filtered from the actual signal and measured and compared at channel end for ensuring the proper interface between detector and electronics. These 4 kHz test serves as continuous online diagnostics for linear safety and log rate safety channels. If 4 kHz signal is not received with adequate strength at the channel then the reactor trip is initiated from that channel. This feature ensures continuity of signal, high-voltage cables and connectors of the detector. Short or open circuit of any of the cables or improper detector connection actuates a trip circuit and "normal" indication on front panel goes off [2].

Apart from this, to ensure the healthy operation of the channel, a reference current is injected into the input of the DC amplifier. This signal provides an on-scale reading by devices used for measuring output signal. In case DC amplifier fails to read this current, a trip signal is initiated and "Normal" indication on front panel goes off. This minimum signal diagnostic is implemented in linear safety channels.

HV supply in Campbell channels, linear and log rate safety channels is continuously monitored by in-built circuits. If HV decreases below set-point then reactor trip is initiated from that channel. Continuous monitoring of DC supply and their ripple content is provided in for RRS. If DC supply varies beyond set-point or ripple content increases beyond then alarm is generated. Apart from these, continuous trend monitoring of nuclear channel and RRS parameters is available for manual comparison among signals, identifying abnormal changes in signal and post-event analysis.

3 Conclusions

By adopting proper methodology and adequate surveillance and diagnostics, required reliability and availability for critical systems such as nuclear instrumentation is achieved. For nuclear detectors, it is very important to identify any developing faults and take remedial actions for ensuring their intended performance. Based on surveillance data, around 16 UICs have been replaced in Dhruva in last three decades before they completely fail to carry out their function. Design improvements in UICs have been implemented based on faults observed during surveillance and their root cause.

Acknowledgements This work is supported by Research Reactor Maintenance Division of Bhabha Atomic Research Centre, India.

References

1. Report on Performance of Electronic Division make Neutron Ion Chambers used in Dhruva
2. Design Manual of Nuclear Instrumentation of Dhruva

Reliability Estimation of Reaction Control System of a Flight Vehicle Using Bayesian Networks with Multilevel Data

S. Muthu Kumar, P. Subhash Chandra Bose, R. A. Srivardhan and C. S. P. Rao

Abstract A Reaction control system (RCS) is a spacecraft system that uses number of thrusters to provide attitude control of flight vehicles. RCS thrusters give small amounts of thrust in any desired direction or combination of directions and capable of providing torque to allow control of rotation (roll, pitch, and yaw). Reaction Control system consists of many flight critical components arranged in series configuration. Even one subsystem failure, leads to total mission failure. Hence RCS need to be highly reliable system. It is very expensive to conduct more number of system level tests to gather reliability information about system. Hence the data drawn from component level, subsystem level along with expert opinion is used for reliability estimation. In this paper, Bayesian network modelling was carried out for estimating the reliability of RCS system considering the dependency between system and components. Bayesian Inference was carried out using Markov Chain Monte Carlo (MCMC) simulations and results are compared with Junction tree based exact inference algorithm. MATLAB code is developed to estimate the reliability of RCS system.

Keywords Reaction control system · Propulsion system reliability · MCMC · Bayesian networks · Bayesian inference · Multilevel data

1 Introduction

Reaction control system is a complex system and consists of multiple components which are arranged in a series configuration. Each component/subsystem of RCS performs the intended function for the specified duration to the specified during flight;

S. Muthu Kumar (✉) · R. A. Srivardhan
DRDO, Hyderabad 500058, India
e-mail: muthu.kumar@rcilab.in

P. Subhash Chandra Bose · C. S. P. Rao
Department of Mechanical Engineering,
National Institute of Technology, Warangal 506004, India

© Springer Nature Singapore Pte Ltd. 2020
P. V. Varde et al. (eds.), *Reliability, Safety and Hazard Assessment for Risk-Based Technologies*, Lecture Notes in Mechanical Engineering,
https://doi.org/10.1007/978-981-13-9008-1_59

otherwise, there is a risk to the mission and its payload. To ensure the failure-free operation of RCS, adequate attention needs to be paid for reliability in the early stage of design and development. Initially, reliability is predicted based on historical data. Later, system reliability is estimated using the field and also system-level test data to compare with predicted reliability. In development phase of RCS system, conducting full-level test is difficult due to cost and time constraints. Hence, the reliability of the system can be computed from other sources of information. Component-level historical data is available with component suppliers at various conditions. Other sources of data are acceptance test and qualification test data, and maintenance data of a similar system can also be used. In addition to above, available expert knowledge, simulation results at component, subsystem and system level can be considered. Real challenge is to combine these data available from various sources to learn about the reliability of the full system. This methodology of using data from various sources to determine system reliability is called multilevel information integration.

Traditionally, reliability block diagram and fault tree methods are the most commonly used tools for system reliability modelling. These tools do not consider the dependences/failure relationship between the system and its components in modelling. The functional interactions are unknown for the newly developed system in which some uncertainties may present. Bayesian network model is a powerful tool used for modelling of the system with uncertainties in the system using probabilistic methods. Conditional probabilities are used, which are capable of giving the failure relationship between the components for a complex system. For example, a Bayesian network can be used to represent the probabilistic relationships between diseases and symptoms. If and only if symptoms are given, the network can be used to compute the probabilities of the presence of various diseases. Hamada et al. [1] discussed fault tree modelling of a complex system using multilevel data and further suggested to use a fully Bayesian approach to model complex system, when considering dependency instead of using fault tree. Guo et al. [2] proposed a novel methodology for reliability estimation of complex multicomponent systems by combining multiple sources of data. Proposed methodology considers functional relationship between components, and the system eliminates the errors naturally during combination of data at multilevel. Zhai et al. [3] described the advantages of Bayesian networks in reliability analysis like modelling flexibility, easy derivation of mathematical formula and accurate reasoning, etc. Liu et al. [4] estimated prior probabilities by synthesizing the expert opinion. Zhou et al. [5] explained a case study in which discrete and continuous scenarios and system with multiple states. Yontay et al. [6] described methodology to find the posterior probabilities for both complete data and incomplete data. The author used MCMC technique to find the posterior to the system, subsystems and components. Wilson et al. [7] proposed a model to estimate the system reliability when non-availability of conditional probabilities and also he proposed another method when additional data is available.

2 Bayesian Network for Reliability Modelling

A Bayesian network (BN), also called as Bayesian belief network (BBN), is a graphical model to characterize the probabilistic failure relationship between the component and systems. They are very useful for modelling most mission and safety-critical systems when the limited information about system is available. It is a valuable tool to study the reliability of systems at the early stage of design. Further BN allows combining various types of data from multiple sources. The input data may be obtained from various experiments, simulations, historical data and expert opinion, etc. BN has provision to update the currently available new information B and efficient in the building of reliability models with less computational effort. Hence, BN is one of the best tools for aiding decision-making process due to its quick response in the computation of results.

A BN model comprises of two main parts, i.e. qualitative part and quantitative part. Qualitative part consists of directed acyclic graph (DAG) where nodes represent the random variables. The random variable could be either discrete or continuous variable. The relationships between nodes are expressed by directed arcs. Nodes without directed arc towards them are called root or parent nodes. Nodes with directed arrows are called child node as shown in Fig. 1.

Quantitative part consists of conditional probability tables for each node. Every parent node will be quantified by the marginal probability distribution. Every child node will be quantified by conditional probability table given the value of parent node. Relational ships between child and parent nodes are expressed by conditional probability tables. *Typical BN* is represented by pair of nodes and edges, i.e. $G = ((V, E), P)$, where V is a set of variables or nodes, $V = \{X_1, X_2,\ldots, X_n\}$ and E is a set of edge (arcs or links) and P is a probability. Let $\text{pa}(X_i)$ to be a set of all parents of variable X_i. The conditional probability distribution attached to variable X_i is represented as $P(X_i|\text{pa}(X_i))$. The joint probability distribution of all variables

Fig. 1 Bayesian network

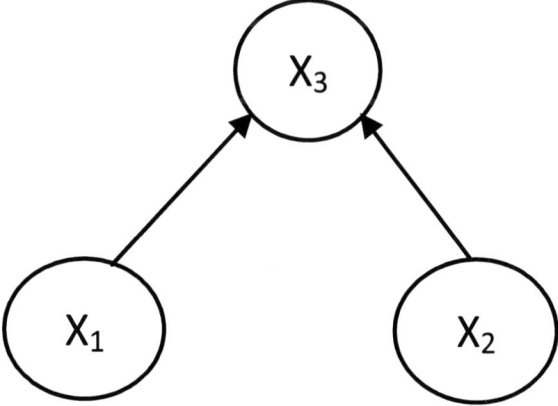

specified in V can be constructed from the conditional probability distributions as shown mathematically below in Eq. (1)

$$P(X_1, X_2, X_3, \ldots, X_n) = \prod_{i=1}^{n} (PX_i/pa(X_i)) \tag{1}$$

Figure 1. Consists of three nodes, namely X_1, X_2 and X_3. The joint probability distribution of three-node BN is given by Eq. (2).

$$P(X_1, X_2, X_3) = P(X_3/X_2, X_3)P(X_2), p(X_3) \tag{2}$$

Where $P(X_1)$, $P(X_2)$ are marginal probability distributions, and $P(X_3|X_1, X_2)$ is a conditional probability distribution. These values are required for computation of joint probability distribution of a three-variable Bayesian network. These values can be inferred from experimental results, simulation results, failure history or expert judgement.

3   Inference in Bayesian Networks

Once the Bayesian network is created, with the algorithms we can check the inference in the Bayesian network. There are two methods of inference which are exact inference and approximate inference. In exact inference, the distribution of conditional probabilities is computed analytically over the interested variables. Sometimes, it becomes too hard to calculate analytically in that case statistical approximation techniques are used to get a sample from the distribution. The most popular approximate inference technique is Markov chain Monte Carlo (MCMC). MCMC is an algorithm for performing sampling of probability distribution and widely used in the fields of Bayesian statistics, computational physics and linguistics, etc. Performing manual integration of complex hierarchical models is very difficult or almost impossible. In such situation, MCMC is very useful for the computation of numerical approximation of multidimensional integrals. Most commonly used MCMC algorithms are Metropolis–Hastings algorithm, Gibbs sampling and slice sampling. The most popularly used sampling algorithm is Gibbs sampling, and samples are obtained in chronological order from complete conditional distributions which will unite to the joint posterior distribution till parameters of distribution parameters are updated constantly. So, after a certain number of preliminary iterations, the samples drawn from simulation chains can be viewed as if they are from the targeted joint posterior distribution. In the reliability literature, Wilson and Huzurbazar and many authors used MCMC for Bayesian inference. To implement MCMC, Win BUGS statistical software for Bayesian inference [8] is used.

Junction tree algorithm [9] is a popularly used tool for performing exact inference Bayesian networks. In exact inference, nodes are collected within the form of junction tree representation. Inference in JTA is carried out in moralisation, triangulation and message passing. JTA is most suitable for performing inference in Bayesian network having less than 10 nodes. If a number node increases more than ten, approximate inferences can be chosen to improve the efficiency and reduce error.

4   Case Study: Reliability Estimation of RCS System Using Bayesian Network

A typical simplified cold gas-based reaction control system is shown in Fig. 2. RCS consists of multiple thrusters and used for pitch, yaw, roll control of flight vehicles. Cold gas RCS systems are mostly used when total impulse needed is below 350 kgf-sec. The major subsystems of RCS are high-pressure gas tank, pyro valve, regulator, valve and series of thrusters, etc. Gas is filled to the gas tank by fill port and gas isolated by electrically operated pyro valve. When the command is issued by flight vehicle computer, pyro valve gets opened and gas from the tank passes through the pyro valve, filter 2, regulator till valve $V2$. On demand, the computer of launch vehicle issues the command to open the valve 2 and operation of thrusters valve which generates required torque for attitude control of flight vehicle.

In Fig. 2, RCS is a series system, and initially reliability model can be build using fault tree (FT). The translation of fault tree to BN is simple, with the basic events that contribute to an intermediate even represented as parents and children. Bayesian network for RCS is given in Fig. 3. To build the system Bayesian network model, a Bayes Net Toolbox for MATLAB is used in this work.

Then, the joint probabilities of RCS reliability are computed from Eq. 3.

$$P(RCS) = \sum_{I=0}^{1}\sum_{J=0}^{1}\sum_{k=0}^{1}\sum_{l=0}^{1}\sum_{m=0}^{1} P\left(S=1 \middle| \begin{array}{l} V1 = i, GB = j, \\ PV = k, F = 1, R = m, \\ V2 = n, \\ V3 = 0 \end{array}\right)$$
$$\times P(V1 = i, GB = j, PV = k, F = 1, R = m, V2 = n, V3 = 0) \tag{3}$$

All the qualification test data, acceptance test data, ground test data and flight test data of RCS were collected systematically from individual component designers and are summarised in Table 1.

4.1   Computation of Posterior Probabilities of RCS Subsystems

Posterior probabilities of RCS subsystems were computed using MCMC. Win BUGS software was used for computation of posterior probabilities. Due to non-availability

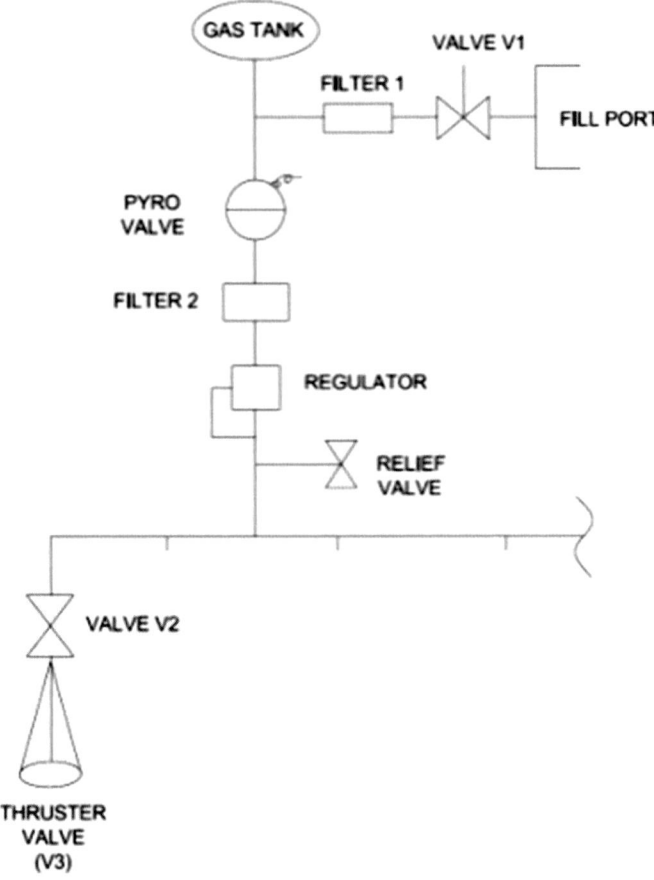

Fig. 2 Typical example of cold-gas RCS

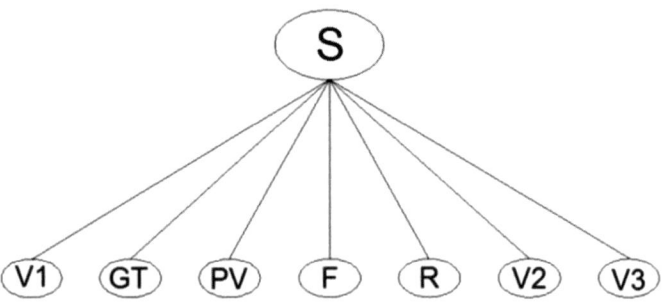

Fig. 3 Bayesian network model of RCS. *V*1—valve, GT—gas tank, PV—pyro valve, *F*—filter, *R*—regulator, *V*2—valve, *V*3—thruster valve

Table 1 Ground-level test data of RCS subsystems

Subsystem	Subsystem tested	No. of failures	No of subsystem passes
Valve (V1)	200	3	197
Gas tank	200	1	199
Pyro valve	200	0	200
Filter	195	0	195
Regulator	200	3	197
Valve (V2)	200	2	198
Thruster valve (V3)	200	0	200

Table 2 Numerical summary of posterior distributions of RCS subsystems

Node	Mean	2.50%	Median	97.50%
$p[1]$-$V1$	0.9809	0.9604	0.9822	0.9949
$p[2]$-GT	0.9899	0.9722	0.9915	0.9987
$p[3]$-PV	0.9952	0.9826	0.9968	0.9999
$p[4]$-F	0.9948	0.981	0.9964	0.9999
$p[5]$-R	0.9797	0.9556	0.9816	0.9944
$p[6]$-$V2$	0.9849	0.9652	0.9863	0.9964
$p[7]$-$V3$	0.9953	0.9813	0.9967	0.9999

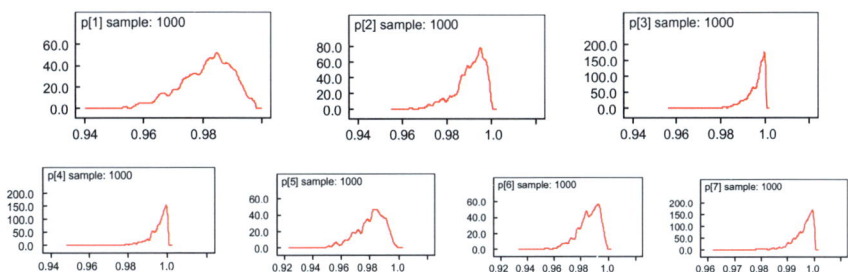

Fig. 4 Computed posterior distribution of RCS subsystems

subsystem prior test information, uniform prior is used for calculations. This prior was combined with test data as shown in Table 1. Summary of subsystem posterior distributions is given in Table 2.

Individual probability plots of subsystems generated by Win BUGS software are shown in Fig. 4.

Table 3 Comparison of results

Sl. No.	Method	Estimated reliability
1	Bayesian inference in BN using Markov chain Monte Carlo algorithm	0.7997
2	Bayesian inference in BN using junction tree algorithm (JTA)	0.8003
3	Bayesian method [10]	0.92

4.2 Elicitation of Conditional Probabilities

Complex relationship between system and components is represented by conditional probabilities. Conditional probabilities are elicited based on expert opinion. During the elicitation process, the team of experts thoroughly reviewed the historical records, simulation results, and experience in similar systems and conditional probabilities were computed.

4.3 Final Reliability Estimation of RCS System

Reliability of RCS is estimated from computed conditional probabilities, subsystem posterior probabilities in previous steps using Eq. (2). To validate the above method, the exact inference was performed using the junction tree algorithm. MATLAB code developed for JTA and results are shown in Table 3.

Table 3 shows the results obtained from two BN methods and Bayesian method. The estimated reliability using MCMC method is 0.7997, whereas from junction tree algorithm is 0.8003. It can be observed that estimated results by MCMC are very close to JTA. Estimated reliability of RCS using Bayesian method [10] values is 0.92. There is a small difference of 0.12 observed between Bayesian networks and Bayesian methods.

5 Conclusions

In this paper, dependency modelling between system and components was carried out using Bayesian networks. Available information about subsystem/component was collected for historical records, simulation results and knowledge of similar systems. Conditional probabilities of each node were obtained from expert opinion. Inference in Bayesian network was carried out using Markov chain Monte Carlo simulations. Code was developed for performing MCMC using Win BUGS software. In case of non-availability of component test level data, uniform distributions are assumed as prior for computation of posterior probabilities of subsystem. Hence, MCMC is very useful tool for quantifying system reliability, when limited, incomplete data. Results obtained from MCMC are validated using junction tree algorithm and Bayesian method. MATLAB code was created for performing computations in JTA.

Elicitation conditional probability in a Bayesian network is a tedious, time consuming and errors may occur in manual elicitation. Hence, it is proposed to develop suitable algorithms, for easier elicitation of conditional probabilities as a part of future work.

References

1. M. Hamada, H. Martz, C.S. Reese, T. Graves, V. Johnson, A. Wilson, A fully Bayesian approach for combining multilevel failure information in fault tree quantification and optimal follow-on resource allocation. Reliab. Eng. Syst. Saf. **86**, 297–305 (2004)
2. J. Guo, A.G. Wilson, Bayesian methods for estimating the reliability of complex systems using Heterogeneous multilevel information. Technometrics, 461–472 (2013)
3. S. Zhai, S. Lin, *Bayesian Networks Application in Reliability Analysis of Multi-State System*, in Proceedings of the 2nd International Symposium on Computer, Communication, Control and Automation (ISCCCA-13). (Atlantis Press, Paris, France, 2013)
4. Y. Liu, P. Lin, Y.-F. Li, H.-Z. Huang, Bayesian reliability and performance assessment for multi-state systems. IEEE Trans. Reliab. **64**(1), 394–409 (Mar 2015)
5. D. Zhou, R. Pan, Chair D. McCarville, M. Zhang, *The Application of Bayesian Networks in System Reliability* (Thesis, Arizona state university, 2014)
6. P. Yontay, R. Pan, A computational bayesian approach to dependency assessment in system reliability. Reliab. Eng. Syst. Saf. **152**, 104–114 (Aug 2016)
7. C.S. Reese, A.G. Wilson, J. Guo, M.S. Hamada, V.E. Johnson, A Bayesian model for integrating multiple sources of lifetime information in system—reliability assessment. J. Qual. Technol. **43**(2), 127–141 (2011)
8. D. Spiegelhalter, A. Thomas, N. Best, D. Lunn, *WinBUGS User Manual MRC Biostatistics Unit* (UK Cambridge, 2005)
9. F.V. Jenson, A.L. Madsen, LAZY propagation; junction tree inference based on lazy evaluation. Artif. Intell. **113**, 203–245 (1999)
10. H.F. Martz, R.A. Waller, E.T. Fickas, Bayesian reliability analysis of series systems of binomial subsystems and components. Techno Metr. **30**(2), 143–154 (1988)
11. A.G. Wilson, C.M. Anderson-Cook, A.V. Huzurbazar, A case study for quantifying system reliability and uncertainity. Reliab. Eng. Syst. Saf. **96**(9), 1076–1084 (2011)

S. Muthu Kumar Scientist, is working in Reliability and Quality Assurance (R&QA), Division of RCI, DRDO, Hyderabad. He has more than 20 years of experience in Quality Assurance and Reliability of Structures, Propulsion systems, Mechanisms of Flight vehicles. He received "Young Scientist of the year" award from DRDO for the year 2007. He is a "Certified Quality Engineer" of American Society of Quality (ASQ). He also received Six Sigma Green Belt certification from Engineering Staff College (ESCI) of India. He is a Life Member of Society for Aerospace Quality and Reliability (SAQR) and National Institution for Quality and Reliability (NIQR). He has published seven papers in National/International Conferences.

Dr. Subhash Chandra Bose currently working as an Associate Professor in Mechanical Engg. Department of National Institute of technology, Warangal, Telangana. He has published more than 20 papers in International Journals. As Principal Co-ordinator, he has handled six Research/Consultancy Projects funded by ISRO and DRDO.

Dr. R. A. Srivardhan Scientist, DRDL, Hyderabad. He is presently working in the area of development of anti-tank guided missile systems. Previous experience includes design and development of aerospace mechanisms, production of propulsion systems and missile system integration for over 24 years. He is a Life Member of High Energy Material Society of India (HEMSI), Indian National Society for Aerospace and Related Mechanisms, Society for Aerospace Quality and Reliability (SAQR), etc., Recipient of AGNI Performance Excellence Award-2008 and Technology Group Award-2012. He has 15 papers published in Journals/Conferences.

Safety Enhancements in Boiling Water Reactor (BWR) Tarapur Atomic Power Station (TAPS-1&2)

Ritesh Raj, Venkata V. Reddy, Sameer Hajela and Mukesh Singhal

Abstract The primary means of preventing and mitigating the consequences of accidents is through 'defence in depth', which is fundamental to the safety of nuclear installations. Defence in depth provides a hierarchical deployment of different independent levels of equipment and procedures in order to maintain effectiveness of physical barriers placed between radioactive materials, the workers, public and the environment, during normal operation and potential accident conditions. The design basis accidents, when postulated unmitigated by engineered safety systems, lead to beyond design basis accidents situations, where core may either degrade or melt due to inadequate core cooling. During these postulated accident scenarios in boiling water reactors if sufficient water is lost from the reactor coolant system (RCS), there could be uncovery and overheating of the core leading to fuel pin cladding oxidation occurs. After accident at Fukushima NPPs, safety review of the design features to handle postulated design extension conditions (DEC) was done for Indian nuclear reactor including TAPS-1&2, and the following safety improvements have been retrofitted.

- Automatic reactor shutdown initiation on sensing seismic activity.
- Pre-inerting of TAPS-1&2 containment.
- Provisions for hookup arrangements for adding cooling water into reactor pressure vessel (RPV) and emergency condenser (EC) system.
- Containment filtered venting system (CFVS) to maintain the containment integrity.

R. Raj (✉) · V. V. Reddy · S. Hajela · M. Singhal
Reactor Safety & Analysis, Nuclear Power Corporation of India Limited,
Mumbai 400094, India
e-mail: riteshraj@npcil.co.in

V. V. Reddy
e-mail: vvreddy@npcil.co.in

S. Hajela
e-mail: shajela@npcil.co.in

M. Singhal
e-mail: singhalm@npcil.co.in

© Springer Nature Singapore Pte Ltd. 2020 707
P. V. Varde et al. (eds.), *Reliability, Safety and Hazard Assessment for Risk-Based Technologies*, Lecture Notes in Mechanical Engineering,
https://doi.org/10.1007/978-981-13-9008-1_60

The above enhanced measures will be used in implementing accident management guidelines to mitigate design extension scenarios. This paper discusses the basis and details of various systems, arrangements and management actions to ensure the effective incorporation of above features.

Keywords DEC · CFVS · Inerting of containment · Management actions

1 Introduction

Tarapur Atomic Power Station (TAPS) is a twin-unit BWR plant with an installed capacity of 2×210 MWe, presently operating at capacity of 2×160 MWe [1, 2]. The reactor is a forced circulation boiling water reactor producing steam for direct use in the steam turbine. The fuel consists of uranium dioxide pellets contained in Zircaloy-2 tubes. Water serves as both the moderator and coolant. Coolant enters from the bottom of the reactor core and flows upwards through the fuel element assemblies where it absorbs heat and gets converted into steam. Steam separators located above the fuel assemblies separate the water particles from the steam–water mixture. This steam is then allowed to pass through a steam dryer to become dry saturated steam before it falls on turbine blades via steam line. During gliding on the turbine blade, this steam loses its energy in moving the turbine. At the end of the series of turbine blades, low-pressure steam is available, which is then passed through the condenser. Low-pressure and low-temperature water is then again allowed to pass through the reactor via two recirculation pumps to the tube side of two secondary steam generators and then to the reactor core. Figure 1 shows the general arrangement of reactor pressure vessel (RPV) and recirculation loops of the TAPS-1&2.

In the era of post-accidents at TMI, Chernobyl and Fukushima, IAEA Safety Guide [3] requires that NPPs should be equipped with independent hardware provisions in order to fulfil the fundamental safety functions, as far as is reasonable for design extension conditions including core melt scenarios (severe accidents), to enhance further defence in depth (DID). This Safety Guide also requires that equipment upgrades aimed at enhancing preventive features of the NPPs should be considered on priority. In the mitigatory domain, equipment upgradation is aimed at preservation of containment function. In order to enhance preventive measures to avoid severe accidents, provisions in terms of hardware and facilities are incorporated in TAPS-1&2.

Following Fukushima event, safety review was carried out for all nuclear power plants in India both by NPCIL and AERB independently. Safety is moving target and is always evolving; with these measures, safety of TAPS-1&2 is enhanced for DEC scenarios.

- Automatic reactor shutdown initiation on sensing seismic activity.
- Pre-inerting of TAPS-1&2 containment.

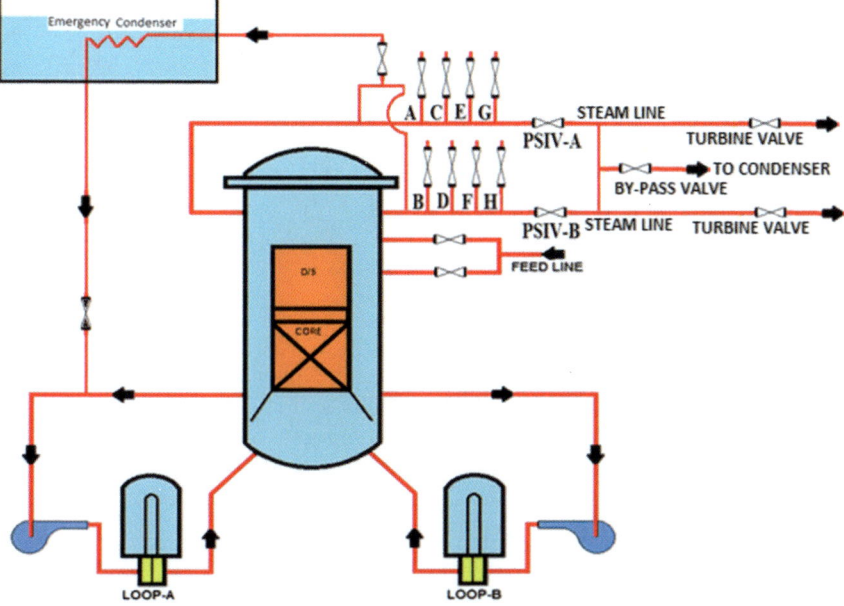

Fig. 1 General arrangement of reactor pressure vessel and recirculation loop

- Provisions for hookups arrangements for adding cooling water into reactor pressure vessel (RPV) and emergency condenser (EC) system.
- Containment filtered venting system (CFVS) to maintain the containment integrity.
- To ensure the reliability of these retrofitted systems, periodic testing and surveillance are carried out. Periodic drills are also performed to train the operator with the functioning of these systems.

2 Improvements in Safety After Accident at Fukushima NPPs

2.1 Provision of Seismic Trip

TAPS-1&2 is located on the west coast of India about 100 km north of Mumbai, and the location is categorized as seismic zone 3 area. By considering low probability of seismic events and to avoid any spurious trip, manual trip on seismic event was earlier considered. Also as per past history of TAPS-1&2, plant did not experience any trip due to seismic event. But after accident at Fukushima NPPs, a trip on seismic event is provided in TAPS-1&2, as a safety enhancement in order to reduce the reactor power to decay heat level as soon as possible after seismic event.

2.2 Nitrogen Inerting System

Primary containment of TAPS-1&2 is provided to contain energy as well as radioactive materials released during an accident. It encloses RPV and other systems connected to it. It comprises of three distinct volumes, namely drywell, suppression pool and common chamber. Each suppression pool contains 1500 m^3 of water. Common chamber is common to both the units and is separated from the suppression pool via rupture diaphragms. Figure 2 shows a schematic diagram of primary containment in TAPS-1&2.

Secondary containment (reactor building) is provided for controlled, elevated release of the building atmosphere under accident conditions, and to provide primary containment during periods when the reactor vessel is open. Secondary containment, also called as reactor building (RB), houses the refuelling and reactor servicing equipment, new and spent fuel storage facilities and other reactor auxiliary or service equipments.

At initial stage of TAPS-1&2 design, primary containment nitrogen inerting was designed as a hydrogen management system for design basis accidents. Considering operational experiences and design basis analysis, containment nitrogen inerting was discontinued after one year of commissioning of plant. During TAPS-1&2, safety upgradation again necessity of nitrogen inerting was felt to ensure safety in design extension scenarios. Also from analysis of severe accident scenarios, without any mitigation, it is found that hydrogen gets generated and hydrogen management system was considered essential to maintain primary containment integrity. By considering scenario similar to accident at Fukushima NPP, it was decided to maintain the primary containment in pre-inerted condition. Therefore, nitrogen inerting system is brought back to prevent hydrogen interaction with oxygen, disallowing hydrogen

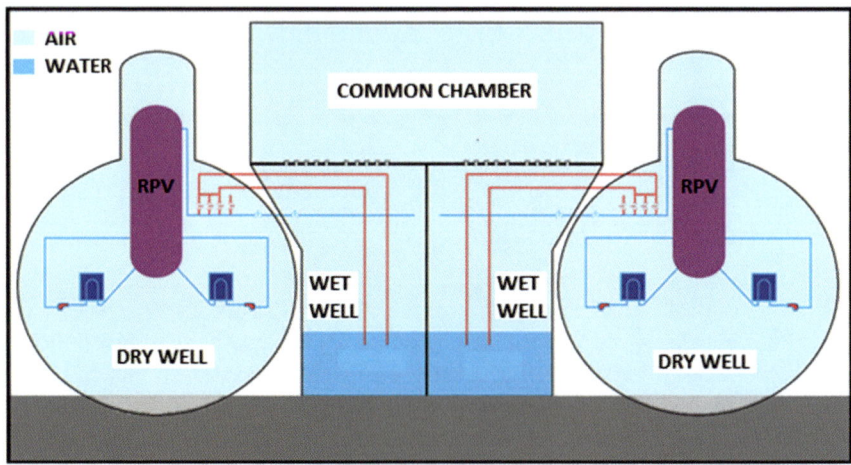

Fig. 2 Schematic diagram of primary containment in TAPS-1&2

burning to protect the containment integrity. The inerting of the primary containment is achieved by filling it with nitrogen such that oxygen concentration is less than 4% by volume, which is the design requirement as per USNRC 10 CFR 50.44 rule on hydrogen management for US NPPs.

2.3 Commissioning of a 200-kVA Mobile DG for Power Supply During Prolonged SBO

Post-accident at Fukushima NPP, it was decided to have diesel-driven electric generator (air-cooled and not requiring external cooling) to cater power needs during extended SBO/BDBA conditions. Accordingly, a 200-kVA mobile DG set has been provided in addition to the emergency DG sets and station blackout DG set. This DG set is self-contained and does not depend on any external resources for starting and operation for the specified duration of minimum 12 h. The set is normally kept as standby at site, and its operation is generally limited to periodic testing.

Complete assembly consisting of diesel generator set, lighting helmets, cables, power supply outlet arrangement, etc., is mounted on truck constituting a mobile DG set.

2.4 Hookup Systems

Even though passive system of emergency condenser is provided in design to remove decay heat during design extension conditions, by considering accident at Fukushima NPPs, need of injecting water from outside was felt necessary to remove decay heat. In response to this, hookup systems are provided to supply water to different systems from outside the reactor building/plant buildings for providing the heat sink [4]. Hookup points are provided for addition of light water to take care of leakages and inventory loss from RPV, evaporation loss from emergency condenser shell and spent fuel pool. These hookup systems are to be used only when designed power and water systems at the NPP become unavailable. Such hookups are provided for injection of water in the following systems:

- Reactor pressure vessel
- Shell side of emergency condenser
- Fuel pool

The schematic diagrams showing the retrofitted scheme of water injection into the RPV, shell side of EC and fuel pool are shown in Figs. 3, 4 and 5, respectively. They are described below:

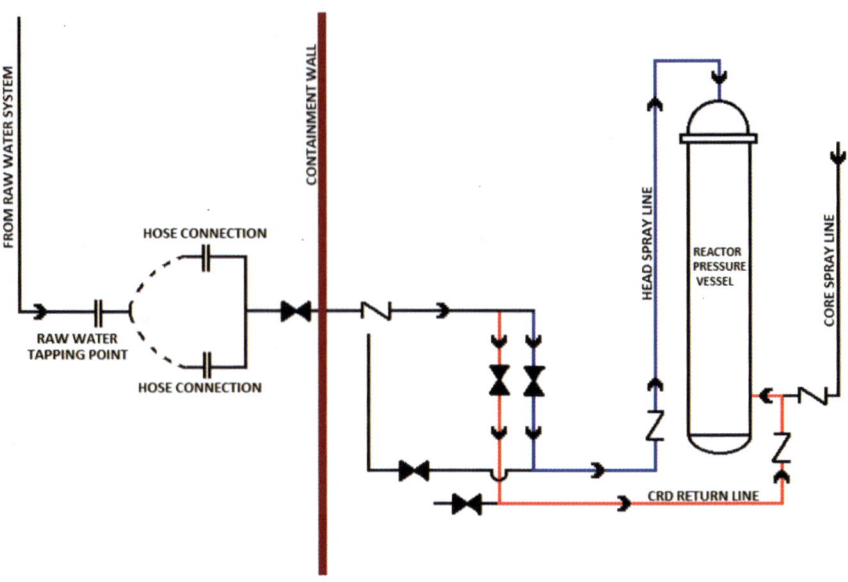

Fig. 3 Schematic diagram for water injection into RPV

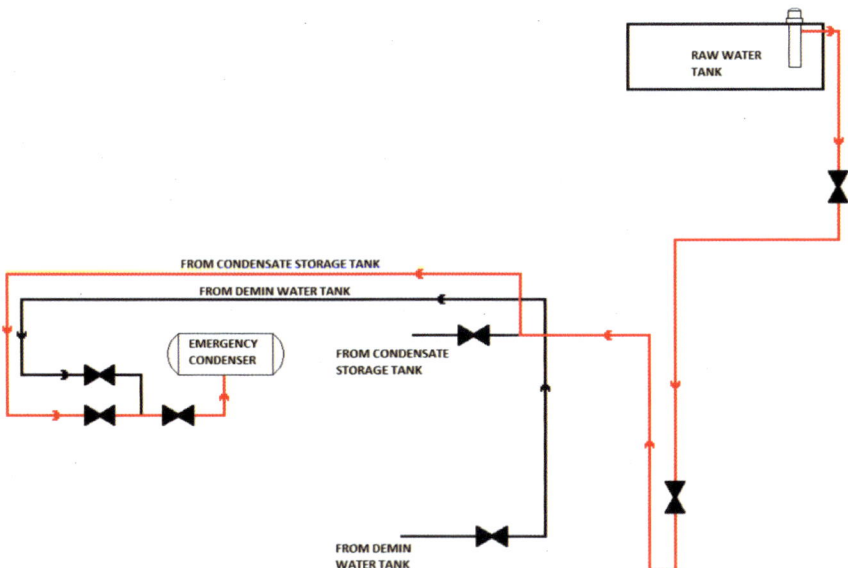

Fig. 4 Water injection into emergency condenser shell side

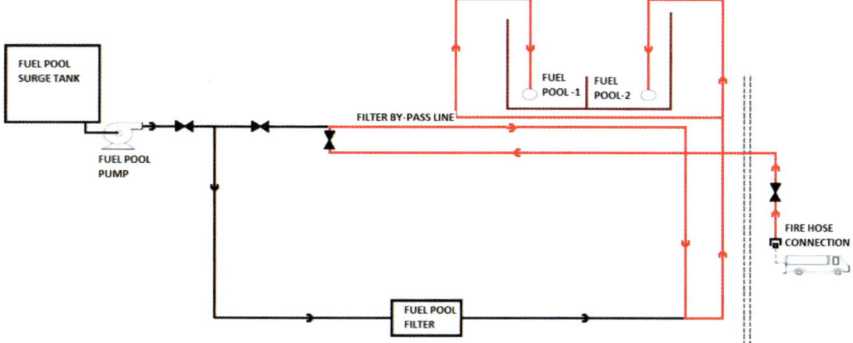

Fig. 5 Injection into fuel pool

Table 1 Designed flow rate for reactor pressure vessel and EC shell

AMG actions	Emergency condenser system available	Emergency condenser system not available
Initial flow rate requirement to RPV (kg/s)	2.0	5.0
Initial flow rate requirement to secondary side of EC (kg/s)	2.0	–

- *Reactor Pressure Vessel (RPV)*

Based on safety analysis, the minimum flow rate required to be injected into the RPV to maintain the reactor core in the cold state for a long duration is given in Table 1.

- *Secondary side of the Emergency Condenser System (EC)*

Based on safety analysis, the minimum flow rate required to be injected into the RPV and secondary side of EC to maintain the reactor core in the cold state for a long duration is given in Table 1.

- *Spent Fuel Pool*

In extended SBO, water make up in spent fuel bays is required not only for cooling but also to maintain effective shielding. Water addition rates were estimated by considering evaporation of water from fuel pools. Required injection rate to fuel pool is given in Table 2.

Table 2 Designed flow rate for spent fuel pool hookup

		Fuel pool
Flow rate requirement to SFSBs (kg/s)	Normal load	2
	Emergency load	3

2.5 Containment Filtered Venting System (CFVS)

The lesson drawn from accident at Fukushima NPPs is that, in the absence of any measure to reduce the containment pressure build-up due to steam and non-condensable gas accumulation during an accident, the venting of the containment should be an essential accident management measure for preservation of its structural integrity. The accident also highlighted the importance of having a filtration capacity to reduce environmental releases for safe and reliable utilization in situations resulting from extremely damaging events. Subsequent to Fukushima nuclear accident, extensive studies have been done worldwide on severe accident to assess the design capability of various safety systems to handle severe accident and mitigate the consequences. The studies pointed out that a system is needed which can operate independent of the state of the reactor itself and avoid catastrophic failure of the containment structure by discharging steam, air and non-condensable gases to the atmosphere. As unfiltered containment venting could result in significant radioactive releases to the environment, the implementation of filtering systems on the venting line can largely reduce the release of radioactive fission products. Hence, the concept of a containment filtered venting system (CFVS) which will be operated during severe accident condition is evolved [5]. The containment filtered venting system consists of a hardened vent pipe, a scrubber tank, chemical storage tanks, air-operated valves, manual valves and piping. Procedurally, the preferred venting route is from the common chamber due to following considerations.

Out of drywell, wet-well and common chamber, the design pressure of common chamber is minimum; thus, venting from the common chamber would ensure the integrity of all three volumes.

Venting through common chamber will give additional advantage of scrubbing of radioactive materials through suppression pool water. The venting scheme, when venting is to be exercised from common chamber, will allow venting from both the units. The schematic diagram of the system is shown in Fig. 6.

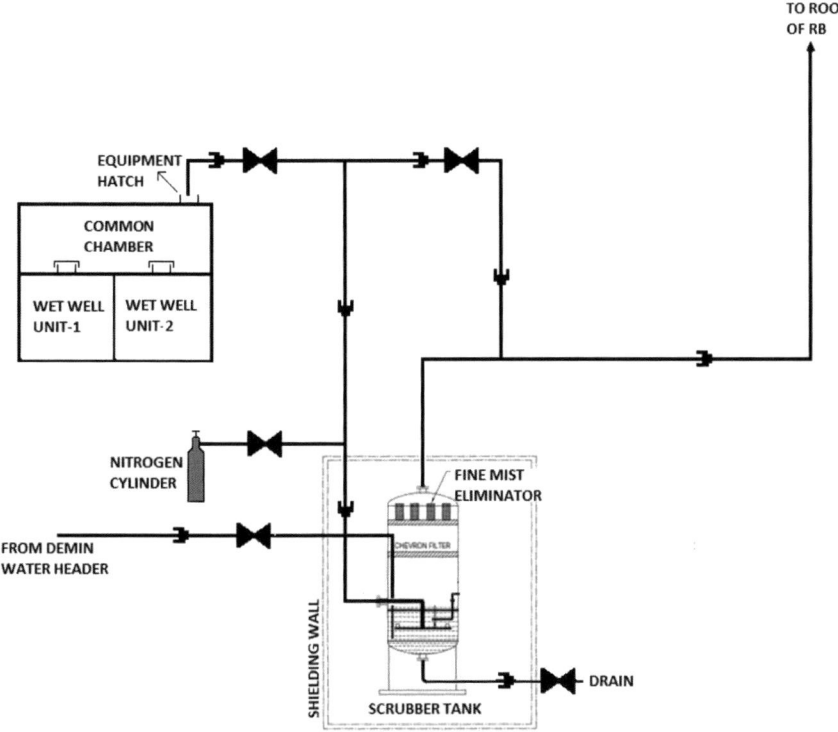

Fig. 6 Layout of CFVS at TAPS-1&2

3 Conclusions

Safety enhancement in nuclear power plant is always important by considering the experience of past events. By implementing above safety enhancement features, now plant is equipped with handling accident at Fukushima NPP type of design extension scenarios with above accident management features.

References

1. S.C. Katiyar, S.S. Bajaj, Tarapur Atomic Power Station Units-1 and 2 design features, operating experience and license renewal. Nucl. Eng. Des. **236**, 881–893 (2006)
2. R. Raj et al., Analysis of anticipated transient without scram for TAPS-1&2, Advances in Reactor Physics (ARP-2017), Mumbai, India, 06–09 Dec. 2017
3. IAEA Safety Standards (Safety Guide NS-G-2.15) (2009)
4. Seismic qualification reports of External Hook up points
5. Design Basis Report TAPS-3&4/DBR/73190/2001 Containment Filtered Venting System (CFVS)

Reliability Analysis of Composite Components Under Low-Velocity Impact

Shivdayal Patel and V. K. Gupta

Abstract Safety guarantee and failure calculation of polymer composites of an auto-mobile industry due to impact is important for the connected probabilistic judgment. It is imperative to account scatters related to Young's modulus and strength properties and subjected to impact. Possibility of this vulnerability causing a succession of failure events plays a vital role in reliability analysis. The Young's modulus and strength properties of polymers typically reveal uncertainties due to their anisotropic characteristics and composite defects. In fact, the likelihood of incident of such a situation is due to huge scatters arising in the structures. Risk analysis of polymer composites due to impact is determined incorporating variabilities of polymer composites and velocity of the impactor. A distinctive failure fissure initiates and propagates further into the boundary causing delamination between different plies. While individual defects in the lamina are complicated to track, the continuum damage model is used in the FE code to overcome these problems. The Gaussian process response surface method is currently performed to estimate reliability. A comparative study is also performed for different arrangement of impactor masses and velocities. The sequence of failure events due to dissimilar modes of failure is adopted to determine the consequences of failure situation. Frequencies of incidence of precise shock hazards yield the estimated risk due to economic loss.

Keywords Composites · Limit state function · Reliability and safety · Uncertainty modeling

S. Patel (✉) · V. K. Gupta
Indian Institute of Information Technology Design and Manufacturing, Jabalpur, India
e-mail: shivdayal@iiitdmj.ac.in

V. K. Gupta
e-mail: vkgupta@iiitdmj.ac.in

© Springer Nature Singapore Pte Ltd. 2020
P. V. Varde et al. (eds.), *Reliability, Safety and Hazard Assessment for Risk-Based Technologies*, Lecture Notes in Mechanical Engineering,
https://doi.org/10.1007/978-981-13-9008-1_61

1 Introduction

Advanced polymer composites have been considered in spacious applications in the automobile industry due to their better strength and stiffness properties. The projectile hit by the composite structure subjected to the low velocity impact (LVI), the structural damaged plays an important role to predict the safety and reliability of polymer laminates. Substantial effort on the failure estimation of FRP under LVI has been investigated . Abrate [1] provided the state-of-the-art in LVI and failure prediction of the polymer composite. The failure mechanisms of composites material incorporated the matrix and fiber breakages. Olsson [2] performed that the impact period robustly changes the result of targets and the impactor plate mass ratio. Small-mass impacts on FRP plates were found to be more crucial than large-mass impacts of the same energy. The deterministic response of polymer laminated plate due to LVI have been performed by Singh et al. [3] and Bandaru et al. [4]. They considered the customized form of Hashin criteria to determine failure initiation in dissimilar modes followed by degradation of the polymer properties due to damage.

Uncertainties in polymer laminate occur because of the variability in volume fractions of fiber and temperature effects, etc. Reliability analysis is essential to consider the scatters in the input parameters. Sriramula and Chyssanthopoulos [5], Patel and Guedes Soares [6] studied the scatters in polymer structure and potted the dissimilar stochastic models. The technique to perform the system failure of the polymer was crucially investigated by allowing for failures of dissimilar ply in total failure. Stefanou and Papadrakakis [7] performed the stochastic technique to estimate the stochastic behavior of polymer structures with the first-order reliability method (FORM). Sutherland and Guedes Soares [8] determined a phenomenological failure criterion to develop the performance function of polymer structures for collapse estimation. Jeong and Shenoi [9] performed a Monte Carlo simulation (MCS) method to evaluate the stochastic failure of polymer structures. However, FORM and MCS methods unsuccessful to evaluate the Pf precisely with nonlinear problems with low Pf [10]. Gaussian response surface method (GRSM) is able to overcome these restrictions and accurately predict the Pf [11–15] presented a failure analysis of polymer plates due impact. On the other hand, some aspect of the stochastic response of energy dissipation of polymer plates under LVI is not broadly investigated and therefore has a plentiful scale of advanced study.

First, the deterministic analysis of polymer laminated plate due to impact load is performed using the mean value of the parameters. If the stresses in the lamina are such that the performance function $(g(x) < 0)$, the probabilistic progressive failure model is adopted to predict the probability of failure (Pf). In polymer composites subjected to impact, failure initiates either due to large transverse shear stresses under the impactor at the top of the laminate or due to bending stresses at the bottom and then spreads inwards. Here, Pf is therefore calculated for the middle (fifth) ply of the laminate using damage mechanics-based finite element simulations combined with GPRSM.

2 Composite Damage Modeling

In engineering domain, to get a solution for simple or complex problems, numerical methods were used by discretizing the entire model into small (finite) elements which are easy to do calculations by using the ordinary differential equations are called finite element method (FEM). A mathematical damage model developed and implemented with certain constraints and assumptions to predict the failure behavior of the composite structure using numerical methods according to application or process. All finite element equations assembled together to frame equations for the entire model to get results of the whole model. A solution of typical problems like crushing analysis, heat transfer and structural analysis can do as well as estimate results with the numerical methods.

2.1 Damage Initiation

Hashin [16] criteria are investigated to calculate the damage initiation for tensile fiber and matrix, while the damage model given as [17] to model the matrix compressive failure. In fiber reinforced plastic composite, damage initiation occurs when satisfying any among the following equations

- Tensile fiber failure for $\sigma_{11} \geq 0$ [16]

$$\left(\frac{\sigma_{11}}{X^{\mathrm{T}}}\right)^2 + \alpha\left(\frac{\tau_{12}}{S^{\mathrm{L}}}\right)^2 = \begin{cases} \geq 1 \text{ failure} \\ < 1 \text{ No failure} \end{cases} \tag{1}$$

- Compressive fiber failure for $\sigma_{11} < 0$ [16]

$$\left(\frac{\sigma_{11}}{X^{\mathrm{C}}}\right)^2 = \begin{cases} \geq 1 \text{ failure} \\ < 1 \text{ No failure} \end{cases} \tag{2}$$

- Tensile matrix failure for $\sigma_{22} > 0$ [16]

$$\left(\frac{\sigma_{22}}{Y^{\mathrm{T}}}\right)^2 + \left(\frac{\tau_{12}}{S^{\mathrm{L}}}\right)^2 = \begin{cases} \geq 1 \text{ failure} \\ < 1 \text{ No failure} \end{cases} \tag{3}$$

where

σ_{11}, σ_{22} and τ_{12} = Stress in longitudinal, transverse, and shear direction
$X^{\mathrm{T}}, X^{\mathrm{C}}$ = Strength properties of fiber in longitudinal and transverse
$Y^{\mathrm{T}}, Y^{\mathrm{C}}$ = Strength properties of matrix in longitudinal and transverse
$S^{\mathrm{L}}, S^{\mathrm{T}}$ = Shear strength properties in longitudinal and transverse
$\alpha = 1$

- Compressive matrix failure for $\sigma_{22} < 0$ [17]

$$F_{22}^{C} = \left(\frac{\sigma_{t}^{n}}{S_{23}^{A} + \mu_{t}^{n}\sigma_{n}^{n}} \right)^{2} + \left(\frac{\sigma_{l}^{n}}{S_{12} + \mu_{l}^{n}\sigma_{n}^{n}} \right)^{2} - 1 = 0 \tag{4}$$

σ_{n}^{n}, σ_{l}^{n}, and σ_{t}^{n} are the stresses on the fracture plane in normal (n), longitudinal (l), and tangential (t) directions.

2.2 Damage Propagation Criteria

The constitutive equation of the material after damage initiation and the relation between stress and strain of the damaged composite material under the impact loadings as follows

$$\sigma = C_{d}\varepsilon \tag{5}$$

$$C_{d} = \frac{1}{D} \begin{bmatrix} (1 - d_{f})E_{1} & (1 - d_{f})(1 - d_{m})\vartheta_{21}E_{1} & 0 \\ (1 - d_{f})(1 - d_{m})\vartheta_{12}E_{2} & (1 - d_{m})E_{2} & 0 \\ 0 & 0 & D(1 - d_{s})G_{12} \end{bmatrix}$$

$D = 1 - (1 - d_{f})(1 - d_{m})\vartheta_{12}\vartheta_{21}$
E_{1}, E_{2}, G_{12}, ϑ_{12}, and ϑ_{21} are undamaged material properties
where d_{f}, d_{m}, and $d_{s} =$ current damage value in fiber, matrix, and delamination after initiation

Fiber and Matrix tensile failure,

$$d_{f,m}^{T} = \frac{\varepsilon_{f,m}^{fT}}{\varepsilon_{f,m}^{fT} - \varepsilon_{f,m}^{0T}} \left(1 - \frac{\varepsilon_{f,m}^{0T}}{\varepsilon_{f,m}} \right) \tag{6}$$

where

$\varepsilon_{f}^{fT} = \frac{2G_{fc}^{T}}{X^{T} \times l^{*}}$ and $\varepsilon_{f}^{0T} = \frac{X^{T}}{E_{1}}$
$\varepsilon_{m}^{fT} = \frac{2G_{mc}^{T}}{Y^{T} \times l^{*}}$ and $\varepsilon_{m}^{0T} = \frac{Y^{T}}{E_{2}}$

Here, ε_{f}^{fT} represents the tensile failure strain for matrix cracking.
Compressive fiber failure,

$$d_{f}^{C} = \frac{\varepsilon_{f}^{fC}}{\varepsilon_{f}^{fC} - \varepsilon_{f}^{0C}} \left(1 - \frac{\varepsilon_{f}^{0C}}{\varepsilon_{f}} \right) \tag{7}$$

where

$$\varepsilon_f^{fC} = \frac{2G_{fc}^C}{X^C \times l^*} \text{ and } \varepsilon_1^{0C} = \frac{X^C}{E_1}$$

T, C represent tension or compression and f, m represent fiber or matrix

$\varepsilon_{f,m}$ = equivalent displacement in between damage initiation and propagation

$\varepsilon_{f,m}^{fT,fC}$ = equivalent displacement at final failure in tension or compression

$\varepsilon_{f,m}^{0T,0C}$ = equivalent displacement at damage initiation in tension or compression

σ_m^{0C} = equivalent stress

$G_{fc,mc}^{T,C}$ = matrix cracking fracture energy

Compressive matrix cracking failure

$$d_m^c = \frac{\gamma_r^{max}}{\gamma_r^{max} - \gamma_r^f}\left(1 - \frac{\gamma_r^f}{\gamma_r}\right) \tag{8}$$

where γ_r^f represents the shear failure strain, γ_r^{max} represents the maximum fracture energy in shear strain.

The above damage initiation and propagation model are used in finite element analysis as VUMAT code.

3   Reliability of Composites Structures

Reliability investigation is robustly suggested by different design codes, particularly to accomplish difficult design targets for acceptable examine and endurance. It is because the deterministic design does not incorporate the randomness of the composite systems. Risk analysis is investigated by allowing the variability for the systems. The uncertainty modeling of composite material properties required to perform experimental tests to determine the statistical characteristics such as standard deviation and distribution of material properties. The stochastic method explicitly determines the level of safety or Pf.

4   Results and Discussions

4.1   Numerical Study of Low-Velocity Impact

In this study, the deformable plate is modeled with eight plies of polymer composites at dissimilar patterns $[0^0/90^0/0^0/90^0]_s$. The plate consists of eight layers in each ply thickness of 0.3175 mm. The impactor was assigned with mass and velocity of 7.66 g, 10 m/s, respectively. The size of the target plate is 200 mm × 200 mm × 2.54 mm. The laminated plate is meshed with 8-noded brick elements. Full integration was used to avoid inaccuracies due to hourglassing. In this investigation, reduced integration scheme, hourglass distortion control, and element deletion were applied using finite element analysis. We have to be careful while assigning the mesh stacking direction in the material property section. Mesh stacking direction have been a lot of importance in the simulation of the composite structure. Size of mesh element has been taking from literature in this study otherwise mesh convergence study gives the suitable element size, as well as it also varies based on the structure of the deformable tube. The coefficient of friction between the impactor and plate was taken as 0.2.

4.2   Validation Study of Contact Force

User-defined material model (VUMAT) is used to validate the experimental study by using explicit FE code of ABAQUS as well as presented data in the reference paper by Krishnamurthy et al. [18]. In this numerical study, essential properties of polymer composite plate with ply orientation of $[0^0/90^0/0^0/90^0]_s$ are taken from an experimental investigation did by Krishnamurthy et al. [18]. The same impactor mass and velocity is applied for numerical study as published research paper [18]; however, the small discrepancy appeared in the present result compared with an experimental result [18] as shown in Fig. 1. This discrepancy appeared due to the damping effect, damage model, and property assumptions from literature which was not given in the research paper.

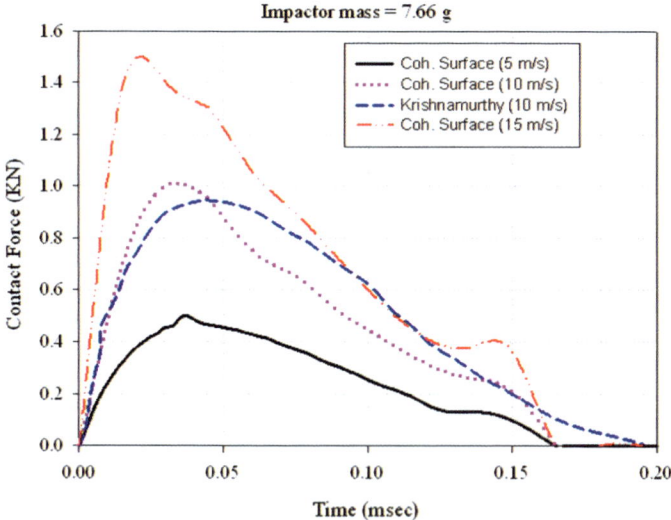

Fig. 1 Validation study of contact force versus time (10 m/s)

4.3 Failure of the Midplane of the Composite Plate Under Impact

In the current study, it is supposed that the polymer plate fails if the midplane of the laminate reaches a failure criterion. The GPRSM is therefore used here to find Pf. The suggested design value of Pf for the polymer plate due to LVI for application under study by Patel et al. [17] lies between 10^{-3} and 10^{-5}. If the Pf is larger than 10^{-3} then the system Pf is known as undesirable Pf.

If the performance function is based on failure initiation in the midplane of the polymer plate, then for the velocity lying between 11.2 m/s and 11.6 m/s, Pf lies between 10^{-3} and 10^{-5} and the plate is considered safe. On the other hand, if the performance function is based on failure evaluation, then the suitable velocity range varies from 12 to 13 m/s. These results for the impactor mass of 7.66 g are shown in Table 1. In Table 2, for impactor velocity of 10 m/s, the acceptable mass is 9.975 g for the performance function based on damage initiation and varies from 11.49 to 15.32 g for limit state based on damage propagation. On plotting, it is also seen in Fig. 2 that both the variations in Tables 1 and 2 are linear in nature and may be used to determine Pf corresponding to a particular velocity or a particular mass. The acceptable Pf for failure initiation based limit state function is achieved for the impactor velocity lying between 11.2 and 11.6 m/s and mass of impactor 7.66 g and undesired Pf failure is found that the velocity lies between 12 and 13 m/s and mass of impactor 7.66 g.

Velocity (m/s)	Damage initiation GRSM failure (Pf)	Damage propagation GRSM failure (Pf)
11.0	0.000	0.000
11.2	7.19E−5	0.000
11.4	2.59E−4	0.000
11.6	4.90E−4	0.000
11.8	0.1106	0.000
12.0	1.000	2.5E−5
13.0	1.000	9.1E−3
14.0	1.000	1.000

Table 1 Effect of impactor velocity on probability of failure due to impactor mass of 7.66 g

Mass (g)	Damage initiation GRSM failure (Pf)	Damage propagation GRSM failure (Pf)
1.0M	0.000	0.000
1.5M	0.205	0.000
2.0M	1.000	4.12E−2
3.0M	1.000	1.000

Table 2 Effect of impactor mass on probability of mid-plane failure due to impactor velocity of 10 m/s

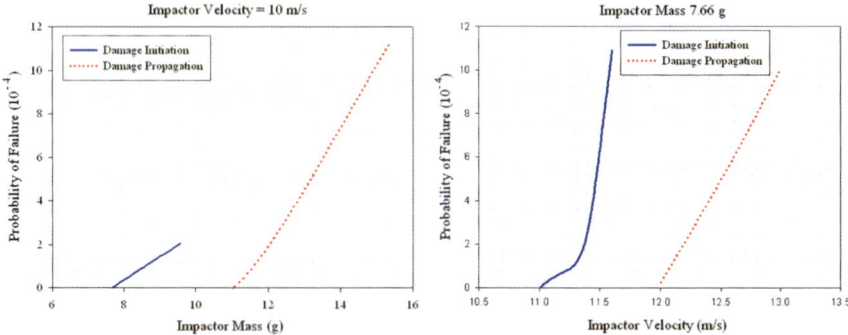

Fig. 2 Midplane probability of failure against variation in velocity

5 Conclusions

A stochastic failure initiation and propagation models are used in the FE code
ABAQUS. Computational responses are verified and obtain to be in good conformity
with accessible in the literature. The peak impact force is a linear function of impactor
velocity for a constant mass. For the masses considered here, the impact force did
not show much variation with mass although the impact duration is significantly
affected. For determining the Pf, two methods MCS and GPRSM are used. GPRSM
is computationally ten times less expensive compared to MCS. For a mass of 7.66 g
with failure initiation as limit state function, this probability is achieved for impactor
velocity lying between 11.2 and 11.6 m/s, whereas with final failure (corresponding
to $d_{22} = 1$) as limit state function, the velocity lies between 12 and 13 m/s. With
impactor velocity fixed at 10 m/s, the safe and reliable design is achieved for a mass
of 9.975 g for performance function based on damage initiation and for a mass lying
between 11.49 and 15.32 g for limit function based on final failure.

References

1. S. Abrate, Modeling of impacts on composite structures. Compos. Struct. **51**, 129–138 (2001)
2. R. Olsson, Mass criterion for wave controlled impact response of composite plates. Compos.
   Part A-Appl. Sci. **31**, 879 (2000)
3. H. Singh, K.K. Namala, P. Mahajan, A damage evolution study of E-glass/epoxy composite
   under low velocity impact. Compos. Part B Eng. **76**, 235–248 (2015)
4. A.K. Bandaru, S. Patel, Y. Sachan, S. Ahmad, R. Alagirusamy, N. Bhatnagar, Low velocity
   impact response of 3D angle-interlock Kevlar/basalt reinforced polypropylene composites.
   Mater. Des. **105**, 323–332 (2016)
5. S. Sriramula, M.K. Chyssanthopoulos, Quantification of uncertainty in stochastic analysis of
   FRP composites. Compos. Part A-Appl Sci. **40**, 1673–1684 (2009)
6. S. Patel, C. Guedes Soares, System probability of failure and sensitivity analyses of composite
   plates under low velocity impact. Compos. Struct. **180**, 1022–1031 (2017)

7. G. Stefanou, M. Papadrakakis, Stochastic finite element analysis of shells with combined random material and geometric properties. Comput. Methods Appl. Mech. Eng. **193**(1–2), 139–160 (2004)

8. L.S. Sutherland, C. Guedes Soares, Review of probabilistic models of the strength of composite materials. Reliab. Eng. Syst. Saf. **56**, 183–196 (1997)

9. H.K. Jeong, R.A. Shenoi, Reliability analysis of mid-plane symmetric laminated plates using direct simulation method. Compos. Struct. **43**(1), 1–13 (1998)

10. M.R. Rajashekhar, B.R. Ellingwood, A new look at the response surface approach for reliability analysis. Struct. Saf. **12**, 205–220 (1993)

11. S. Patel, S. Ahmad, P. Mahajan, Reliability analysis of a composite plate under low velocity impact using the Gaussian response surface method. Int. J. Comput. Methods Eng. Sci. Mech. **15**(3), 218–226 (2014)

12. S. Patel, S. Ahmad, P. Mahajan, Probabilistic finite element analysis of S2-glass epoxy composite beams for damage initiation due to high velocity impact. ASME J. Risk Uncertain. Part B **2**(4) (2016). 044504-044504-3

13. S. Patel, S. Ahmad, Probabilistic failure of graphite epoxy composite plates due to low velocity impact. ASME J. Mech. Des. **139**(4), 044501 (2017)

14. S. Patel, C. Guedes Soares, Reliability assessment of glass epoxy composite plates due to low velocity impact. Compos.Struct. **200**, 659–668 (2018)

15. S. Patel, S. Ahmad, P. Mahajan, Stochastic finite element analysis of composite body armor. In: V. Matsagar (ed.), *Advance in Structural Engineering Mechanics* (Springer, India, 2015), pp. 259–272

16. Z. Hashin, Failure criteria for unidirectional composites. J. Appl. Mech. Eng. **47**, 329–334 (1980)

17. A. Puck, H. Schurmann, Failure analysis of FRP laminates by means of physically based phenomenological models. Compos. Sci. Technol. **62**, 1633–1662 (2002)

18. K.S. Krishnamurthy, P. Mahajan, R.K. Mittal, Impact response and damage in laminated composite cylindrical shells. Compos. Struct. **59**, 15–36 (2003)

Failure Analysis

A Study on Emission Pattern of Semiconductor ICs Using Photon Emission Microscopy

Debasish Nath, Arihant Jain, V. K. Tapas, N. S. Joshi and P. V. Varde

Abstract The electronics industry has achieved a phenomenal growth over the last few decades mainly due to the rapid advances in integration technologies of semiconductor devices. At the same time, fault localization and failure identification of defects in complex integrated circuits (ICs) has become more difficult with time due to the increasing number of devices per chip. Photon emission microscopy (PEM) is a useful failure analysis technique for localizing defects by detecting photon emissions generated as a result of an electroluminescence process. The emission pattern of a faulty IC can be compared to that of a healthy one to have an idea about fault localization. The objective of this work is to generate a reference database by recording the emission patterns of various ICs under specified electrical biasing. Emission patterns of used ICs have been observed in this work using PEM, which can be used as a signature and can be compared with that of faulty ones. The procedure followed and the results of the experiment are discussed in detail.

Keywords Emission pattern · Fault identification · Fault localization · Photon emission microscopy

D. Nath
National Institute of Technology, Silchar 788010, Assam, India
e-mail: debasish.nath15@gmail.com

A. Jain (✉) · V. K. Tapas · N. S. Joshi · P. V. Varde
Research Reactor Services Division, Reactor Group,
Bhabha Atomic Research Center, Mumbai, India
e-mail: arihantj@barc.gov.in

V. K. Tapas
e-mail: vktapas@barc.gov.in

N. S. Joshi
e-mail: nsjoshi@barc.gov.in

P. V. Varde
e-mail: varde@barc.gov.in

© Springer Nature Singapore Pte Ltd. 2020
P. V. Varde et al. (eds.), *Reliability, Safety and Hazard Assessment
for Risk-Based Technologies*, Lecture Notes in Mechanical Engineering,
https://doi.org/10.1007/978-981-13-9008-1_62

1 Introduction

Semiconductor chip technology has produced dramatic advances over the past years enabling the manufacture of high-density integrated circuits (ICs), with millions of devices on a single die [1]. In the early 1970s, design innovations propelled advancement from small-scale integration (SSI) to large-scale integration (LSI) [2]. The decade from mid-1970 to 1980 was characterized by significant advances in silicon technology which led to very-large-scale integration (VLSI) with 2^{16}–2^{21} devices on a single die [3]. In today's arena of submicron technologies for ultra-large scale integration (ULSI) and wafer-scale integration (WSI), failure analysis has progressed to a level of sophistication and complexity matching the components that must be analyzed [4]. Emergence of flip-flop packaging has further complicated the job by making front-side analysis of the circuit almost impossible without removing the chip bounds physically.

Near-infrared photon emission microscopy (PEM) is an established fault localization technique for microelectronic failure analysis [5]. It is a backside analysis technique which uses photon wavelengths in the infrared (IR) and near-infrared regions to exploit the relative transparency of the silicon die to such wavelengths. It is useful for localization of defects in devices that exhibit leakage of electric current. Various electric current failures are classified into four categories of photon emission situations [6]. In some cases, photon emission represents a defect like leaky junction, silicon mechanical damage, or contact spiking, etc., and in some, it may be because of design or test conditions. There are various defects which do not produce photon emission (ex-ohmic shorts, surface inversion, etc.), and hence, they cannot be detected and localized by PEM. Parekh and Milburn [7] studied on the effectiveness of PEM as a fault localization tool in CMOS ICs and showed that about 66% of the chip-related failures can be detected and localized by PEM. The use of PEM for analyzing IC failures caused by electrical overstress and electrostatic discharge (ESD) was explored by Wills [8]. The study resulted that about 80% of the failure sites were found out through the combination of photon emission and curve tracer testing. All these studies show that PEM is a very effective technique for defect localization in complex ICs.

Photon emission spectrum of IC can be served as its 'fingerprint' or 'signature.' Spectrum of healthy one acts as a reference to be compared against that of a faulty IC for defect localization. Also, the emission spectrums of faulty ICs with known faults can be used as 'defect fingerprint' for fault identification.

A reference database for comparison of emission patterns generated by various ICs is required for fault localization and identification. The objective of this study is to generate the reference database by recording the emission patterns of various ICs under specified electrical biasing. Emission pattern of four ICs has been observed in this work using INSCOPE labs PEM 1200. The emission patterns observed are discussed in Sect. 2 of the paper. Conclusions inferred from the work and future scopes are discussed in Sects. 3 and 4, respectively.

2 Emission Patterns of Tested Integrated Circuits (ICs)

Four different ICs were tested in this work. The ICs were biased in line with the specifications mentioned in the respective datasheets. The used ICs are:

- HD74LS367A (non-inverted hex bus driver)
- SN74LS06N (high-voltage open collector inverter buffer)
- LM3900 (quadruple Norton operational amplifier)
- SN74LS365A (non-inverting buffer with NOR gate output control)

The detailed observations are presented in this section. A brief description of each IC is presented followed by details about the electrical biasing conditions. Emission spectrum of the IC for each biasing is then presented followed by a discussion on observations.

2.1 HD74LS367A (Non-inverted Hex Bus Driver)

The IC HD74LS367A has 16 pins with non-inverted data outputs with a three-state outputs hex bus driver IC [9]. This device contains six independent gates each of which performs a non-inverting buffer function. When enabled, outputs exhibit the low impedance characteristics with additional drive capability to permit the driving of bus lines without external resistors. When disabled, both the output transistors are turned off presenting a high impedance state to the bus line. Thus, the output will act neither as significant load nor as a driver. To minimize the possibility that two outputs will attempt to take a common bus to opposite logic levels, the disable time is shorter than enable time of the outputs. The 1st buffer having input 1A, output 1Y, and gate control pin no-1 is used for experimentation. Pin out diagram of the IC is given in Fig. 1. Recommended supply voltage V_{cc} varies from 4.75 to 5.25 V at room temperature.

2.1.1 Functional Table of HD74LS367A

See Table 1.

Two cases are considered for the study. In the first case, the input A is high, and in the second case, it is low.

2.1.2 Case 1 (Input at 5 V)

The inputs applied to the IC are:

- Pin 1 ($\overline{G1}$): 0 V
- Pin 2 (1A): 5 V

Fig. 1 Pin out diagram of
HD74LS367A

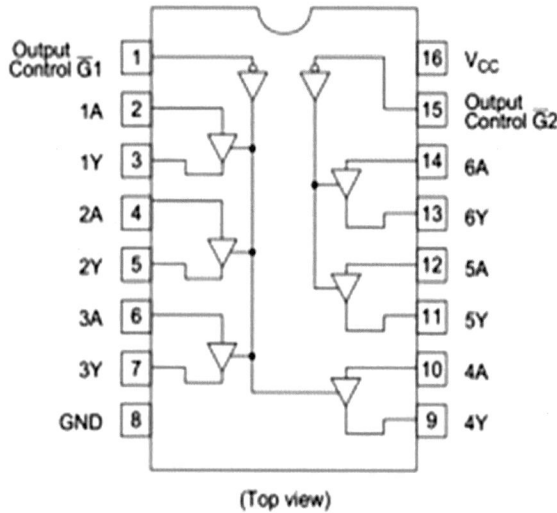

(Top view)

Table 1 Functional table of
HD74LS367A [9]

$\overline{G}$ (output control)	A (input)	Y (output)
H	X	Z
L	L	L
L	H	H

Here, 'H' represents high level, 'L' represents low level, 'X' represents irrelevant, and 'Z' represents high impedance state of a three-state output

- Pin 16 (V_{CC}): 5 V
- Pin 8 (GND): 0 V

The output in this case is high (5 V), which is observed at pin 3(1Y). The emission pattern observed is given in Fig. 2 (left).

2.1.3   Case 2 (Input at 0 V)

The inputs applied to the IC are:

- Pin 1 ($\overline{G1}$): 0 V
- Pin 2 (1A): 0 V
- Pin 16 (V_{CC}): 5 V
- Pin 8 (GND): 0 V

The output in this case is low (0 V), which is observed at pin 3(1Y). The emission pattern observed is given in Fig. 2 (right).

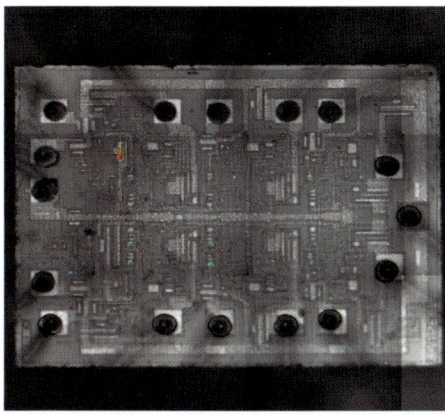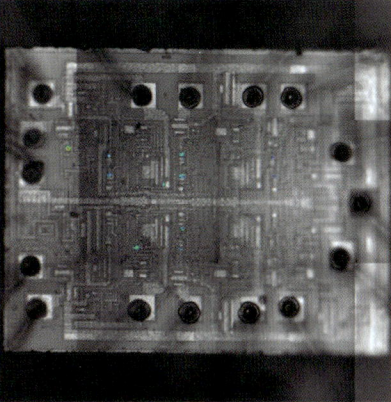

Fig. 2 Observed emission pattern of HD74LS367A for mentioned case 1 (left). Observed emission pattern of IC for case 2 (right)

2.1.4 Observation

The IC HD74LS367A has six non-inverting buffers with NOT gate output control. One buffer is used for the experiment which has input pin 1A and output gate control pin $\overline{G1}$. The expected outputs are observed in pin 1Y. For case 1, the input is kept at high logic (5 V), and for case 2, it is kept at low logic (0 V). Photon emission spectrum is observed under PEM. Red pattern indicates high-intensity photon emission while blue indicates faint emission.

2.2 SN74LS06N (High-Voltage, Open Collector Inverter Buffer)

These devices feature high-voltage, open collector outputs to interface with high-level circuits (such as MOS), or for driving high-current loads and also used as inverter buffers for driving TTL devices [10]. The open collector device is suitable for high-voltage translation applications. Pin out diagram of the IC is given in Fig. 3. The functional table is given in Table 2. It has six buffers. For our experiment, we have chosen the buffer having 1A as input and 1Y as output.

Two cases are considered for the study. In the first case, the input A is high and in the second case it is low.

2.2.1 Case 1 (Input at 0 V)

The inputs applied to the IC are:

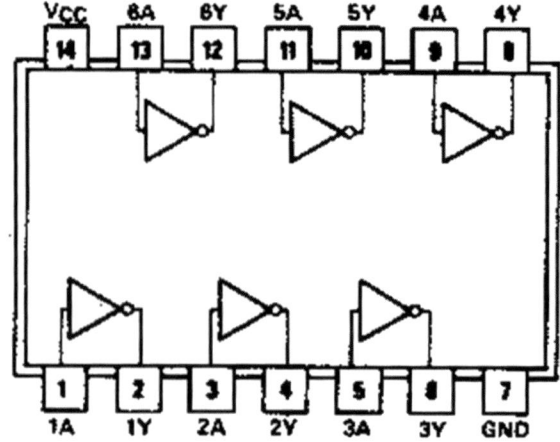

Fig. 3 Pin out diagram of SN74LS06N

- Pin 1: 0 V
- Pin 14 (V_{CC}): 5 V
- Pin 7 (GND): 0 V

For this case, output observed at pin 2 (1Y) is low (high impedance state). The observed emission pattern is shown in Fig. 4 (left).

2.2.2 Case 2 (Input at 5 V)

The inputs applied to the IC are:

- Pin 1: 5 V
- Pin 14 (V_{CC}): 5 V
- Pin 7 (GND): 0 V

For this case, output observed at pin 2 (1Y) is at logic low (0 V). The observed emission pattern is shown in Fig. 4 (right).

Table 2 Functional table of SN74LS06N

A (Input)	Y (Output)
H	L
L	Z

Here, 'H' represents high level, 'L' represents low level, and 'Z' represents high impedance state output

Fig. 4 Observed emission pattern of SN74LS06N for mentioned case 1 (left). Observed emission pattern of IC for case 2 (right)

2.2.3 Observation

The buffer used for our experiment has input pin 1A and output 1Y. For case 1, high emission pattern is observed while for case 2 no detectable photon emission is observed. Even when current flow exists through the IC, emission is not observed. It can be concluded that IC SN74LS06N does not generate any detectable emission pattern when biased under case 2 as specified.

2.3   LM3900 (Quadruple Norton Op-Amp)

LM3900 is a 14 pin DIP containing four identical, independent high gain frequency compensated op-amps [11]. These devices provide wide bandwidth and large output voltage swing. The pin out diagram is given in Fig. 5. For our experiment, we have chosen the op-amp having pin 1 and 6 as input and pin 5 as output.

Two different input cases are considered for the study as below:

2.3.1   Case 1

The inputs applied to the IC are:

- Pin 1: 0 V
- Pin 16 (V_{CC}): 5 V
- Pin 6: 0 V
- Pin 7 (GND): 0 V

For this case, the observed emission pattern is shown in Fig. 6 (left).

2.3.2 Case 2

The inputs applied to the IC are:

- Pin 1: 5 V
- Pin 16 (V_{CC}): 5 V
- Pin 6: 0 V
- Pin 7 (GND): 0 V

For this case, the observed emission pattern is shown in Fig. 6 (right).

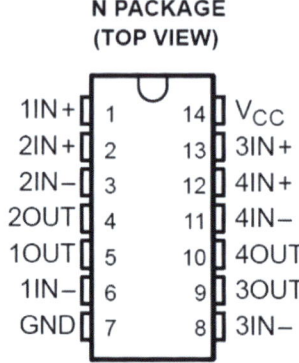

Fig. 5 Pin out diagram of LM3900

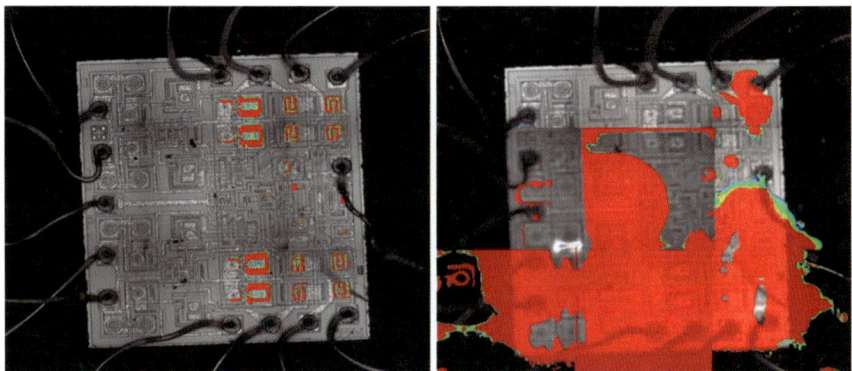

Fig. 6 Observed emission pattern of LM3900 for mentioned case 1 (left). Observed emission pattern of IC for case 2 (right)

2.3.3 Observation

LM3900 has four OP-Amps out of which one is used for our experiment and biased as per both cases. Logically correct analog outputs are observed across pin 5 which indicates healthiness of the IC. For both cases, photon emission spectrum is observed under PEM. In case 2, comparatively high photon emission is observed than case 1.

2.4 SN74LS365A (Non-inverting Buffer with nOR Gate Output Control)

The IC SN74LS365A has high-speed hex buffers with three-state outputs. They are organized with non-inverting data-path. It has a common 2-input NOR enable [12]. When the output enable is high, the outputs are forced to high impedance 'off' state. The pin out diagram of the IC is given in Fig. 7, and its functional table is given in Table 3. For our work, we have used the buffer having input pin 2 and output pin 3.

Fig. 7 Pin out diagram of SN74LS365A

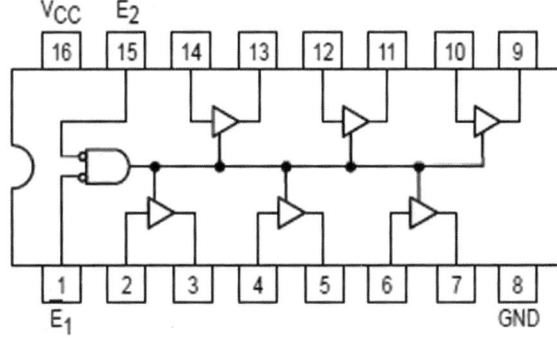

Table 3 Functional table of SN74LS365A

Control enable		Input	Output
$\overline{E1}$	$\overline{E2}$		
L	L	L	L
L	L	H	H
H	X	X	Z
X	H	X	Z

Here, 'H' represents high level, 'L' represents low level, 'X' represents irrelevant, and 'Z' represents high impedance state of a three-state output

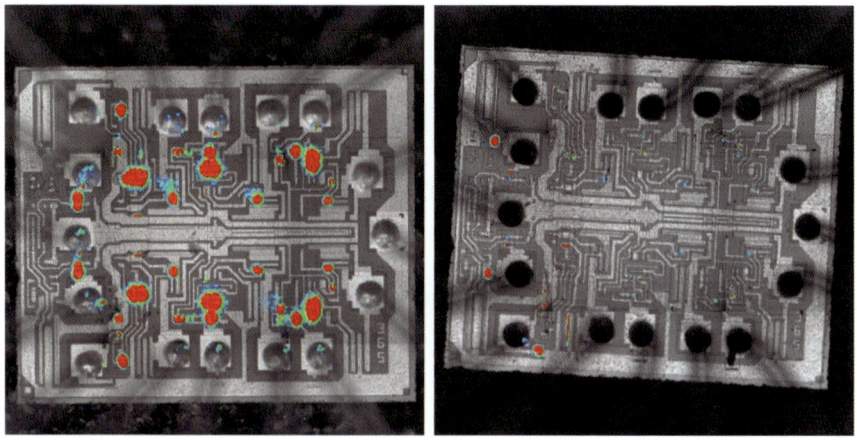

Fig. 8 Observed emission pattern of SN74LS365A for mentioned case 1 (left). Observed emission pattern of IC for case 2 (right)

Two cases are considered for the study. In the first case, the input A is high and in the second case it is low.

2.4.1   Case 1 (Input at 5 V)

The inputs applied to the IC are:

- Pin 1 ($\overline{E1}$): 0 V
- Pin 15 ($\overline{E2}$): 0 V
- Pin 16 (V_{CC}): 5 V
- Pin 2: 5 V
- Pin 8 (GND): 0 V

For this case, output observed at pin 3 is high (5 V). The observed emission pattern is shown in Fig. 8 (left).

2.4.2   Case 2 (Input at 0 V)

The inputs applied to the IC are:

- Pin 1 ($\overline{E1}$): 0 V
- Pin 15 ($\overline{E2}$): 0 V
- Pin 16 (V_{CC}): 5 V
- Pin 2: 0 V
- Pin 8 (GND): 0 V

For this case, output observed at pin 3 is low (0 V). The observed emission pattern is shown in Fig. 8 (right).

2.4.3 Observation

We have used one out of six buffers having input at pin 2 and output is observed at pin 3. Logically correct output indicates the healthiness of used IC. In both the cases mentioned, photon emission spectrum is observed under PEM and both are different.

3 Conclusions

PEM is a useful analysis technique for localizing defect sites in devices exhibiting electrical current leakages. In this study, experiments are performed on four ICs under specified biasing conditions and emission spectrum is observed. Very high-intensity emission spreading over a wide area is observed for LM3900 whereas no emission is observed for SN74LS06N case 2 biasing mode. Localized low-intensity emission is observed for SN74LS365A and HD74LS367A.

It is clear from the observations that photon emission of various ICs differs from each other. The observed patterns serve as 'signature' or 'fingerprint' of IC which can be used as a reference for comparison with that of a defect one.

4 Future Scope

Use of PEM for defect localization requires extensive database of emission patterns in order to serve as reference for comparison. Results obtained from this study will be part of the database. Further experiments need to be performed to include the emission pattern of other ICs in the database. Also, emission patterns of ICs with known faults will serve as defect fingerprint for fault identification and those should be included in the database. Further experiments are required for the same.

Acknowledgements We thank Shri S. Bhattacharya, Associate Director, Reactor Group for his encouragement and support during the work.

References

1. R.B. Fair, Challenges to manufacturing submicron, ultra-large scale integrated circuits. Proc. IEEE **78**(11), 1687 (1990)
2. C.T. Sah, Evolution of the MOS transistor-from conception to VLSI. Proc. IEEE **76**(10), 1280–1326 (1988)

3. A. Reisman, Device, circuit and technology scaling to micron and submicron dimensions. Proc. IEEE **71**(5), 550 (1983)
4. J.M. Soden, R.E. Anderson, IC failure analysis: techniques and tools for quality and reliability improvement. Proc. IEEE **81**(5), 703–715 (1993)
5. J.C.H. Phang, D.S.H. Chan, S.L. Tan, W.B. Len, K.H. Yim, L.S. Koh, C.M. Chua, L.J. Balk, A review of near infrared photon emission microscopy and spectroscopy, in *Proceedings of the 12th International Symposium on the Physical and Failure Analysis of Integrated Circuits* (2005), pp. 275–281
6. G. Shade, Physical mechanisms for light emission microscopy, in *Proceedings of International Symposium for Testing and Failure Analysis* (1990), p. 121
7. K. Parekh, R. Milburn, Effectiveness of emission microscopy in the analysis of CMOS ASIC devices, in *Proceedings of International Symposium for Testing and Failure Analysis* (1997), p. 299
8. K. Wills, S. Vaughan, C. Duvvury, O. Adams, J. Bartlett, Photoemission testing for EOS/ESD failures in VLSI devices: advantages and limitations, in *Proceedings of International Symposium for Testing and Failure Analysis* (1989), p. 183
9. Hitachi Semiconductor, [Online]. Available: http://www.alldatasheet.com/datasheet-pdf/pdf/64049/HITACHI/HD74LS367A.html
10. Texas Instruments, [Online]. Available: http://www.ti.com/lit/ds/symlink/sn74ls06.pdf
11. Texas Instruments, [Online]. Available: http://www.ti.com/product/LM3900
12. On Semiconductor, [Online]. Available: https://www.alldatasheet.com/datasheet-pdf/pdf/12650/ONSEMI/SN74LS365A.html

Human Reliability Analysis

CQB—A New Approach for Human Factor Modeling

P. V. Varde and Arihant Jain

Abstract Even though human factor has been found to be one of the major contributors to accidents, the state of the art shows that there is no single approach to model human reliability. It is well-recognized fact that integrating human in engineering system model is a complex and challenging task, particularly for emergency scenario. Hence, a robust human model is required to have a dependable human factor modeling approach. The cognition, consciousness, conscience, and brain (C^3B)-based approach is presented in this paper. This approach has a comprehensive framework that enables capturing of human elements that were not addressed in earlier approaches. The traditional approaches human stress employing indirect methods, like available time window, design of man-machine interface, etc. In CQB model, stress is considered as measurable parameter, where experiments are performed to ensure that most an effective framework for stress modeling can be developed. This paper will discuss the R&D efforts and results of the analysis performed in support of risk analysis.

Keywords CQB · Human factor modeling · Human error · Reliability

1 Introduction

Human interaction is an integral and essential part of most of the engineering systems being used today. Human-induced error forms a major contributing factor for industrial accidents across the world [1]. Analysis of major nuclear accidents like Three Mile Accident, Chernobyl, and Fukushima revealed that the accidents can be prevented by proper decision making and mitigation actions, failure of which can be

P. V. Varde · A. Jain (✉)
RRSD, Bhabha Atomic Research Center, Mumbai, India
e-mail: arihantj@barc.gov.in

P. V. Varde
e-mail: varde@barc.gov.in

© Springer Nature Singapore Pte Ltd. 2020
P. V. Varde et al. (eds.), *Reliability, Safety and Hazard Assessment for Risk-Based Technologies*, Lecture Notes in Mechanical Engineering, https://doi.org/10.1007/978-981-13-9008-1_63

743

attributed to human factors [2]. This makes it very important to model the human reliability in a proper way to address the possibility of human-induced failures.

The work on human reliability was started by Sandia National Laboratories in 1952 [3]. USNRC-sponsored technique for human error-rate prediction was first presented in 1963 [4] followed by a later version in the 1980s [5]. Rasmussen developed WASH-1400 [6] in 1975 as part of PRA for boiling water reactors and pressurized water reactors. ASEP [7] was developed to determine the effect of human error on risk (core damage frequency). Human cognitive reliability (HCR) [8], success likelihood index [9], human error assessment and reduction technique (HEART) [10], time reliability correlation (TRC) [11], and systematic human action reliability procedure (SHARP) [8] are other first-generation techniques developed in the 1980s. Due to the unavailability of human error data, these techniques were based on expert opinion and analysis using cognitive aspects of human behavior.

Various second-generation techniques developed in the 1990s are the cognitive reliability evaluation and assessment methodology (CREAM) [12] and technique for human error analysis (ATHENA), and success plant risk analysis-human reliability SPAR-H [13]. These techniques, in addition to cognitive aspects, also attempt to model dynamic aspects of human behavior. Nuclear action reliability assessment (NARA) [14], a modified version of human error assessment and reduction technique (HEART), is a third-generation technique that also incorporates the plant experience in form of human reliability data.

USNRC study on human error contribution to risk in nuclear plants [15] indicates that the contribution of human error is significant in all the analyzed events. The study concluded that around 81% of events analyzed had design changes and work practice errors while 76% had maintenance-related errors, both of which can be attributed to human factors. PRA results also indicate that the contribution of human error is significate to the risk especially when human error induces common cause failure in redundant systems. The possibilities and effects of human-induced error can be reduced either by reducing the human interaction with the systems or by addressing the human reliability in a better way. Development of automated and advanced reactor control and mitigation systems is a step forward in addressing the former option. The latter one needs a human factor model that can model the human behavior taking into account various aspects of cognition, consciousness, and conscience. This paper discusses the proposed comprehensive and robust human reliability approach (CQB) that aims at modeling the human behavior by extending cognition to consciousness and conscience.

In this paper, the next section describes the definition and importance of consciousness, cognition, and conscience on the performance of an individual. Various parameters that can be used to quantify the effect of human performance are also described. A mathematical model is proposed to determine human error probability based on stress level (cognitive parameter); consciousness and other extrinsic parameters are described in the third section. The detailed procedure to be followed for the development of CQB model is described in Sect. 4. This paper concludes by summarizing the discussion and further work being done in the development of CQB model in Sect. 5.

2 CQB (Consciousness, Cognition, Conscience, and Brain)

Significant contribution of human-induced errors in nuclear accidents emphasizes the need of human and organizational factors modeling as an attempt to improve safety. There exists a need of a comprehensive and robust model for estimation of reliability associated with various actions. The models that exist currently are mainly based on expert opinion and were developed in 1980–2000. These models do not incorporate the operating experience of the nuclear industry and hence need to be modified taking into account the plant-specific data and simulator data. Also, the existing models mainly take into account the contribution arising due to cognition, but other human aspects like consciousness, conscience, and brain are ignored.

CQB stands for cognition, consciousness, conscience, and brain (C^3B). It is a comprehensive human model that also includes the contribution from consciousness (attention, awareness) and conscience (ethics). It proposes an approach where various physiological and psychological measures of stress can be used for reliability quantification.

The four elements that form the CQB approach are consciousness, cognition, conscience, and brain. Consciousness can be defined as a state of being aware. It can also be interpreted as attention. In CQB, consciousness is related to the awareness for cognition and hence is considered a prerequisite for cognition modeling. Consciousness is not limited to an individual but is also the communication that links various individuals of an organization.

Consciousness can be classified into type I and II. Level of awareness, awakened state, and attention are some of the attributes of type I consciousness. It can be simplified as the ability to respond to any situation based on some sensory input received from various sources. It is also termed as 'lower state of consciousness.' Type II consciousness termed as 'higher state of consciousness' is related to spirituality. It is related to the culture and behavior response at higher levels (individual, plant, and national level). As reported by the National Diet of Japan in findings of Fukushima accident, the cultural and behavioral shortcomings lead to a man-made event in Fukushima [1]. Type II consciousness takes into account the existing culture of the workplace and institution that affects the human performance.

Some of the measures to quantify the consciousness may be qualification level, supervisor's rating, past service record, institutions failure record, etc. These parameters basically attempt to quantify awareness of the surroundings by taking knowledge as the basic parameter. The measures can be extended to include some other measures that directly or indirectly relate to the individual's level of awareness, alertness, and attention.

Cognition refers to *knowing*. It includes memory, perception, and judgment. It indicates the capability to take correct judgment under any stressful situation. It forms a very important part of human model and can be evaluated quantitatively through simulator experiments, qualification program and by analyzing the past service record of any individual. Stress quantification forms an important part in incorporating cognition into human model. Various parameters that are normally used for stress quantification are time window available, effect of failure on safety, etc. Apart from these physiological parameters like pulse rate, heart response (ECG), brain response (EEG), saliva testing, blood pressure, etc., can also be used for the purpose. In CQB, human error probability is determined taking into account the individual's response to any external stress. Traditional approaches lack the ability to tackle the human's behavior as an individual and assume that the response of all individuals will be similar under a given condition. Quantification of stress together with individual's consciousness level gives the results valid for the individual in question. Various simulator experiments and other available data can be used for characterization.

Conscience is related to ethics, to know right and wrong. NG-T-1.2 [16] published by IAEA established code of ethics for all working personnel in nuclear institutions. Individual's and organization's ethics play a major role in ensuring safety and security. Conscience quotient of the individual is used to incorporate conscience in CQB model. Conscience quotient can be defined as the probability of an individual to follow practices that are in public interest. This is modeled as the probability that the individual wants to take corrective action in case of any emergency or not. Detailed modelling of conscience-quotient is not in the scope of present work and hence is not discussed further.

Brain is the central processing unit for all human actions, and all the consciousness, cognitive response, and conscience come directly from it. Study of brainwaves can lead to better understanding of brain operation and process. It gives an indication of the activities going on in the brain. Human brain plays a central role in the modeling of consciousness and cognition. The sensory inputs from various sensors like eyes, ears, etc., go into the brain and are affected by the consciousness of the individual. Based on the inputs, decision is taken that again is dependent on cognitive response. Various parameters that can be related to internal or external stress affect the function.

3   Mathematical Modeling

A mathematical model incorporating the effects of consciousness, cognition, and conscience is developed to determine human error probability for an individual performing any action. Following points were considered for formulation of the model:

- Model should incorporate considerations of consciousness, cognition, and conscience.
- Available data should be incorporated in the model. This can be done in the form of base failure probability.
- Stress is considered as cognition parameter. Model should be able to relate stress level with failure.
- External parameters that may include the type of job to be done, environment, etc., must form part of the model.

Taking into account above-mentioned points, the human error probability with stress ratio (S_R) is:

$$P(S_R) = \left[1 - C_{SC}(1 - P_o) \exp\left(-\frac{S_R}{B} A \right) \right]$$

P_o is reference (basic) human error probability. It comes from simulator experiments. S_R is the stress ratio.

If multiple stress quantification parameters are available, weighted average can be taken:

$$S_R = \sum_i W_i p_i$$

W_i is the weighting factor and p_i is the measured value of stress parameter.

B is the stress normalizing parameter. It quantifies the consciousness level. Simulator results can be used for the determination of B using the relation:

$$\ln(1 - P) = \ln(C_{SC}) + \ln(1 - P_o) - \frac{S_R}{B} A$$

A incorporates the effects of extrinsic parameters like type of job, working condition, etc.
C_{SC} is the conscience quotient with 1 signifying good ethical decision making and 0 signifying bad ethics.

The variation of human error probability with stress ratio is shown graphically in Fig. 1.

The proposed model takes into account the major parameters affecting human performance, viz. consciousness, cognition, and conscience along with their integration in Brain to give a human model that can be used to model the performance of an

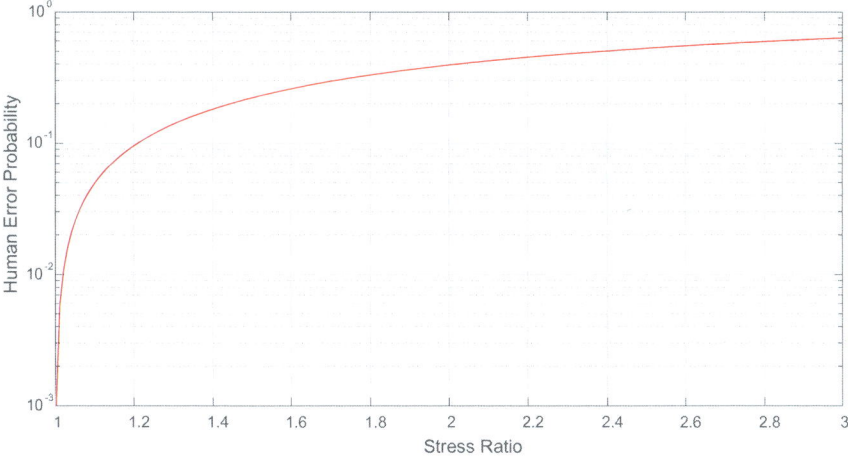

Fig. 1 Human error probability versus stress ratio ($A = 1, B = 2, P_o = 1e{-}3, C_{SC} = 1$)

individual under any condition. The model has various parameters like S_R, P_o, A, B, and C_{SC} which are to be determined using simulator experiments, analysis of human failure data, measurement of various stress parameters like ECG, EEG, saliva, pulse rate, etc.

4 CQB Methodology

CQB methodology proposes a holistic approach aimed toward modeling the human behavior taking into account various aspects of consciousness, cognition, and conscience. Probability of human error can be estimated using the human factor model and will hence help in preventing the errors. The complex process of human model development and error assessment can be carried out in acceptable and credible manner following the well-laid procedural steps shown in Fig. 2.

Plant familiarization involves developing an understanding of the plant and the organization with an aim to understand the major human-related activities carried out in the plant and the organizational setup that affects the working and performance of individual. It gives an insight of the type II consciousness in the organization, qualification level and training methodology of the institute, conscience quotient of the individuals, and the various activities taking place in the plant. Familiarization

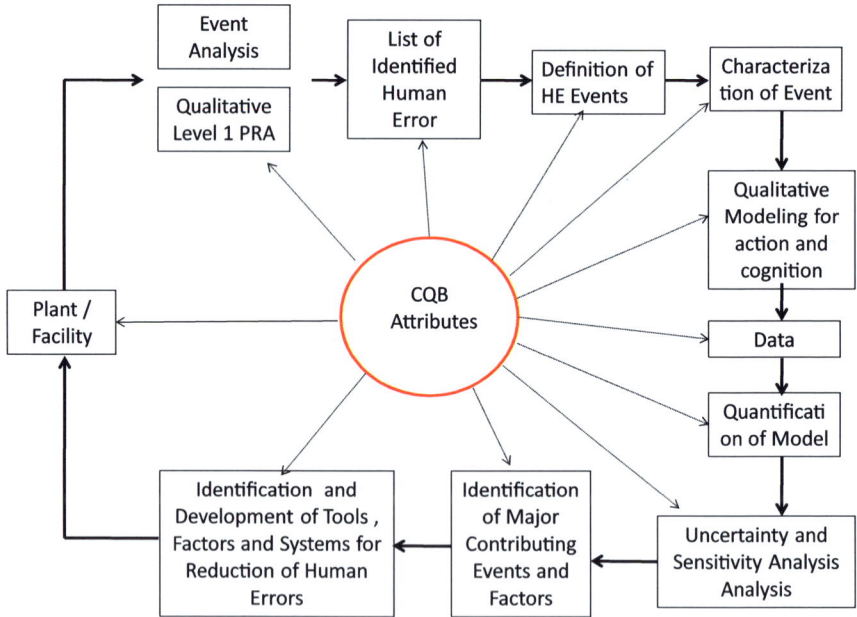

Fig. 2 Basic procedural steps in CQB methodology

also gives an idea about the risk associated with the processes and the impact of human reliability on safety. Important factors are communication framework, operator training level, human-machine interface, etc.

Identification of human error events gives the risk associated with various human errors and their impact on plant and individual's safety. Various sources like plant reports, operational procedures, PRA reports, failure records, etc. can be referred in order to obtain human error data. Analysis of past failure data also gives an understanding of various deficiencies that might exist in the form of human-machine interface or inadequate EOPs.

The identified events are then to be screened and characterized among various categories based on error type, safety impact, stress level, work type, etc. This helps in determining the various factors necessary for quantification of human error probability for other events. The various parameters used in the proposed model can be determined using the characterized human error data.

The next step is qualitative assessment of the identified events. Qualitative assessment gives a logical model to be incorporated for the analysis. Some of the methods that can be used for the analysis are operator action tree, HRA tree, and confusion matrix. It gives a cognitive understanding of the event under consideration. The logical model obtained can be used for quantification in the next step of CQB analysis.

The next step is to collect the data from various sources that may include plant failure reports, generic sources, simulator results, etc. In case the data sources are not available, expert opinion can be used to qualitatively generate data for analysis. Data collection results along with the logical model available from the previous step are used for quantification of the human error probability. The understanding of cognitive and consciousness parameters can be used to determine the values of parameters required in the mathematical model. To carry out initial assessment available sources and models like THERP, ASEP, HCR, etc., can be used. The understanding developed on the parameters of the CQB model can then be used to evaluate human error probability in other cases too. It will form an integral part of risk assessment studies.

5 Conclusion

The CQB human factor model incorporates the effect of consciousness, cognition, and conscience to determine the human error probability of an individual performing any action. The mathematical model developed takes into account the stress acting on the individual, consciousness level, the effect of extrinsic factors like environment and type of job to predict the human error probability. Basic procedure described for the development of CQB model outlines the major steps required, the final result being a model that can predict the human performance taking into account the individual's ability to handle stress, the level of knowledge and skill. Conscience generating from ethical inclination of the individual is also captured in the model.

The proposed model requires parameters that can be derived by analysis of past failures, generic data sources, simulator test results, and expert opinion. The work on data analysis and parameter estimation is currently going on with an aim to employ the model in risk assessment of Indian research reactors being operated by BARC. The model once developed and validated will then be used in place of currently available models like HCR, THERP, etc.

References

1. M. Stringfellow, Accident analysis and hazard analysis for human and organizational factors, Ph.D. thesis, Massachusetts Institute of Technology, 2010
2. J. Vucicevic, Human Error—Crucial Factor in Nuclear Accidents (2016)
3. A. Swain, Human reliability analysis: need, status, trends and limitations. Reliab. Eng. Syst. Saf. **29**, 301–313 (1990)
4. A. Swain, A Method for Performing a Human Factors Reliability Analysis, Monograph SCR-685 (Sandia Laboratories, USA, 1963)
5. A. Swain, H. Guttmann, Handbook of Human Reliability Analysis with Emphasis on Nuclear Power Plant Applications, NUREG-CR-1278 (U.S. Nuclear Regulatory Commission, Washington, 1983)
6. N. Rasmussen, "WASH-1400," Nuclear Safety Study (1975)
7. A. Swain, *Accident Sequence Evaluation Program-Human Reliability Analysis Procedure, NUREG/CR-4772* (US Nuclear Regulatory Commission, Washington, DC, 1987)
8. G. Hannaman, A. Spurgin, Y. Lukie, *Human Cognitive Reliability (HCR) Model for PRA Analysis, EPRI NUS-4531* (Electric Power Research Institute, USA, 1984)
9. D. Embrey et al., SLIM-MAUD (Success Likelihood Index Method–Multi Attribute Utility Decomposition): an approach to accessing human error probability using structural expert judgment, NUREG/CR-3518, 1984
10. J. Williams, HEART (Human Error Assessment and Reduction Technique)—a proposed method for assessing and reducing human error, in *9th Advances in Reliability Technology Symposium*, London, UK (1984)
11. R. Hall, J. Fragola, J. Wreathall, Post Event Human Decision Errors: Operator Action Tree/Time Reliability Correlation, NUREG-CR-3010, BNL-NUREG-51601 (U.S. Nuclear Regulatory Commission, 1983)
12. E. Hollnagel, Cognitive Reliability and Error Analysis Method: CREAM (Elsevier, Amsterdam, 1998)
13. G. Gertman, H. Blackman, J. Marble, J. Byres, C. Smith, The SPAR-H Human Reliability Analysis Method, NUREG/CR-6883 (U.S. Nuclear Regulatory Commission, Washington, DC, 2005)
14. B. Kirwan, Nuclear Action Reliability Assessment (NARA)—A data-based HRA tool. Saf. Reliab. **25**(2) (2005)
15. U.S. Nuclear Regulatory Commission, Review of Findings for Human Error Contribution to Risk in Operating Plants, Job Code E8238, INEEL/EXT/-01-01166, NUREG, Office of Nuclear Regulatory Research (U.S. Regulatory Commission, Washington, DC, 2001)
16. International Atomic Energy Agency, Establishing a Code of Ethics for Nuclear Operating Organizations, IAEA Nuclear Energy Series No NG-T-1.2 (IAEA, Vienna, 2007)

Effects of Human and Organizational Factors on the Reliability and Maintainability of CNC Turning Center

Rajkumar B. Patil, Basavraj S. Kothavale and Rajendra S. Powar

Abstract Human and Organizational Factors (HOFs) play an important role in the safe, reliable, and maintainable operation of the CNC turning center (CNCTC). Several human performance influencing factors (PIFs) and organizational factors (OFs) influence the human reliability. In this paper, some human PIFs and OFs which may affect the human reliability during maintenance phase are defined and considered for the prioritization according to their criticality using the expert judgments. It is observed that experience is the most important human performance influencing factor (PIF) and safety culture is the most critical organizational factor (OF) affecting the human reliability. The time-between-failure (TBF) and time-to-repair (TTR) data significantly influenced by HOFs are analyzed using the techniques of reliability and maintainability, and the results of the analysis are compared with those of the TBF and TTR data which are not significantly affected by HOFs. The field failure and repair data were sorted considering the influence of hardware, software, and HOFs using expert judgments and outcomes of reliability and maintainability analysis. It has been observed that 16.33% of the total failures and 15.49% of total repairs are significantly influenced by HOFs. Nearly 66% of the total failures and repairs are due to hardware system. The reliability and maintainability of the CNCTC are greatly influenced by HOFs. The HOFs can reduce the expected life of the components or sub-systems of the CNCTC by 33%.

Keywords Human and organizational factors · Human reliability · Maintainability analysis · Performance influencing factors · Reliability analysis

R. B. Patil (✉)
Annasaheb Dange College of Engineering and Technology, Ashta, Sangli, India
e-mail: rajkumarpatil2009@gmail.com

B. S. Kothavale
MAEER's MIT College of Engineering, Kothrud, Pune, India
e-mail: basavraj.kothavale@mitcoe.edu.in

R. S. Powar
Faculty of Engineering, D. Y. Patil Technical Campus, Talsande, Kolhapur, India
e-mail: rspowar68@gmail.com

© Springer Nature Singapore Pte Ltd. 2020   751
P. V. Varde et al. (eds.), *Reliability, Safety and Hazard Assessment for Risk-Based Technologies*, Lecture Notes in Mechanical Engineering,
https://doi.org/10.1007/978-981-13-9008-1_64

1   Introduction

CNC turning center (CNCTC) is one of numerically controlled machine tools used in manufacturing industries. Human and Organizational Factors (HOFs) play an important role in the safe, reliable, and maintainable operation of the CNCTC. Several human performance influencing factors (PIFs) such as training, experience, stress, work memory, physical capacity, access to the equipment and organizational factors (OFs) such as safety culture, staffing, problem-solving resources, communication influence the human reliability in the maintenance phase. Most of the times the human errors occur during maintenance phase. Therefore, there is a need to identify the significant HOFs occurring. The expert judgments are collected for prioritizing these HOFs according to their criticality. Furthermore, the time between failure (TBF) and time to repair (TTR) data influenced by HOFs are identified and sorted. Furthermore, the TBF and TTR data significantly influenced by HOFs are analyzed using the techniques of reliability and maintainability, and the results of the analysis are compared with those of the TBF and TTR data which are not significantly affected by HOFs. The present work is supported by SPM Tools, Ichalkaranji, Maharashtra (India), who is one of the leading manufacturers of special purpose machine tools. The required data for reliability and maintainability analysis are collected from SPM Tools and several industries from FIE Group, Ichalkaranji.

2   Literature Survey

Human faults and human failures are the two key aspects of human reliability [1]. Human fault is the error caused by intentional or negligent behavior, often punishable. Human failure is the common and massive errors or mistakes that have several consequences. These mistakes can be easily identified, diagnosed, and generally excusable. It is expected that the human errors or mistakes are the part of every functioning. These errors or mistakes are considered as the most undesirable and unacceptable aspects. Although several errors are often identified, and corrected before leading to harm and are often a way of predicting forthcoming problems [2].

Human error is one of the sources of accidents and possible negative consequences in areas such as aerospace, nuclear, process, manufacturing, and other complex industries. The probability of human error or mistake has to be monitored very closely for its possible effect on system malfunction or failure. The critical human errors need to be managed effectively in order to reduce failure rates. In this context, it is essential to take into account the available failure and repair data on complex systems and their relationships with human errors in order to estimate their probabilities. New methods need to be recommended based on error state exclusion and systematic learning to find out how to appropriately manage the occurrence of human error [3].

Several industries developed techniques based on expert judgment such as Success Likelihood Index Method (SLIM), Justified Human Error Data Information (JHEDI), technique for human error-rate prediction (THERP), and human error assessment and reduction technique (HEART) for preliminary assessment of Human Error Probability (HEP) [4]. Incorporating HEP in the development of design and operational procedures can considerably improve the overall system reliability [3, 5].

Human Reliability Assessment (HRA) techniques have been applied in the field of transportation and engineering [5, 6]. The HRA can be carried out by different approaches such as: Cognitive Reliability Error Analysis Method (CREAM), and A Technique for Human Error Analysis (ATHENA) [7–9]. Since maximum activities in industries take in human participation in terms of management, decision making, supervision, inspection, monitoring, and maintenance, human errors appear unavoidable. Errors can occur at any stage due to the failure to take a necessary action. There are many sources for human error, including poor system design, lack of training, poor work layout, inadequate lighting, improper tools, loud noise, and poor working procedures [8]. Human errors can be classified into six categories as design, assembly, installation, inspection, operation, and maintenance [9].

There may be many reasons why error occurs during maintenance and include inadequate training, inadequately written maintenance procedures, obsolete maintenance handbooks, experience, complex maintenance tasks, poor work layout, harsh environments, and fatigue [10, 11]. The need of experience and proper training to minimize maintenance mistakes were discussed by several researchers. The employees with necessary knowledge, experience, higher skill level, better emotional stability, higher moralities, and greater gratification with the workgroup have less likelihood of making errors.

3 Prioritization of Human PIFs and OFs of the CNCTC According to Their Criticality

3.1 Prioritizing Human PIFs

PIFs can be defined as basic human mistake tendencies and the making of error-likely situations. Various PIFs such as well-designed programs, maintenance procedures, and proper training will vary from the best practicable to worst possible. PIFs and performance are closely linked to each other, if PIFs are optimal, performance will be optimal and the probability of occurrence of error will be minimized. The significant PIFs are to be identified and allied with the problem areas that will increase the chances of occurrence of human errors. In this view, twelve human PIFs which may increase the human error probability are taken for prioritization. Table 1 gives the human PIFs under considerations.

As discussed above, 12 human PIFs are chosen for the CNCTC reliability analysis. The severity rank data are collected from experts in the field of CNCTC. The most critical human PIF is assigned with rank '1' and the least critical human PIF with

Table 1 Human PIFs under consideration

Human PIFs	PIF description
Training	Training is related to an ability of employee to recognize each maintenance action most effectively and do the required activities to complete the maintenance task
Experience	Experience is linked to how an employee will complete the maintenance activities effectively
Stress	Stress is related to the incapability of an employee to complete the task successfully due to nervousness and pressure
Work memory	Work memory is associated with short- and long-term memory of an employee to execute the maintenance program.
Physical capabilities and condition	This is related to the functional skill, ability, and condition of the employee to maintain the system under the given working environment
Work environment	This is related to how an operator identifies the conditions of the workplace used for maintenance
Access to equipment	Access to equipment or tools refers to the authority of the worker to easily access the equipment or tools
Distraction	Distraction is a thing that prevents someone from concentrating on something else
Behavior	Morale, motivation, and attitude decide the readiness to complete maintenance tasks, the amount of efforts an employee devote to various activities and the state of mind of the employee
Fatigue	Fatigue is related to the mental and physical weariness resulting from too little rest, high demands, or overexertion
Time pressure	Time pressure is like task pressure; however, it adds the component of perception of time to the number of tasks; this time perception can affect the stress level of the employee beyond the stress of having too many tasks
Task difficulty	Task difficulty is defined as the physical and mental demands of the task at hand

rank '12.' Table 2 shows the list of human PIFs and corresponding severity ranks given by each expert.

The severity rank given by each of the experts is generally influenced by several parameters such as experience, number of CNCTC under supervision, number of employees or workers under supervision, knowledge, operating conditions of the CNCTC systems, organizational culture, experience in various organizations, and so on. However, in this work, three major parameters are considered, viz. 'experience,' 'number of CNCTC systems under supervision,' and 'number of employees under supervision' in order to assign weight to each expert for normalizing the severity

Table 2 Human performance influencing factors and severity ranks

S. No.	Human PIFs	Severity rank given by the experts											
		1	2	3	4	5	6	7	8	9	10	11	12
1	Training	3	2	3	4	2	3	2	1	4	1	2	2
2	Experience	2	1	1	1	1	2	1	2	1	2	1	1
3	Stress	11	11	12	10	11	11	3	6	5	13	12	10
4	Work memory	6	5	7	6	5	7	4	3	11	7	8	6
5	Physical capacity and condition	4	3	5	5	6	4	7	7	2	6	6	5
6	Work environment	5	4	4	7	7	6	6	5	6	5	5	4
7	Access to equipment	8	6	6	8	4	5	11	11	7	8	4	7
8	Distraction	12	8	11	11	10	12	12	9	12	10	11	12
9	Behavior	1	7	2	3	3	1	5	4	3	4	3	3
10	Fatigue	7	10	9	9	8	8	10	8	9	9	10	9
11	Time pressure	10	9	8	2	12	9	9	10	8	2	7	8
12	Task difficulty	9	12	10	12	9	10	8	12	10	11	9	11

Table 3 Scale for assigning the weight for each of the experts

Experience in years	Number of CNCTC systems under supervision	Number of employees under supervision	Scale
>16	>40	>24	1
14–16	35–40	21–24	2
12–14	30–35	18–21	3
10–12	25–30	15–18	4
8–10	20–25	12–15	5
6–8	15–20	9–12	6
4–6	10–15	6–9	7
2–4	5–10	3–6	8
<2	<5	<3	9

ranks given by each expert. Various classes for each of the selected parameters are formed. Minimum scale of '1' is assigned for the maximum value of the selected factors and maximum scale of '9' is assigned for the minimum value of the selected parameters. Table 3 gives the scale for assigning the weight for each of the experts for the selected parameters.

Table 4 Estimation of overall weight value of each expert

Expert	Experience in years		Number of CNCTC systems under supervision		Number of employees under supervision		Overall weight of expert
	Weight	Relative weight	Weight	Relative weight	Weight	Relative weight	
1	1	0.0161	2	0.0270	2	0.0308	0.0739
2	6	0.0968	7	0.0946	7	0.1077	0.2991
3	3	0.0484	4	0.0541	3	0.0462	0.1486
4	5	0.0806	8	0.1081	5	0.0769	0.2657
5	6	0.0968	8	0.1081	8	0.1231	0.3280
6	4	0.0645	6	0.0811	4	0.0615	0.2071
7	8	0.1290	8	0.1081	7	0.1077	0.3448
8	5	0.0806	6	0.0811	8	0.1231	0.2848
9	1	0.0161	3	0.0405	1	0.0154	0.0721
10	7	0.1129	6	0.0811	6	0.0923	0.2863
11	8	0.1290	8	0.1081	6	0.0923	0.3294
12	8	0.1290	8	0.1081	8	0.1231	0.3602

The weight values are assigned for each of the experts for all the selected parameter using scale developed in Table 3. The relative weight value of each expert for each of the parameters is estimated. The sum of relative weight values of all parameters gives the overall weight of the expert. Table 4 shows the estimation of overall weight value of each expert.

The severity rank given by each expert for each of the PIFs (Table 2) is multiplied by the overall weight of the expert value of the corresponding expert (Table 4) to estimate normalized severity rank. The overall severity value of each PIF is calculated by summing all the corresponding normalized severity rank. The severity rank of all PIFs is decided based on the overall severity value. The most critical PIF is one having minimum overall severity value and the least critical PIF having maximum overall severity value. Table 5 shows the estimation of severity rank for human PIFs. Experience, training, and behavior are the most critical human PIFs which affect the reliability and maintainability of the CNCTC.

3.2 Prioritizing OFs

The organizational factors (OFs) are nothing but the factors that are defined by or under the control of the organization. Further, the management is defined as the personnel at the upper layers of the organization. In case of small- and medium-scale industries, the role of organization and management is not differentiable. Therefore,

Table 5 Estimation of overall severity rank of human PIFs

S. No.	Human PIFs	Normalized severity rank given by expert												Overall severity value	Severity rank
		1	2	3	4	5	6	7	8	9	10	11	12		
1	Training	0.2	0.6	0.4	1.1	0.7	0.6	0.7	0.3	0.3	0.3	0.7	0.7	6.5	2
2	Experience	0.1	0.3	0.1	0.3	0.3	0.4	0.3	0.6	0.1	0.6	0.3	0.4	3.9	1
3	Stress	0.8	3.3	1.8	2.7	3.6	2.3	1.0	1.7	0.4	3.7	4.0	3.6	28.8	10
4	Work memory	0.4	1.5	1.0	1.6	1.6	1.4	1.4	0.9	0.8	2.0	2.6	2.2	17.5	6
5	Physical capacity and condition	0.3	0.9	0.7	1.3	2.0	0.8	2.4	2.0	0.1	1.7	2.0	1.8	16.1	5
6	Work environment	0.4	1.2	0.6	1.9	2.3	1.2	2.1	1.4	0.4	1.4	1.6	1.4	16.0	4
7	Access to equipment	0.6	1.8	0.9	2.1	1.3	1.0	3.8	3.1	0.5	2.3	1.3	2.5	21.3	7
8	Distraction	0.9	2.4	1.6	2.9	3.3	2.5	4.1	2.6	0.9	2.9	3.6	4.3	32.0	12
9	Behavior	0.1	2.1	0.3	0.8	1.0	0.2	1.7	1.1	0.2	1.1	1.0	1.1	10.7	3
10	Fatigue	0.5	3.0	1.3	2.4	2.6	1.7	3.4	2.3	0.6	2.6	3.3	3.2	27.0	9
11	Time pressure	0.7	2.7	1.2	0.5	3.9	1.9	3.1	2.8	0.6	0.6	2.3	2.9	23.2	8
12	Task difficulty	0.7	3.6	1.5	3.2	3.0	2.1	2.8	3.4	0.7	3.1	3.0	4.0	30.9	11

Table 6 Organizational factors of the CNCTC under consideration

Organizational factors	Description
Safety culture	Safety culture is the values, attitude, and beliefs of the organization towards the safety of the employee and public
Staffing	Staffing is the way that the organization hires employees and allocates responsibilities
Problem-solving resources	The problem-solving resources comprise of the necessary information, procedures, and tools required to perform a task. Operators and maintenance personnel are typically limited to use the resources at hand to complete tasks
Training program	Related to the type and quality of training program organized by the organization to develop the ability of the employees
Communication	Communication is the ability of an employee to pass the information to other employees effectively
Team-based factors	A team is defined as any group of employees expected to work together to accomplish a maintenance activity. Team members are expected to interact directly, either in person or in writing
Task assigning and scheduling	Task assigning and scheduling is defined as how the organization plans and allocates tasks to employees. It includes maintenance activity planning and scheduling, resource allocation, and prioritization of planned activities
Corrective action plan	The corrective action program is the organizational approach to correcting known deficiencies
Motivation	Motivation is given by offering different levels of incentives for successful task completion on the task
Direct supervision	Direct supervision serves as the link between employees and management. Sometimes direct supervision can be separated from management because direct supervisor works with and assign tasks to the employees whereas the management is indirectly involved in various activities through the supervisor

in this analysis, the organization and management are assumed to be same and are playing same role. The management can be interpreted as the authorized person or group of persons, lead a group without being directly involved in all aspects of day-to-day operations. The OFs include the attitude and behavior of a certain organization that influence the performance of employees which affect the system reliability and maintainability. The OFs taken for the analysis are summarized in Table 6.

The severity rank data of these 10 OFs are collected. The most critical OF is assigned with rank '1' and the least critical OF with rank '10.' As in case of human PIFs, the severity rank given for each OF by each expert is also influenced by several parameters and the parameters, viz. 'experience,' 'number of CNCTC systems under supervision,' and 'number of employees under supervision' are taken as it is to assign weight value for the experts. Table 7 shows the estimation of overall severity value and rank for the selected OFs. It reveals that safety culture, problem-solving resources, corrective action program, and training program are the most critical OFs which affect the reliability and maintainability of the CNCTC.

4 Effects of HOFs on the Reliability and Maintainability of the CNCTC

The HOFs influence the reliability and maintainability of the CNCTC. If the maintenance is greatly influenced by HOFs, then it may increase the time required for maintenance of a component and the next failure of the same component may take place. As such, if the time required for maintenance is far greater than the mean time to repair (MTTR) and the failure of the component takes place far before the mean time between failure (MTBF), then repairs or failures are assumed to be influenced by HOFs. In this context, the failure and repair times influenced by HOFs can be identified in consultation with experts in the field. The failure and repair data are sorted according to the pure failures and repairs due to hardware, software, and HOFs.

Figure 1 gives failure frequency analysis of hardware, software, human, and organizational factors. It is seen that the hardware failures dominated the total failures of the CNCTC. Nearly 66% of the total failures are purely hardware failures. Software failures contributed nearly 17.27% of the total failures whereas 16.33% of the total failures were influenced by HOFs. Failure frequency analysis reveals that 16 failures out 100 are directly influenced by HOFs.

4.1 Best-Fit Distribution

The failure data of 'hardware,' 'software,' and 'HOFs' are further analyzed, and best-fit distribution is estimated. Table 8 shows the estimation of best-fit distribution for TBF data of hardware, software, and HOFs using Kolmogorov–Smirnov (K–S) test and correlation coefficient (CC). It is seen that Weibull 2P is the best-fit distribution

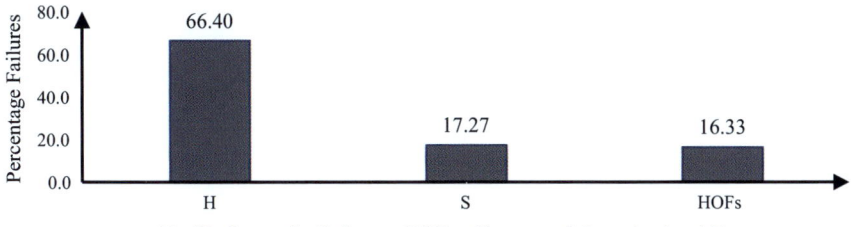

Fig. 1 Failure frequency of hardware, software, and human and organizational factors

Table 7 Estimation of overall severity rank of OFs

S. No.	Organizational factor	Normalized severity rank given by expert												Overall severity value	Severity rank
		1	2	3	4	5	6	7	8	9	10	11	12		
1	Safety culture	0.5	1.8	0.1	0.8	0.7	0.8	1.0	0.3	0.2	0.6	0.3	0.4	7.5	1
2	Staffing	0.7	1.5	0.3	2.1	2.0	0.4	0.3	0.9	0.7	2.0	3.0	2.2	16.0	6
3	Problem-solving resources	0.1	0.3	0.6	0.5	1.6	0.2	1.7	0.6	0.1	1.1	1.6	0.7	9.4	2
4	Training program	0.1	2.4	0.4	1.1	0.3	0.6	2.1	1.1	0.3	2.3	0.7	1.1	12.5	4
5	Communication	0.3	2.1	0.7	1.3	2.3	1.0	0.7	1.7	0.4	1.7	1.0	1.8	15.1	5
6	Team-based factors	0.4	3.0	1.2	2.4	2.6	1.2	2.4	2.3	0.4	1.4	2.3	3.6	23.3	8
7	Task assigning and scheduling	0.4	2.7	1.3	2.7	3.0	2.1	3.4	2.6	0.1	0.9	2.0	2.5	23.6	10
8	Corrective action program	0.2	0.6	0.9	0.3	1.0	1.4	1.4	1.4	0.5	0.3	2.6	1.4	12.1	3
9	Motivation	0.6	0.9	1.5	1.9	3.3	1.9	3.1	2.8	0.6	2.9	1.3	2.9	23.6	9
10	Direct supervision	0.7	1.2	1.0	1.6	1.3	1.7	2.8	2.0	0.6	2.6	3.3	3.2	22.1	7

Table 8 Estimation of best-fit distribution for TBF data of hardware, software, and HOFs

Factors	Test	Exp. 1P	Exp. 2P	Log-normal	Normal	Weibull 2P	Weibull 3P	Best-fit distribution
Hardware	K-S	0.99	1	0.9941	1	0.3398	0.3761	Weibull 2P $\beta = 0.9254$, $\theta = 2181$
	CC	−95.8	−97.6	97.61	82.88	99.45	99.52	
Software	K-S	0.19	0.17	0.8143	0.9850	0.0049	0.001	Weibull 2P $\beta = 0.9704$, $\theta = 2226$
	CC	−96.5	−96.9	96.13	93.65	99.57	99.37	
HOFs	K-S	0.37	0.07	0.5817	0.9986	0.1620	0.0089	Weibull 3P, $\beta = 0.88$, $\theta = 1386$, $\gamma = 10$
	CC	−98.9	−99.1	97.95	91.18	99.45	99.58	

for hardware and software failure data whereas Weibull 3P is the best-fit distribution for the failure data influenced by HOFs. The shape parameter (β) of all the factors under consideration is very close to 1 and therefore it is concluded that the failure rates are nearly constant. The scale parameter (θ) for HOFs is 1386 h which is very less than that for hardware ($\theta = 2181$ h) and for software ($\theta = 2226$ h). Therefore, it can be concluded that the percentage of failures due to HOFs is only 16.33% however the failure impact of HOFs is more.

The serial configuration is taken for hardware, software, and HOFs by assuming the situation of the CNCTC. Therefore, the reliability of the CNCTC can be estimated using Eq. (1).

$$R_{\text{CNCTC}}(t) = R_{\text{Hardware}}(t) \times R_{\text{Software}}(t) \times R_{\text{HOFs}}(t)$$

$$R_{\text{CNCTC}(t)} = e^{-(t/2181)^{0.9254}} \times e^{-(t/2226)^{0.9704}} \times e^{-(t-10/1386)^{0.8782}} \tag{1}$$

Reliability characteristics of CNCTC such as reliable life, reliability, and MTBF have been estimated in order to understand the failure characteristics of the hardware, software, and HOFs in detail and given in Table 9. Reliability at different time has been estimated using Eq. (1). It is seen that MTBF for HOFs is nearly 1500 h whereas the MTBF for hardware and software is nearly 2250 h. Therefore, it can be concluded that the reliability of the CNCTC is greatly influenced by HOFs. HOFs can reduce the expected life of the components or sub-systems of the CNCTC by 33%.

Table 9 Estimation of reliability, reliable life, and MTBF

Factors	Reliable life in hours for a given reliability						Reliability at a given time				MTBF
	99%	95%	90%	80%	75%	50%	2190	4380	6570	8760	
Hardware	15	88	192	431	568	1468	0.37	0.15	0.06	0.03	2261
Software	19	104	219	475	616	1526	0.37	0.15	0.06	0.02	2256
HOFs	17	57	117	261	345	923	0.23	0.06	0.02	0.00	1488

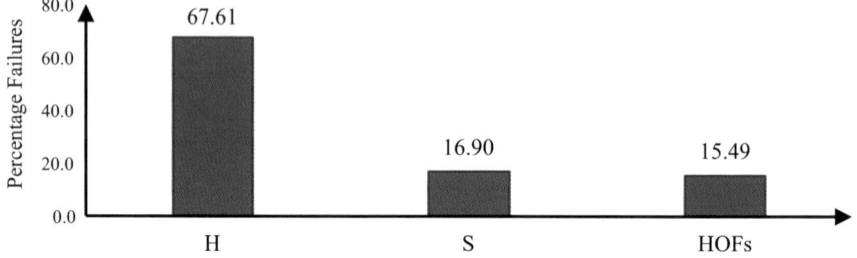

Fig. 2 Repair frequency of hardware, software, and human and organizational factors

Figure 2 shows the repair frequency analysis of hardware and software systems and the repair activity influenced by HOFs. It is seen that the hardware repairs dominated the total repairs of the CNCTC. Nearly 67.61% of the total repairs are purely due to hardware repairs. Software repairs contributed nearly 16.90% of the total repair. There are 15.49% of the total repairs activities which are influenced by HOFs. Repair frequency analysis reveals that nearly 16 repairs out 100 are directly influenced by HOFs.

Table 10 shows the best-fit distribution for TTR data of hardware and software systems and the TTR data influenced by HOFs. Log-normal distribution is the best-fit distribution for hardware, software, and HOFs for all the factors under considerations. The MTTR for hardware and software is 10 h whereas for HOFs is 16.5 h. It shows that influence of HOFs can increase the time required for maintenance activity by nearly 65%.

5   Conclusions

The analysis carried out in this paper reveals that the Human and Organizational Factors (HOFs) play an important role in the reliable, maintainable, and safe operation of the CNCTC. Twelve human PIFs and ten OFs were considered for the analysis. It has been shown that experience is the most important human PIF followed by training, behavior, work environment, physical capacity and condition, work memory, access to equipment, time pressure, fatigue, stress, task difficulty, and distraction.

Table 10 Estimation of best-fit distribution for TTR data of hardware, software, and HOFs

Class	Test	Exp. 1P	Exp. 2P	Log-normal	Normal	Weibull 2P	Weibull 3P	Best-fit distribution	MTTR
Hardware	K-S	1	1	0.9999	1	1	0.9999	Log-normal $\sigma' = 1.4918$, $\mu' = 1.2895$	10
	CC	–	−69.5	95.40	46.2	86.57	93.56		
Software	K-S	1	1	0.9635	1	0.9999	0.9751	Log-normal $\sigma' = 1.5045$, $\mu' = 1.2678$	10
	CC	–	−84.3	96.58	62.0	88.81	94.14		
HOFs	K-S	1	1	0.9896	1	0.9999	0.9931	Log-normal $\sigma' = 1.7415$, $\mu' = 1.4584$	16.5
	CC	–	−80.3	97.07	57.8	89.86	95.44		

Safety culture is the most critical OF affecting the human reliability followed by problem-solving resources, corrective action program, training program, communication, staffing, direct supervision, team-based factors, motivation, and task assigning and scheduling. Most of the times, the human errors occur during maintenance phase. Therefore, it is necessary to take corrective action on the significant HOFs to improve reliability and maintainability of CNCTC. Most of the failures of the CNCTC are due to hardware system and contribute nearly 66% of the total failures whereas nearly 17% of the total failures are typically due to software system. It has been also observed that 16.33% of the total failures and 15.49% of total repairs are significantly influenced by HOFs. The HOFs can reduce the expected life of the components of the CNCTC by 33%. The HOFs can increase the mean time to repair (MTTR) of the CNCTC by nearly 65%. Therefore, it is concluded that the reliability and maintainability of the CNCTC are significantly influenced by HOFs.

Acknowledgements This work is supported by SPM Tools, Ichalkaranji and HK Group, Ichalkaranji. Authors also would like to thank Dr. S. G. Joshi, Former Professor, Walchand College of Engineering, Sangli.

References

1. S. French, T. Bedford, S.J.T. Pollard, E. Soane, Human reliability analysis: a critique and review for managers. Saf. Sci. **49**(6), 753–763 (2011)
2. B. Reer, Review of an advances in human reliability analysis of errors of commission part 1: EOC identification. Reliab. Eng. Syst. Saf. **93**, 1091–1104 (2008)
3. A. Noroozi, N. Khazad, F. Khan, S. Mackinnon, R. Abbassi, The role of human error in risk analysis: application to pre and post maintenance procedures of process facilities. Reliab. Eng. Syst. Saf. **119**, 251–258 (2013)
4. O. Strater, H. Bubb, Assessment of human reliability based on evaluation of plant experience: requirements and implementation. Reliab. Eng. Syst. Saf. **63**, 199–219 (1999)
5. R.L. Boring, S.M.L. Hendrickson, J.A. Forester, T.Q. Tran, E. Lois, Issues in benchmarking human reliability analysis method: a literature review. Reliab. Eng. Syst. Saf. **95**(6), 591–605 (2010)
6. T. Reiman, P. Oedewald, Assessment of complex sociotechnical systems—theoretical issues concerning the use of organizational culture and organizational core task concepts. Saf. Sci. **45**, 745–768 (2007)
7. G.A.L. Kennedy, C.E. Siemieniuch, M.A. Sinclair, B.A. Kirwan, W.H. Gibson, Proposal for a sustainable framework process for the generation, validation, and application of human reliability assessment within the engineering design lifecycle. Reliab. Eng. Syst. Saf. **92**, 755–770 (2007)
8. K.M. Groth, A. Mosleh, A data-informed hierarchy for model-based human reliability analysis. Reliab. Eng. Syst. Saf. **108**, 154–174 (2012)
9. G.I.J. Zwetsloot, L. Drupsteen, E.M.M. de Vroome, Safety, reliability and worker satisfaction during organizational change. J. Loss Prev. Process Ind. **27**, 1–7 (2014)
10. S.M. Asadzadeh, A. Azadeh, An integrated systematic model for optimization of condition-based maintenance with human error. Reliab. Eng. Syst. Saf. **124**, 117–131 (2014)
11. M. Voirin, S. Pierlet, M. Llory, Availability organizational analysis: is it a hazard for safety? Saf. Sci. **50**, 1438–1444 (2012)

Investigation of Human Factors Using HFACS Framework—A Case Study for Unintended Reactor Trip Events in NPP

M. Karthick, C. Senthil Kumar and T. Paul Robert

Abstract Safety remains to be the principal concern in nuclear industry, and layered protective measures are implemented to ensure safety to public and environment. Nevertheless, accidents in a nuclear plant do occur, and human factors are confirmed to be the prime cause of these accidents. Several studies have been carried out for human factor analysis, and still, there is a need for additional methods to model human errors. In this paper, human factor analysis and classification system (HFACS) is used to analyze the causal factors for an accident/incident. HFACS is a validated and reliable model for human factor analysis and provides a simple way to evaluate the causal factors in an accident. Although HFACS framework can be applied to analyze any accident events, in this paper, it is applied to review eighteen unintended reactor trip events due to human error from 2000 to 2006. In this study, experts analyze and rate the events against each hierarchal level in HFACS. Expert judgment is sought as the uncertainty and complexity involved in modeling human and organizational factors are very high. Homogeneity of experts is assessed using inter-rated reliability and Cohen's kappa value. Expert evaluation on human factors is quantified and based on the correlation in rating, most significant factor is identified. Further, based on the information on the reported events and the analysis, statistical tests are employed to confirm the associations between the hierarchal levels in the HFACS framework for reactor trip events. The results indicate four pairs of factors, viz. resource management, planned inappropriate actions, adverse mental states, and skill-based errors have strong associations between adjacent categories. Based on these associations, adverse mental states are deemed the most potent for accident occurrences. Further analysis indicated that slips of attention, memory failures for events such a failure to open or close the valves, and perception failures for visual

M. Karthick (✉) · C. Senthil Kumar
Safety Research Institute, Atomic Energy Regulatory Board, Kalpakkam, India
e-mail: mkarthik@igcar.gov.in

C. Senthil Kumar
e-mail: cskumar@igcar.gov.in

T. Paul Robert
Department of Industrial Engineering, Anna University, Chennai, India
e-mail: prpaul@annauniv.edu

© Springer Nature Singapore Pte Ltd. 2020
P. V. Varde et al. (eds.), *Reliability, Safety and Hazard Assessment for Risk-Based Technologies*, Lecture Notes in Mechanical Engineering,
https://doi.org/10.1007/978-981-13-9008-1_65

illumination (i.e., 'skilled based errors') are the most common 'unsafe acts' committed and appear more frequently along with nuclear control room actions. The results reveal that human errors are caused mainly due to potential human factors associated with cognition. Few active failures are also identified. Thus, the analysis highlights the importance of attention toward possible human factors among others. Overall, HFACS tool proves useful in categorizing operator errors from reported events. HFACS framework and the quantification method help to determine which factors are more dominant and influential in the accident sequence and provide additional insight for the analysts into the significance and relative importance of each of the human factors.

Keywords Cognitive factor · Human error identification · Human factor analysis and classification systems · Reliability measurement · Reactor trip event

1 Introduction

While the nuclear industry has witnessed tremendous successes in safety over the last two decades, it remains as one of the risk professions worldwide from the viewpoint of safety. Safety of a system may be affected by technical difficulties, design parameters, or by human activities. The most challenging work is to improve safety by minimizing human errors in such complex systems. It has been reported in the literature that Human and Organizational Factors (HOFs) are the main root cause to large number of industrial accidents. In nuclear industry, contribution of error due to HOFs is 50–70% in large accidents [9]. The importance of human factors study in nuclear safety was highlighted by the Three Mile Island in 1979 and Chernobyl in 1986 which led the nuclear community to acknowledge the role of non-technical aspects of operations [24]. These two accidents stimulated research in psychological mechanism of human errors. Recently, the accident at the Fukushima power plant in 2011 drew more attention to the critical role of safety culture, management, and organizational factors [1, 11]. Although technology has advanced quickly in the last two decades, HOFs still remain as a major concern. Thus, there is a greater willingness among nuclear safety researchers and analysts to seek alternative, proactive solutions to model human errors. Several techniques have been developed, viz. ACCI-MAP, STAMP, and Activity Theory model to analyze the human error systematically. However, these are lacking in model representation, simplification of active and latent failure classification, domain-specific application, etc. To overcome the above limitations, researchers have applied human factors analysis and classification system (HFACS) in a human factors accident analysis framework [26]. Though HFACS was successfully utilized in different applications, this model relies on expert judgment to identify the cause of errors. Opinion from two or more experts helps to determine the potential cause. In such cases, the reliability of different judgments should also be assessed to identify the real cause.

There are many statistical methods available to measure the homogeneity and degree of agreement among the assessors. Cohen's Kappa, inter-rater reliability (IRR), and Chi-square are some methods widely employed. For example, Li et al. [16] did IRR on civil accidents; Madigan et al. [17] applied Chi-square on rail accidents; Theophilus et al. [25] employed Chi-square on Oil industry accidents, etc. However, there is no HFACS assessment studies carried out to recognize and identify human error causes for the nuclear industry. The present study therefore critically assesses expert opinions using IRR, Cohen's Kappa and Fisher's exact test on HFACS for nuclear industry to identify the potential cause.

The rest of this paper is organized as follows: The present section discusses the significance of human errors in nuclear industry. In Sect. 2, human error and the state-of-the-art review of the application of HFACS is discussed. The model, data collection, and the coding process are explained in Sect. 3. Results obtained using the models are discussed in Sect. 4. In the final Sect. 5, research outcomes and potential contributions through nuclear safety are extensively discussed.

2 Literature Survey

Literature survey has been done on HFACS and detailed in the following sub-sections.

2.1 Human Error

Human errors can be classified into two types: active human errors and latent human errors. Active errors are committed by operators in direct contact with the nuclear operations, latent errors arise in organizational and managerial spheres and their adverse effects may take a long time to become evident [20]. The active and latent failure is commonly referred to as HOFs. While these factors may very well play a separate and significant role in an operational failure, it is often a combination of several humans, organizational, and technological factors that lead to events and accidents in complex industries [8]. Prevention and reduction in the number of accidents and incidents require a reduction in human error. To minimize the human error in such complex industries, human errors need to be appropriately identified and systematically classified. The first step in this process is error identification. To determine human errors in the occurrence of incidents and accidents leads to the development of appropriate prevention and alleviation strategies. Several models are available, viz. 'Wheel of Misfortune' taxonomy [18], HFACS [22], The Technique for the Retrospective and Predictive Analysis of Cognitive Errors [23], Systems Theoretic Accident Modelling and Processes [14, 15], Human Factors Investigation Tool [10], ACCI-MAP [2], etc., to critically investigate and recognize the significant factors contributing to human error. One of the most widely used models of HFACS is adopted in this study to identify the potential human factors.

2.2   HFACS Review in a Nutshell

HFACS is a robust accident analysis and investigation tool developed by [22]. HFACS system was initially designed as an evaluation framework to examine operator errors in aviation accidents [22]. HFACS appears to be more suitable for aviation accident analysis as observed in the literature [3, 16, 21, 22]. Although this method was developed for the aviation industry, it has been widely applied to other disciplines such as railways [4], mining [19], marine [5], health care [6], nuclear industry [12], and oil and gas industry [25]. Expert judgment plays a crucial role in the application of HFACS to analyze and identify the potential causes of human error. Opinions from two or more experts could help to determine the cause correctly. However, a different view from multiple experts might also lead to wrong error identification due to non-homogeneity. It is necessary to check the consensus of the opinion for the HFACS level. Several studies have adopted different techniques to measure the inter-rater reliability and intra-rater reliability to evaluate the coding process across the various HFACS levels. For example, Li et al. [16] used Chi-square method and Ergai et al. [7] measured the inter-rater and intra-rater analysis of HFACS levels using Krippendorff's Alpha.

This study examines eighteen nuclear near-miss events, i.e., human performance-related events occurred at Indian NPPs using the HFACS framework.

3   The Methodology

This study adopts HFACS to examine potential causes of human error. The calculation of inter-rater reliability on various levels of HFACS is obtained using percentage of agreement method and reliability measurement is carried out by Chi-square and Cohen' Kappa method. Details of the model are explained below.

3.1   Human Factor Analysis and Classification System (HFACS)

The framework of HFACS is improved from the Swiss cheese model. This model consists of four levels: (a) organizational factor, (b) unsafe supervision, (c) preconditions for unsafe acts, and (d) actual unsafe acts. Organizational factor is the first level consisting of three sub-factors, viz. resource management, organizational influence, and organizational process. Unsafe supervision consists of inadequate supervision, planned inappropriate operations, failed to correct problem and supervisory violations. Environmental factors, conditions of operators, and personal factors are classified under precondition of unsafe act. Errors and violations are categorized under unsafe acts. The framework of HFACS is given in Fig. 1 [22].

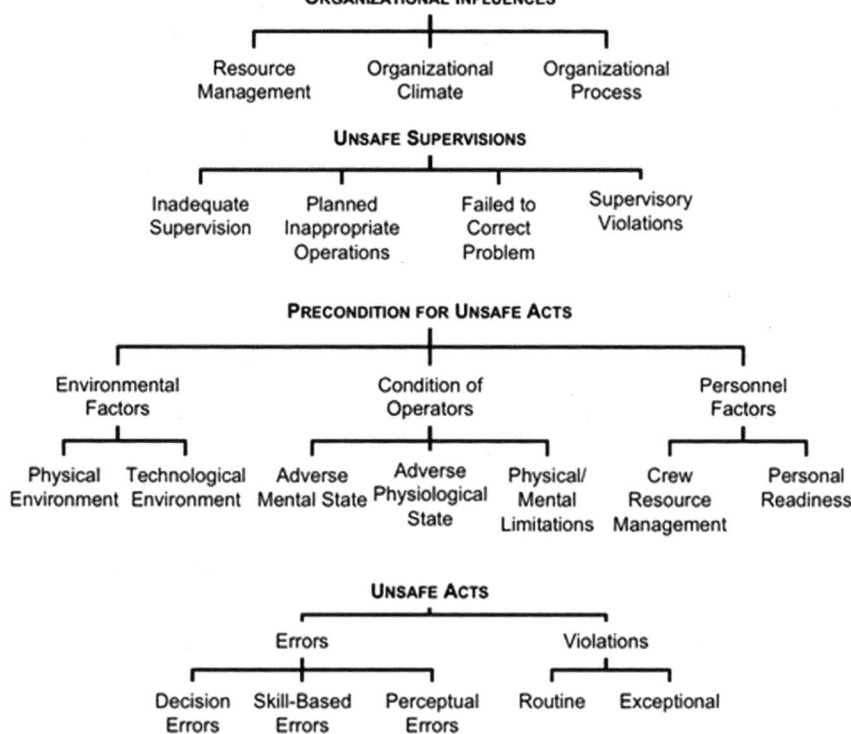

Fig. 1 HFACS framework

3.2 Data Collection

In this study, eighteen human-related performance events reported in a sample study are considered. These events include reactor trips, pump failures, bus failures, turbine trip, etc. in a NPP. The operator actions required the possible errors and causes are elicited from each of the events.

3.3 Coding Process

Two independent raters (PSA Experts) are requested to code each accident separately. Both experts have excellent knowledge in the reference domain and are well-aware about the events. Adequate time is given to both experts to ascertain about HFACS model and to analyze each event regarding human factor aspect. Two excel sheets are prepared with HFACS levels for each event. The presence or absence of each category is evaluated from the narrative of each accident report. Each HFACS level is assessed only once so that this figure indicates the existence or absence of each category in 18 different categories within an accident. If there are contradictions in the classification of an accident, the raters' convened and addressed their observations.

4 Results and Discussion

4.1 Inter-rater Reliability

At the inception, the analyses of accidents were performed using frequency counts. Before the resolution of any inconsistencies in coding between the two raters, the index of concordance (IoC) was used to assess the inter-rater reliability (IRR). The ratio of the agreeing pairs out of all pair using IoC is as follows [Agreements/(Agreements + Disagreements)]. A simple statistical analysis of Cohen's Kappa (κ) is also performed to measure IRR for all categories [13]. Equation (1) shows the computational method for Kappa.

$$\kappa = \frac{P_\text{o} - P_\text{e}}{1 - P_\text{e}} \tag{1}$$

where P_o is a relative observed agreement among raters and P_e is conditional probability of chance agreement. According to Landis and Koch [13], Kappa value scale is commonly between 0 and 1, in which 0 represents complete independence and 1 indicates extreme dependence. Generally, Kappa value greater than 0.40 considered as acceptable for strong agreement between the raters. The computation of frequency counts, Kappa, and IRR is shown in Table 1.

From the results, it is observed that the Kappa values are more than 0.40 in half of the categories, which is acceptable [13]. Kappa values are less than 0.40 in some of the categories which resulted from low observed frequencies of rater's opinion. These low-observed frequencies are the result of disagreement between raters in the category. However, in view of the fact that Kappa values are more than 0.40 in more than half of the categories and IRR range between 67 and 91%, the Cohen Kappa method is found suitable for this study.

Furthermore, causal factors are recognized at all levels. As can be seen in Fig. 2, HFACS factors concerning nuclear accidents, unsafe act errors contributed to 19.04% cases. Preconditions for unsafe acts are associated with 43.82% of the cases while unsafe leadership was identified in 20% of the cases, whereas organizational influences are identified in 17.14%. As anticipated, a large number of conditions of operator factor were identified within precondition of unsafe acts, particularly adverse mental state (70%) such as memory, attention, and workload. It is found that, in the unsafe act level, skill-based errors were associated with the most significant percentage of accidents. Approximately 11.33% of all events were associated with at least one skill-based error in this category followed by knowledge-based 4.7% and

Table 1 Computation of IRR and Kappa for each category of HFACS for all 18 events

HFACS category	Frequency	Percentage	Cohen's Kappa	Inter-rater reliability
Level-4: Organizational influences				
Resource management	5	27.7	0.78	0.78
Organizational climate	3	16.6	1	1
Organizational process	10	55.5	0.88	0.91
Level-3: Unsafe supervision				
Inadequate supervision	8	44.4	0.5	0.71
Planned inappropriate actions	5	27.7	0.76	0.75
Fail to correct a problem	3	16.6	0.76	0.67
Leadership violations	5	27.7	0.85	0.80
Level-2: Preconditions of unsafe acts				
Physical environment	8	44.4	0.11	0.70
Technological environment	5	27.7	0.64	0.76
Adverse mental state	14	77.7	0.82	0.82
Adverse physiological state	2	11.1	0.86	0.83
Adverse psychological states	3	16.6	0.67	0.73
Physical/mental limitations	1	0.05	0.76	0.67
Crew resource management	9	0.5	0.47	0.81
Personal readiness	4	22.2	0.75	0.72
Level-1: Unsafe acts				
Knowledge-based errors	5	27.7	1	1
Skill-based errors	12	66.6	0.72	0.86
Rule-based errors	1	0.05	0.85	0.8
Routine	2	11.1	1	1
Exceptional	0	0.0	–	–

Fig. 2 Distribution of causal
factors

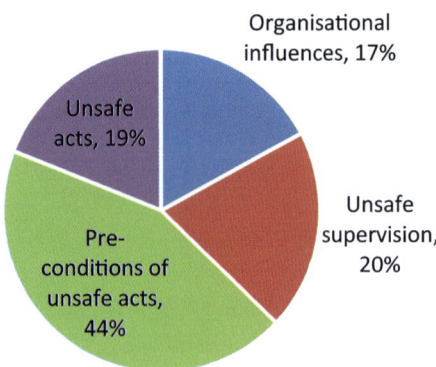

rule-based 0.9%. Finally, the occurrences of accidents related to rule-based errors were relatively low. However, an operator may perform a task incorrectly merely because the correct procedure is not known either due to lack of training or retention of information. Regardless, these types of errors suggest that additional training or procedural aids like checklists may prove useful. Similar to accidents associated with violations, those attributable at least in part to violations of rules and regulations are associated with 1.9% of the accidents examined.

4.2 Statistical Analysis of HFACS

A Chi-square test was carried out to examine the null hypotheses of homogeneity and measure the statistical significance between factors in HFACS using 2×2 table. Chi-square test is further compared with Fisher's test as shown in Table 2. Fisher's exact test measures the strength of the relationship between categories that showed significant relationship in Chi-square. One of the notable findings is that there are no significant differences between at $\alpha = 0.05$ in both tests. p-value less than 0.001 is considered as significant relationship between categories. For example, errors and violations are influenced by condition of operators, supervisory violations, etc.

The significant association between the two factors is shown in Table 2. It is observed from Table 2 that there is a link between resource management and inadequate supervision, χ^2 (1, 18 = 10.08, $p < 0.001$) as well as between resource management and supervisory violations, χ^2 (1, 18 = 9.791, $p < 0.001$). Both factors are more often involved in nuclear events occurring in reactor trip than in pump failure. This indicates that majority of causal factors at this level fell into the 'resource management' category and the most often cited example involved training, inadequate procedure, procedure unavailability, procedure not clear. Confusion of roles, shift scheduling, and task schedule was identified as problems associated with worker schedule in the organizational process. There is also a strong relationship between adverse mental state and skill-based error, χ^2 (1, 18 = 11.49, $p < 0.001$) as well as

Table 2 Significance test results

Factor association	Chi-square test		Fishers exact test
	Calculated	p-value	p-value
Resource management × inadequate supervision	10.087	0.001	0.0025
Resource Management × supervisory violations	9.791	0.001	0.0033
Organizational process versus inadequate supervision	11.49	0.001	0.0011
Organizational climate versus supervisory violations	11.10	0.001	0.0014
Inadequate supervision versus adverse mental state	10.83	0.0009	0.0052
Adverse mental state versus skill-based errors	11.49	0.0001	0.0028
Adverse mental state versus knowledge-based errors	10.83	0.0009	0.0085

between adverse mental state and knowledge-based errors, χ^2 (1, 18 = 10.83, $p <$ 0.001). This reveals that 'conditions of operators' such as memory, decision making, perception, attention deficit, and attention influences more. At this level of HFACS framework, attention and memory errors are both significantly associated with incident type. Besides, analysis of this level reveals that cognitive domain is the topmost human error contributing 12.26% (frequency analysis). This indicates that operators are mostly affected by cognitive factors, viz. attention, memory, and perception.

5 Conclusion

In this study, we examined countermeasures against human errors by dealing with active and latent factors such as unsafe acts, preconditions of unsafe acts, unsafe supervisions, and organizational influences. The countermeasures are computed by analyzing associations between the HFACS levels statistically. This study provides detailed information regarding the role of human factors, particularly human error, in adverse incidents in the nuclear industry. The main focus of this paper is to analyze human error data and identify important contributing factors to strengthen the human safety barriers in nuclear industry. HFACS proved to be a useful taxonomy for classifying the causal factors associated with human errors. A higher percentage is classified as skill-based errors as compared to decision errors. Overall, the HFACS framework is effective in categorizing errors from existing investigation reports and proves useful in capturing a wide range of errors, particularly organizational errors, influencing to an incident. However, this study has limitations to be noted and further studied. Firstly, it is not easy to address some kinds of causal factors, such as organizational climate, crew resource management, and violations. Second, it is very tough to find the errors of 'Condition of operators' as it is common that operators hide the actual truth to protect themselves and therefore true cognitive errors are not revealed. Due to this, most of the investigation reports do not represent the cognitive errors realistically. For future work, the HFACS framework and classifications may be validated through more field studies.

Acknowledgements This work was supported and funded by Safety Research Institute, Atomic Energy Regulatory Board, India. The authors would like to express sincere thanks to Director, NSARG, SRI-AERB for his continuous support to do this research.

References

1. M. Aoki, G. Rothwell, A comparative institutional analysis of the Fukushima nuclear disaster: lessons and policy implications. Energy Policy **53**, 240–247 (2013)
2. K. Branford, N. Naikar, A. Hopkins, Guidelines for AcciMap analysis. In: *Learning from High Reliability Organisations* (2009 Jan), pp. 193–212
3. C.G. Bryan, *An Analysis of Helicopter EMS Accidents Using HFACS: 2000–2012* (2014)
4. M. Celik, S. Cebi, Analytical HFACS for investigating human errors in shipping accidents. Accid. Anal. Prev. **41**(1), 66–75 (2009)
5. C. Chauvin, S. Lardjane, G. Morel, J.P. Clostermann, B. Langard, Human and organisational factors in maritime accidents: analysis of collisions at sea using the HFACS. Accid. Anal. Prev. **59**, 26–37 (2013)
6. T. Diller, G. Helmrich, S. Dunning, S. Cox, A. Buchanan, S. Shappell, The human factors analysis classification system (HFACS) applied to health care. Am. J. Med. Qual. **29**(3), 181–190 (2013)
7. A. Ergai, T. Cohen, J. Sharp, D. Wiegmann, A. Gramopadhye, S. Shappell, Assessment of the human factors analysis and classification system (HFACS): intra-rater and inter-rater reliability. Saf. Sci. **82**, 393–398 (2016)
8. S.F. Galan, A. Mosleh, J.M. Izquierdo, Incorporating organizational factors into probabilistic safety assessment of nuclear power plants through canonical probabilistic models. Reliab. Eng. Syst. Saf. **92**(8), 1131–1138 (2007)
9. D.I. Gertman, B.P. Halbert, M.W. Parrish, M.B. Sattison, D. Brownson, J.P. Tortorelli, *Review of Findings for Human Error Contribution to Risk in Operating Events* (NUREG/CR-6753) (Idaho National Engineering and Environmental Laboratory, 2001)
10. R. Gordon, R. Flin, K. Mearns, Designing and evaluating a human factors investigation tool (HFIT) for accident analysis. Saf. Sci. **43**(3), 147–171 (2005)
11. E. Hollnagel, Y. Fujita, The Fukushima disaster-systemic failures as the lack of resilience. Nucl. Eng. Technol. **45**(1), 13–20 (2013)
12. S.K. Kim, Y.H. Lee, T.I. Jang, Y.J. Oh, K.H. Shin, An investigation on unintended reactor trip events in terms of human error hazards of Korean nuclear power plants. Ann. Nucl. Energy **65**, 223–231 (2014)
13. J.R. Landis, G.G. Koch, The measurement of observer agreement for categorical data. Biometrics **33**(1), 159–174 (1977)
14. N. Leveson, M. Daouk, N. Dulac, K. Marais, *A Systems Theoretic Approach to Safety Engineering* (2003)
15. N. Leveson, A new accident model for engineering safer systems. Saf. Sci. **42**, 237–270 (2004)
16. W.-C. Li, D. Harris, C.-S. Yu, Routes to failure: analysis of 41 civil aviation accidents from the Republic of China using the human factors analysis and classification system. Accid. Anal. Prev. **40**(2), 426–434 (2008)
17. R. Madigan, D. Golightly, R. Madders, Application of human factors analysis and classification system (HFACS) to UK rail safety of the line incidents. Accid. Anal. Prev. **97**, 122–131 (2016)

18. D. O'Hare, The 'Wheel of Misfortune': a taxonomic approach to human factors in accident investigation and analysis in aviation and other complex systems. Ergonomics **43**, 2001–2019 (March 2013) (2000)
19. J.M. Patterson, S.A. Shappell, Operator error and system deficiencies: analysis of 508 mining incidents and accidents from Queensland, Australia using HFACS. Accid. Anal. Prev. **42**(4), 1379–1385 (2010)
20. S.K. Sharma, Human reliability analysis : a compendium of methods, data and event studies for nuclear power plants (TEC. DOC. NO. AERB/NPP/TD/O-2). Atomic Energy Regulatory Board Niyamak Bhavan Anushaktinagar Mumbai—400 094 India (2008)
21. A. Scarborough, L. Bailey, J. Pounds, *Examining ATC Operational Errors Using the Human Factors Analysis and Classification System* (Federal Aviation Administration, 2005)
22. S.A. Shappell, D.A. Wiegmann, *The Human Factors Analysis and Classification System— HFACS* (U.S. Department of Transportation Federal Aviation Administration, 2000)
23. S.T. Shorrock, B. Kirwan, Development and application of a human error identification tool for air traffic control. Appl. Ergon. **33**(4), 319–336 (2002)
24. G. Steinhauser, A. Brandl, T.E. Johnson, Comparison of the Chernobyl and Fukushima nuclear accidents: a review of the environmental impacts. Sci. Total Environ. **470–471**, 800–817 (2014)
25. S.C. Theophilus, V.N. Esenowo, A.O. Arewa, A.O. Ifelebuegu, E.O. Nnadi, F.U. Mbanaso, Human factors analysis and classification system for the oil and gas industry (HFACS-OGI). Reliab. Eng. Syst. Saf. **167**, 168–176 (2017)
26. D.A. Wiegmann, S.A. Shappell, A human error analysis of commercial aviation accidents: application of the human factors analysis and classification system (HFACS). Aviat. Space Environ. Med. **72**, 1006–1016 (2001)

Experience in Development of a Research Reactor Simulator

Narendra Joshi, Sparsh Sharma, Jainendra Kumar, P. Y. Bhosale, Parag Punekar and P. V. Varde

Abstract Nuclear safety not only depends on the reliability and safety of the structures, systems, and components (SSCs) but also largely depends on human factors and interaction of operators with machines. Simulators greatly support in the training and transfer of knowledge to the plant operators. This paper shares an experience in indigenous development of full scope, non-replica simulator for a more than 30 years old, 100 MW research reactor at BARC Trombay. The simulator was so designed that over and above training of the plant personnel, it will be used for safety studies in the area of reactor physics and thermal hydraulics. The interface was designed to support the interaction of the operators with the simulator through virtual panels supported by touch screens with high-fidelity graphic displays. Important postulated initiating events (PIEs) like loss of offsite power, loss of coolant accident, loss of regulation incident, etc. have been modeled. This information will provide an insight to verify and validate the plant emergency operating procedures (EOPs). The simulator will also be used for the development of human factor models as well as the design of future visual display unit (VDU)-based control rooms. This will also enable the operators to align themselves from the traditional panel-based human–machine interface (HMI) to personal computers.

N. Joshi · S. Sharma (✉) · J. Kumar · P. Y. Bhosale · P. Punekar · P. V. Varde
Bhabha Atomic Research Center, Mumbai, India
e-mail: sparsh@barc.gov.in

N. Joshi
e-mail: nsjoshi@barc.gov.in

J. Kumar
e-mail: jaink@barc.gov.in

P. Y. Bhosale
e-mail: pyb@barc.gov.in

P. Punekar
e-mail: punekar@barc.gov.in

P. V. Varde
e-mail: varde@barc.gov.in

© Springer Nature Singapore Pte Ltd. 2020
P. V. Varde et al. (eds.), *Reliability, Safety and Hazard Assessment for Risk-Based Technologies*, Lecture Notes in Mechanical Engineering,
https://doi.org/10.1007/978-981-13-9008-1_66

Keywords Human factors · Human–machine interface · Research reactor · Simulator · Training

1 Introduction

If we look at the past history of nuclear plant operations world over, most of the accidents that have taken place can be attributed to human errors. These accidents have changed the public perception about nuclear energy though the nuclear reactors are one of the safest plants with incorporation of defence-in-depth philosophy; multiple barriers for accidental release of radiation and thorough training and re-training of plant personnel make use of the latest available technologies. The operation of a complex system like a nuclear reactor requires thoroughly trained and licensed staff. One of the key elements that determine the operational safety of a nuclear reactor or nuclear power plant (NPP) is the training and technical knowledge of its operators about the behavior of their plant under normal, transient, and accidental conditions. Though the occurrence of an accident is a very low probability event, the operator has to be sufficiently trained to handle such situations effectively. It is widely recognized that simulators play an essential and extremely important role in establishing the training programs for NPPs. The full-scope training simulators provide such in-depth training tools for reactor operators. Apart from improved quality of training, the other objectives of the simulator are 'performance of safety studies,' 'validation of emergency operating procedures (EOP),' 'development of specifications for state-of-the-art modern control rooms using VDUs' and 'development of human factors,' considering available the plant-specific data and simulator data. A full-scope simulator has been developed first time for a research reactor, Dhruva at BARC. The details about the development activities are covered in this paper.

2 Simulator Development

2.1 Selection of Systems

It was proposed to develop a simulator for an old research reactor Dhruva at BARC. Dhruva is a 100 MWth research reactor at BARC. The tank-type reactor is fuelled by natural uranium, cooled, moderated, and reflected by heavy water which is in operation since year 1985. Generally, the simulators are constructed along with the plant, and for NPPs, it is a regulatory requirement. However, for a research reactor such regulatory requirement doesn't exist. However, considering the enhancement in the training quality and performance of safety studies it was decided to develop a simulator for Dhruva after its operation for several years. In view of the limitation on available space, it was decided to develop a full-scope, partial replica simulator.

Table 1 Reactor systems selected for modeling

S. No.	Module/system	Development platform
Process systems		
1	Main coolant system and sub-systems like shutdown cooling system moderator system	PFTool
2	Emergency core cooling system	PFTool
3	Process water system	PFTool
4	Sea water system	PFTool
5	Vault and shield cooling system	PFTool
6	Reactor ventilation system	PFTool
C & I and safety systems		
7	Emergency cooling signal control logic	LCTool
8	Reactor start-up and shutdown module	LCTool
9	Reactor protection system	LCTool
10	Reactor regulating system	LCTool
11	Primary shutdown system	LCTool
Specialized modules		
12	Core neutronics	RTTool
13	Core thermal hydraulics	RTTool
Power supply system		
14	Power supply and distribution system	ELTool

It was decided to retain the dimensions of the main control console, as it is without any change, from which all the operations are performed and the hardwired graphic panels were reproduced in the form of soft touch screens [1].

The details of reactor systems selected for modeling and simulation development platform of the simulation software are provided in Table 1. The details about the simulation software are provided in subsequent section. The development activities were completed in a time-bound manner. The development contract was awarded to a vendor with whom the requisite expertise and indigenously developed simulation software were available. The relevant data on design and operation parameters, details of interlocks, alarm and trip settings, process & instrumentation (P&I) diagrams, etc. pertaining to each system were provided for model development along with the details of steady-state simulations, transients, and initiating events, if any associated with the system.

2.2 Simulator Architecture and Simulation Tool [2]

The simulator hardware consists of nine PCs and a data server PC interconnected on a fast ethernet network. Figure 1 shows the simulator architecture. Brief description of the hardware is provided below:

Server: The simulator process and control models generated using simulation suite will be running in this system. Plant database will be maintained in this system.

Console: The replica of main control console of Dhruva. It consists of various equipments to display and operate the safety-related devices. The equipments include recorders, switches, push buttons, analog meters, digital meters, demand power switch, etc.

IO System: This system consists of IO software used to drive the equipment of console. IO software sends inputs (equipment commands like switch positions, push button, set power, etc.) to simulation server and receives model outputs (like reactor power, moderator level, pump and valve status, etc.) from simulation server periodically.

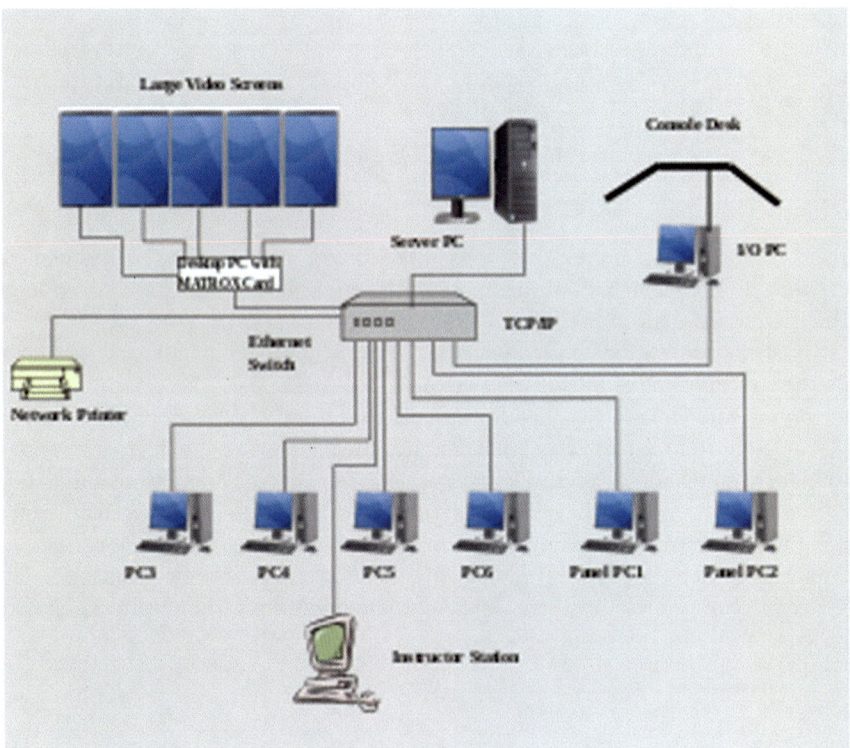

Fig. 1 Schematic representation of the simulator architecture

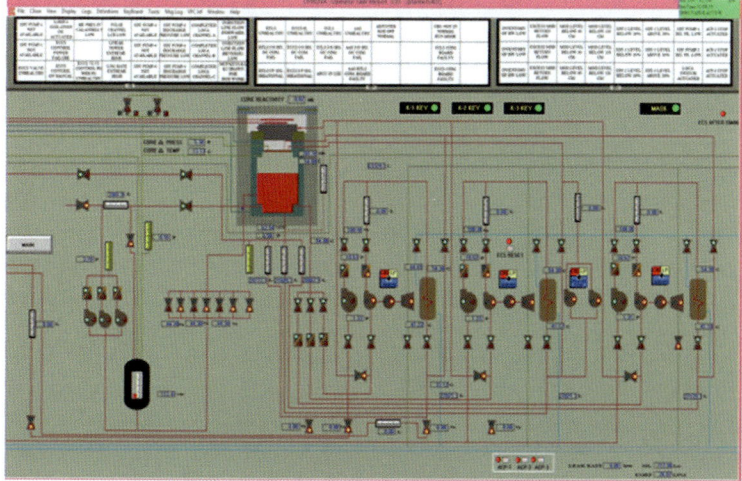

Fig. 2 Mimic of main coolant and moderator system

HMI Server: HMI software will be running on this PC. The plant mimics with dynamic data will be displayed on the LV screens. Data can be displayed in various formats like graphical, tabular, and matrix formats. Data logging can be done periodically.

HMI Clients: HMI client software will be running on these systems. The plant mimics with dynamic data will be displayed on the large video (LV) screens. Initially, each client shows different screens. But any schematic can be viewed in any of the clients. Operator can operate from any of the HMI systems. The mimic of reactor process systems is shown in Figs. 2 and 3, and the same for power supply system is shown in Fig. 4.

Instructor Station: Instructor station software will be running on this system. Instructor can control the simulation by using this software. He can save and load initial conditions (IC), introduce malfunctions, and run/freeze the simulation. Instructor can conduct training sessions to evaluate the operator.

Process Simulation Suite is an indigenously developed software package. The package provides integrated configuration and simulation environment, which allows engineers who are familiar with power plant operations, schemes, and equipment to efficiently develop high-quality, high-fidelity models. It is a PC-based simulation tool running on windows operating system that facilitates to build simulation load using different modules without writing any code. Facility is provided with the tool to develop user modules written in FORTRAN and interface to third-party software. Efficient and elaborate GUI interface is provided which allows users to operate with ease various control operations while keeping an eye on the process dynamics.

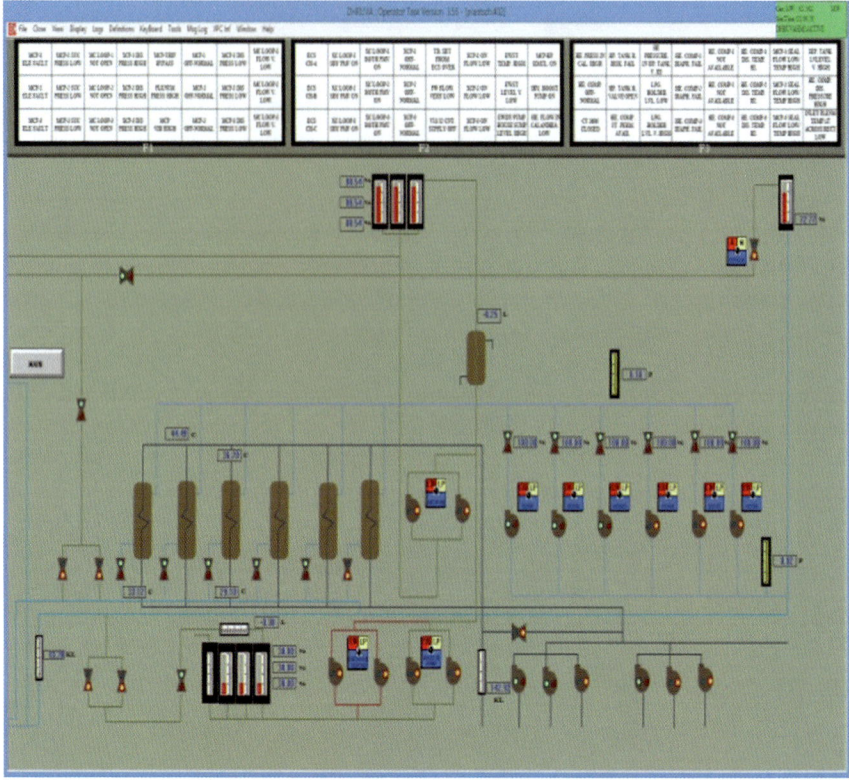

Fig. 3 Mimic of ventilation, vault and shield, ECCS, system

The various modules in the simulation suite are:

PFTool—Process flow network solver for calculating pressures, flows, and temperatures of a given network generated using plant P&I diagrams.

LCTool—Logic and control network generator used for drawing control and logic circuits like PID controllers, drive modules, group sequence modules, etc. The typical screenshots of the PFTool and LCTool are shown in Fig. 5a and b, respectively.

ELTool—Electrical network tool is used to generate the power plant electrical distribution network and calculate the node voltages and currents for the same.

DBTool—Database management tool for handling the huge database comprising of signal database and module database. The back end for this tool is MS SQL.

RTTool—Real-time executive tool used for the execution of the models periodically at selectable cycle rates. It allows the user to view the plant data, save IC, reset IC, and backtrack, etc. User can control the simulation session using step, run, and freeze commands.

ISTool—Instruction station tool is used to display the model variables, adding malfunctions, backtracking, trending, etc. User can control the simulation session using step, run, and freeze commands.

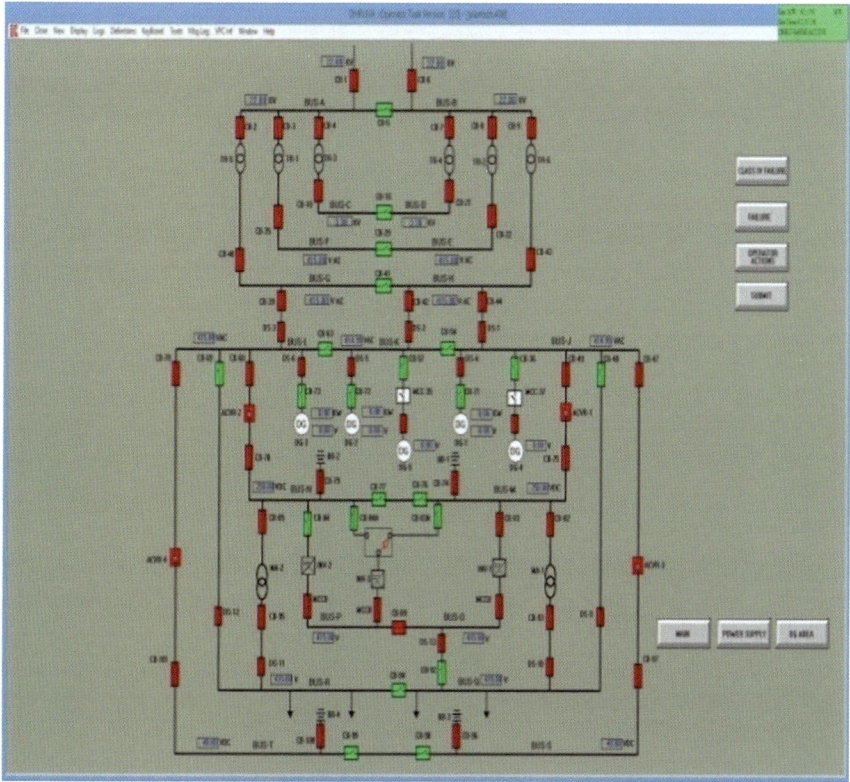

Fig. 4 Network of power supply and distribution system

(a) **(b)**

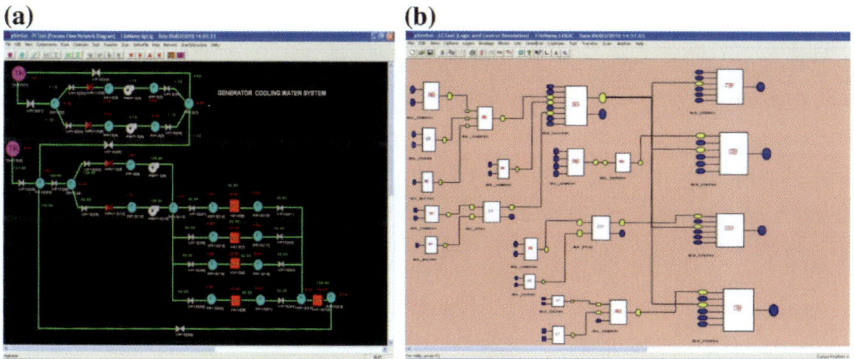

Fig. 5 Screenshots of **a** PF tool and **b** LC tool

(a)   **(b)**

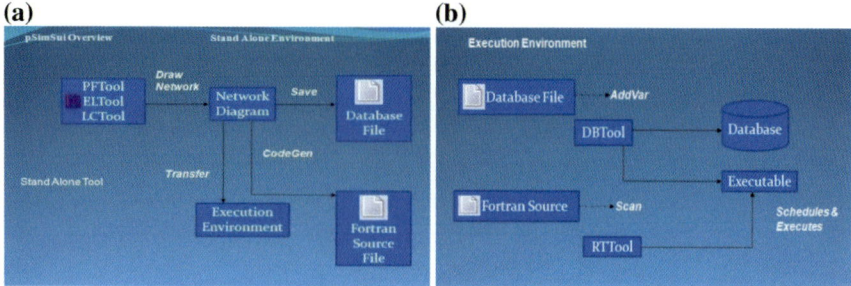

Fig. 6 a, b Overview of simulation tool

GMS Tool (Picture Editor)—GMS tool is a graphic tool that provides facility to generate plant schematics, control schemes, and soft panels. The tool includes all the required objects/icons/components for generating the schemes and assigning dynamics.

Development of process, control, or electrical subsystem involves generation of module in standalone environment as shown in Fig. 6a. Here, the network diagram of module is drawn using any one of the above-mentioned tools. Subsequently, a FORTRAN code corresponding to this network is generated along with a database file containing the variables used in the code. The generated module is then included in the run-time environment of the simulation suite as per Fig. 6b. Here, the generated FORTRAN code is executed as per the user-defined cycle time. During each cyclic execution, the variables (specified by the DB file) are updated and stored in the database using the DB tool.

2.3 Selection of Hardware [3, 4]

Few factors determined the final configuration and type of the simulator. To conduct effective training, the facility should resemble as much possible to the actual control room. Also, the space available to accommodate the simulator systems was limited, and the entire project had to be completed within the sanctioned budget. Keeping these factors in mind, the main control console was designed and fabricated similar to one in control room with identical meters and switches to have a feel of control room. All the equipment on the control console is hardwired and retained as it is. To accommodate the control room graphic panels, large amount of space was needed and that was the major constraint, so the graphic panels were replaced with LV touch screens. However, the mimic details on the graphic panels were retained and produced on the soft screens. The screens were installed in such a manner that the operator sitting at the console can view the entire mimic. In case of actual control room, the entire panel is not visible from the console. As the control console was large in size, it was not possible to shift the console in the room in a single piece or

Fig. 7 Simulator control console

Fig. 8 Development work in progress

two pieces. So a wire frame model was made to check the transportation from the lab corridors and doors, and based on the results, the console was fabricated in four parts and later these were joined at site. The simulator control console is shown in Figs. 7 and 8.

2.4 Model Development

As shown in Table 1, the reactor systems were selected for modeling. The development of process system models was done using the PF tool. The models of core neutronics, reactor regulating system (RRS), reactor trip logic system, start-up logic system, alarm annunciation system, and coolant activity monitoring system were developed in house and interfaced with the main module. Here, a brief description about development of core neutronics and RRS model is provided.

2.4.1 Core Neutronics Point Kinetics Model

Dhruva being a strong neutronically coupled core, point kinetics model is adequate to simulate various reactivity-initiated transient's conditions. Point reactor kinetics code is based on point kinetics model where a finite sized reactor is assumed to be a point reactor; i.e., change in neutron density under reactivity insertion reflects uniformly at all points within a reactor. Rate of change in neutron density and precursor concentration is described by a set of coupled linear ordinary differential equations given below.

$$\frac{dn(t)}{dt} = \frac{\rho(t) - \beta}{\Lambda} n(t) + \sum_{i=1}^{g} \lambda_i C_i(t) + S(t) \tag{1}$$

$$\frac{dC_i(t)}{dt} = \frac{\beta_i}{\Lambda} n(t) - \lambda_i C_i(t) \tag{2}$$

where

n = core average thermal neutron density,
C_i = core average delayed neutron precursor density,
S = external neutron source,
Λ = prompt neutron generation time,
ρ = reactivity,
β_i = fraction of ith delayed neutron,
g = number of precursor groups.

Fourth-order Runge–Kutta method is used to solve these equations.

Reactivity calculations include reactivity control and safety devices, Xenon, voiding in channels and power level changes. The parameters displayed are:

- Neutron power (% full power)
- Neutron power rate (%/second)
- Xenon load
- Core reactivity.

2.4.2 Modeling of Reactor Regulating System

The RRS block interacts with the core neutronics module so as to control the reactor power at the desired level. Figure 9 shows the block diagram of RRS of Dhruva.

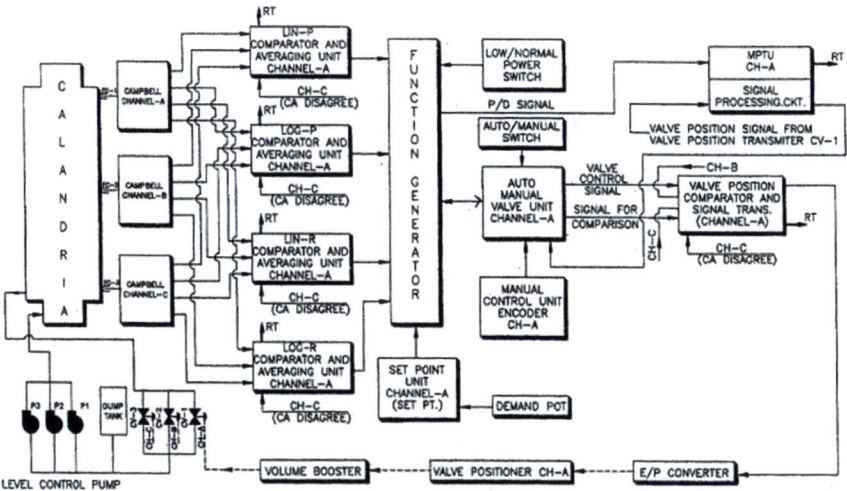

Fig. 9 Block diagram of single-channel reactor power control loop of Dhruva

In Fig. 9, the calandria is the reactor core which is modeled mathematically as core neutronics module and from Campbell Channel onwards it is the RRS module.

The instantaneous power signal is fed to RRS module through Campbell Unit. It consists of a linear amplifier, log amplifier, and differentiator circuits for generating the voltages corresponding to linear power, log power, linear rate, and log rate. These four signals are then fed to comparator units so as to generate representative average signals and to reject the faulty signal. Average values of each of the four signals are then fed to function generator which computes the error signal between actual reactor power and the set power. Based on this, it generates the corrective signal for the control valve so that actual reactor power is controlled at the desired value, i.e., set power.

Mathematical modeling of RRS involved computation of transfer function for each block (using the actual circuits deployed in Dhruva) and then converting into time-domain differential equations. The set of equations so obtained were converted into a FORTRAN code and were coupled with the core neutronics module.

Subsequently, extensive real-time integrated testing of the system was carried out. During this phase, observed steady-state error between the set power and actual reactor power and fluctuations in the control valve signal leading to hunting of position of control valves were the main problems encountered which were overcome by systematic fine tuning of the RRS module.

3   Simulator Features

The simulator has been developed to meet most of the training requirements as well as verification of plant EOPs. The only variation is the physical fidelity of the simulator. Soft panels offer all the functionality of a full-scope simulator except that the operator interaction is through a mouse click while sitting on the console or through tactile displays of images over the screen. This is the only difference between the control room panels and simulator. However, this can be easily overcome by the operators as for most of the time they are working in the control room.

All the control room graphic panels of process systems have been accommodated on the LV screens as it is as shown earlier. Mimic of any process system can be displayed on any of the screens. Upon right click of the cursor on any main equipment, its major specifications are displayed. Only there is a change in the relative position of the components as compared to main control room. All the trip and alarm windows have been positioned on the top of each screen with similar buzzer sound, and one screen is devoted for safety parameter display system. Also, the information related to fuel channel temperatures, channel flows, channel output power, etc. has been made available. Real-time trends of reactor power, moderator level, Xenon poison level, etc. are displayed. A reactivity meter has also been incorporated in the reactor physics model which is not the part of actual plant.

The simulator also incorporates additional features that can only be provided by a simulator with soft panels. Some of them serve to simplify the course of training sessions. For instance, every time the instructor set a new scenario or reset the IC, all the devices are adjusted automatically to their new correct positions and number of such ICs can be pre-defined.

4   Simulator Performance

Considering the space constraints, the whole simulator panels display a reduced scale image of the control room. At the commissioning stages, models of the individual systems were thoroughly checked for process parameters, interlocks, logics, etc. Acceptance procedures were prepared for each system, and the readings of different parameters were noted at different power levels and compared with the actual reading and found to be satisfactory. Subsequently, integrated testing of the simulator was carried out. This stage involved complete interlock and logic checks, checking of trips and alarms, and functioning of equipment. All the modeling errors were immediately identified and corrected as this would have adversely affected the training programme. Also, tuning of the RRS was also done so as to match the simulation results with that of the plant data. The performance of the simulator was also checked, and the initiating events like CL-IV power failure, loss of coolant accident, and loss of regulation incident were simulated, and the results were found to be in agreement with the theoretical calculations. The fact has to be accepted that in case of soft panels there are no physical handles or switches/push buttons and the size of the object on the screen may also vary from the actual. The operations from the hardwired console were also smooth.

5   Conclusion

The simulator with soft panels interface is being used to train the operators at Dhruva, BARC. This demonstrates that hybrid systems containing an appropriate mix of hardwired devices and use of software driven soft panels can perform the same tasks that were performed by expensive systems as the classic full-scope replica simulator of the control room.

Furthermore, the soft panels approach provides additional benefits that facilitate its use and expand its functionality and range of application.

For other more specific applications, such as the training on the auxiliary panels that are not included in the scope of traditional replica simulators, a soft panel interface appears as the most suitable solution.

6   Future Scope

Apart from training, the simulator will be extensively used for safety studies in the area of reactor physics, thermal hydraulics, validation of EOPs, design of future hybrid control rooms with optimum mix of soft and hard panels. The simulator will also be used for development of a human reliability database for research reactors using the available plant-specific data of human errors and experiments performed on simulator.

Acknowledgements   The authors would like to thank Mr. Subba Ram, Ms. Pratima, Mr. Pradeep, Ms. Prashanthi, and Mr. Praveen of ECIL, Hyderabad, who were involved in development, installation, and commissioning activities of the simulator.

References

1. International Atomic Energy Agency, Selection, specification, design and use of various nuclear power plant training simulators, IAEA-TECDOC-995
2. ECIL manual on simulation suite pSimSui
3. P.A. Corcuera, A full scope nuclear power plant training simulator: design and implementation experiences. Applied Mathematics and Computer Science Department, University of Cantabria
4. J. Tavira-Mondragon, R. Cruz-Cruz, Development of modern power plant simulators for operator training center, in *Proceedings of the World Congress on Engineering and Computer Science 2010 Vol IWCECS 2010*, Oct 20–22, 2010, San Francisco, USA

HAMSTER: Human Action Modelling—Standardized Tool for Editing and Recording

Vincent Bonelli and Jean-François Enjolras

Abstract EDF has developed a new HRA-type C method called Human Actions Modelling—Standardized Tool for Editing and Recording (HAMSTER). This method aims to address many constraints, among which are:

- A cost-effective, industrial and reproducible method: minimization of "authors effects" as far as possible,
- A user-friendly method for engineers:
  - The expertise is supported by the method itself (through embedded quantification rules) and its initial settings which mainly depend on the design and the nature of the guidelines,
  - The analysis is focused on a minimum of essential tasks, based on available support studies,
  - The production of results is computerized as far as possible,
- HFE assessments consistent with the staff organization and the type of implemented EOPs/SAMGs,
- A large scope of applicability: internal events as well as hazards, Level 1 PSA as well as Level 2 PSA, MCR actions as well as local actions, reactor building as well as spent fuel pool initiating events,
- Especially for hazards, once the internal events HFE probabilities have been evaluated, the HAMSTER method is very practical to assess the results in hazards contexts. The HAMSTER method provides also homogeneous results on all human missions which are credited in the internal events PSA model, instead of cliff edge effects induced by "scoping" methods usually used in hazards HFE assessments,
- An accurate treatment, if necessary, of dependencies between actions and HMI failures, dependencies between actions themselves and inopportune actions,
- An optional function, if necessary, allowing a quick and conservative assessment of HFE probabilities.

V. Bonelli (✉) · J.-F. Enjolras
Electricité de France—Design and Technology Branch, Lyon, France
e-mail: vincent.bonelli@edf.fr

J.-F. Enjolras
e-mail: jean-francois.enjolras@edf.fr

© Springer Nature Singapore Pte Ltd. 2020
P. V. Varde et al. (eds.), *Reliability, Safety and Hazard Assessment for Risk-Based Technologies*, Lecture Notes in Mechanical Engineering,
https://doi.org/10.1007/978-981-13-9008-1_67

791

This method was used to assess:

- The HFE of an EPR PSA model: more than 300 mitigation actions, taking into account HFE/HFE and HFE/I&C dependencies,
- Some HFE specific to hazards initiating events,
- Some HFE performed by the FARN (Nuclear Rapid Action Force of EDF in case of severe accident),
- The internal events HRA of the 1300 MWe French plants PSA in the frame of an impact study.

These first applications have highlighted the HAMSTER ability to identify the need for optimizations of the operating procedures. In the future, this method will be extended to the PSA models of the whole French fleet in the frame of the next periodic safety reassessments.

Keywords Human failure event (HFE) · Method · Real-time applications · Severe accident management guideline (SAMG) · State-oriented approach (SOA)

1 Introduction

Most utilities and vendors are facing big issues with the HRA topic, mainly due to the extension of the scope of the PSAs to internal and external hazards as well as the implementation of new SSC (mobile or not) and/or specific staffs decided after the accident of Fukushima.

In this context, EDF has developed a new HRA-type C method called Human Actions Modelling—Standardized Tool for Editing and Recording (HAMSTER).

This method aims to address many constraints, among which are:

- A cost-effective, industrial and reproducible method,
- A user-friendly method for engineers, the expertise being supported by the method itself (through embedded quantification rules) and its initial settings which mainly depend on the design and the nature of the guidelines,
- A large scope of applicability: internal events as well as hazards, Level 1 PSA as well as Level 2 PSA, MCR actions as well as local actions, reactor building as well as spent fuel pool initiating events,
- An accurate treatment, if necessary, of dependencies between actions and HMI failures, dependencies between actions themselves,
- An optional function, if necessary, allowing a quick assessment of HFE, which can be very useful to address actions dealing with system recoveries (usually embedded in the system fault trees) or specific actions credited in real-time applications (e.g. risk-informed technical specifications).

2 HAMSTER Principles

The HAMSTER method enables users to determine probabilities of failure of operator actions credited after a PSA initiating event. This can concern either:

- Actions prescribed in the operating procedures (both EOPs and SAMGs),
- Actions requested by the local or national emergency teams,
- Actions required to mitigate a hazard, before or after occurrence of a plant initiating event induced by it,
- Execution of inopportune actions likely to occur after a plant initiating event.

2.1 Systemic Approach of Operating

When the HRA is set in the reality of an accident, the course of the actions emerges from multiple interactions:

- Between the members of the operating team and/or the emergency teams, whose roles are defined by the organization,
- Between individuals and the various human-system interfaces: instrumentation and control, computerized or non-computerized I&C, automatic messages to national emergency teams, etc.

Operation is finally produced by this complex "system", whose failure cannot stem solely from execution of erroneous actions by isolated persons. In this "systemic approach", the "system" would prepare and define the operating actions to be implemented, and the role of each operator would be that of executing those actions.

It is hence necessary to define the operating system as the location of multiple but structured interactions, between members of the operating team, plant emergency director (PED), local and national emergency teams, procedures, HMI and processes, and search for the failure mechanisms within that unit.

The HAMSTER method thus takes into account this notion of "operating system" in order to model it as effectively as possible (see Fig. 1).

2.2 Functional Model of the Operating System

Within the framework of PSA modelling, the end purpose of the operating system is the execution of an action (or a series of actions) aimed, directly or indirectly, at restoring or maintaining one of the three safety functions:

- Control of reactivity (Level 1 PSA),
- Evacuation of residual heat (Level 1 PSA),
- Control of containment (Level 2 PSA).

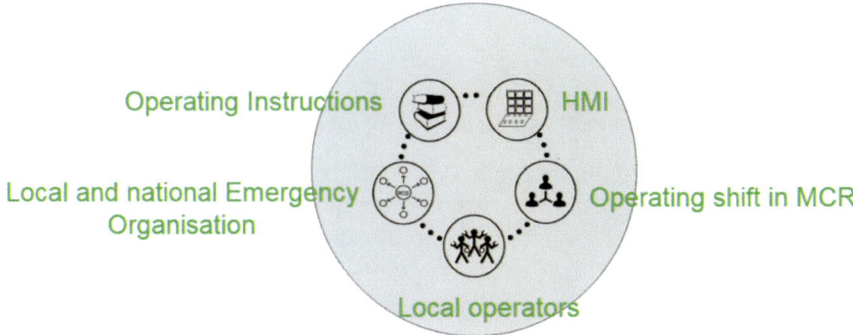

Fig. 1 The "operating system"

The "operating system" is described via four main functions that participate in the process of execution of function events, i.e. diagnosis, strategy, prognosis and action (DSPA).

In so doing, this moves well away from the simplified models of operation in accidental situations, which put forward two "traditional" sequential functions:

- Diagnosis of an accident,
- Followed by execution of actions triggered directly after the diagnosis.

Such a model is considered too simplistic, due to its reactive nature associating response to a stimulus, and due to the fact that it does not feature to the "operating system" any capacities of anticipation, or examination of alternative or complex action plans on the basis of the diagnosis.

That's why two other functions are added on these grounds:

- Strategy, which defines the operating objectives on the basis of the diagnosis covering the overall state of the facility and plans the action prior to its execution,
- Prognosis, once the National Emergency Organization (NEO) has been set up, enabling all concerned to envisage the various possible evolutions in the state of the facility and anticipate implementation of complementary or alternative strategies to the strategy initially adopted and applied.

These four functions (diagnosis, strategy, prognosis and action—DSPA) which described the "operating system" are envisaged as relatively independent processes leading to results that can be reviewed.

Failure of one of the functions leads to the failure of any operator action modelled in the PSA (see Fig. 2).

Fig. 2 The four "DSPA" functions modelling the "operating system"

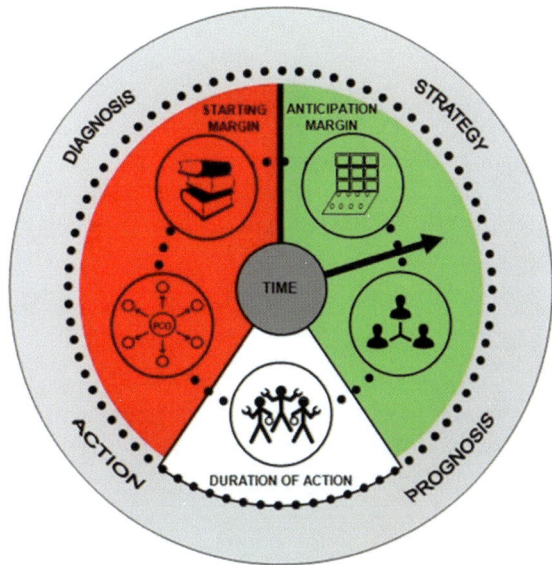

2.3   Notion of Failure Scenarios

The HAMSTER method relies on the identification and quantification, for each of the four functions (diagnosis, strategy, prognosis and action), of the various scenarios that could lead to the failure of an operator action.

A failure scenario for an operator action results from:

- A "situation property" characterizing the emergence of an operating system configuration that, if it lasts more than a certain length of time, will lead to failure of the operator action. Two types of properties are to be considered in light of the scenarios envisaged:

  - The behavioural properties, which characterize collective and individual behaviour patterns of the various staff members—on and outside the site—involved in performing the action,
  - The HMI properties, which characterize unexpected operation of the HMI likely to lead to possible failure scenarios: failure of a measurement chain, failure of a command chain or even failure of a complete computerized platform, for example.

- The failure to reconfigure the operating system within the allotted time: "system non-reconfiguration" thus enables to assess the relevancy of the scenario over time or the probability that the operating system configuration is maintained long enough to lead to failure of the operator action.

2.4 Characteristic Times, Action Durations and Time Margins

Characteristic times and grace periods are defined to credit the available time for the operating system to reconfigure itself.

They usually come from PSA support studies, whatever they are (simulators, safety reports…). Nevertheless, and as much as possible, it is recommended to use realistic support studies.

2.4.1 Characteristic Times

t_C: Compatibility of the operating system with execution of the action
The time t_C corresponds to the time at which the operating system reaches a configuration that is compatible with execution of the operator action (e.g. first representative alarm of the PIE).

t_P: Time of reaching the criterion or criteria on which the action is Prescribed
The time t_P corresponds to the time of reaching the criterion or criteria on which the action is prescribed in the operating instructions.

t_U: Ultimate time for execution of the action
The time t_U corresponds to the time after which the execution of the action no longer meets its objective.

2.4.2 Action Durations

Definition of the duration of an action

The duration of an action d_A corresponds, in practice, to the sum of the durations necessary for:

- Execution of the action itself (i.e. not including execution of the diagnosis, or drawing up a strategy or a prognosis): each elementary operating gesture is assessed (especially in case of local actions, when several actuators and/or sensors have to be manipulated),
- And, in case of actions carried out locally, movements inside the different areas of the NPP prior to local implementation of the action itself.

Penalization of the action durations in the case of a facility that has been deteriorated

Table 1 sets out the penalizing factors to be applied to the action durations, in the case of some facilities that have been deteriorated:

Table 1 Penalization of the action duration for an action for a facility that has deteriorated

Conditions of intervention	Penalization of the duration of action	Justification
No deterioration	1	
Deteriorated ambience (pressure, temperature, hygrometry, dose rate) and/or deteriorated lighting	1.4	Necessity of extra PPE Slower movements Difficulty in identifying the organs to be operated
Areas "physically" deteriorated (flooding, earthquake)	2	Difficult movements
Inaccessible areas or action that cannot be executed (flooding, earthquake, fire)	If the state of the facility prevents execution of the action, the probability of failure of the function event is forced to 1	

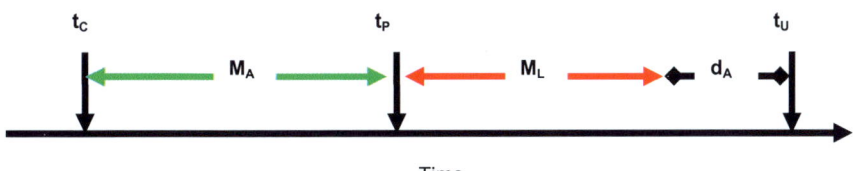

Anticipation margin: $M_A = t_P - t_C$

Launching margin: $M_L = t_U - t_P - d_A$

Fig. 3 HAMSTER time margins

2.4.3 Time Margins

Using the characteristic times and the action duration defined previously, it is possible to define two separate time margins to characterize the maximum time period during which an operating system configuration can continue over time without leading to the failure of the operator action:

- The anticipation margin (M_A) corresponding to the time available for mental preparation concerning execution of an action before it is actually required,
- The launching margin (M_L) corresponding to the time made available by the system to start (i.e. give an order covering) an action after the instructions to do so have been issued through operating procedures (see Fig. 3).

2.5 Notion of Context

The context is defined as a set of functional data enabling precise characterization of the state of deterioration of the facility.

Table 2 Levels of complexity of the context

Unavailable support and safety systems	Internal event	Internal or external hazard
No unavailable support system and 1unavailable safety system at the most	Nominal	Medium
1unavailable support system and 2 unavailable safety systems at the most	Medium	High
All other cases	High	Very high

The level of complexity of the context is defined thanks to:

- The number of unavailable support and safety systems, because it makes application of the procedures more complicated and can result in failure of I&C chains,
- The origin of the transient, either an internal event or a hazard (internal or external), because a hazard leads to a "physical" deterioration in the facility, and increases the workload of the operating and emergency teams and can affect the whole site.

The complexity of the context is discretized into four levels in HAMSTER (see Table 2).

2.6 Processing of Human Failure Events in Hazards PSA

In the case of a hazard, the internal events HFE probabilities are penalized:

- By a systematic increment of the level of the complexity of the context.
- And, on a case by case analysis, by:
  - The development of specific failure scenarios induced by the hazard,
  - An additional increment of the level of the complexity of the context if the hazard induces additional losses of support or safety systems,
  - An increase of the action duration (see Sect. 2.4.2),
  - Taking into account of a possible multi-units impact on the operating team organization.

Once the internal events HFE probabilities have been assessed, the HAMSTER method is very practical to adapt the results in hazards contexts (see Fig. 4).

The HAMSTER method provides also homogeneous HFE results, instead of cliff edge effects induced by "scoping" methods usually used in hazards HFE assessments. There is no need to re-examine in detail all the human events of the internal events PSA model in a hazard context.

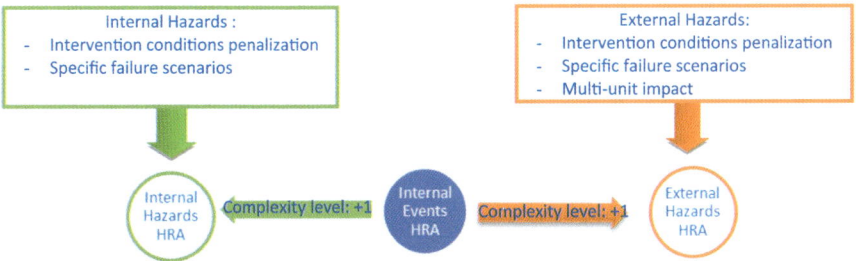

Fig. 4 The treatment of hazards by the HAMSTER method

2.7 Quantification of the HFE Probability

The HFE probability corresponds to the underlined Boolean sum of the set of N failure scenarios identified and can be written as follows:

$$P_{\text{HFE}} = \sum_{\text{Boolean } n=1}^{N} [P_{\text{PS}} \times P_{\text{NRS}}]_n + P_{\text{residual}} = 1 - \prod_{n=1}^{N}(1 - [P_{\text{PS}} \times P_{\text{NRS}}]_n) + P_{\text{residual}}$$

where

N Number of failure scenarios.
P_{PS} Probability of the situation property for a scenario.
P_{NRS} Probability of system non-reconfiguration for a scenario.

The residual probability aims to take into account the probability of the occurrence of any possible undefined scenarios.

2.7.1 Identification and Quantification of the Situation Properties (P_{PS})

To identify situation properties that can affect an operation that a team can undertake, it is advisable to:

- List the objectives, the instruction criteria and the route in the procedures for the action to be executed, together with the concurrent operating objectives if relevant,
- Deduce the necessary and sufficient conditions for the failure of the mission,
- Try to put oneself "in the shoes" of the operating team:

  – What reasons would the operators have for proceeding in that way?
  – What could cause them to fail to meet the conditions for success of the mission?

The analyst or HRA expert has to:

- Define the situation properties which are necessary and sufficient for the emergence of the corresponding failure scenario in order to justify the probability

Table 3 HAMSTER discrete scale of probabilities for PPS

Event qualification	Corresponding probability	Comments
Certain	1	Observable
Quasi-certain	0.99	These events can be observed (and refuted) by a series of tests
Highly probable	0.9	
Probable	0.7	
Indeterminate	0.5	
Fairly probable	0.3	
Not very probable	0.1	
Improbable	10^{-2}	Theoretical
Quasi-impossible	10^{-3}	Events stemming from a theoretical approach, representing the "reliability" of the procedures and the operating system

assigned to the situation property, referred to as P_{PS}. P_{PS} is selected from among the probabilities put forward in the scale of discrete values in Table 3.

- Identify the situation properties requiring specific treatment in the PSA modelling tool in order to benefit from Boolean fusion with other function events. This involves:

  – HMI failures concerning the instrumentation and control chains necessary for the DSPA functions, and especially diagnosis and action,
  – Possible scenarios involving failure of a beforehand HFE that could lead to the failure of the mission examined.

Based on this discrete scale of probabilities, EDF has identified generic "situation properties" for each EDF plant series with specific quantification rules in order to avoid "author effects.

As an example, the chapter 4 presents different "situation properties" established for EPR series.

2.7.2 Quantification of the Failure to Reconfigure the Operating System (P_{NRS})

The probability of system non-reconfiguration, P_{NRS}, quantifies the operating system's failure to reconfigure itself to execute a given operator action.

Two models of $P_{NRS} = f$ (context, time) are defined below, depending on the organization of the operating system:

- Normal operation,
- Applying EOPs or SAMGs instructions.

Table 4 $P_{NRS}(M_A)$ and $P_{NRS}(M_L)$ for the actions performed as part of normal operation of the plant

$M_{A/L}$	Number of independent controls/verifications during $M_{A/L}$	$P_{NRS}(M_A)$	$P_{NRS}(M_L)$
$M_{A/L} \leq 8$ h	0	1	1
8 h $< M_{A/L} \leq 16$ h	1	5×10^{-1}	10^{-1}
16 h $< M_{A/L} \leq 24$ h	2	10^{-1}	10^{-2}
24 h $< M_{A/L}$	3	5×10^{-2}	10^{-3}

Model suitable for the actions performed as part of normal operation of the plant

$$P_{NRS} = P_{NRS}(M_A) \times P_{NRS}(M_L)$$

with

$P_{NRS}(M_A)$: Non-reconfiguration probability crediting the controls/verifications which could reasonably be performed during the anticipation margin M_A. Indeed, this margin can be "credited" to the operating system, which can make mental preparations concerning execution of the action (or even anticipate it).

$P_{NRS}(M_L)$: Non-reconfiguration probability crediting the controls/verifications which could reasonably be performed during the launching margin M_L (see Table 4).

Model suitable for the actions performed as part of incidental or accidental operation of the plant

In these situations, EOPs or SAMGs instructions have to be applied in a "loyal and efficient" way, with a continuous control of the shift supervisor and a permanent monitoring of the facility by the shift manager or the safety engineer.

P_{NRS} are also based on parameters which are representative of the operating system's understanding of the plant state.

French EOPs are based on the state-oriented approach (similar to the symptom-based approach). Specific post-accidental parameters are used to determine the proper strategy to be applied and are periodically checked in order to change of strategy, if necessary.

Each member of the operating system working in the MCR applies his own post-accidental procedure, each strategy being cut into different phases. Phases and strategies are structured in order to be applied regularly, in the form of "loop operation".

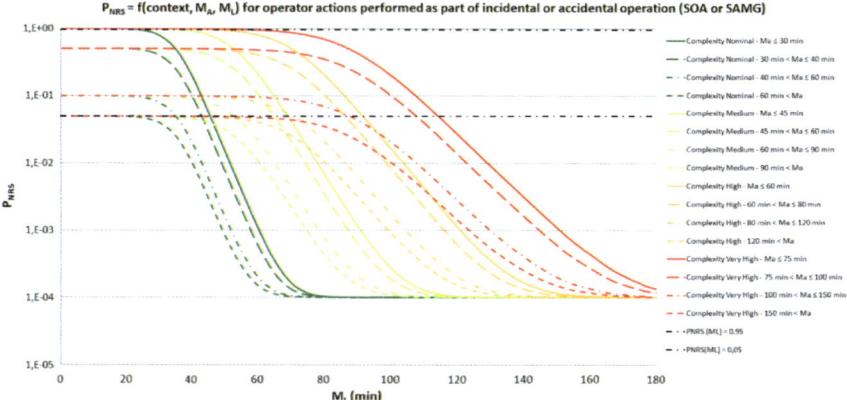

Fig. 5 The HAMSTER approach for modelling non-reconfiguration of the "operating system" in case of actions precsribed in EOP's or SAMG

The shift supervisor procedure (gathering the two operator's procedures) and the specific safety engineer[1] procedure (called SPE—permanent survey of the plant state) are dedicated to manage and survey the two operator's actions.

In the HAMSTER method, it is considered that the number of loops performed in the procedures by the shift supervisor and the safety engineer (or the shift manager, see Footnote 1) is a good indicator of the understanding of the plant state. Indeed, after a first loop in the procedure, the "operating system" is aware of the operating objectives, the systems unavailability and the main actions prescribed.

It is considered that the average duration of a shift supervisor loop increases with the complexity of the context while the average duration of a SPE loop is less time dependent: the SPE application is not affected by the systems unavailability. However, the duration includes the time for the shift manager and the safety engineer arrivals (20 and 40 min at most, respectively).

As an example, Fig. 5 gives the P_{NRS} curves obtained for an EPR Plant, whose EOP's present the particularity to be based on an Automatic (and permanent) diagnosis (of the proper strategy).

The curves use different parameters to be compared to the EOP's loops, as M_A (anticipation margin), M_L (launching margin) and finally the level of complexity of the context.

They are based on a "logistic function", easier to manipulate than a log-normal statistic function, and nevertheless very close to the modelling objective.

---

[1]In case of EOPs or SAMGs application, the shift manager, or, when arrived in MCR, the safety engineer has to apply a specific procedure called "SPE" leading to a dedicated and specific permanent monitoring of the plant state.

Table 5 Time history of the residual probability

	$M_L < 240$ min	$240 \leq M_L < 480$ min	$M_L \geq 480$ min
Residual probability	10^{-4}	Exponential decay from 10^{-4} to 10^{-5}	10^{-5}

2.7.3 Residual Probability

The residual probability ($P_{residual}$) is added to the probabilities of the failure for the various scenarios involving a given operator action to take into account the probability of occurrence of possible unrecognized scenarios (see Table 5).

2.8 Dependencies

The dependencies between HFE and I&C failures and between two HFE are taken into account through specific treatment in the PSA modelling tool of situation properties, in order to benefit from Boolean fusion with other function events.

The operator action is thus modelled in the PSA tool as a fault tree which identifies:

- HMI failures concerning the required instrumentation and control for diagnosis and action: these situation properties are modelled through a fault tree linked to the I&C components of the PSA model,
- Possible scenarios involving failure of a beforehand operator action function event that could lead to the failure of the mission: the situation properties of these scenarios are modelled through a specific basic event called in both beforehand and afterward operator actions fault trees.

2.9 Simplified and Conservative Assessment of HFE Probabilities

If an order of magnitude of the HFE probability is seen as being acceptable with regard to the objective of the PSA model or its applications, HAMSTER can be used to execute a simplified quantification of the function event.

For all the four "operating system" functions (diagnosis, strategy, action, prognosis) playing a part in the execution of an operator action, just one failure scenario is adopted, whose situation property probability (P_{PS}) is set to 1, and whose system non-reconfiguration depends on the context and the time lapse.

Indeed, it has been noticed that the sum of the P_{PS} of a detailed analysis is very rarely superior to 1.

The expression of the probability of failure for the operator action thus takes the following form:

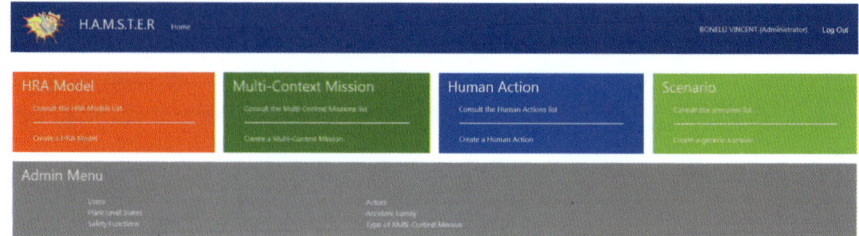

Fig. 6 Screenshot of the home screen

$$P_{\text{HFE}} = P_{\text{NRS}}$$

3 HAMSTER Software

A HAMSTER software has been developed in English language to assist the implementation of the method, to edit and record the HFE assessments and their justifications.

This software also leads to calculate HFE probabilities and to facilitate sensitivity studies, hazards PSA processing's, and implementation of the HFE values in the PSA tool (see Figs. 6 and 7).

4 Implementation of HFE in the CO-PSA Model

4.1 The CO-PSA Model

The CO-PSA (common operating PSA) was developed by EDF for both Flamanville 3 and Hinkley Point C EPR.

This integrated L1/L2 PSA addresses internal events for reactor building as well as spent fuel pool.

All the 300 type C operator actions of the CO-PSA were assessed with HAMSTER, taking into account HFE/HFE and HFE/I&C dependencies.

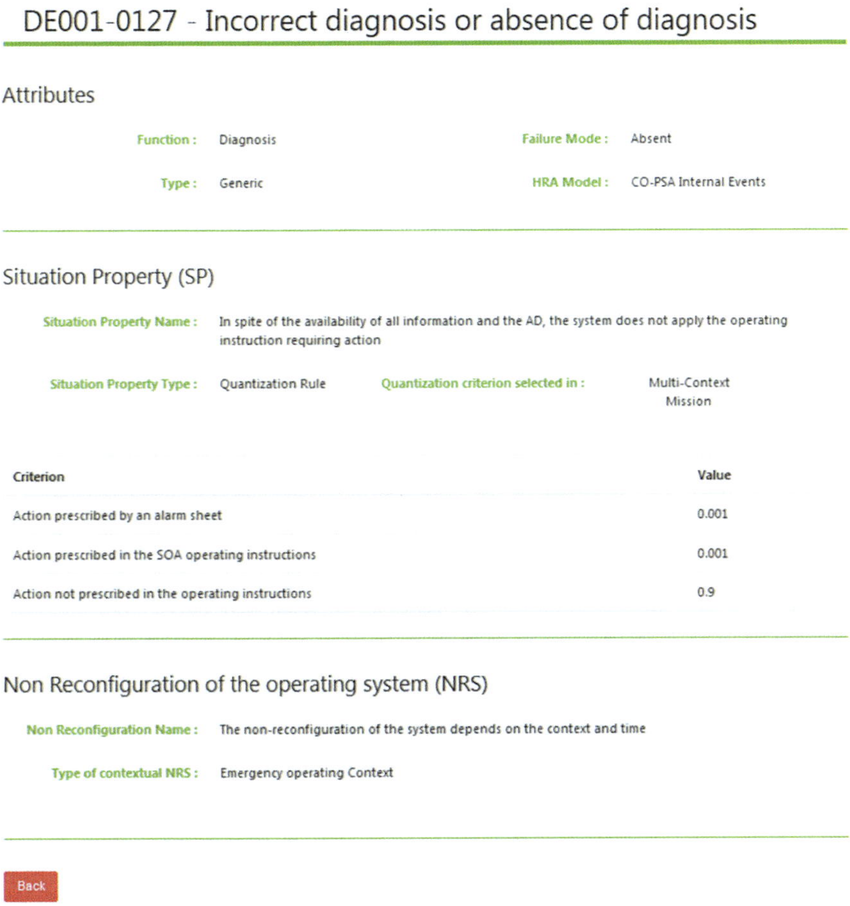

DE001-0127 - Incorrect diagnosis or absence of diagnosis

Attributes

Function :	Diagnosis	Failure Mode :	Absent
Type :	Generic	HRA Model :	CO-PSA Internal Events

Situation Property (SP)

Situation Property Name : In spite of the availability of all information and the AD, the system does not apply the operating instruction requiring action

Situation Property Type : Quantization Rule   Quantization criterion selected in :   Multi-Context Mission

Criterion	Value
Action prescribed by an alarm sheet	0.001
Action prescribed in the SOA operating instructions	0.001
Action not prescribed in the operating instructions	0.9

Non Reconfiguration of the operating system (NRS)

Non Reconfiguration Name : The non-reconfiguration of the system depends on the context and time

Type of contextual NRS : Emergency operating Context

Back

Fig. 7 Screenshot of the failure scenario viewer

4.2 Generic Failure Scenarios

Most of the operator actions credited in the CO-PSA model are performed during incidental or accidental operation with the state-oriented approach EOPs.

In these situations, the organization of the operating team is clearly defined: SOA instructions have to be applied in a "loyal and efficient way", with a formalized control of the shift supervisor and an independent monitoring of the plant by the shift manager or the safety engineer.

The mechanism of the operating system failures can be translated into generic failure scenarios for each of the functions diagnosis, strategy and action. The applicability of each of these failure scenarios is analysed for each operator action.

For each generic failure scenario, specific-quantification rules are defined to avoid "authors effects" and to ensure results reproducibility: **the expert judgements are embedded in the HAMSTER method (once for all analyses), not by the analysts during its industrial application** (see Table 6).

Two examples of quantification rules are given hereafter.

4.2.1   Example No. 1 of Generic Failure Scenario: "Incorrect Action"

Action is the function that switches the facility from one state to another.

The HAMSTER method considers that the failure scenario "incorrect action" depends on several parameters:

- The number of components to be operated,
- The operating mode,
- The risk of confusing the components to be operated,
- The technical difficulty of the actions to be carried out,
- The familiarity of operators with the action to be carried out,
- The self-checking carried out during the action,
- The supervisor monitoring the proper execution of the action,
- For local actions, the communication between the operator and the field operator.

Table 7 proposes a use of the table of discrete values to quantify the probability of the "incorrect action" situation property.

4.2.2   Example No. 2 of Generic Failure Scenario: "Application of Operating Instructions at an Inappropriate Rate"

In general, the application of operating instructions at an inappropriate rate is necessary for the emergence of this situation property: the operators and the supervisor are slow or delayed in their operating instructions and no strategic decision is taken by the operating system to accelerate.

The probability of this situation property depends on the route in the operating instructions before reaching the action request as well as on the time available.

In certain situations of accumulations, a step-by-step monitoring of operating instructions without specific strategy is sufficient for the emergence of this situation property.

For the EPR series, Table 8 proposes a probability of this situation property by considering that the slowness of application of operating instructions is partly "erased" by the reorientations of the automatic diagnosis (which is EPR specificity in the French NPP fleet).

It will thus be supposed for EPR that, at the latest, when the required criteria are physically achieved, the automatic diagnosis directs the team to the correct operating instructions.

Table 6 Generic failure scenarios of the CO-PSA model

Label	Situation Property type	P_{PS}	P_{NRS}
Failure or inaccessibility of the main information	HMI Situation property Operating from Process Information and Control System (PICS)	PSA fault tree	Quantification Rule 1
Failure or inaccessibility of the main control		PSA fault tree	$P_{NRS}=f(context, M_L)$
Undetectable failure of the Automatic Diagnosis		PSA fault tree	Quantification Rule 2
Partial detectable failure of the Automatic Diagnosis		PSA fault tree	$P_{NRS}=f(context, M_A, M_L)$
Failure of the " Protection System "		PSA fault tree	1 if the operator action is not stipulated in a "Loss of Protection System" EOP $P_{NRS}=f(context, M_A, M_L)$ otherwise

Label	Situation Property type	P_{PS}	P_{NRS}
Failure of the PICS	HMI Situation property Operating from Safety Information and Control System (SICS)	PSA fault tree	$P_{NRS}=f(context, M_A, M_L)$
Failure of the "Process System"		PSA fault tree	1 if the operator action is not stipulated in a "Loss of Process System" EOP $P_{NRS}=f(high context, M_A, M_L)$ otherwise
Total loss of computerized I&C		PSA fault tree	1 if the operator action is impossible at NCSS $P_{NRS}=f(high context, M_A, M_L)$ otherwise
Incorrect diagnosis	Behavioral Situation Property	Quantification Rule 3	$P_{NRS}=f(context, M_A, M_L)$
Incorrect operating instructions application		Quantification Rule 4	
Application of operating instructions at an inappropriate rate		Quantification Rule 5	
Prioritization of a rival strategy		Quantification Rule 6	
Incorrect action		Quantification Rule 7	
Local action not executed		Quantification Rule 8	

Table 7 Value of the failure probability for the failure scenario "incorrect action"

Action type			Formalized in an operating instruction	Not formalized in an operating instruction
Action in control room	$\leq$10 switch commands only		10^{-3} (quasi-impossible)	10^{-2} (improbable)
	At least one adjustment command or more than 10 switch commands		10^{-2} (highly improbable)	
Local action	Switch command	$\leq$2 commands	10^{-2} (highly improbable)	10^{-1} (not very probable)
		From 3 to 5 commands	10^{-1} (not very probable)	0.3 (fairly probable)
		>5 commands	0.3 (fairly probable)	0.5 (indeterminate)
	Adjustment command	1 command	10^{-1} (not very probable)	
		$\geq$2 commands	0.3 (fairly probable)	

Table 8 Value of the failure probability for the failure scenario "application of operating instructions at an inappropriate rate"

Number of applied pages/M_L ratio expressed in page/minute	Probability of the situation property "The operating system applies the operating instructions too slowly"
More than two pages/min of operating instructions or specific situation for which a "step by step" application of operating instructions leads to the failure of the function event	1 (certain)
$1 \leq$ page/min < 2	0.5 (indeterminate)
$0.5 \leq$ page/min < 1	0.3 (fairly probable)
page/min < 0.5	10^{-1} (not very probable)

As a result, for EPR, the slowness in the application of operating instructions may be measured from the ratio between the number of pages of the operating instructions to cover before reaching the action request and the starting margin, deducted from a possible waiting time required by the operating instructions.

Note: Some operating instructions may require waiting times from operators which must be taken into account to estimate the operating instruction coverage time. These times may be assigned as credit to the system to prepare itself to carry out the action: they must therefore be taken into account in the anticipation margin and not in the starting margin: as a result, the definition of time t_P must take into account the waiting times required in the operating instructions.

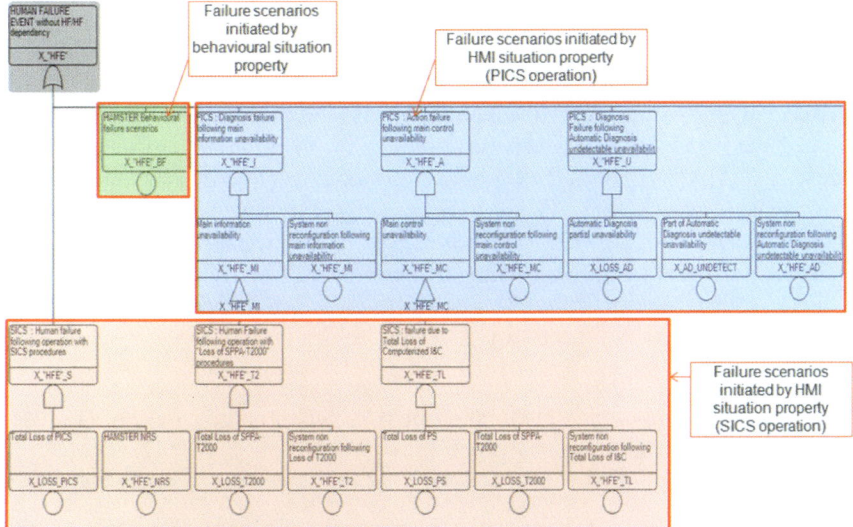

Fig. 8 Generic fault tree of operator actions in the CO-PSA model

4.3 Implementation in the PSA Using Risk Spectrum

Each CO-PSA operator action is thus modelled as a fault tree in the PSA model (see Fig. 8).

5 Conclusion

EDF has developed a new HRA-type C method called Human Actions Modelling—Standardized Tool for Editing and Recording (HAMSTER).

This method aims to address many constraints, among which are:

- A cost-effective, industrial and reproducible method: minimization of "authors effects" as far as possible,
- A user-friendly method for engineers:

  - The expertise is supported by the method itself (through embedded quantification rules) and its initial settings which mainly depend on the design and the nature of the guidelines,
  - The analysis is focused on a minimum of essential tasks, based on available support studies,
  - The production of results is computerized as far as possible,

- HFE assessments consistent with the staff organization and the type of implemented EOPs/SAMGs,

- A large scope of applicability: internal events as well as hazards, Level 1 PSA as well as Level 2 PSA, MCR actions as well as local actions, reactor building as well as spent fuel pool initiating events,
- Especially for hazards, once the internal events HFE probabilities have been evaluated, the HAMSTER method is very practical to assess the results in hazards contexts. The HAMSTER method provides also homogeneous results on all human missions which are credited in the internal events PSA model, instead of cliff edge effects induced by "scoping" methods usually used in hazards HFE assessments,
- An accurate treatment, if necessary, of dependencies between actions and HMI failures, dependencies between actions themselves and inopportune actions,
- An optional function, if necessary, allowing a quick and conservative assessment of HFE probabilities.

The first applications have highlighted the HAMSTER ability to identify the need for optimizations of the operating procedures.

The method was presented to the IRSN (support of the French safety authority).

Reliability Methods

Markov Probabilistic Approach-Based Availability Simulation Modeling and Performance Evaluation of Coal Supply System of Thermal Power Plant

Hanumant P. Jagtap and A. K. Bewoor

Abstract The high demand of electricity from the society can be fulfilled by various sources, and thermal power plant is one of the largest sources of power generation in India. Availability of thermal power plant is dependent upon its subsystem and equipment in use. The equipment of thermal power plant can be maintained highly reliable if suitable maintenance is performed at defined time interval. This paper presents availability simulation modeling for coal supply subsystem using Markov birth–death probabilistic approach for thermal power plant generating 500 MW from Unit 1 located in western region of India. The equipment considered for availability analysis is coal mill, stacker reclaimer, and wagon tippler. For this reason, the differential equations are generated. These equations are solved using normalizing condition so as to find out the steady-state availability of coal supply system. The effects for occurrence of failures activity as well as availability of repair facilities on system performance are investigated. This study revealed that stacker reclaimer is the most critical equipment of coal supply system which needs more attention for form maintenance point of view and followed by coal mill and wagon tippler. The results show that availability simulation modeling based on Markov birth–death probabilistic approach is very effective tool finding critical equipment of thermal power plant. Further, particle swarm optimization method is used for finding the optimized availability parameter which in turn helps to select maintenance strategy.

Keywords Availability analysis · Critical equipment · Markov approach · Particle swarm optimization · Thermal power plant

H. P. Jagtap (✉)
Zeal College of Engineering and Research, Narhe, Pune 411041, Maharashtra, India
e-mail: jagtaphp@gmail.com

A. K. Bewoor
Cummins College of Engineering for Women, Pune 411052, Maharashtra, India
e-mail: dranandbewoor@gmail.com

© Springer Nature Singapore Pte Ltd. 2020
P. V. Varde et al. (eds.), *Reliability, Safety and Hazard Assessment for Risk-Based Technologies*, Lecture Notes in Mechanical Engineering, https://doi.org/10.1007/978-981-13-9008-1_68

1 Introduction

The coal-based thermal power plant is the largest source of power generation in India. A coal-fired thermal power plant is a complex engineering system comprising of various system, viz. coal handling, steam generation, cooling water, crushing ash, ash handling, power generation, feed water, steam and water analysis system, and flue gas and air system [1]. These systems are connected in complex configuration, i.e., in either series configuration or parallel configuration. The reliability and sensitive analysis of coal handling unit of thermal power plant is investigated by Kumar et al. [2]. The coal handling unit consisting conveyor and wagon tippler should function continuously without any obstacle for power generation and identified as critical subsystem. Likewise, availability simulation model of coal handling unit consisting of five subsystem, viz. screener, feeder, hooper, wagon tippler, and conveyor is developed and analyzed for performance analysis by Gupta et al. [3]. Conradie et al. [4] address the operational scheduling of coal extraction, stacking, and reclaimer using practical approaches for modeling and solving the coal scheduling problem. Song et al. [5] studied the configuration methods of coaling, unloading, the coal yard, and its equipment. The early detection of fault of crusher rotor of coal handling unit is studied and diagnosed by Jha et al. [6] for the improvement in reliability as well as availability. Further, Kajal et al. [7] used genetic algorithm-based Markov birth–death approach for optimization of failure and repair rates of coal handling unit. Further, Zhao et al. [8] studied operation process of coal handling system in thermal power plant and analyzed technical characteristics of coal handling system and also propose suitable maintenance strategy for the safe operation of coal handling unit.

The root cause of failures of coal handling plant for reduction in availability is studied by Ojha et al. [9]. Gupta et al. [10] studied performance analysis of coal handling unit of thermal power plant using Markov probabilistic approach and the equipment consider for analysis is wagon tippler and conveyor subsystem. The opportunities for availability simulation modeling and maintenance decision making for steam and water analysis system in thermal power plant are presented by Gupta et al. [11]. Kumar et al. [12] developed Markov model for performance evaluation of coal handling unit of thermal power plant using probabilistic approach. The results revealed that the maintenance priority should be as per the order of (1) conveyor, (2) wagon tippler, (3) hooper, (4) feeder, and (5) side arm changer. Therefore, probabilistic analysis of system under given operative condition is helpful in forecasting equipment behavior which further helps in design to achieve minimum failure in system, i.e., to optimize the system working [13].

With critical literature survey and detail discussion with domain expert at DTPP, coal supply system is considered for performance evaluation. This paper proposes Markov probabilistic approach for performance-based availability analysis of coal supply system of Dahanu Thermal Power Plant (DTPP). The historical data corresponding to failure rate as well as repair rate of last eighteen years is collected and analyzed for availability analysis. The detailed analysis is discussed in next section.

2 System Description

Coal supply system of thermal power plant consists of three subsystem, viz. coal mill, stacker reclaimer, and wagon tippler. (a) The coal mill 'A' subsystem consists of three units. Failure of any one unit leads to run at reduced capacity, (b) the stacker reclaimer 'B' subsystem is single unit, failure of which leads to unit failure, and (c) the wagon tippler 'C' subsystem consists of two units, failure of any one unit leads to run at reduced capacity.

2.1 Assumptions

(a) Failure and repair rates of each subsystem are constant and statistically independent; (b) not more than one subsystem fails at a time; (c) a repaired unit is as good as new performance wise; (d) the standby units are of the same nature and capacity as the active units.

2.2 Nomenclature

⬤ : Good capacity state   ⬭ : Reduced capacity state   ▭ : Failed state

A, B, C: Subsystems are in good operating state
a, b, c: Indicating the failed state of A, B, C
$\overline{A}$ and $\overline{C}$: Indicating reduced capacity state of A and C
λ_i: Mean constant failure rate
μ_i: Mean constant repair rate
$'$: Derivative w.r.t. 't'
$P_i(t)$: Probability that at time 't' the subsystem is in ith state

3 Availability Simulation Model for Performance Analysis of Coal Supply System in Thermal Power Plant

The proposed availability simulation model for coal supply system of thermal power plant is based on Markov birth–death probabilistic approach. Figure 1 gives the transition diagram of coal supply system of thermal power plant with three different working states, viz. working at full capacity, reduced capacity, and failed state. In contains total 12 states ('0' to '11') out of which state '0' represents working of subsystem with full capacity, states '1,' '2,' and '3' designate the working of subsystem

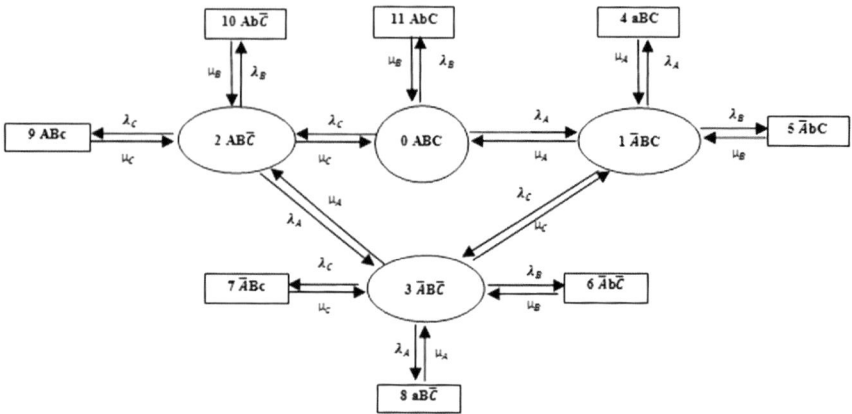

Fig. 1 Transition diagram of coal supply system

with reduced capacity, and remaining states from '4' to '11' represent to failed state in transition diagram.

The probability consideration gives the following differential equations, i.e., Eqs. (1–11) using Laplace transformation technique associated with the transition diagram (Fig. 1) which are given as follows

$$P_0'(t) + (\lambda_A + \lambda_B + \lambda_C) P_0(t) = \mu_A P_1(t) + \mu_B P_2(t) + \mu_C P_3(t) \tag{1}$$

$$P_1'(t) + (\lambda_A + \lambda_B + \lambda_C + \mu_A) P_1(t) = \mu_A P_4(t) + \mu_B P_5(t) + \mu_C P_3(t) + \lambda_A P_0(t) \tag{2}$$

$$P_2'(t) + (\lambda_A + \lambda_B + \lambda_C + \mu_C) P_2(t) = \mu_A P_3(t) + \mu_B P_{10}(t) + \mu_C P_9(t) + \lambda_C P_0(t) \tag{3}$$

$$P_3'(t) + (\lambda_A + \lambda_B + \lambda_C + \mu_A + \mu_C) P_3(t) = \mu_A P_8(t) + \mu_B P_6(t) + \mu_C P_7(t)$$
$$+ \lambda_A P_2(t) + \lambda_C P_1(t) \tag{4}$$

$$P_4'(t) + \mu_A P_4(t) = \lambda_A P_1(t) \tag{5}$$

$$P_5'(t) + \mu_B P_5(t) = \lambda_B P_1(t) \tag{6}$$

$$P_6'(t) + \mu_B P_6(t) = \lambda_B P_3(t) \tag{7}$$

$$P_8'(t) + \mu_A P_8(t) = \lambda_A P_3(t) \tag{8}$$

$$P_9'(t) + \mu_C P_9(t) = \lambda_C P_2(t) \tag{9}$$

$$P_{10}'(t) + \mu_B P_{10}(t) = \lambda_B P_2(t) \tag{10}$$

$$P'_{11}(t) + \mu_B P_{11}(t) = \lambda_B P_0(t) \tag{11}$$

Initial conditions at time $t = 0$, $Pi(t) = 1$ for $i = 0$ otherwise, $Pi(t) = 0$. For long-run availability steady state, the system can be analyzed by setting $\frac{d}{dt} \to 0$ and $t \to \infty$. The limiting probabilities from Eqs. (1) to (11) are

$$(\lambda_A + \lambda_B + \lambda_C)P_0 = \mu_A P_1 + \mu_B P_2 + \mu_C P_3 \tag{12}$$

$$(\lambda_A + \lambda_B + \lambda_C + \mu_A)P_1 = \mu_A P_4 + \mu_B P_5 + \mu_C P_3 + \lambda_A P_0 \tag{13}$$

$$(\lambda_A + \lambda_B + \lambda_C + \mu_C)P_2 = \mu_A P_3 + \mu_B P_{10} + \mu_C P_9 + \lambda_C P_0 \tag{14}$$

$$(\lambda_A + \lambda_B + \lambda_C + \mu_A + \mu_C)P_3 = \mu_A P_8 + \mu_B P_6 + \mu_C P_7 + \lambda_A P_2 + \lambda_C P_1 \tag{15}$$

$$\mu_A P_4 = \lambda_A P_1 \tag{16}$$

$$\mu_B P_5 = \lambda_B P_1 \tag{17}$$

$$\mu_B P_6 = \lambda_B P_3 \tag{18}$$

$$\mu_A P_8 = \lambda_A P_3 \tag{19}$$

$$\mu_C P_9 = \lambda_C P_2 \tag{20}$$

$$\mu_B P_{10} = \lambda_B P_2 \tag{21}$$

$$\mu_B P_{11} = \lambda_B P_0 \tag{22}$$

Let us assume,

$$P_{1-3} = L_{1-3} P_0, \ P_4 = K_A P_1, \ P_5 = K_B P_1, \ P_6 = K_B P_3, \ P_7 = K_C P_3,$$
$$P_8 = K_A P_3, \ P_9 = K_C P_2, \ P_{10} = K_B P_2, \ P_{11} = K_B P_0$$
$$\text{where,} \quad K_A = \frac{\lambda_A}{\mu_A}, \quad K_B = \frac{\lambda_B}{\mu_B}, \quad K_C = \frac{\lambda_C}{\mu_C}$$

Solving Eqs. (12) and (22) using matrix method of equations, we get $\sum_{i=0}^{11} P_i = 1$

$$P_0 = [1 + L_1 + L_2 + L_3 + K_A L_1 + K_B L_1 + K_B L_3 + K_C L_3$$
$$+ K_A L_3 + K_C L_2 + K_2 L_2 + K_B]^{-1}$$

Now, the simulation model for steady-state availability of the power generation system is obtained as the summation of all the working state probabilities,

$$A_V = [P_0 + P_1 + P_2 + P_3] \tag{23}$$

$$A_V = [1 + L_1 + L_2 + L_3]P_0 \tag{24}$$

4 Result and Discussions for Markov Approach

The performance analysis is base for all activities related to performance improvement such as modification in design, implementation new maintenance activities, etc. The performance of coal supply unit is majorly affected by failure and repair rate of subsystem used. Hence, the failure and repair rate of coal mill, stacker reclaimer, and wagon tippler are taken from maintenance history using Eq. (23) or Eq. (24), and then, availability values are plotted from Tables 1, 2, and 3.

Likewise, Figs. 2, 3, and 4 represent effect of failure and repair rates of subsystems of coal supply system on system availability. The various availability levels are computed for different possible combination of failure and repair rates of coal supply system for unit 1 of DTPP, which are discussed next.

(1) It is observed from Table 1 and Fig. 2 that, for some constant values of failure and repair rates of other subsystems, as the failure rate λ of coal mill of unit 1 increases from 0.00016253 to 0.00033361, the system availability decreases by about 0.08%. Similarly, as repair rate 'μ' of coal mill increases from 0.009577 to 0.019658, the system availability increased by about 0.02%.

Table 1 Availability matrix for coal mill of unit 1

μ_1	λ_1					
	0.00016	0.00017	0.00022	0.00027	0.00033	Constant values
0.009577	0.9931	0.9931	0.9928	0.9925	0.9922	$\lambda_2 = 0.0000583, \mu_2 =$
0.010081	0.9931	0.9931	0.9929	0.9926	0.9923	0.008798
0.013273	0.9932	0.9932	0.9931	0.993	0.9928	$\lambda_3 = 0.0000516,$
0.016466	0.9933	0.9933	0.9932	0.9931	0.993	$\mu_3 = 0.010471$
0.019658	0.9933	0.9933	0.9933	0.9932	0.9931	

Table 2 Availability matrix for stacker reclaimer of unit 1

μ_2	λ_2					
	0.000055	0.000058	0.000076	0.000095	0.000113	Constant
0.008358	0.9931	0.9928	0.9906	0.9884	0.9863	$\lambda_1 = 0.0001711, \mu_1$
0.008798	0.9934	0.9931	0.991	0.989	0.9869	$= 0.010081$
0.011584	0.9949	0.9947	0.9931	0.9915	0.99	$\lambda_3 = 0.0000516,$
0.01437	0.9958	0.9956	0.9944	0.9931	0.9918	$\mu_3 = 0.010471$
0.017155	0.9965	0.9963	0.9952	0.9942	0.9931	

Table 3 Availability matrix for wagon tippler of unit 1

μ_3	λ_3					
	0.000049	0.000051	0.000067	0.000084	0.000100	Constant
0.009948	0.9931	0.9931	0.9931	0.993	0.9928	$\lambda_1 = 0.0001711, \mu_1$
0.010471	0.9931	0.9931	0.9931	0.993	0.9928	$= 0.010081$
0.013787	0.9931	0.9931	0.9931	0.9931	0.993	$\lambda_2 = 0.0000583, \mu_2$
0.017103	0.9931	0.9931	0.9931	0.9931	0.9931	$= 0.008798$
0.020419	0.9931	0.9931	0.9931	0.9931	0.9931	

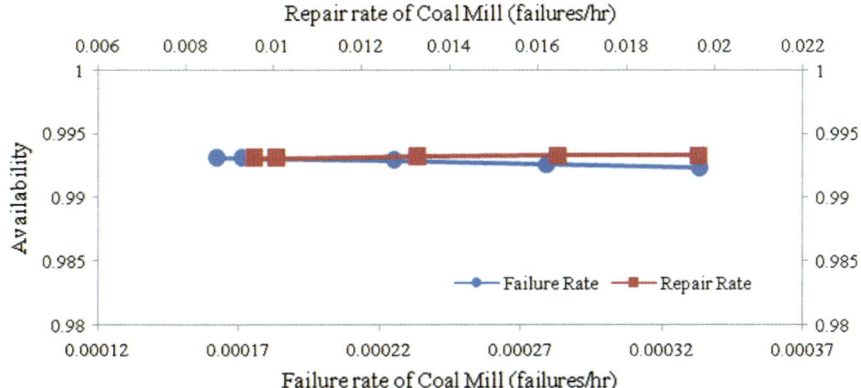

Fig. 2 Effect of failure and repair rates of coal mill on system availability

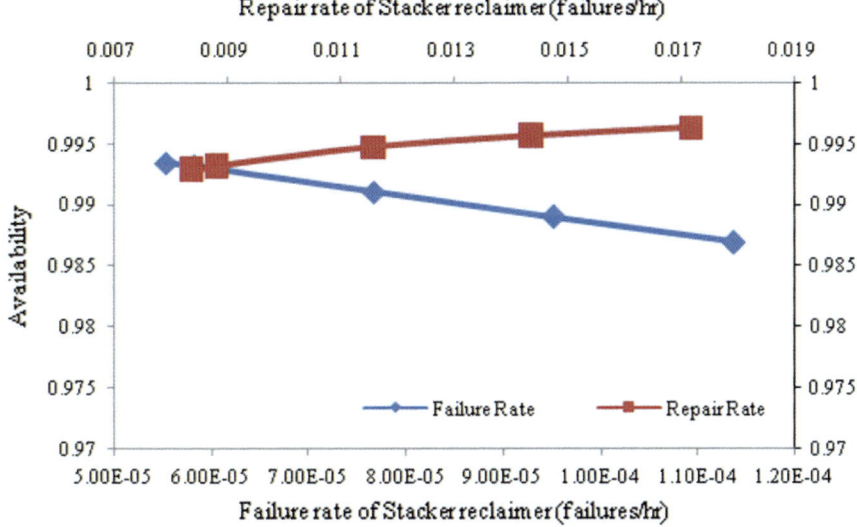

Fig. 3 Effect of failure and repair rates of stacker reclaimer on system availability

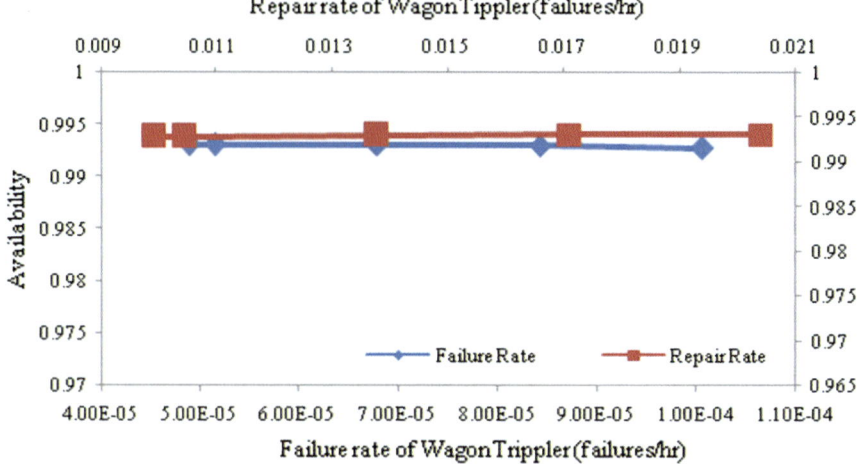

Fig. 4 Effect of failure and repair rates of wagon tippler on system availability

(2) It is observed from Table 2 and Fig. 3 that, for some constant values of failure and repair rates of other subsystems, as the failure rate 'λ' of stacker reclaimer of unit 1 increased from 0.000055419 to 0.00011375, the system availability decreased by about 0.65%. Similarly, as repair rates μ of stacker reclaimer increased from 0.008358 to 0.017155, the system availability increased by about 0.35%.

(3) It is observed from Table 4 and Fig. 4 that, for some constant values of failure and repair rates of other subsystems, as the failure rate 'λ' of wagon tippler of unit 1 increased from 0.000049002 to 0.00010058, the system availability decreased by about 0.03%. Similarly, as repair rates 'μ' of wagon tippler increased from 0.009948 to 0.020419, the system availability increased by about 0.03%. The next study is extended to optimize the availability parameter using particle swarm optimization technique which is discussed next.

5   Particle Swarm Optimization for Coal Supply System of Thermal Power Plant

The modern optimization methods such as genetic algorithm, simulated annealing, ant colony optimization, and neural network-based methods are developed and adopted by earlier researchers. As complex engineering problem can be solved by using modern optimization methods, its applicability in case of thermal power plant for availability analysis is not reported in published literature. The study reported here is an attempt to fulfill the said research gap. Particle swarm optimization method (PSOM) has advantage over the other method, as it is not largely affected by the size and non-linearity of the problem [14]. PSOM is used to optimize the availability

Table 4 PSO parameter of CSS

S. No.	Parameter	Value	Remark
1	Inertia weight	0.9	Lies between 0 and 1
2	Cognitive component $c1$	1.5	Randomly selected between 0 and 2
3	Social component $c2$	1.5	Randomly selected between 0 and 2
4	Number of particles	10–100	To find optimum performance

of coal supply system (CSS) of DTPP. In case of CSS of TPP application, PSOM allocates randomly generated particles and random velocity to each particle which propagates in search space toward optima over number of iterations. Each particle has past memory attaining best position P_{best} and particle with best value of fitness G_{best} (refer Eqs. 25–26)

Let,

$$X_i = \{X_i\} \text{ and } V_i = \{V_i\} \tag{25}$$

for $i = 1$ to n, updating the PSO rules for velocity and position,

$$V_i = W \times V_i + c_1 r_1 (P_{\text{best}} - X_i) + c_2 r_2 (G_{\text{best}} - X_i) \tag{26}$$

$$X_i = X_i + V_i \tag{27}$$

where r_1 and r_2 are random numbers (0–1) as well as c_1 and c_2 are acceleration constant for P_{best} and G_{best}. The inertia weigh is W [15]. The PSO algorithm code is generated for CSS, and further optimized availability parameters results are obtained which is discussed next.

5.1 Optimization Modeling

In this study, all possible failure and repair parameters combination for equipment of CSS are considered for optimization for availability of the system. The numbers of parameters are 6 [three values of failure rate 'λ' as $\lambda_1 \in (0.00016253, 0.00033361)$, $\lambda_2 \in (0.000055, 0.00011375)$, $\lambda_3 \in (0.000049, 0.00010058)$, and three values of repair rate 'μ' as $\mu_1 \in (0.009577, 0.019658)$, $\mu_2 \in (0.008358, 0.017155)$, $\mu_3 \in (0.009948, 0.020419)$]. The performance of the system has been determined by imposing constraints on failure and repair parameters, i.e., minimum and maximum values of 'λ' and 'μ' which are obtained from Markov analysis. The PSO parameters used in this work are tabulated in Table 4.

Fig. 5 Effect of number of particles on system availability

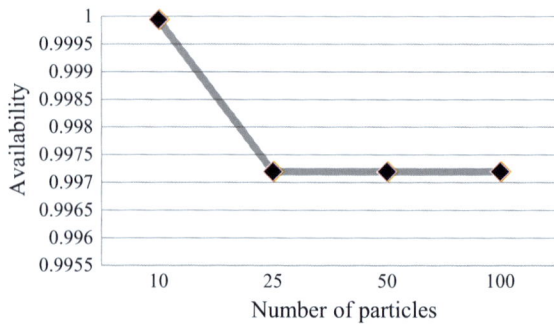

Table 5 Effect of number of particles on system availability

Parameters	No of particles			
	10	25	50	100
λ_1	0.000069	0.000334	0.000334	0.000334
λ_2	0.000045	0.000114	0.000114	0.000114
λ_3	0.000058	0.000101	0.000101	0.000101
μ_1	0.018086	0.019658	0.009577	0.009577
μ_2	0.003343	0.017155	0.017155	0.008358
μ_3	0.023031	0.009948	0.020419	0.009948
Availability	0.999938	0.997194	0.997194	0.997194

5.2 Result and Discussion for PSO

The PSO method is applied successfully for CSS of DTPP. The optimum availability parameters are obtained. The effect of number of particle on system availability is shown in Fig. 5. The optimum value for availability of CSS performance is 98.05%, for which the best possible combination of failure and repair parameters are $\lambda_1 = 0.000069$, $\lambda_2 = 0.000045$, $\lambda_3 = 0.000058$, $\mu_1 = 0.018086$, $\mu_2 = 0.003343$, $\mu_3 = 0.023031$ which is also tabulated in Table 5.

The results obtained from PSO method are used for determination of optimum value for MTTF and MTTR for CSS of DTPP which intern used to selected suitable maintenance strategy. Table 6 represents comparison of result for availability parameters with PSO approach. The availability is defined as fraction of time a system perform its function under given operating conditions. The availability of system is reliable on system performance parameters such as mean time to failure (MTTF) and mean time to repair (MTTR). In Table 6, the availability parameters viz. MTTF and MTTR has been optimized using particle swarm optimization method (PSO) and improvement in availability of 0.85% is obtained for coal supply system (CSS) of thermal power plant.

The result obtained from PSO revealed that system availability is improved by 0.85% with optimized availability parameters. These parameters are used to suggest

Table 6 Result comparison for availability parameters for CSS subsystems of DTPP

Equipment	Availability parameters			Optimized availability parameters			Improved availability (%)
	MTTF	MTTR	Availability	MTTF	MTTR	Availability	
Coal mill	5845	99	**97.20%**	14,566	55	**98.05%**	**0.85%**
Stacker reclaimer	17,142	114		22,231	299		
Wagon tippler	19,387	96		17,162	43		

Table 7 Optimum values failure and repair rates of Unit 1

Equipment	Failure rate (λ_i)	Repair rate (μ_i)	Decrease in Av. due to (λ_i) (%)	Increase in Av. due to (μ_i) (%)	Maximum availability (%)
Coal mill	0.0001711	0.010081	0.08	0.02	99.33
Stacker reclaimer	0.0000583	0.008798	0.65	0.35	99.65
Wagon tippler	0.0000516	0.010471	0.03	0.03	99.31

suitable maintenance strategy for CSS of DTPP and in turn to scheduled frequency of monitoring. Further work can be extended to validate optimized maintenance schedule by the application of other optimization techniques such as neural network and genetic algorithm.

6 Conclusions

The availability simulation model is used to implement maintenance strategies of coal supply system in thermal power plant. The availability simulation model helps to analyze the performance of coal handling system in thermal power plant which will ensure the maximum overall system availability. The optimum values of failure and repair rates for each subsystem of coal supply system with maximum availability level are tabulated in Table 7 for CSS of DTPP. On the basis of best suitable combination of failure and rates, the optimum maintenance strategy is suggested.

It is concluded that stacker reclaimer is identified as most critical subsystem as its failure will affect system performance rapidly. On the basis of failure and repair rates, the maintenance priority of coal supply system should be given as per order: (1) stacker reclaimer, (2) coal mill, and (3) wagon tippler.

Further, the results obtained from PSO revealed that system availability of CSS is improved by **0.85%** with optimized availability parameters. These parameters are used to suggest suitable maintenance strategy for subsystem at DTPP.

The findings of this study are discussed with domain expert at thermal power plant, and results are found to be beneficial to prioritize the coal supply system according to its effect on overall availability. The further work can be extended to implement optimum maintenance strategy on the basis of best suitable combination of failure and rates of thermal power plant.

Acknowledgements The author would like to express thanks to Head of the Department Mr. Suhas Patil, Domain Experts Mr. Atul Deshpande and Mr. Hemant Bari, Dahanu Thermal Power Plant for their valuable guidance and comments. Also, the authors would like to acknowledge especially maintenance and planning department as a facilitator for data collection.

References

1. R. Kumar, A.K. Sharma, P.C. Tewari, Performance evaluation of a coal-fired power plant. Int. J. Perform. Eng. **9**(4), 455–461 (2013)
2. A. Kumar, M. Ram, Reliability measures improvement and sensitivity analysis of a coal handling unit for thermal power plant. Int. J. Eng. **26**(9), 1059–1066, (2013)
3. S. Gupta, A.K. Sharma, P.C. Tewari, An availability simulation model and performance analysis of a coal handling unit of a typical thermal plant. S. Afr. J. Ind. Eng. **20**, 159–171 (2009)
4. D.G. Conradie, L.E. Morison, J.W. Joubert, Scheduling at coal handling facilities using simulated annealing. Math. Methods Oper. Res. **68**(2), 277–293 (2008)
5. F. Song, Analysis of optimization configuration for coal handling system in large power plant. *IEEE* (2012), pp. 2–5
6. P.K. Jha, R. Rajora, Fault diagnosis of coal ring crusher in thermal power plant: a case study. *International Conference on Automatic Control and Dynamic Optimization Techniques*, *ICACDOT* (2016), pp. 355–360
7. S. Kajal, P.C. Tewari, P. Saini, Availability optimization for coal handling system using genetic algorithm. Int. J. Perform. Eng. **9**(1), 109–116 (2013)
8. R. Zhai, M. Zhao, K. Tan, Y. Yang, Optimizing operation of a solar-aided coal-fired power system based on the solar contribution evaluation method. Appl. Energy **146**, 328–334 (2015)
9. S. Ojha, B.K. Pal, B.B. Biswal, Minimising the breakdown in belt conveyor system of coal handling plant. Int. J. Mech. Eng. **2**(9), 3–6 (2015)
10. S. Gupta, P. Tewari, A. Sharma, A Markov model for performance evaluation of coal handling unit of a thermal power plant. J. Ind. Syst. Eng. **3**(2), 85–96 (2009)
11. S. Gupta, P.C. Tewari, Simulation model for coal crushing system of a typical thermal power plant. Int. J. Eng. Tech. **1**(2), 156–164 (2009)
12. Y. Kumar, S. Kumar, R. Singh, A markov model for reliability analysis of coal handling unit of Badarpur thermal power plant. Int. J. Inno. Sci. Res. **1**, 17–22 (2013)
13. S. Gupta, P.C. Tewari, Simulation modeling and analysis of a complex system of a thermal power plant. J. Ind. Eng. Manag. **2**(2), 387–406 (2009)
14. S. Pant, D. Anand, A. Kishor, B. Singh, A particle swarm algorithm for optimization of complex system reliability. Int. J. Perform. Eng. **11**(1), 33–42 (2015)
15. P. Kumar, P.C. Tewari, Performance analysis and optimization for CSDGB filling system of a beverage plant using particle swarm optimization. Int. J. Ind. Eng. Comput. **8**, 303–314 (2017)

Aerospace Power Supply Modules Reliability Estimation Using ARINC Reliability Apportionment Technique

Khilawan Choudhary and S. Monisha

Abstract Reliability assures the proper functioning of the system for specified period. Development of a new system design requires certain specifications to be fulfilled to achieve the target reliability. Reliability Apportionment technique is a useful tool to achieve the target reliability by assigning the reliability to be met by each of the subsystems by consideration of risk involved, development cost and time. Reliability Apportionment helps in determining the weakest subsystem and guides the designer to focus on this subsystem to improve the reliability. The simplest reliability apportionment technique is equal apportionment which assigns equal reliability to all the subsystems. But equal apportionment assigns the reliability without consideration of risk involved in the development of each subsystem. Hence, reliability apportionment technique has to be selected which apportions the reliability to the subsystems based on the configuration, mission time and risk involved in development. In this paper, ARINC reliability apportionment technique has been adopted in assigning the reliability goals for the modules of aerospace power supply. Power supply consists of modules connected in series, components having constant failure rate, and the system follows exponential failure distribution which makes ARINC method suitable for apportionment. In ARINC method, weighting factor is calculated based on the failure rate predicted, which indicates the weakest card of the system and guides the designer to focus on that card for improvement of reliability.

Keywords Reliability · Reliability apportionment · ARINC · Weighted factor

1 Introduction

The reliability of the system is set to achieve the required performance for the specific period without any failures. The reliability to be assigned to each of the subsystems has to be estimated to achieve the target reliability. Reliability estimation

K. Choudhary (✉) · S. Monisha
Microwave Tube Research & Development Centre, DRDO, Bangalore, India
e-mail: khilawan@mtrdc.drdo.in

© Springer Nature Singapore Pte Ltd. 2020   825
P. V. Varde et al. (eds.), *Reliability, Safety and Hazard Assessment for Risk-Based Technologies*, Lecture Notes in Mechanical Engineering, https://doi.org/10.1007/978-981-13-9008-1_69

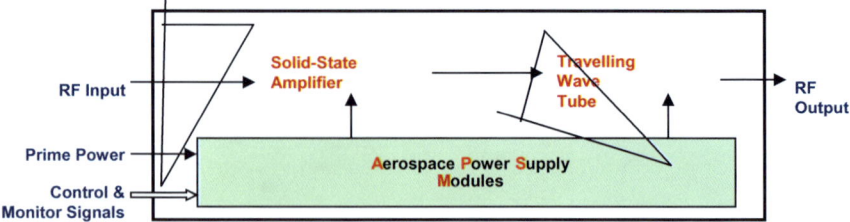

Fig. 1 Block diagram of microwave power module

is performed using reliability apportionment technique. Reliability apportionment technique apportions the target reliability to the subsystems considering various factors. There are various methods of performing reliability apportionment, and the appropriate method is selected based on the system characteristics.

The equal reliability apportionment is the simplest technique used when no proper details of the system are available. This apportionment equally divides the reliability to all the subsystems without considering the importance of each subsystem. The development risk, time and cost of the subsystem are not considered in this technique. This may result in improper reliability allocation to the subsystems.

Reliability allocation is usually done based on weighting factors or optimal reliability allocation is used. The weighting factors method is concerned with allocation of subsystem reliability to achieve target reliability. The optimal reliability allocation is concerned with the allocation of individual component reliability to meet some desired level of system reliability or cost [1].

ARINC reliability apportionment method is based on the weighting factors which assigns improved failure rates to subsystems using criteria that are considered important for system performance [1].

In this paper, ARINC reliability apportionment is used to determine the reliability to be allocated for the aerospace power supply modules in Microwave Power Module (MPM).

Typical block diagram of MPM consisting of power supply modules is shown in Fig. 1. The power supply modules cater the power supply needs for the solid state amplifier and travelling wave tube in the MPM.

Power supply consists of five modules: modulator card, high voltage card, housekeeping power supply card, SSPA power supply card and logic card. These modules are connected in series, the components in the modules have constant failure rate and the system follows exponential failure distribution which makes ARINC reliability apportionment technique suitable for reliability allocation. The predicted failure rate and reliability of the system are used to determine the weighting factors and improved failure rate of the modules using ARINC apportionment technique.

Based on the value of the weighting factors determined for the subsystems of the system the designer decides which card has to be focused and given importance to achieve the target reliability of the system.

2 ARINC Reliability Apportionment

ARINC (Aeronautical Radio Inc.) reliability apportionment technique allocates the reliability based on the predicted failure rate of the subsystems. Improved system failure rates are derived from the predicted failure rate of the subsystems based on the weighting factors [2]. Weighting factor w_i is the probability that the system will fail if the subsystem fails. Weighting factor of '1' implies that the subsystem is essential for successful system operation. Weighting factor of '0' implies that the subsystem has no impact on system performance. This method is applied for the subsystems that are connected in series and independent, components having constant failure rate and for the system whose mission time equals each subsystem's mission time [1].

Weighting factor w_i is calculated from the predicted failure rate λ_i as follows:

$$w_i = \frac{\lambda_i}{\sum_{i=1}^{n} \lambda_i} \tag{1}$$

Improved failure rate to be allocated to the subsystem is calculated as follows:

$$\lambda_i' = w_i * \lambda_s \tag{2}$$

where w_i is the weighting factor, λ_i is the predicted failure rate of the ith subsystem, λ_i' is the improved failure rate allocated to the ith subsystem, λ_s is the target system failure rate and n is the total number of subsystems.

3 System Failure Rate and Reliability Prediction

MIL-STD-217F (part stress method) is used for prediction of failure rate of power supply modules. Reliability for these modules is calculated based on this predicted failure rate considering mission time to be 5 years. Predicted failure rate and reliability values for each module of power supply are given in Table 1.

Table 1 Predicted failure rate and reliability

Subsystem name	Failure rate λ_i (FIT)	Reliability (%)
Modulator card	174.2	99.24
High voltage (HV) card	91.0	99.60
Housekeeping power supply (HKPS) card	44.5	99.80
SSPA power supply card	39.3	99.83
Logic card	12.2	99.95
System level	**361.2**	**98.43**

Target power supply reliability to be achieved is 98%. Target system failure rate is calculated as follows:

$$\lambda_s = \frac{-\ln(R_s)}{t} = \frac{-\ln(0.98)}{(5 * 365 * 24)} = 461 \text{ FIT} \tag{3}$$

4 Weighting Factor and Improved Failure Rate Calculations

Power supply modules' weighting factors and improved failure rate to be allocated are calculated using Eqs. (1–3).

4.1 Modulator Card

(i) Weighting factor:

$$w_1 = \frac{\lambda_1}{\sum_{i=1}^{5} \lambda_i} = \frac{174.2}{361.2} = 0.48$$

(ii) Improved failure rate:

$$\lambda_1' = \lambda_s * w_1 = 461 \times 10^{-9} * 0.48 = 221 \text{ FIT}$$

4.2 High Voltage (HV) Card

(i) Weighting factor:

$$w_2 = \frac{\lambda_2}{\sum_{i=1}^{5} \lambda_i} = \frac{91.0}{361.2} = 0.25$$

(ii) Improved failure rate:

$$\lambda_2' = \lambda_s * w_2 = 461 \times 10^{-9} * 0.25 = 115 \text{ FIT}$$

4.3 Housekeeping Power Supply (HKPS) Card

(i) Weighting factor:

$$w_3 = \frac{\lambda_3}{\sum_{i=1}^{5} \lambda_i} = \frac{44.5}{361.2} = 0.12$$

(ii) Improved failure rate:

$$\lambda_3' = \lambda_s * w_3 = 461 \times 10^{-9} * 0.12 = 55 \, \text{FIT}$$

4.4 SSPA Power Supply Card

(i) Weighting factor:

$$w_4 = \frac{\lambda_4}{\sum_{i=1}^{5} \lambda_i} = \frac{39.3}{361.2} = 0.11$$

(ii) Improved failure rate:

$$\lambda_4' = \lambda_s * w_4 = 461 \times 10^{-9} * 0.11 = 51 \, \text{FIT}$$

4.5 Logic Card

(i) Weighting factor:

$$w_5 = \frac{\lambda_5}{\sum_{i=1}^{5} \lambda_i} = \frac{12.2}{361.2} = 0.034$$

(ii) Improved failure rate:

$$\lambda_5' = \lambda_s * w_5 = 461 \times 10^{-9} * 0.034 = 16 \, \text{FIT}$$

5 Results and Discussions

5.1 Results

In order to compare and identify the weakest subsystem, ARINC reliability apportionment calculations for power supply modules are tabulated in Table 2.

The comparison of ARINC reliability apportionment results with equal apportionment results for power supply modules is brought about in Table 3.

Table 2 ARINC reliability apportionment calculations for power supply modules

Subsystem name	Weighting factor w_i	Failure rate λ_i (FIT)	Reliability (%)
Modulator card	0.48	221	99.04
High voltage (HV) card	0.25	115	99.49
Housekeeping power supply (HKPS) card	0.12	55	99.76
SSPA power supply card	0.11	51	99.78
Logic card	0.034	16	99.93

Table 3 Reliability apportionment results for power supply modules

Module name	Predicted reliability (%)	Equal apportioned reliability (%)	ARINC apportioned reliability (%)
Modulator card	99.24	99.6	99.04
High voltage (HV) card	99.60	99.6	99.49
Housekeeping power supply (HKPS) card	99.80	99.6	99.76
SSPA power supply card	99.83	99.6	99.78
Logic card	99.95	99.6	99.93

5.2 Discussions

The modulator card design has the highest predicted failure rate because of switching power MOSFETs [3] which is the highest failure contributing components and the logic card design has the least predicted failure rate because of the lowest failure contributing components. Based on these predicted failure rate values and target power supply failure rate, the weighting factors for each the module is calculated and represented graphically in Fig. 2. It is seen that modulator card has the highest weighting factor which implies that this card has to be given more importance. Similarly based on the weighting factors other modules of the system are given importance.

The predicted reliability, equal apportionment reliability estimation and ARINC apportionment reliability estimation values given in Table 3 for power supply modules are represented graphically in Fig. 3. This graph depicts that the equal apportionment technique allocates reliability value of 99.6% to all the modules based on the target power supply reliability without consideration of predicted component failure rate and design of the modules. This allocation implies that the designer must modify the card's design to achieve the target reliability.

ARINC apportionment values imply that the highest failure contributing modulator card has to be allocated the least reliability value of 99.04% based on the highest

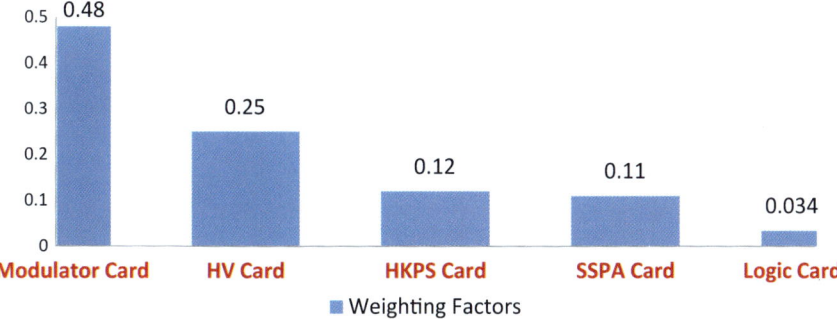

Fig. 2 Weighting factors of power supply modules

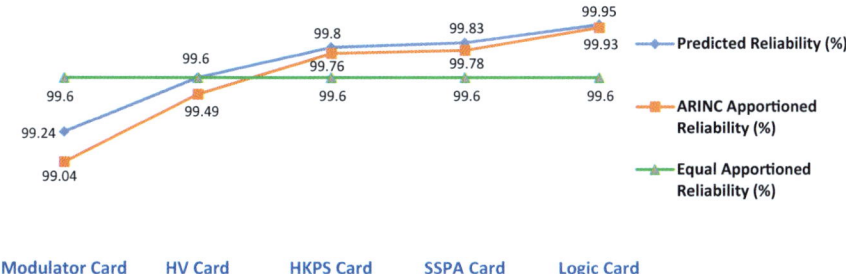

Fig. 3 Predicted reliability, ARINC apportioned reliability and equal apportioned reliability

weighting factor of 0.48 and the least failure contributing logic card has to be allocated the highest reliability value of 99.93% based on the least weighting factor of 0.034. It is seen that ARINC apportionment reliability values are already met by the power supply cards which indicates that the power supply modules meet the target reliability.

6 Conclusion

Aerospace power supply modules are apportioned using ARINC reliability apportionment method. Weighting factor and improved failure rate to be allocated to each of the subsystems are determined. The modulator card has the highest weighting factor, and the logic card has the least weighting factor. Based on these weighting factors, reliability values to be allocated to power supply modules are calculated. The designer must give more importance to modulator card and less importance to logic card if the allocated reliability values are higher than the predicted reliability values. Since the reliability allocation using ARINC apportionment method is lesser than the predicted reliability values there is no need for the designer to modify the design as these modules already meet the reliability requirements to achieve the target power supply reliability.

Acknowledgements The authors are thankful to Director, MTRDC, for permitting to publish this work and all MTRDC officers, supporting staff for their contributions to complete the above activities.

References

1. J.A. Cruz, *Applicability and Limitations of Reliability Allocation Methods* (NASA, Glenn Research Center, Cleveland, Ohio)
2. Reliability Allocation Using Lambda Predict—Weibull.com
3. K. Choudhary, N. Kumar, S. Monisha, *High Switching Beam Focusing Electrode Circuit MOS-FETs Reliability Analysis* (Microwave Tube Research & Development Centre (MTRDC), Bangalore, India)
4. PTC Windchill Quality Solutions 10.2 documents and software
5. MIL-HDBK-338, Electronic Reliability Design Handbook
6. P.D.T. O'Connor, *Practical Reliability Engineering,* 2nd edn.
7. Restricted Internal Design Reports on EPC, Microwave Tube Research and Development Centre, DRDO, B E Complex, Bangalore

Prediction and Assessment of LHD Machine Breakdowns Using Failure Mode Effect Analysis (FMEA)

J. Balaraju, M. Govinda Raj and Ch. S. N. Murthy

Abstract Across the world, production industries are always searching for enhancement of productivity by producing the targeted level of production. In the mining industry, Load Haul Dumper (LHD) is one of the major production equipments generally utilized as an intermediate level technology-based transportation system. LHDs are prone to uneven modes of multiple failures/breakdowns due to harsh operating environmental conditions. This leads to a decrease in the performance of the equipment and increases the maintenance cost, the number of unplanned outages (downtime), as well as loss of production levels. This can be controlled by adequate prediction of machine failures through root cause analysis (RCA). In the present investigation, a well-known fault prediction technique, i.e., failure mode effect analysis (FMEA) was utilized to identify the modes of potential failure, causative factors and recognize the effects of these failures on performance and safety. The risk-based numerical assessment was made by prioritizing the failure modes through the risk priority number (RPN) model. The criticality of failure was estimated using RPN values. They are calculated by the product of risk indexed/ruled parameters [severity (S), occurrence (O) and detection (D)]. Further, an attempt has been made to suggest suitable remedial actions to reduce or eliminate the various potential failures.

Keywords RCA · FMEA · RPN · LHD · Severity · Occurrence and detection

J. Balaraju (✉) · M. Govinda Raj · Ch. S. N. Murthy
Mining Engineering, NITK, Surathkal 575025, India
e-mail: jakkulabalraj@gmail.com

M. Govinda Raj
e-mail: mandelaraj88@gmail.com

Ch. S. N. Murthy
e-mail: chsn58@gmail.com

© Springer Nature Singapore Pte Ltd. 2020
P. V. Varde et al. (eds.), *Reliability, Safety and Hazard Assessment for Risk-Based Technologies*, Lecture Notes in Mechanical Engineering,
https://doi.org/10.1007/978-981-13-9008-1_70

1 Introduction

Faced with tough global competition and to enhance the expected targets of production and productivity every industry continuously looks for enhancement of mechanization performance. Failure mode effect analysis (FMEA) is such a performance measurement metric which will indicate the performance rate with very simple calculations. This is an engineering technique used to describe the root cause of failure and control actions. The initial development of FMEA was done by American Army in the year 1949 to estimate the effect of machine and component failures on the achievement of operation and protection of employees [1]. FMEA can be defined as "an engineering technique of quality and reliability study anticipated to recognize breakdowns affecting the performance of a machine/equipment and facilitate measures for suitable action to be place" [2]. FMEA technique is a qualitative risk assessment tool, mostly counting on the decision made by the risk analyst for a failure mode [3]. Risk assessment is a procedure to recognize the severity of the risk and comparing the level of hazard against predetermined values, aspects including the extent of exposure and the number of persons exposed and the risk of that hazard being found out. A wide variety of methods are used to identify the overall risk of the failure from a basic calculation with the utilization of low, medium, and high priorities. Risk management process involves different steps such as identification of risk, risk assessment and analysis, development of action plan and its implementation and evaluate, control, and measure of effectiveness [4] (Fig. 1).

Fig. 1 Risk management process

2 Failure Mode Effect Analysis (FMEA)

FMEA is a methodology for investigating the potential failures of reliability systems occurring during the design and development phase. It helps to recommend suitable preventive actions to overcome these problems and thus enhance the system reliability [5]. By performing FMEAs, failure modes are identified. Failure modes are the ways, or modes, in which an asset can fail. The severity, probability of occurrence, and risk of non-detection are estimated and used to rate the risk associated with each failure mode. In usual practice, level of risk of occurred failure can be measured by the computed metrics of "Risk Priority Number (RPN)" [6]. Three elements are generally taken into account while comparing the hazard of breakdown: the severity; the possibility of occurrence; and the detection probability of a failure [6, 7].

From the perspective view of the point of failure investigation, application of FMEA can be of many types [8]:

- Concept FMEA: used to analyze the concepts for systems and subsystems in the early stages.
- Design FMEA: used to analyze the products, high-standard machines, components, standard production tools, etc. before the products are delivered into the market.
- Process FMEA: used to analyze the small-scale machines or tools that allow for customized selection of component parts, machine structure, tooling, bearings, coolants, etc.
- Service FMEA: used to analyze manufacturing and assembly processes.

3 Risk Indexed/Ruled Parameters

FMEA searches for prioritizing the breakdown modes of a system with a specific end goal to allocate the accessible and constrained assets to the most genuine hazard things. In general, priority is determined through RPN, which is achieved by multiplying the indexes of O, S, and D of each failure.

- **Severity (S)**
  Severity evaluates the significance of the effect of the potential risk occur. S score is evaluated against the effect of the impact caused by the disappointment mode is shown in Table 1.
- **Occurrence (O)**
  Occurrence estimates the frequency that potential risk(s) will occur for a given situation or a system. Likelihood score is rated against the likelihood that the impact happens because of a disappointment mode is shown in Table 2.
- **Detection (D)**
  Detectability is the probability of the breakdown being detected before the impact of the breakdown to the process or system being assessed is detected. The D score

Table 1 Severity criteria for FMEA

Severity of a failure		
S. No.	Severity description	Risk scale
1	No effect	1
2	Very minor effect	2
3	Minor effect	3
4	Very low effect	4
5	Low effect	5
6	Moderate effect	6
7	High effect	7
8	Very high effect	8
9	Hazardous warning	9
10	Hazardous no warning	10

Table 2 Probability of occurrence criteria for FMEA

Occurrence of a failure		
Probability of occurrence	Failure	Rating
Remotely occurrence	1E−06	1
Very less chance of occurrence	1E−05	2
Less chance of occurrence	0.0001	3
Moderate probability: 1 in 2000	0.0005	4
Moderate probability: 1 in 400	0.0025	5
Moderate probability: 1 in 80	0.0125	6
High chance of occurrence: 1 in 20	0.05	7
High chance of occurrence: 1 in 8	0.125	8
Very high chance of occurrence: 1 in 3	0.25	9
Very high chance of occurrence: 1 in 2	0.5	10

is rated against the capability to detect the consequence of the breakdown mode is shown in Table 3.

- **Risk Priority Number (RPN)**
  RPN is the result of three data sources (severity, occurrence and detect) rating utilized when assessing risk to identify failure modes.

$$RPN = \text{Severity } (S) \times \text{Occurrence}(O) \times \text{Detect}(D)$$

The three risk factors of S, O, and D are evaluated using a ten-point scale. Failure modes with higher RPN values are assumed to be more hazardous, and there was an urgent requirement to resolve them than those with lower RPN values [9].

Table 3 Delectability criteria for FMEA

Detection of a failure		
Probability of detection	Detection scale	Detection scale
Almost certain	1	1
Very high	0.5	2
High	0.25	3
Moderately high	0.125	4
Moderate	0.05	5
Low	0.0125	6
Very low	0.0025	7
Remote	0.0005	8
Very remote	0.0001	9
Absolute uncertainty	0	10

(a) (b)

Fig. 2 **a** LHD machine at the maintenance workshop. **b** LHD machine during the repair

4   Field Investigation

Field investigation for the present analysis has been carried out in the underground metal mine of Hindustan Zinc Limited (HZL) located in the northeast region in India. This is India's largest and second largest zinc producing mining industry in worldwide. Sustainable underground mining operations are being carried out to extract the valuable minerals. Ore extraction is done by drilling and blasting, and LHD is used as the main workhorse for ore handling and transportation. LHDs are used to scoop the extracted ore, load it into the bucket, and dump it in the bottom of mine to undergo primary crushing before being hoisted to the surface out of the mine. Figure 2a shows the typical LHD equipment at the maintenance workshop kept for performing schedule maintenance, and Fig. 2b shows the replacement of the LHD's bucket boom.

5 Results of FMEA Analysis

In the present investigation, five numbers of LHD machines coded by 17LH21, 17LH22, 17LH24, 17LH25, and 17LH26 were considered to perform the analysis. The failure data, such as scheduled service hours (SSH), breakdown hours (BDH), spare parts waiting hours (SWH), idle hours of operator shift change, ore delay, and environmental factors (IDH) and in addition scheduled working hours (SWH) were collected from the field investigation. The data were collected from the data sheet and real-time monitoring in the field for financial year 2016–17 (Table 4).

Key performance indicators (KPI) of equipment such as availability percentages (% AVL), utilization percentage (% UTL) factors were calculated from the values of machine downtime hours (MDH), machine idle hours (MIH), machine available hours (MAH), and machine operating hours (MOH). In order to predict the LHD performance, the failure rate (FR) of each machine was computed with the measures of mean time between failure (MTBF) and mean time to repair (MTTR). MTBF is the ratio of uptime of LHD and its number of failures, and similarly, MTTR is the ratio of downtime of LHD and its failure frequency. Reliability (R) percentages were also calculated to identify the probability of LHD failure (Table 5). The utilized time of the machine was compared with its available time (Fig. 3) and computed FR metrics were also compared to identify the variations, shown in Fig. 4.

Table 4 Collected data from field visit (Mar 2016–Apr 2017)	Machine ID	SWH	SSH	BDH	SWH	IDH
	17LH21	8760.00	167.00	2180.75	0.00	674.00
	17LH22	8760.00	417.75	1716.83	10.50	434.00
	17LH24	8760.00	273.67	2308.17	0.00	456.00
	17LH25	8760.00	526.50	1756.67	0.00	600.00
	17LH26	8760.00	340.17	1931.58	0.00	575.00

Fig. 3 Comparison of availability and utilization in percentage

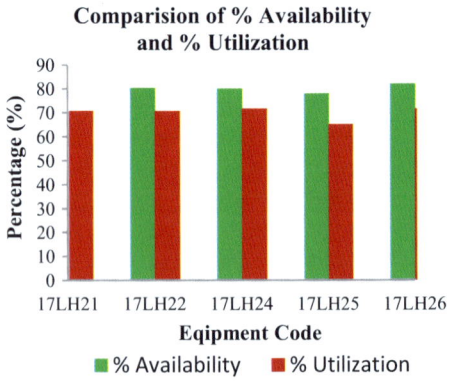

Table 5 Computed values of machine parameters

Machine ID	MDH	MIH	MAH	MOH	% AVL	% UTL	MTBF	MTTR	FR	R
17LH21	2180.75	841.00	7032.67	6191.7	80.28	70.68	351.67	151.08	0.0285	43.85
17LH22	1727.33	851.75	7032.67	6180.9	80.28	70.56	351.67	128.95	0.0285	43.85
17LH24	2308.17	729.67	7003.33	6273.7	79.95	71.62	411.96	179.69	0.0242	57.92
17LH25	1756.67	1126.50	6828.42	5701.9	77.95	65.09	359.39	151.74	0.0278	44.58
17LH26	1931.58	915.17	7177.98	6262.8	81.94	71.49	422.23	167.45	0.0236	61.15

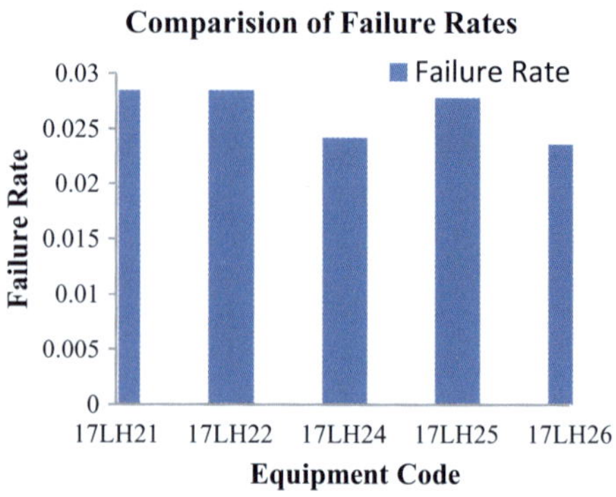

Fig. 4 Comparison of failure rate for various equipments

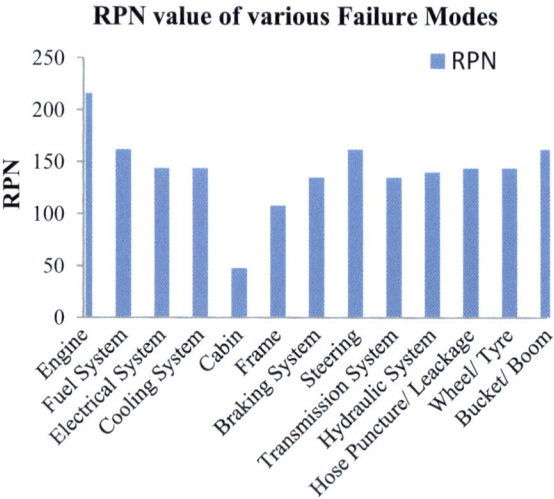

Comparision of Failure Rates

Fig. 5 Risk indexed RPN values of various failure modes

RPN value of various Failure Modes

A prospective risk analysis technique (FMEA) has been implemented for the present investigation. FMEA attempts to detect the potential failure modes of the product. The technique is used to anticipate the causes of failure and prevent them from happening. FMEA uses, occurrence (O) and detection (D) probability criteria in conjunction with severity (S) criteria to develop risk priority numbers (RPN) for prioritization of corrective action considerations [10]. Ranking criteria of S, O, and D were selected appropriately by analyzing the failure data of the equipment. The values of RPN were computed with the utilization of S, O, and D parameters (Table 6). Risk indexed RPN values, action results of RPN and its comparison with various failure modes are shown in Figs. 5, 6, and 7.

Table 6 FMEA worksheet (Process FMEA) of LHDs

Component description	Potential failure mode	Effects description	Potential cause(s)	Risk index				Recommended actions	Actions taken	Action results			
				S	O	D	RPN			S	O	D	RPN
Engine	Accelerator pedal stuck	Dangerous to the operator and the opposite person	Dirty air in the throttle body and cable in the air intake system	9	4	6	216	Cleaning the throttle body	Repair action should release this symptom	5	2	6	60
	Exhaust manifold gasket problem	Performance issues such as decrease in power, accelerator, and fuel efficiency	Engine wiring and heat from the exhaust gases	8	3	4	96	Fit the new gasket by pushing it into the cylinder head	Replacement action should improve the performance of the engine	5	3	4	60
Fuel System	Fuel airlock	Complete stoppage of fuel flow	Air leaking into the fuel delivery line	9	6	3	162	Eliminated by turning the engine over for a time using the starter motor	Replacement of modern diesel injection systems reduce the air locking	7	5	3	105
	Fuel top up damage	Decrease the fuel efficiency	A faulty pump with low pressure	7	3	7	147	Replacement of the fuel pump with a new one	Improves the performance of the machine	5	3	7	105
Electrical System	Starting problem	The machine won't start	Starter broken, bad ignition system	9	3	5	135	Replacement of modified starter and ignition plug	Replacement of parts should operate the machine	5	3	5	75

(continued)

Table 6 (continued)

Component description	Potential failure mode	Effects description	Potential cause(s)	Risk index				Recommended actions	Actions taken	Action results			
				S	O	D	RPN			S	O	D	RPN
	Water insert in the electrical panel	Damage of wires and short circuit	Wires can mold or corrosion	6	4	6	144	Replace the wires and protect with necessary provisions	Replacement action should operate the machine	3	4	6	72
Engine Cooling System	Overheating problem	Radiator damage	A small leak or evaporation of coolant oil	7	6	2	84	Repairing of a leak and provide coolant oil	Repairing of the parts will control the leakage	5	6	2	60
	Fan belt and pulley problem	Catastrophic engine damage	Misalignment forces the belt to kink or twist	8	6	3	144	Provide premature wear resistant and proper alignment	Repair action should reduce the fan belt breakages	5	6	3	90
Cabin	Door glass sliding problem	Bind at the corners	Adjusting screw fully tightened	3	6	2	36	Provide proper alignment between glass and frame	Alignment action reduces the sliding problem	2	6	2	24
	Door glass came out	The operator gets affected by the environmental issues	Misadjustment between glass and frame	6	4	2	48	Adjust the glass with a roller adjustment screw	Adjustment action improves the alignment	4	4	2	32

(continued)

Table 6 (continued)

Component description	Potential failure mode	Effects description	Potential cause(s)	Risk index				Recommended actions	Actions taken	Action results			
				S	O	D	RPN			S	O	D	RP N
	Rear wiper motor connection failure	Hard functioning of the wiper blades	Noisy in operation and stuck in operation	6	3	2	36	Replacement of new wiper set up	Replacement action provides the smooth operation	4	3	2	24
Frame	Rear light frame mounting broken	Impacts on vehicle operation	Vibration due to the harsh working environment	9	7	2	126	Replacement of light frame mounting	Replacement action should provide the machine operable	6	7	2	84
	Front axle stud broken	Uneven wear on the tire	Aging factor of the stud	9	4	3	108	Replace with a new stud for the axle	Replacement action helps to machine operable	6	4	3	72
Braking system	Parking problem	Traffic congestion causes to brake failure	Lack of sufficient parking at the event site	7	6	3	126	Recommended actions suggest to the operator	Skill of the operator should improve the brake life	5	6	3	90
	Brake oil leakage	Serious safety concerns	Fluid level in the reservoir is low	9	5	3	135	Maintain the fluid level by proper arrangements	Repair actions could control the oil leakage	7	5	3	105

(continued)

Table 6 (continued)

Component description	Potential failure mode	Effects description	Potential cause(s)	Risk index				Recommended actions	Actions taken	Action results			
				S	O	D	RPN			S	O	D	RP N
Steering	Gear shifting problem	Overall damage of the transmission system	A clutch that fails to disengage from the fly wheel	8	5	4	160	Replacement of the clutch plates with a new one	Replacement action should reduce the gear shifting problem	6	5	4	120
	Articulation problem	Slurring of speech, hearing loss	Difficulties in articulating sounds	9	6	3	162	Recommend suggestions to the operator about machine operation	Skill of the operator should minimize the problem	7	6	3	126
Transmission System	Torque converter lockup problem	Won't be able to transfer the power from engine to the transmission system	A strange noise, higher stall speeds and slipping of gears	8	4	4	128	Repair the lockup problem	Repair action should reduce the problem	6	4	4	96
	Drive shaft belt broken	Stops working of the machine	Belt breaks, slips, and wear out	9	5	3	135	Replace with a new belt	Replacement action should help to machine operable	7	5	3	105
Hydraulic System	Cylinder damage	Cracks on the cylinder cover	Improper lubrication and wear and tear	7	4	5	140	Provide sufficient lubrication	Proper lubrication system should increase the cylinder life	5	4	5	100

(continued)

Table 6 (continued)

Component description	Potential failure mode	Effects description	Potential cause(s)	Risk index				Recommended actions	Actions taken	Action results			
				S	O	D	RPN			S	O	D	RPN
	Hydraulic pump failure	Failure chain reaction in the system	Poor system design and low contamination control	6	4	4	96	Replacement of a pump with modified design	Replacement action should contribute to control the failures	4	4	4	64
Hose puncture/leakage	Charge air cooler hose open	Damage the intercooler parts	Drop in the pressure of the compressed air	5	5	4	100	Replace the new intercooler parts	Replacement action controls the damages	3	5	4	60
	Coolant leakage	Hoses get hard and brittle	Coolant join with water pump, radiator, and heater core	6	4	6	144	Recommend provisions to hose get smooth	Repair action reduces the hose problems	4	4	6	96
	Hydraulic oil leakage	Machine stoppages and component failures	Contaminated lubricants	5	3	7	105	Replace the hydraulic actuating valve	Replacement action should control the machine stoppages	3	3	7	63
Wheel/Tire	Wheel stud broken	Wheel and tire to separate them from vehicle	Over-torquing and under-torquing the lug nuts	9	5	3	135	Recommend suggestion to the maintenance personnel	Recommended suggestions should improve the stud life	7	5	3	105

(continued)

Table 6 (continued)

Component description	Potential failure mode	Effects description	Potential cause(s)	Risk index				Recommended actions	Actions taken	Action results			
				S	O	D	RPN			S	O	D	RPN
	Tire puncture	Machine stopped for repair	Poor underfoot condition and improper inflation	9	4	4	144	Remove and send to maintenance workplace	Maintenance of the machine helps to machine operable	7	4	4	112
	Tire nozzle lock	Cover for the tire valve breakage	Harsh road condition	8	5	3	120	Replace with a new cover plate	Replacement action should help to machine runs smooth	6	5	3	90
Bucket/Boom	Dump box welding work	Production stoppages	Poor welding at the joints	8	5	2	80	To be weld at the required portions	Repaired portions should continuous the production	6	5	2	60
	Bucket attachment system	Production stoppages	Broken due to poor design, week strength, and overload	9	6	3	162	Replace with a modified design and strengthened one	Replacement action should continue the work	7	6	3	126
	Front axle replacement	Hard operation of equipment	Overloading, lifting cylinder breakage					Replace the cylinder breakage with a modified design	Replacement of the component improves the operation				

(continued)

Table 6 (continued)

Component description	Potential failure mode	Effects description	Potential cause(s)	Risk index				Recommended actions	Actions taken	Action results			
				S	O	D	RPN			S	O	D	RP N
	Boom functioning slow	Reduce the production levels	Feed assembly damage	9	5	3	135	Remove and replace the feed assembly	Replacement action improves the production levels	7	5	3	105
	Boom broken	Production stoppages	Overloading, lifting cylinder breakage, and poor welding at the joints	9	4	4	144	Remove and replace with a strengthened one	Replacement action should continue the production	7	4	4	112

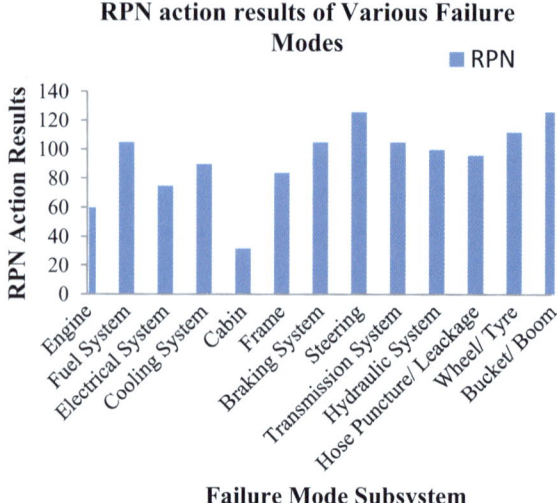

Fig. 6 RPN action results of various failure modes

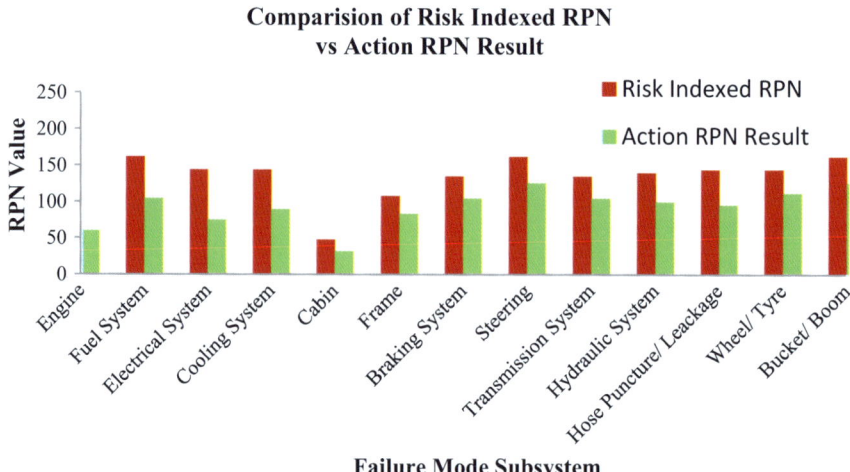

Fig. 7 Comparison of risk indexed RPN versus action RPN result

6 Discussion and Conclusion

Reliability and safety are the fundamental attributes of any modern technological systems. Focusing on system efficiency and safety, the reliability prediction and analysis aimed at the quantification of the probability of failure of the system. One of the sophisticated techniques used to predict the reliability parameters is FMEA. This is a well-situated tool for root cause analysis (RCA) of the failure. This easy and accurate method identifies the potential failure mode and its consequential impacts on machine operation and performance. In the present study, the performance of LHD machines was analyzed by utilizing the FMEA technique.

From the KPI prediction, it was identified that available percentage of all the equipment was on an average 80.08% (for 17LH21, 17LH22, 17LH24, 17LH25, and 17LH26). The remaining 19.92% time has been loosed due to continuous and randomly accruing breakdowns in a system could be the main cause for reduction of machine available hours. From the computed results of percentage utilization, it was understood that the utilization of the equipment was also not under the satisfactory level (69.88%) due to the increase of MIH. In addition to the KPI factors, the percentages of reliability were also evaluated based on the failure rate. The least rate of reliability was observed for the machines 17LH21 and 17LH22 (43.85%) with its corresponding failure rate of 0.0285 and the maximum was 17LH26 (61.15%) with a failure rate 0.0236. Increase of FR is mainly caused due to maintenance and operational related failures. The modes of failure, the reasons for the occurrence, and risk impact on subsequent components were evaluated by the utilization of FMEA.

Component wise detailed analysis was carried out in this paper. The scales of S, O, and D are allocated for each potential failure mode to evaluate the risk indexed RPN value. From the results of (Table 5), it was noticed that the higher RPN value was obtained from the engine (216), bucket/boom (162), steering (162), and fuel system (162). In FMEA investigation, the component with high RPN value results in more risk hazardous in the system. The S and O of risk hazardous failure can be reduced/eliminated by taking suitable preventive actions. After successful implementation of suggested actions, RPN values were reduced to 60 (engine), 126 (bucket/boom), 126 (steering), and 105 (fuel system). It was concluded that the RPN will give guidance for ranking the potential failures and it is easy to identify or recommend the necessary actions for the design or process changes to lower S or O. If the RPN value is high, then the effect of failure mode is more critical. A critical mode of the failure in a system causes to reduce the overall performance. This can be improved by conducting the scheduled maintenance within time intervals for the equipment, by giving training to the maintenance personnel and operating crew to enhance their skills.

References

1. P.C. Teoh, K. Case, An evaluation of failure modes and effects analysis generation method for conceptual design. Int. J. Comput. Integr. Manuf. **18**(4), 279–293 (2005)
2. BS5760, 2009. Part 5, page 3. *BS5760.* British Standards Institution
3. J. Moubray, *Reliability-Centered Maintenance* (Industrial Press Inc., New York, 1992)
4. Risk Management Handbook for the Mining Industry, (1997), http://www.dpi.nsw.gov.au/__data/assets/pdf_file/0005/116726/MDG-1010-Risk-Mgt-Handbook-290806-website.pdf
5. P. Gudale, D.V. Naik, Use of FMEA methodology for development of semiautomatic averaging fixture for engine cylinder block. Int. J. Innov. Res. Sci. Eng. Technol. **3**(5), 12452–12462 (2014)

6. G. Dieter, *Engineering Design a Materials and Processing Approach* (McGraw-Hill, New York, 2000)
7. D.H. Stamatis, *Failure Mode and Effect Analysis: FMEA from Theory to Execution* (Quality Press, 2003)
8. D.J. Lazor, Failure Mode and Effects Analysis (FMEA) and Fault Tree Analysis (FTA) (Success Tree Analysis—STA). *Handbook of Reliability Engineering and Management* (McGraw-Hill, New York, 1995), pp. 6.1–6.46
9. Y.M. Wang, K.S. Chin, G.K. Poon, J.B. Yang, Risk evaluation in failure mode and effects analysis using fuzzy weighted geometric mean. Expert Syst. Appl. **36**(2), 1195–1207 (2009)
10. R. Rakesh, B.C. Jos, G. Mathew, FMEA analysis for reducing breakdowns of a sub system in the life care product manufacturing industry. Int. J. Eng. Sci. Innov. Technol. (IJESIT) **2**(2), 218–225 (2013)

Investigation of Reliability of DC–DC Converters Using Reliability Block Diagram and Markov Chain Analysis

M. Sujatha and A. K. Parvathy

Abstract In this paper, the reliability of DC–DC Luo converter is analyzed for its failure rate by considering the quality of the components and the internal parameter variation, as power electronics components are always prone to failure. Methods of improving the reliability of converter are discussed. By using redundancy technique, reliability is improved and it is proved mathematically. The results can be applied to basic as well as improved converters. The results are verified using reliability block diagram approach and markov analysis by using SHARPE tool.

Keywords DC–DC converters · Mean time between failures · MIL-HDBK-217 · SHARPE

1 Introduction

Reliability Engineering is the procedural discipline of assessing, controlling, and handling the probability of failure in devices and systems [1]. This plays an increasing role in all engineering disciplines. A quite standard procedure for assessing the reliability of a system is to decompose it into its integral components, estimate the reliability of each of these components and finally combine the component reliabilities using arithmetic methods to evaluate the reliability of the whole system.

A solar PV system would require a reliable DC–DC converter for its operation. This paper covers the reliability analysis and methods of improving the reliability of DC–DC Luo converter. This paper includes the analysis of DC–DC Luo converter for components of different qualities and for different internal parameter variation. Also redundancy technique is used to improve the reliability of the converter.

M. Sujatha (✉) · A. K. Parvathy
Hindustan Institute of Technology and Science, Chennai, India
e-mail: msujatha@hindustanuniv.ac.in

A. K. Parvathy
e-mail: akparvathy@hindustanuniv.ac.in

© Springer Nature Singapore Pte Ltd. 2020
P. V. Varde et al. (eds.), *Reliability, Safety and Hazard Assessment
for Risk-Based Technologies*, Lecture Notes in Mechanical Engineering,
https://doi.org/10.1007/978-981-13-9008-1_71

2 Reliability Computation of DC–DC LUO Converter

2.1 DC–DC Luo Converter

Luo converter is an advanced DC–DC converter, which is derived from Buck–Boost converter [2]. Depending upon the duty ratio (D), it can either step-up or step-down the voltage and current. Due to its simple structure and high voltage applications, it has high usage. PWM switching signal is used to drive the switch S, with frequency fs and duty ratio D. The energy storage elements are inductors L_1 and L_2 and capacitors C_1 and C_2, and R is the load resistance. Diode D represents the freewheeling diode in the circuit. V_{in} and V_o are the input supply voltage and output load voltage.

The reliability of converter is estimated by calculating the reliability of individual components. The circuit diagram of Luo converter is as shown in Fig. 1.

The converter is designed for the following values of components: $L_1 = L_2 = 1$ mH; $C_1 = C_0 = 20 \, \mu F$; $R = 20 \, \Omega$; $V_{in} = 12$ V; $V_o = 15$ V; frequency $= 5$ kHz. The output voltage of the converter is given in Fig. 2.

The waveforms of current, voltage, power across the switch are given in Fig. 3.

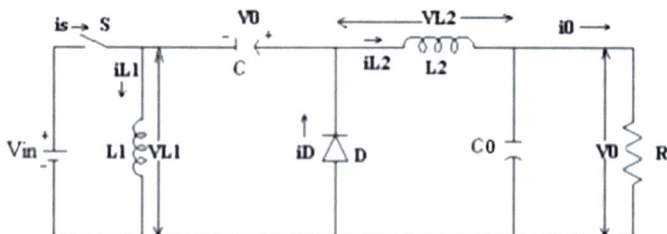

Fig. 1 Circuit diagram of Luo converter

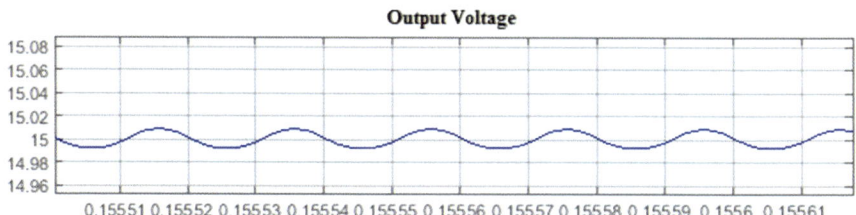

Fig. 2 Output voltage of the converter

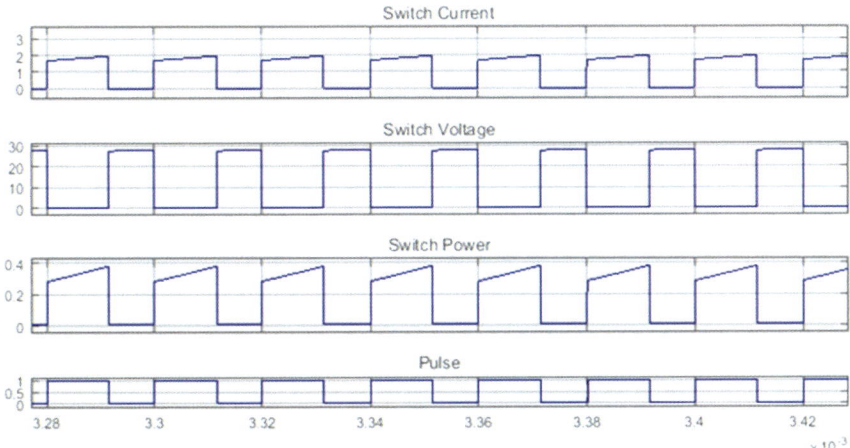

Fig. 3 Switch current, voltage, power waveforms with the given pulse

2.2 Failure Rate Calculation of Individual Components of Converter

For MOSFET switch and diode, the failure rate calculation is given by Eqs. (1) and (2) [3],

$$\lambda_p = \lambda_b \pi_T \pi_A \pi_Q \pi_E \text{ Failures}/10^6 \text{ h(for switch)} \tag{1}$$

$$\lambda_p = \lambda_b \pi_T \pi_C \pi_Q \pi_S \pi_E \text{ Failures}/10^6 \text{ h(for diode)} \tag{2}$$

λ_b = Base failure rate = 0.012(for Switch); λ_b = Base failure rate = 0.064 (for diode);

π_q = Quality factor = 5.5; π_E = Environmental factor = 1; π_c = Construction factor = 1, π_s = Stress factor;

π_A = Application factor = 10; π_T = Temperature factor = $\exp[-1925 * (\frac{1}{T_j+273} + \frac{1}{298}]$ (for switch);

$\pi_T = \exp[-3091 * (\frac{1}{T_j+273} + \frac{1}{293}]$ (for diode);

$T_c = T_a + \theta_{ca} * P_{loss}; T_j = T_c + \theta_{jc} * P_{loss}; \theta_{jc} = 0.25$ (for switch); $\theta_{jc} = 1.6$ (for diode), $T_a = 27, \theta_{ca} = 1.$

For capacitor, failure rate is, $\lambda_c = \lambda_b \pi_{cv} \pi_Q \pi_E$ where $\pi_{cv} = 0.34C^{0.18}, \pi_q = 5.5, \pi_E = 1$ and $\lambda_b = 0.0013.$

For inductor, failure rate is given by, $\lambda_L = \lambda_b \pi_c \pi_Q \pi_E$ where $\pi_c = 1, \pi_q = 4, \pi_E = 1$ and $\lambda_b = 0.00053.$

For resistor, failure rate is given by, $\lambda_{resistor} = \lambda_b \pi_R \pi_Q \pi_E$ where $\pi_R = 1, \pi_q = 5.5, \pi_E = 1$ and $\lambda_b = 0.012.$

Table 1 Failure rate of the components

S. No.	Component	Failure rate (failures/10^6 h)
1	Switch	0.77448
2	Diode	1.56965
3	Capacitor	3.467×10^{-4}
4	Inductor	2.12×10^{-3}
5	Resistor	0.066

Under normal operating conditions, λ_p (switch) $= 0.77448$; λ_p (diode) $= 1.56965$; $\lambda_C = 3.467 * 10^{-4}$; $\lambda_L = 2.12 \times 10^{-3}$; $\lambda_{resistor} = 0.066$.

Failure rate of the system is given by Eq. (3),

$$\lambda_{sys} = \lambda_p(\text{switch}) + \lambda_p(\text{diode}) + (2 * \lambda_C) + (2 * \lambda_L) + \lambda_{resistor} \tag{3}$$

$\lambda_{sys} = 2.415$; MTBF (mean time between failures) of the system $= (1/\lambda_{sys}) \times 10^6$ h $= 0.41407 \times 10^6$ h.

Reliability is given by Eq. (4) (Table 1)

$$R = e{-}^{\lambda t} \tag{4}$$

3 Reliability Analysis

3.1 Parts Selection According to Quality

The components of the DC–DC converter may be procured to a variety of different quality grades like commercial, industrial, and military grade [4]. The usage of highest-quality components will improve the reliability of the system. Table 2 gives the reliability of Luo converter with different quality grade components.

As the quality of the component increases, the failure rate decreases, and hence, the MTBF increases. Table 3 shows the quality and mean time between failures of the system. As per MIL HDBK, the quality of a component with S grade is greater than R grade and so on, i.e., S > R> P > M.

3.2 Comparison of Reliability Indices for Variation in System Internal Parameter

Switches are the most failure prone component in converters. In MOSFET switch, the on-state resistance has been described as the most significant aging factor. With

Table 2 Failure rate of the system for different grade components

Component	S		R		P		M		Standard	
	π_q	λ_p	π_q	λ_p	π_q	λ_p	π_q	λ_p	π_q	λ_p
Switch	0.03	4.2×10^{-3}	0.1	0.1408	5.5	0.77448	1.0	0.1408	0.3	4.2×10^{-2}
Diode	0.03	8.56×10^{-3}	0.1	0.0285	5.5	1.5696	1.0	0.2854	0.3	8.56×10^{-2}
Capacitor	0.03	1.89×10^{-6}	0.1	6.3×10^{-6}	5.5	3.46×10^{-4}	3.0	1.89×10^{-4}	0.3	1.89×10^{-5}
Resistor	0.03	3.6×10^{-4}	0.1	0.0012	5.5	0.066	5.0	0.06	0.3	3.6×10^{-3}
Inductor	0.03	1.2×10^{-5}	0.1	4.01×10^{-5}	4	2.21×10^{-3}	4.0	1.6×10^{-3}	0.3	1.20×10^{-4}
Failure rate		0.0131		0.0438		2.4126		0.48798		0.1313

Table 3 Converter system failure rate and MTBF

S. No.	Quality	λ_{sys}	MTBF (h)
1	S	0.01313	76,160,000
2	R	0.0438	22,830,000
3	P	0.1313	7,616,000
4	M	0.4386	2,279,000
5	Standard	2.4126	414,400

Table 4 Failure rate of system for different $R_{DS(on)}$

S. No.	$R_{DS(on)}$	T (°C)	Switch		Diode		λ_{System}	MTBF(h)
			π_T	λ_p (Switch)	π_T	λ_p (Diode)		
1	0.034	27	1.168	0.7708	1.079	1.369	2.2107	452,345
2	0.036	34	1.352	0.89232	1.365	1.7319	2.69512	371,040
3	0.038	41	1.555	1.0263	2.040	2.5883	3.6855	271,333
4	0.040	49	1.811	1.1952	2.183	2.7697	4.0358	247,782
5	0.042	56	2.056	1.35696	2.6778	3.3975	4.82536	207,238

increase in age, on-state resistance increases. It will also increase because of the stress in the device, to maintain required voltages and currents when operated in closed loop [5]. As the value of $R_{DS(on)}$ increases, the thermal stress on the switch increases, which increases the junction temperature and changes the operating point of the converter. This will eventually reduce the reliability of the converter.

$R_{DS(on)}$ increases with the temperature because the mobility of the hole and electron decreases as the temperature rises [6]. The $R_{DS(on)}$, at a given temperature of a P/N-channel power MOSFET can be estimated with Eq. (5) [7].

$$R_{DS(on)}(T) = R_{DS(on)}(25\,°C)\frac{T}{300} \tag{5}$$

where T = absolute temperature. By using the above equation, the temperature for different $R_{DS(on)}$ is calculated as given in Table 4. As $R_{DS(on)}$ increases, the thermal stress (temperature) also increases. For different $R_{DS(on)}$, by calculating the switch and diode failure rate, the system failure rate is calculated. It is represented in Table 4.

From the table, it is evident that as $R_{DS(on)}$ increases, the failure rate of the system increases and the time between failure decreases. The graphical representation of the same is given in Fig. 4.

Fig. 4 $R_{DS(on)}$ versus MTBF

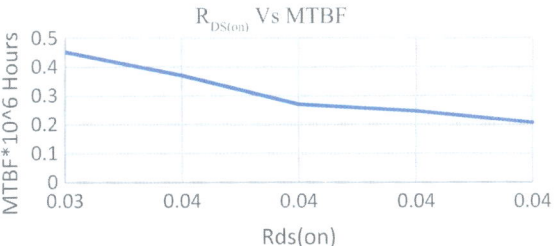

4 Reliability Improvement by Redundancy Technique

Redundancy is widely used tool for reliability enhancement. It involves designing of one or more alternative signal paths in the system through addition of series or parallel elements [8]. However, the incorporation of redundancy in a system results in increase in mass, volume, and power. For non-repairable system like DC–DC converter requires the availability of optimal redundancy for critical elements to avoid single point failures. To enhance the reliability of DC–DC converter, passive redundancy is provided.

4.1 Series System

For a series system consisting of k components, if any one component fails in an open circuit mode, the whole system will fail; whereas a malfunction in all components of the system in short circuit mode will lead to whole system failure. Many switches can be connected in parallel to improve the reliability during open circuit fault and for short circuit fault many switches should be connected in series.

The probabilities of system fails in open circuit mode and fails in short circuit mode are

$$F_o(k) = 1 - (1 - q_o)^k \tag{6}$$

and

$$F_s(n) = q_s^k \tag{7}$$

where

q_o = open circuit mode failure probability of each component;
q_s = short circuit mode failure probability of each component.

The system reliability is given by (8),

$$R_s(k) = (1 - q_o)^k - q_s^k \tag{8}$$

There exists an optimum number of components, say k^*, which maximizes the system reliability. If we define

$$K_o = \frac{\log\left(\frac{q_s}{1-q_o}\right)}{\log\left(\frac{q_o}{1-q_s}\right)} \tag{9}$$

then the system reliability $R_s(k^*)$ is maximum for

$$k^* = \lfloor k_0 \rfloor + 1 \tag{10}$$

if k_0 is not an integer;

$$k^* = k_0 \text{ or } k_0 + 1 \tag{11}$$

if k_0 is an integer.

For the Luo converter designed the loss in switch and the failure rate of switch are given as,

under normal operating condition, $P_{loss} = 0.2491; \lambda_p(\text{switch}) = 0.77448$
under open circuit condition, $P_{loss} = 0.00144; \lambda_p (\text{switch}) = 0.76938$
under short circuit condition, $P_{loss} = 0; \lambda_p (\text{switch}) = 0.769344$

The optimal number of switches in series

$$\frac{\log\left(\frac{0.76938}{1-0.769344}\right)}{\log\left(\frac{0.769344}{1-0.76938}\right)} = 0.9999; \quad k^* = [0.9999] + 1 = 1$$

4.2 Parallel System

Let a parallel system consists of k components. For a parallel configuration to fail completely, all the components in the system must fail in open circuit mode or at least one component must fail in short circuit mode.

$$\text{The system reliability is } R_p(k) = (1 - q_s)^k - q_o^k \tag{12}$$

where k is the number of components connected in parallel. The probability that no components fail in short circuit mode is represented by $(1 - q_s)^k$, and the probability that all components fail in open circuit mode is represented by q_o^k. If we define

$$k_o = \frac{\log\left(\frac{q_s}{1-q_o}\right)}{\log\left(\frac{q_o}{1-q_s}\right)} \tag{13}$$

then the system reliability $R_p(k^*)$ is maximum for

$$k^* = \lfloor k_0 \rfloor + 1 \tag{14}$$

if k_0 is not an integer;

$$k^* = k_0 \text{ or } k_0 + 1 \tag{15}$$

if k_0 is an integer.

The optimal number of switches in parallel for the designed Luo converter is,

$$\frac{\log\left(\frac{0.769344}{1-0.76938}\right)}{\log\left(\frac{0.76938}{1-0.769344}\right)} = 1.00009$$

$k^* = [1.00009] + 1 = 2$. System will be more reliable if two switches are put in parallel.

5 Reliability Prediction by Reliability Block Diagram Method

Combinational model such as reliability block diagram is commonly used to predict the reliability and maintainability of systems. Using SHARPE tool, the model is prepared for the converter system with and without redundancy and is checked for the reliability. Figure 5 shows the block diagram of non-redundant converter circuit.

The analysis from SHARPE tool for the non-redundant system is given by,

$$t = 0.000000 \text{ Reliability}(t): \ 1.00000000e + 000;$$
$$t = 5.000000 \text{ Reliability}(t): 5.69844630e - 006$$

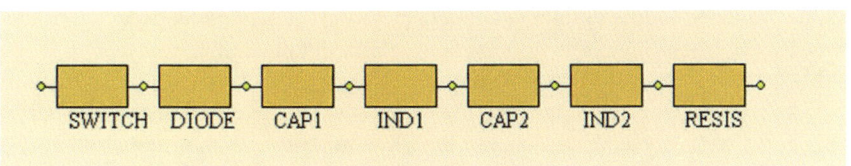

Fig. 5 Block diagram of non-redundant converter

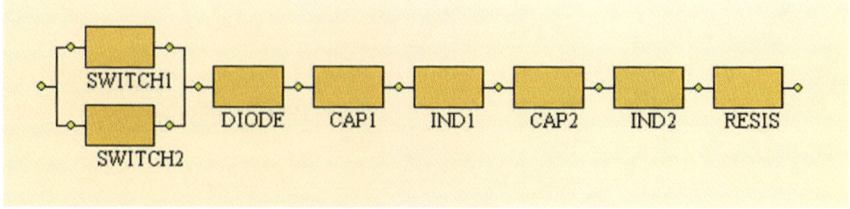

Fig. 6 Block diagram of redundant converter

$$t = 10.000000 \text{ Reliability}(t) : 3.24722471e - 011;$$
$$t = 15.000000 \text{ Reliability}(t) : 0.00000000e + 000$$
$$\text{Mean time to failure MTTFval: } 4.14067805e - 001 \left(*10^6 \text{ h}\right)$$

The converter system with redundancy in switch circuit is shown in Fig. 6 using SHARPE tool.

The analysis from SHARPE tool for the non-redundant system is given by,

$$t = 0.000000 \text{ Reliability}(t) : 1.00000000e + 000;$$
$$t = 5.000000 \text{ Reliability}(t) : 1.14963157e - 005$$
$$t = 10.000000 \text{ Reliability}(t) : 6.74920120e - 011;$$
$$t = 15.000000 \text{ Reliability}(t) : 0.00000000e + 000$$
$$\text{Mean time to failure MTTFval: } 5.15177604e - 001 \left(*10^6 \text{ h}\right)$$

From the analysis, it is evident that in the redundant system, the mean time between failures increases and so the system is more reliable than non-redundant system. The results are also shown by the graphical representation in Fig. 7.

6 Reliability Prediction by Markov Model Analysis

Markov analysis is a powerful tool in the reliability and maintainability analysis of dynamic system. In Markov analysis, the system configuration is divided into number of states. Transition rates are used to connect the states. Transition matrix can be evaluated for the system under consideration. Transition matrices are used to evaluate the reliability of the systems, through matrix manipulation.

Markov chain model of two switches with no repair facility will be analyzed for the converter system [9]. Two identical switches which are connected in parallel are considered for analysis. The switches are prone to failure with no repair facility is available. The state space is given as $S = \{0, 1, 2\}$, where $\{0, 1, 2\}$ is the number of switches failed.

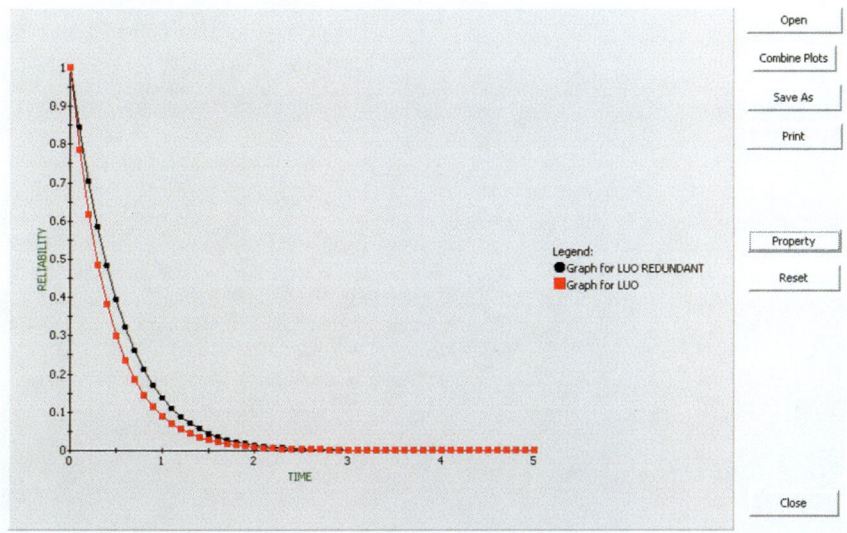

Fig. 7 Reliability comparison of non-redundant and redundant Luo converter

Let 'a' be the constant probability of failure of the switch at a given time interval. The transition probability matrix (TPM) is given by,

$$P = \begin{bmatrix} P_{00} & P_{01} & P_{02} \\ P_{10} & P_{11} & P_{12} \\ P_{20} & P_{21} & P_{22} \end{bmatrix}$$

Let us examine state 0, where no switches failed and compute the transition probabilities P_{00}, P_{01}, and P_{02}

$P_{00} = P$ {none of the switches fail in the given time} $= (1 - a)(1 - a) = (1 - a)^2$

$P_{01} = P$ {exactly one switch fails in the given time} $= a(1 - a) + (1 - a)a = 2a(1 - a)$

$P_{02} = P$ {both switches fail in a given time} $= a * a = a^2$

$$P_{00} + P_{01} + P_{02} = 1; \quad \text{TPM} = \begin{bmatrix} (1 - a)^2 & 2a(1 - a) & a^2 \\ 0 & (1 - a) & a \\ 0 & 0 & 1 \end{bmatrix}$$

The chain has three states {0}, {1}, and {2}. In state {0}, none of the switch fails. In state {1}, one of the two switches fail where as in state {2} both the switches fail. Here, state {1} is accessible from state {0} but not vice versa. Hence, state {0} and {1} do not communicate. Similarly, state {2} is accessible from state {1}, but they

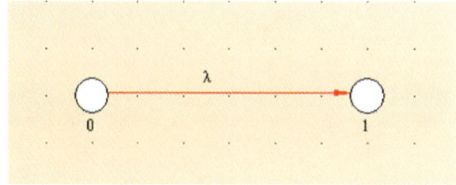

The analysis by SHARPE tool is given by,

Input parameters values: lambda= 2.414

MTTAb: 4.14250207e-001(*10^6 Hours)

Fig. 8 Converter with two states of operation

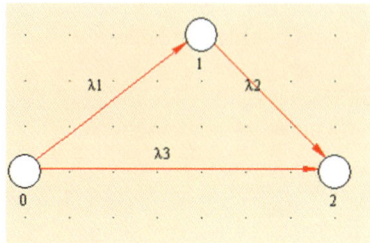

The analysis by SHARPE tool is given by,

Input parameters values: lambda1=2.414, lambda2=2.414, lambda3=1.941

MTTAb_0: 6.43871332e-001(*10^6 Hours)
MTTAb: 4.59242250e-001(*10^6 Hours)

Fig. 9 Converter with three states of operation

do not communicate. The state {0} and {1} are open, while the class {2} is closed. State {0} and {1} are transient states, where as state {2} is an absorbing state.

Markovian analysis of non-redundant Luo converter, which is having only two states of operation, working state and failure state, is given by SHARPE tool as shown in Fig. 8.

Markovian analysis of redundant Luo converter, which is having three states of operation, working state, degraded state, and failure state, is given by SHARPE tool as shown in Fig. 9.

The Markov analysis shows that the redundant system is more reliable than non-redundant system.

7 Conclusion

In this work, the reliability of converter system for different quality of components is analyzed and it is evident that converter using S grade (military grade) components are having the highest reliability. The effect of $R_{DS(on)}$ of the MOSFET switch is considered for performance degradation study. It is evident that switch with higher $R_{DS(on)}$ is less reliable. Also attempt has been made to predict the reliability based on reliability block diagram model and Markov model. Both analyses reveal that the redundant system is more reliable than non-redundant system.

References

1. K.K. Aggarwal, *Reliability Engineering* (Springer-Science+Business Media, 1993)
2. M. Sujatha and A.K. Parvathy, Chaos control of Luo converter for improving the reliability. Int. J. Appl. Eng. Res. **10**(77). ISSN 0973-4562 (2015)
3. U.S. Department of Defense, *Reliability Prediction of Electronic Equipment*, Military Handbook 217F, Feb. 28, 1995
4. Song Yantao, Wang Bingsen, Survey on reliability of power electronic systems. IEEE Trans. Power Electron. **28**(1), 591–604 (2013)
5. D. Umarani, R. Seyezhai, Reliability assessment of two-phase interleaved boost converter. Life Cycle Reliab. Saf. Eng. **7**(1), 43–52 (2018)
6. M. Sultana Nasrin, F.H. Khan, Real time monitoring of aging process in power converters using the SSTDR generated impedance matrix, in *Twenty-Eighth Annual IEEE Applied Power Electronics Conference and Exposition (APEC)* (2013)
7. AN-9010 MOSFET Basics, Fairchild Semiconductor Corporation, www.onsemi.com
8. H. Pham, *Handbook of Reliability Engineering* (Springer international Edition, 2003)
9. N. Viswanadham, Y. Narahari, *Performance Modeling of Automated Manufacturing Systems* (PHI Learning Private Limited, 2015)

Maximum Entropy Principle in Reliability Analysis

G. P. Tripathi

Abstract The principle of maximum entropy (PME) is a very powerful tool for analysis in a situation, where partial information about the system in terms of simple statistical moments is available. Various useful interpretations of this principle enable researchers to apply it in problems of its own interest. In the present paper, the principle of maximum entropy has been applied to determine the probability of failure of proportions of components of a network (system), when the proportion of these components is subjected to different levels of stress and the failure probability distribution of entire network is known. It is shown that the obtained probability distribution is a continuous variate version of a popular Maxwell–Boltzmann distribution widely used in statistical mechanics. This provides the safe range of the stress to the system for its safe use. Interesting properties of the obtained probability distribution are discussed. The application of the proposed model in other areas as in Network Centric Combat Force (NCCF), social, organizational, urban and defence is also presented.

Keywords Maximum entropy · Stress distribution · Reliability · Failure distribution

1 Introduction

Reliability and safety analysis, particularly of a complex and high-risk systems like nuclear plants, large chemical plants, space vehicle, etc. has assumed ever increasing importance in a recent year. A number of techniques for reliability and safety analysis of system/network are available in a literature [2]. The principle of maximum entropy (PME) is a powerful tool to solve a problem in a situation when information available is partial or it is expressed in terms of simple statistical moments. This technique

G. P. Tripathi (✉)
Institute for Systems Studies and Analyses, Defence Research
and Development Organization, Metcalfe House Complex, Delhi 110054, India
e-mail: gptr2007@rediffmail.com

© Springer Nature Singapore Pte Ltd. 2020
P. V. Varde et al. (eds.), *Reliability, Safety and Hazard Assessment
for Risk-Based Technologies*, Lecture Notes in Mechanical Engineering,
https://doi.org/10.1007/978-981-13-9008-1_72

has already been applied in different field of science and technology [1, 4–6]. In the present paper, an effort has been made to exploit the powerful tool of PME in reliability analysis by maximizing Shannon [3] measure of entropy. A risk can be defined as the expected value of loss due to various outcomes or in terms of uncertainty and damage. In either case, we have to estimate a probability distribution, and we have to do it when only partial information is available. If this information is available in the form of some expected or average values, we can use the principle of maximum entropy (PME) to estimate the probability distribution.

In the present paper, we shall find the probability of failure of a component when it is subjected to a stress up to a value of 'x', when we know the expected value of the probability of failure and when different proportions of a component are subjected to different stresses. As we shall show in this case, we shall get the continuous version of Maxwell–Boltzmann distribution which we find so useful in statistical mechanics, transportation and population distribution in cities, regional and urban planning and econodynamics. This can also be applied if we know the average time to failure of an equipment or average stress at which a certain effect appears or average loss due to hazards or an average loss of lives lost in accidents. This also shows its connection with reliability theory. The reliability of systems depends on the reliability of its components and the manner in which these components are connected in a network, in series or in parallel.

In making a prediction about the reliability of products, we want to (i) be as unbiased as possible (ii) be consistent with whatever is known and (iii) make use of the a priori information we may have, even though this a priori information may not be in the form of a frequency distribution.

The MEP can enable us to generate hypothesis about a priori probability distribution, e.g. that this distribution must be exponential if the mean time to failure is known. We can then make an observation to get the posterior probability distribution by making use of Baye's theorem. From this distribution, we can find expected life, etc. of interest.

If a number of components are connected in various paths, which may have common components and if the proportion of non-failure of components of different types is known, we can use MaxEnt (maximum entropy) to find probabilities of non-failures of individual components.

2 The Problem

The problem is concerned with the reliability of proper functioning of systems of the risk of their failure. A system is a network of components, and it is useful to know the probability of failure $p_F(x)dx$ when a component is subjected to stress. The stress (may be thermal or mechanical in terms of temperature and load, respectively.) between $x - \frac{dx}{2}$ and $x + \frac{dx}{2}$. Alternatively, we may like to find the cumulative probability of failure defined by

$$P_F(x) = \int_0^x p_F(x)dx, \quad P_F(x) = \int_0^X p_F(x)dx = 1, \quad \frac{dP_F(x)}{dx} = p_F(x) \quad (1)$$

So that 'X' is the maximum stress, which the component can take and beyond which it will fail.

To find $p_F(x)$, we take the network having number of components and subject them to different degrees of stresses. Let $p_S(x)dx$ be the proportion of components subjected to stress lying between $x - \frac{dx}{2}$ and $x + \frac{dx}{2}$ and let

$$P_s(x) = \int_0^x p_s(x)dx \quad (2)$$

We assume that the company, which manufactures the components, gives us information about $p_S(x)$ and about the overall probability of failure

$$p_{\text{fail}} = \int_0^X P_F(x)p_S(x)dx = \int_0^X p_S(x)dx \int_0^X p_F(x)dx \quad (3)$$

The only knowledge about $p_F(x)$ we have is given by (1) and (3), and on the basis of only these two pieces of information, we have to find an estimate of $p_F(x)$.

3 Methodology

Since we have to find $p_F(x)$, we first convert (3) to a form involving a single integral. Integrating (3) by parts

$$p_{\text{fail}} = [P_S(x)P_F(x)]_0^X - \int_0^X P_S(x)p_F(x)dx$$

$$= 1 - \int_0^X P_S(x)p_F(x)dx$$

probability of non-failure

$$1 - p_{\text{fail}} = \int_0^X P_S(x)p_F(x)dx \quad (4)$$

This knowledge of p_{fail} is equivalent to the knowledge of the expected value of the known function $P_S(x)$ with respect to the unknown distribution with density function $p_F(x)$. We also know that

$$\int_0^X p_F(x)dx = 1 \tag{5}$$

For the unknown density function $p_F(x)$, (5) is a natural constraint and (4) gives a constraint in the form of the moment. To get the most unbiased estimate for $p_F(x)$, we maximize the Shannon's measure of entropy function

$$S = -\int_0^X p_F(x) \ln p_F(x)dx \tag{6}$$

subject to (4) and (5). Using Lagrange's method, we get

$$p_F(x) = Ae^{-\beta P_S(x)} \tag{7}$$

where A is determined by using (5), so that

$$p_F(x) = \frac{e^{-\beta P_S(x)}}{\int_0^X e^{-\beta P_S(x)}dx} \tag{8}$$

And then β is determined by using (4) so that

$$1 - p_{\text{fail}} = \frac{\int_0^X P_S(x)e^{-\beta P_S(x)}dx}{\int_0^X e^{-\beta P_S(x)}dx} \tag{9}$$

From (1) and (8),

$$P_F(x) = \frac{\int_0^x e^{-\beta P_S(x)}dx}{\int_0^X e^{-\beta P_S(x)}dx} \tag{10}$$

Thus, we have obtained maximum entropy estimates for both $p_F(x)$ and $P_F(x)$ in terms of the given function $P_S(x)$. Now, in the next section, we discuss the properties of the solution. In order to do that we note that

(i) Equation (8) may be regarded as the continuous variate version of the Maxwell–Boltzmann distribution.
(ii) The function $P_S(x)$ is a monotonic increasing function of x since

$$\frac{dP_S(x)}{dx} = p_S(x) \geq 0 \tag{11}$$

If we impose stress up to a maximum limit of X_0, we get

$$\frac{dP_S(x)}{dx} = p_S(x) = 0, \quad \text{when } X_0 < x < X.$$

4 Properties of the Solution

In this section, we investigate the behaviour of $P_S(x)$, $p_F(x)$ and $P_F(x)$ with respect to the parameter β in order to find the range of stress within which the network will function without failure.

4.1 β *Is a Monotonic Increasing Function of* p_{fail}

From (9), we get

$$-\frac{dp_{\text{fail}}}{d\beta} = -\frac{\int_0^X e^{-\beta P_S(x)} \int_0^X P_S^2(x)e^{-\beta P_S(x)}dx + \left[\int_0^X P_S(x)e^{-\beta P_S(x)}dx\right]^2}{\left[\int_0^X e^{-\beta P_S(x)}dx\right]^2} \tag{13}$$

So that using Cauchy–Schwartz inequality according to which

$$\int_0^X f^2(x)dx \int_0^X g^2(x)dx \geq \left[\int_0^X f(x)g(x)dx\right]^2 \tag{14}$$

And the equality sign holds if $f(x) = g(x)$ almost everywhere.
We get

$$\frac{dp_{\text{fail}}}{d\beta} > 0 \tag{15}$$

Since

$$P_S(x) = 1 \text{ only when } X_0 \leq x \leq X \tag{16}$$

From (15), we get

$$\frac{d\beta}{dp_{\text{fail}}} > 0 \tag{17}$$

So that β is a monotonic increasing function of p_{fail}. We may call β is a failure temperature and say that this temperature increases as p_{fail}.

Now when $\beta = 0$, from (9)

$$1 - p_{\text{fail}} = \frac{\int_0^X P_S(x)\mathrm{d}x}{X} \tag{18}$$

Or

$$(p_{\text{fail}})_0 = 1 - \frac{\int_0^X P_S(x)\mathrm{d}x}{X} \tag{19}$$

If given p_{fail} is greater than $(p_{\text{fail}})_0$ $\beta > 0$ and if given p_{fail} is less than $(p_{\text{fail}})_0$ $\beta < 0$. This follows since β increases as p_{fail} increases.

4.2 $P_F(x)$ Increases with X, β and p_{fail}

It is easily seen that $P_F(x)$ increases with x, since

$$\frac{\mathrm{d}P_F(x)}{\mathrm{d}x} = p_F(x) \geq 0 \tag{20}$$

4.3   Variation of $P_F(x)$ with Respect to x, β and p_{fail}

From (8), $P_F(x)$ decreases with x if $\beta > 0$ and increases with x if $\beta < 0$.

Also,

$$\left[\int_0^X e^{-\beta P_S(x)}\mathrm{d}x\right]^2 \frac{\mathrm{d}P_F}{\mathrm{d}\beta} = \left[\int_0^X e^{-\beta P_S(x)}\mathrm{d}x\right]\int_0^x e^{-\beta P_S(x)}[-P_S(x)]$$

$$+ \int_0^x e^{-\beta P_S(x)}\int_0^X p_s(x)e^{-\beta P_S(x)}\mathrm{d}x$$

$$= \int_0^x e^{-\beta P_S(x)}\int_0^X e^{-\beta P_S(x)}\mathrm{d}x\left[\frac{\int_0^X P_S(x)e^{-\beta P_S(x)}\mathrm{d}x}{\int_0^X e^{-\beta P_S(x)}\mathrm{d}x} - P_S(x)\right] \tag{21}$$

Using (9), we get

$$\frac{\mathrm{d} P_F(x)}{\mathrm{d}\beta} = \frac{\int_0^x e^{-\beta P_S(x)}}{\int_0^X e^{-\beta P_S(x)}\mathrm{d}x}[1 - p_{\text{fail}} - P_S(x)]$$

$$= P_F(x)[1 - p_{\text{fail}} - P_S(x)] \tag{22}$$

Now, we define $P_S(x)$ by

$$P_S(x_0) = 1 - p_{\text{fail}}, \tag{23}$$

So that

$$\frac{\mathrm{d} P_F(x)}{\mathrm{d}\beta} = p_F(x)[P_S(x_0) - P_S(x)] \tag{24}$$

Since $P_S(x)$ is a monotonic increasing function of x

$$\frac{\mathrm{d} P_F(x)}{\mathrm{d}\beta} \mathop{\gtrless}\limits_{<}^{\geq} 0 \text{ according as } x \mathop{\gtrless}\limits_{<}^{\geq} x_0 \tag{25}$$

Using (17),

$$\frac{\mathrm{d} P_F(x)}{\mathrm{d} p_{\text{fail}}} \mathop{\gtrless}\limits_{<}^{\geq} 0 \text{ according as } x \mathop{\gtrless}\limits_{<}^{\geq} x_0$$

$$\frac{\mathrm{d} P_F(x)}{\mathrm{d}\beta} \geq 0$$

$$\frac{\mathrm{d} P_F(x)}{\mathrm{d}\beta} = P_F(x)[1 - p_{\text{fail}} - P_S(x)]$$

$$\frac{\mathrm{d} P_F(x)/\mathrm{d}\beta}{P_F(x)} = [1 - p_{\text{fail}} - P_S(x)]$$

$$\frac{\mathrm{d}}{\mathrm{d}\beta} \log P_F(x) = (1 - p_{\text{fail}} - P_S(x))$$

$$P_F(x) = \exp \int (1 - p_{\text{fail}} - P_S(x))\mathrm{d}\beta$$

So that $P_F(x)$ increases with β also. From (17),

$$\frac{\mathrm{d} P_F(x)}{\mathrm{d} p_{\text{fail}}} \geq 0 \tag{26}$$

So that $P_F(x)$ increases with p_{fail} for every value of x.
Further, when $\beta = 0$

$$P_F(x) = \frac{x}{X} \tag{27}$$

So that

$$P_F(x) \underset{<}{\overset{\geq}{=}} \frac{x}{X} \qquad (28)$$

$$\text{according } \beta \underset{<}{\overset{\geq}{=}} 0 \qquad (29)$$

Thus, we find that

(a) $P_F(x)$ decreases with x if $\beta > 0$ and increases with x if $\beta < 0$.
(b) $P_F(x)$ increases with β if $x < x_0$ and decreases with β if $x > x_0$.
(c) $P_F(x)$ increases with p_{fail} if $x < x_0$ and decreases with p_{fail} if $x > x_0$.

From (1), it follows that
$P_F(x)$ is a concave function of x if $\beta < 0$. $P_F(x)$ is represented by straight line if $\beta = 0$.

4.4   Graphical Representation of $P_F(x)$

From the above discussion, the graph of $P_F(x)$ is shown in Fig. 1.

When $\beta > 0$, $P_F(x)$ is a concave function from 0 to X_0 and is linear from X_0 to X.
When $\beta = 0$, $P_F(x)$ is a linear function with slope $1/s$.
When $\beta < 0$, $P_F(x)$ is a convex function from 0 to X_0 and is linear from X_0 to X.
When $\beta = \infty$, $P_F(x)$ is unity almost everywhere, $P_F(x) = \delta(x)$.
When $\beta = -\infty$, $P_F(x)$ is almost zero everywhere, $P_F(x) = \delta(x - X)$.
The tangent to all curves at $x - x_0$ has slope $1/x$.

The graph of $P_F(x)$ is as shown in Fig. 2.

-   When $p_{fail} = 1$, $\beta = \infty$, $P_F(x) = \delta(x)$ and even very small stresses can lead to failure of system/network means those systems which have an increasing failure rate lead to failure if they have to function even with very small stress.
-   When $p_{fail} = 0$, $\beta = -\infty$, $P_F(x) = \delta(x - X)$ and the path/component can be used safely almost up to the maximum load.

Fig. 1   Graph of failure distribution w.r.t β

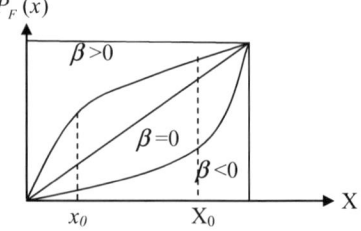

Fig. 2 Variation of $P_F(x)$ w.r.t. x (stress)

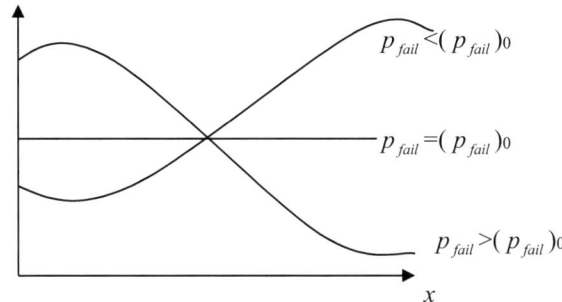

- As p_{fail} increases from 0 to $(p_{fail})_0$, β increases from $-\infty$ to 0.
- $P_F(x)$ increases for every value of x, but remains below the value x/X and approaches it.
- As p_{fail} increases from $(p_{fail})_0$ to unity, β increases from 0 to ∞, $P_F(x)$ continues to increase for every value of x is always greater than x/X and approaches a value unity as p_{fail} approaches unity.

4.5 S_{max} Is a Concave Function of p_{fail}

From (6) and (8),

$$S_{max} = -\int_0^X P_F(x) \ln P_F(x)\,dx$$

$$= -\int_0^X P_F(x)(\ln A - \beta P_S(x))\,dx$$

$$= -\ln A + \beta(1 - p_{fail}) = \ln \int_0^X e^{-\beta P_S(x)}\,dx + \beta(1 - p_{fail}) \tag{30}$$

$$\frac{dS_{max}}{d\beta} = \frac{-\int_0^X P_S(x)e^{-\beta P_S(x)}\,dx}{\int_0^X e^{-\beta P_S(x)}\,dx} + 1 - p_{fail} - \beta\frac{dp_{fail}}{d\beta}$$

$$= -\beta\frac{dp_{fail}}{d\beta}$$

$$\text{Or}\quad \frac{dS_{max}}{dp_{fail}} = -\beta \tag{31}$$

Fig. 3 Variation of maximum entropy w.r.t. overall network failure prob (p_{fail})

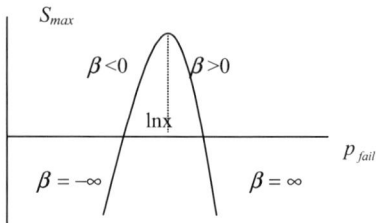

So that the maximum entropy increases (decreases) with a probability of failure according as $\beta < (>)\,0$, i.e. according as $p_{\text{fail}} < (>)(p_{\text{fail}})_0$. Again differentiating (31) and using (13) and (17)

$$\frac{\mathrm{d}^2 S_{\max}}{\mathrm{d}p_{\text{fail}}^2} = -\frac{\mathrm{d}\beta}{\mathrm{d}p_{\text{fail}}} < 0. \tag{32}$$

So that $S_{\max}$ is a concave function of the probability of failure. Its graph is given in Fig. 3.

$S_{\max}$ is maximum when $\beta \to 0$, $p_{\text{fail}} = (p_{\text{fail}})_0$, $p_F(x) = 1/X$

And

$$(S_{\max})_{\max} = -\int_0^X \frac{1}{X} \ln \frac{1}{x} \mathrm{d}x = \ln X. \tag{33}$$

When $\beta \to -\infty$ or ∞, $p_F(x)$ is a delta function and $S_{\max}$ approaches $-\infty$. When $\beta > 0$, $S_{\max}$ decreases as β increases.

Suppose we are given average stress at failure, i.e. we are given

$$\hat{x} = \int_0^Y x p_F(x)\mathrm{d}x \tag{34}$$

In this case, maximizing S subject to (4), (5) and (34), we get

$$p_F(x) = \frac{e^{-\mu x - \beta P_S(x)}}{\int_0^X e^{-\mu x - \beta P_S(x)}\mathrm{d}x} \tag{35}$$

where μ and β are determined by using

$$\frac{\int_0^X P_S(x) e^{-\mu x - \beta P_S(x)}\mathrm{d}x}{\int_0^X e^{-\mu x - \beta P_S(x)}\mathrm{d}x} = 1 - p_{\text{fail}} \tag{36}$$

and

$$\frac{\int_0^X x e^{-\mu x - \beta P_{\mathrm{S}}(x)} \mathrm{d}x}{\int_0^X e^{-\mu x - \beta P_{\mathrm{S}}(x)} \mathrm{d}x} = \hat{x} \tag{37}$$

If we know the average stress at failure and probability of non-failure of the proportion of the components, then (35) gives failure probability of components.

5 Conclusions and Applications of Proposed Methodology in Other Area

The proposed methodology can also be applied in other areas with appropriate interpretation. Principle of maximum entropy is a powerful tool to solve problems in the presence of partial information. Shannon's measure satisfied most of the essential properties and there are various faces (interpretations) of entropy as a measure of uniformity, measure of equality, measure of randomness, flexibility, interaction, etc. [1] In the present paper, measure of entropy has been used as a measure of equality. The entropy failure probability distribution is maximized subject to a prescribed constraint so that the probability of failure of the proportion of components becomes as equal as possible. Using this interpretation, interesting results are obtained.

The methodology developed in the present paper can be applied in an organization, social system, urban planning, defence and agriculture, etc. with the proper interpretation of the involved parameter. Some of the applications are as follows:

- The principle of maximum entropy has been applied to estimate the probability of failure of various proportions of components, which are affected by different stresses, when the overall failure of network is known. We could also consider proportion of path sets instead of components under stress. Alternatively, cut sets of systems can also be determined for reliability evaluation.
- Consider a Network Centric Combat Force (NCCF) in which the components are information gathering processes, data fusion, command and control, combat, logistics and battle damage assessment connected with highly reliable links. There is a threat (stress) to this network which can be measured in terms of force strength and combat support systems effectiveness of friendly and enemy forces. Knowing the overall or average threat (stress), we could estimate the threat distribution of this network. This may help in deploying the forces to defend the network or in other words to compare the network of different countries.
- In order to assess the damage, consider some area (land) to be defended is divided into a specified number of sub areas and assume fragments of warheads falls in this area. Now knowing the average energy of fragments and number of fragments in different sub areas, we could estimate the damage distribution in terms of kill probability of area.
- If we know proportions of different categories of scientist of an organization, working under physical or mental stress and the overall performance of the orga-

nizational system, then by using this model we can estimate the performance distribution of scientist.

- If we know proportion of road crossings (components)/roads (path sets) having vehicular stress (traffic density) at specified time in traffic flow and we also know the overall probability of traffic jam (failure probability) in city, then this methodology may help us in estimating the traffic flow distribution in all the crossings/roads.
- If we know the proportion of agricultural land affected by the stress like rain, fertilizer, etc. also if we know the overall production of the crop in this agricultural land, then we can in position to estimate the production distribution of different proportions of agricultural land.
- Similarly, one can formulate the problem of his own interest, and the proposed models can be applied analogically to get the required results. In the above problem, only one type of stress has been considered. We could consider more than one type of stress as thermal, mechanical, etc. This requires evaluation of multiple integral or simulation techniques. The data is a basic requirement to implement the proposed models.

Acknowledgements The author acknowledges the peer reviewer of ICRESH-2019 for valuable suggestions.

References

1. J.N. Kapur, *Maximum Entropy Models in Science &Engineering* (Wiley Eastern Ltd., Delhi, 1989)
2. F.M. Martz, R.A. Waller, *Bayesian Reliability Analysis* (Wiley, U.S.A., 1982)
3. C.E. Shannon, A mathematical theory of communication. Bell Syst. Tech. J. **27**, 379–423 & 623–659 (1948)
4. G.P. Tripathi, Weighted generalized directed divergence measure to assess military requirements. Def. Sci. J. **15**(1), 101–106 (2000 Jan)
5. G.P. Tripathi, The need for an alternative measure of information in electronic warfare. Presented & published in *Proceedings in Experience Workshop in Third International Conference on Pattern Recognition and Machine Intelligence (PReMI'09)*, 16–20 Dec 2009 in IIT Delhi (2009)
6. G.P. Tripathi, Maximum entropy principle to estimate threat intensity distribution in network centric combat force. Presented and published in *Proceedings of International Congress on Productivity, Quality, Reliability, Optimization & Modelling 2011*, Delhi (2011)

Dr. G. P. Tripathi obtained his M.Sc. (Statistics) from IIT Kanpur in 1985 and Ph.D. in 1989. He was SRF (CSIR) at IIT Delhi and School of Computer and Systems Sciences, JNU, New Delhi. He joined DRDO as a scientist at the Institute of Systems Studies and Analyses, Delhi in 1991. His area of research includes applications of maximum entropy principle in science and engineering, reliability analysis, network centric warfare, systems analysis, simulation and modelling. He is a Certified Reliability Engineer (CRE) (American Society for Quality, 2016) and published number of research papers and technical reports in national/international journals. He is a member of Indian society for information theory and its applications (ISITA).

Nuclear Safety

Development and Validation of Direct Method-Based Space-Time Kinetics Code for Hexagonal Geometry

Tej Singh, Tanay Mazumdar, Paritosh Pandey, P. V. Varde and S. Bhattacharya

Abstract A 3D, multigroup, neutron kinetics code DINHEX (DIrect method-based Nodal kinetics code for HEXagonal geometry) is written to simulate reactivity-driven transient events in loosely coupled reactor cores with hexagonal lattice. In this code, space-time-based neutron kinetics equations are solved by direct method to obtain the spatial as well as temporal variation of neutron flux. In direct method, space and time variables are discretized to form a space-time grid and the kinetics equations are solved at these discretized grid points. Nodal Expansion Method (NEM) is used to obtain the spatial variation of neutron flux, and implicit method is used to obtain the temporal variation of neutron flux at each grid point. A number of benchmark problems are analyzed to test the performance of the code. The validation study shows satisfactory agreement with other international codes.

Keywords Benchmark · Direct method · Nodal expansion method · Kinetics · Transient analysis

T. Singh (✉) · T. Mazumdar · P. Pandey
Reactor Physics and Nuclear Engineering Section, Research Reactor
Services Division, Reactor Group, Bhabha Atomic Research Centre, Mumbai, India
e-mail: t_singh@barc.gov.in

T. Mazumdar
e-mail: tanay@barc.gov.in

P. Pandey
e-mail: paritosh@barc.gov.in

P. V. Varde
Reactor Group, Bhabha Atomic Research Centre, Mumbai, India
e-mail: varde@barc.gov.in

S. Bhattacharya
Reactor Projects Group, Bhabha Atomic Research Centre, Mumbai, India
e-mail: subhatt@barc.gov.in

© Springer Nature Singapore Pte Ltd. 2020
P. V. Varde et al. (eds.), *Reliability, Safety and Hazard Assessment for Risk-Based Technologies*, Lecture Notes in Mechanical Engineering,
https://doi.org/10.1007/978-981-13-9008-1_73

1 Introduction

Point kinetics model is found suitable for any local reactivity perturbation (e.g., withdrawal of control rod) in research reactors, which are usually small in size. On the contrary, power reactors are large in size, and therefore, neutron flux is mostly affected in the vicinity of the perturbation site, while the distant parts seem to be "unaware" about the perturbation. Such "out of phase" dynamics of large reactors can only be explained by solving space-time kinetics equations. Space-time factorization and direct integration are two methods to solve these equations. In the factorization method, neutron flux is factorized in such a manner that the "Amplitude" function contains only the strong dependence of time, while the "Shape" function contains the entire space dependence and the weak dependence of time. Therefore, amplitude and shape equations are solved in micro- and macro-time step, respectively. This method has already been implemented in one of our codes, IQSHEX [1]. In direct method, space and time are discretized simultaneously. Space discretization is based on Nodal Expansion Method (NEM) for which the reactor core is divided into a number of assembly-sized nodes. Thereafter, the neutron flux in each mesh and energy group is expressed in terms of polynomial up to third order in x-y plane and fourth order in z-direction, which is one order higher than what proposed by Lawrence [2]. The partially integrated leakage terms are expressed in terms of polynomial up to second order. Thus, the spatially discretized kinetics equations form a coupled set of equations, which are first order in time. Time discretization is performed using implicit method due to its better stability to the solution. The final inhomogeneous response matrix equation is solved using conventional fission source iteration. NEM-based space discretization and implicit method-based time discretization are combined together to make a computer code DINHEX. Performance of the kinetics code is verified against number of 2D and 3D benchmark problems including a prompt super criticality case, a reactivity-driven transient study in the presence of adiabatic temperature feedback, etc. Results of DINHEX are found in good agreement with our own code IQSHEX as well as other international codes like DIF3D, KIKO3D, DYN3D, BIPR8, etc.

2 Solution of Space-Time Kinetics Equations

The multigroup, time-dependent diffusion equations are given below.

$$
\frac{1}{v_g} \frac{\partial \phi_g(r, t)}{\partial t} = \left[\vec{\nabla}. D_g(r, t) \vec{\nabla} - \sum_{rg}(r, t) \right] \phi_g(r, t)
$$

$$
+ \left[(1 - \beta) \chi_g^p \sum_{g'=1}^{G} v \sum_{fg'}(r, t) + \sum_{g'=1}^{G} \sum_{\substack{sg' \to g \\ g' \neq g}}(r, t) \right] \phi_{g'}(r, t) + \sum_{d=1}^{D} \chi_g^d \lambda_d C_d(r, t) \qquad (1)
$$

$$\frac{\partial C_d(r, t)}{\partial t} = \beta_d \sum_{g'=1}^{G} v \sum_{fg'} (r, t)\phi_{g'}(r, t) - \lambda_d C_d(r, t) \tag{2}$$

where the symbols have their usual meaning. Since direct method requires simultaneous discretization of space and time variables, NEM is chosen to obtain the space dependence of neutron flux in a reactor core, while implicit method is chosen to find out how the space dependence of flux changes with time due to reactivity addition. An advanced NEM for hexagonal geometry is adopted for spatial discretization. In this method, reactor core is divided into a number of nodes. The 3D, multigroup, steady-state neutron diffusion equation in k-th node and g-th energy group is integrated over two transverse directions to obtain 1D equation. In order to solve the 1D equation, the partially integrated radial flux ($\phi_{gx}^{k}(x, t)$) and axial flux ($\phi_{gz}^{k}(z, t)$) are expressed in terms of third- and fourth-order polynomial, respectively.

$$\text{(i) } \phi_{gx}^{k}(x, t) \approx 2y_s(x)\left[\overline{\phi}_g^{k}(t) f_0(x) + \sum_{i=1}^{5} a_{gxi}^{k}(t) f_i(x)\right]\Delta z_k, \text{ (ii) } \phi_{gz}^{k}(z, t) \approx \overline{\phi}_g^{k}(t)g_0(z)$$

$$+ \sum_{i=1}^{4} b_{gzi}^{k}(t)g_i(z) \tag{3}$$

where $a_{gxi}^{k}(t)$ and $b_{gzi}^{k}(t)$ are expansion coefficients, $f_i(x)$ is polynomial of x defined as $f_0(x) = 1, f_1(x) = x/h = \xi, f_2(x) = 36\xi^2/13 - 5/26, f_3(x) = 10\xi^2/13 - |\xi|/2 + 3/52, f_4(x) = \xi(|\xi|-1/2), f_5(x) = 15\xi^3/14 - 11\xi|\xi|/14 + \xi/8$ and $g_i(z)$ is polynomial of z defined as $g_0(z) = 1, g_1(z) = z/\Delta z_k = \eta, g_2(z) = 3\eta^2 - 1/4, g_3(z) = \eta(\eta^2 - 1/4), g_4(z) = (\eta^2 - 1/20)(\eta^2 - 1/4)$. Before calculating the expansion coefficients, it is necessary to discretize the time variable at a number of points on the time axis, along which the reactor core evolves under a reactivity transient. If Δt is the time step, then first-order time derivative of $\phi_{gx}^{k}(x, t)$ can be discretized in the following manner.

$$\frac{\partial \phi_{gx}^{k}(x, t)}{\partial t} \approx \frac{\phi_{gx}^{k}(x, t + \Delta t) - \phi_{gx}^{k}(x, t)}{\Delta t} = \frac{\phi_{gx}^{k,n+1}(x) - \phi_{gx}^{k,n}(x)}{\Delta t} \tag{4}$$

If we assume $t = n\Delta t$, then $(t + \Delta t)$ will be $(n + 1)\Delta t$. Therefore, time "t" is denoted by the superscript "n" (and "$(t + \Delta t)$" by "$(n + 1)$"). Since implicit method is used for time discretization, rest of the terms of the 1D diffusion equation are evaluated at $(n + 1)$, instead of n. $a_{gx1}^{k}(t)$ and $a_{gx2}^{k}(t)$ are calculated by equating surface-averaged flux along $\pm x$-direction to $\overline{\phi}_{gx\pm}^{k}$. $a_{gx3}^{k}(t)$ is calculated by equating the discontinuity of first derivative of flux at $x = 0$ to average y-directed leakage. $a_{gx4}^{k}(t)$ and $a_{gx5}^{k}(t)$ are calculated using weighted residual method to the transverse integrated equation with $sgn(x)$ and $f_1(x)$ as weighting functions. $b_{gz1}^{k}(t)$ and $b_{gz2}^{k}(t)$ are obtained in a similar way as followed for $a_{gx1}^{k}(t)$ and $a_{gx2}^{k}(t)$. $b_{gz3}^{k}(t)$ and $b_{gz4}^{k}(t)$ are calculated using weighted residual method with weights $g_1(z)$ and $g_2(z)$, respectively. Finally, outgoing partial currents of the k-th node are combined to form following matrix equation.

$$\left[J_{g,out}^{k,n+1}\right]_{8\times1} = \left[P_g^k\right]_{8\times9}\left[\hat{Q}_g^{k,n,n+1} - \overline{L}_g^{k,n+1}\right]_{9\times1} + \left[R_g^k\right]_{8\times8}\left[J_{g,in}^{k,n+1}\right]_{8\times1} \quad (5)$$

Assumption of constant axial leakage term $L_{gz}^k(x, y, t)$ makes all leakage moments in x-y plane zero. In z-direction, transverse leakage $L_{gxy}^k(z, t)$ is expressed in terms of polynomials up to second order. The coefficients are calculated from the average leakage of two neighboring nodes, and leakage moments are represented in terms of these coefficients.

3 Results and Discussion

In the present paper, one 2D and two 3D benchmark problems are solved to demonstrate the performance of the code DINHEX. For all the problems, the number of energy groups of prompt neutron is two and the number of delayed neutron precursor groups is six (except the problem in Sect. 3.3, in which there is only one delayed neutron precursor group). Two sets of calculations, namely without (DINHEX-I) and with (DINHEX-II) the inclusion of $f_5(x)$ in the flux expansion on x-y plane, are carried out for each of the problems. In the nodal calculation, size of a node is taken to be equal to the size of an assembly. Starting with some guess value of multiplication factor (k_{eff}) and neutron flux, convergence with the final solution is obtained in an iterative manner. For this purpose, convergence criterion for neutron flux (or fission source) in each node and multiplication factor are set at 1.0×10^{-5} and 1.0×10^{-7}, respectively. Incoming current zero boundary condition is applied at the outermost periphery of the core. The initial state of transient event is supposed to be critical, i.e., $k_{\text{eff}} = 1.0$. If it is not so, then $\nu\Sigma_f$ cross section will be modified by dividing with the multiplication factor calculated ($\nu\Sigma_f^{\text{new}} = \nu\Sigma_f^{\text{old}}/k_{\text{eff}}$), and with the modified $\nu\Sigma_f$ cross section, k_{eff} will be calculated again. Arrangement of various materials in core is pictorially represented by DISLIN [3]. The benchmark problems and their results are given below.

3.1 2D, 60° Symmetric, 2 Energy Group Benchmark Problem [4]

This 2D, 60° symmetric, hexagonal benchmark core consists of 133 Fuel-1 assemblies (including 19 control assemblies), 126 Fuel-2 assemblies (including 18 control assemblies), 168 Fuel-3 assemblies (including 24 control assemblies), 72 fuel assemblies in buckled zone, 168 lithium target, 6 vacancy, and 246 core exterior assemblies. Lattice pitch of the arrangement is 17.78 cm. Two energy group homogenized lattice parameters of the core materials and kinetics parameters of the delayed neutron precursors are available in [4]. At $t = 0.0$ s, steady-state calculation is performed and the results obtained are compared with the reference solution [5]. k_{eff} value, as com-

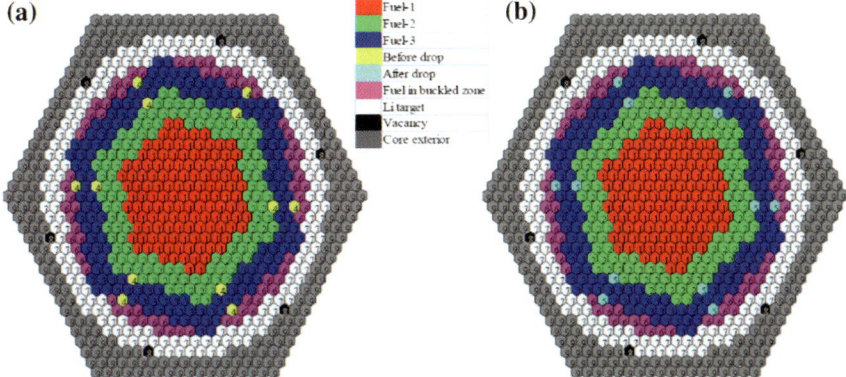

Fig. 1 Horizontal cross section of 2D core **a** before and **b** after dropping of fuel rods at designated locations

puted by DINHEX-I and DINHEX-II, is 0.991792 (error: −17.6 pcm) and 0.991894 (error: −7.2 pcm), respectively, while the reference value is 0.991965. Maximum relative percentage error for assembly power is 1.2% for DINHEX-I and 2.3% for DINHEX-II.

Three transient cases are analyzed. In first case, two fuel bearing rods per 60° sector, i.e., total twelve rods, are dropped into the core, causing 0.755 \$ positive reactivity addition. Figure 1a shows cross section of the reactor core at $t = 0.0$ s (before the drop). In this figure, the drop locations are indicated by the number "4". Since duration of the drop is 0.205 s, material at "4" is replaced with fuel material "5" at $t = 0.205$ s and the corresponding cross section of the reactor core is shown in Fig. 1b. In second case, 0.744 \$ positive reactivity is inserted into the core by decreasing the thermal absorption cross section of material "1" in central zone by 4.5% in 0.2 s. Introduction of perturbation into the core in third case is similar to that in second case with a difference in duration of transient (at $t = 0.0$ s) and amount of reduction in thermal absorption cross section, which is 6.0%, causing 1.17 \$ (i.e., prompt supercritical case) positive reactivity addition [6]. Temporal variation of relative power in all the cases is given in Fig. 2a–c. Results of DINHEX are compared with the results of our own code IQSHEX and other codes like DIF3D, FX2-TH, etc. In first two cases, results of FX2-TH are considered as reference solution [4]. As shown in Fig. 2a, b, prediction of reactor power by DINHEX is comparable with IQSHEX, DIF3D, and FX2-TH. As shown in Fig. 2c, result of DINHEX is in good agreement with the reference solution of HENKO.

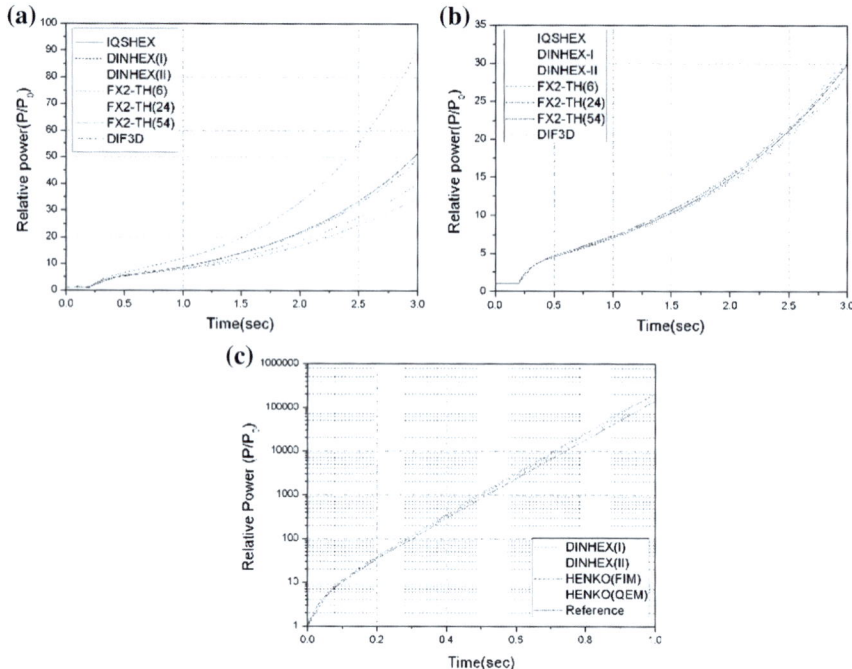

Fig. 2 Variation of relative reactor power with time in 2D benchmark core for three transient cases

3.2 AER-DYN-002 [7]

Transient in "AER-DYN-002" is introduced by ejecting a control rod from a 180°
symmetric reactor core. Horizontal cross section of half of the core is shown in
Fig. 3a. In the core map, "1", "2", and "3" are fuel assemblies. "4/2" is partially
inserted control assembly in which there is absorber material ("4") from 50 to 250 cm
and fuel follower (type "2") from 0.0 to 50 cm, as shown in Fig. 3b. The hexagonal
assemblies are arranged in a pitch of 14.7 cm. Height of active core is 250 cm.
Two energy group homogenized lattice parameters of the core materials and kinetics
parameters of the delayed neutron precursors are available in [7].

The transient starts due to withdrawal of one control rod from its location
(shown with hatched hexagon in Fig. 3a in the reactor core, which is operated
at 1.375 KW power at 0.0 s. The rod is withdrawn with a speed of 1250 cm/s.
So, in 0.16 s time, the rod goes out of the core. Uniqueness of the problem is
the power surge, caused due to the withdrawal, will not be arrested by SCRAM,
rather we will rely on Doppler temperature feedback developed due to increase in
fuel temperature. Simple adiabatic model is considered in the simulation to calculate the fuel temperature. The adiabatic feedback is determined by $\Sigma_{f2}(t) =$
$\Sigma_{f2}^0 \left[1 - 7.228 \times 10^{-4} \left(\sqrt{T_{\text{fuel}}(t)} - \sqrt{T_{\text{fuel}}^0} \right) \right]$ where fuel temperature at 0.0 s

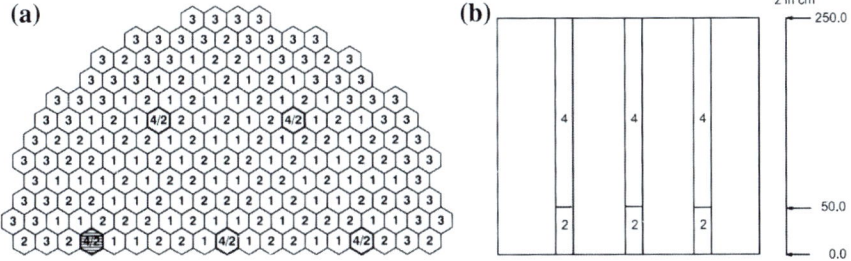

Fig. 3 **a** Horizontal (half reactor core), **b** Initial vertical cross section of AER-DYN-002 benchmark problem

(T_{fuel}^0) is 260 °C and Σ_{f2}^0 is the thermal fission cross section of fuel at 0.0 s. The transient is studied up to 2.0 s. In the simulation by DINHEX, one hexagonal assembly is taken as one mesh in x-y plane and the active height of the core is divided into 12 meshes. Temporal variation of power fuel temperature and peaking factor is shown in Fig. 4. Good agreement is observed with that of KIKO3D, DYN3D/M2, BIPR8 ("Dyn002.doc" in [8]), and IQSHEX [1]. From Fig. 4a, it is evident that the control rod ejection induces the power surge, which continues till 0.23 s, and after that, the Doppler feedback, caused due to increase in fuel temperature, dominates over the power surge and subsequently reactor power decreases.

3.3 3D, 60° Symmetric, 2 Energy Group Benchmark Problem [9]

In another 3D benchmark problem, the reactor core consists of 7, 252, 168, 210, and 252 number of assemblies in five annular regions, as shown in Fig. 5a. All hexagonal assemblies are arranged in a pitch of 17.78 cm. Height of active core is 500 cm (Fig. 5b). Two energy group homogenized lattice parameters of the core materials and kinetics parameters of the delayed neutron precursors are available in [9]. In this benchmark core, 87% positive reactivity is inserted by reducing the thermal absorption cross section of central zone by ~15%. In the simulation, velocity of thermal neutron is taken as 1775 m/s, as suggested in [4], though in the description of the problem, the velocity is given as "∞". In the simulation by DINHEX, one hexagonal assembly is taken as one mesh in x-y plane and the active height of the core is divided into 40 meshes. Figure 6 shows the normalized radial thermal flux variation at $z = 250$ cm at different instant of the transient. The reference value is calculated by DIF3D-nodal [4]. The comparison shows that the results of DINHEX ($\Delta t = 5$ ms) are in good agreement with IQSHEX and reference values.

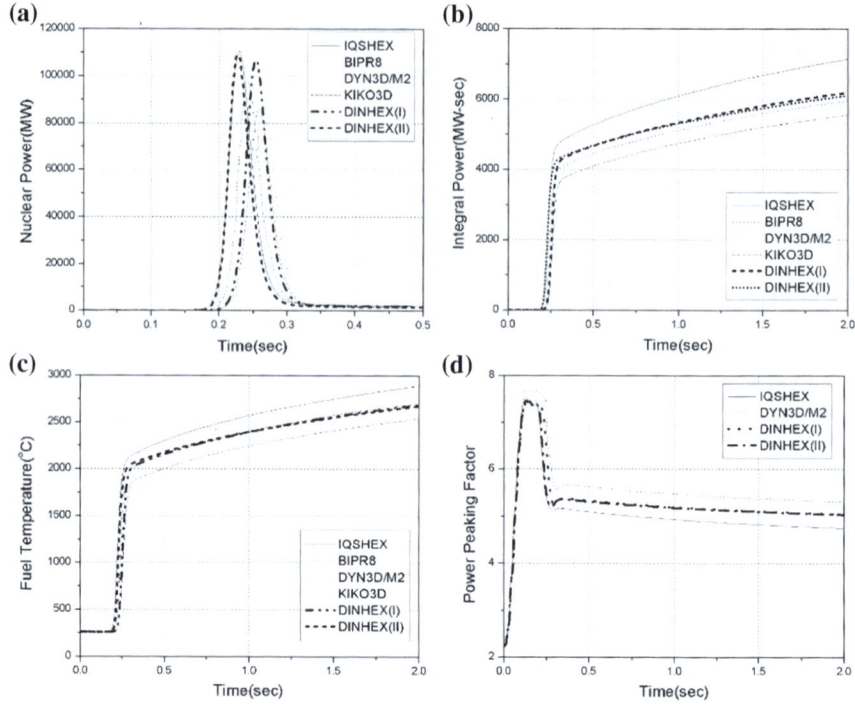

Fig. 4 Response to the transient caused by one control rod withdrawal in 3D VVER-440 core without SCRAM

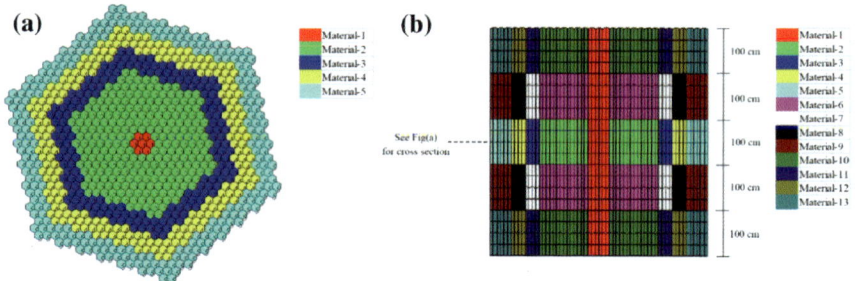

Fig. 5 **a** Horizontal cross section [200 cm < z < 300 cm], **b** vertical cross section at mid-elevation of 3D, 60° symmetric benchmark problem

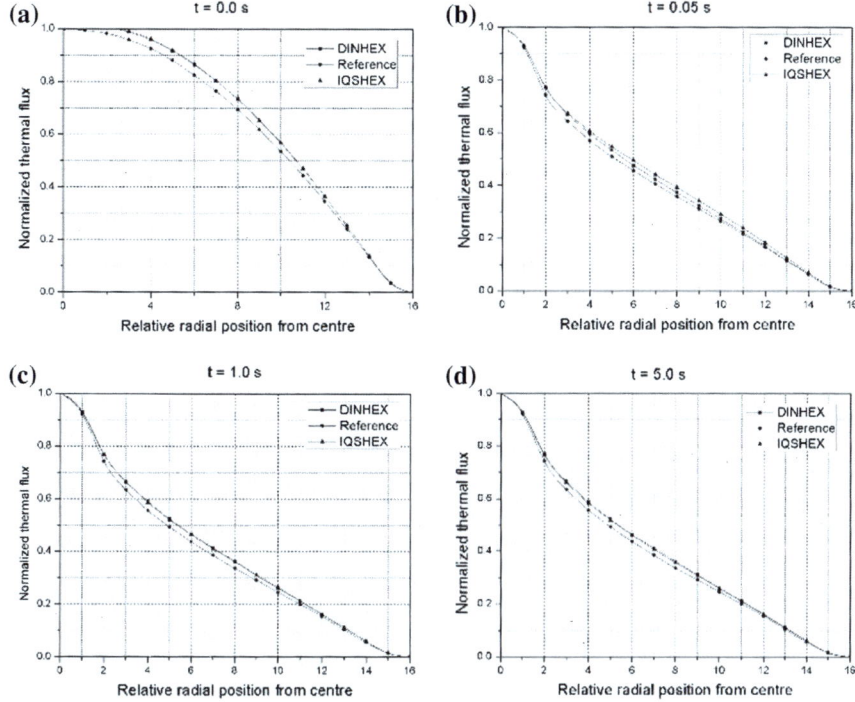

Fig. 6 Thermal flux variation in radial direction at different time of transient in 3D core

4 Conclusion

DINHEX is verified against number of benchmark problems with perturbation of different variety. There are problems of prompt super criticality where 1.17 $ positive reactivity is added instantly into the core, AER-DYN-002, where control rod is ejected causing fuel temperature to increase and thereby adiabatic temperature feedback to arrest the power surge. The results of DINHEX agree well with the results of our own code IQSHEX and other codes like FX2-TH, DIF3D, HENKO, KIKO3D, DYN3D, BIPR8, etc. While simulating the drop of control rod into the core, lattice parameters of meshes, encountering mixture of materials, are homogenized using flux weighting technique to avoid the cusp problem in transient power variation. Coupling of the kinetics code with thermal hydraulics of two-phase coolant will make the code more versatile and accurate.

References

1. T. Singh, T. Mazumdar, Development of a Nodal Expansion Method based Space time kinetics code for hexagonal geometry. Ann. Nucl. Energy **110**, 584–602 (2017)
2. R.D. Lawrence, *The DIF3D Nodal Neutronics Option for Two-and Three-Dimensional Diffusion Theory Calculations In Hexagonal Geometry* (Argonne National Laboratory, Lemont, Ill, USA, 1983)
3. H. Michels, *DISLIN Manual* (Max-Planck-Institut fur Aeronomie, Katlenburg-Lindau, 1997), p. 288
4. T.A. Taiwo, H.S. Khalil, *The DIF3D Nodal Kinetics Capability in Hexagonal-Z Geometry-Formulation and Preliminary Tests.* In Int. Top. Meeting, Pittsburgh (1991)
5. Y.A. Chao, Y.A. Shatilla, Conformal mapping and hexagonal nodal methods—II: implementation in the ANC-H code. Nucl. Sci. Eng. **121**(2), 210–225 (1995)
6. S. Zhang, Z. Xie, Numerical solution of space-time kinetics equation in hexagonal geometry. Nucl. Power Eng. **18**(3), 226–232 (1997)
7. U. Grundmann, U. Rohde, Definition of the second kinetic benchmark of AER. In: *Proceeding of the Third Symposium of AER* (1993), p. 325
8. AER benchmark. (Available at http://aerbench.kfki.hu/aerbench/)
9. M.R. Buckner, J.W. Stewart, Multidimensional Space-Time Nuclear-Reactor Kinetics Studies—Part I: Theoretical. Nucl. Sci. Eng. **59**(4), 289–297 (1976)

Reactivity Initiated Transient Analysis of 2 MW Upgraded Apsara Reactor

Tanay Mazumdar, Jainendra Kumar, Tej Singh, P. V. Varde
and S. Bhattacharya

Abstract The 2 MW upgraded Apsara reactor is an Indian, open pool-type, light water research reactor which is commissioned recently. U_3Si_2-Al dispersion-type low-enriched uranium (LEU) (17.0 wt%) fuel has been used for this reactor. Expected maximum in-core thermal neutron flux is 6.1×10^{13} n/cm^2/s. In-house computer codes are used to carry out the reactivity initiated transient analysis of the reactor. The analysis shows that the maximum temperature of fuel, clad and coolant in the enveloping scenario is well within the stipulated limit. In the present work, safety of the core, under the transient events postulated for the reactor, is demonstrated based on detailed safety analysis.

Keywords Upgraded Apsara · LORI · Safety · Start-up accident · Transient

1 Introduction

Before commissioning a reactor, regulatory body demands demonstration of the safety of the reactor, which deals with the analysis whether the reactor will remain

T. Mazumdar · J. Kumar · T. Singh (✉)
Reactor Physics and Nuclear Engineering Section, Research Reactor Services Division, Reactor Group, Bhabha Atomic Research Center, Mumbai, India
e-mail: t_singh@barc.gov.in

T. Mazumdar
e-mail: tanay@barc.gov.in

J. Kumar
e-mail: jaink@barc.gov.in

P. V. Varde
Reactor Group, Bhabha Atomic Research Center, Mumbai, India
e-mail: varde@barc.gov.in

S. Bhattacharya
Reactor Projects Group, Bhabha Atomic Research Center, Mumbai, India
e-mail: subhatt@barc.gov.in

© Springer Nature Singapore Pte Ltd. 2020
P. V. Varde et al. (eds.), *Reliability, Safety and Hazard Assessment for Risk-Based Technologies*, Lecture Notes in Mechanical Engineering, https://doi.org/10.1007/978-981-13-9008-1_74

safe or become vulnerable under any postulated initiating events (PIEs). The term 'postulated initiating event' refers to an unintended event including an operating error, equipment failure or external influence that directly or indirectly challenges basic safety functions of a reactor. Typically, such an event necessitates protective actions (automatic or manual) to prevent or mitigate undesired consequences for reactor equipment, reactor personnel, the public or the environment. Postulated initiating events are identified based on engineering evaluation, logical analysis and past operating experience. Listing of the initiating event as applicable to upgraded Apsara reactor has been done based on the guidelines given in IAEA Safety Standard Series No. SSR-3 [1]. The PIEs considered here are mainly responsible for reactivity insertion accidents (RIAs) like reactor start-up accident, loss of regulation incident (LORI), etc. In order to assess the severity of such accidents, we have already developed computer codes—RITAC [2] and SACRIT [3] which solve point kinetics equation as well as thermal hydraulics equations in a coupled manner for plate-type fuel geometry to analyze reactivity-induced transients. In RIAs, positive reactivity insertion in a given manner makes neutron density and subsequently fuel, clad and coolant temperatures increasing with time, and this rise in fuel and coolant temperatures causes reactivity feedback which perturbs reactivity and ultimately the neutron density. This paper is written in the following way. In Sect. 2, upgraded Apsara is briefly introduced. Then, input data and underlying assumptions of the analysis are given in Sect. 3. Subsequent sections are dedicated to steady-state thermal hydraulic analysis followed by results of all PIEs, namely start-up accident, LORI, etc. Finally, conclusion is drawn based on the analysis carried out.

2   2 MW Upgraded Apsara Reactor

Upgraded Apsara reactor is a 2 MW swimming pool-type reactor. It uses low-enriched uranium (LEU)-based U3Si2 fuel dispersed in aluminum matrix (17 wt% U235 enrichment). The reactor core consists of 11 standard fuel assemblies (SFAs) and 4 control fuel assemblies (CFAs) as shown in Fig. 1. A standard fuel assembly consists of 17 fuel-bearing plates and two aluminum inert plates swaged into grooves provided in aluminum side plates. A water gap of 2.5 mm is maintained between two successive fuel/inert plates. Each fuel plate has a meat thickness of 0.7 mm and the fuel meat is clad with 0.4-mm-thick aluminum sheet. Control fuel assembly is basically a standard fuel assembly except the fact that there are 12 fuel plates with a water gap of 2.4 mm between them. At either end of the assembly, 9.45-mm-wide recess is provided for movement of fork-type twin-absorber blade. Out of four control fuel assemblies, two will work as control cum shutoff rods (CSRs) and other two will work as shutoff rods (SORs). The reactor core is cooled by demineralized water flowing from top to bottom through the core. Primary coolant pumps draw water from pool outlet line at a rate of 4900 lpm, and it is passed through a heat exchanger where heat is rejected to secondary coolant. Calculation of core physics parameters of the reactor is given in [4].

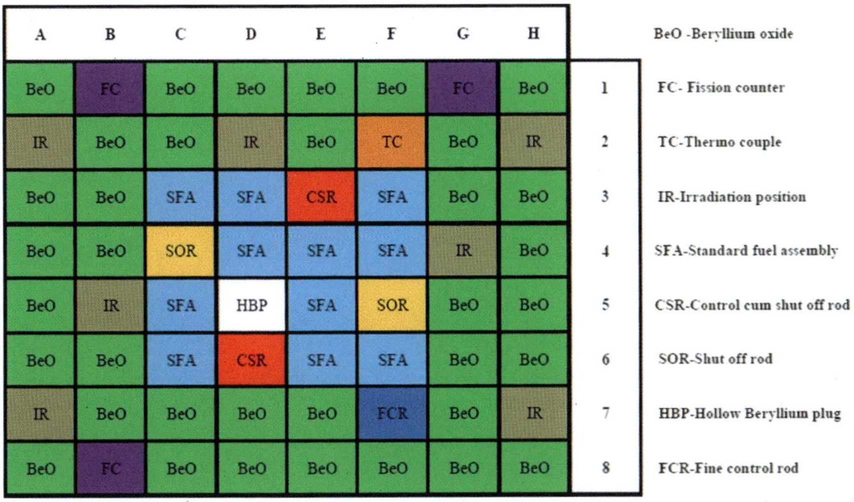

	A	B	C	D	E	F	G	H		
	BeO	FC	BeO	BeO	BeO	BeO	FC	BeO	1	
	IR	BeO	BeO	IR	BeO	TC	BeO	IR	2	
	BeO	BeO	SFA	SFA	CSR	SFA	BeO	BeO	3	
	BeO	BeO	SOR	SFA	SFA	SFA	IR	BeO	4	
	BeO	IR	SFA	HBP	SFA	SOR	BeO	BeO	5	
	BeO	BeO	SFA	CSR	SFA	SFA	BeO	BeO	6	
	IR	BeO	BeO	BeO	BeO	FCR	BeO	IR	7	
	BeO	FC	BeO	BeO	BeO	BeO	BeO	BeO	8	

BeO -Beryllium oxide

FC- Fission counter

TC-Thermo couple

IR-Irradiation position

SFA-Standard fuel assembly

CSR-Control cum shut off rod

SOR-Shut off rod

HBP-Hollow Beryllium plug

FCR-Fine control rod

Fig. 1 Equilibrium core configuration of upgraded Apsara

3 Input Data and Assumption

Based on steady-state analysis, it is found that 50% IN condition of the control blades is the most conservative for thermal hydraulic calculation. Hence, the same is considered for carrying out the transient analysis. Computer codes RITAC and SACRIT are used to carry out the transient calculation, and results, shown in subsequent sections, are the conservative one. Underlying assumptions of the analysis are given below.

(i) Transients, caused due to uncontrolled addition of positive reactivity, lead to rise in reactor power. In normal operation scenario, this power surge is arrested by tripping the reactor on regulation/protection parameters.

(ii) However, in the analysis, no credit has been given to regulation system trips. In addition to this, it is assumed that the first trip is ignored and the reactor is tripped on the next available trip to assess the severity of the transients.

(iii) Reactor trip is manifested by dropping two control rods (partly inserted into the core) and two shutoff rods (parked outside the core) into the core though in the analysis a typical combination of one control rod and one shutoff rod, which adds least negative reactivity worth (~69 mk) among all the possible combinations of two out of four rods, has been considered to simulate the worst possible case.

(iv) The analysis has been performed considering maximum drop time of SORs and CSRs as 0.5 s for 90% drop and 2.5 s. for the 100% drop.

(v) A time delay, caused due to signal processing, before the rods fall into the core is taken to be about 500 ms in the calculation.

(vi) The reactivity feedback coefficients used in the calculation are: (a) Doppler coefficient: responsible for increased neutron absorption in U^{238} epithermal

Table 1 Steady-state thermal hydraulic analysis for SFA, CFA-CSR and CFA-SOR

Parameters	SFA	CFA-CSR	CFA-SOR
Fuel centerline temperature	90.3	84.1	87.1
Fuel–clad interface temperature	89.1	82.9	85.8
Clad–coolant interface temperature	88.4	82.3	85.2
Coolant outlet temperature	58.5	53.4	57.1
ONBR	1.37	1.47	1.42

resonance regime as the temperature increases in fuel (~ -0.014 mk/°C), (b) water temperature coefficient: responsible for hardening of neutron spectrum (~ -0.07 mk/°C) and (c) water void/density coefficient: responsible for neutron leakage due to change in water density as the water is heated up or boiled (~ -1.6 mk/°C). Core averaged values of the reactivity feedback coefficients are considered for both average and hot channel.

(vii) Six group effective delayed neutron fractions (β_{eff}) are used in the calculation. Their values have been determined with the help of a computer code NEMSQR [5].

(viii) The nominal flow rates for control and standard fuel assemblies are taken to be 190 and 240 lpm, respectively. At 2 MW, maximum assembly power is 173 KW for SFA, 124 KW for CFA-CSR and 140 KW for CFA-SOR. Axial peaking factors for SFA, CFA-CSR and CFA-SOR are 1.36, 1.63 and 1.33, respectively, while local peaking factors are 1.45, 1.18 and 1.3, respectively.

(ix) Coolant inlet temperature remains at 38 °C throughout the transient.

(x) In order to carry out conservative analysis incorporating the uncertainty in single-phase heat transfer coefficient correlation, heat transfer coefficient value is taken 15% lesser than the value obtained from the correlation [6].

4 Results and Discussion

4.1 Steady-State Thermal Hydraulic Analysis

In order to carry out steady-state thermal hydraulic analysis for standard (SFA) and control (CFA-CSR and CFA-SOR) fuel assemblies, radially fuel meat is divided into 10 meshes and clad into 3 meshes while axially entire fuel plate is divided into 10 axial meshes. Table 1 shows comparison of results between hottest SFA, CFA-CSR and CFA-SOR.

It is to be noted that plate power is maximum for CFA-SOR, followed by SFA, whereas coolant velocity for SFA is lesser as compared to CFA-SOR. As a combined effect, fuel centerline, fuel–clad interface, clad–coolant interface and coolant outlet temperatures are found to be maximum for SFA. These temperatures are 90.3, 89.1,

Table 2 Steady-state thermal hydraulic analysis for SFA, CFA-CSR and CFA-SOR (85% flow and 115% power)

Parameters	SFA	CFA-CSR	CFA-SOR
Fuel centerline temperature	105.1	97.4	101.0
Fuel–clad interface temperature	103.6	96.0	99.6
Clad–coolant interface temperature	102.9	95.2	98.9
Coolant outlet temperature	65.7	58.8	63.8
ONBR	1.18	1.28	1.23

88.4 and 58.5 °C, respectively. Onset of nucleate boiling ratio (ONBR), which is a ratio of maximum clad–coolant (or clad outer surface) temperature to onset of nucleate boiling temperature (T_{ONB}), is found to be minimum for SFA among all assemblies though its value (1.37) implies sufficient margin of maximum clad outer surface temperature from T_{ONB}. In view of the fact that at any point of time during reactor operation, reactor power and coolant flow may mismatch, an analysis is carried out considering reactor operation at: (i) 115% power and nominal flow, (ii) nominal power and 85% flow and (iii) 115% power and 85% nominal flow. Case (iii) is considered to be the most conservative steady-state thermal hydraulic analysis which shows no melting of fuel–clad or boiling of coolant for the present reactor (Table 2).

4.2 Start-up Accident

In this accident analysis, it is postulated that due to circuit malfunction, control rods are withdrawn continuously from its most sensitive position at a maximum rate of travel with reactor initially critical at 1 W. However, reactor power at critical level except first approach to criticality will remain more than 1 W due to a buildup of photo-neutron source. Uncontrolled withdrawal of control rods during reactor start-up leads to uncontrolled reactivity addition into the core. In the analysis, only one control rod withdrawal case is considered. Maximum rod withdrawal speed will be 1.5 mm/s and the corresponding maximum reactivity insertion rate will be ~0.22 mk/s. This uncontrolled withdrawal of control rod will be terminated as soon as reactor trip is registered. Though control rods are normally kept at partially IN condition, they are considered to be in fully OUT condition in the present analysis since the former is less severe than latter. It is found that maximum reactivity of 6.26 mk is inserted into the core and log rate reaches 224%/s. Reactor power is peaked at 3.89 MW, and maximum fuel centerline, clad–coolant interface and coolant outlet temperatures of SFA are estimated to be about 113.9, 111.0 and 65.8 °C, respectively. Variation of reactor power, reactivity, log rate and fuel centerline, clad–coolant interface and coolant outlet temperature of SFA with time is given in Fig. 2.

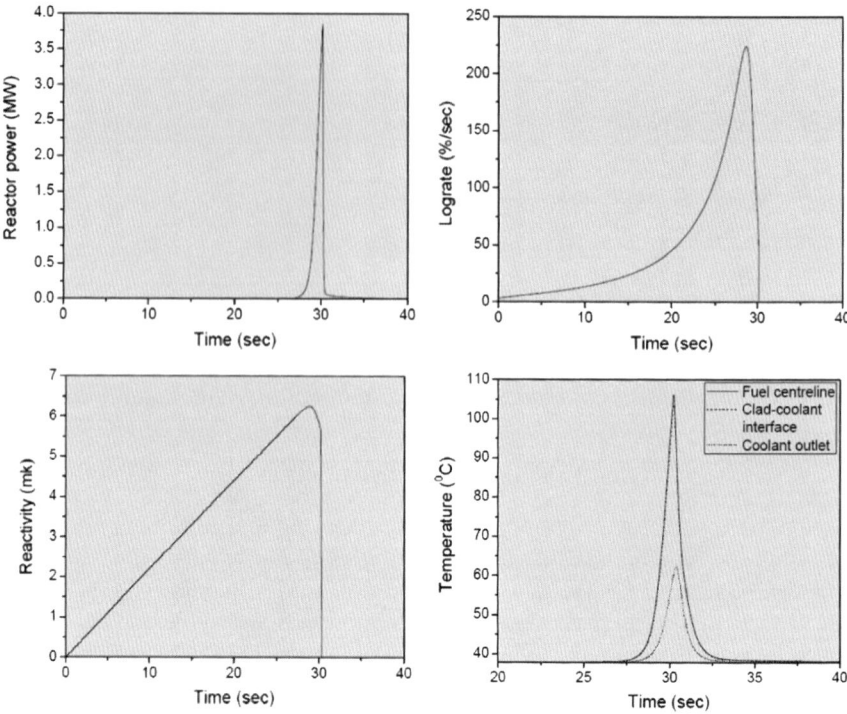

Fig. 2 Results of start-up accident

4.3 Inadvertent Withdrawal of CSR at Full-Power Operating Condition

In this accident analysis, it is postulated that during normal reactor operation at full power, i.e., 2 MW with normal coolant flow through the core, one of the CSRs is withdrawn in uncontrolled manner with maximum permissible speed of 1.5 mm/s. The corresponding maximum reactivity insertion rate will be ~0.22 mk/s. In this analysis, first trip, which is overpower trip at 2.3 MW, has been ignored and second trip, which is log rate at 6%/s, is taken into account. It is assumed that uncontrolled withdrawal of control rod will be terminated as soon as the trip is registered. As a consequence, maximum reactivity of 1.04 mk is inserted into the core and log rate reaches up to 6.24%/s. During the transient, the peak power is reached up to 2.61 MW and maximum fuel, clad and coolant temperatures of SFA are estimated to be about 103.7, 101.2 and 64.0 °C, respectively. Variation of reactor power, reactivity, log rate and fuel centerline, clad–coolant interface and coolant outlet temperature of SFA with time is given in Fig. 3.

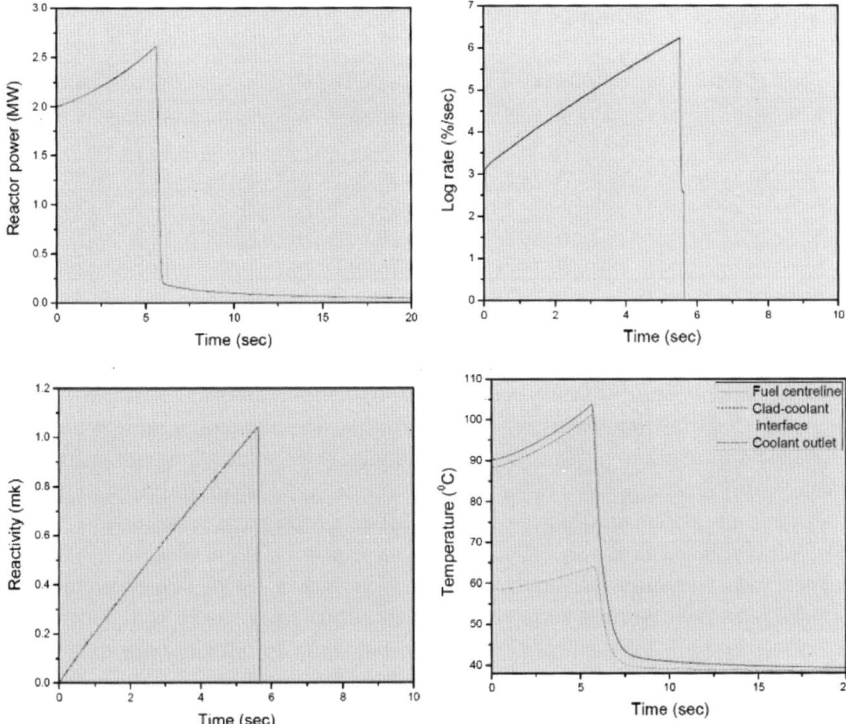

Fig. 3 Results of inadvertent withdrawal of CSR at full-power operating condition

4.4 Loss of Regulation Incident (LORI)

Reactivity control for start-up and power regulation is achieved by controlled movement of fine control rod and control cum shutoff rods in the reactor core. The fine control rod is used to adjust reactivity in smaller step on auto/manual mode while the control cum shutoff rods are meant for coarse adjustment of reactivity on manual mode. The regulating system employs three independent channels and works on 2/3 coincidence logic. The system design is such that a fault/defect beyond a pre-set level in any channel automatically leads to its rejection and the control action is taken over by the rest two channels without affecting reactor operation. Further malfunction in anyone of the remaining two channels before the rejected channel which is rectified and restored would cause a reactor trip. In loss of regulation incident (LORI) analysis, enveloping scenario that positive reactivity getting added into the system at the maximum design rate by the way of continuous and uncontrolled withdrawal of (i) Fine Control Rod (FCR) alone and (ii) FCR and one of the CSRs together, are considered which in turn increases the reactor power.

4.4.1 Only FCR Withdrawal Case

The maximum withdrawal speed of FCR is limited to 16 mm/s and the corresponding maximum addition of reactivity in the core is about 0.33 mk/s. Reactivity worth of FCR is chosen to be less than the effective delayed neutron fraction ($\beta_{eff} \sim 7.1$ mk) of the core to restrict the consequences of a loss of regulation scenario. Worth of the FCR in equilibrium core is about 5 mk. However, in the analysis, reactivity worth of the FCR has been conservatively assumed to be 10% higher than 5 mk considering the fact that the worth varies while the core evolves from fresh to equilibrium phase. In order to analyze the severity of reactivity initiated transients at different operating conditions, the transients are assumed to be initiated from five different initial power levels, namely 1, 25, 100, 500 and 2 MW. Consequence of the transient initiated from 100 KW (or less) is more severe than those initiated from higher initial powers. This is expected because at higher initial powers, the overpower trip level of 2.3 MW is reached earlier, thereby restricting further addition of positive reactivity. It is important to note that if credit is taken for the termination of the transients by the high log rate trip at 6%/s instead of the overpower trip at 2.3 MW, the consequences of the LORI would be hardly of any concern.

Maximum reactivity inserted into the core is found to be about 5.46 mk and log rate reaches up to 108%/s. During the transient, the peak power is reached up to 2.85 MW. Maximum fuel, clad and coolant temperatures are observed to be about 102.9, 100.5 and 62.7 °C, respectively, for SFA. All the temperatures are well within their limiting value and no boiling occurred in the core. Figure 4 shows the variation of reactor power, reactivity, log rate and temperatures of fuel centerline, clad–coolant interface and coolant outlet with time for all the above-mentioned cases in SFA. It is evident from the figure that higher the initial power, the curves are shifted toward left as the time available between the initiation of the transient and its arrest due to protection system decreases with the increase in initial power.

4.4.2 FCR and One of the CSRs Withdrawal Case

Withdrawal of FCR causes maximum reactivity addition at a rate of 0.33 mk/s while withdrawal of one of the CSRs causes maximum reactivity addition at a rate of 0.22 mk/s. If FCR and one CSR are inadvertently withdrawn together, which is considered to be extremely rare event, then positive reactivity will be added into the core at a rate of $(0.33 + 0.22) = 0.55$ mk/s. Similar to the analysis described in previous section, the transient is assumed to be initiated from five different initial power levels, namely 1, 25, 100, 500 and 2 MW. Consequence of the transient initiated from 1 KW is more severe than those initiated from higher initial powers.

Maximum reactivity inserted into the core is found to be about 6.27 mk and log rate reaches up to 237%/s. During the transient, the peak power is reached up to 4.76 MW. Maximum fuel, clad and coolant temperatures are observed to be about 125.3, 121.9 and 69.7 °C, respectively, for SFA. Though the bulk coolant temperature

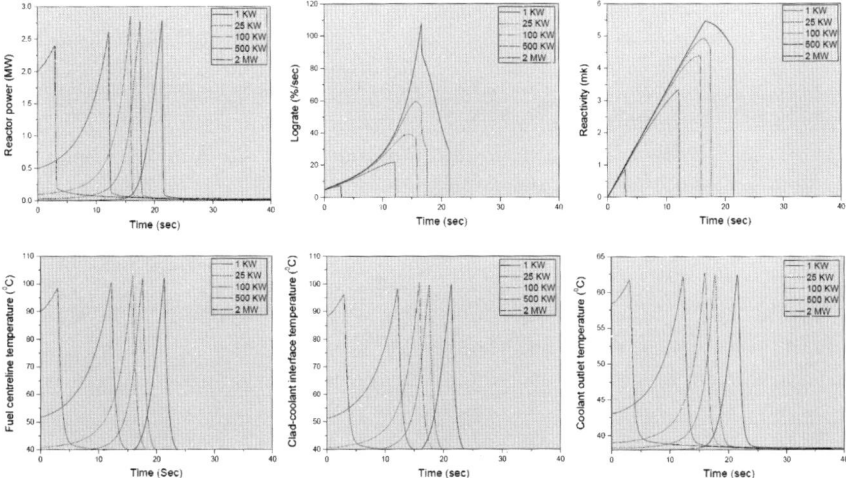

Fig. 4 Results of LORI due to withdrawal of FCR alone

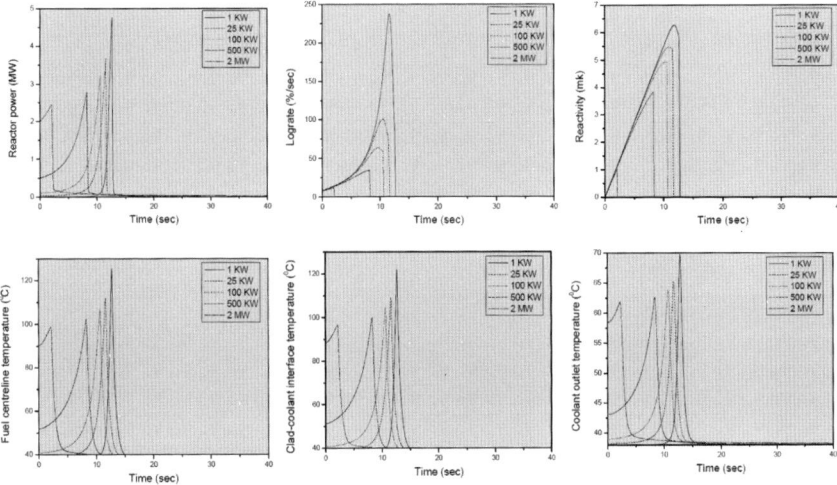

Fig. 5 Results of LORI due to withdrawal of FCR and one of the CSRs

remains much below the saturation temperature (~115 °C), the clad–coolant interface temperature exceeds saturation temperature momentarily for about 0.18 s.

Figure 5 shows the variation of reactor power, reactivity, log rate and temperatures of fuel centerline, clad–coolant interface and coolant outlet with time for all the above-mentioned cases in SFA.

4.5   Removal of in-Core Experimental Assembly

There is one position inside core where an experimental assembly can be loaded. Though this assembly will be removed from core only when the reactor is in shutdown state, a hypothetical scenario has been thought of where this assembly will be removed from a critical core of initial power 2 MW. In this analysis, reactivity load of the experimental assembly is restricted to be less than or equal to 2/3 of β_{eff}, i.e., 5 mk which is added instantly into the core. As a consequence, peak power is reached up to 6.64 MW. Maximum fuel, clad and coolant temperatures are observed to be about 133.9, 128.8 and 84.8 °C, respectively, for SFA. Though the bulk coolant temperature remains much below the saturation temperature (~115 °C, in this case), the clad–coolant interface temperature exceeds saturation temperature momentarily for about 0.6 s.

5   Conclusion

The Analysis of various reactivity initiated transient events for 2 MW upgraded Apsara reactor is carried out using two in-house computer codes—RITAC and SAC-RIT with a good degree of accuracy. Based on the detailed analysis, it is concluded that the core is safe under the reactivity initiated transient events postulated for the reactor. It is important to note that the maximum estimated clad temperature, 133.9 °C in the event of removal of in-core experimental assembly, is far below the melting point. In the same event, subcooled boiling of coolant takes place at the hot spot which has negligible effect due to its short duration (~ 0.6 s).

Acknowledgements   Authors express their special gratitude to Shri P. Mukherjee, Project Manager, Apsara Upgradation Project, and entire project team of Apsara upgradation for their support to carry out this work.

References

1. International Atomic Energy Agency. Safety of Research Reactors—Safety Requirements, *IAEA Safety Standards Series No. SSR-3* (2016)
2. T. Mazumdar, T. Singh, H.P. Gupta, K. Singh, RITAC: reactivity initiated transient analysis code–an overview. Ann. Nucl. Energy **43**, 192–207 (2012)
3. T. Singh, J. Kumar, T. Mazumdar, V.K. Raina, Development of neutronics and thermal hydraulics coupled code–SAC-RIT for plate type fuel and its application to reactivity initiated transient analysis. Ann. Nucl. Energy **62**, 61–80 (2013)
4. T. Singh, P. Pandey, T. Mazumdar, K. Singh, V.K. Raina, Physics design of 2 MW upgraded Apsara research reactor. Ann. Nucl. Energy **60**, 141–156 (2013)
5. T. Singh, T. Mazumdar, P. Pandey, NEMSQR: A 3-D multi group diffusion theory code based on nodal expansion method for square geometry. Ann. Nucl. Energy **64**, 230–243 (2014)
6. M. Kristof, I. Vojtek, Uncertainty analyses for LB LOCA in VVER-440/213 (2003)

Reactivity Initiated Transient Analysis of 30 MW High Flux Research Reactor

Tanay Mazumdar, Tej Singh, P. V. Varde and S. Bhattacharya

Abstract High Flux Research Reactor (HFRR) is a 30 MW pool-type research reactor which has been proposed to be built in India. It has a benefit of free access from pool top and a large inventory of ultimate heat sink. The core is compact in nature and is fueled with 19.75 wt% (U^{235}) enriched U_3Si_2 fuel, dispersed in aluminum matrix. The reactor has a maximum in core thermal and fast neutron flux levels of about 5.3×10^{14} and 1.7×10^{14} n/cm^2/s, respectively. Based on the detailed safety analysis, it is concluded that the core is safe under the reactivity initiated transient events postulated for the reactor.

Keywords HFRR · LORI · Safety · Start-up accident · Transient

1 Introduction

In safety analysis, PIE or postulated initiated event is a common term which refers to an unintended event including an operating error, equipment failure or external influence that directly or indirectly challenges the basic safety functions of a reactor. Typically, such an event necessitates protective actions to mitigate the undesired consequences to reactor equipment, reactor personnel, public and environment. The

T. Mazumdar · T. Singh (✉)
Reactor Physics and Nuclear Engineering Section, Research Reactor
Services Division, Reactor Group, Bhabha Atomic Research Center, Mumbai, India
e-mail: t_singh@barc.gov.in

T. Mazumdar
e-mail: tanay@barc.gov.in

P. V. Varde
Reactor Group, Bhabha Atomic Research Center, Mumbai, India
e-mail: varde@barc.gov.in

S. Bhattacharya
Reactor Projects Group, Bhabha Atomic Research Center, Mumbai, India
e-mail: subhatt@barc.gov.in

© Springer Nature Singapore Pte Ltd. 2020
P. V. Varde et al. (eds.), *Reliability, Safety and Hazard Assessment for Risk-Based Technologies*, Lecture Notes in Mechanical Engineering,
https://doi.org/10.1007/978-981-13-9008-1_75

PIEs, identified from the list given in IAEA safety standard series no. SSR-3 [1] based on engineering evaluation, logical analysis and past operating experience for the proposed design of 30 MW High Flux Research Reactor (HFRR), are reactor start-up accident and loss of regulation incident (LORI), which are mainly responsible for reactivity insertion accidents (RIA). In order to assess the severity of such accidents, numerous computer codes have been developed so far, and this development process is still very much going on. In RIAs, positive reactivity insertion increases reactor power and subsequently fuel, clad and coolant temperatures with time. This rise in fuel and coolant temperature causes reactivity feedback which perturbs the reactivity and ultimately the reactor power. So, it is necessary to solve the equations of neutron kinetics and thermal hydraulics in a coupled manner to carry out the safety analysis of the reactor. The present analysis has been carried out using such an in-house computer code (RITAC) reactivity initiated transient analysis code [2]. This paper is written in the following way. In Sect. 2, HFRR is briefly introduced. Then, modeling and underlying assumptions of the analysis are given in Sect. 3. Subsequent sections are dedicated to steady state thermal hydraulic analysis followed by results of all PIEs, namely, start-up accident, LORI, etc. Finally, a conclusion is drawn based on the analysis carried out.

2   30 MW High Flux Research Reactor (HFRR)

High Flux Research Reactor (HFRR) is a 30 MW pool-type reactor. It uses low-enriched uranium (LEU; 19.75 wt% U^{235} as enrichment)-based U_3Si_2 fuel, dispersed in aluminum matrix. The core of HFRR is mainly loaded with two types of plate-type fuel assemblies, namely, standard fuel assemblies (SFA) and control fuel assemblies (CFA). An SFA consists of 20 fuel-bearing plates and two aluminum end plates swaged into the grooves provided in the side plates. Each fuel plate has a meat thickness of 0.6 mm, and the fuel meat is cladded with 0.4 mm thick aluminum sheet. The meat of fuel plate contains LEU fuel with a loading density of 4.5 gm/cc. A water gap of 2.3 mm is provided between the fuel plates as well as between a fuel plate and an aluminum end plate. CFA is similar to SFA except the fact that it has 13 fuel plates with fuel loading density of 3.5 gm/cc. 11.55 mm wide recess is provided at the extreme ends of CFA to accommodate the movement of fork type twin Hafnium (Hf) blades of dimension $5.5 \times 68 \times 660$ mm. Each recess is created between the end plate and aluminum guide plate, which are swaged into the side plates of CFA. Apart from these two types of assemblies, there is fine control rod (FCR) assembly, which is made by removing four fuel plates from the middle portion of SFA in order to provide a gap of 9.7 mm in between two aluminum guide plates of thickness 1.2 mm for the vertical movement of a Hf blade of dimension $4 \times 68 \times 650$ mm. Cadmium wire, which works as a burnable poison, is embedded in the side plates of all assemblies (near to both the edges of each fuel plate). Details of all these assemblies are available in [3]. The reactor core consists of 17 SFAs, six CFAs, one FCR and one central water hole as shown in Fig. 1. Out of six CFAs, four are control

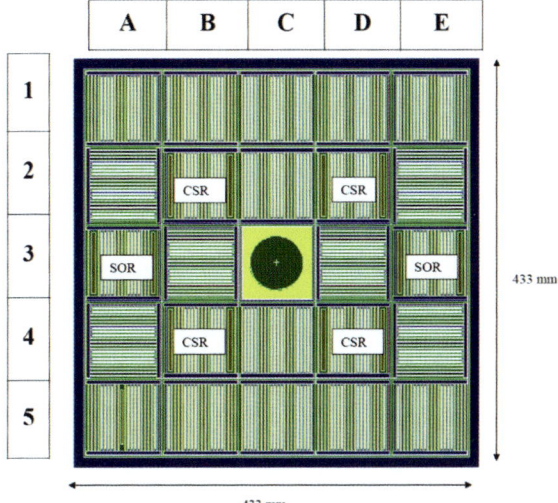

Fig. 1 Layout of 30 MW HFRR core

cum shut off rods (CFA-CSRs) and two shut off rods (CFA-SORs). The reactor core is cooled by demineralized water flowing from bottom to top through the core.

3 Modeling and Assumption

Computer code RITAC is used to carry out the transient calculation. Underlying assumptions of the analysis are given below.

(i) Transients caused due to uncontrolled addition of positive reactivity lead to a rise in reactor power. In normal operation scenario, this power surge is arrested by tripping the reactor on regulation/protection parameters.

(ii) However, in the analysis, no credit is given to the regulation system trips. In addition to this, it is assumed that the first trip is ignored and the reactor is tripped on the next available trip to assess the severity of the transients.

(iii) Reactor trip is manifested by dropping four CSRs (partly inserted into the core) and two SORs (parked outside the core) into the core though, in the analysis, a typical combination of three CSRs, which adds least negative reactivity worth (~90 mk) among all the possible combinations of three out of four CSRs, is considered to simulate the worst possible case. The CSR completes its 90% drop into the core within 0.6 s.

(iv) A time delay of 500 ms, before the CSRs fall into the core, is taken in the present calculation. This is caused due to signal processing (total 450 ms out of which 300 ms delay in electronic circuit and 150 ms delay in clutch de-energization) and distance travel between the parking elevation and the core top (50 ms).

(v) The reactivity feedback coefficients used in these calculations are: (i) Doppler coefficient—responsible for increased neutron absorption in U^{238} epithermal resonances as the temperature increases in fuel, (ii) Water temperature coefficient—responsible for hardening of neutron spectrum and (iii) Water void/density coefficient—responsible for leakage of neutrons due to change in density of water as the water is heated up or boiled. Core averaged values of the reactivity feedback coefficients are considered for both the average and hot channels.

(vi) Values of six group delayed neutron fractions (β_i) and effective delayed neutron fraction (β_{eff}), used in the calculation, are determined with the help of an in-house computer code NEMSQR [4].

(vii) The nominal flow rate through one SFA and one CFA is taken to be 1900 and 1500 lpm, respectively. An analysis is performed to study the effect of reactor operation under an enveloping scenario that the channel flow is equal to 98% of the nominal flow through the fuel assemblies. Maximum assembly power at 30 MW reactor power is taken to be 1670 KW for SFA, 1060 KW for CFA-CSR and 1100 KW for CFA-SOR. Axial peaking factors for SFA, CFA-CSR and CFA-SOR are 1.60, 1.70 and 1.45, respectively, while local peaking factors are 1.25, 1.19 and 1.25, respectively.

(viii) The analysis is performed considering the maximum drop time of CSRs as 0.6 s for 90% drop and 2.5 s for 100% drop.

(ix) Coolant inlet temperature remains at 40 °C throughout the transient. All input parameters discussed above are listed in Sect. 1.4. Two channel analysis-based computer codes RITAC and SACRIT are used to carry out the transient calculation.

(x) In order to carry out conservative analysis incorporating the uncertainty in single-phase heat transfer coefficient correlation, the coefficient value is taken 15% lesser than the value obtained from the correlation [5].

4 Results and Discussion

4.1 Start-Up Accident

In this accident analysis, it is postulated that due to circuit malfunction CSR is withdrawn continuously from its most sensitive position at a maximum rate of travel with reactor initially critical at 1 W. However, reactor power at critical level except the first approach to criticality (FAC) will remain more than 1 W due to buildup of photo-neutron source. Uncontrolled withdrawal of CSR during reactor start-up leads to uncontrolled reactivity insertion to the core. Maximum withdrawal speed of one CSR is 1.5 mm/s, which corresponds to the maximum reactivity addition rate of ~0.25 mk/s. Since the transient starts from 1 W, enough time is available before the reactor reaches the trip level. In this analysis, it is assumed that the uncontrolled

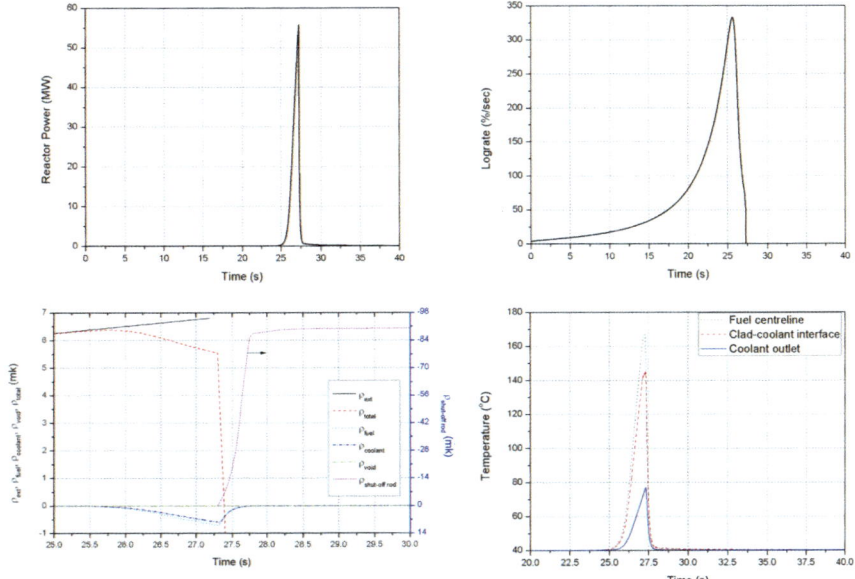

Fig. 2 Results of start-up accident

withdrawal of CSR will be terminated as soon as the trip is registered. As a consequence, maximum reactivity of 6.37 mk is inserted into the core and the log rate reaches 332.8%/s. Reactor power is peaked at 55.77 MW and maximum fuel, clad (clad-coolant interface, to be precise) and coolant temperatures are estimated to be about 167.7 °C (SFA), 145.1 °C (SFA) and 77.5 °C (CFA-SOR), respectively. Variation of reactor power, log rate, reactivity and fuel centerline, clad-coolant interface and coolant outlet temperature of SFA with time are given in Fig. 2.

4.2 Loss of Regulation Incident (LORI)

Reactivity control for start-up and power regulation is achieved by the controlled movement of FCR and CSRs in the reactor core. The FCR is used to adjust the reactivity of smaller magnitude on auto/manual mode, while the CSRs are meant for coarse adjustment of reactivity on manual mode. The regulating system employs three independent channels and works on 2/3 coincidence logic. The system is designed in such a way that a fault/defect beyond a pre-set level in any channel automatically leads to its rejection and the control action will be taken over by the other two channels without affecting the reactor operation. Further malfunction in one of the remaining two channels, prior to the rectification and restoration of the rejected channel, would cause a reactor trip.

In loss of regulation incident (LORI) analysis, an enveloping scenario, that positive reactivity getting added into the system at the maximum design rate by the way of continuous and uncontrolled withdrawal of FCR, is considered which in turn increases the reactor power. The maximum withdrawal speed of FCR is limited to 16 mm/s and the corresponding maximum addition of reactivity in the core is about 0.33 mk/s. Reactivity worth of FCR is chosen to be less than the β_{eff} (~ 7.0 mk) of the core in order to restrict the consequences of a loss of regulation scenario. Worth of the FCR in equilibrium core is about 5 mk. However, in the analysis, reactivity worth of the FCR has been conservatively assumed to be 10% higher than 5 mk considering the fact that the worth varies while the core evolves from fresh to equilibrium condition. Uncontrolled reactivity addition rate of 0.33 mk/s corresponding to the maximum design rate is considered for the analysis. In order to analyze the severity of reactivity initiated transients at different operating conditions, the transients are assumed to be initiated from three different initial power levels, namely, 1 KW, 1 and 30 MW.

It is clear from the analysis that the consequences of the transients initiated from 1 MW are more severe than those initiated from higher initial powers. This is expected because at higher initial powers the overpower trip level of 33.0 MW is reached earlier and thereby restricting further addition of positive reactivity. It is important to note that if credit is taken for the termination of the transients by the trips on reactor period (such as the high log rate trip at 5%/s as provided in the log rate safety channels) instead of the overpower trip at 33.0 MW, the consequences of the LORI would be hardly of any concern. Maximum fuel, clad (clad-coolant interface, to be precise) and coolant outlet temperatures are observed to be about 145.8 °C (SFA), 128.8 °C (SFA) and 68.2 °C (CFA-SOR), respectively. All the temperatures are well within their limiting values and no boiling occurs in the core. Figure 3 shows the variations of reactor power, log rate, reactivity and fuel.

5   Conclusion

Analysis of reactivity initiated transients, i.e., start-up accident and LORI are carried out for the 30 MW HFRR. In the analysis, maximum fuel, clad and coolant temperatures are found to be about 167.7 °C (SFA), 145.1 °C (SFA) and 77.5 °C (CFA-SOR), respectively, during the start-up accident. The results clearly indicate that the estimated peak temperatures of fuel, clad and coolant during the transients are well within the prescribed limit, i.e., the maximum fuel temperature remains less than 400 °C.

Acknowledgements Authors express their special gratitude to Shri S. Mammen, Project Manager, HFRR and entire project team of HFRR for their support to carry out this work.

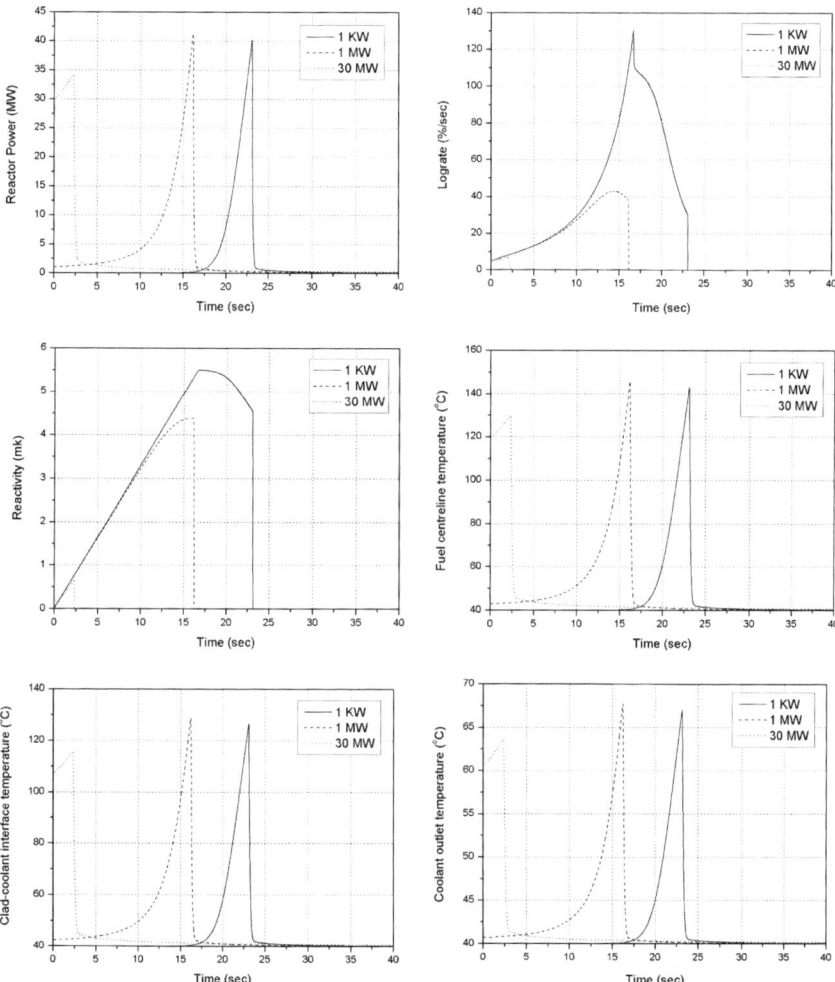

Fig. 3 Results of LORI due to withdrawal of FCR

References

1. International Atomic Energy Agency, *Safety of Research Reactors—Safety Requirements* (IAEA Safety Standards Series No. SSR-3, 2016)
2. T. Mazumdar, T. Singh, H.P. Gupta, K. Singh, RITAC: reactivity initiated transient analysis code–an overview. Ann. Nucl. Energy **43**, 192–207 (2012)
3. *Preliminary safety analysis report—reactor core physics, high flux research reactor*, Revision-1, USI No. 0111, Reactor Physics and Nuclear Engineering Section (RPNES), Research Reactor Services Division (RRSD), Reactor Group (RG), March, 2016
4. T. Singh, T. Mazumdar, P. Pandey, NEMSQR: A 3-D multi group diffusion theory code based on nodal expansion method for square geometry. Ann. Nucl. Energy **64**, 230–243 (2014)
5. M. Kristof, I. Vojtek, Uncertainty analyses for LB LOCA in VVER-440/213 (2003)

Development of a Computer Code for Reactivity Initiated Transient Analysis of Research Reactor with Pin-Type Fuel

Tanay Mazumdar, Tej Singh, P. V. Varde and S. Bhattacharya

Abstract RITAC (reactivity initiated transient analysis code), a coupled code of point reactor kinetics and thermal hydraulics, was developed for analyzing reactivity initiated transient (RIT) events of research reactors with plate-type fuel. Now, the code has been modified for pin-type fuel also. The code solves the coupled system of equations of neutron kinetics based on point reactor model and equations of thermal hydraulics for pin-type fuel to carry out safety analysis of fuel assembly. Point kinetics equations are solved numerically by piecewise constant approximation (PCA) method. In thermal hydraulics, mass, momentum and energy conservation equations for coolant region and energy conservation equations for pin-type fuel and clad regions are solved by Crank–Nicolson-based finite difference method. For the calculation of heat transfer coefficient from clad to coolant in different boiling regimes, a number of correlations are used in RITAC. Finally, results of the code are validated against the known results of a pin-type-fuel-based research reactor.

Keywords Point kinetics · Transient · Thermal hydraulic · Pin-type fuel

T. Mazumdar · T. Singh (✉)
Reactor Physics and Nuclear Engineering Section, Research Reactor Services Division,
Reactor Group, Bhabha Atomic Research Centre, Mumbai, India
e-mail: t_singh@barc.gov.in

T. Mazumdar
e-mail: tanay@barc.gov.in

P. V. Varde
Reactor Group, Bhabha Atomic Research Centre, Mumbai, India
e-mail: varde@barc.gov.in

S. Bhattacharya
Reactor Projects Group, Bhabha Atomic Research Centre, Mumbai, India
e-mail: subhatt@barc.gov.in

© Springer Nature Singapore Pte Ltd. 2020
P. V. Varde et al. (eds.), *Reliability, Safety and Hazard Assessment for Risk-Based Technologies*, Lecture Notes in Mechanical Engineering,
https://doi.org/10.1007/978-981-13-9008-1_76

1 Introduction

Nuclear reactors are meant for safe and sustained production of energy from nuclear fission reaction. Hence, it is very important to study various safety aspects of reactors which help us to figure out whether reactors will remain safe or become vulnerable under any undesirable reactivity insertion accidents (RIA) like reactor start-up accident, loss of regulation accident (LORA), etc. In order to assess the severity of such accidents, numerous computer codes have been developed so far and this development process is still very much going on. In reactivity initiated transients, positive reactivity insertion in a given manner makes neutron density and subsequently fuel, clad and coolant temperatures increasing with time, and the rise in fuel and coolant temperatures causes reactivity feedback which perturbs reactivity and ultimately the neutron density. So, it is necessary to solve the coupled system of equations of neutron kinetics and thermal hydraulics to carry out safety analysis of fuel assembly. Our work presented in this paper covers the development of a computer code RITAC [1], having two modules—one module for neutron kinetics and other module for thermal hydraulics of pin-type fuel. In kinetics module, point kinetics equations are solved numerically by PCA approximation [2]. In thermal hydraulics module, energy conservation equations for fuel and clad regions and mass, momentum and energy conservation equations for coolant region are solved by finite difference method along with Crank–Nicolson technique, which is a semi-implicit method and provides unconditionally stable solution. In RITAC, a number of correlations of heat transfer coefficient are used to calculate the heat transfer from clad to coolant in different boiling regimes. The code is validated against thermal hydraulic results of Dhruva [3], which is 100 MW heavy water research reactor with pin-type fuel.

2 Method of Calculation

2.1 Point Kinetics Module

The point kinetics equations for g number of delayed neutron precursor groups are as follows [4].

$$\frac{dn(t)}{dt} = \frac{\rho(t) - \beta}{\Lambda} n(t) + \sum_{i=1}^{g} \lambda_i C_i(t) \tag{1}$$

$$\frac{dC_i(t)}{dt} = \frac{\beta_i}{\Lambda} n(t) - \lambda_i C_i(t) \tag{2}$$

where $1 < i < g$, $n(t)$ is neutron density at time t, $C_i(t)$ is density of ith group delayed neutron precursor at time t, $\rho(t)$ is reactivity at time t, β_i is delayed neutron fraction corresponding to ith group precursor, β is total delayed neutron fraction, Λ is prompt neutron generation time and λ_i is decay constant of ith group precursor. In Eqs. 1 and 2, there are $(g + 1)$ coupled first-order ordinary differential equations. All these equations, with the PCA approximation that $\rho(t)$ is constant over a time interval $\Delta t = (t_{i+1} - t_i)$, i.e., $\rho(t) \sim \rho((t_{i+1} + t_i)/2) = \rho_i$, if $t_i \leq t \leq t_{i+1}$, are converted into a first-order linear ordinary differential matrix equation which has the following kind of solution.

$$\overrightarrow{x}(t_{i+1}) = e^{(A+B_i)\Delta t}\overrightarrow{x}(t_i) \tag{3}$$

With the help of matrix diagonalization, Eq. 3 reduces to

$$\overrightarrow{x}(t_{i+1}) = X_i e^{D_i \Delta t} X_i^{-1}\overrightarrow{x}(t_i) \tag{4}$$

where X_i contains eigenvectors of $(A + B_i)$ and $e^{D_i \Delta t}$ is a diagonal matrix whose nonzero elements are $e^{w1\Delta t}, e^{w2\Delta t}, \ldots, e^{wi\Delta t}, \ldots, e^{wg+1\Delta t}$ (w_i s' are eigenvalues of $(A + B_i)$). This solution requires initial conditions $n(0) = n_0$ and $C_i(0) = (\beta_i n_0)/(\lambda_i \Lambda)$. In general, $\rho(t)$ is sum of external reactivity $(\rho_{ext}(t))$, which will be added into the core if reactivity devices, moderator level control, etc. malfunctions and reactivity feedback developed within the core due to change in temperature of fuel, coolant and moderator $(\rho_{feed}(t))$. Step, ramp and sinusoidal are three types of $\rho_{ext}(t)$ currently available in the code.

2.2 Thermal Hydraulics Module

The original version of RITAC, which was developed for plate-type fuel, is modified to simulate reactivity initiated transient in pin-type-fueled reactor. The transient begins as positive reactivity (due to uncontrolled withdrawal of control rod or moderator pump up, etc.) is added into a core, which is assumed to be in equilibrium condition, i.e., the reactor power is constant. This addition of positive reactivity drives the temperatures of all fuel and non-fuel materials to rise. In order to obtain the spatial and temporal distribution of temperature, steady-state calculation is performed at $t = 0$ s and based on the steady-state temperature distribution, temperatures in subsequent time are calculated solving the thermal hydraulics equations. As shown in Fig. 1, the entire pin is divided into a number of axial meshes (on left), and then, each of these axial meshes is further divided into few radial meshes (on right). Temperatures are calculated at the midpoints of all these meshes. In a radial mesh i in fuel, the energy conservation equation is discretized using Crank–Nicolson technique [5] to get temperature of ith mesh at time $t + \Delta t$

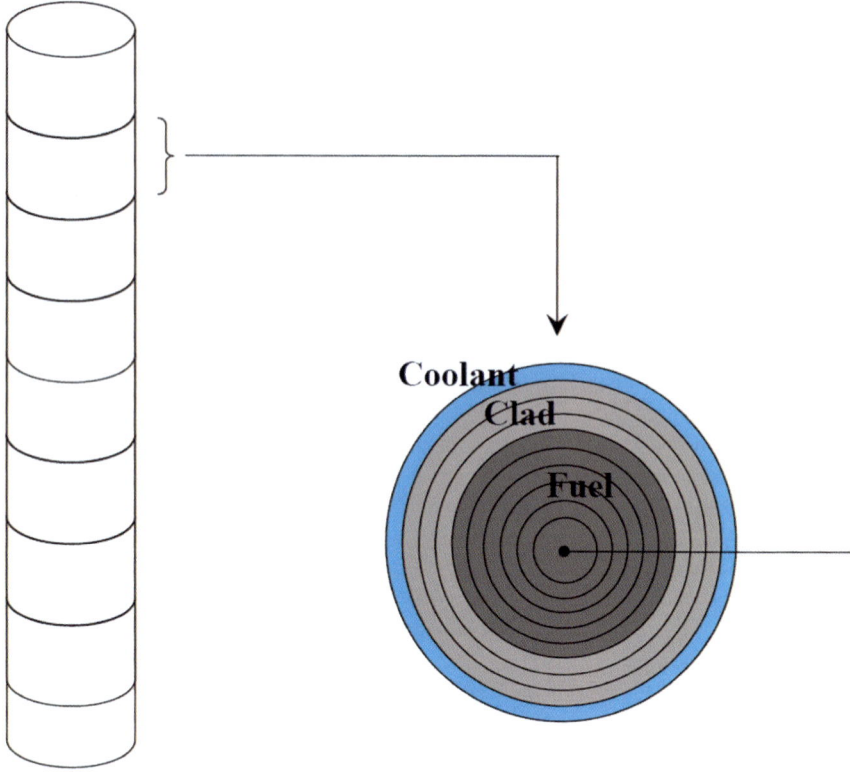

Fig. 1 Axial and radial mesh divisions in pin-type-fuel geometry

$$T_i(t + \Delta t) = F_1 \delta_{i-1} T_{i-1}(t + \Delta t) - F_1(\delta_{i-1} + \delta_{i+1}) T_i(t + \Delta t)$$
$$+ F_1 \delta_{i+1} T_{i+1}(t + \Delta t) + F_1 \delta_{i-1} T_{i-1}(t) + (1 - F_1(\delta_{i-1} + \delta_{i+1})) T_i(t)$$
$$+ F_1 \delta_{i+1} T_{i+1}(t) + \frac{Q(t)\Delta t}{2D_f C_f} + \frac{Q(t + \Delta t)\Delta t}{2D_f C_f} \tag{5}$$

where $F_1 = (K_f \Delta t)/(2D_f C_f \Delta r_f)$, $\Delta r_f = d_f/(2n_f)$, $\delta_{i-1} = [r_i \ln(r_i/r_{i-1})]^{-1}$, $\delta_{i+1} = [r_i \ln(r_{i+1}/r_i)]^{-1}$, K_f, D_f and C_f are thermal conductivity, density and specific heat of fuel, d_f is fuel pin diameter, n_f is total number of radial meshes considered in fuel region, Δr_f is thickness of a radial mesh in fuel, r_i is distance of ith mesh from center of the fuel pin and Q is heat generation rate per unit volume, which is proportional to $n(t)$ of Eq. 1 and 2. At fuel centerline, it is assumed that the temperature distribution will be symmetric, which acts as a boundary condition. In a radial section i in clad region, the energy conservation equation is discretized in a similar fashion to get

$$T_i(t + \Delta t) = F_3 \delta_{i-1} T_{i-1}(t + \Delta t) - F_3(\delta_{i-1} + \delta_{i+1}) T_i(t + \Delta t) + F_3 \delta_{i+1} T_{i+1}(t + \Delta t)$$

$$+ F_3\delta_{i-1}T_{i-1}(t) + (1 - F_3(\delta_{i-1} + \delta_{i+1}))T_i(t) + F_3\delta_{i+1}T_{i+1}(t) \qquad (6)$$

where $F_3 = (K_{cl}\Delta t)/(2D_{cl}C_{cl}\Delta r_{cl})$ and $\Delta r_{cl} = d_{cl}/n_{cl}$, K_{cl}, D_{cl} and C_{cl} are thermal conductivity, density and specific heat of clad, d_{cl} is thickness of clad, n_{cl} is total number of radial meshes considered in clad region and Δr_{cl} is thickness of a radial mesh in clad. The energy conservation equation is also discretized in coolant region to obtain

$$
\begin{aligned}
T_{n_f+n_{cl}+1}(t + \Delta t) = {} & h F_{51} T_{n_f+n_{cl}}(t + \Delta t) - (h F_{51} + F_{52})T_{n_f+n_{cl}+1}(t + \Delta t) \\
& + h F_{51} T_{n_f+n_{cl}}(t) + (1 - (h F_{51} + F_{52}))T_{n_f+n_{cl}+1}(t) \\
& + F_{52}(T_{in}(t + \Delta t) + T_{in}(t)) \qquad (7)
\end{aligned}
$$

where $F_{51} = \Delta t/[2D_m C_m \Delta r_c\{1 + (\Delta r_c/(d_f + 2d_{cl} + \Delta r_{cl}))\}]$, $F_{52} = v\Delta t/\Delta z$. During transient caused by power surge, air bubble may be formed in coolant depending on the severity of reactivity insertion. Hence, a homogeneous mixture of liquid and gas phase of coolant is considered while calculating the coolant temperature. D_m and C_m are density and specific heat of coolant where subscript "m" is used for homogeneous mixture of liquid and gas phase of coolant. Δr_c is thickness of the only radial mesh considered in coolant, h is coefficient of heat transfer from clad to coolant, v is coolant velocity and Δz is height of axial mesh. Equations 5–7 give temperatures of all radial meshes at height z. All these equations are clubbed together to form the following matrix equation.

$$\vec{T}(t + \Delta t) = A\vec{T}(t + \Delta t) + B\vec{T}(t) + \vec{C}(t) + \vec{C}(t + \Delta t) \qquad (8)$$

where $T(t)$ contains temperatures of all radial meshes at a height z, A is a tridiagonal matrix which contains thermophysical parameters and mesh size of fuel, clad and coolant, $B = I + A$, where I is an identity matrix, $C(t)$ contains heat source terms. Apart from Eq. 8, mass and momentum conservation equations are solved for coolant region. Coolant mass conservation equation is given as [6]

$$\frac{\partial D_m}{\partial t} + \frac{\partial G_m}{\partial z} = 0 \qquad (9)$$

where G_m is coolant mass flux. It is clear from the above equation that variation of coolant mass flux along the height of a fuel pin is influenced by temporal change of coolant density. After calculating the coolant temperature at height z, corresponding coolant density is first calculated, and then based on the rate of change of coolant density at height z, change of coolant mass flux across Δz (i.e., height of an axial mesh) is determined applying the following equation.

$$G_m(z + \Delta z, t + \Delta t) = G_m(z, t + \Delta t) + \frac{\Delta z}{\Delta t}(D_m(z, t) - D_m(z, t + \Delta t)) \qquad (10)$$

Momentum conservation equation of coolant is given below [6].

$$\frac{\partial G_m}{\partial t} + \frac{\partial}{\partial z}\left(\frac{G_m^2}{D_m^+}\right) = -\frac{\partial p}{\partial z} - \frac{f G_m |G_m|}{2 D_H D_m} - D_m g \tag{11}$$

where $D_m^+ = [\{(1-x)^2/((1-\alpha)\,D_l)\} + \{x^2/(\alpha D_g)\}]^{-1}$, p is coolant pressure, f is friction factor, D_H is hydraulic diameter of coolant channel and it is $2(R^2 - Nr^2)/(R - Nr)$ for pin-type geometry where R is radius of coolant channel, N is number of fuel pins per assembly and $r\,[= (d_f/2) + d_{cl}]$ is fuel pin radius and g is acceleration due to gravity. Integrating Eq. 11 over z with a limit from channel inlet $(z = 0)$ to a height z and rearranging the terms gives

$$p(z) = p(0) - D_m g z - \frac{f G_m |G_m|}{2 D_H D_m}\overline{\phi^2} z - \left\{\left(\frac{G_m^2}{D_m^+}\right)_z - \left(\frac{G_m^2}{D_m^+}\right)_0\right\} - \frac{\partial}{\partial t}\overline{G_m} z$$

$$= p_{in} - \Delta p_{grav} - \Delta p_{fric} - \Delta p_{acc} - \Delta p_{tra} \tag{12}$$

where $p(0)$ is inlet coolant pressure (p_{in}), Δp_{grav} is gravitational pressure drop, Δp_{fric} is frictional pressure drop, $\overline{\phi^2}$ is homogeneous multiplication factor $= ((D_l/D_g) - 1)\,x + 1$, subscript "l" and "g" stand for liquid and gas phases of coolant and x is steam quality. Friction factor f is calculated based on Reynolds number (Re).

$$f = \begin{cases} 64/Re, & Re < 2100 \\ 0.316\,Re^{-0.25}, & 2100 < Re < 3 \times 10^4 \\ 0.184\,Re^{-0.2}, & 3 \times 10^4 < Re < 10^6 \end{cases} \tag{13}$$

Δp_{acc} is pressure drop due to acceleration, Δp_{tra} is transient acceleration pressure drop and $\overline{G_m} = (1/z)\int_0^z G_m dz'$. Considering the fact that pressure drop may occur due to abrupt change in geometry, e.g., spacer, etc., an extra term, written as $\Delta p_{form} = \sum_i \left(\overline{\phi^2}\,K_i\,\frac{G_m |G_m|}{2 D_l}\right)_i$, where K_i is single-phase pressure loss coefficient and D_l is density of liquid phase of coolant and is added on right-hand side of Eq. 12. Therefore, total pressure exerted by coolant at a height z is

$$p(z) = p_{in} - p_{grav} - p_{fric} - p_{acc} - p_{tra} - p_{form} \tag{14}$$

Total coolant pressure, calculated using Eq. 14, is used to find out the saturation temperature of coolant and subsequently identify the boiling regime of coolant which is the prerequisite of finding out the heat transfer coefficient. Size of mesh depends on the error introduced due to conversion of the differential equation into the difference equation. The error is proportional to the mesh size to the power of the order of the differential equation. At the beginning of the simulation, temperatures of all radial meshes at coolant inlet are calculated. Then, we move to the next elevation to calculate the temperatures of all radial meshes and so on till we reach the coolant outlet. It is assumed that the coolant at outlet takes larger time to reach the inlet via heat exchanger as compared to the residence time of the transient. Hence, inlet temperature is kept

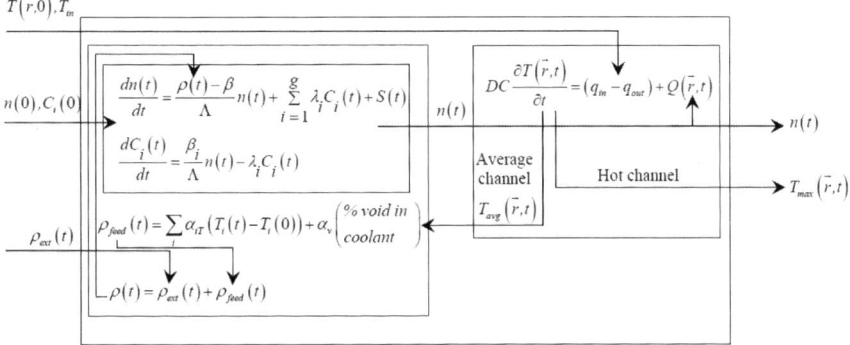

Fig. 2 Block diagram of RITAC

constant throughout the calculation, which serves as another boundary condition. Two-channel analysis is considered to be sufficient for research reactors, which are, in general, small in size. In the analysis, temperatures are calculated for two channels. One is hottest channel to find out the maximum fuel, clad, coolant temperatures and another is average powered channel to find out the average fuel, clad, coolant temperatures. After calculating the fuel and coolant temperatures in all meshes of the average powered channel, temperature feedback of reactivity is calculated by weighting average fuel and coolant temperatures at each elevation with the square of corresponding fuel mesh power. This feedback modifies the net reactivity $\rho(t)$ in Eqs. 1 and 2 and hence the power, which is proportional to $n(t)$. The coupling between the point kinetics and the thermal hydraulics module of the code is shown in Fig. 2.

Coefficient of heat transfer from clad to coolant in single-phase and nucleate boiling regimes are calculated using a number of correlations. For single-phase liquid regime with forced convection, Dittus-Boelter correlation is used for turbulent flow (i.e., Reynolds number $(Re) > 10^4$).

$$h_{FC} = 0.023 \frac{K_1}{D_H} (Re)^{0.8} (Pr)^{0.4} \tag{15}$$

where K_1 is thermal conductivity of coolant (W/m/K), D_H is hydraulic diameter (m), Re is Reynolds number of coolant, Pr is Prandtl number of coolant. If the flow is laminar (i.e., $Re < 2100$), Roshenow and Choi correlation is used.

$$h_L = 4.0 \frac{K_1}{D_H} \tag{16}$$

In the code, max(h_{FC}, h_L) is used as single-phase heat transfer coefficient as it maintains continuity between two flow regimes. Nucleate boiling starts in coolant

Table 1 Results of steady-state calculations

Channel power (KW)	Channel flow (lpm)	Computer codes used	Max fuel temp (°C)	Max clad–coolant interface temp (°C)	Max fuel–clad interface temp (°C)
1125	480	RITAC	371.8	122.9	138.0
		COBRA	377.3	127.0	138.4
1294	375	RITAC	435.6	149.1	166.4
		COBRA	434.7	159.5	172.6

when wall temperature reaches onset of nucleate boiling temperature (T_{ONB}) which is calculated using Bergles and Rohsenow correlation as given by

$$T_{\text{ONB}} = T_{\text{sat}} + 0.556\left(\frac{\Phi}{1082 p^{1.156}}\right)^{0.463 p^{0.0234}} \tag{17}$$

where Φ is heat flux (W/m^2) and p is coolant pressure (bar). For nucleate boiling, Chen correlation, as given below, is used.

$$q'' = h(T_w - T_c) = h_{NB}(T_w - T_{\text{sat}}) + h_{FC}(T_w - T_c)$$
$$\Rightarrow h = h_{NB}\frac{T_w - T_{\text{sat}}}{T_w - T_c} + h_{FC} \tag{18}$$

3 Results and Discussion

Dhruva is a heavy water cooled, moderated and reflected 100 MWth vertical tank-type, pin-type-fueled research reactor. Fuel assembly consists of seven natural uranium metallic pins cladded with aluminum. Diameter of the fuel pin is 12.0 mm and thickness of the aluminum clad is 1.35 mm. Steady-state calculations for two cases, viz. nominal fuel channel power (1125 KW)—nominal coolant flow (480 lpm), and safety limit power of fuel channel (1294 KW)—trip set coolant flow (375 lpm), are carried out using RITAC and COBRA [7]. Results are in good agreement and are given in Table 1. Under prediction of RITAC with respect to COBRA may be due to differences in the underlying assumptions taken.

For validation of transient calculation, loss of regulation incident (LORI) is simulated for Dhruva, where reactivity control for reactor start-up and power regulation is achieved by controlling moderator level in the reactor vessel (RV) via constant inflow and variable outflow principle. This control function is driven by the reactor regulating system (RRS) which has three independent channels working on 2/3 coincidence logic. The regulating system generates absolute trips on high log rate

(6/6/6%/s), overpower (110/110/110 MW) and power more than demand (10/10/10% above the desired power level).

If the regulation system fails, then reactivity will be added into the system at the maximum design rate by continuous and uncontrolled moderator pump up in RV. Under most conservative scenario, all three control valves will be fully closed and all three level control pumps will pump moderator with its rated capacity of 2070 lpm. This will cause maximum reactivity addition rate of ~0.25 mk/s (at 200 cm moderator height). It is worthy to mention here that reactor cannot be operated on automatic control below 1 KW reactor power. Hence, the power level at the onset of LORI is to be kept $\geq$ 1 KW. In the present analysis, LORI is initiated from 1 KW power, which is considered to be the most severe scenario. For this power level, initial critical moderator height is 200 cm and coolant inlet temperature is 35 °C. Due to uncontrolled reactivity addition, as discussed above, reactor power increases rapidly with time and there are trips from regulation and protection system, which are capable to arrest this power surge by tripping the reactor. In order to make conservative estimate, no credit has been given to regulation system trips and the LORI transient is terminated only by the second available trip (on overpower) from protection system ignoring the first one (on high log rate) from the same. Termination of the transient has been assumed to be achieved by the fast insertion of only seven out of the nine shutoff rods assuming two strongest shutoff rods being unavailable for shutdown action (total worth of the 7 rods = 60 mk). No credit has been given to moderator dump as it is much slower compared to the former. The sluggish profile (M-91) of shutoff rod with an actuation delay of 0.5 s has been used so as to achieve the worst accident scenario. Effective delayed neutron fraction (β) is 0.00804 because of 6 groups of delayed neutrons and 8 groups of photoneutrons. Computer codes RITAC and SECMOD [8] have been used for carrying out the analysis. Only fuel temperature coefficient (~ −0.013 mk/°C) is considered in order to calculate the reactivity feedback due to Doppler effect in fuel. In the analysis, a combination of nominal fuel channel power to coolant, which is 1125 KW, and trip set coolant flow, which is 375 lpm, is considered. In Table 2, results are compared with that of SECMOD and it shows good agreement. Thus, safety of the fuel assembly is ensured from the results of the analysis as the fuel temperature, even with conservative approximations, remains well below the temperature required for alpha-to-beta phase transition of uranium metal fuel (i.e., 668 °C).

4 Conclusion

For a practical problem like LORI analysis in Dhruva, good agreement is observed between the results obtained from RITAC and other internationally recognized codes. Accuracy of the code depends on how wisely the spatial mesh size and time steps are chosen. The choice depends on the kind of temperature or reactivity variation is expected across the mesh or time step. There are scopes of work in future on analysis of self-limited transient, loss of flow accident, etc. Two-channel analysis, mentioned

Table 2 Results of LORI initiated from 1 KW reactor power with existing assembly power 1125 KW and coolant channel flow 375 lpm

Parameters (unit)	Value	
	SECMOD	RITAC
Trip actuated at (s)	32.8	32.4
Peak reactor power (MW)	146	136
Maximum reactivity inserted (mk)	7	7.05
Maximum log rate (%/s)	94	99
Maximum fuel temperature (°C)	584	614
Maximum clad temperature (°C)	223	203[a]/174[b]
Maximum coolant temperature (°C)	97	94
Total energy released in 100 s (MJ)	461	457

[a]Fuel–clad interface temperature
[b]Clad–coolant interface temperature

in the paper, is adequate for small-size or strongly coupled reactors. However, for large or loosely coupled reactors, space-time kinetics equations need to be solved in a coupled manner with multichannel analysis in thermal hydraulics.

Acknowledgements Authors would like to thank Dr. H. P. Gupta, ThPD (now retired), Shri Abhishek Tripathi, LWRD, and Shri V. K. Kotak, RRDPD, for spending their valuable time with fruitful suggestions.

References

1. T. Mazumdar, T. Singh, H.P. Gupta, K. Singh, RITAC: reactivity initiated transient analysis code —an overview. Ann. Nucl. Energy **43**, 192–207 (2012)
2. M. Kinard, E.J. Allen, Efficient numerical solution of the point kinetics equations in nuclear reactor dynamics. Ann. Nucl. Energy **31**, 1039–1051 (2004)
3. S.K. Agarwal, C.G. Karhadkar, A.K. Zope, K. Singh, Dhruva: main design features, operational experience and utilization. Nucl. Eng. Des. **236**(7–8), 747–757 (2006)
4. D.L. Hetrick, *Dynamics of Nuclear Reactors* (The University of Chicago Press, Chicago, 1971)
5. J. Crank, P. Nicolson, A practical method for numerical evaluation of solutions of partial differential equations of the heat conduction type. Proc. Camb. Phil. Soc. **43**, 50–67 (1947)
6. N.E. Todreas, M.S. Kazimi, *Nuclear Systems Volume I: Thermal Hydraulic Fundamentals*, second ed. (Taylor and Francis, Pennsylvania, 1993)
7. C.L. Wheeler, C.W. Stewart, R.J. Cena, D.S. Rowe, A.M. Sutey, *COBRA-IV-I: An interim version of COBRA for thermal-hydraulic analysis of rod bundle nuclear fuel elements and cores (No. BNWL-1962)* (Battelle Pacific Northwest Labs., Richland, Wash, USA, 1976)
8. SECMOD-A point kinetics-thermal hydraulics coupled computer code based on "*Numerical integration of dynamic nuclear systems equations by optimum integrating factors,*" Ph.D. thesis by Phillip Allen Seeker, Jr., The University of Arizona, (1969)

Reactor Power Variation Due to Tripping of All Three Level Control Pumps

Archana Sharma, P. Y. Bhosale, N. S. Joshi and P. V. Varde

Abstract Dhruva is a 100 MW heavy water moderated and cooled vertical tank-type research reactor. Reactor power is controlled through controlled addition or removal of heavy water in the calandria tank. There are three level control pumps which facilitate pumping of heavy water from dump tank to the calandria. To control the reactor power, the moderator level in the reactor vessel is regulated automatically as demanded by the reactor control system. During the steady state operation of the reactor when the moderator level variation is zero, all the three level control pumps operate at a constant speed and feed the moderator at the corresponding rate and the control valves in the equilibrium position will bleed the moderator at the same rate as the feed rate. Hence, the net inflow/outflow from the dump tank or calandria will be zero, and the system operates at steady equilibrium condition. When there is a demand for the moderator level variation, the opening of the control valves is automatically adjusted and accordingly there will be variation of outflow from the reactor vessel to dump tank. The automatic control of the control valve opening is carried out in proportion to the error signal generated from the reactor power regulating system. The transient introduced due to tripping of any one of the operating LCPs does not affect reactor operation. The regulating system generates the signal to close the control valves suitable to offset reduction in flow due to LCP trip. Similarly, the regulating system copes up by a slight reduction in reactor power even in case of tripping of two LCPs. However, in case of tripping of all the LCPs, the reactor power reduces gradually due to the continuous drop in moderator level and eventually the reactor reaches in shutdown state. The experiment was conducted on Dhruva simulator to establish the above statement. The paper describes the experiment details and discusses the results of the experiment.

A. Sharma (✉) · P. Y. Bhosale · N. S. Joshi · P. V. Varde
Research Reactor Services Division, Bhabha Atomic Research Center, Mumbai, India
e-mail: archphy@barc.gov.in

N. S. Joshi
e-mail: nsjoshi@barc.gov.in

P. V. Varde
e-mail: varde@barc.gov.in

© Springer Nature Singapore Pte Ltd. 2020
P. V. Varde et al. (eds.), *Reliability, Safety and Hazard Assessment for Risk-Based Technologies*, Lecture Notes in Mechanical Engineering, https://doi.org/10.1007/978-981-13-9008-1_77

Keywords Dhruva · Level control pump · Reactivity · Simulator

1 Introduction

Dhruva is natural uranium fueled, heavy water moderated and heavy water cooled thermal research reactor having a thermal power output of 100 MW and a maximum thermal neutron flux of 1.8×10^{14} n/cm 2/s. The reactor is controlled by varying the reactor size by controlling the moderator level in the reactor vessel (controlling the neutron leakage). In Dhruva reactor, there are three level control pumps (three canned motor pumps, 3231-P1, 3231-P2 and 3231-P3) which provide constant inflow of heavy water from dump tank to the calandria. Similarly, there are three control valves connected in parallel whose openings are varied to control the moderator outflow from the reactor vessel. By matching moderator outflow through control valves (CVs) opening to the moderator inflow by moderator level control pumps (LCPs), reactor power is controlled automatically as demanded by the reactor control system. There are three dump valves also, which remain fully closed during normal reactor operation and fly open on receiving reactor trip signal. Whereas, the basic function of the control valves is to control the moderator outflow from the reactor vessel, which results in a change in the moderator level (and therefore a change in core reactivity); the dump valves are on–off type valves required for fast draining of moderator from the calandria, when desired.

During steady state operation of the reactor when the moderator level variation is zero, all the three level control pumps operate at constant speed and feed the moderator at the corresponding rate and the control valves in the equilibrium position will bleed the moderator at the same rate as the feed rate. Hence, the net inflow/outflow from the dump tank or calandria will be zero, and the system operates at steady equilibrium condition.

When there is a demand for the moderator level variation, the opening of the control valves is automatically adjusted and accordingly there will be a variation of outflow from the reactor vessel to dump tank. The automatic control of the control valve opening is carried out in proportion to the error signal generated from the reactor power regulating system. The error signal is the difference between the set power and the actual power. For the complete range of error signal, i.e., ±100%, the control valve position will be controlled with the speed of the level control pumps constant at 2875 rpm and a flow of 700 lpm. The (+) 100% error signal will correspond to maximum reactivity insertion at a rate of 1/3 mk/s, while the (−) 100% error signal corresponds to the maximum reactivity removal rate of 1/3 mk/s. Hence, for (+) 100% error signal, the control valves and dump valves remain fully closed [1]. As the error signal reduces the control valves start opening and attain equilibrium position when the error is ~0% corresponds to equalization of inlet and outlet flow.

The permissive for LCP operation is available only after raising the safety bank. However, the level rise in the reactor vessel is achieved only after reactor start-up permissive is available which enables closing of CV and DV. The rate of level rise is controlled manually from the control console or on AUTO. In manual mode, the opening of the CVs is adjusted to get the desired rate keeping a watch on the log rate signal. In AUTO mode, the error feedback signal controls the rate of level rise in the reactor. The maximum rate of level rise is controlled by the pre-set log rate limit in regulating system. The LCPs take suction from the D2O dump tank. The three pumps operating in parallel deliver a net flow of 2000 lpm. The moderator inlet line has a check valve V-3249 which allows the forward flow from the pumps to reactor vessel; however, it prevents heavy water flowing back from the reactor vessel to dump tank in case of stoppage of level control pumps. The inventory in dump tank is controlled administratively such that even if the entire inventory, up to DT low-level trip, is pumped up the reactor vessel, the level will not cross the maximum pumpable height. There is a trip on level control pumps at high level (365 cms) in reactor vessel and low level (24 cm/4%) in dump tank [1].

The transient introduced due to tripping of any one of the operating LCPs does not affect reactor operation. The regulating system generates the signal to close the control valves suitable to offset reduction in flow due to LCP trip. Similarly, the regulating system copes up by slight reduction in reactor power even in case of tripping of two LCPs. However, in case of tripping of all the LCPs, the reactor power reduces gradually due to continuous drop in moderator level and eventually the reactor reaches in shutdown state. The paper describes the experiment conducted on Dhruva simulator to validate the above statement. Two sets of experiment were planned. In both experiments, all the three LCPs were shut down, but in the first experiment the NP/LP switch was changed to LP as soon as the power reached in KW and in the second experiment NP/LP switch was set at NP throughout the experiment.

2 Experiment-1

2.1 Details of the Experiment

The experiment was conducted at Dhruva simulator. Following steps were carried out:

Step 1   Reactor was operating at 100 MW, with a set power of 100 MW power, moderator height 311 cm and adjuster rods fully IN. Coolant inlet temperature was 480C. Xenon and Iodine loads were 197.6 and 27.23 mk, respectively.

Step 2   All the three LCPs were stopped intentionally, and it was noticed that within 40 s all the CVs were closed and moderator level came down to 305 cm due to some leakage.

Step 3   Reactor power, Xenon load, Iodine load, core reactivity were noted down at regular time intervals.

Step 4   When reactor power reduces to KW level, NP/LP switch is changed to LP.

2.2 *Observations*

As soon as LCPs were shutdown, to maintain the reactor at the desired power level, all the control valves close. During this time, there is a leakage of moderator from CVs and reactor power reduces to 84.5 MW from 100 MW. Reduction in reactor power causes Xenon load to increase with time, which reflected in further reduction in power, because there is no source to combat the Xenon load. Within 60 min, reactor power was reduced to 1 MW. At this time, the control console switch was changed from NP to LP. After 10.2 h, Xenon load reached at a peak value of 96 mk corresponding to reactor power of 8.7 KW. Now the Xenon load started decreasing with time, resulting in an increase in reactor power. The decrease in Xenon load introduces positive reactivity into the system. Figure 1 gives the variation of Xenon load and Iodine load with time after LCPs were shutdown.

When Xenon concentration starts decreasing, it introduces positive reactivity into the system, eventually reactor power increases. Reactor regulation system controls the rise on reactor power and does not allow power to rise above LP, as the set power switch is at LP. Figure 2 gives the variation of reactor power with time.

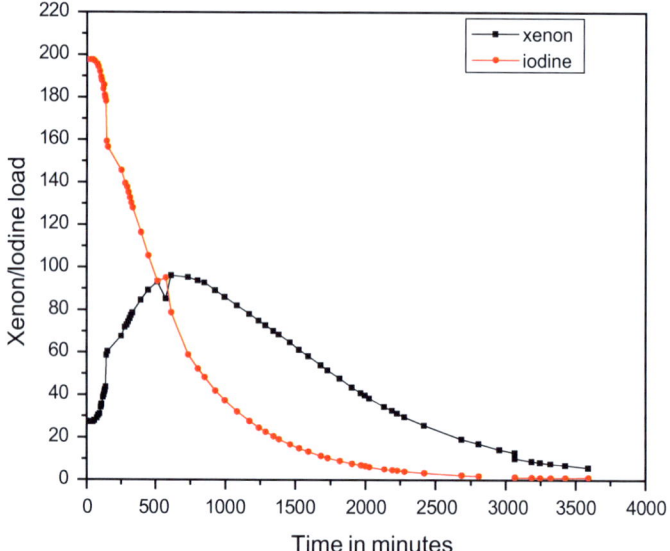

Fig. 1 Variation of Xenon load and Iodine load with time after LCPs were shutdown

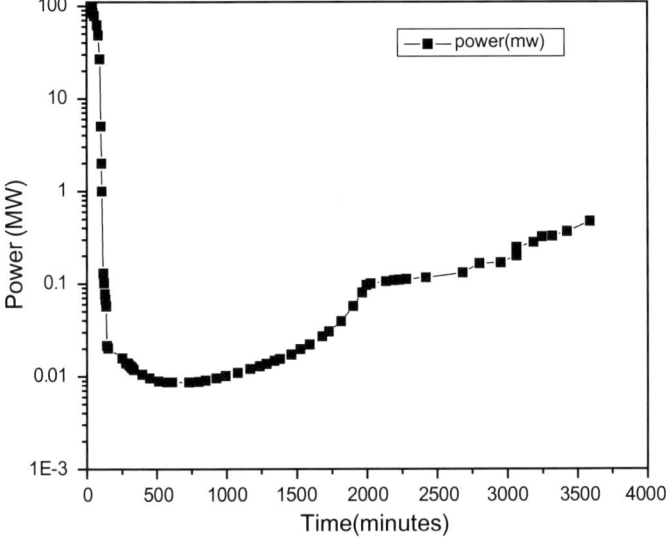

Fig. 2 Change of Reactor power with Time

3 Experiment-2

3.1 Details of the Experiment

The experiment was conducted at Dhruva simulator. Following steps were carried out:

Step 1   Reactor was operating at 100 MW, with a set power of 100 MW power, moderator height 311 cm and adjuster rods fully IN. Coolant inlet temperature was 480C. Xenon and Iodine loads were 197.6 and 27.23 mk, respectively.

Step 2   All the three LCPs were stopped intentionally, and it was noticed that within 40 s all the CVs were closed and moderator level came down to 305 cm due to some leakage.

Step 3   Reactor power, Xenon load, Iodine load, core reactivity were noted down at regular time intervals.

Step 4   When reactor power reduces to KW level, NP/LP switch is kept at NP with no change.

3.2 Observations

Observations were the same as they were in experiment 1. As soon as LCPs were shutdown, to maintain the reactor at the desired power level, all the control valves

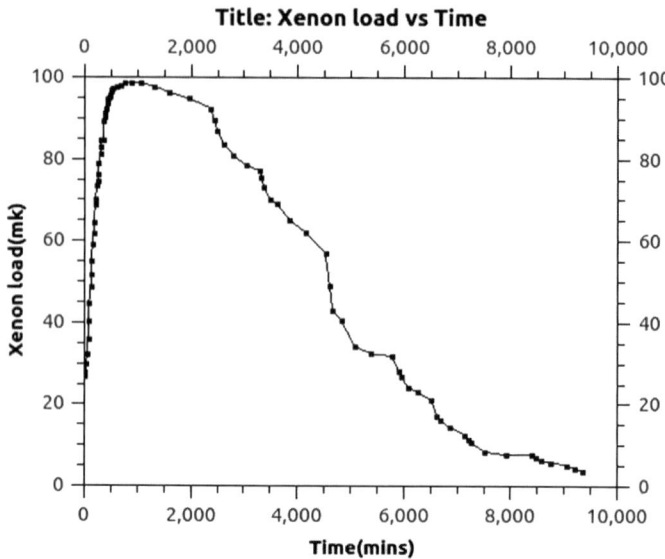

Fig. 3 Variations of Xenon load and Iodine load with time after LCPs were shutdown

close. It takes around 25 s for the control valve to change to fully close condition. During this time, there is a leakage of moderator from CVs and reactor power reduces to 87.2 MW from 100 MW. Reduction in reactor power causes Xenon load to increase with time, which reflected in further reduction in power, because there is no source to combat the Xenon load. Within 60 min, reactor power was reduced to 436 KW. After 14 h, Xenon load reached at a peak value of 97 mk corresponding to reactor power of 8.5 KW. Now the Xenon load started decreasing with time, resulting in an increase in reactor power. The decrease in Xenon load introduces positive reactivity into the system. Because RRS is enabled, so it controls the reactivity addition rate by opening CVs such that the log rate will not exceed the trip value. The reactor power cannot reach set power because some heavy water moderator is drained out of the vessel to control the reactivity addition rate into the system. Figures 3 and 4 give the variation of Xenon load and Iodine load with time after LCPs were shutdown. Figures 5 and 6 give the variation of reactor power and core reactivity with time after LCPs were shutdown.

4   Conclusions

The Dhruva simulator provides efficient hands-on learning and a clear understanding of the physics and engineering design of the reactor. Those experiments can be conducted on Dhruva simulator which we cannot afford to do on real reactor because of safety limitations. In this experiment, we conclude that, in case of tripping of all

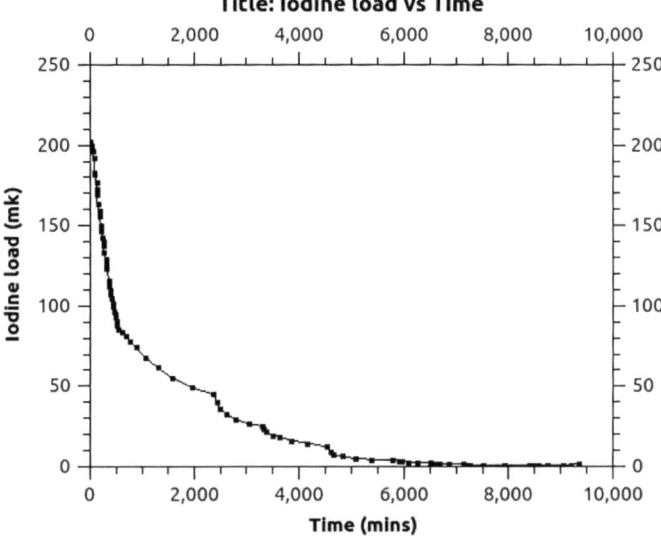

Fig. 4 Variations of Iodine load with time after LCPs were shutdown

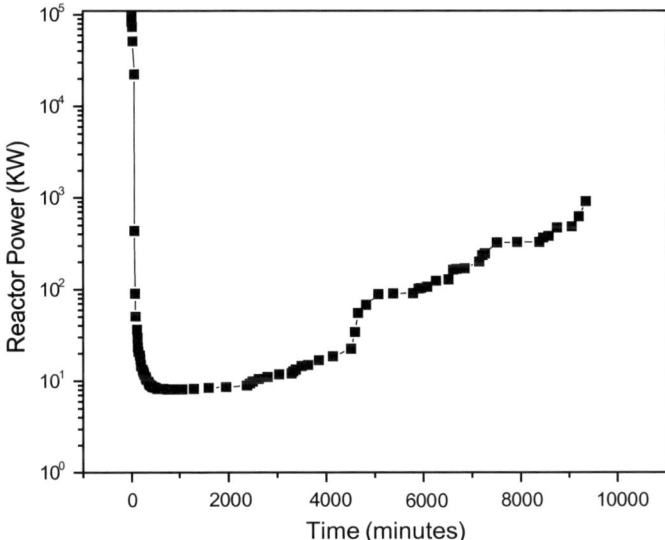

Fig. 5 Change of reactor power with time

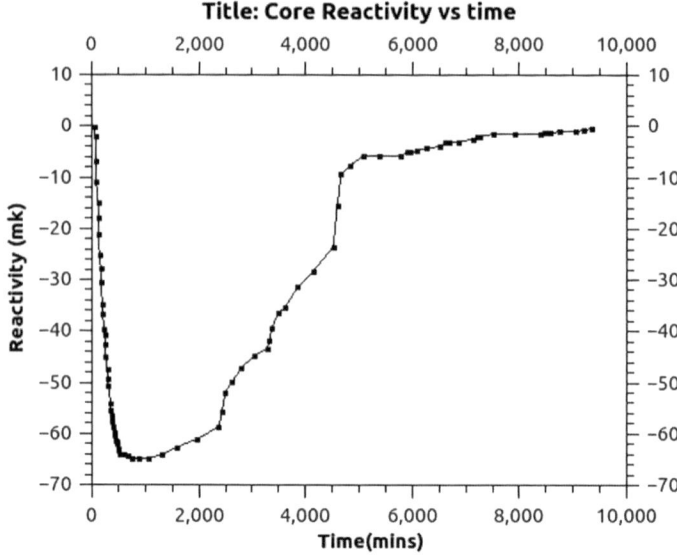

Fig. 6 Change of core reactivity with time

the LCPs while the reactor is not shutdown, the reactor power reduces gradually due to the continuous drop in moderator level and will attain an equilibrium value lower than the set power level.

Acknowledgements Authors are grateful to RRSD and ROD colleagues for their support and technical discussions, due to which this experiment could be successfully completed.

Reference

1. Design Manual for Heavy Water System, *Dhruva Design Manuals*, 27 May 2008

Risk Based

Fuzzy FMEA Analysis of Induction Motor and Overview of Diagnostic Techniques to Reduce Risk of Failure

Swapnil K. Gundewar and Prasad V. Kane

Abstract Induction motors are one of the most popular and widespread electrome-chanical units commissioned in the industry. This paper includes a comprehensive overview of mechanical and electrical failure mode causes for the induction motor-driven unit. The reliability factor plays an important role in the motor-driven system where a failure will affect the huge loss, especially in power production industry. Hence, to enhance the reliability, failure mode and effects analysis (FMEA), fuzzy FMEA is used to analyze the most important factors responsible for the failure. The possibilities of identifying the fault by different techniques at the incipient level are reported in this paper. This analysis would reduce the risk priority number (RPN) as the detectability would be increased by selecting appropriate diagnostic techniques for induction motor-driven system.

Keywords Condition-based maintenance (CBM) · Failure mode and effects analysis (FMEA) · Induction motor (IM) · Risk priority number (RPN)

1 Introduction

Induction motors are at the heart of almost all the rotating machine. The horsepower ratings of a motor vary from 1/12 to 2000 HP depending upon the requirement of a machine. Almost 90% of the industry uses an induction motor as a prime mover; hence, these are kept at the focus of this study. As the induction motor is a critical component in the industry, it is necessary to maintain it in good condition to avoid the breakdown. Failure mode and effects analysis [FMEA] plays an important role in the reliability-centered maintenance [1]. Failure mode and effects analysis is a bottom-up approach used to relate the failures to the root causes. It helps to address

S. K. Gundewar (✉) · P. V. Kane
Department of Mechanical Engineering, VNIT, Nagpur, India
e-mail: swapnilgundewar32@gmail.com

P. V. Kane
e-mail: prasadkane20@gmail.com

© Springer Nature Singapore Pte Ltd. 2020
P. V. Varde et al. (eds.), *Reliability, Safety and Hazard Assessment for Risk-Based Technologies*, Lecture Notes in Mechanical Engineering,
https://doi.org/10.1007/978-981-13-9008-1_78

the issue of identifying failure modes based on the severity, non-detectability and frequency of occurrence. A detailed FMEA and fuzzy FMEA analysis of induction motor is carried out to improve the maintenance policies and avoid the breakdown in this work.

FMEA is a well-known failure analysis technique which is used worldwide for systematic failure reporting and analysis of a system to obtain the risk priority number (RPN). As it is an opinion-based technique, the subjectivity is involved in assigning the scores for the occurrence (O), detection (D) and severity (S) [2]. Therefore, in this work, fuzzy FMEA is also implemented. To obtain fuzzy rules, which are Gaussian in nature, ANFIS technique is used which is a hybrid of artificial neural network (ANN) and fuzzy logic. The inputs to it are the all possible combinations of S, O and D in the conventional FMEA. The results of both techniques are compared, i.e., traditional RPN calculation and RPN by fuzzy logic. The FMEA analysis is done using the data collected from the power plant and other industries which is based on the questionnaire. In this paper, diagnostic techniques for induction motor faults are summarized. Good diagnostic techniques would enhance the detectability so that the RPN would be reduced.

2   Literature Survey

Industries are seriously adopting the concept of reliability-centered maintenance technique to identify the risk of damage for the modification of maintenance policies (International Standard, ISO 14224, 1999). In order to avoid the breakdown condition, every plant needs to implement the predictive maintenance on the periodic basis. Predictive maintenance is an important factor for any industry to avoid the high cost of a breakdown. During the maintenance activity, different types of faults need to be prioritized on the basis of their criticality. FMEA helps to prioritize different fault on the basis of RPN. Christopher [3] described an automated FMEA assistant that promises to reduce significantly the amount of effort that an engineer has to expend in order to produce an FMEA of a designed system while enabling FMEA reports to be produced in a much more timely fashion. In the traditional FMEA approach, each of the parameters like severity, occurrence and non-detectability has equal importance. Sometimes, though the severity is less, the RPN values get increased on the basis of the other two factors like non-detectability and occurrence rate. K XU [4] in 2001 used the fuzzy system for the FMEA analysis of engine system, in which the failure information is described as a fuzzy variable which indicates the more realistic and flexible reflection of a real-life system. A. Azad [5] in 2009 used a fuzzy interface system for a pump failure diagnosis in order to improve the maintenance process. Fuzzy FMEA has the provision to provide weightage to each of the three parameters like severity, occurrence and non-detectability. Each of the terms in the traditional FMEA is explained in the next section.

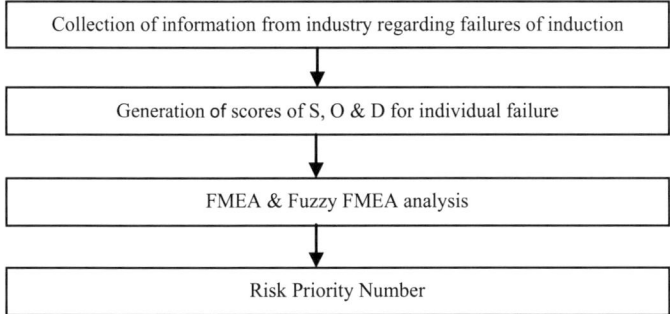

Fig. 1 Methodology for FMEA and fuzzy FMEA

3 Methodology

In an induction motor, the faults are classified as mechanical and electrical. Each type of fault has its own probability of occurrence. Depending on the probability of occurrence, information is collected from industry about the type of failures of an induction motor. In FMEA, each fault is assigned with the rating of S, O and D. The analysis is carried on the basis of scores of S, O and D using FMEA and fuzzy FMEA. In traditional FMEA, RPN is obtained simply by the product of S, O and D, while fuzzy FMEA does the fuzzification and defuzzification to get the RPN. On the basis of the RPN number, ranks are allotted to the faults of an induction motor. The methodology adopted for FMEA analysis is shown in Fig. 1.

4 Failure Mode and Effects Analysis

For the first time, the formal methodology of FMEA was proposed by NASA in 1963 for their reliability requirement. From that time, it is widely used in different types of industry such as nuclear, aerospace and automobile related to safety and reliability analysis of the processes and products [6]. Ford Motor Company in 1977 promoted and adopted the FMEA. The steps involved in calculating the risk priority number are given as follows:

1. Identify the purpose of a system design, process or service. Divide the system into a number of a subsystem such as item, part or assembly and define its objective.
2. Identify the number of ways in which the failure can occur which will give the idea about the failure modes.
3. For each individual failure mode, identify its effects on the system, connected systems, process, connected processes, which indicate the potential effects of failure.

4. Determine the severity of each effect which is known as severity rating in traditional FMEA which is indicated by S. The rating of the severity will vary from 1 to 10 in which 1 will indicate negligible and 10 will indicate catastrophic.
5. Related to each failure mode, identify the root causes and list them in the FMEA table. For each cause define the occurrence rating in the scale of 1 to 10 which is indicated by the symbol O in the traditional FMEA. In the occurrence rating, 1 will indicate remote and 10 will indicate the very high occurrence rate.
6. For non-detectability assign 1 for failure mode, which can be easily detected, while 10 for the failure mode which is difficult to detect.
7. Calculate the risk priority number which is the product of O, S and D. This will guide to rank the problems in order to decide the sequence in which they need to be addressed.

In the traditional FMEA, to calculate RPN, each of the three factors such as severity, occurrence and non- detectability has given equal importance, so sometimes it is possible to have equal RPN in two different conditions. For instance, consider two different conditions having values $S = 5$, $O = 4$ and $D = 3$, and $S = 10$, $O = 3$ and $D = 2$, respectively. Both the conditions will give the same RPN value of 60; however, the severity of both these conditions is different. From this condition, though the second condition has higher severity as compared to the first one, both are showing the same RPN by the traditional FMEA method. Also, in traditional FMEA, equal weightage is given to all the terms like severity, non-detectability and occurrence which is not good for practical applications. Suitable weighting factor needs to be assigned to make the decision more reliable and precise.

Tables 1, 2 and 3 show the scales used for deciding the rating of an occurrence, severity and non-detectability.

Table 1 Occurrence rating scale [7]

Rating	Description	Potential failure rate
10	Certain probability of occurrence	Failure occurs at least once a day, or failure occurs almost every time
9	Failure is almost inevitable	Failure occurs predictably, or failure occurs every 3–4 days
8	Very high probability of occurrence	Failure occurs frequently, or failure occurs about once per week
6–7	Moderately high probability of occurrence	Failure occurs approximately once per month
4–5	Moderate probability of occurrence	Failure occurs occasionally, or failure occurs once every 3 months
2–3	Low probability of occurrence	Failure occurs rarely, or failure occurs about once per year
1	Remote probability of occurrence	Failure almost never occurs; no one remembers the last failure

Table 2 Severity rating scale [7]

Rating	Description	Definition
10	Extremely dangerous	Failure could cause the death of a customer (patient, visitor, employee, staff member, business partner) and/or total system breakdown, without any prior warning
8–9	Very dangerous	Failure could cause a major or permanent injury and/or serious system disruption with interruption in service, with prior warning
7	Dangerous	Failure could cause a minor to moderate injury with a high degree of customer dissatisfaction and/or major system problems requiring major repairs or significant rework
5–6	Moderate danger	Failure could cause a minor injury with some customer dissatisfaction and/or major system problems
3–4	Low to moderate danger	Failure could cause a very minor or no injury but annoys customers and/or results in minor system problems that can be overcome with minor modifications to the system or process
2	Slight danger	Failure could cause no injury and the customer is unaware of the problem; however, the potential for minor injury exists. There is little or no effect on the system
1	No danger	Failure causes no injury and has no impact on the system

Table 3 Detection rating scale [7]

Rating	Description	Definition
10	No chance of detection	There is no known mechanism for detecting the failure
8–9	Very remote/unreliable chance of detection	The failure can be detected only with a thorough inspection, and this is not feasible or cannot be readily performed
7	Remote chance of detection	The error can be detected with a manual inspection, but no process is in place, so that detection left to chance
5–6	Moderate chance of detection	There is a process for double-checks or inspections, but it is not automated and/or is applied only to a sample and/or relies on vigilance
3–4	High chance of detection	There is 100% inspection or review of the process, but it is not automated
2	Very high chance of detection	There is 100% inspection of the process, and it is automated
1	Almost certain chance of detection	There are automatic "shut-offs" or constraints that prevent failure

Table 4 RPN Calculation by traditional FMEA

Component	Potential Failure Mode	Failure Effect	S	Failure Causes	O	Current Control Method	D	RPN
Bearing	High vibration	Overload and heat	8	Insufficient lubricant or incorrect lubricant	5	Weekly inspection	8	320
Stator	Stator fault	Motor inefficiency	8	High temperature, corrosion, contamination	5	Check air gap between rotor and stator	7	280
Shaft and Coupling	Misalignment	Improper installation and corrosion	7	Aging, abd maintenance	2	Misalignment check	5	70
Rotor	Rotor defect	Bearing damage	8	Imbalance, thermal stresses	3	Check air gap between rotor and stator	3	72
Winding	Winding fault	Motor failure	9	Overheat, moisture, insulation breakdown	5	Normal measurement including voltage	8	360
Fan	Fan failure	Overheat	8	Environmental issues, lack of cleaning	1	Visual inspection and cleaning	2	16

In Table 4, different types of failure modes for induction motor are stated, and the rating for those failure modes is specified on the basis of the scale mentioned in Tables 1, 2 and 3. The highest RPN rating is obtained for winding fault, while the lowest RPN is for fan failure. The winding fault is the most critical fault in the induction motor, whereas fan fault is the least critical fault in an induction motor.

5   Fuzzy Failure Mode and Effects Analysis

As reported by various authors [8–10], fuzzy FMEA is a better option for FMEA analysis since it takes into consideration the fuzziness of assigning different scores of S, O and D. While applying the fuzzy FMEA, the most important step is writing fuzzy rule base. To obtain this rule base, ANFIS is used in this work. ANFIS is a hybrid of ANN and fuzzy logic, which is based on the Takagi–Sugeno fuzzy inference system. The input to this is S, O and D, while the output is RPN number. In MATLAB, a Genfis 2 function is used to model a fuzzy inference system (FIS) using the input and output data. Five-layered ANFIS model structure is shown in Fig. 2. The first layer is of inputs assigning the values of S, O and D. In the second layer, the database defines the membership functions of the fuzzy sets used in the fuzzy rules. A decision-making unit which performs the inference operation on the rules is completed in the third layer. The fourth layer transforms a fuzzification inference which converts crisp input degrees of a match with the linguistic variables. A defuzzification inference which transforms the fuzzy results of the inference into crisp input is accomplished in the fifth layer.

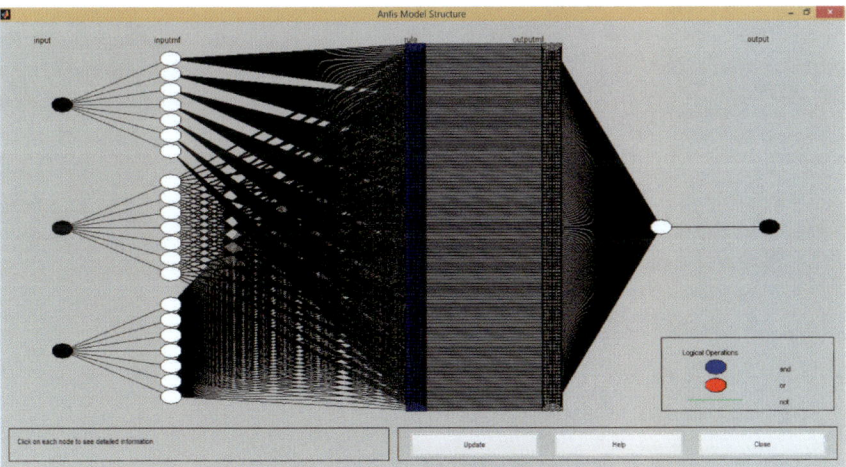

Fig. 2 Five-layer ANFIS model structure in fuzzy

Table 5 Comparison between traditional RPN and ANFIS (ANN and Fuzzy) RPN

Fault	Occurrence	Severity	Non-detectability	Traditional RPN	Rank	Fuzzy RPN	Rank
Bearing	5	8	8	320	2	327	2
Stator	5	8	7	280	3	285	3
Misalignment	2	7	5	70	5	72.3	4
Rotor	3	8	3	72	4	71	5
Winding	5	9	8	360	1	364	1
Fan	1	8	2	16	6	16.2	6

ANFIS info:
Number of nodes: 734
Number of linear parameters: 343
Number of nonlinear parameters: 42
Total number of parameters: 385
Number of training data pairs: 1000
Number of fuzzy rules: 343.

In Table 5, RPN obtained by traditional FMEA and fuzzy FMEA is compared for the different types of faults in an induction motor. On the basis of RPN, these faults are ranked on the scale of 1–6. In traditional FMEA, RPN is obtained by the multiplication of the rating assigned to S, O and D, while fuzzy FMEA used fuzzy rule base generated using ANFIS to obtain RPN. According to traditional FMEA, rotor fault has more RPN than misalignment, while in fuzzy, misalignment fault needs to be attended on a priori basis as compared to rotor fault. As misalignment leads to the generation of fault in other components like bearing, the whole assembly will be affected. Thus, in fuzzy FMEA, RPN calculation is more precise as compared with traditional FMEA. Figure 3 shows the fuzzy rule base. In the rule base, each parameter of RPN is divided into 7 IMF as per Tables 1, 2 and 3. The obtained output is shown in fuzzy RPN box. Total 343 rules are defined in the rule base. Figure 4 shows the RPN calculator using fuzzy logic toolbox. Based on the values of S, O and D, the corresponding fuzzy RPN value is generated. The input parameter values need to be put in the input box separated by space, and the fuzzy RPN value is obtained at the top portion of this calculator.

6 Overview of the Diagnostic Technique to Reduce the Risk of Failure for Induction Motor

To avoid the condition of catastrophic failure, maintenance is needed on periodic basis with the different diagnostic techniques to get the idea about the condition of the equipment under testing. Each diagnostic technique uses a particular parameter

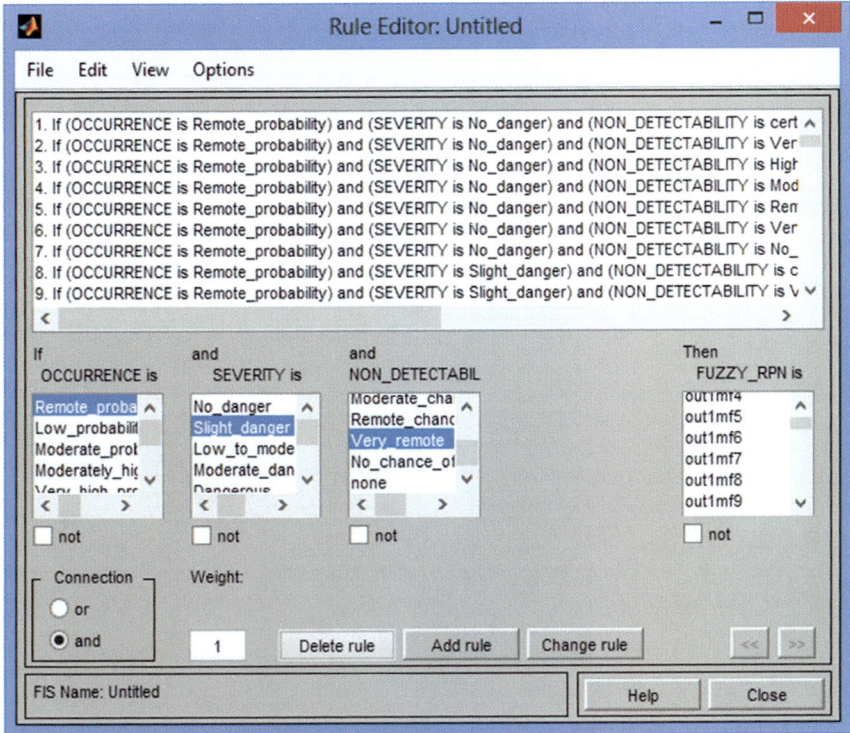

Fig. 3 RPN Calculator using Fuzzy Logic Toolbox

to identify health conditions. The parameters used for the condition monitoring are vibration, acoustics, motor current and temperature depending upon the suitability for a particular condition. These are normally used on rotating equipment, auxiliary systems and other machinery (compressors, pumps, internal combustion engines, presses) [11]. Condition-based maintenance helps to diagnose the fault, predict the fault and improve the machine life and proper management of spares inventory. Depending on the parameter used, the condition monitoring techniques are presented in the following section.

6.1 Vibration Analysis

It is the most convenient and reliable test for a maintenance engineer to get the idea about the condition of a machine. For a healthy condition of an induction motor, the amplitude of vibration will be normal, while in case of any fault, the amplitude of vibration gets increased rapidly or it will show an impulse in signal [12]. In induction motor, bearing faults can be better analyzed using vibration monitoring [13]. In fact,

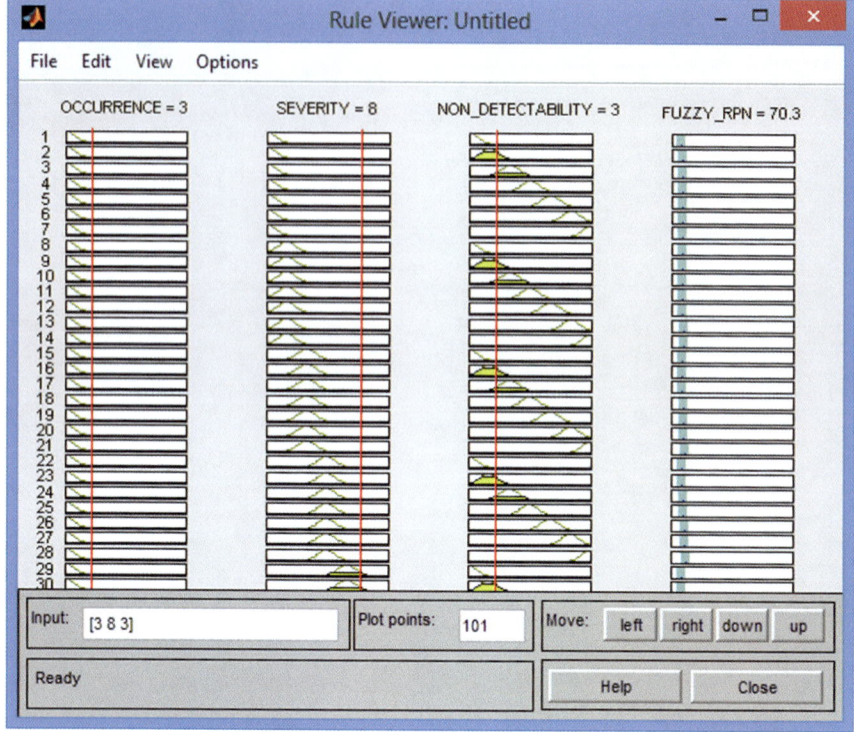

Fig. 4 Rule Base for RPN calculation in Fuzzy Logic Toolbox

it is the best method to identify the bearing fault in induction motor [14]. Using the accelerometer, the radial or axial vibration signals can be captured. Vibration monitoring is also useful for misalignment and rotor unbalance fault [13].

6.2 Acoustic Emission Analysis

In this technique, acoustic signals are used for condition monitoring purpose having a magnitude in the order of 2 MHz or more than that which can be sensed by a photoelectric sensor provided on the surface creating it. Acoustic method is useful for dealing the rotor and bearing faults in induction motor, but its accuracy depends on the surrounding environment [12]. Ultrasonic waves are also useful in case stator bar faults [15]. Doubling the speed of induction motor will result in the rise of noise level up to 12 dB [16]. The limitation to acoustic analysis is of the surrounding environment, which needs to be free from noise.

6.3 Motor Current Signature Analysis [MCSA]

In case of the remote location or an underground machine, it is not possible to mount the accelerometer on a machine, in that case with the current drawn from the component the condition can be identified to generate a power spectrum. In MCSA for an induction motor, stator current is processed to produce an power spectrum profile [17]. In asymmetric rotor bars, the nonzero rotating backward fields are formed because of which harmonics pertaining to stator winding currents are induced [18]. Stator current can be used to identify the number of broken rotor bar in induction motor [19]

6.4 Thermography

Anybody above absolute zero temperature will emit the radiation. The radiations emitted by the body are used to diagnose the condition of a machine whether it is healthy or faulty [20]. In the case of the stator or bearing of an induction motor, it is not convenient to mount the thermocouple over it. In that case, infrared thermal imaging plays an important role. A thermal imaging device is also used to diagnose the bearing fault in induction motor [21].

6.5 Surge Test

Stator winding condition monitoring in induction can be performed using a surge test [22]. In this test, two high-voltage, high-frequency pulses are imposed while grouping is applied to the remaining phase of motor winding [23]. Reflected pulses are used to detect the insulation faults between windings and coils in an induction motor. Surge test is also useful to detect eccentricity problems in rotor [24].

These diagnosis techniques when judiciously selected would identify faults in its early stage. It will help to reduce non-detectability score in FMEA leading to reduce RPN. Hence, a detailed literature review has been carried out to study development in motor fault diagnosis.

7 Conclusion

The critical analysis of different faults of induction motor has been carried out using FMEA and fuzzy FMEA. From the analysis, it can be concluded that electrical faults like stator winding fault have higher RPN compared to a mechanical fault like bearing and misalignment. It is observed from the literature review that vibration-based technique is more important in bearing fault, while MCSA will give better results in case of an electrical fault. Using these diagnostic techniques, the detectability can be improved to reduce the RPN. The focus of this work was on single component (induction motor) failure modes; however, many multiple failure modes are completely or partially diagnosed and all possibilities are available to extend the analysis into multiple failure modes if required. This analysis is the first step to go into the detailed failure analysis and fault diagnosis of an induction motor.

References

1. H. Arabian-Hoseynabadi, H. Oraee, P. Tavner, Failure modes and effects analysis (FMEA) for wind turbines. Int. J. Electr. Power Energy Syst. **32**, 817–824 (2010)
2. K.-S. Chin, A. Chan, J.-B. Yang, Development of a fuzzy FMEA based product design system. Int. J. Adv. Manuf. Technol. **36**, 633–649 (2008)
3. C.J. Price, D.R. Pugh, M.S. Wilson, N. Snooke, The flame system: automating electrical failure mode and effects analysis (FMEA), in *Reliability and Maintainability Symposium, 1995. Proceedings., Annual*, 1995, pp. 90–95
4. K. Xu, L.C. Tang, M. Xie, S.L. Ho, M. Zhu, Fuzzy assessment of FMEA for engine systems. Reliab. Eng. Syst. Saf. **75**, 17–29 (2002)
5. A. Azadeh, V. Ebrahimipour, P. Bavar, A fuzzy inference system for pump failure diagnosis to improve maintenance process: The case of a petrochemical industry. Expert Syst. Appl. **37**, 627–639 (2010)
6. B. Splavski, V. Šišljagić, L. Perić, D. Vranković, Z. Ebling, Intracranial infection as a common complication following war missile skull base injury. Injury **31**, 233–237 (2000)
7. M.M. Silva, A.P.H. de Gusmão, T. Poleto, L.C. e Silva, A.P.C.S. Costa, A multidimensional approach to information security risk management using FMEA and fuzzy theory. Int. J. Inf. Manag. **34**, 733–740 (2014)
8. L.-H. Chen, W.-C. Ko, Fuzzy linear programming models for new product design using QFD with FMEA. Appl. Math. Model. **33**, 633–647 (2009)
9. M. Kumru, P.Y. Kumru, Fuzzy FMEA application to improve purchasing process in a public hospital. Appl. Soft Comput. **13**, 721–733 (2013)
10. K. Meng Tay, C. Peng Lim, Fuzzy FMEA with a guided rules reduction system for prioritization of failures. Int. J. Qual. Reliab. Manag. **23**, 1047–1066 (2006)
11. M. Montanari, S.M. Peresada, C. Rossi, A. Tilli, Speed sensorless control of induction motors based on a reduced order adaptive observer. IEEE Trans. Control Syst. Technol. **15**, 1049–1064 (2007)
12. W. Li, C.K. Mechefske, Detection of induction motor faults: a comparison of stator current, vibration and acoustic methods. J. Vib. Control **12**, 165–188 (2006)
13. C. Kral, T. Habetler, R. Harley, F. Pirker, G. Pascoli, H. Oberguggenberger et al., A comparison of rotor fault detection techniques with respect to the assessment of fault severity. in *Diagnostics for Electric Machines, Power Electronics and Drives, 2003. SDEMPED 2003. 4th IEEE International Symposium on*, 2003, pp. 265–270
14. V. Rai, A. Mohanty, Bearing fault diagnosis using FFT of intrinsic mode functions in Hilbert-Huang transform. Mech. Syst. Signal Process. **21**, 2607–2615 (2007)
15. Y.-S. Lee, J. Nelson, H. Scarton, D. Teng, S. Azizi-Ghannad, An acoustic diagnostic technique for use with electric machine insulation. IEEE Trans. Dielectr. Electr. Insul. **1**, 1186–1193 (1994)
16. R. Singal, K. Williams, S. Verma, Vibration behaviour of stators of electrical machines, part II: experimental study. J. Sound Vib. **115**, 13–23 (1987)

17. A. Siddique, G. Yadava, B. Singh, A review of stator fault monitoring techniques of induction motors. IEEE Trans. Energy Convers. **20**, 106–114 (2005)
18. M.R. Mehrjou, N. Mariun, M.H. Marhaban, N. Misron, Rotor fault condition monitoring techniques for squirrelcage induction machine—a review. Mech. Syst. Signal Process. **25**, 2827–2848 (2011)
19. J. Siau, A. Graff, W. Soong, N. Ertugrul, Broken bar detection in induction motors using current and flux spectral analysis. Aust. J. Electr. Electron. Eng. **1**, 171–178 (2004)
20. S. Bagavathiappan, T. Saravanan, N. George, J. Philip, T. Jayakumar, B. Raj, Condition monitoring of exhaustsystem blowers using infrared thermography. Insight-Non-Destructive Test. Condition Monit. **50**, 512–515 (2008)
21. J.J. Seo, H. Yoon, H. Ha, D.P. Hong, W. Kim, Infrared thermographic diagnosis mechnism for fault detection of ball bearing under dynamic loading conditions, in *Advanced materials research*, 2011, pp. 1544–1547
22. J. L. Kohler, J. Sottile, and F. C. Trutt, Condition-based maintenance of electrical machines, in *Industry Applications Conference, 1999. Thirty-Fourth IAS Annual Meeting. Conference Record of the 1999 IEEE*, 1999, pp. 205–211
23. O. Thorsen, M. Dalva, Condition monitoring methods, failure identification and analysis for high voltage motors in petrochemical industry (1997)
24. Y.-R. Hwang, K.-K. Jen, Y.-T. Shen, Application of cepstrum and neural network to bearing fault detection. J. Mech. Sci. Technol. **23**, 2730 (2009)

Risk-Based Analysis of Electro-Explosive Devices Switching System: A Case Study of Aerospace System

Mukesh Kumar, G. Kamalakar and S. Giridhar Rao

Abstract The conventional design approach used to focus on functionality and performance requirements while safety was treated as merely rule compliance. Risk-based design involves satisfying design goals and meeting safety and reliability objectives also at the design stage itself. In an aerospace system, various ground and in-flight events occur in the planned sequence leading to meeting the functional and performance requirements of the mission. Switching circuits are used to control and execute the events by firing the electro-explosive devices (EEDs). The design of the switching circuits must cater for the highest order of safety requirements considering the catastrophic consequences of inadvertent initiation of the EED as it may cause huge potential loss to human lives, resources, environment and reputation of the nation. However, meeting the safety objectives may result in compromising with the reliability, thus increasing the probability of failure. This paper gives a case study of risk-based analysis of various redundancy configurations of switching circuit consisting of electromagnetic relays and relay drivers for the initiation of EED in an aerospace system. The analysis also involves reliability block diagrams (RBDs) modelling to arrive at an optimized design of switching system to meet the dual goals of safety as well as reliability.

Keywords Electromagnetic relay · Redundancy configuration · Relay driver · RBD · Risk-based design · EED · Switching system

M. Kumar (✉) · G. Kamalakar
ASQ Certified Quality Engineer & Scientist-E, Advanced Systems Laboratory, DRDO, Hyderabad, Telangana 500058, India
e-mail: mukeshkumar@asl.drdo.in

G. Kamalakar
e-mail: kamalakarg@asl.drdo.in

S. Giridhar Rao
Scientist-G, Advanced Systems Laboratory, DRDO, Hyderabad, Telangana 500058, India
e-mail: sgiridhar@asl.drdo.in

© Springer Nature Singapore Pte Ltd. 2020
P. V. Varde et al. (eds.), *Reliability, Safety and Hazard Assessment for Risk-Based Technologies*, Lecture Notes in Mechanical Engineering,
https://doi.org/10.1007/978-981-13-9008-1_79

941

1 Introduction

There has been a vast change in design approach with time. Before World War II, the market was content with mere quality aspects. But, later reliability took the centre stage in product development and product's warranty clauses were driven by the product's reliability figures. But, today mere quality, reliability and warranty are no longer able to satisfy the consumers. The durability, robustness, fault tolerance, fail safety, etc., are the focus now other than the quality and reliability. In case of aerospace system, the safety aspects play a crucial role and hence must be considered at the design level itself. Risk and safety are related to each other: the higher the risk, the lower the safety. Risk assessment is also referred as safety assessment with practically no difference in engineering applications [1]. With the fast pace advancement in technology, increasing complexity and miniaturization the conventional design approach is not adequate today particularly in case of aerospace applications. Hence, the erstwhile design approach is being replaced by risk-based design approach. The old approach is called rule-based design as safety is treated as just rule compliance and hence was a passive approach as far as the safety issues were concerned. The design engineers, therefore, catered for functionality, and performance and mere compliance to safety rules. However, in the new trend called risk-based design, safety too is treated as major design objective on par with functionality and performance goals and all these three must be considered and taken care of at the design stage itself. In fact, the risk-based design needs to put additional efforts as the design output must cater for reliability, safety, durability, robustness, fault tolerance, etc. This paper explains with a case study the risk-based analysis of EED switching system for an aerospace application. In particular, this paper also carries out detailed analysis of various redundancy configurations of the switching circuit consisting of electromagnetic relays and relay drivers. The analysis also involves reliability block diagrams (RBD) modelling and reliability calculation to finally arrive at an optimized design of switching system to meet the goals of highest order of safety as well as reliability.

2 EED and EED Switching Scheme

2.1 EED

An electro-explosive device (EED) also called a pyrotechnic device is a device which is activated by the application of electrical energy and then initiates an explosive, burning, electrical or mechanical train. EEDs are used to provide highly reliable critical functions. The basic mechanism of an EED is that it converts electrical energy into chemical energy providing a gas or heat output.

EEDs are used to accomplish different functions such as ignition, initiation, release, severance, fracture, cutting, jettisoning, switching, time delay and actuation.

Fig. 1 Basic EED switching
scheme

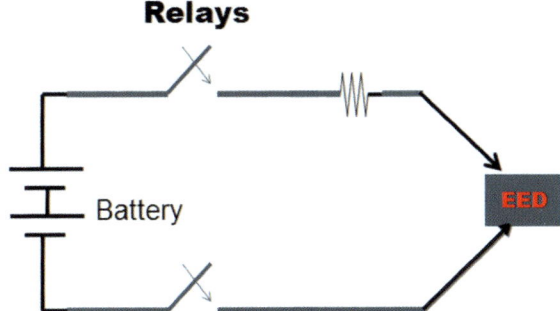

EEDs have got its applications in many areas, viz. aircraft, oil rigs, civil construction, defence and satellite applications. In applications related to defence and satellite launch vehicles, the EEDs are used for activation of various critical events such as battery activation, piston actuation, solenoid valve activation, pyro bolt separation, propulsion system ignition, stage separation, retro motor firing, thrust termination for flight safety, skin separation, satellite injection into orbits, etc.

2.2 Basic EED Switching Scheme

In an aerospace system, various ground and in-flight events occur in the planned sequence leading to meeting the functional and performance requirements of the mission. Switching circuits are used to control and execute the events by firing the electro-explosive devices (EEDs). An EED firing or switching circuit consists mainly of an EED device and an electrical relay which switches the battery power to the EED (Fig. 1).

The relays are operated with the help of relay drivers (Fig. 2). These drivers are basically transistors, FET or MOSFET devices and switch the relays ON or OFF according to the digital control signals or commands received from a controller. The controller generally is microcontroller or FPGA-based system. In aerospace system mostly Electromagnetic relays are used due to its various crucial features such as multipole and multi-throw contacts, no Off-state leakage current, very low contact voltage drop, etc. Hence, electromagnetic relays have an edge over solid state relays and are the first choice in critical applications. However, SSRs are also used in few cases but mainly for redundancy purposes. Figure 3 illustrates a typical circuit for 2PDT non-latch relay and driver. Figures 4 and 5 illustrate typical electromagnetic relay and MOSFET-based driver microcircuit.

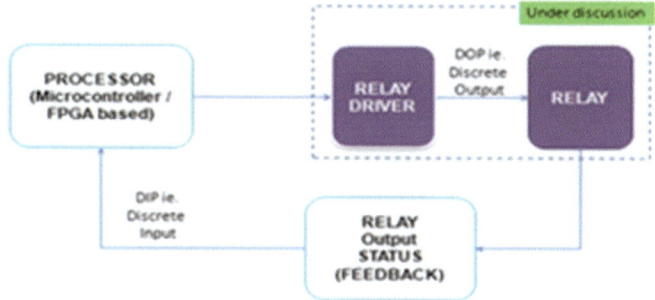

Fig. 2 Block Diagram of EED switching circuit

Fig. 3 A typical circuit
illustration for 2PDT
non-latch relay and driver

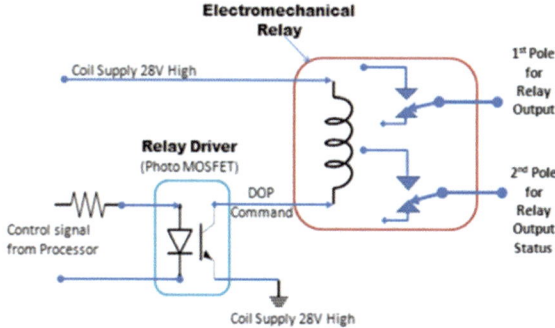

Fig. 4 A typical EMR

Fig. 5 A typical photograph
of MOSFET

2.3 Risk-Based Analysis for EED Switching

The design of the EED switching circuits must cater for the highest order of reliability and safety requirements considering the catastrophic consequences of failure in operation of the switching system resulting in EED initiation failure, as it may cause huge potential loss to human lives, resources, environment and reputation of the nation. Table 1 shows two instances related to EED initiated events of two aerospace missions of NASA, USA. In one case, the payload fairing separation failure resulted in the mission failure (Orbiting Carbon Observatory satellite launch mission aboard the Taurus-XL launch vehicle in 2009), and in the second case, remotely operated EED was fired by the Range Safety Officer from the ground station to destroy the rocket to avoid its dangerous fall back to the earth after a mid-flight fault caused the rocket's first stage engine to prematurely shut down (GOES-G satellite aboard Delta-3914 rocket in 1986). These two are two classical examples which show two different and complementary aspects or use of EED. In the former case, the EED was used to prevent the system become unsafe after one subsystem's failure. In the later case, as an EED-based separation system did not function and resulted in the mission failure.

It may be noted that in a mission critical aerospace system, both reliability and safety are important. They are closely related but have different meanings [2]. If a system failure causes loss of property, injury or loss of life, it is a safety problem. On the other hand, if a system failure results in premature loss of the system say a rocket, or causes the mission to fail, it is a reliability problem. However, common denominator of the two can be unexpected failures. The goal of risk-based analysis is to ensure that the design must meet the highest order of both safety and reliability. Meeting or improving, one should not be at the cost of the other. Extra efforts may be required if one aspect seems to be getting compromised while improving the other one. In fact, this paper's case study is a typical example of such a situation as will be explained in further sections.

2.4 Redundancy

Redundancy is the duplication of critical components or functions of a system with the intention of increasing reliability of the system, usually in the form of a backup or fail safe, or to improve actual system performance. Figure 6 illustrates an example of redundancy in subsystem B as three such subsystems are configured in parallel in the overall system consisting of the five subsystems. The application of redundancy is not without penalties. It will increase weight, space requirements, complexity, cost and time to design. The increase in complexity results in an increase in unscheduled maintenance. Thus, safety and mission reliability is gained at the expense of adding an item(s) in the unscheduled maintenance chain [3]. Redundancy, duplication or replication in association with (lesser) complexity and excess strength are the three

Table 1 Mission failures related to EED initiated events

Year	Launch vehicle	Satellite	Mission purpose	Mission failure details
03 May 1986	Delta-3914	GOES-G	A weather satellite for (National Oceanic & Atmospheric Administration)	Only 71 s after lift-off, an electrical fault (due to lightning) caused rocket's 1st stage engine to shut down prematurely NASA then destroyed the rocket 20 s later using remotely operated EED to avoid having it falls dangerously back on to the earth
2009 February 24	Taurus-XL	Orbiting Carbon Observatory (OCO)	A NASA satellite mission designed to map sinks and sources of carbon in earth's atmosphere, and for understanding of carbon cycle	Failed to reach orbit after launch as the Payload fairing (a protective clamshell covering the satellite) of the rocket failed to separate The extra weight prevented the satellite from reaching orbit

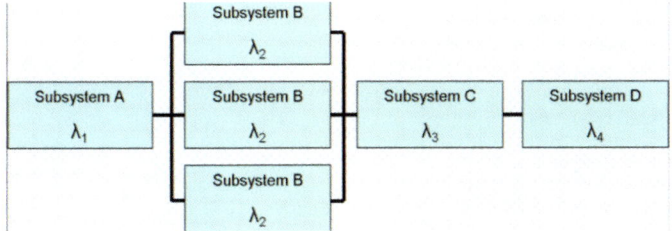

Fig. 6 An illustration of redundancy: redundancy in subsystem B

crucial factors which are at the core of design or construction which have secured their longevity [4].

3 A Case Study of Risk-Based Analysis of EED Switching Circuit for an Aerospace Application

3.1 Risks Associated with EED Firing System

The first step for a risk-based design is to understand and identify the **potential risks**. There is a large number of safety and reliability-related risks associated with EED firing system. However, they can be categorized under few major headings. These are *sabotage, inadvertent operation, uncertainties, random failures, severe in-flight induced environmental stresses* (such as vibration, stage separation shocks, linear acceleration, dynamic pressure), *inefficiency of switching circuits and fault intolerance* of the system. Most of the mentioned risks are not uncommon to any given system; be it aerospace, industrial or commercial applications. But it is emphasized hereby that though the risks seem to be similar, but the consequences and the severity of consequences of the risks in an aerospace system are of catastrophic nature, and hence, the risk treatment must be of highest priority and must be addressed at the design stage itself. In fact, the risk-based design approach should consider and carry out a comprehensive exercise from risk identification and assessment to planning of appropriate risk mitigation measures using engineering judgment.

3.2 Objectives of Risk-Based Design

After identifying and assessing the risks, it is required to list out the design objectives so that the potential risks could be avoided or mitigated. The following are the major design objectives to avoid the mentioned risks.

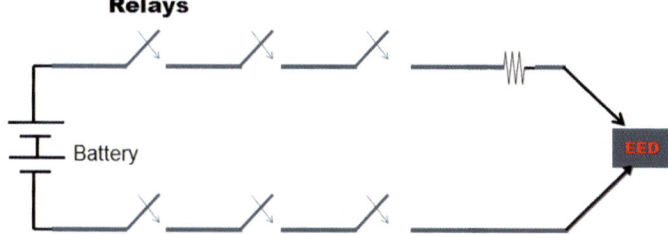

Fig. 7 Use of multiple breaks (relay units) in EED firing circuit

a. No single-point control: to avoid safety issues caused by inadvertent operation of EED switching system.
b. Multiple breaks instead of single break: to avoid inadvertent operation.
c. Both electrical (relay breaks) and physical breaks: for safety.
d. Distributed architecture with the various relay breaks and relay drivers distributed across multiple sections and packages: Hence, it would not be possible to fire the EED without powering ON a no. of subsystems. This helps to address safety issues caused by sabotage like factors and also to avoid inadvertent operation.
e. Multiple level redundancy: to ensure desired level of reliability and for fault tolerance.
f. Indeterminate states of digital circuit at Power-ON should not result into a fault: to avoid uncertainty related faults and failures. It is achieved by using two drivers with complementary logics to drive a single relay.
g. Selection of preferred and reliable parts, derating and mounting: Should be fault tolerant to take care of uncertainties.
h. No single-point failure: to ensure high system reliability as the system would be tolerant of single failure.
i. Must have mechanism to preventing it to become unsafe: for having a fail-safe system, i.e. to ensure safety even in case there is premature system failure.

3.3 Mechanism Implemented

a. **Multiple point control**: *A no. of relay units and relay driver units* located at different locations for EED-based events execution (refer Fig. 7):
b. **Redundancy** at multiple levels:

- Relays: *No redundancy, dual modular redundancy (DMR), triple modular redundancy (TMR)*
- Drivers: Two or more drivers *steering* a relay
- Driver units or switching chains: *DU(M) & DU(S), i.e. driver units' main and standby*

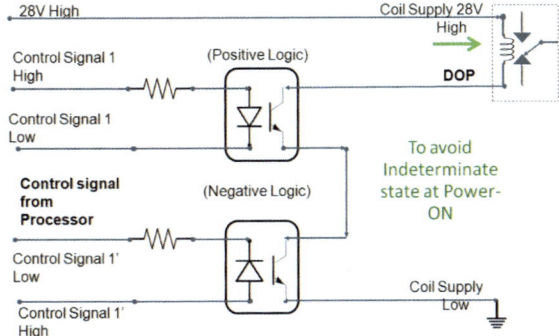

Fig. 8 Use of two drivers in series with complementary logic to drive a relay

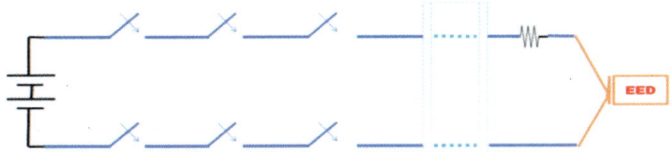

Fig. 9 Use of physical break in the circuit along with electrical breaks (the dotted line indicates the safety patch cable harness)

- Control unit: two sets of *launch PC (while checkout) or onboard computer (for inflight purposes).*

c. **Complementary logics in Relay drivers**: To avoid faults due to Power-ON indeterminate states of driver outputs. Figure 8 illustrates one typical schematic of using complementary logics in relay driving circuit.

d. **Assembly**: *Orthogonal mounting of relays and/or drivers, i.e. being electrically parallel but mounted orthogonal to each other in the electronic packages to ensure undue in-flight environmental stresses experienced in one particular direction don't hurt and damage both of them.*

e. **Use of Physical break in series with electrical breaks**: This is a small 'Cable Harness' with mating connectors at the terminal ends. It is basically used as a physical break which is connected in the Switching circuit while the "arming" process is carried out. This is called "Safety Patch" as it is used to ensure safety related to inadvertent/unintended operation of the EEDs (refer Fig. 9).

f. **Combination of latch and non-latch relays**: *For power economization, it is required to use latch relays where continuous power supply required.*

g. **Grounding of relays till execution of final command**: *Use of bleeder resistance to ground to avoid risk of inadvertent firing of the EEDs due to static charges.*

h. **Current limiting resistors**: *To limit the firing current and avoiding damage to the driver circuit.*

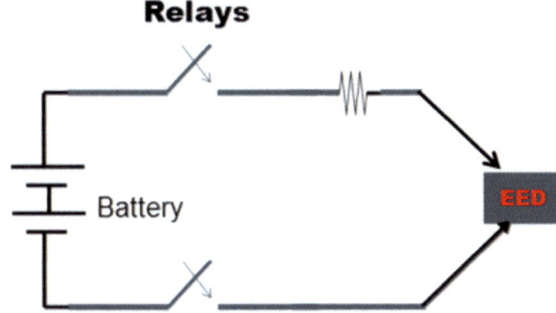

Fig. 10 EED firing circuit with a single (pair of) break

Fig. 11 EED firing circuit with several breaks

i.   ***Cross coupling of outputs***: *Two or more chains used and the power supplies are cross coupled to the EEDs for further redundancy.*

3.4   *Redundancy in EED Switching Circuit*

A single relay break between the power supply and the EED may be technically sufficient to initiate the EED firing as shown in Fig. 10 below.

However, as it was mentioned earlier, while planning the EED switching circuit design, there are several breaks used (refer the figure below) instead of just one (refer Fig. 11) for addressing the safety and other similar issues.

But, in this case, the reliability of the circuit significantly comes down due to more no. of relays. For example, three relays in series will have combined reliability of 73% even if each having individual reliability of 90%. *So, the reliability is getting compromised in order to meet the safety requirements.* It is hence necessary to relook and put extra effort to ensure the reliability is not compromised.

Here, three *redundancy configurations* for one relay break are being analysed to understand their operational and reliability issues.

(1) *Single relay configuration—no or without redundancy*:

Fig. 12 Single relay with no redundancy

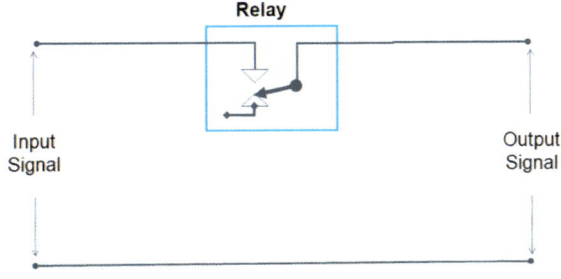

Table 2 Truth table for a single relay without redundancy

S. No.	Relay state	Output
1	OFF	OFF
2	ON	ON

Fig. 13 DMR configuration

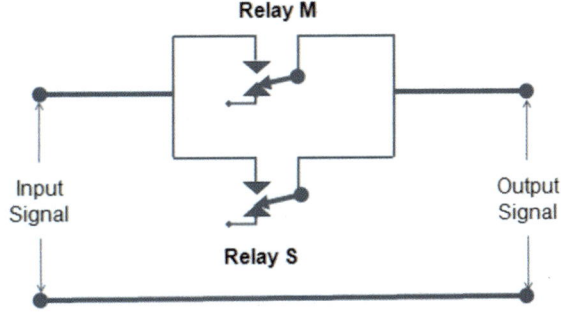

The simplest configuration is with a single relay break (refer Fig. 12) and hence without any redundancy at the relay level. The typical figure and its truth table are as given below:

It may be noted that if the relay fails, it will not be possible to control the break as it is clear from the truth table in Table 2.

There are two possible configurations when the relays are considered with the drivers: (1) the relay being operated by only one driver unit, so there is no redundancy at either the relay or the driver level; (2) two driver units in parallel steering the relay. So, in the second case, there is redundancy at driver unit level but no relay level redundancy.

(2) *Dual Modular Redundancy*

The DMR configuration is with two relay breaks in parallel. The typical figure and its truth table are as given below in Fig. 13 and Table 3, respectively:

It may be noted that though this configuration is superior than the first configuration (with single relay without any redundancy) mentioned above, this DMR too is having one *serious weakness*. If one relay fails in OFF (open) condition, it would be still possible to control the output through the other relay. However, suppose one relay

Table 3 Truth table of DMR configuration

S. No	Main relay state	Standby relay state	Output
1	OFF	OFF	OFF
2	OFF	ON	ON
3	ON	OFF	ON
4	ON	ON	ON

Fig. 14 TMR configuration

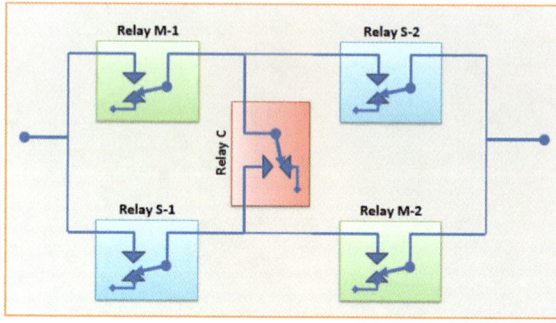

(say the Main relay) fails in On (i.e. closed) condition and remains permanently On (closed), there would always be a conductive path between the input and the output through the Main relay M. Hence, the standby relay would have no control over the output. Therefore, the reliability of the DMR configuration assuming a single-point failure (of one relay failure in ON condition) is quite low.

There are three possible relay driver configurations in case of DMR: (1) one separate driver unit for each of the two relays, i.e. total two driver chain and one chain for each relay. Driver unit DU-M operating the main relay M and the second driver unit DU-S operating the standby relay S (2) DMR with drivers steering and with cross-connection; i.e. both the relays are being operated by both the driver units (or chains). Here, thus both DU-M and DU-S operating both relays M and R due to the cross-connection. (3) DMR with steering but without cross-connection. In this case, there are independent main and standby driver units are there for each of the two relays. Thus, there are total four driver units, two in each chain, i.e. driver units (M1) and S1 steering main relay M and the driver units (M2) and S2 steering the standby relay S. It is clear that the second and third configurations of DMR are more reliable than the first one. But, the third configuration needs more resources (drivers, i.e. MOSFETs) vis-a-vis the second configuration.

(3) *Triple Modular Redundancy configuration*

The TMR configuration shown here is having three relays and two driver chains. The typical figure and its truth table are as given below in Fig. 14 and Table 4, respectively. Each chain with one driver unit thus total two relay driver units DU(M) and DU(S).

Table 4 Truth table of TMR configuration

S. No	Main relay state	Standby relay state	Common relay state	Output
1	OFF	OFF	OFF	OFF
2	OFF	OFF	ON	OFF
3	OFF	ON	OFF	OFF
4	OFF	ON	ON	ON
5	ON	OFF	OFF	OFF
6	ON	OFF	ON	ON
7	ON	ON	OFF	ON
8	ON	ON	ON	ON

- Relay M-1 and Relay M-2 represent first and second poles of main relay M operated by main driver unit, i.e. DU(M).
- Relay S-1 and Relay S-2 represent first and second poles of standby relay S operated by standby driver unit, i.e. DU(S).
- Relay C represents a common relay and is operated by both DU(M) and DU(S).

From the truth table, the following can be inferred:

- There are four possible signal flow paths from input to output. (1) Relay M-1—Relay S-2, (2) Relay S-1—Relay M-2, (3) Relay M-1—Relay C—Relay M-2 and (4) Relay S-1—Relay C—Relay S-2.
- At least two relays must operate for proper function.
- This TMR configuration can tolerate single-point failure. That is, even if a relay fails in either ON or OFF condition, output can still be controlled using the remaining two relays.
- In case of ON command DOP failure of a DU, output can be switched OFF.
- This configuration is especially useful where it is mandatory to withdraw power/command at some point during the mission sequence.

3.5 Reliability Analysis of the Various Redundancy Configuration Using RBD

Reliability calculation was carried out for all the mentioned redundancy configurations of relays and separately for both non-latch- and latch-type relays. The reliability was calculated in two conditions. (1) reliability when no failure assumed and (2) reliability assuming a single-point failure, i.e. assuming failure of any of the drivers or relays in the relay driver circuit. The RBD diagram for each case is involved and hence has not been covered in this paper. However, for illustration the RBD of a DMR configuration of non-latch relays with two driver chains has been given below

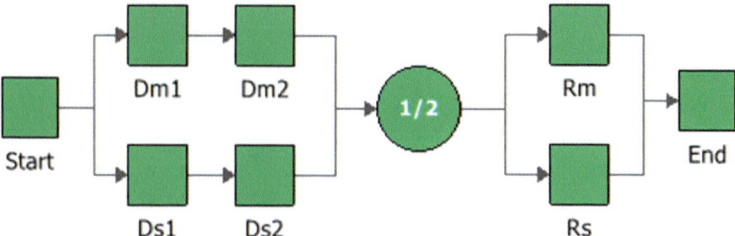

Fig. 15 DMR configuration of two non-latch relays with steering drivers and cross-connection between the driver units and the two relays: *RBD for OFF-to-ON switching*

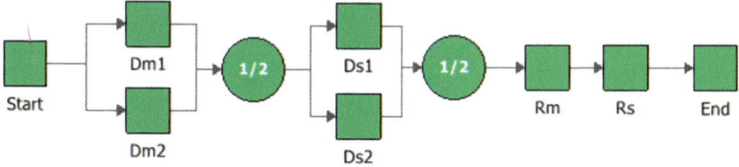

Fig. 16 DMR configuration of two non-latch relays with steering drivers and cross-connection between the driver units and the two relays: *RBD for ON-to-OFF switching*

in Figs. 15 and 16. As mentioned earlier, the two driving MOSFETs have been considered in complementary logic (to avoid indeterminate output conditions of the drivers at Power-ON) for driving each relay. Thus, Dm1 and Dm2 are having complementary logic and are inside driver unit main (DU-M), and similarly, Ds1 and Ds2 are having complementary logic and are inside standby driver unit (DU-S). Figure 15 shows the RBD for OFF-to-ON switching, and Fig. 16 shows the RBD for ON-to-OFF switching.

The reliability of the drivers and the relays has been assumed to be **0.90 and 0.98, respectively, while carrying out the reliability analysis using RBD.**

3.6 Outcome

After carrying out the risk-based analysis and estimating the reliability through the RBD exercise, TMR configuration was found to be the most reliable one. This is the only configuration out of the three which has no single-point failure, and hence the circuit and hence the mission will survive without losing its functionality even in the event of failure of one driver, driver unit or relay. This TMR can thus be used even if one of the relays fails in ON (CLOSED) condition. The summary of the reliability values calculated for TMR is given in the Table 5.

It can be observed from the above table that the TMR configuration involves a lot of complexity and highest number components (relays and drivers) vis-a-vis the other two configurations. Hence, it is used for critical signals, power lines or events

Table 5 Reliability calculation for TMR configuration

S No.	Relay type	If no failure assumed		If single point failure assumed		Qty of components
		OFF→ON	ON→OFF	OFF→ON	ON→OFF	
1.	Non latch relays	0.9769	0.9998	**0.6301–0.7498** (Rc Fails: **0.6301** Rm or Rs Fails: **0.7498** Dm/Ds Fails: **0.6301)**	**0.9413–0.9967** (Rc Fails: **0.9413** Rrn or Rs Fails: **0.9413** Dm/Ds Fails: **0.9967)**	Driver: 8 Relay: 3
2.	Latch relays	0.9574	0.9574	D_on fails: **0.6176** D_off Fails: **0.8704** R Fails: **0.7349**	D onfails: **0.8704** D off Fails: **0.6176** R Fails: **07340**	Driver: 16 Relay: 3

Symbols used *(in Table 5 above)*
ON → OFF
All relays are in ON or closed state (implying the end-to-end circuit is ON, i.e. closed) and the circuit between input and output needs to be made OFF, i.e. open
OFF → ON
All relays are in OFF or open state (implying the end-to-end circuit is OFF, i.e. open) and the circuit needs to be made ON, i.e. closed
D: Driver R: Relay Dm: Main Driver Ds: Standby driver Dc: Common Driver
D_on: Driver for ON coil D_Off: Driver for OFF coil Rm: Main relay Rs: Standby relay Rc: Common relay

such as emergency thrust termination. DMR has been selected for use for non-critical events. The scheme has been implemented across a number of systems over time, and the same have been successfully used for aerospace applications.

4 Conclusions

A risk analysis is always a proactive approach in the sense that it deals exclusively with potential accidents [5]. Risk-based design analysis is at the heart of designing electrical and electronic circuits of an aerospace system which are mission critical and severity of consequences of any fault or failure are catastrophic in nature. The design needs utmost care in addressing, analysing and meeting the highest order of safety and reliability issues. Dealing with safety issues cannot be treated as just rule compliance as was usually the case in traditional design approach. Further, equal emphasis is required to be put on both the safety and the reliability issues and at the design stage itself. In this paper, risk-based analysis of an EED switching system has been discussed in detail. From a general introduction of the basic components of an EED firing circuit through various considerations or objectives, etc., have been discussed. In the case study of an aerospace application, it has been explained what mechanisms were employed to meet those objectives. Various configurations of redundancy were discussed and RBD-based reliability analysis was discussed briefly. Finally, it was also shown that the reliability of TMR configuration is found to be the best and that the design have been implemented across many systems and have been successfully used in various missions.

References

1. A.K. Verma, S. Ajit, D.R. Karanki, *Reliability and Safety Engineering* (2010)
2. E.L. Keith, *Fundamentals of Rockets and Missiles* (2002)
3. MIL-HDBK-338B Electronic Reliability Design Handbook, U.S. Department of Defense (1 Oct 1998)
4. D.J. Smith, *Reliability, Maintainability and Risk—Practical methods for Engineers including Reliability Centred Maintenance and Safety-Related Systems,* Eighth edition (2011)
5. M. Rausand, *Risk Assessment: Theory, Methods, and Applications*, John Wiley & Sons (2013)

Value Tree Analysis Approach for Integrated Risk-Informed Decision Making: Revision of Allowed Outage Time

Poorva P. Kaushik and SokChul Kim

Abstract This study puts forward a methodology of integrated risk-informed decision making (IRIDM) for extension of allowed outage time (AOT) of safety system through Value Tree Analysis (VTA) approach. The purpose of regulatory decision making is to demonstrate compliance with regulatory requirements. Objectivity, transparency, and auditability are the foremost requirements for decisions on nuclear safety. Practical implementation of the IRIDM process may be very difficult because of the problems with the inputs prioritization and the decision options evaluation. An important element of any IRIDM process should be the explicit consideration of all effects because improvements in one area may adversely affect another area and to achieve balance in the overall safety measures. VTA replaces simple ranking of the inputs based on the engineering judgement by a selection of appropriate value functions, as much as possible, basing on the quantitative criteria. In this study, Value Tree for extension of AOT has been developed considering deterministic, probabilistic, and operating experience and economy as inputs. The case study has also been conducted to demonstrate the proposed approach for the extension of AOT of a single inoperable emergency diesel generator from 3 days to 7 days. It is observed that the benefits of the proposed methodology are: transparency, consistency, and reproducibility. This approach significantly makes the IRIDM process well-structured and easier to apply. The VTA complements the existing IRIDM framework proposed by International Atomic Energy Agency. It also increases the accountability and auditability of decisions.

Keywords Allowed outage time · Integrated risk-informed decision making · Multi-attribute decision making · Value tree analysis

P. P. Kaushik (✉)
Atomic Energy Regulatory Board, Mumbai, India
e-mail: poorvakaushik@gmail.com

S. Kim
Korea Institute of Nuclear Safety, 62 Kwahak-Ro, Yuseong, Daejeon 34142, Korea
e-mail: jupiter@kins.re.kr

© Springer Nature Singapore Pte Ltd. 2020
P. V. Varde et al. (eds.), *Reliability, Safety and Hazard Assessment for Risk-Based Technologies*, Lecture Notes in Mechanical Engineering, https://doi.org/10.1007/978-981-13-9008-1_80

1 Introduction

When a safety system is unavailable in the nuclear power plants (NPPs), there is an increase in risk due to loss of its safety function. The Allowed Outage Time (AOT) of a safety system is the time period it may remain unavailable during power operation before a plant shutdown is required [1]. AOTs are usually determined using the traditional deterministic approach. However, deterministic requirements are sometimes overestimated due to conservative analysis. Experience with plant operation has indicated that AOT may require revision to optimize the safe plant operation. Probabilistic analysis, on the other hand, uses realistic data in the risk models. However, due to the inherent uncertainties in current risk models of NPPs, the probabilistic approach is also not sufficient in itself.

Because of these inevitable gaps, making a decision only on the basis of deterministic analysis or probabilistic analysis can lead to unoptimized decisions. In addition to the above, factors such as operating experience (OE), economic implications, and implementation complexity can also have a pivotal role in decision making. The decision making that involves the integration of such a wide variety of information, insights, and perspectives has been termed as integrated risk-informed decision making (IRIDM) [2].

The International Atomic Energy Agency (IAEA) has issued various technical documents related to IRIDM. The IAEA has outlined basic concepts for the use of risk information for decision making on NPP safety issues or regulatory activities [2]. It has also identified the basic framework of IRIDM and defined key inputs and principles of application of IRIDM [3].

IRIDM is a multi-attribute problem that considers a wide variety of inputs. Quantitative determination of the relative significance of these inputs and their impact on the final decision is difficult. Decision makers usually rely on their subjective decree to evaluate inputs and there is no strategy commonly applied to deal with this issue.

The purpose of regulatory decision making is to demonstrate compliance with regulatory requirements. It involves making an informed judgment. In the nuclear field, where risk is high, good reasoning is as important as the decision itself. Objectivity, transparency and auditability are the foremost requirements for decisions on nuclear safety.

Mieczyslaw Borysiewicz et al. have suggested the application of Value Tree Analysis (VTA) method in decision making as an improvement and further extension of the framework recommended by IAEA [4]. VTA method replaces subjective judgments with value functions. The gains of application of the VTA methods within a multi-attribute decision making process is proven in varied industries. In the nuclear field, VTA was used successfully as one of the alternate methods for making decisions on the fuel conversion of the research reactor MARIA, Poland [4].

The present work proposes a new approach to IRIDM input evaluation for AOT optimization based on VTA methodology resulting in objective and transparent decision making.

2　Methodology

VTA method is used in multiple criteria decision making in which objectives/inputs are arranged hierarchically. Each objective is defined by attributes. Attributes are the measure of objectives. There can be several layers of objectives. Attributes are added to the lowest level of objectives to construct the value tree (see Fig. 1). A value tree outlines the hierarchical relationship between multiple layers of objectives and attributes [5].

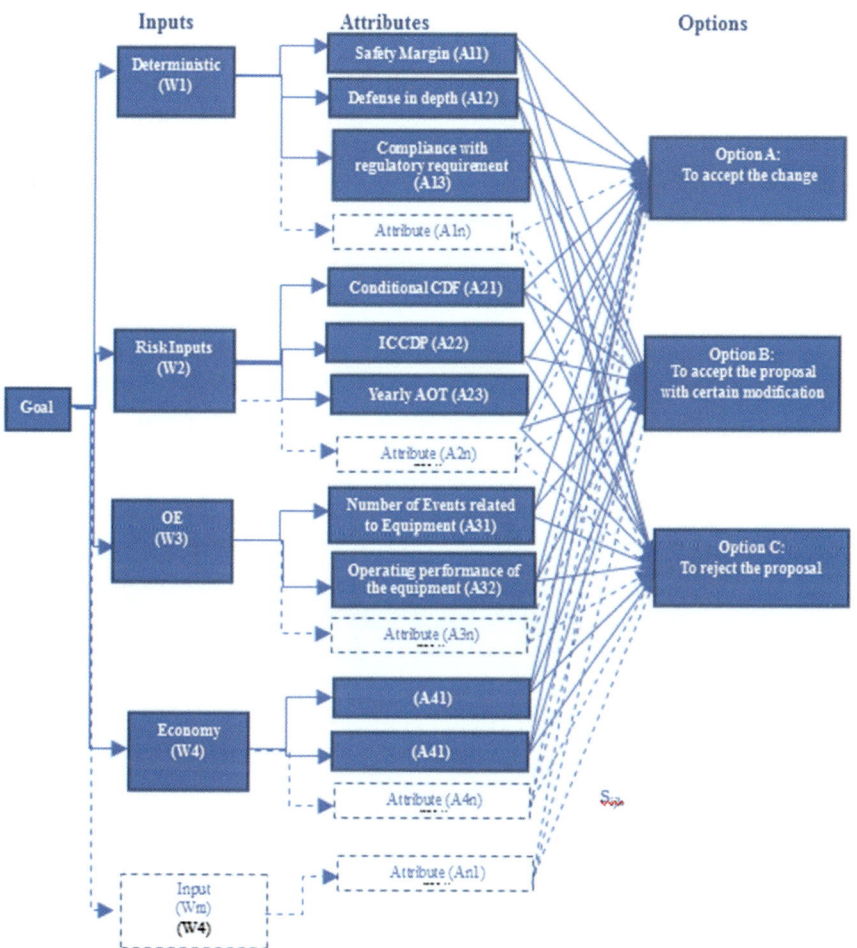

Fig. 1　Value tree model for allowed outage time

2.1 Framework for Proposed Methodology

VTA methodology comprises of following steps:

I. Problem Structuring: The first step in problem structuring is a clear definition of an issue for which decision has to be made and identification of various decision alternatives. The second step involves careful selection of inputs that need consideration for making the decision. These inputs will be specific to the issue under consideration. Various inputs that can be considered by the regulatory body are deterministic insights, probabilistic insights, cost benefit, and OE. The third step is the identification of attributes (quantitative or qualitative) for respective inputs.

II. Preference Elicitation: The aim of this step is to measure and estimate the preferences for various inputs and attributes. This step is to set up the hierarchical order between various inputs and attributes to construct the value tree. It is carried out in the following two steps:

    i. Weightage elicitation: It involves assigning priorities among various inputs and their attributes. The relative importance of ith input is given by W_i. The relative importance of the jth attribute for the ith input is given by A_{ij}.

    ii. Value elicitation: It describes the importance and desirability of achieving different performance levels of the given attribute for each alternative. This is achieved through the evaluation of consequence factor S_{ijk}. This factor describes how the implementation of the kth option would affect the jth attribute of ith input.

    iii. Evaluation of decision options: Assuming the independence among the attributes and additive model, once the values of all attributes for each input are determined, the best option with the highest score can be identified by using the following equation:

$$S_k = \sum_i W_i \sum_i A_{ij}.S_{ijk} \tag{1}$$

2.2 Methods for Preference Elicitation

Various prioritization and value evaluation methods can be employed within the VTA method. Following are the commonly used methods:

2.2.1 Methods for Weightage Elicitation

a. SMART—Simple multi-attribute rating technique.
b. SWING—SMART with swing weighting.

c. SMARTER—Simple multi-attribute rating technique exploiting rank is based
    on a formally justifiable weighting procedure developed by Barron and Barrett
    for multi-attribute utility measurement. In this method, inputs or attributes are
    ranked first, and then the weight W_i or A_{ij} is determined [6]. Edward and Baron
    have derived the following equation for the weights (where $W_1 > W_2 > W_3....$
    Wn):

$$W_i = \frac{1}{N} \sum_{n=i}^{N} \frac{1}{n}$$

(2)

It has been shown that without requiring any difficult subjective judgments (a
prerequisite for SMART), the SMARTER is an improvement to SMART and
performs about 98% accurately as SMART [6].

2.2.2 Methods for Value Elicitation

For value elicitation, the end points of the range of an attribute have to be first fixed.
The range should be optimized for the application under consideration. Once the
range of an attribute is fixed, the following methods of value elicitation can be used:

a. Direct Rating: It is most appropriate when performance levels of the attribute
    can be judged only by subjective measures. This value judgment is carried out
    by experts. In this method, first the worst and the best alternative are identified
    and a score of 0 and 100 are assigned, respectively. The value of the remaining
    alternatives is then considered to reflect the strength of the preferences for one
    alternative over another.
b. Value Function Form Assessment: A value function of different shape can be
    applied to each measurable attribute considered during the IRIDM process. Value
    function can be obtained as a function of any parameter X, the variation of which
    will decide the performance level of an attribute. The form of the value function
    should be specified in order to describe the relation between the value of X and the
    S_{ijk}. The shape of the curve is decided according to the importance of parameter
    X within the given range. This is a preferred method for quantitative attributes.

3 Results and Discussions

The IAEA has provided general guidance for making regulatory decisions using
IRIDM [2, 3]. In addition, The United States Nuclear Regulatory Commission (US-
NRC) has also developed guidelines [7, 8] for using risk-informed approach in AOT
revision. However, the challenge is to develop a model that systematically gives
weights to each of these IRIDM inputs to make a transparent and auditable decision. A

value tree model has been developed for IRIDM of AOT based on the considerations outlined in the above references by following the steps below.

3.1 Problem Structuring for Allowed Outage Time

Major decision alternatives identified for this case are: accepting the change as it is, denying the change, or accepting the change after additional modifications. Inputs and attributes to be considered for making the decision for AOT will vary from case to case. In this study, four major inputs are identified for decision making of AOTs. Deterministic assessment and probabilistic assessment are the two major inputs recognized in the literature. These two inputs are also considered in this study. Experience gained from construction, commissioning, operation, and decommissioning of NPP is termed as OE. It includes events, precursors, deviations, good practices, lessons learned, and corrective actions. OE is valuable information for improving nuclear and radiological safety and hence has been identified as the third input. It is important to assure that NPPs are competitive with respect to high availability. The factors, which directly influence availability and costs, are the outage frequency, outage duration, and resources used. Hence, economy has been considered as the fourth input.

For evaluation of the above inputs, qualitative or quantitative attributes are identified for each of them. Each attribute has to be assessed to identify the best decision option that satisfies the goal.

a. Attributes for Deterministic Input: Implementation of a proposed change should ensure that the existing regulatory requirements are complied with an adequate safety margins as envisaged are maintained. In addition to above, defense in depth (DID) which deals with independent multiple layers of prevention, protection, and mitigation, shall always be maintained to prevent any radiological consequences. Thus, deterministic inputs for AOT are sufficiency of the safety margin, compliance to regulatory standards and adequacy of DID. As all these attributes are qualitative in nature, value elicitation can be done by direct rating.

b. Attributes for Probabilistic Input: Three risk measures are identified as attributes for the evaluation of risk impact because of extended AOT. Instantaneous core damage frequency (ICDF) is the increased risk level when the component is known to be unavailable and is the first attribute. ICDF is calculated by setting the safety system down event to a true state in the probabilistic safety assessment and recalculating the core damage frequency (CDF). The second attribute identified is the cumulative (integrated) risk over the AOT period or incremental conditional core damage probability (ICCDP). It is the single downtime risk [4].

$$ICCDP = [ICDF - Base\ CDF] \times AOT \tag{3}$$

The third attribute is yearly AOT risk. This is the integrated risk over the duration of repair or AOT period. If the same component is undergoing the AOT 'n' number of times in a year, the yearly AOT risk would be

$$\text{Yearly AOT risk} = n \times \text{ICCDP} \tag{4}$$

c. Attributes for OE Input: The past performance of the safety system and events related to it must be considered while reviewing the AOT. Thus, attributes identified to assess the OE input can be past significant failure of the safety system and related international events [8].

d. Attributes for economy Input: The shutdowns of a plant due to conservative AOT are mostly unplanned thus may not lead to optimized resource utilization. Sometimes maintenance carried out under pressure due to short AOT may result in human error and thus also have a potential to reduce the safety of an operating NPP. Thus, a major attribute that can be used to assess the economy factor is the prevention of unplanned outage and in some cases; it can also be performance improvement of the safety system due to better maintenance. The following can be various measures:

   i. Prevention of unplanned outage: Expected increased plant availability or increased load factor or unplanned load reductions or Unplanned capability loss factor [9]:

   ii. Expected improvement in plant safety system availability/performance or reduced rate of human error.

3.2 Preference Elicitation for Allowed Outage Time

a. Weightage elicitation of inputs and attributes: As discussed weightage elicitation of inputs can be done by one of the various prioritization methods. In this study, the inputs are proposed to be weighted through SMARTER method as it is less subjective and yet easily applied and accurate. In ranking the inputs, the deterministic input is considered as the most important input followed by the probabilistic input. This is because of uncertainties present in risk models. OE is ranked third and economy as fourth. For the significance order: Deterministic (W_1) > Probabilistic (W_2) > OE (W_3) > Economy (W_4), the SMARTER method (Eq. 2) would produce the weights as follows: $W_1 = 0.521$, $W_2 = 0.271$, $W_3 = 0.146$, and $W_4 = 0.063$.

Similarly, the weightage elicitation for all the attributes can be done by various prioritization methods. SMARTER can be used to weigh the attributes for probabilistic, OE, and economy inputs. Deterministic attributes can be weighed equally through direct rating since all three attributes have equal priority (see Fig. 1).

b. Value elicitation for the evaluation of consequence factor: For value elicitation, the performance level of all the attributes has to be measured for each decision option. Qualitative attributes for deterministic input can be measured by direct rating. Direct rating of safety margin can be carried out by engineering judgment with respect to the compliance and consequences of exceeding the acceptable values of the corresponding safety parameters. In the case of evaluation for AOT, it can be considered enough if assumption made in final safety analysis are complied with. To assess the adequacy of DID, various elements have been identified by US-NRC and the fulfillment of these can be the basis of rating [10]. Quantitative attributes, in the case of probabilistic input, can be evaluated through identification of value function. Value function can be identified as a function X given by the following equation [4]

$$X = \frac{xa - xf}{xa - xi} \tag{5}$$

where xa = acceptable value of considerable parameter which cannot be exceeded, xi = initial value of parameter, and xf = final value of parameter.

Thus, different decision option can be obtained by different values of xf. The shape of the curve should also be specified in order to describe the relation between the X-value and the S_{ijk}. For example, when any changes in the lower region of the parameter X space are more important to the decision makers, then the changes of the same size in the upper region the concave curve should be chosen. Acceptance criteria for these attributes are given in Table 1. Attributes for OE and economy similarly can be assessed either by identification of value function or direct assessment.

3.3 Evaluation of Decision Options

Once the values of all attributes for each input are determined, the best option with the highest score can be identified by Eq. (1).

Table 1 Acceptance criteria for probabilistic measures

Risk measure	Risk Measure value	Acceptable action
ICDF or ΔCDF	ICDF < E−06 per reactor year	AOT extension will be considered regardless of total CDF
	E-06 per reactor year < ICDF < E−05 per reactor year	AOT extension will be considered only if total CDF < E−04 per reactor year
	ICDF > E−05 per reactor year	AOT extension will not be considered
ICCDP	ICCDP < E−06 per reactor year	AOT extension is allowed

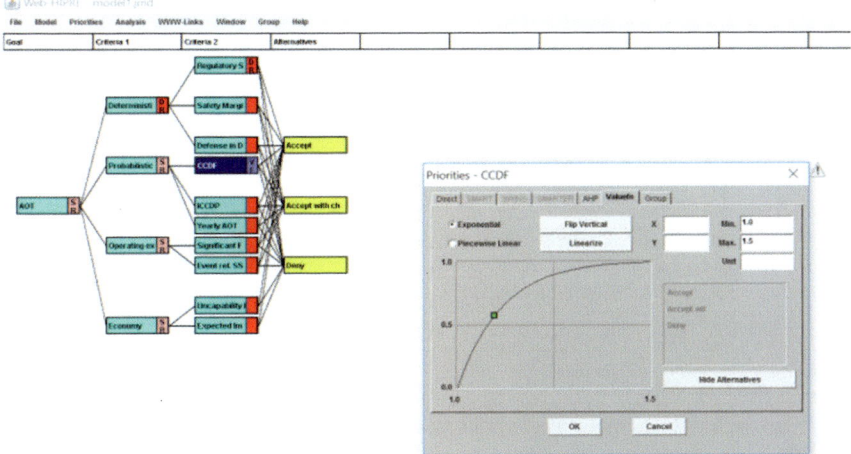

Fig. 2 Example for value elicitation for "conditional core damage" attribute through value function in Web-HIPRE

3.4 Software Tool

HIerarchial PREference (HIPRE) is a software tool that can be used by decision makers for multi-criteria decision analysis. It has a visual graphical interface which is easy to understand (see Fig. 2). The prioritization methods available in HIPRE are based on multi-attribute value theory. A decision problem is visually structured into a value tree of objectives/attributes. Each decision alternative is assessed in a performance matrix.

4 Conclusions

A systematic approach has been developed for the input evaluation and for weight assignment to each input and attribute. This approach significantly makes the IRIDM process well-structured and easier to apply. Present work puts forward a methodology of risk-informed decision making for extension of allowed outage time (AOT) of safety system. The value tree approach complements the existing IRIDM framework proposed by IAEA. It also increases the accountability and auditability of decisions.

Acknowledgements This research paper was made possible with the support of the Korea Institute of Nuclear Safety (KINS), Republic of Korea. I express my sincere gratitude to the Atomic Energy Regulatory Board (AERB) for giving me the opportunity to be a part of KINS-KAIST MS International program.

References

1. NUREG, *Handbook of methods for risk based Analyzes of Technical Specification*, CR141
2. International Nuclear Safety Group, *A framework for an Integrated Risk informed decision Making Process*, INSAG-25
3. International Atomic Energy Agency, *Risk informed regulation of nuclear facilities: Overview of the current status*, TECDOC-1436
4. M. Borysiewicz, K. Kowal, S. Potempski, An application of the value tree analysis methodology within the integrated risk informed decision making for the nuclear facilities. Reliab. Eng. Syst. Saf. **139**, 113–119 (2015)
5. Multiple Criteria Decision Analysis E learning Helsinki University of Technology Systems Analysis Laboratory
6. Edwards, F.H. Barron, SMARTS and SMARTER: improved methods for multi-attribute utility measurement, Organizational Behaviour and Human decision Processes, 1994
7. United States Nuclear Regulatory Commission, An approach for using Probabilistic Risk Assessment in Risk-Informed decisions on Plant -Specific Changes to the licensing Basis, Regulatory guide 1.174
8. United States Nuclear Regulatory Commission, An approach for plant -specific, Risk informed decision making: Technical Specifications. Regulatory guide 1.177
9. International Nuclear Safety Group, Guidance for optimizing nuclear power plant maintenance programmes, TECDOC1383
10. J. Mustajoki, R.P. Hamalainen, Web-HIPRE: global decision support by value tree and AHP analysis INFOR2000 **38**(3), 208 (2016)

A Deterministic Approach to Quantify the Risk in Propellant Processing Unit Using Bow Tie Analysis

K. Sreejesh, V. R. Renjith and S. C. Bhattacharyya

Abstract The energy requirements of the rocket propulsion systems in defense and space applications are largely fulfilled by the use of micron-sized metal powders and oxidizers embedded in polymer binders forming a solid crystalline matrix. These are developed and produced through a series of unit chemical operations, with critical operational limits for process parameters and differing reactant combinations including aluminum metal powders. Explosive processing unit accident histories prove the potential explosive characteristics of micron-sized aluminum metal powder. Some of the raw materials and the product involved in such a process have inherent explosive potential. Hence in most of the operations, involving these materials requires to be carried out with utmost safety compliance for personnel, plant and environment. Many catastrophic failures reported from the propellant processing plants were attributed to the potential explosive characteristic of aluminum metal powders and their complex interactions with the processed raw materials and equipment. This study is intended to assess the risk values associated with the propellant mixing unit using a bow tie method. The mixer unit is susceptible to failures arising from abnormal conditions resulting from process deviations. The catastrophic failures leading to a mass explosion in the mixing units are considered as the critical event for developing a bow tie. The credible top event scenarios were selected following a detailed preliminary hazard analysis using HAZOP and fire and explosion indices calculation of the propellant processing plant. This work intended to identify the failure probabilities and outcome frequencies of identified cases. The failure probabilities of the components and respective outcome frequencies are calculated by deterministic

K. Sreejesh (✉) · S. C. Bhattacharyya
Defence Research & Development Organisation, Delhi, India
e-mail: safesree@cusat.ac.in

S. C. Bhattacharyya
e-mail: sc.bhattacharyya@acem.drdo.in

V. R. Renjith
Division of Safety & Fire Engineering, Cochin University of Science and Technology, Kerala, India
e-mail: renjithvr75@gmail.com

© Springer Nature Singapore Pte Ltd. 2020
P. V. Varde et al. (eds.), *Reliability, Safety and Hazard Assessment for Risk-Based Technologies*, Lecture Notes in Mechanical Engineering,
https://doi.org/10.1007/978-981-13-9008-1_81

methods, and the failure probabilities of safety barriers are used for a probabilistic prediction of the outcome case in bow tie model.

Keywords Composite propellant · Deterministic BT · ETA · FTA · MCS

1 Introduction

The explosive processing industry deals with a variety of hazardous materials which possess both toxic and explosive threats to human life and environment. Normally, the operations in such plants are operated remotely through programmable logic controller (PLC) and supervisory control and data acquisition (SCADA), thus avoiding human exposure to hazardous processes. The propellant class explosives normally fall under United Nations (UN) explosive class Hazard Division 1.3. Processing of composite propellant involves various processes, viz. raw material preparation, mixing, material feeding, casting, hot water circulation, slurry flow control and TDI addition. Considering the inherent hazard potential and operational criticality of an involved process, quality aspects and safety, the logic, sequence and redundant systems should be designed on a fail-safe concept with a remotely operated processing unit with centralized monitoring and high-reliability control systems. Mixing is one of the unit operations by which the solid propellant is prepared from liquid binders, oxidizers and metallic powders. Due to the complexities associated with the process, this operation is classified under UN Hazard Division 1.1, i.e., a process with a mass explosion hazard. Many accidents were reported in propellant processing facilities around the globe in both defense and space programs. Though these accidents are of low probability, they have high-intensity outcomes. The propellant mixer units normally range from subscale (60 kg) to larger quantities more than 5000 kg for preparing solid propellant with varying ranges.

The propellant processing plants were normally operationalized after detailed quantitative risk assessment (QRA) studies [1, 2]. Since inception in 1980, QRA is proved to be the best available, analytic and predictive tool in the complex chemical process system. Any deviation from normal operation in such industries can lead to catastrophic accidents. The importance of hazard analysis and risk assessment comes at this stage which helps in identification, analysis and evaluation of risk involved in a particular process. There are two main types of risk assessment methods commonly followed, qualitative and quantitative risk assessment methods. Qualitative method is a descriptive one which gives cause and consequence as output. The bow tie method is one of the methods that become popular in high hazard industries like oil and gas, aviation and mining [3]. A bow tie method is a combination of fault tree (FT) on the left-hand side and an event tree (ET) on the right-hand side which are interconnected by a common critical event. The explosive processing plant being one of the highly hazardous areas, the application of bow tie will help in identifying a cause–consequence relationship in one single pictorial representation. Bow tie representation will enable risk analyst to reach multiple interpretations for a particular outcome scenario.

2   Review of the Literature

Aluminum metal powder of micron size is a widely used constituent of ammonium perchlorate–hydroxyl-terminated polybutadiene (AP-HTPB)-based composite solid propellant compositions. By decreasing the particle size of the materials, the researchers have achieved an increased rate of energy release from composite propellant formulations. Research data says that the risks associated with ultrafine and micron-sized metal powders with respect to their fire, explosion and health characterization have been tremendously altered with the reduction in the particle size. The 1973 accident at the premix batch plant of a slurry explosive factory in Norway figures out the dust explosion susceptibility of micro- as well as nano-scaled metal powders used for explosives and propellant processing with minimum ignition energy(MIE) as low as 1 mJ [4].

Risk is traditionally defined as a combination of the probability or likelihood and consequence of the undesired event. As reported by the Center for Chemical Process Safety (CCPS), several approaches are enlisted to identification of hazard and the related risk associated, viz, what-if analysis, hazard and operability studies (HAZOP) and fault tree analysis. Estimation of potential accident frequencies and evaluation of event consequences are considered as the central steps for the whole QRA. Consequence estimation is used to determine the potential for damage or injury from specific unwanted events. This is normally performed using a number of physical–mathematical and empirical models.

HAZOP is a systematic approach to investigating each node in a process to identify all of the ways in which parameters can deviate from the intended design conditions and lead to hazards or operability problems [5]. The HAZOP is carried out by a multidisciplinary team, with the help of predesigned guide words to review the process parameters and nodes for any potential hazards and operability problems.

2.1   Fault Tree Analysis

A fault tree is a deductive logic analysis used to identify the event sequences which could lead to the unwanted (top) event. All combinations of failure of equipment or operators that can lead to the top event are shown in a systematic and logical format in the form of fault tree. It uses Boolean logic to construct a tree of possible failure paths leading to a single top event at the end (usually a critical failure or loss of control) [3]. The fault tree is developed with graphic symbols and logic operators.

The basic process in the technique of FTA is to identify a particular effect or outcome from the system and trace backward into the system by the logical sequence to prime cause(s) of this effect [6]. Fault tree symbols are divided into four categories: primary event, intermediate event, gates and transfer. Primary events are considered as the end events. Whereas intermediate events are events which propagates to gates, gates are otherwise called as graphic operators or logic gates. The solution of a fault tree gives a list of combination of primary events that are sufficient to result in an accident. These combinations of failures are known as minimal cut sets [2].

A cut set in an FT is considered as any group of PE which, if all occur, will lead to the occurrence of the TE. But there is a least group of such combination of PE or basic events which, if all occur, will cause the top event to occur. In this particular approach, it is followed each MCS turns out to be a causal combination for TE [1].

2.2 Event Tree Analysis

An event tree analysis (ETA) is a procedure based on inductive logic that shows all possible outcomes resulting from an initiating event or critical event, taking into account whether installed safety barriers are functioning or not, and additional events and factors. An event tree relates an abnormal event to all its known consequences [7]. By studying all relevant accidental events (that have been identified by the HAZOP), the ETA can be used to identify all potential accident scenarios and sequences in a complex system. This method can be used to quantify the outcome probabilities. In a bow tie, the ET is displayed on the right-hand side of the diagram.

The *TE* probability thus calculated is the initiating event probability of the ET constructed on the right-hand side. The end-state probability can be determined by following the branches of the event tree to that particular state. The deterministic value of each end state is calculated by multiplying the failure probabilities of each branch related to the state.

3 Methodology

3.1 Scenario Identification and BT Development

The application of hazard identification technique, such as a bow tie analysis, is performed at this step to provide a visual representation of consequences, causes and related safety barriers in place to mitigate or control the hazards. The "knot" of the bow tie diagram known as "critical event (CE)" is the same top event as in the FTA. Consequences in left-hand part of the diagram and causes in the right-hand part of the diagram have been, respectively, indicated as "intermediate events" (IEs) and "outcome events" (OEs).

A bow tie diagram is a simple and effective graphical tool for connecting the primary event (PE), intermediate event (IE) leading to the critical event and its final

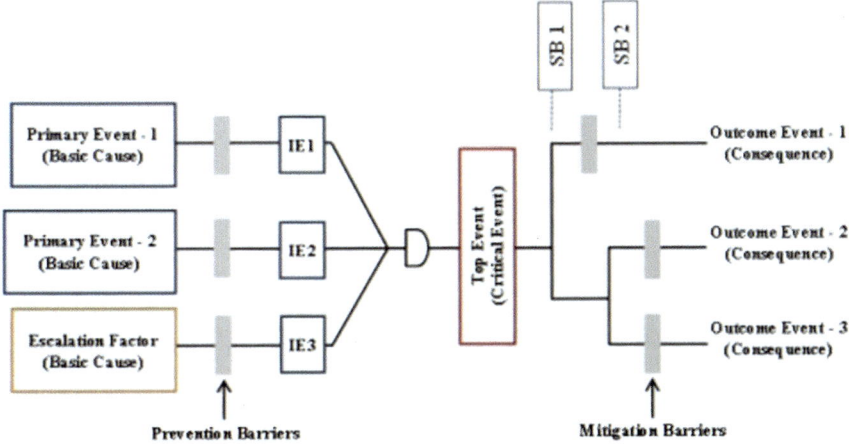

Fig. 1 Bow tie representation

outcome event (OE). The diagrams clearly display the links between the potential causes, preventative and mitigative barriers and consequences of a major incident. Bow tie diagrams are prepared as a combination of a fault tree (FT) on the left-hand side with possible primary events leading to a critical event and an event tree with outcome cases on the right-hand side. A bow tie combines a cause–consequence diagram and merges it with barriers into a single diagram [3]. The general structure of a bow tie diagram is represented in the diagram (Fig. 1). Bow tie diagrams may also be integrated with semiquantitative analysis techniques such as layers of protection analysis (LOPA) depending on the level of complexity required.

3.2 Calculation of Failure Probabilities—Deterministic Estimates

The failure function is identified for each barrier in the BT diagram using a deterministic method. The top event probabilities are calculated from the FT part using MCS methods. The MCS directly links the top event to the primary events or basic events. The probability of TE can be calculated using the following equations

$$P(T) = P(M_1 + M_2 + \ldots + M_N) \text{ where } ``+" = \text{Logical OR} \qquad (1)$$

$$P(T) = \Sigma P(M_k)\text{Sum of MCS Probabilities (Rare Event Approximation)} \quad (2)$$

$$P(M) = P(E_1)P(E_2) \ldots P(E_M) \text{ Product of Independent } PE \text{ Probabilities} \quad (3)$$

where T = top event, M = MCS and E = primary event

The TE probability data is used to solve the ET part of the bow tie diagram to estimate the outcome frequencies. Consider $P(T)$ as the frequency of the top event calculated out by FT analysis. Let $Pr(Bi)$ denote the probability of event $B(i)$. The ET branches out to various accidental outcome events with deterministic probabilities depending on the success and failure rates of each safety barrier in the ET path. Here, all the probabilities are conditional given the result of the process until "barrier i" is reached. The probability of "outcome i" can be calculated as follows:

$$Pr(\text{Outcome } i) = Pr(B_1 \cap B_2 \cap \ldots \cap B_n) \tag{4}$$

$$Pr(\text{Outcome } i) = Pr(B_1) \cdot Pr(B_2|B_1) \cdot Pr(B_3|B_1 \cap B_2) Pr(B_n|B_1 \cap B_2 \cap B_3 \ldots B_{n-1}) \tag{5}$$

3.3 Probabilistic Estimates of Outcome Cases

The end-state probability can be determined by following the branches of the event tree to that particular end state. The ET follows a cumulative distribution function describing the failure probability of each barrier and represents the prior knowledge about the barriers. The values are obtained from the generic equipment reliability data available with the IAEA, CCPS, OREDA and MIL-HDBK. Although this method is very convenient and simple to use, it does not give a precise answer to the problem since the failure probability of a safety system tends to follow a distribution [8]. It is assumed that all the safety barrier failure probabilities follow a beta distribution.

$$FP = \mu = \frac{\alpha}{\alpha + \beta} \tag{6}$$

where α and β are the parameters by which the beta distribution is defined. These parameters are estimated using plant-specific data from the maintenance log books and safety expert feedback from the processing unit.

4   System Description

The propellant mixer considered for study is a vertical planetary mixer with 3 ton capacity. The mixer system includes subsystems like feeding system, propellant slurry temperature control system, PLC-based control system, compressed air system for operating the bowl, spill tray collection system for collecting the remnant propellant on the agitators and bowl lid lifting system.

The vertical mix has three helical-shaped blades mounted vertically on mixer head. The blades are arranged and termed as two outer agitators and one central agitator. The mixer is operated by hydraulic power with suitable reduction gears for rotation of the central agitator and rotation and revolution of the outer agitators in a specified ratio. The mixing operation is performed in a bowl aligned through a cradle assembly and gravity locked to the shroud assembly. Mixing bowl is aligned to the mixer head using hydraulic power, and raw materials are fed to the mixer through feed lines from the feeder room located above the mixer room level. The pictorial representation of overall assembly of the mixer with the detachable bowl and different feed line layouts are shown in Fig. 2. As a safety precaution, all the raw materials are screened through sieves before loading into the feed bins and prior to addition preventing foreign material entry into the mixer.

Though the propellants are grouped under hazard category 1.3, considering the process variables and critical parameters the mixing plant is classified under Hazard Division 1.1. The ammonium perchlorate (AP)-based propellant, which is the most widely used solid propellant, aluminum powder is used as the fuel. AP powder is the oxidizer and contains AP particles with three average diameters and aluminum particles (Al). The binder is hydroxyl-terminated polybutadiene (HTPB) [9]. The mixer system is facilitated by other subsystems like feeding system, propellant slurry temperature control system, PLC-based control system, compressed air system for operating the bowl, spill tray collection system for collecting the remnant propellant on the agitators and bowl lid lifting system.

Since the mixer is susceptible to varying load and torque conditions during charging of powders and unusual viscosity buildup during mixing operations, it has been designed and provided with overload or over torque protection devices in gearboxes. These barriers are known as shear pin assembly and tension bar mechanisms. An automatic bowl drop mechanism in bowl lift hydraulic circuit will help to relieve the unusual pressure buildup during mixing operations. This mechanism is activated through a rupture disk (RD) placed in the hydraulic subsystem of the mixer unit. The chemical unit operation mixing is performed in batch process using this unit to mix the propellant into a homogenized crystal matrix.

5 Results and Discussion

The bow tie analysis for the propellant mixing plant is conducted for three different failure cases, viz. overpressure, overload and excessive torque. These deviations from normal operation are identified after a detailed HAZOP and fire and explosion index (F&EI) calculations at plant. The final outcome events are categorized into four major outcome cases, viz. plant shutdown (safe), emergency shutdown (near miss), loss of containment (accident) and fire/toxic/detonation (major accident).

5.1 Case 1: Overload Condition

The overload condition in the bowl is identified as one of the failure cases in the vertical propellant mixer unit. This condition can be developed during mixing oper-

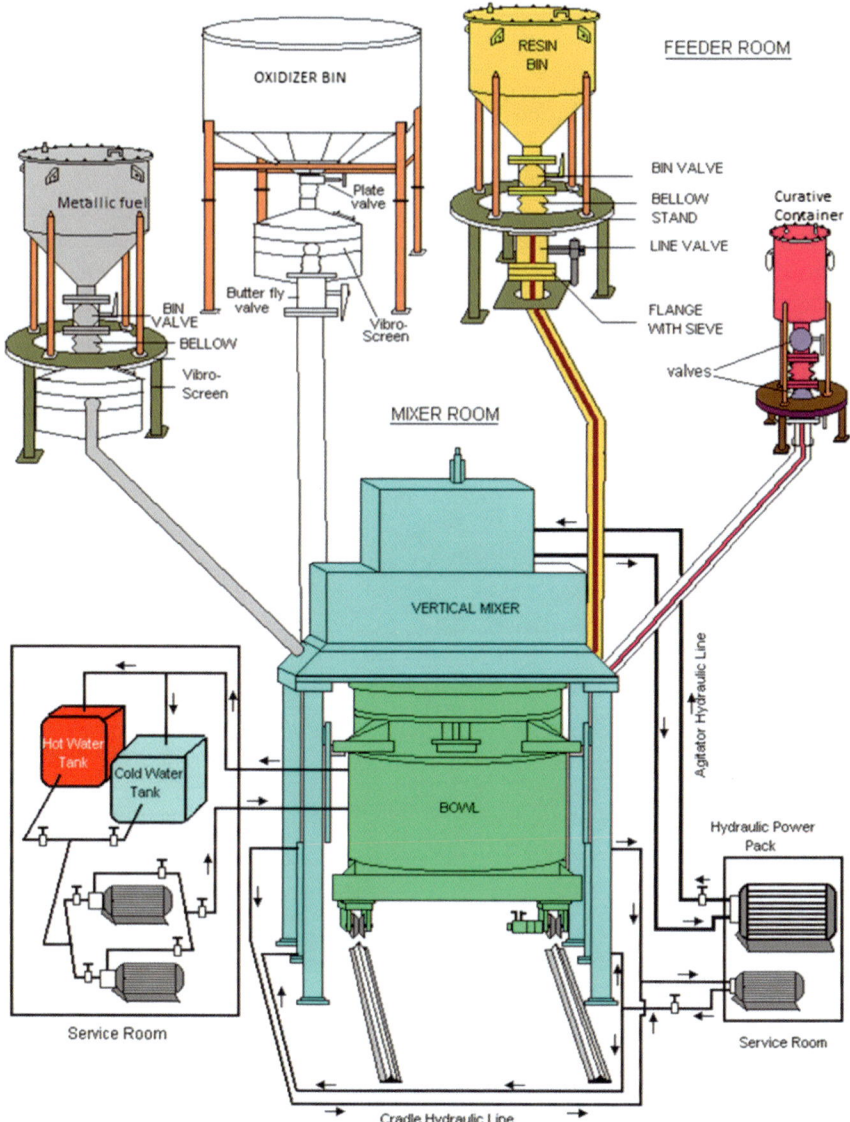

Fig. 2 Vertical mixer facility mixing arrangement [10]

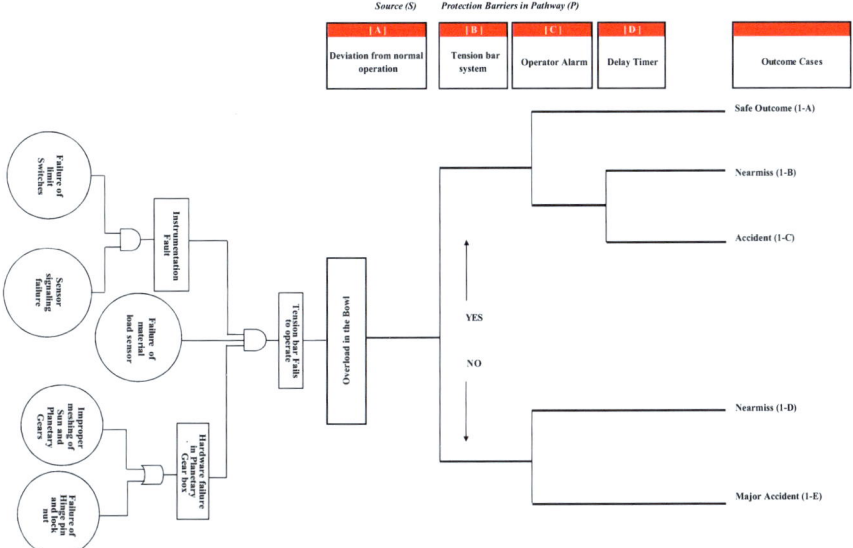

Fig. 3 Bow tie for overload condition

ation due to the increased amount of reactants in the bowl. The basic/primary events are identified using FT and the final outcome by ET. The bow tie representation is shown in Fig. 3. In order to protect the system and occupants from such a condition, three safety barriers are provided in the system, viz.

(a) Tension Bar (TB) System: It is a protection system in the planetary gearbox (PGB) which connects the sun gear with stationary gear housing. The failure of TB during overload condition stops the planetary motion, but all three agitators will be rotating about their own axis.

(b) Operator Alarm: The TB failure is sensed by the sensor circuit, and an indicator both visual and audio alarm is indicated in the control room. The operator intervenes to stop the power supply and takes the plant to a shutdown mode during such alarms.

(c) Delay Timer: Two sensors positioned at 180° apart on the bottom disk of the PGB activate two limit switches mounted on the shroud body to monitor the speed of the planetary motion. When the sensor does not get signal for 110 s., it recognizes the failure of TB, creates an emergency condition by switching off the power supply to drive motor and stops agitator rotation.

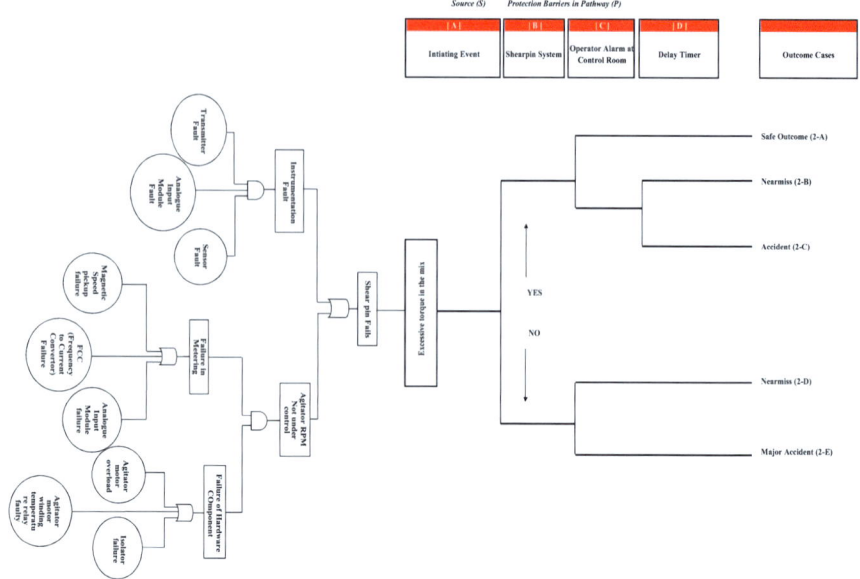

Fig. 4 Bow tie for excessive torque

5.2 Case 2: Excessive Torque

The upper gearbox (UGB) is designed to transmit full torque to agitator shaft and reduces the speed from 1750 to 10 rpm in three stages. During operation, conditions with high load demand a higher torque from the UGB. Hence, the excessive torque is identified as another major condition with high-impact outcome. The bow tie generated for a particular condition is shown in Fig. 4. The safety barriers are also provided to protect the mixer unit during such condition, viz.

(a) Shear Pin System: The shear pin is connected with the UGB. When the excessive torque is experienced in the mixing zone, the shear pin fails and the agitator and planetary rotations are stopped.
(b) Operator Alarm: The shear pin failure is detected by the sensors in the UGB, and audiovisual alarm is initiated in the control room for operator intervention.
(c) Delay Timer: This system has a circular disk mounted on the top of central axis, whose rotation is sensed by proximity sensors. When the sensors do not get signal for 12 s, it recognizes the failure of shear pin, creates an emergency condition by switching off the power supply to drive motor and stops agitator rotation.

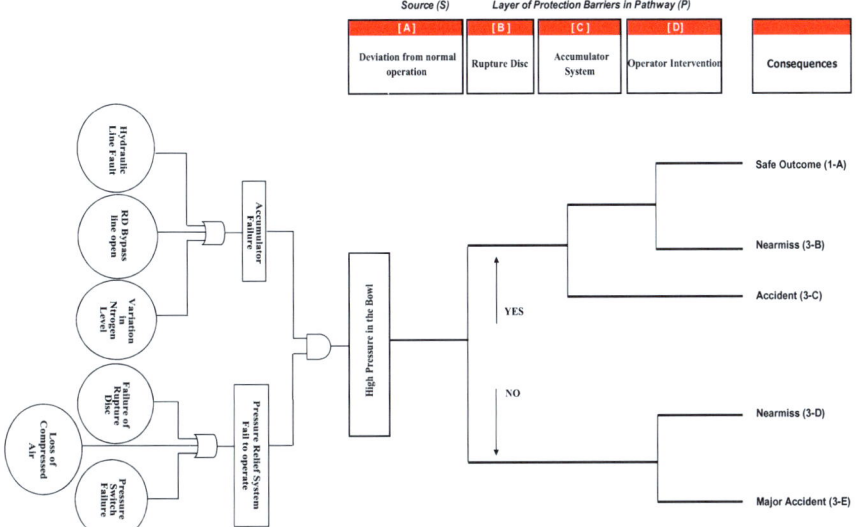

Fig. 5 Bow tie for overpressure scenario

5.3   Case 3: Overpressure Condition

Due to any unforeseen circumstances, if the pressure inside the bowl during mixing increases beyond a stipulated value of 5 psi, an overpressure scenario is created and it warrants an emergency bowl drop. The *BT* developed for the event is presented in Fig. 5. The safety barriers provided in the system to avert the dangerous condition are:

(a)   Rupture Disk (RD): The hydraulic line is provided with an RD at a 5 psi increment from working pressure. If this fails, there is a redundant RD on the line which fails at 170 bar and routes the hydraulic supply to accumulator.

(b)   Accumulator Mechanism: It is a tank pre-charged to 40 bar with nitrogen. This mechanism helps in slow bowl drop.

(c)   Pressure Switch: The pressure switch is provided in the actuator line. By detecting the flow in accumulator line, this switch cuts the power supply to hydraulic pump.

The study adopted a deterministic method to quantify the risk figures associated with the *BT*. The failure probabilities for each safety system and component were adopted from process equipment reliability data of CCPS, component reliability data for use in probabilistic safety assessment of International Atomic Energy Agency (IAEA) and MIL-HDBK-217E and plant-specific failure case histories refer Table 1.

The end-state probability of the *TE* and outcome probabilities are calculated by the MCS method mentioned in the earlier session. The occurrence data for primary

Table 1 Failure probabilities of safety systems

Safety systems	Failure probabilities
Tension bar system	0.02
Shear pin	0.05
Rupture disk	0.02
Accumulator system	0.01
Delay timer mechanism	0.03
Operator alarm (For RD system)	0.15 (0.2)

Table 2 Probabilities of primary events in FT

Index	Components	Probability
Case 1		
PE1	Limit switch	0.0089
PE2	Sensor signaling circuit	0.0045
PE3	Material load sensor	0.00434
PE4	PGB meshing fault	0.0019
PE5	Hinge pin and locknut	0.0043
Case 2		
PE1	Transmitter	0.000572
PE2	Analog input module	0.000836
PE3	Proximity sensor	0.000642
PE4	Magnetic speed pickup	0.00273
PE5	Frequency current convertor	0.0009
PE6	Agitator motor overload	0.0084
PE7	Motor winding temperature fault	0.000572
PE8	Isolator fault	0.000836
Case 3		
PE1	Hydraulic line failure	0.00045
PE2	RD line bypass open state	0.00006
PE3	Variation in nitrogen level in accumulator	0.029
PE4	Failure of RD	0.000034
PE5	Loss of compressed air	0.0035
PE6	Pressure switch failure	0.000021

events for each FT part was adopted from IAEA, and CCPS data and MIL-STD are mentioned in Table 2.

The risk quantification associated with the cause–consequence chain was calculated by considering the failure probability data of each component from the FT side and the safety barriers at ET side. Table 3 brings out both the deterministic and prob-

Table 3 Result summary of occurrence probability

Sl. No.	Outcome scenario	Occurrence probabilities (deterministic value)	Occurrence probabilities (probabilistic value)
1	1-A	7.08E−09	8.03E−09
2	1-B	3.19E−09	2.46E−09
3	1-C	2.04E−10	1.57E−10
4	1-D	2.19E−10	9.78E−11
5	1-E	1.05E−10	4.69E−11
6	2-A	9.01E−05	7.53E−05
7	2-B	2.18E−05	3.39E−05
8	2-C	6.76E−07	2.17E−06
9	2-D	1.37E−06	1.93E−06
10	2-E	9.01E−05	7.53E−05
11	3-A	8.66 E−05	7.49 E−05
12	3-B	1.53 E−05	2.63 E−05
13	3-C	1.03 E−06	6.11 E−07
14	3-D	1.79 E−06	2.33 E−06
15	3-E	3.15 E−07	8.19 E−07
16	3-F	8.66 E−05	7.49 E−05

abilistic outcome frequencies of each outcome case calculated as per the methods mentioned in the earlier section. It is also evident that the outcome frequencies sum to that of the initial event [11].

6 Conclusions

The worst-case outcome scenarios in an explosive processing unit propagate through microsecond interval, and such a process needs a quantitative judgment of risk figures associated with each activity for the selection of protective devices and mitigation methods. Bow ties facilitate effective risk conversations by providing a visual aid to prompt conversation rather than relying on sometimes cumbersome text-based policies, procedures and checklists [12]. The output can be used for the representation of risk associated with a process and a platform to take risk control decision. The bow tie model directly establishes a cause–consequence sequence giving the sequences of events which might lead to an adverse consequence. Here, the FT part of the diagram brings out the scope of prevention barriers which will help in controlling a primary event developing into a top event, whereas the ET parts help in identifying the mitigation barriers which can reduce the severity of a top event once it has already occurred. Though the *BT* provides deterministic values for risks associated, there is a lack of data/failure in updating data related to the particular event sequences. However, in some cases the probabilistic values obtained are more than those of deterministic values. The occurrence probabilities obtained through beta distribution are found to be affected by the nature and number of failure cases surveyed. This can be addressed and further refined by a dynamic updation of abnormal failure data using Bayesian analysis. However, the present study only quantifies the occurrence probability of particular outcome cases in a propellant mixer unit.

References

1. P.K. Roy, A. Bhatt, C. Rajagopal, Quantitative risk assessment in titanium sponge plant. Defence Sci. J. **54**, 549–562 (2004)
2. C. Rajagopal, A.K. Jain, Risk assessment in a process plants. Defence Sci. J. **47**(2), 197–205 (1997)
3. Ruijter A. De, F. Guldenmund, The bowtie method: a review. Saf. Sci. **88**, 211–218 (2016). https://doi.org/10.1016/j.ssci.2016.03.001
4. R.K. Eckhoff, *Dust Explosion in the Process Industries*. vol. i. 3rd ed. (Gulf Professional Publishing, Elsevier Science, 2003)
5. V.S. Kumar, L. Ramanathan, Generalized hazop analysis for process plant. Int. J. Sci. Res. Dev. **3**, 1281–1284 (2015)
6. V.R. Renjith, G. Madhu, V.L.G. Nayagam, A.B. Bhasi, Two-dimensional fuzzy fault tree analysis for chlorine release from a chlor-alkali industry using expert elicitation. J. Hazardouse Mater. **183**, 103–110 (2010). https://doi.org/10.1016/j.jhazmat.2010.06.116
7. M. Kalantarnia, F. Khan, K. Hawboldt, Dynamic risk assessment using failure assessment and Bayesian theory. J. Loss Prev. Process. Ind. **22**, 600–606 (2009). https://doi.org/10.1016/j.jlp.2009.04.006
8. M. Kalantarnia, F. Khan, K. Hawboldt, Dynamic risk assessment using failure assessment and Bayesian theory. J. Loss Prev. Process Ind. **22**(5), 600–606 (2009)
9. B.A.I. Wasaki, K.M. Atsumoto, R.B. An, S.Y. Oshihama, T.N. Akamura, H.H. Abu. The continuous mixing process of composite solid propellant slurry by an artificial muscle actuator. **14**, 107–110 (2016)
10. V. Paliwal et al., Safety management in propellant mixing area-vertical mixer experience, in *International Conference on Safety and Fire Engineering*, (2017), p. 20
11. F.P. Lees, *Loss Prevension in Process Industries* (Butterworth and Hienemann, Chapter 9, 1996), pp. 9/31–9/36
12. B. Mary, B. Mulcahy, C. Boylan, S. Sigmann, Using bowtie methodology to support laboratory hazard identification, risk management, and incident analysis. J. Chem. Health Saf. **24**(3), 14–20 (2017)

Risk-Based Approach for Thermal Profiling of Sodium Vessel and Pipelines of Prototype Fast Breeder Reactor

P. Chithira, N. Kanagalakshmi, P. Rajavelu, R. Prasanna, P. Sreenivasa Rao and V. Magesh Mari Raj

Abstract Monitoring of the temperature profile and accurate assessment of temperature data from the measured values in main vessel (MV), primary and secondary sodium piping and related pipe-mounted equipment, during the commissioning phase of a sodium-cooled fast reactor, is extremely important, since a fall in temperature below the plugging temperature value of sodium could result in freezing of sodium within the pipes or in various crevasses in valves, pumps, etc., or also in the MV. While it is possible to augment the heating of the piping segments for re-melting the sodium, the same will not be possible, if the large volume of sodium in the MV (of the order of 1000 tonnes) is to either freeze completely or even agglomerate and settle down in the various crevasses of in-pool equipment. This paper deals with the various failure modes of heating equipment and the necessary precautions to be taken for ensuring the required temperature by monitoring the temperature data along with the uncertainties during commissioning phase of PFBR. It is essential to ensure that the temperature profiles of the MV are well above the plugging temperature of the

P. Chithira (✉) · N. Kanagalakshmi · P. Rajavelu · R. Prasanna · P. Sreenivasa Rao · V. Magesh Mari Raj
Department of Atomic Energy, Bharatiya Nabhikiya Vidyut Nigam Ltd, Kalpakkam, India
e-mail: pchithira_bhavini@igcar.gov.in

N. Kanagalakshmi
e-mail: kanaga@igcar.gov.in

P. Rajavelu
e-mail: rajavelu_bhavini@igcar.gov.in

R. Prasanna
e-mail: rprasanna@igcar.gov.in

P. Sreenivasa Rao
e-mail: psrao@igcar.gov.in

V. Magesh Mari Raj
e-mail: magesh_bhavini@igcar.gov.in

© Springer Nature Singapore Pte Ltd. 2020   981
P. V. Varde et al. (eds.), *Reliability, Safety and Hazard Assessment for Risk-Based Technologies*, Lecture Notes in Mechanical Engineering,
https://doi.org/10.1007/978-981-13-9008-1_82

sodium, and for this, a detailed failure mode and effect analysis (FMEA) is carried out. Also, based on the computed reliability numbers from the mechanistic models of the nitrogen blowers and heaters and the rotor dynamic equipment of the secondary loops, viz. main sodium pumps and electromagnetic pumps, a risk-based model is proposed.

Keywords Fast breeder reactor · FMEA · RLSE · Sodium

1 Introduction

Prototype Fast Breeder Reactor (PFBR) is a 500 MWe (1250 MWt) power, (U-Pu) O_2-fuelled, sodium-cooled, fast reactor. It is a pool-type reactor, in the sense that the entire primary system is housed within a single vessel called main vessel (MV). The reactor components are characterized by large size thin-walled shell structures, such as main vessel, safety vessel (guard vessel kept surrounding the main vessel to prevent complete loss of coolant), inner vessel that separates the hot and cold sodium pools and thermal baffles to protect the main vessel from the heat emanating radially from the core. There are three major systems in PFBR, viz. primary, secondary and steam water circuits. The heat generated in the core is removed by circulating primary sodium from the cold pool to the hot pool. The sodium from hot pool after transporting its heat to four intermediate heat exchangers (IHX) mixes with the cold pool. The heat from IHX is in turn transported to eight steam generators (SG) by sodium flowing in the secondary circuit.

The commissioning process essentially requires filling of the auxiliary secondary loops, where sodium is circulated with the help of electromagnetic pumps, through a purification loop, in order to bring down the plugging temperature, as close to its melting point (98 °C). Subsequently, the main loop is filled and the main pumps are run, so as to circulate the sodium through the intermediate heat exchangers (IHX), housed within the MV. As a part of commissioning activities, the vessels and pipelines need to be preheated before filling hot sodium, to avoid thermal shocks to structures and components and solidification of sodium. The temperature to which the structure must be preheated is dictated by melting point of pure sodium, level of impurity of sodium and cleanliness of the internals of vessel and pipelines. Preheating of MV is achieved by purging hot nitrogen. Preheating of pipelines and equipment is by mineral insulated surface heaters (MI heaters). Temperature is measured by thermocouples fixed on the walls of MV, surface of pipelines and equipment.

2 Preheating System of Main Vessel and Pipelines

2.1 Main Vessel Preheating

MV is a cylindrical shell of size Ø12.9 × 12.79 m with a specially shaped tori-spherical dished end weighing 135 T. MV houses the reactor assembly components like core catcher, core support structure, grid plate, core, transfer arm, control plug, inner vessel, primary sodium pumps, intermediate heat exchangers and decay heat exchangers with large volume of sodium of the order of 1000 MT. Inadequate or insufficient preheating of MV increases the risk of agglomeration of sodium in the crevasses or even a risk of freezing sodium in the MV during filling. To achieve a uniform temperature distribution across the MV internals, hot nitrogen gas is purged in the interspace between MV and safety vessel (SV) and through the internals of MV (initially). Once it is ensured that the temperature is stabilised above 150 °C, sodium can be filled in the MV. Keeping the temperature of the vessel in the expected temperature range is very much essential, as the filling of sodium in the MV is the point of no return. Secondary sodium system through IHX imparts heat to the MV internals in addition to the heat input by hot nitrogen gas. Preheating nitrogen circuit consists of blowers, heaters, filters, inlet and outlet headers, associated pipes/ducts and valves (Fig. 1). During preheating, data on temperature observed at various locations of MV and internals were collected.

In order to maintain the temperature profiles of the MV well above the plugging temperature of the sodium, a detailed failure mode and effect analysis (FMEA) is carried out for the blowers and heaters of the nitrogen system. Based on the computed reliability numbers from the mechanistic models of the nitrogen blowers and heaters and the rotor dynamic equipment of the secondary loops, probability of both the secondary sodium loops failing during the sodium filling exercise in MV is very low. This ensures that the possibility of sodium freezing/agglomeration within the MV is extremely low. The reliability of the model is further enhanced by ensuring class III power supply to the blowers and heaters and also by making the availability of adequate spares of parts most likely to fail based on the component FMEA of each of the said equipment.

2.2 Preheating of Pipelines

Total length of sodium and auxiliary piping system of PFBR is in the order of 25,000 m, and these pipelines are varying from a size of 15 to 800 NB. After laying of the MI heating cables on the outer surface of the pipeline, valves and components, they are covered with thermal insulation and outer metallic clad. These heating

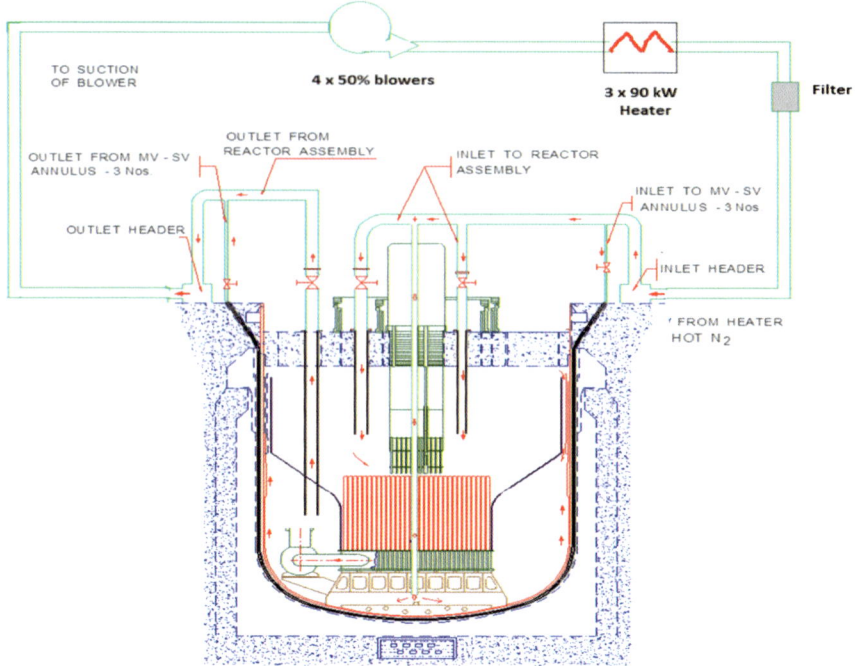

Fig. 1 Main vessel preheating circuit

cables are to raise the temperature of empty pipelines/equipment from ambient to a minimum of 150 °C, for the same reason as that of MV. Preheating of pipelines is required as one of the important activities during commissioning phase of the sodium system before sodium filling operation. Subsequently, preheating of the pipelines is required before refilling with sodium if draining of sodium from the loop is essential for any reason. Any repair/replacement of either the heaters or thermocouples warrants, switching off the heaters, removal of insulation, cooling the lines for effective checking of the heaters and thermocouples, replacement of long coiled lengths of the same and subsequent normalization. This requires a close study of the temperature profiles in the various pipe segments, through the measurements available and corresponding control of the ON/OFF cycles of the heaters.

3 Data Analysis

3.1 MV Preheating

3.1.1 Estimation by Least Square Approximation Method

Temperature at various locations of the MV monitored using thermocouples was obtained from data acquisition system of PFBR during commissioning. Thermal profiling of MV is carried out based on selected data by removing outliers. From this, the temperature of MV (T_{MV}) during preheating is found to be a function of inlet nitrogen temperature (T_{N2}) and secondary sodium temperature (T_{Na}).

$$T_{MV} = f\left(T_{N2}, T_{Na}\right) \tag{1}$$

Heat addition through the IHX to the MV is achieved through the churning effect of secondary sodium pump (SSP) and the pipeline heaters. Availability of the components of the secondary sodium system, mainly SSP and associated piping system, is playing a vital role in the heat addition process. Failure rate of SSP is 7e−5/ry [1], sodium pipelines and welds are 1e−8/ry [1], and that of valves are 1.9 e−5/ry [1]. Thus, the failure probability of both the secondary system is very low. Also, from the data analysis, effect of secondary sodium system on T_{MV} is found to be less significant compared to the nitrogen heating system. From the commissioning data, rate of change of MV temperature due to secondary sodium system is found to be sluggish compared to the rate of change of nitrogen heating system. Hence, the role of nitrogen heating system on MV temperature is studied in detail compared to secondary sodium system. From the trend analysis of MV preheating system, the coldest point is identified, which is the crown of MV. Correlation of thermocouple readings at crown portion and nitrogen inlet temperature is found out using least square approximation method, and the dominant contributor for the preheating is found to be nitrogen inlet temperature. Equation for straight line based on least square approximation can be written as [2]

$$y \, \nabla k_0 \lessgtr k_1 x \tag{2}$$

where 'y' is the temperature at crown of MV and 'x' is the nitrogen inlet temperature. From the collected data sets, k_0 and k_1 are calculated as explained below and the best fit is reported here.

Sl. No	x	y	$x - \bar{x}$	$y - \bar{y}$	$(x - \bar{x}) \times (y - \bar{y})$	$(x - \bar{x})^2$	$(y - \bar{y})^2$	x^2	y^2	xy
1	97	70.45	−35.7	−22.97	820.9	1276.9	527.7	9409	4963.2	6833
2	105	76.65	−27.7	−16.77	465.2	769.1	281.3	11,025	5875.2	8048
3	107	77.63	−25.7	−15.79	406.4	662.2	249.4	11,449	6026.4	8306
4	111	79.58	−21.7	−13.84	300.8	472.3	191.6	12,321	6332.9	8833
5	112	81.53	−20.7	−11.89	246.6	429.9	141.4	12,544	6647.1	9131
6	106	82	−26.7	−11.42	305.4	714.7	130.5	11,236	6724	8692
7	111	81.5	−21.7	−11.92	259.1	472.3	142.1	12321	6642.3	9046
8	116	83.5	−16.7	−9.92	166	280	98.5	13,456	6972.3	9686
9	120	86.5	−12.7	−6.92	88.1	162.1	47.9	14,400	7482.3	10,380
10	124	84	−8.7	−9.42	82.3	76.3	88.8	15,376	7056	10,416
11	160	103	27.27	9.57	261.1	743.5	91.7	25,600	10,609	16,480
12	170	112	37.27	18.57	692.3	1388.8	345.1	28,900	12,544	19,040
13	182	123	49.27	29.57	1457.2	2427.2	874.8	33,124	15,129	22,386
14	183	127	50.27	33.58	1687.8	2526.7	1127.4	33,489	16,129	23,241
15	187	133	54.27	39.58	2147.7	2944.9	1566.4	34,969	17,689	24,871
Sum	1991	1401.3	5E−05	−5E−05	9387	15,346	5904.79	279,619	136,821	195,391

$$\text{Regression coefficient, } k1 = \frac{S_{xy}}{S_x^2} \tag{3}$$

S_{xy} is the sample covariance which is found out using

$$S_{xy} = \frac{1}{(n-1)} \sum_{j=1}^{n} x_j y_j - \frac{1}{n} \left(\sum_{i=1}^{n} x_i \right) \left(\sum_{i=1}^{n} y_i \right) \tag{4}$$

$$S_x^2 = \frac{1}{n-1} \left[\sum_{j=1}^{n} x_j^2 - \frac{1}{n} \left(\sum_{j=1}^{n} x_j \right)^2 \right] \tag{5}$$

$$S_y^2 = \frac{1}{n-1} \left[\sum_{j=1}^{n} y_j^2 - \frac{1}{n} \left(\sum_{j=1}^{n} y_j \right)^2 \right] \tag{6}$$

For the best fit, 'n' is 15

Therefore, $S_{xy} = \frac{1}{(15-1)} \left[195,391 - \frac{1}{15}(1991 \times 1401.3) \right] = 670$

$$S_x^2 = \frac{1}{15-1} \left[279,619 - \frac{1}{15}(1991^2) \right] = 1096$$

I.e., $k_1 = 0.611$

$$\bar{x} = \frac{1991}{15} = 132.73$$

$$\bar{y} = \frac{1401.3}{15} = 93.42$$

From Eq. (2), $k_0 = \bar{y} - k_1 \bar{x}$

$$k_0 = 93.42 - 132.73 \times 0.611 = 12.32.$$

$$\text{Correlation coefficient, } r = \frac{S_{xy}}{S_x S_y}$$

$$r = \frac{670}{\sqrt{1096}\sqrt{422}} = 0.985 \tag{7}$$

$r^2 = 0.97$, which shows that the best-fit equation is

$$y = 12.32 + 0.611x \tag{8}$$

The predicted temperature based on the above correlation is verified with the observed data for a period of two months and is found to be within ±10 °C (Fig. 2).

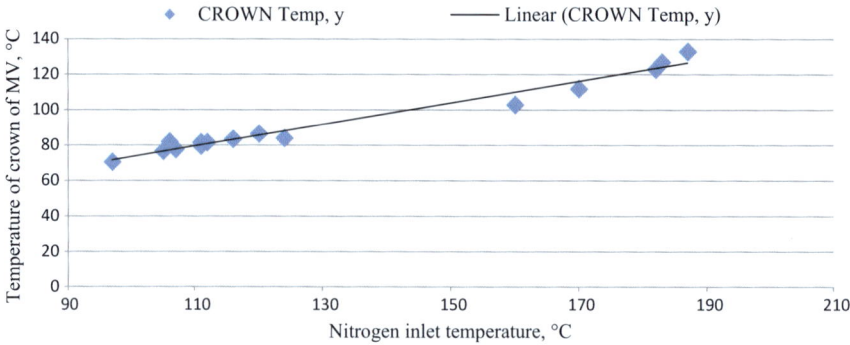

Fig. 2 Trend of nitrogen inlet temperature and crown temperature

3.1.2 Estimation by Recursive Least Square Approximation Method

Once the preheating is stabilised, the sodium will be filled in the MV. At that point of time, the number of measurement becomes more with multivariable dependency. This is applicable when the plant is operating also. In such cases, we need a model that can work with online data which changes its model parameters to reduce the error between the real data and the estimated data. Thus, recursive least square estimate (RLSE) becomes the best suitable tool for this.

The recursive least squares (RLSs) algorithm is the recursive application of the well-known least squares (LSs) regression algorithm. Recursive least squares is an arrangement of the least squares solution in which each new measurement is used to update the previous least squares estimate that was based on the previous measurement. Instead of processing all of the measurement data at once, the measurements are processed one at a time, with each new measurement causing a modification in the current estimate. The method allows for the dynamical application of LS to time series acquired in real-time. A linear recursive estimator can be written in the form [3]

$$z_{k+1} = H_{k+1}x + \vartheta_{k+1} \tag{9}$$

$$\hat{x}_{k+1} = \hat{x}_k + K_{k+1}\left[Z_{k+1} - h_{k+1}^T\hat{x}_k\right] \tag{10}$$

and,

$$\hat{x}_k = \left[H_k^T.H_k\right]^{-1} H_k^T.Z_k \tag{11}$$

where, $K_{k+1} = P_k.h_{k+1}.\delta_{k+1}^{-1}$, $P_k = \left[H_k^T.H_k\right]^{-1}$ and $\delta_{k+1} = 1 + h_{k+1}^T.P_k.h_{k+1}$

That is, we compute x_{k+1} on the basis of the previous estimate x_k and the new measurement z_{k+1}. K_{k+1} is a matrix to be determined called the estimator gain matrix. The quantity $\left[Z_{k+1} - h_{k+1}^T\hat{x}_k\right]$ is called the correction term. Note that if the correction term is zero, or if the gain matrix is zero, then the estimate does not change from time step k to $k+1$ and the estimate is converging.

3.2 Preheating of Pipelines

During the trial preheating, the temperature of few pipeline segments was not raising up to 150 °C causing a heat unbalance in the piping system. The epistemic uncertainties in the estimation of heater power requirement for huge length of pipelines contributed to this situation. These circumstances are called for augmentation of heater capacity to the system. Thermocouple measurements taken during trial preheating are scrutinized to remove the outliers. Mechanistic modelling of the temperature rise in the pipe segments from ambient to 150 °C under the site conditions is carried out using the refined data. Temperature expected at each pipe segments is calculated based on the heater power delivered and by considering all heat loses at particular pipe segments using Eq. (13).

$$\Delta T = \frac{Q}{2\pi} \left[\frac{\ln \frac{r_2}{r_1}}{k_1} + \frac{\ln \frac{r_3}{r_2}}{k_2} + \frac{1}{r_3 h_0} \right] \tag{13}$$

where Q is the power delivered by the laid heater, r_1 is the radius of pipe, r_2 is the radius with inner layer of insulation, r_3 is the radius with outer layer of insulation, k_1 and k_2 are the thermal conductivity of inner and outer layer insulation, and h_0 is the outer heat transfer coefficient.

Based on the analysis, increase in insulation thickness, change in voltage across the heaters, increase/decrease of pitch of heater laying, energising duplicate heater, etc., were proposed. Further, the expected temperature rise for various sodium systems of PFBR is validated with the achieved temperature during preheating. The error is in ±10 °C, in general. Validation of the calculated temperature of pipeline w.r.t preheating data is a prudent forethought to minimize the risk of sodium freezing during sodium filling in the pipelines.

4 Future Works

PFBR is under advanced stage of commissioning. Filling of sodium in the pipelines is completed, and the commissioning phase is marching towards filling of MV with sodium. During the sodium filling activities, data collection on the temperature profile of the MV and its internals, secondary sodium system and the operation of steam water system will go along with commissioning. Refining of the data sets by removing the uncertainty using *ARMAX* modelling and casting in a *RLSE* framework in a real-time window is the future study area for understanding the temperature profile of the PFBR reactor assembly components during sodium filling and plant operation.

5 Epilogue

Engineering decisions could be taken to operate and maintain the preheating parameters with the identified prognostic parameters and help to identify/maintain the various components. On the basis of trend forecasting, the lead time could also enable full preparations at the places of concern to handle the potential risk of sodium freezing or agglomeration. Statistical analysis of data collected during pre-commissioning and commissioning phase of PFBR is well utilized to establish risk-based methodologies to avoid the risk of freezing of sodium or agglomeration of sodium in the crevasses. Predicted values validated during the commissioning phase give confidence on filling of MV and pipelines with sodium.

Ensuring minimum temperature at pipelines and vessel is established by providing separate class III power supply to the heating system and keeping adequate spares management to avoid the risk of sodium freezing/agglomeration. Precautions such as continuous monitoring of the thermocouple readings and availability of nitrogen are important for maintaining required temperature on the pipelines and vessel, and it is ensured during commissioning phase of PFBR sodium systems.

Acknowledgements We thank the guidance provided by Dr. Kallol Roy, Shri V. Rajan Babu and Shri B. K. Chennakeshava for carrying out the studies based on PFBR commissioning experience.

References

1. *Level 1 Probabilistic safety analysis of PFBR for internal events at full power* –June 2017
2. E. Kreyszig, *Advanced Engineering Mathematics*, Tenth Ed - John Wiley & Sons, Inc
3. G.H. Hostetter, *Digital control system design,* Holt, Rinehart and Winston, Inc

Risk Estimation of a Typical Uranium Metal Production Plant

Bikram Roy, Arihant Jain, Subhankar Manna, Santosh K. Satpati, N. S. Joshi and Manharan L. Sahu

Abstract Risk quantification requires estimation of accident frequency and associated consequences. HAZOP analysis is carried out to identify various hazards associated with the plant operation. The principal hazards associated with handling and storage of uranium compounds and different chemicals in solid, solution and vapour forms are radioactivity, chemical and industrial hazard and fire hazard. Probabilistic safety assessment is carried out for quantification of accident occurrence frequency. Starting from the identification of initiating events, event progression is modelled to determine the accident scenarios that may lead to undesirable state in the plant using event tree approach. Fault tree approach is used to model safety systems employed for mitigation of accident scenarios and estimation of failure probability on demand. Human error probability for various human actions involved for accident mitigation is determined using IAEA guidelines. This paper discusses the hazards associated, event progressions, accident frequency and risk estimation of the plant. Insights obtained from the study are also discussed in the paper.

Keywords PSA · Uranium metal production plant · HAZOP · Risk

B. Roy (✉) · A. Jain · S. Manna · S. K. Satpati · N. S. Joshi · M. L. Sahu
Bhabha Atomic Research Center, Mumbai, India
e-mail: bikram@barc.gov.in

A. Jain
e-mail: arihantj@barc.gov.in

S. Manna
e-mail: smanna@barc.gov.in

S. K. Satpati
e-mail: sksatp@barc.gov.in

N. S. Joshi
e-mail: nsjoshi@barc.gov.in

M. L. Sahu
e-mail: mlsahu@barc.gov.in

© Springer Nature Singapore Pte Ltd. 2020
P. V. Varde et al. (eds.), *Reliability, Safety and Hazard Assessment for Risk-Based Technologies*, Lecture Notes in Mechanical Engineering,
https://doi.org/10.1007/978-981-13-9008-1_83

1 Introduction

The main goal of industrial safety in a uranium metal production plant is to keep the exposure of radioactive product and harmful chemical to members of the public and plant personnel as low as reasonably achievable (ALARA) during normal operational states and in the event of accident. In order to adhere to the safety goal, safety analysis has become almost mandatory for all industrial facilities handling toxic and radioactive material. Safety analysis of uranium metal plant is aimed at determination of risk and ensuring compliance to the regulatory requirement. It is based on probabilistic safety assessment (PSA) which includes all possible accident scenarios and their quantification in terms of release frequency and consequences.

Probabilistic safety assessment (PSA) is an analytical method for deriving numerical estimates of risk associated with a facility; it is aimed at protecting the public health and safety [1]. The major steps to carry out PSA are [2]: hazard identification, initiating event frequency quantification, consequence analysis and risk quantification. Event progression modelling is carried out for all initiating events to determine the undesirable scenarios that may occur and its likelihood (frequency). Fault tree and event tree methods are the most widely used techniques for incident frequency quantification and accident sequence analysis leading to undesirable consequences.

2 Brief Description of a Typical Uranium Metal Plant

Impure uranium materials (mainly in the form of uranates, oxides and peroxide) are dissolved in nitric acid. The crude uranyl nitrate (CUN) solution, obtained after dissolution, is purified by solvent extraction using diluted tributyl phosphate (TBP) as solvent. The uranyl nitrate pure solution (UNPS) is neutralized with ammonia gas to precipitate uranium as ammonium diuranate (ADU). Pure ADU is calcined in calcination furnaces to obtain uranium trioxide (UO_3). UO_3 powder is reduced to UO_2 by using cracked ammonia gas. UO_2 is then converted to UF_4 (green salt) using anhydrous hydrogen fluoride (AHF). Green salt is finally reduced to uranium metal using magnesium chips during magnesio-thermic reaction (MTR).

3 Hazard Associated with Uranium Metal Plant

Various unit operations and processes are involved in the uranium metal plant for handling both radioactive and non-radioactive materials. The principal hazards associated with handling and storage of uranium compounds and different chemicals in solid, solution and vapour forms are radioactivity, chemical and industrial hazard and fire hazard. All the hazards have been taken into consideration in the safety requirements and in the design and operation of the processing facility as per pre-

Table 1 PIE and frequency of occurence (/year) [5]

Sl. No.	PIEs	Frequency (/year)
1	Spillage of radioactive liquid	4.037e−4
2	Leakage of ammonia from equipment and pipeline	2.0e−3
3	Leakage of HF from pipeline	2.0e−3
5	Leakage of UF_6 gas from equipment or pipeline	1.0e−3
6	Contact of uranium metal with air during machining	1.0e−4
7	Non-availability of class IV power	1.2
8	Vigorous reaction inside dissolution vessel	1.1e−3

vailing standards followed by BARC. Various design and administrative control mechanisms have been incorporated in the facility as built-in safety features.

Some of the techniques used for hazard identification include HAZOP analysis, failure mode and effects analysis (FMEA), 'What If' Analysis, Preliminary Hazard Analysis (PHA) and Checklist Analysis. The AIChE-CCPS Guidelines for Hazard Evaluation Procedures [3] describe these techniques in detail. Among all these methods, HAZOP is widely popular method for hazard study due to its simplicity and multidisciplinary approach. A HAZOP study is a structured and systematic examination of a planned or existing process or operation in order to identify and evaluate problems that may represent risks to personnel or equipment, or prevent efficient operation. A preliminary HAZOP [4] study for uranium metal production facility—Vizag (UMPF-V)—has been carried out and initiating events are identified based on final consequences and listed accordingly in Table 1.

The chance of occurrence of several events is very rare considering the inherent safety as well as administrative measures. Taking into account the 58 years of operating experience, the events with very low occurrence frequency are not considered for probabilistic safety assessment [5].

4 Safety Mitigation Systems, Human Action and Reliability Analysis

The details of process instrumentation and control system are divided into three categories:

- Field instruments
- Logical interlocks
- Data acquisition and alarm annunciation system.

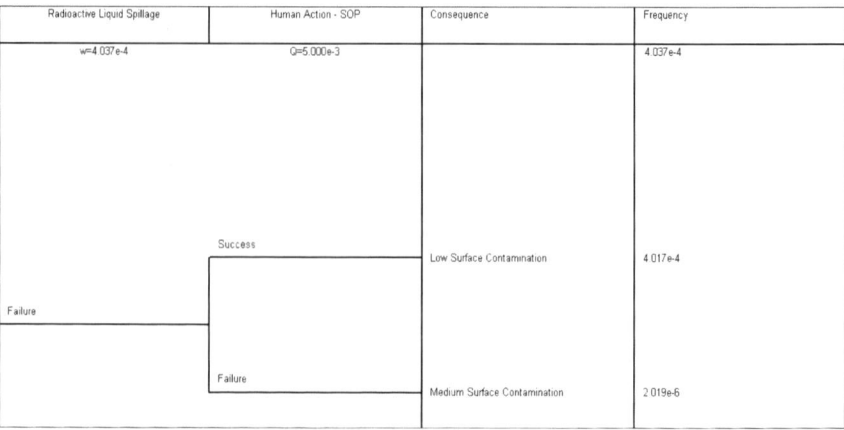

Radioactive Liquid Spillage	Human Action - SOP	Consequence	Frequency
w=4.037e-4	Q=5.000e-3		4.037e-4
	Success	Low Surface Contamination	4.017e-4
Failure			
	Failure	Medium Surface Contamination	2.019e-6

Fig. 1 Event tree diagram of spillage of radioactive liquid

Brief details regarding the purpose of these systems are given in Table 2 along with different safety instrumentation system being employed.

Reliability analysis of various safety systems was carried out using fault tree approach to determine failure probability on demand. The results of reliability analysis are summarised in Table 3. Human error probability has been addressed for various human actions before and during the abnormal conditions [6].

5   Event Progression and Quantification of Consequence Frequency

- Spillage of radioactive liquid can be caused by tank overfilling. The tank overfilling can be caused by several factors such as error in level transmitter, operation error. The consequences of such hazard initiating event are discussed in the form of event tree (Fig. 1).
- Leakage of ammonia from equipment is a result of loose fitting or coupling. Any leakage which results in build-up of ammonia concentration beyond the TLV-TWA (25 ppm) is considered as hazard. The consequences of such hazard initiating event are discussed in the form of event tree diagram in Fig. 2.
- Leakage of HF from equipment is a result of loose/corroded fitting or coupling. Any leakage which results in build of HF vapour concentration beyond the TLV-TWA (0.5 ppm) is considered as hazard. The consequences of such hazard initiating event are discussed in the form of event tree diagram in Fig. 3.
- The consequences leakage of UF6 from pipeline and flanges is discussed in the form of event tree diagram in Fig. 4.

Table 2 Instrumentation and safety system

Type of instrumentation	Components	Purpose	Design and performance codes
Field instruments	Process transmitter	Indicates and transmit different process values Application: Level transmitters (LT), pressure transmitter (PT), temperature transmitter (TT) on process equipments	2004/108/EC, EN 61326
	Switches	Indicates high/low values of defined process variables Application: Level switch (LS) on tanks Pressure switch (PS) on vessel	IEC 61058 and equivalent
	Actuator	To maintain the value of manipulated variable as per instruction from the controller. The actuator can be a Thyristor module for temperature control or control valve for flow control application	ISA 75 and equivalent
	Annunciator	Creates audio visual indication at field and control room to intimate abnormal process situation Application: HL alarm in target tank, LS alarm in target tank, empty alarm in source tank, gas monitoring system alarm, fire alarm etc.	IEC 62682

(continued)

Table 2 (continued)

Type of instrumentation	Components	Purpose	Design and performance codes
Logical interlocks	Process controller, programmable logic controller (PLC) and switchgears	Implementation of process interlocks and control strategy. Takes input from field instruments and controls the output variable to maintain the set point within pre-defined limit. It is often used to activate local alarm modules and necessary control action. Application: Pump interlock, fail-safe operation by PLC, etc.	IEC 61131-1-4, IEC 61439 etc., IEC-61000
Data acquisition system	Supervisory control and data acquisition system (SCADA)	The centralised data acquisition system based on PLC and SCADA acquires all field data and display them in the legible form to the operator. The display also specifies process value range corresponding to particular variable intimate operator about any alarm situation. Centralised control strategy may also by programmed on this system	IEC 61131
Safety instrumentation	Gas monitoring system	Gas detectors will be provided for different gaseous component like NH3, HF, NOx, H2 and organic. The detection system will raise an alarm in case the concentration of toxic builds beyond the set permissible value	IEC 60079-29-3
	Fire alarm system	Smoke and heat base fire detectors with audible and visible alarm system will be provided at proper place to detect fire	NFPA 72

(continued)

Table 2 (continued)

Type of instrumentation	Components	Purpose	Design and performance codes
	Radiation monitoring system	Surface contamination, radiation field and air contamination in the plant will be monitored using available suitable instruments. Plant exhaust will be monitored using online discharge monitoring system/manual monitoring equipment. TLDs and electronic personnel dosimeters (EPD) will be used for the external dose assessment of plant personnel	IEC 61504:2017, IEC 60325:2002 etc.

Table 3 Reliability data of process and safety systems

Systems	Unavailability (per demand) [3, 7]
High-level alarm in target tank	9.1e−3
Low-level alarm in source tank	9.1e−3
Level switch alarm in target tank	7.1e−3
Pump interlock	3.9e−4
Gas monitoring system alarm	2.4e−4
Fail-safe operation by controller	3.5e−2
High-temperature interlock to disable feeding	7.1e−3
Pressure transmitter alarm	1.2e−3

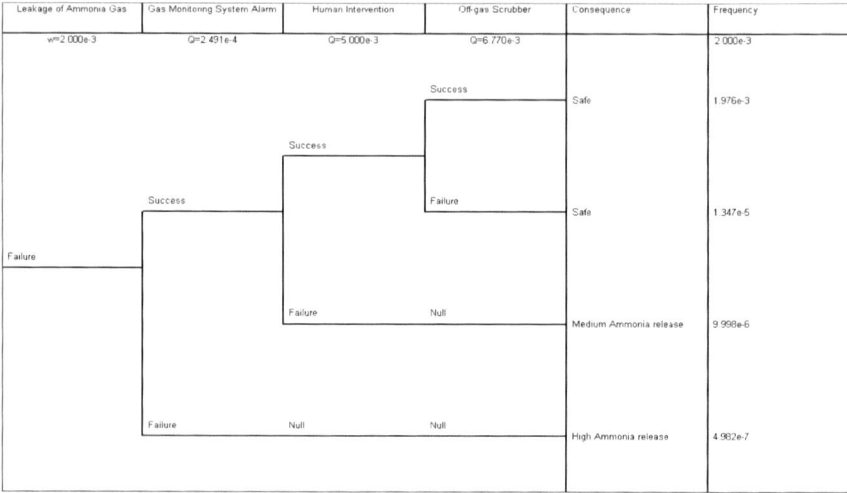

Fig. 2 Event tree diagram of leakage of ammonia gas

- Uranium metal may start oxidation in contact with air which may result in fire. The consequences of such hazard initiating event are discussed in the form of event tree diagram in Fig. 5.
- In case of grid power failure, the class III power is made available through captive power plant which is necessary for safe shutdown of process as well as prevents any damage to heated equipment. The consequences of such hazard initiating event are discussed in the form of event tree diagram in Fig. 6. Figure 7 represents the consequence of uncontrolled reaction at dissolution vessel which may arise due to failure of feeding system due to power failure of any other reason.

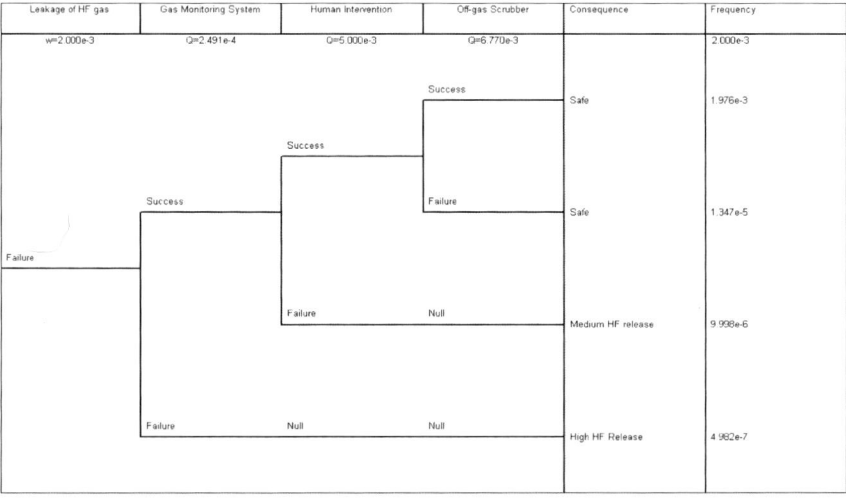

Fig. 3 Event tree diagram of leakage of HF gas

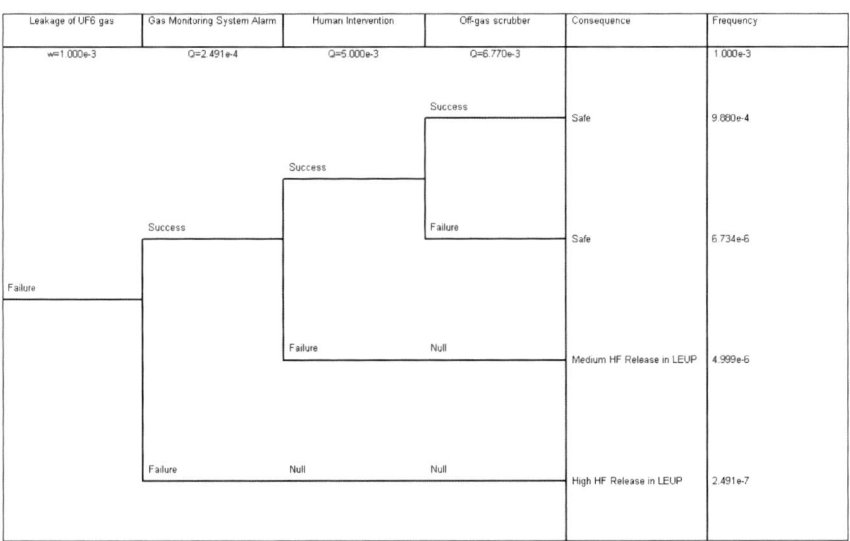

Fig. 4 Event tree diagram of leakage of UF6 gas

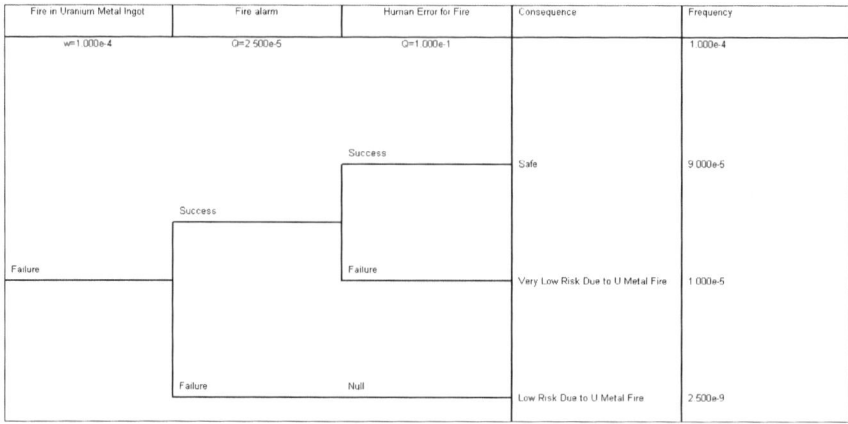

Fig. 5 Event tree diagram of uranium metal fire

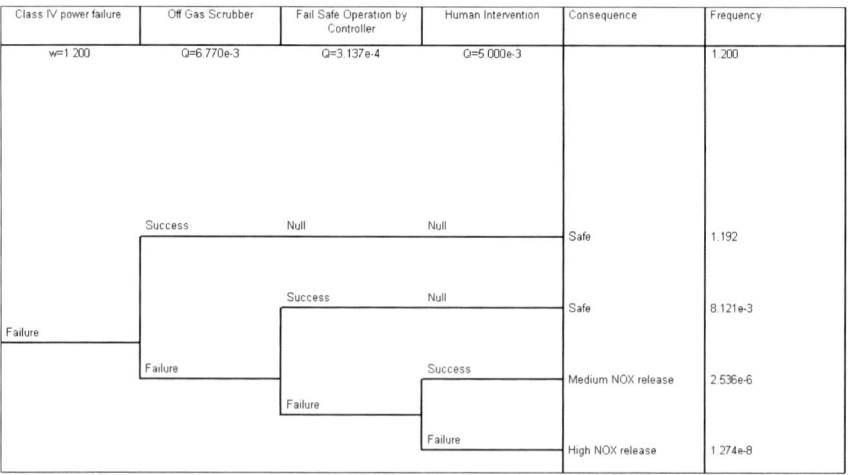

Fig. 6 Event tree diagram of leakage class IV power failure

6 Results

Event tree analysis gives a likelihood of occurrence of various consequences origi-
nating from initiating events. The results are given in Table 4. The major consequence
is surface contamination (low and medium) which is limited to plant personnel only.
The effects associated with uranium metal fire are also limited within the plant only.

Other consequences do not pose much threat to the society (environment and
public) due to very low occurrence frequency. Radiation hazard to the environment
and public due to the plant is negligible.

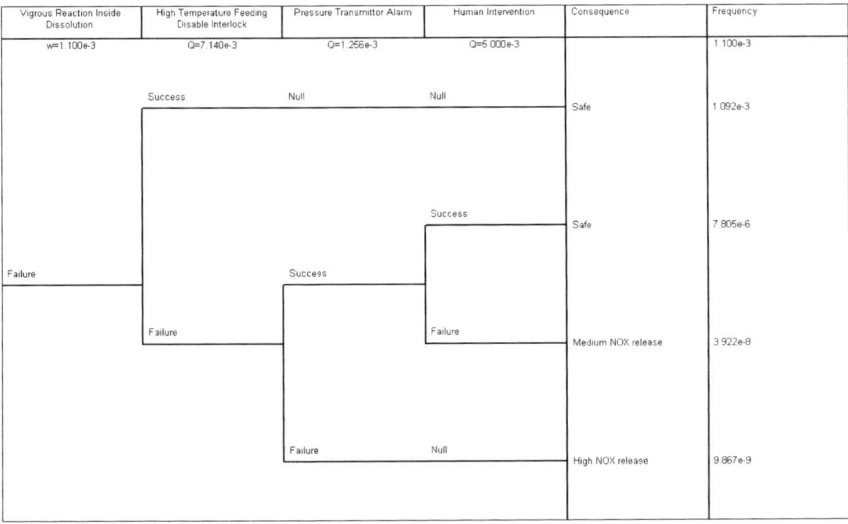

Fig. 7 Event tree diagram of vigorous reaction in dissolution vessel

Table 4 Hazard frequency due to various initiating events

Release state	Consequence frequency (/yr)	
	Confidence mean	95% confidence value
Low surface contamination	4.866e−4	1.069e−3
Medium surface contamination	2.445e−6	5.373e−6
Medium ammonia release	9.997e−6	9.997e−6
High ammonia release	6.215e−7	1.071e−6
Medium HF release	9.997e−6	9.999e−6
High HF release	6.215e−7	1.071e−6
Very low risk due to uranium fire	1.229e−5	2.809e−5
Low risk due to uranium fire	3.156e−9	6.758e−9
Medium NOx release	3.741e−6	8.734e−6
High NOx release	3.796e−8	8.373e−8

In order to consider the uncertainty associated with the failure data, uncertainty analysis is carried out. Ninety-five per cent confidence values are reported in Table 4.

7 Conclusion

The probabilistic safety analysis has been carried out for a typical uranium metal plant after identification of different hazard initiating events. Event tree and fault tree analysis techniques have been adopted to calculate the frequency of hazard. The final consequences, in terms of frequency/year, have been listed with reference to data obtained from event tree analysis. The highest mean accident frequency among all the states is found to be 4.017e−4/year for "low surface contamination". The consequence reported here is only related to working personnel inside plant. The effect of these consequences to the public outside plant boundary is very less.

References

1. R.R. Fullwood, *Probabilistic Safety Assessment in the Chemical and Nuclear Industries* (Butterworth Heinemann, 1998)
2. G.R Van Sciver, Quantitative risk analysis in chemical process industry, Reliab. Eng. Syst. Saf. **29**, 55–68 (1990)
3. Guidelines for Hazard Evaluation Procedures, *Prepared by Battelle Columbus Division for CCPS/AIChE* (AIChE, New York, 1985)
4. Preliminary Hazard and Operability (Pre-HAZOP) of Uranium Metal Production Facility at Vizag (UMPF-V), UMPF-V/PreHAZOP/R0 Dated 12/03/2018
5. B. Roy, S. Manna, A. Jain, *Probabilistic Safety Assessment of Uranium Metal Production Facility* (Vizag, 2018)
6. A.D. Swain, H.E. Guttmann, Handbook of HRA with Emphasis on Nuclear Power Plant Application, Final Report: NUREG\CR. 1278, Sandia National Laboratories, Aug 1989. NUREG/CR-5695
7. International Atomic Energy Agency, *Reliability Data for Probabilistic Safety Assessment of NPP*, (IAEA-TECDOC-478, Vienna 1986)

Radiation Levels in Structural Components of Cirus Reactor—Estimation and Measurement

Y. S. Rana, Tej Singh and P. V. Varde

Abstract Estimation of radioactivities and radiation levels is carried out for structural components of the Cirus reactor. The estimated results of radiation levels are compared with in situ measurements. The data will be useful for future decommissioning of Cirus reactor from the viewpoint of decontamination techniques, dose budgeting and waste management. Neutron fluxes were estimated in different structural components by considering the operational history of the reactor. The neutron fluxes in core and reflector regions were estimated by using computer code FINHEX. Point depletion computer code ORIGEN2 was used for carrying out activity calculations. The activities estimated by ORIGEN2 were used as input in point kernel computer code IGSHIELD for estimating radiation level.

Keywords Cirus · ORIGEN2 · FINHEX · IGSHIELD · Neutron flux · Activity

1 Introduction

Cirus reactor was commissioned in 1960. It is a 40 MW_t tank type research reactor with natural uranium as fuel, heavy water as moderator and light water as coolant. Graphite, consisting of two concentric cylinders of 9 and 24.5 in. thickness, is used as a reflector. The maximum thermal neutron flux available in the reactor is 6.7×10^{13} n/cm^2/s. The reactor core is housed in a vertical cylindrical vessel made of aluminium. The fuel assemblies are located in lattice tubes which are rolled to the top and bottom tube sheets of the reactor vessel. In radial and axial directions, the core is covered by thermal and biological shields. The top thermal shield in the axial

Y. S. Rana (✉) · T. Singh · P. V. Varde
Bhabha Atomic Research Center, Mumbai, India
e-mail: yps@barc.gov.in

T. Singh
e-mail: t_singh@barc.gov.in

P. V. Varde
e-mail: varde@barc.gov.in

© Springer Nature Singapore Pte Ltd. 2020
P. V. Varde et al. (eds.), *Reliability, Safety and Hazard Assessment for Risk-Based Technologies*, Lecture Notes in Mechanical Engineering,
https://doi.org/10.1007/978-981-13-9008-1_84

direction consists of one aluminium and two carbon steel tube sheets with water sandwiched in between. The bottom thermal shield consists of one aluminium and three carbon steel tube sheets with water sandwiched in between. The thicknesses of the top and bottom thermal shields are 3 and 4 ft., respectively. The thermal shield in the radial direction is known as a side thermal shield and consists of two annular rings of cast iron with a thickness of 6 in. each. The biological shield is made of high-density concrete with a thickness of 8 ft. Figures 1a, b give a schematic of the reactor structure.

The reactor operated successfully for about 50 years (except for a period of about 6 years from 1997 to 2003, during which the reactor remained shut down for undergoing major refurbishment jobs for its life extension). The average availability factor until 1992 was 61% at 40 MW power. From 1992 to 1997, the reactor operated at 20 MW with an availability factor of about 46%. After refurbishment, the reactor operated at 40 MW during the year 2004 with an availability factor of about 68%. From the year 2005 to 2010, the average availability factor was about 81% at 20 MW reactor power. The reactor was permanently shut down on 31 December 2010. The core has been unloaded completely and the heavy water in the reactor vessel has been drained.

As part of preparation towards decommissioning of Cirus, it is planned to bring the reactor to a state wherein minimum surveillance is required. In this respect, neutron-induced activities and radiation levels in different reactor structure components have been estimated. Neutron fluxes in core and reflector regions were estimated by using computer code FINHEX [1] considering the operational history of the reactor. Point depletion computer code ORIGEN2 [2] was used for carrying out activity calculations. The activities estimated by ORIGEN2 were used as input in point kernel computer code IGSHIELD [3] for estimating radiation level.

The present paper gives a detailed account of the calculations for activities and radiation levels in different structural components of Cirus reactor. The results are compared with measured values. The data will be useful for future decommissioning of Cirus reactor from the viewpoint of decontamination techniques, dose budgeting and waste management.

2 Calculation Methodology

Neutron fluxes were estimated in different structural components (reactor vessel, reactor vessel tubes and reflector) considering three energy groups, viz. thermal (<0.625 eV), epithermal (0.625 eV−821 keV) and fast (>821 keV). The neutron fluxes in core and reflector regions were estimated by using computer codes WIMSD and NEMHEX. The thermal neutron flux so obtained was used in point depletion computer code ORIGEN2 for carrying out activity calculations. For this purpose, the actual operational history of the reactor was taken into account. PHWR spectrum of neutrons was used in the code. The activities estimated by ORIGEN2 were used as input in point kernel computer code IGSHIELD for estimating the radiation level.

(a)

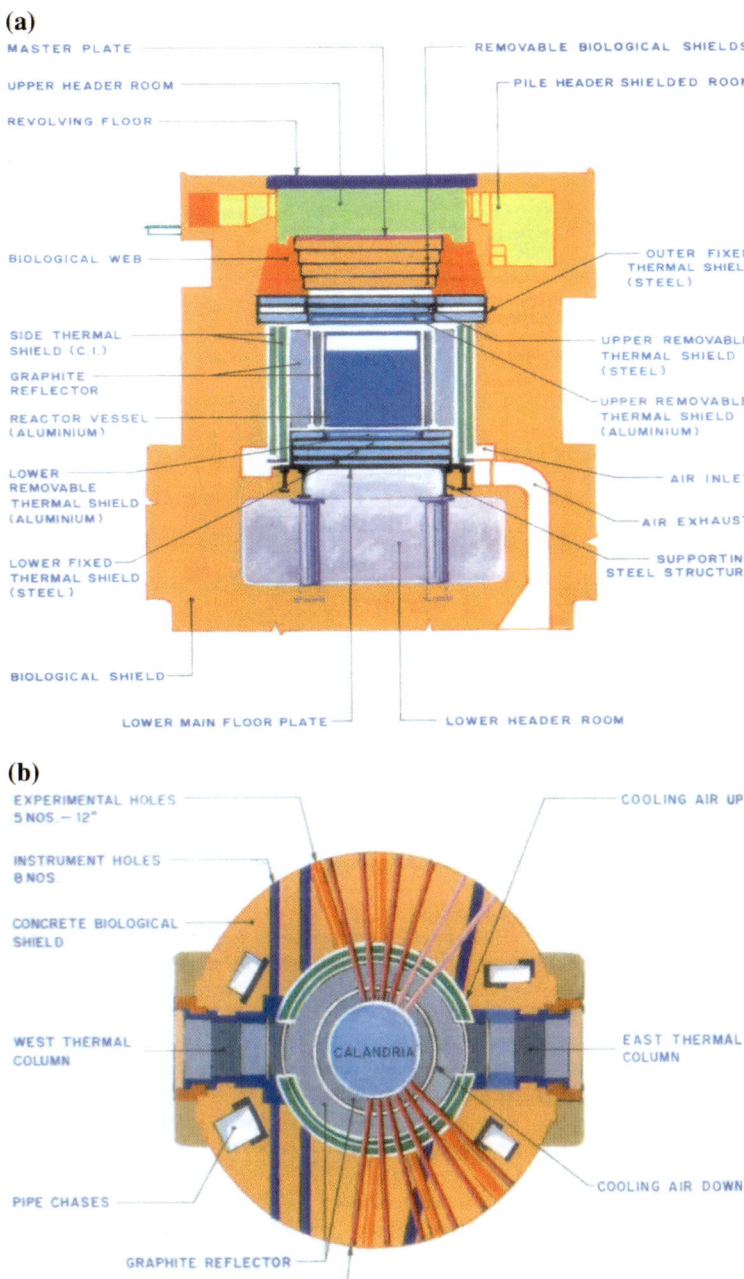

(b)

Fig. 1 **a** Schematic of the reactor structure—elevation. **b** Schematic of the reactor structure—Plan

3   Radionuclide Inventory and Radiation Levels

3.1   Reactor Vessel

The reactor vessel (RV) is made of aluminium (ALCAN6056) and its weight is around 500 kg. The impurity contents (wt%) in ALCAN6056 are given in Table 1. The average thermal neutron flux on RV is about 1.0×10^{13} n/cm²/s. The activities of different nuclides as a function of S/D time are given in Table 2. It can be seen from the table that the activity is mainly on account of Fe^{55} and Co^{60} radionuclides. However, due to its beta activity, Fe^{55} is not a concern for radiation level. Thus, the radiation level on RV is mainly due to Co^{60}. The radiation level on RV as a function of S/D time is given in Table 3. It can be seen from the table that the radiation level on contact of RV after 4 years of shut down is about 540 mGy/h which reduces to about 130 mGy/h after 15 years. Thus, the reduction in radiation level follows the decay pattern of Co^{60} (Half-life 5.2 years).

Table 1   Chemical composition of ALCAN-6056 [4]

Element	wt%
Cu	0.01
Fe	0.4
Si	0.2
Mn	0.01
Ti	0.01
Ni	0.005
Zn	0.005
Cr	0.005
Ga	0.025
V	0.025
B	0.002
Cd	0.002
Co	0.001
Mg	0.01

Table 2   Activity of RV as a function of shut down time

Nuclide	Activity (MBq) after shut down							
	4 years	5 years	6 years	7 years	8 years	9 years	10 years	15 years
Mn^{54}	8.18E+02	3.64E+02	1.62E+02	7.22E+01	3.20E+01	1.42E+01	6.33E+00	1.10E−01
Fe^{55}	1.25E+06	9.55E+05	7.33E+05	5.62E+05	4.29E+05	3.29E+05	2.52E+05	6.66E+04
Co^{60}	3.38E+06	2.96E+06	2.60E+06	2.28E+06	2.00E+06	1.75E+06	1.54E+06	7.96E+05
Zn^{65}	2.22E+03	7.84E+02	2.78E+02	9.84E+01	3.48E+01	1.23E+01	4.37E+00	2.43E−02

Table 3 Radiation level on RV

S/D time (years)	Radiation level (mGy/h)
4	540
5	470
6	415
7	364
8	319
9	280
10	245
15	127

Table 4 (a) Activity of central RV tubes as a function of shut down time. (b) Activity of peripheral RV tubes as a function of shut down time

Nuclide	Activity (MBq) after shut down							
	4 years	5 years	6 years	7 years	8 years	9 years	10 years	15 years
(a)								
Mn^{54}	1.49E+01	6.66E+00	2.96E+00	1.32E+00	5.85E−01	2.60E−01	1.16E−01	2.02E−03
Fe^{55}	2.29E+04	1.75E+04	1.34E+04	1.03E+04	7.88E+03	6.03E+03	4.63E+03	1.22E+03
Co^{60}	3.31E+04	2.90E+04	2.55E+04	2.23E+04	1.96E+04	1.72E+04	1.51E+04	7.81E+03
Zn^{65}	4.22E+01	1.49E+01	5.29E+00	1.87E+00	6.62E−01	2.35E−01	8.33E−02	4.63E−04
(b)								
Mn^{54}	8.62E+00	3.85E+00	1.71E+00	7.59E−01	3.38E−01	1.50E−01	6.70E−02	1.17E−03
Fe^{55}	1.32E+04	1.01E+04	7.73E+03	5.92E+03	4.55E+03	3.47E+03	2.66E+03	7.03E+02
Co^{60}	2.55E+04	2.24E+04	1.96E+04	1.72E+04	1.51E+04	1.32E+04	1.16E+04	6.03E+03
Zn^{65}	2.37E+01	8.40E+00	2.97E+00	1.05E+00	3.74E−01	1.32E−01	4.66E−02	2.60E−04

3.2 Reactor Vessel (RV) Tubes

The RV tubes are also made of aluminium (ALCAN6056) and weight of a single tube in the active core region is about 2.2 kg. The average thermal neutron fluxes on RV tubes in central and peripheral regions are 4.3×10^{13} and 1.0×10^{13} n/cm^2/s, respectively. The activities of different nuclides in the RV tubes as a function of S/D time are given in Tables 4a, b. The radiation levels on RV tubes as a function of S/D time are given in Table 5. It can be seen from the table that the radiation levels on contact of central and peripheral RV tubes after 4 years of shut down are about 170 and 130 mGy/h, respectively. After 15 years of cooling, these values reduce to about 40 and 30 mGy/h, respectively. Again, it can be seen that the reduction in radiation level follows the decay pattern of Co^{60}.

Table 5 Radiation level on RV tubes

S/D time (years)	Radiation level (mGy/h)	
	Central region	Peripheral region
4	169	130
5	148	114
6	130	100
7	114	88
8	100	77
9	88	68
10	77	59
15	40	31

3.3   Top and Bottom Tube Sheets

The top and bottom tube sheets are made of 3-in. thick Al alloy ALCAN6056 annealed plates. Both tube sheets are water cooled from the primary cooling water system. Water passages are formed by rows of 2-in. holes bored diagonally. The 2-in. holes are sealed by ALCAN 61S machined plugs fitted with diamond-shaped stainless steel gaskets. The upper tube sheet is provided with 16 water passages while there are 8 water passages in the lower tube sheet since it is partially cooled by heavy water. The average thermal neutron fluxes on the bottom and top aluminium tube sheets at 40 MW reactor power are estimated to be about 4×10^{12} and 2×10^{12} n/cm^2/s, respectively. The activities of different nuclides in the top and bottom tube sheets are given in Tables 6a, b, respectively. The corresponding radiation levels are given in Table 7. It can be seen from the table that the radiation levels on top and bottom tube sheets are about 710 and 1390 mGy/h, respectively, after 4 years of cooling. After 15 years of cooling, these values reduce to about 170 and 330 mGy/h, respectively. As in the case of RV and RV tubes, the reduction in radiation level follows the decay pattern of Co^{60} as other major radionuclides such as Fe^{55} and Ni^{63} are beta emitters.

3.4   SS-316 Fittings on Tube Sheets

The water passages in upper and lower tube sheets are connected to separate individual common inlet and outlets of ½ in. OD SS-316 pipes through swagelock fittings made of SS-316 and Helium passages are also provided in both tube sheets. The mass of fitting is about 70 gm. Activities produced in the fittings on top and bottom tube sheets are given in Table 8a, b, respectively. The corresponding radiation levels are given in Table 9. It can be seen from the table that the radiation levels on fittings on top and bottom tube sheets are about 2550 and 4920 mGy/h, respectively, after

Table 6 (a) Activity of top tube sheet. (b) Activity of bottom tube sheet

Nuclide	Activity (MBq) after shut down							
	4 years	5 years	6 years	7 years	8 years	9 years	10 years	15 years
(a)								
Mn^{54}	3.81E+02	1.69E+02	7.51E+01	3.34E+01	1.49E+01	6.62E+00	2.94E+00	5.11E−02
Fe^{55}	5.77E+05	4.44E+05	3.39E+05	2.60E+05	1.99E+05	1.52E+05	1.17E+05	3.08E+04
Co^{60}	1.79E+06	1.57E+06	1.37E+06	1.21E+06	1.05E+06	9.25E+05	8.10E+05	4.22E+05
Ni^{59}	2.28E+02	2.28E+02	2.28E+02	2.28E+02	2.28E+02	2.28E+02	2.28E+02	2.28E+02
Ni^{63}	2.75E+04	2.72E+04	2.70E+04	2.68E+04	2.66E+04	2.64E+04	2.62E+04	2.53E+04
Zn^{65}	1.01E+03	3.59E+02	1.27E+02	4.51E+01	1.59E+01	5.66E+00	2.00E+00	1.11E−02
(b)								
Mn^{54}	7.62E+02	3.38E+02	1.51E+02	6.70E+01	2.98E+01	1.32E+01	5.88E+00	1.02E−01
Fe^{55}	1.16E+06	8.88E+05	6.81E+05	5.22E+05	4.00E+05	3.05E+05	2.34E+05	6.18E+04
Co^{60}	3.48E+06	3.05E+06	2.68E+06	2.35E+06	2.06E+06	1.80E+06	1.58E+06	8.18E+05
Ni^{59}	4.44E+02	4.44E+02	4.44E+02	4.44E+02	4.44E+02	4.44E+02	4.44E+02	4.44E+02
Ni^{63}	5.48E+04	5.44E+04	5.37E+04	5.33E+04	5.29E+04	5.25E+04	5.22E+04	5.03E+04
Zn^{65}	2.04E+03	7.22E+02	2.56E+02	9.07E+01	3.21E+01	1.14E+01	4.03E+00	2.24E−02

Table 7 Radiation level on tube sheets

S/D time (years)	Radiation level (mGy/h)	
	Top tube sheet	Bottom tube sheet
4	710	1390
5	620	1210
6	550	1080
7	480	940
8	420	820
9	370	720
10	320	630
15	170	330

4 years of cooling. After 15 years of cooling, these values reduce (mainly following the decay pattern of Co^{60}) to about 600 and 1160 mGy/h, respectively.

4 Comparison with Measured Experimental Data

After one year of shut down, measurement was carried out for radiation level in the core region and the value was found to be in the range of 2000–3000 mGy/h. Based on the estimations for radiation levels on the structural components, the radiation level in core region after four years of cooling was estimated to be about 1680 mGy/h.

Table 8 (a) Activity of SS-316 fittings on top tube sheet. (b) Activity of SS-316 fittings on bottom tube sheet

Nuclide	Activity (MBq) after shut down							
	4 years	5 years	6 years	7 years	8 years	9 years	10 years	15 years
(a)								
Mn^{54}	3.55E+00	1.58E+00	7.03E−01	3.13E−01	1.39E−01	6.18E−02	2.75E−02	4.77E−04
Fe^{55}	5.40E+03	4.14E+03	3.17E+03	2.43E+03	1.86E+03	1.43E+03	1.09E+03	2.88E+02
Co^{60}	4.37E+03	3.81E+03	3.35E+03	2.94E+03	2.58E+03	2.26E+03	1.98E+03	1.02E+03
Ni^{59}	3.85E+01	3.85E+01	3.85E+01	3.85E+01	3.85E+01	3.85E+01	3.85E+01	3.85E+01
Ni^{63}	4.66E+03	4.63E+03	4.59E+03	4.55E+03	4.51E+03	4.48E+03	4.44E+03	4.29E+03
(b)								
Mn^{54}	7.10E+00	3.16E+00	1.41E+00	6.25E−01	2.78E−01	1.24E−01	5.51E−02	9.58E−04
Fe^{55}	1.08E+04	8.29E+03	6.36E+03	4.85E+03	3.74E+03	2.85E+03	2.18E+03	5.77E+02
Co^{60}	8.44E+03	7.40E+03	6.48E+03	5.70E+03	5.00E+03	4.37E+03	3.85E+03	1.98E+03
Ni^{59}	7.55E+01	7.55E+01	7.55E+01	7.55E+01	7.55E+01	7.55E+01	7.55E+01	7.55E+01
Ni^{63}	9.21E+03	9.14E+03	9.07E+03	8.99E+03	8.95E+03	8.88E+03	8.81E+03	8.47E+03

Table 9 Radiation level on SS-316 fittings on tube sheets

S/D time (years)	Radiation level (mGy/h)	
	Top tube sheet	Bottom tube sheet
4	2550	4920
5	2230	4320
6	1960	3780
7	1720	3320
8	1500	2910
9	1320	2550
10	1160	2240
15	600	1160

Assuming the decay pattern of Co^{60}, the value comes out to be about 2500 mGy/h which is in agreement with the measurement.

In another set, measurements were carried out for radiation levels at different pile positions in the gap between the tube sheet and Al thermal shield after six years of shut down. The measured value of radiation level at different lattice positions between top tube sheet and Al thermal shield varied from 1200 to 9000 mGy/h. The corresponding value between the bottom tube sheet and Al thermal shield was 1000–16500 mGy/h. The radiation level at any lattice position on the tube sheet is mainly due to the SS fitting and the tube sheet itself. It can be seen from Tables 7 and 9 that the resultant radiation level on the top tube sheet will be about 2500 mGy/h. The corresponding value for the bottom tube sheet is about 4000 mGy/h. Thus, the theoretically estimated values are in agreement with the measurements.

5 Summary and Conclusions

Reactor physics modelling of Cirus reactor has been carried out by deterministic method based computer codes. The models have been used for estimation of activities and radiation levels in various structural components of the reactor. The activity and radiation level in structural components is mainly on account of Co^{60} radionuclide. The results are compared with measurements of radiation levels observed during radiation mapping of reactor structure. A good agreement between calculated and measured values is observed. The data will be useful for future decommissioning of Cirus reactor for deciding on important issues such as decontamination techniques, dose budgeting and waste management.

Acknowledgements The authors are grateful to Shri Rakesh Ranjan, Decommissioning Superintendent, Cirus and the entire team of Cirus operations for their inputs during the work.

References

1. K. Singh, V. Kumar, Solution of the multi-group diffusion equation in hex-z geometry by finite Fourier transformation. Ann. Nucl. Energy **20**(3), 153–161 (1993)
2. A.G. Croff, ORIGEN2: a versatile computer code for calculating the nuclide compositions and characteristics of nuclear materials. Nucl. Technol. **62**(3), 335–352 (1983)
3. K.V. Subbaiah, R. Sarangapani, IGSHIELD: a new interactive point kernel gamma ray shielding code. Ann. Nucl. Energy **35**(12), 2234–2242 (2008)
4. T Singh et al., *Estimation of Radiation Levels in Structural Components of Cirus Reactor*, BARC/2004/E/032 (2004)

Author Index